Lecture Notes in Mechanical Engineering

Lecture Notes in Mechanical Engineering (LNME) publishes the latest developments in Mechanical Engineering—quickly, informally and with high quality. Original research reported in proceedings and post-proceedings represents the core of LNME. Volumes published in LNME embrace all aspects, subfields and new challenges of mechanical engineering.

To submit a proposal or request further information, please contact the Springer Editor of your location:

Europe, USA, Africa: Leontina Di Cecco at Leontina.dicecco@springer.com
China: Ella Zhang at ella.zhang@springer.com
India: Priya Vyas at priya.vyas@springer.com
Rest of Asia, Australia, New Zealand: Swati Meherishi
at swati.meherishi@springer.com

Topics in the series include:

- Engineering Design
- Machinery and Machine Elements
- Mechanical Structures and Stress Analysis
- Automotive Engineering
- Engine Technology
- Aerospace Technology and Astronautics
- Nanotechnology and Microengineering
- Control, Robotics, Mechatronics
- MEMS
- Theoretical and Applied Mechanics
- Dynamical Systems, Control
- Fluid Mechanics
- Engineering Thermodynamics, Heat and Mass Transfer
- Manufacturing Engineering and Smart Manufacturing
- Precision Engineering, Instrumentation, Measurement
- Materials Engineering
- Tribology and Surface Technology

Indexed by SCOPUS, EI Compendex, and INSPEC.

All books published in the series are evaluated by Web of Science for the Conference Proceedings Citation Index (CPCI).

To submit a proposal for a monograph, please check our Springer Tracts in Mechanical Engineering at https://link.springer.com/bookseries/11693.

Sudev Das · Narasimha Mangadoddy ·
Jaap Hoffmann
Editors

Proceedings of the 1st International Conference on Fluid, Thermal and Energy Systems

ICFTES 2022

Springer

Editors
Sudev Das
Department of Chemical Engineering
National Institute of Technology Calicut
Calicut, Kerala, India

Narasimha Mangadoddy
Department of Chemical Engineering
Indian Institute of Technology Hyderabad
Hyderabad, Telangana, India

Jaap Hoffmann
Department of Mechanical Engineering
Stellenbosch University
Stellenbosch, Western Cape, South Africa

ISSN 2195-4356 ISSN 2195-4364 (electronic)
Lecture Notes in Mechanical Engineering
ISBN 978-981-99-5989-1 ISBN 978-981-99-5990-7 (eBook)
https://doi.org/10.1007/978-981-99-5990-7

This Springer imprint is published by the registered company Springer Nature Singapore Pte Ltd.
The registered company address is: 152 Beach Road, #21-01/04 Gateway East, Singapore 189721, Singapore

Preface

This book comprises the select *Proceedings of the 1st International Conference on Fluid, Thermal and Energy Systems (ICFTES 22)* organized by the National Institute of Technology Calicut in technical collaboration with Stellenbosch University, South Africa, and in association with the Indian Institute of Chemical Engineers (IIChE) and Indian Society for Heat and Mass Transfer (ISHMT), Asian Union of Thermal Science and Engineering (AUTSE). This conference witnessed more than 220 papers presented across three days (June 9th–11th, 2022) in three parallel sessions chaired by eminent academic researchers from around the globe. It is encouraging to see that many of the contributions we received were from young researchers and academics. Through our screening process, we did our utmost to guarantee that only high-quality papers were chosen to publish in this proceedings.

We expect this publication, "Select Proceedings of *ICFTES 22*," to be helpful to researchers from both academics and industries. This conference discussed fluid, thermal, and energy systems, with an emphasis on the applications of heat and work to engineering challenges, fluid behavior under the influence of temperature and pressure gradients, and energy systems. This book is also anticipated to highlight the current scenario on new and innovative technologies in fluid, thermal, and energy systems worldwide, which will assist researchers, faculty, and practising engineers in finding solutions to complex problems.

Calicut, India — Dr. Sudev Das
Hyderabad, India — Prof. Narasimha Mangadoddy
Stellenbosch, South Africa — Prof. Jaap Hoffmann

Contents

About the Editors

Dr. Sudev Das is currently working as an assistant professor in the Department of Chemical Engineering at the National Institute of Technology (NIT) Calicut. He obtained his Ph.D. in heat transfer engineering from the NIT Agartala. His research interests involve phase change heat transfer, solar thermal energy storage, and novel surface engineering. He has multiple collaborative research works with Kyushu University, Japan, and University of Trento, Italy. He has published 18 research papers in SCI journals and 15 international conferences to his credit. His research group has established a state-of-the-art two-phase flow test rig established with a sponsored research project. He is also a reviewer of journals of national and international repute. He has organized various programs such as faculty development programs, workshops, and seminars in the capacity of coordinator.

Dr. Narasimha Mangadoddy is an associate professor in the Department of Chemical Engineering at the Indian Institute of Technology (IIT) Hyderabad, India. He was awarded the Ph.D. in Mining Engineering in 2010 from the University of Queensland, Australia. Before joining IIT, he worked at TATA Steel India, JKMRC-Australia, on various mineral processing projects. His areas of interest include computational modeling ranging from CFD, DEM, CFD, and DEM for multi-phase flows, particulate technology, mineral processing, slurry rheology, and transport processes of separation devices. He has two international patents and has authored about 75 referred publications. His research is supported by grants (~600 Lakhs) from SERB-DST, Ministry of Mines, Ministry of Education, Hindustan Zinc Limited, R&D TATA Steel, R&D NMDC, Clairs Engineers Private Limited, and Tega Industries Private Limited. In December 2020, he was inducted as the International Advisory Committee member for IMPC from India, representing Academic Research.

Dr. Jaap Hoffmann is an associate professor in the thermofluids division, Stellenbosch University. He did his Ph.D. on performance prediction on natural draft cooling towers under the influence of atmospheric temperature inversions. During the 90's, his attention turned to numerical simulation of dispersed two-phase flows. Recently, his research focus has shifted to solar thermal energy. He was registered as a professional engineer with ECSA (reg. no. 990334) in 1998. He authored about 49 referred publications, including journals and international conferences. He is also a member of several reputed professional organizations.

Mixing Performance of an Electroosmotic Micromixer: Effect of Baffle

Biswajit Gayen, Nirmal K. Manna, and Nirmalendu Biswas

Abstract In recent years, electrokinetic effects, viz., electroosmotic flow and electrophoretic phenomenon finds applications in the field of micromixing. To address such an application, we present a novel approach to micromixing. In our present study, simulation has been performed for Mixing Quality (MQ) of two different concentration fluids in one ring-shaped micromixer using electroosmotic flow. The micromixer consists of two pairs of electrodes at the outer surface of the micromixer at specific locations, also two baffles are considered at prescribed distances from the inlets of the micromixer. The two electrode pairs are excited by an alternating current (AC) and can generate micro-vortices near the electrodes even for a highly ordered laminar flow, as a result of efficient mixing. In the present study, two baffles are also considered to enhance the mixing quality. Comsol Multiphysics solver has been employed to carry out the simulations. We have considered Navier–Stokes equation to know the flow field and the convection–diffusion equation to know the concentration field. The proposed model can achieve 95.61% mixing quality at the inlet velocity of 0.05 mm/s. A comparison study has also been performed between one ring-shaped micromixer with and without baffles.

Keywords Micromixing · Electroosmotic flow · Baffle · Mixing quality (MQ)

B. Gayen · N. K. Manna
Department of Mechanical Engineering, Jadavpur University, Kolkata 700032, India

N. Biswas (✉)
Department of Power Engineering, Jadavpur University, Salt Lake, Kolkata 700106, India
e-mail: biswas.nirmalendu@gmail.com

B. Gayen
Department of Mechanical Engineering, MCKV Institute of Engineering, Howrah 711204, India

S. Das et al. (eds.), *Proceedings of the 1st International Conference on Fluid, Thermal and Energy Systems*, Lecture Notes in Mechanical Engineering,
https://doi.org/10.1007/978-981-99-5990-7_1

1 Introduction

Over the last two decades, microfluidic devices such as Lab-on-a-chip (LOC) and Micro-total analysis systems (μTAS) have emerged as efficient tools in the field of medical sciences, biotechnology, and some engineering applications [1]. The dimensions of the mentioned miniaturization systems can vary from ten micrometers to a few hundred micrometers. Distinguished advantages such as low consumption of sample, quick response time, high throughput, portability, simple geometry, and ease to fabricate can be availed. Now-a-days microfluidic systems can be applied to different fields for different purposes such as blood separations [2], drug delivery devices [3], disease diagnostic [4], and micromixing [5, 6]. Among these different applications of the microfluidic device, micromixing requires in the fields like DNA hybridization [7, 8], point-of-care testing [9, 10], chemical synthesis [11], and biological analysis [12]. Electrokinetics refers to the movement of fluid/particles under the action of an externally applied electric field [13, 14]. Recently, researchers attracted by the concept of electrokinetic effects to implement on micromixing, even for a simple geometry one can achieve high mixing quality.

In general, micromixing can be classified into two categories; active micromixing and passive micromixing [6, 12, 15]. In the case of active micromixing, mixing is produced by an externally applied excitation (force). The nature of the excitation may be in the form of electric, magnetic, acoustic, or thermal. The mixing quality achieved from the micromixer is quite high as compared to passive micromixers. On the contrary, passive micromixing refers to the mixing obtained only by considering the shape of the geometry. In a microfluidic device, the hydraulic diameter is in the range of micrometers and the fluid velocity encountered here in terms of mm/s is extremely low. Therefore, the nature of flow exhibits here highly laminar type. In such a situation, it is difficult to achieve high efficient mixing quality as the mixing process primarily depends on molecular diffusion. Consequently, some effective methods should be applied by which mixing can be achieved quickly and also better mixing quality [16, 17]. To encounter this type of problem some researchers considered specific types of geometry to achieve better results in terms of mixing quality [18, 19]. Some researchers applied consequent grooves at the base of their model [20], also some applied their innovative ideas by incorporating a non-conductive plate that may be flexible or nonflexible [21].

Now-a-days, AC microelectrodes have been efficiently used in the application of micromixing. When microelectrodes are excited by an externally applied electric field it can produce electrokinetic forces such as electroosmotic and die-electrophoretic forces and due to these forces micro-vortices generate near the microelectrodes which provide a rapid mixing [22]. Also, this technique can be applied to simple geometry and is easy to fabricate. Inspired by the concept, some researchers investigated the dependency of the mentioned electrokinetic forces on different parameters, viz., AC voltage, frequency of the alternating current, and the electrical properties of fluid medium [23]. One major difficulty they experienced that the fabrication of their huge length model but they reported the mixing quality as 86.5%.

Fig. 1 Schematic diagram of model geometry (all dimensions are in μm)

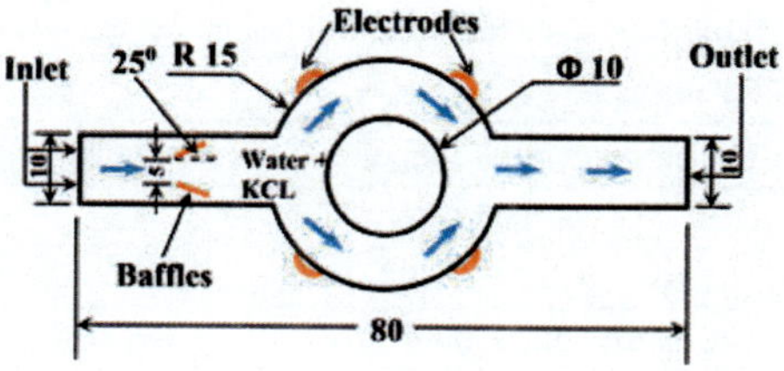

Motivated by the work in enhancing the mixing quality considering the electrokinetic effects, in our present study we have considered a novel geometry that could produce 95.61% mixing quality in 0.8 s with a set of parameters, viz., inlet fluid velocity, frequency of alternating current (AC) and maximum voltage amplitude of AC.

2 Problem Description

In Fig. 1 the novel geometry consists of two identical pairs of electrodes that are located at the angular positions of 45°, −45°, 135°, −135° and two baffles at the inlet side (keeping clearance from the upper and lower wall). The computational domain is filled with diluted KCL and water as a base fluid. The electrical properties of the fluid are disclosed in Table 1. In our present study, we have simulated our model by considering the inlet velocity of 0.05 mm/s, the voltage amplitude of 0.2 V, and the frequency of the alternating current of 4 Hz. A comparison study has also been performed between the one ring-shaped micromixer with and without the baffles. Results show that better mixing quality can be obtained with baffles.

3 Mathematical Modeling

The mathematical model of an electric field, flow field, and concentration field with boundary conditions are presented as follows.

3.1 Electric Field

We apply the Poisson equation [24] to know about the distribution of potential in the computational domain, as follows:

$$\nabla^2\varphi = \frac{\rho_e}{\epsilon\epsilon_0} \tag{1}$$

Table 1 Numerical value of different parameters used in the simulation

Parameters	Value
The permittivity of vacuum (ϵ_0)	8.854×10^{-12} C/V s
Dielectric constant (ϵ)	80.2
The viscosity of the fluid (μ)	0.9×10^{-3} kg/ms
The density of the fluid (ρ)	1000 kg/m^3
Zeta potential of the channel wall (ζ_w)	−0.1 V
Mean inflow velocity (U_0)	0.05 mm/s
The conductivity of the ionic solution (σ)	0.11846 S/m
The maximum value of the AC potential (V_0)	0.2 V
Frequency of the AC (f)	4 Hz
Start time (t)	0 s
Diffusion coefficient of the solution (D)	0.6×10^{-11} m^2/s
Initial concentration (C_0)	1 mol/m^3
Maximum concentration (C_{max})	0.5 mol/m^3

where, ϵ_0 and ϵ represent the electrical permittivity of the vacuum and dielectric constant of the medium, respectively. Moreover, φ and ρ_e demonstrate the distribution of electric potential and free electric charge density in the domain, respectively. As far away from the solid surfaces, the net electric charge density is zero ($\rho_e = 0$) therefore, the Poisson equation takes the form Laplace equation [24] as follows:

$$\nabla^2\varphi = 0 \tag{2}$$

The boundary conditions that are imposed to solve the electric field are as follows:

- Wall of the microchannel and two baffles:

$$\mathbf{n} \cdot \nabla\varphi = 0 \tag{3}$$

- Two electrode pairs:

$$V = V_0 \sin(2\pi f t) \tag{4}$$

- Outlet of the microchannel:

$$\varphi = 0 \tag{5}$$

3.2 Flow Field

We have considered continuity and Navier–Stokes equation to solve the flow field. The fluid is assumed as Newtonian and incompressible. The mentioned two equations can be represented as follows:

- Continuity equation [25]:

$$\nabla \cdot \mathbf{u} = 0 \tag{6}$$

- Navier-Stokes equation:

$$\rho\left[\frac{\partial \mathrm{u}}{\partial t} + \mathbf{u} \cdot \nabla \mathbf{u}\right] = -\nabla P + \mu \nabla^2 \mathbf{u} + \mathbf{E}\rho_e \tag{7}$$

In Eq. (7) $\mathbf{u}$ and $\mathbf{E}$ demonstrate the velocity vector and electric field vector, respectively. Furthermore, ρ and μ designate fluid mass density and fluid's dynamic viscosity, respectively, and ∇P represents the pressure gradient. The last term is on the right side in Eq. (7) representing the electrostatic body force in the thin electric double layer (EDL). The electric field and potential are related by the expression $\mathbf{E} = -\nabla \varphi$.

From the fact that the most solid surfaces acquire a surface charge when brought into contact with an electrolyte, as a result, counterions are attracted by these charged surfaces, and a thin layer is formed, where the net charge is not zero, at the vicinity of the solid surfaces of the order of nanometers known as an electric double layer (EDL). When an electric field is applied due to electroosmotic flow the charged solution displaces in the electric double layer. This scheme generates a force acting on the positively charged ions and the liquid flows in the direction of the applied electric field. The electric double layer can be expressed as follows:

$$\lambda_D = \sqrt{\frac{\epsilon \epsilon_0 k_B T}{2 n_0 (ze)^2}} \tag{8}$$

In Eq. (8) k_B, T, n_0, z, and e demonstrate the Boltzmann constant, absolute temperature, free ionic charge density, valence, and elementary charge, respectively. In our present study, we have neglected the last term $\mathbf{E}\rho_e$ in Eq. (7) outside the electric double layer (EDL) and considered Helmholtz-Smoluchowski slip velocity boundary condition expressed in Eq. (9) for simulating the electroosmotic flow outside EDL.

$$\mathbf{u} = -\frac{\epsilon \epsilon_0 \zeta_w}{\mu}\mathbf{E} \tag{9}$$

where ζ_w represents the zeta potential at the channel wall.

The boundary conditions that are imposed to solve the flow field are as follows [27]:

- Inlet of the microchannel: Fully developed flow.
- Outlet of the microchannel: $P = P_a$ (local atmospheric pressure) with suppressing backflow.
- Wall of the microchannel: Electroosmotic slip velocity.
- Wall of the baffles: No-slip condition.

3.3 Concentration Field

The Convection–Diffusion equation [25] is:

$$\frac{\partial C_i}{\partial t} + u \cdot \nabla C_i = D_i \nabla^2 C_i \tag{10}$$

where, C_i and D_i demonstrate the concentration and diffusion coefficient of ith fluid, respectively. To solve the Eq. (10) we have applied the boundary conditions as follows:

- Inlets of the microchannel: $C = 1$ mol/m^3 (for upper half) and $C = 0$ (for lower half)
- Wall of the microchannel and baffles:

$$\frac{\partial C_i}{\partial t} = 0 \ (t \geq 0) \tag{11}$$

3.4 Mixing Quality Factor

To verify the performance of our proposed model, we have defined a Mixing Quality (MQ) factor in Eq. (12).

$$MQ = 1 - \sqrt{\frac{\sigma^2_{\text{mean}}}{\sigma^2_{\text{max}}}} \tag{12}$$

$$\sigma^2_{\text{mean}} = \int (C_i - C_{\text{max}})^2 dy \tag{13}$$

In Eq. (13) the integration has been performed at the outlet of the microchannel. Whereas, C_{max} (=0.5 mol/m^3) is the ideal concentration after mixing at the outlet, and σ_{max} is the maximum variation in the mixture. According to Eq. (12), a value of 1 of MQ signifies 100% mixing quality and a value of 0 indicates 0% mixing quality.

4 Methodology

Comsol Multiphysics® 5.6 software [26] has been considered to perform the simulations of our proposed computational model domain. The mentioned software utilizes its solver, the Backward Euler method to carry out the simulation. In the current simulation, we have considered triangular elements. Discretization of fluid has been performed by considering $P_2 + P_1$ efficient elements. PARDISO linear solver has been selected to solve the fluid flow variables. PARDISO linear solver can reuse the factorization of the nonsymmetric type system matrix, which is solved for. As a result, the mentioned solver can solve faster than other available solvers. The constant (Newton) method has been imposed to take care of nonlinearity terms and is considered a method of termination. The tolerance factor has been imposed as the criterion of termination. The absolute tolerance is considered 1×10^{-5} for the convergence. The simulation workflow has been disclosed as follows.

4.1 *Model Physics*

To test our proposed model, three physics, viz., electric current, laminar flow, and transport of diluted species have been selected. Thereafter, three physics have been coupled by imposing proper boundary conditions. Coupling between the different physics has been performed thoroughly to modify the mixing analysis in our proposed model.

4.2 *Model Design*

In Fig. 1 one ring-shaped micromixer with two baffles at the inlet side has been depicted. The two corners at the leading edge are separated by a distance of 5 μm and maintain a distance of 13 μm from the inlets. The rotation angle has been considered 25°. Two fluids with different concentrations of 1 mol/m^3 and 0 mol/m^3 enter the upper half and lower half, respectively. In the present study, we have considered one outlet to achieve the desired concentration of 0.5 mol/m^3.

4.3 *Model Parameters*

To simulate our proposed model we have considered different parameters. The numerical values are presented in Table 1. The different parameters govern the fluid flow variables, fluid compositions, electrical properties, and physiochemical properties of the fluid.

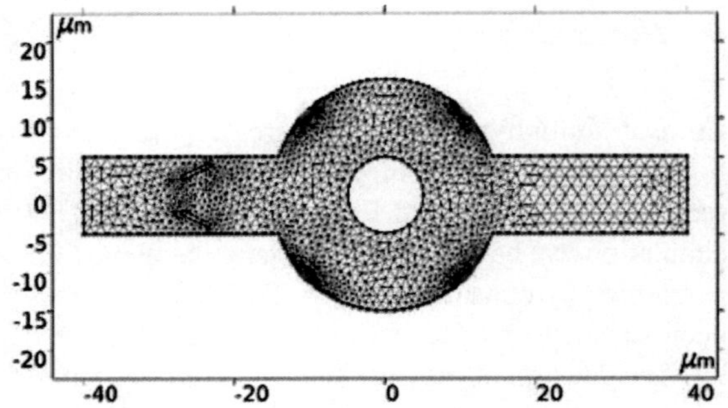

Fig. 2 Triangular meshing system

Table 2 Mesh quality in the computational domain

Number of elements	7424
Average mesh quality	0.8975
Minimum element quality	6024
Element area ratio	0.02462
Mesh area	1130 μm^2

4.4 *Model Meshing*

Free triangular mesh elements with predefined extremely fine mesh have been considered for the entire computational domain and have been shown in Fig. 2. Mesh refinement for triangular elements has been performed by considering different element size parameters such as maximum element size 0.2, minimum element size 0.006, and maximum element growth rate 1.1. The average mesh quality for the entire 2D geometry environment is 0.8975 which is quite acceptable for getting accurate results. The details of mesh quality are presented in Table 2.

4.5 *Time-Dependent Approach*

In our present simulation, we have simulated our proposed model in two steps. First, we have determined the voltage amplitude of the electric current considering a stationary study. After that, we computed the flow field and concentration field by considering the transient steps. At the start time ($t = 0$) of the transient step, the initial state is the solution of the stationary study. This technique considerably reduces the simulation time as well as the memory requirement. Besides, we apply a step function for smoothing the concentration differences and also we disregard the discontinuity of the Convection–Diffusion equation. Because most biological molecules are highly fragile and unstable in an inappropriate environment and the threshold value can be considered 0.1 s. Therefore, it is quite reasonable to consider simulation time 0.8 s in our present study.

5 Results and Discussion

Figures 3 and 4 show instantaneous streamline velocity patterns at $t = 0.0375$ s and surface velocity magnitude at $t = 0.43542$ s, respectively. From the figures, it has been seen that near the four electrodes two micro-vortices are generated due to electroosmotic flow.

In the present study, as we have applied an AC electric field, the resulting electroosmotic flow perturbs the highly ordered laminar flow, such that it continuously pushes up and down two different concentration fluids from the two opposite directions at the beginning of the mixing chamber, causing extensive stretching and folding of the fluid elements. Effective mixing mostly relies on vigorously stirring the fluid elements due to electroosmotic flow and a small contribution to diffusion. In addition to this consideration, two baffles at the inlet side act as a diffuser and there will be a pressure drop and can generate vortices, which enhance the mixing.

The mixing phenomenon is further exemplified by considering Figs. 5 and 6. From Fig. 5 it can be seen that the two fluid streams are well separated also at the outlet when no AC field is applied. Figure 6 represents the concentration field when an AC field is applied at a time, $t = 0.46875$ s. The mixing increases considerably owing to the alternating swirling in the flow.

It should be mentioned that the diagram presented in Fig. 7 is the result of integration at the outlet boundary of the criterion, mixing quality as presented in Eq. (12), derived from the concept of square deviation from the desired concentration (0.5 mol/

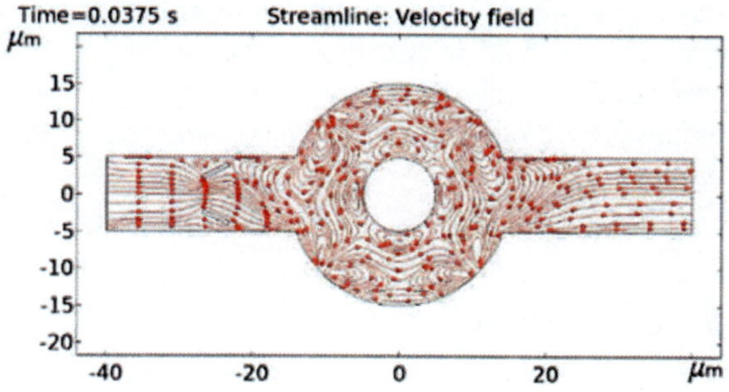

Fig. 3 Streamline velocity field

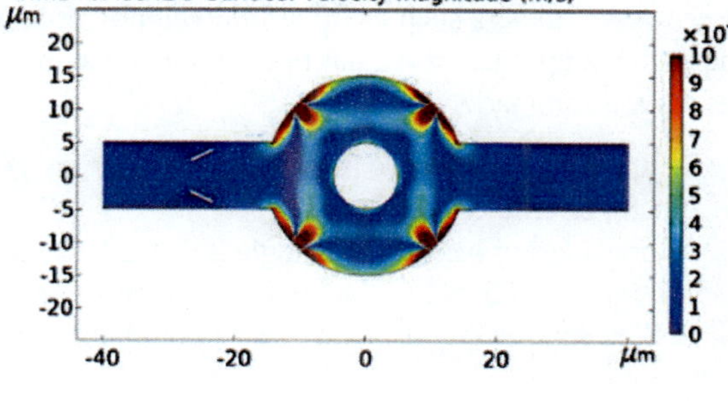

Fig. 4 Surface velocity field

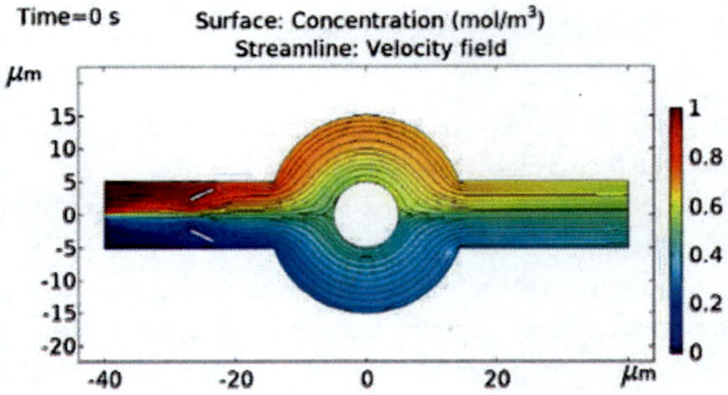

Fig. 5 Concentration field at $t = 0$ s

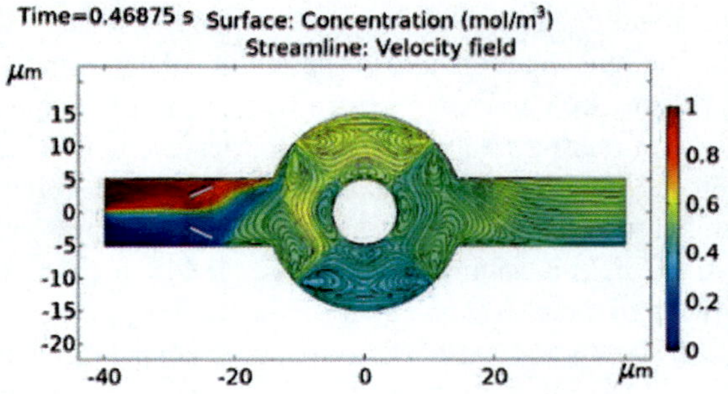

Fig. 6 Concentration field at $t = 0.46875$ s

m^3). Initially, the mixing quality is not good as can be depicted in Fig. 7. For example, more fluid with a concentration of 1 mol/m^3 passes across the outlet boundary as compared to fluid with 0 mol/m^3 concentration. Moreover, we have performed a comparative study of mixing quality between one ring-shaped with and without baffles geometries. Figure 7 depicts an increase in mixing quality by considering the baffles.

6 Conclusions

In this study, an electroosmotic micromixer in one ring-shaped microchamber has been simulated. The mixing quality was further investigated by considering baffles. A set of parameters such as inlet fluid velocity, frequency of AC electric field, and maximum amplitude of AC has been employed to investigate the mixing quality. The results are summarized as follows:

- By applying AC electric field micro-vortices are generated near the electrodes as a result of efficient mixing.
- By considering baffles at an angle 25°, the MQ can further be increased.
- For $U_0 = 0.05\,\text{mm/s}$, $V_0 = 0.2\,\text{V}$, and $f = 4\,\text{Hz}$ we achieved 95.61% and 94.85% mixing quality with and without baffles, respectively.

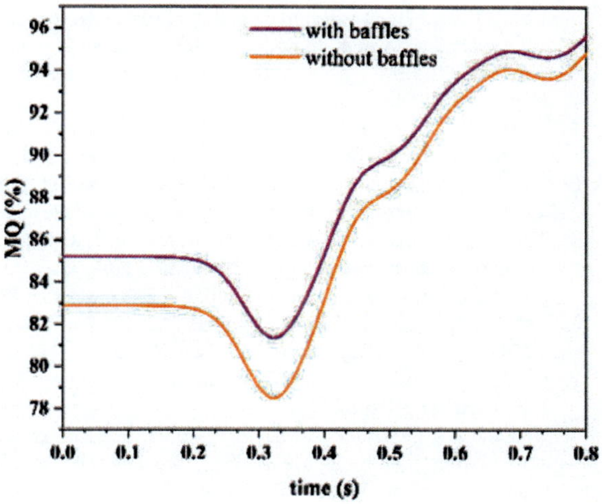

Fig. 7 Mixing quality for two different geometries

Nomenclature

$\boldsymbol{u}$	Velocity vector (m/s)
$\boldsymbol{E}$	Electric field vector (V/m)
P	Pressure (Pa)
V	Potential (V)
ρ	Density of fluid (kg/m^3)
φ	Electric potential distribution (V)
ρ_e	Free charge density (C/m^3)
μ	Dynamic viscosity of the fluid (Pa s)
ϵ_0	Permittivity of vacuum (F/m)
ζ_w	Zeta potential at the channel wall (V)

References

1. Cheri MS, Latifi H, Moghaddam MS, Shahraki H (2013) Simulation and experimental investigation of planar micromixer with short-mixing length. Chem Eng J 234:247–255
2. Di Carlo D, Irimia D, Tompkins RG, Toner M (2007) Continuous inertial focusing, ordering, and separation of particles in microchannels. Proc Natl Acad Sci 104:18892–18897
3. Chen Y, Gao D, Wang Y, Lin S, Jiang Y (2018) A novel 3D breast-cancer-on-chip platform for therapeutic evaluation of drug delivery systems. Anal Chim Acta
4. Su W, Gao X, Jiang L, Qin J (2015) Microfluidic platform towards point-of-care diagnostics in infectious diseases. J Chromatogr A 1377:13–26
5. Ryu S-P, Park J-Y, Han S-Y (2011) Optimum design of an active micro-mixer using successive Kriging method. Int J Precis Eng Manuf 12:849–855
6. Bayareh M, Ashani MN, Usefian A (2020) Active and passive micromixers: a comprehensive review. Chem Eng Process Intensif 147:107771

7. Kang D-K et al (2014) Rapid detection of single bacteria in unprocessed blood using integrated comprehensive droplet digital detection. Nat Commun 5:5427
8. Tian J et al (2004) Accurate multiplex gene synthesis from programmable DNA microchips. Nature 432(7020):1050–1054
9. Alijani H, Özbey A, Karimzadehkhouei M, Kosar A (2019) Inertial micromixing in curved serpentine micromixers with different curve angles. Fluids 4:204
10. Shi H, Nie K, Dong B, Chao L, Gao F, Ma M, Long M, Liu Z (2020) Mixing enhancement via a serpentine micromixer for real-time activation of carboxyl. Chem Eng J 392:123642
11. Thiele M, Knauer A, Malsch D, Csáki A, Henkel T, Köhler JM, Fritzsche W (2017) Combination of microfluidic high-throughput production and parameter screening for efficient shaping of gold nanocubes using Dean-flow mixing. Lab Chip 17:1487–1495
12. Lee TY, Hyun K-A, Kim S-I, Jung H-I (2017) An integrated microfluidic chip for one-step isolation of circulating tumor cells. Sens Actuators B Chem 238:1144–1150
13. Chang CC, Yang RJ (2007) Electrokinetic mixing in microfluidic systems. Microfluid Nanofluid 3(5):501–525
14. Delgado AV, Gonzalez-Caballero F, Hunter RJ, Koopal LK, Lyklema J (2007) Measurement and interpretation of electrokinetic phenomena. J Colloid Interface Sci 309(2):194–224
15. Capretto L, Cheng W, Hill M, Zhang X (2011) Micromixing within microfluidic devices. In: Microfluidics. Springer, pp 27–68
16. Ahmed D, Mao X, Shi J et al (2009) A millisecond micromixer via single bubble-based acoustic streaming. Lab Chip 9(18):2738–2742
17. Li J, Kleinstreuer C (2009) Microfluidics analysis of nanoparticle mixing in a microchannel system. Microfluid Nanofluid 6(6):661–668
18. Okuducu MB, Aral MM (2019) Novel 3–D T–shaped passive micromixer design with helicoidal flows. Processes 7(9):637
19. Stroock AD, Dertinger SK, Whitesides GM, Ajdari A (2002) Patterning flows using grooved surfaces. Anal Chem 74:5306–5312
20. Stroock AD, Dertinger SK, Ajdari A, Mezic I, Stone HA, Whitesides GM (2002) Chaotic mixer for microchannels. Science 295:647–651
21. Cho C-C (2007) Electrokinetically-driven flow mixing in microchannels with the wavy surface. J Colloid Interface Sci 312:470–480
22. Biddiss E, Erickson D, Li D (2004) Heterogeneous surface charge enhanced micromixing for electrokinetic flows. Anal Chem 76:3208–3213
23. Green NG, Ramos A, González A, Morgan H, Castellanos A (2000) Fluid flow induced by nonuniform ac electric fields in electrolytes on micromicroelectrodes. I. Experimental measurements. Phys Rev E 61 4011
24. Kazemi Z, Rashidi S, Esfahani JA (2017) Effect of flap installation on improving the homogeneity of the mixture in an induced-charge electrokinetic micro-mixer. Chem Eng Process Process Intensif 121:188–197
25. Wu Z, Li D (2008) Micromixing using induced-charge electrokinetic flow. Electrochim Acta 53(19):5827–5835
26. Yoshimura M, Shimoyama K, Misaka T, Obayashi S (2019) Optimization of passive grooved micromixers based on genetic algorithm and graph theory. Microfluid Nanofluid 23(3):1–21
27. Gayen B, Manna NK, Biswas N, Chakrabarty A (2022) Assessment of micro-mixing process of electroosmotic flow in a microchannel. Adv Mech Eng 247–256. ISBN: 978-81-952903-6-9

MHD Nanofluidic Convective Behavior in a Hexagonal-Shaped Thermal System

Husain Tamim Bamboowala, Niloy Sen, Soutrik Nag, Nirmal K. Manna, Nirmalendu Biswas, and Dipak Kumar Mandal

Abstract Nanofluidic thermal flow characteristics of magnetohydrodynamic convection in a complex hexagonal cavity-like thermal system are investigated numerically utilizing CuO-water nanofluid. The bottom portion of the complex geometry is partially heated to a uniform temperature T_h while the two inclined upper sidewalls of the cavity are kept at a lower temperature T_c. The rest portion of the enclosure is insulated. In order to analyze the heat flow dynamics within the geometry, the finite element method-based solver is utilized to solve the transport equations. The obtained results are illustrated using local as well as global quantities. The study is conducted for the appropriate range of involved parameters, Rayleigh number ($10^3 \leq \text{Ra} \leq 10^5$), Hartmann number ($0 \leq \text{Ha} \leq 70$), and angle of the magnetic field ($0 \leq \gamma \leq 70$). Furthermore, the heat flow dynamics from the heater to the cooler are scrutinized using the heatlines. The flow structures and temperature characteristics reveal many interesting distribution patterns. Nanofluidic thermal-fluid phenomena are heavily influenced by the Rayleigh and Hartmann number variations with the appearance of multi-vortex flow structures.

Keywords Hexagonal cavity · Differential heating · Heat transfer · Magneto-hydrodynamic (MHD) flow · Nanofluid

H. T. Bamboowala · N. Sen · S. Nag · N. K. Manna
Department of Mechanical Engineering, Jadavpur University, Kolkata 700032, India

N. Biswas (✉)
Department of Power Engineering, Jadavpur University, Salt Lake, Kolkata 700106, India
e-mail: biswas.nirmalendu@gmail.com

D. K. Mandal
Department of Mechanical Engineering, Government Engineering College, Samastipur 848127, India

S. Das et al. (eds.), *Proceedings of the 1st International Conference on Fluid, Thermal and Energy Systems*, Lecture Notes in Mechanical Engineering,
https://doi.org/10.1007/978-981-99-5990-7_2

1 Introduction

In recent studies, the thermo-hydraulic phenomenon' pertaining to the buoyant convective flow in different geometrical enclosures has been widely investigated owing to its diverse industrial solicitations such as heat exchanging devices, micromixing, microelectronic equipment cooling, material processing, etc. The use of nanofluid in a carrier fluid improves the base heat transfer rate, which helps improve the performance and efficiency of industrial systems. The analysis of free-convective heat transfer for various geometries has become a matter of considerable attention worldwide and some have been reflected in the present literature review. Free and forced convective processes and heat transfer in cavities have prospective applications in technological fields such as room ventilation, heat exchangers, electronic device cooling, food processing, solar collectors and ponds, and so on.

Hashemi-Tilehnoee et al. [1] investigated the buoyant convection flow in a permeable cavity containing Al_2O_3-H_2O nanofluid. In this study, it was discovered that the heat transfer and irreversibility increase with the growth of the Rayleigh number and Darcy number. However, they diminish as the Hartmann number increases. Mahapatra et al. [2] studied the free-convective process in a partially heated enclosure regarding entropy generation. The impact of active heater-cooler walls location on the heat flow dynamics and irreversibility production has been thoroughly examined, taking into account six different factors. Izadi et al. [3] numerically studied the impact of geometrical as well as flow controlling parameters such as cavity and heat source aspect ratio and Rayleigh number inside a C-shaped enclosure. Flow structures and thermal aspects in a square cavity comprising a protruded heater body are inspected by Biswas et al. [4] for two common fluids numerically for different Rayleigh numbers. MHD transport phenomena in a differentially heated double-driven enclosure were studied by Sen et al. [5] using the finite Galerkin method to solve coupled governing equations in an analytical way. MHD mixed thermal convection in Cu-water-based nanofluidic flow was studied by Mandal et al. [6]. Other studies [7–9] on MHD free-convective phenomenon in a porous substance and nanofluid packed geometry and an applied magnetic field have been consulted. Several researchers have analyzed the thermo-convective dynamics in a complex geometry like hexagonal-shaped cavity with or with various shaped inner obstacles under multi-physical scenarios [10–14]. Seyyedi et al. [15] studied the buoyant convective phenomenon and irreversibility production in a hexagon-shaped geometry containing Cu-H_2O nanoliquid under the influence of an oriented magnetic field. All these studies have stated that thermal behavior is markedly influenced by the thermal boundary conditions. Biswas et al. [7, 16] and Nag et al. [17] studied different variations in thermal convective systems for varied geometries. However, still, there are many unexplored areas of thermo-flow physics in such complex geometry, which motivates us to fundamentally study the hexagonal cavity under the multi-physical scenarios.

In light of the above-mentioned literature survey, we have considered a regular hexagonal enclosure heated partly from the lower walls and cooled partially from the upper-oriented sidewalls. The cavity is packed with CuO-water nanofluid under the influence of the oriented magnetic field. The evolved heat flow dynamics and flow field will be very useful in the engineering fields. Therefore, the proposed problem is scrutinized extensively for controlling variables like Rayleigh number (Ra), Hartmann number (Ha), magnetic field inclination (γ), and nanoparticle concentrations (φ). The obtained outcomes are demonstrated by streamlines, isotherms, heatlines, and average Nusselt numbers.

2 Mathematical Modeling

A differentially heated regular hexagon is considered in a two-dimensional plane assuming unit depth. The diameter of the circumscribing circle of the hexagon is taken to be unity to simplify the geometry, thus making the side length 0.5. The bottom wall and half lengths of both lower inclined walls are kept constant at an ambient temperature of Th, thus making the heated length one. Similarly, the inclined walls on either side of the top wall, as depicted in Fig. 1, are maintained at cold temperature T_c, also intentionally keeping the cooling length to be one, i.e., identical to the heating length. The remaining walls of the cavity are adiabatic in nature.

The hexagonal cavity is filled with water which is infused with CuO nanoparticles. The concentration of this CuO-water nanofluid is variable, and a subject of our study. The hexagonal cavity is subjected to a uniform intensity (B) based oriented (at an angle, γ) magnetizing field. We vary the orientation of this uniform magnetic field from 0 to 90 degrees and study the effects of this variation. The magnitude of magnetic field strength is also different using the Hartmann number (Ha).

Except for the fluid density, the thermophysical properties of CuO-water nanofluid are regarded to be fixed using the Boussinesq technique. Furthermore, the nanofluid flow is supposed to be 2D and incompressible, making it steady, laminar, and Newtonian. The walls are non-permeable and adhere to a no-slip criterion, which means that all components of velocity at the wall must be ignored. We also ignore the loss of heat from the walls owing to radiative heat transfer.

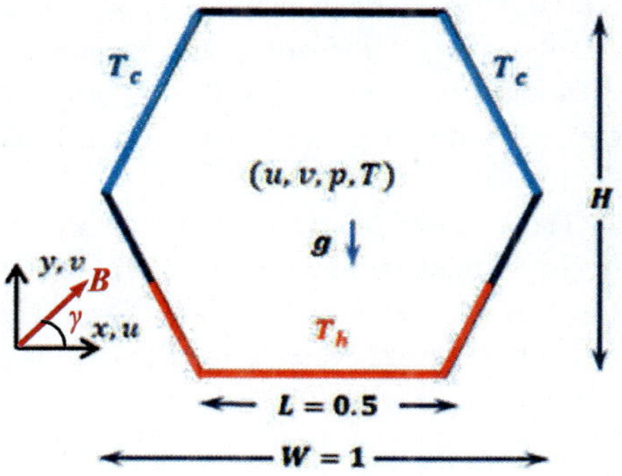

Fig. 1 Schematic geometry of the hexagonal cavity flow model

The Boussinesq approximation addresses the Buoyancy force (which in turn includes gravity force). Due to the limitation of this study to low Hartmann Number values, only Lorentz force is sufficient to be taken into account. For the purpose of applying the single-phase model, we need to limit the concentration of the nanofluid to a 3% concentration of CuO nanoparticles. Thus, the conservation equations in the dimensionless form are

$$\frac{\partial U}{\partial X}+\frac{\partial V}{\partial Y}=0 \tag{1}$$

$$U\frac{\partial U}{\partial X}+V\frac{\partial U}{\partial Y}=-\frac{\rho_f}{\rho_{nf}}\frac{\partial P}{\partial X}+\frac{\nu_{nf}}{\nu_f}\mathrm{Pr}\left(\frac{\partial^2 U}{\partial X^2}+\frac{\partial^2 U}{\partial Y^2}\right) \tag{2}$$

$$U\frac{\partial V}{\partial X}+V\frac{\partial V}{\partial Y}=-\frac{\rho_f}{\rho_{nf}}\frac{\partial P}{\partial Y}+\frac{\nu_{nf}}{\nu_f}\mathrm{Pr}\left(\frac{\partial^2 V}{\partial X^2}+\frac{\partial^2 V}{\partial Y^2}\right) - \frac{\sigma_{nf}}{\sigma_f}\frac{\rho_f}{\rho_{nf}}\mathrm{PrHa}^2 V+\frac{(\rho\beta)_{nf}}{\rho_{nf}\beta_f}\mathrm{RaPr}\theta \tag{3}$$

$$U\frac{\partial \theta}{\partial X}+V\frac{\partial \theta}{\partial Y}=\frac{\alpha_{nf}}{\alpha_f}\left(\frac{\partial^2 \theta}{\partial X^2}+\frac{\partial^2 \theta}{\partial Y^2}\right) \tag{4}$$

Here, the notation '*nf*' indicates the property of the nanoliquid. In fact, the basic properties like *u*, *v*, *p*, *T* transformed into nondimensional variables introducing the scale factors as:

$$(X,Y)=(x,y)/H,\ (U,V)=H(u,v)/\alpha_f$$

$$\theta=(T-T_c)/(T_h-T_c),\ P=pH^2/\rho_f\alpha_f^2 \tag{5a}$$

$$\mathrm{Pr}=\frac{\nu_f}{\alpha_f},\ \mathrm{Ra}=\beta_f H^3(T_h-T_c)\frac{g}{\nu_f\alpha_f},\ \mathrm{Ha}=BH\sqrt{\frac{\sigma_f}{\mu_f}} \tag{5b}$$

Here, the dimensionless numbers are Prandtl (Pr), Rayleigh (Ra), and Hartmann (Ha) numbers. The dynamic (μ) and kinematic (ν) viscosity, electrical conductivity (σ), density (ρ), diffusivity, and expansion coefficient (β) are expressed. The actual thermodynamic properties of CuO-water (through Pr = 5.83) are taken from using the standard relationships (as listed in Table 1), which are available in the literature [7, 16–19].

The imposed boundary circumstances are $\theta = 0$ for the cold wall, $\theta = 1$ for hot walls, normal gradient $\partial\theta/\partial n = 0$ for the rest portion of the walls (adiabatic), and U, $V = 0$ for all the bounding walls. Evaluating the global heat transfer over the active walls, the average Nusselt number (Nu) is considered as

Table 1 Properties of CuO-water nanofluid

Properties	H_2O	CuO
ρ (kg/m^3)	997.1	6500
C_p (J/kg K)	4179	540
k (W/m K)	0.631	18
dp (nm)	–	29
$\beta \times 10^{-5}$ (1/K)	21	0.85
σ (Ω m)$^{-1}$	0.05	2.7×10^{-8}

$$\text{Nu} = k_r \int_0^1 \left(-\frac{\partial \theta}{\partial X} |_{X=0} \right) dY \tag{6}$$

Furthermore, the heat flow dynamics are evaluated using the heatlines contours, which is obtained by solving the heat function equations as expressed by Biswas et al. [4].

$$-\frac{\partial \Pi}{\partial X} = V\theta - \frac{k}{k_f}\frac{\partial \theta}{\partial Y} \text{ and } \frac{\partial \Pi}{\partial Y} = U\theta - \frac{k}{k_f}\frac{\partial \theta}{\partial X} \tag{7}$$

3 Solution Methodology

The dimensionless transport Eqs. (1)–(4) are solved using the finite element [19] based computing solver. The solution domain is structured with a suitable grid distribution. Therefore, the discretized equations are solved iteratively minimizing residuals for achieving the converged results (residuals $<10^{-8}$). Finally, the solved data for temperature and velocity fields are utilized to generate thermo-fluid flow structures. The same solver has already been utilized during the validation of several works [20]. In fact, before conducting the extensive simulation proper mesh structure checking for this problem has been tested through a grid independence study. All these details are not included for brevity.

4 Results and Discussion

This work aims to explore the heat flow dynamics in a complex hexagonal-shaped cavity, occupied by CuO-water nanofluid, partially bottom heated and partially cooled from the upper sidewalls under the influence of magnetic field at varying orientation. The involved flow controlling parameters are Rayleigh number (Ra = 10^3–10^5), Hartmann number (Ha = 0–70), and magnetic field orientation (0°–90°).

The outcome of the results is illustrated using streamlines, isotherms, heatlines, and average Nusselt numbers.

4.1 Influence of Rayleigh Number (Ra)

Figure 2 presents the effects of Rayleigh number, Ra, (10^3, 10^4, 10^5) and the magnetic field orientation (0°–90°) of the differentially heated hexagonal cavity on the streamline behavior in the flow domain of the enclosure assuming the Hartmann number and nanofluid concentration as 30 and 1% respectively. The heated plume rises up due to density differences and is cooled after impacting the top wall owing to the base and surrounding walls' isothermal heating and chilling at the upper inclined walls. This increase in density results in the plume sinking. Due to the hexagonal cavity's symmetry, two oppositely oriented vortices emerge. The increase in Ra clearly reflects the shift in the streamlining pattern.

The rise of the Rayleigh Number visually represents an enhancement of the magnitude of the convective currents due to the augmentation of the heat transfer. At the lower Ra $= 10^3$, due to the low flow strength, the nanofluid flow is mostly governed by thermal conduction. Higher numbers Ra ($>10^3$) cause an increase in flow velocity, which increases convective heat transfer while also reducing the tiny circulation at the bottom right. The influence of convective heat transfer is clearly visible in the main circulation that forms within the cavity. Upon increasing Ra to 10^4 and 10^5, the increase in velocity causes better thermal convection which in turn results in the vortices being stretched toward the periphery of the cavity. Significant flow strength is revealed with a rising Rayleigh Number due to a surge in buoyancy forces. A smaller recirculation zone is visible due to the asymmetry caused by magnetic field orientation, but it vanishes completely when the angle of orientation is 90°, as shown in Fig. 2. Changes in magnetic field orientation cause the nanofluid flow vortices to reorient visually. The angular variation, when combined with Rayleigh number increments, can be used to intelligently modulate the magnitude and orientation of nanofluid fluid flow in confined cavities.

Figure 2 depicts the corresponding isotherm plots. The isotherm lines are equidistant and mostly directed along the cavity's periphery at a lesser value of Ra ($=10^3$). The conduction mode governs heat transfer in this case. When Ra hits 10^5, the isotherm lines along the cavity's active walls become more concentrated. As the Rayleigh number approaches 10^5, the isotherm lines get increasingly warped. This happens for the formation of a thermal boundary layer (which is thin) and higher temperature slope, that in turn transfers a higher amount of heat energy from the source to the sink surface. Thus, when Ra rises, the natural convective mechanism of heat transmission takes precedence over the conductive mode of heat transfer.

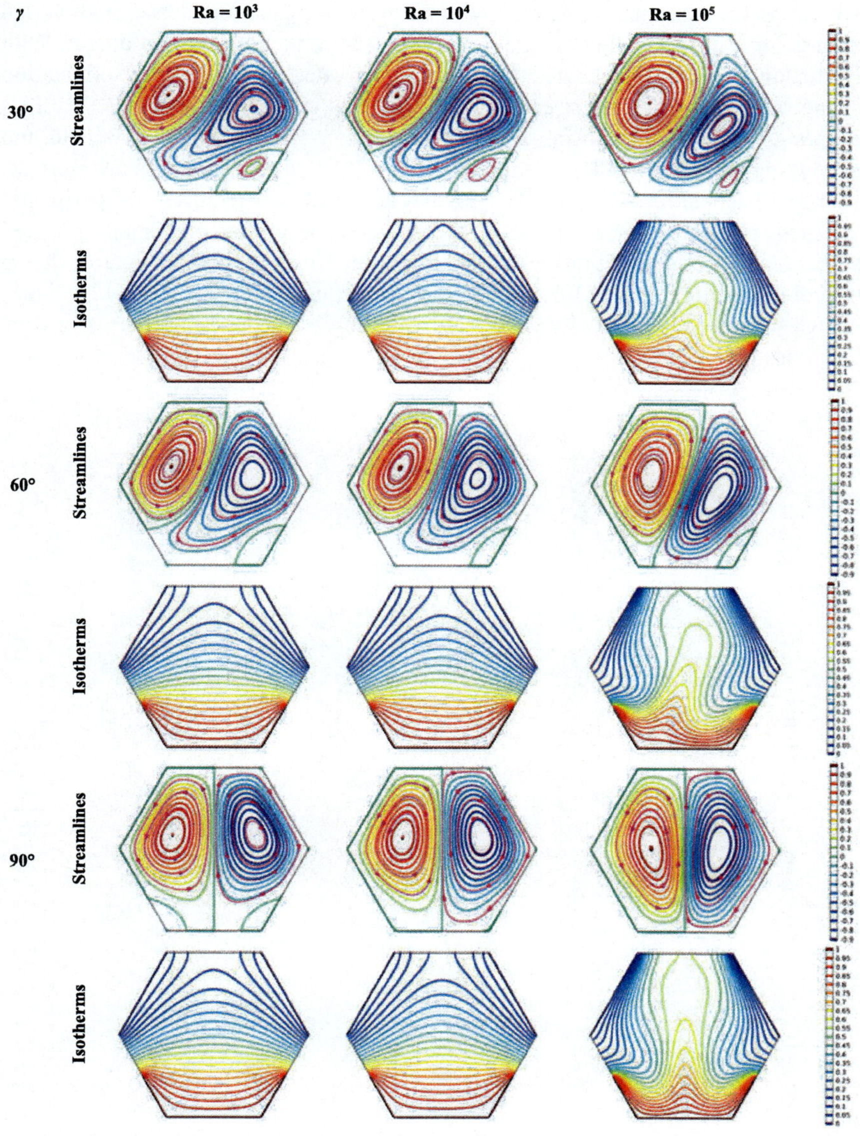

Fig. 2 Streamlines and isotherms patterns for Ha = 30 and $\varphi = 1\%$

4.2 Effect of Hartmann Number

Figure 3 demonstrates how the streamline and isotherm plots are affected by the Hartmann number (Ha) = 10, 70. The shown streamlines are double circulation and concentrated in the core and periphery regions of the hexagonal cavity when Ha =

10, but as Ha increases to 70, they become more stretched and dense, with varied orientations. In fact, a tiny circulation also exists in the bottom-right corner. With Hartmann number increment to 70, the flow velocity decreases along with orientation change and waviness in the streamlines, which is not visible in case of Ha = 10. This occurs as a result of opposing Lorentz forces interfering with the flow field and attempting to reduce the flow strength. As a result, circulation strength deteriorates.

Now for isotherms, when the Ha is 10, they are mostly clustered near the periphery. When the Ha value is large, such as 70, the isotherms become smoother and stretched as the intensity of convection current decreases and heat transport is impaired due to the counteracting effect of Lorentz force. As a result of the magnetic field's dampening effect, the flow velocity decreases, resulting in a decrease in thermal energy transport.

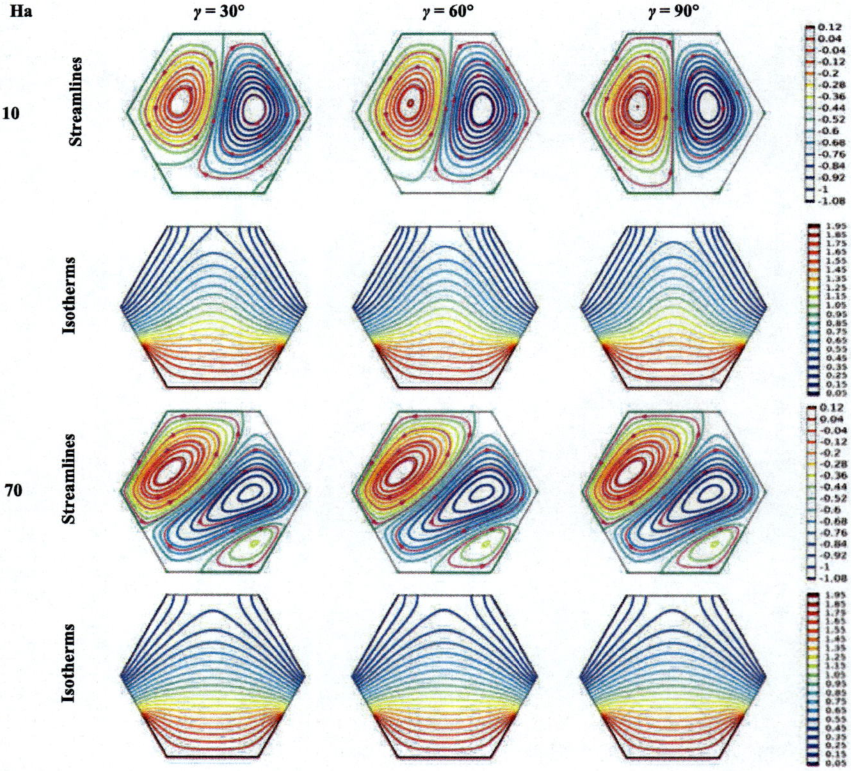

Fig. 3 Streamlines and isotherm patterns for Ra $= 10^4$ and $\varphi = 1\%$

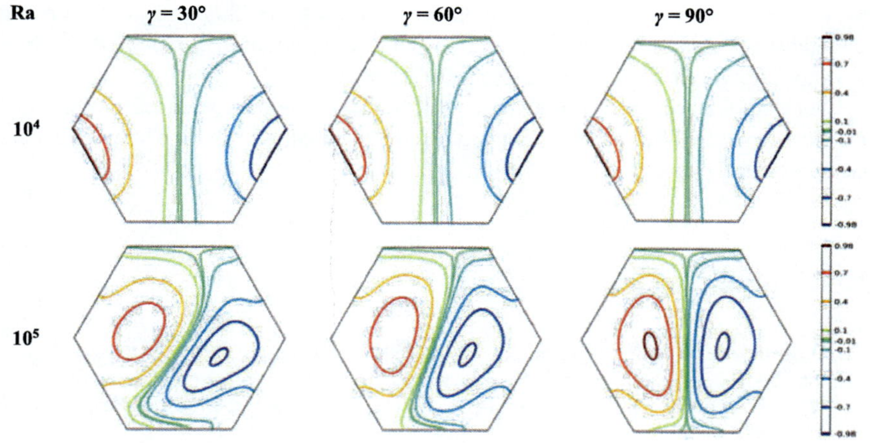

Fig. 4 Plot of heatlines at Ha = 10

4.3 Heatline Visualization

In this section, Fig. 4 illustrates the heat flow dynamics using heatlines contours. Basically, heatlines begin at the bottom heater and terminate at the sink cold walls. In general, the width of the heatlines signifies the intense heat energy transport. At the lower Ra values width of the heatlines corridors is wider, whereas at the higher Ra values thinner corridor of the heatlines is observed, which corresponds to the higher intensity of the heat energy transportation. In fact, at the lower inclination (<90°) and higher Ra values, asymmetric features of the contours are observed. This is due to the unstable convection patterns in the cavity. However, at $\gamma = 90°$ contours become symmetric patterns.

5 Conclusions

The current study focuses to scrutinize the results of partially heated walls of a regular hexagonal enclosure that is undergoing a buoyancy-driven flow induced by the magnetic field on CuO nanoparticles. The parameters under scrutiny are Rayleigh number, Hartmann number, CuO Nanofluid concentration (φ), and angle of inclination (γ) of the hexagonal enclosure.

The outcome of the results clearly illustrates that the heat transfer is dictated by the Rayleigh and Hartmann Numbers. Strong convective heat currents enhance the overall heat transfer in the Hexagonal cavity, directly as a consequence of increasing the Rayleigh number. On the contrary, a rise in Hartmann number thwarts the buoyancy forces that support natural convection by growing the Lorentz force. Consequently, both the intensity of streamlines and heat transfer decrease.

With the superior thermal conductivity of the carrier fluid, the inclusion of nanoparticles always improves heat transfer. The concentration of nanoparticles has a minor effect on velocity and temperature changes.

Our data suggest that variations in heat transfer characteristics are largely due to alterations in flow regime and associated features when the aforementioned components are changed.

References

1. Hashemi-Tilehnoee M, Dogonchi AS, Seyyedi SM, Chamkha AJ, Ganji DD (2020) Magnetohydrodynamic natural convection and entropy generation analyses inside a nanofluid-filled incinerator-shaped porous cavity with wavy heater block. J Therm Anal Calorim 141:2033–2045
2. Mahapatra PS, Manna NK, Ghosh K (2015) Effect of active wall location in a partially heated enclosure. Int Commun Heat Mass Transf 61:69–77
3. Izadi M, Mohebbi R, Chamkha AJ, Pop I (2018) Effects of cavity and heat source aspect ratios on natural convection of a nanofluid in a C-shaped cavity using Lattice Boltzmann method. Int J Numer Methods Heat Fluid Flow 28(8):1930–1955
4. Biswas N, Mahapatra PS, Manna NK (2016) Buoyancy-driven fluid and energy flow in protruded heater enclosure. Meccanica 51(9):2159–2184
5. Sen K, Manna NK, Biswas N (2021) Thermo-fluidic transport process in a double-driven cavity with triangular adiabatic obstacles. Mater Today Proc 52(3):524–531
6. Mandal DK, Biswas N, Manna NK, Gorla RSR, Chamkha AJ (2021) Role of surface undulation during mixed bioconvective nanofluid flow in porous media in presence of oxytactic bacteria and magnetic fields. Int J Mech Sci 211:106778
7. Biswas N, Manna NK, Chamkha AJ, Mandal DK (2021) Effect of surface waviness on MHD thermo-gravitational convection of $Cu-Al_2O_3-$water hybrid nanofluid in a porous oblique enclosure. Phys Scr 96:105002
8. Biswas N, Sarkar UK, Chamkha AJ, Manna NK (2021) Magneto-hydrodynamic thermal convection of Cu–Al_2O_3/water hybrid nanofluid saturated with porous media subjected to half-sinusoidal nonuniform heating. J Therm Anal Calorim 143:1727–1753
9. Khan ZH, Khan WA, Sheremet MA, Hamid M, Du M (2021) Irreversibilities in natural convection inside a right-angled trapezoidal cavity with sinusoidal wall temperature. Phys Fluids 33(8):083612
10. Ghalambaz M, Sabour M, Sazgara S, Pop I, Trâmbiţaş R (2020) Insight into the dynamics of ferrohydrodynamic (FHD) and magnetohydrodynamic (MHD) nanofluids inside a hexagonal cavity in the presence of a non-uniform magnetic field. J Magn Magn Mater 497:166024
11. Haq RU, Soomro FA, Wang X, Tlili I (2020) Partially heated lid-driven flow in a hexagonal cavity with inner circular obstacle via FEM. Int Commun Heat Mass Transf 117:104732
12. Acharya N, Chamkha AJ (2022) On the magnetohydrodynamic Al_2O_3-water nanofluid flow through parallel fins enclosed inside a partially heated hexagonal cavity. Int Commun Heat Mass Transfer 132:105885
13. Ali MM, Alim MdA, Akhter R, Ahmed SS (2017) MHD natural convection flow of cuo/water nanofluid in a differentially heated hexagonal enclosure with a tilted square block. Int J Appl Comput Math 3(Suppl 1):S1047–S1069
14. Asako Y, Yamaguchi Y, Chen L, Faghri M (1996) Combined natural convection and radiation heat transfer in a vertical air cavity with hexagonal honeycomb core of negligible thickness. Numer Heat Transf A 30(1):73–85
15. Seyyedi SM, Dogonchi AS, Hashemi-Tilehnoee M, Ganji DD, Chamkha AJ (2020) Second law analysis of magneto-natural convection in a nanofluid filled wavy-hexagonal porous enclosure. Int J Numer Meth Heat Fluid Flow 30(11):4811–4836

16. Biswas N, Mahapatra PS, Manna NK (2016) Merit of non-uniform over uniform heating in a porous cavity. Int Commun Heat Mass Transfer 78:135–144
17. Nag S, Sen N, Bamboowala HT, Manna NK, Biswas N, Mandal DK (2022) MHD nanofluid heat transport in a corner-heated triangular enclosure at different inclinations. Mater Today Proc 63:141–148
18. Saha A, Manna NK, Ghosh K, Biswas N (2022) Design analysis of geometrical shape impact on thermal management of practical fluids using square and circular cavities. Eur Phys J Spec Top 231:2509–2537
19. Mliki B, Abbassi MA, Omri A, Zeghmati B (2016) Effects of nanoparticles Brownian motion in a linearly/sinusoidally heated cavity with MHD natural convection in the presence of uniform heat generation/absorption. Powder Technol 295:69–83
20. Chatterjee D, Biswas N, Manna NK, Sarkar S (2023) Effect of discrete heating-cooling on magneto-thermalhybrid nanofluidic convection in cylindrical system. Int J Mech Sci 238:107852

Design and Modeling of CO_2 Absorption Column for Carbon Sequestration

P. Vishal Reddy and Praveen Kumar Ghodke

Abstract Carbon dioxide (CO_2) mitigation technologies are emerging faster due to the impact of global warming. The study aimed to sequestrate CO_2 from a thermochemical conversion process of biomass by absorption in a specially designed parallel plate column. The absorption column was designed in Gambit software, while simulations were performed in Fluent software. Simulations were analyzed for different liquid hold-up volumes with variable liquid flow rates at a constant mass transfer area and gas flow rate. The absorption of CO_2 in water was increased with decreasing plate spacing in the column. It was also discovered that liquid and gas flow rates greatly influenced efficiency. The simulations primarily focused on optimizing the liquid flow rate and plate spacing for maximal CO_2 recovery with the least energy and water requirement.

Keywords Carbon dioxide · Absorption column · CFD simulations · Gambit · Fluent

1 Introduction

CO_2 is a significant ozone-damaging gas. It contributes to a 60% abnormal weather change. Growing awareness of the potential effects of ozone-depleting compounds on global environmental change has led to recent efforts to develop ways to reduce CO_2 emissions. Carbon capture and sequestration is one such methodology that has received much attention. CO_2 can be captured using various strategies such as pre-processing technologies, in-situ capturing technologies, and post-processing technologies. In the current study, post-processing technology has been chosen to address the carbon sequestration process [1].

P. Vishal Reddy · P. K. Ghodke (✉)
Department of Chemical Engineering, National Institute of Technology Calicut, Kozhikode 673601, Kerala, India
e-mail: praveenkg@nitc.ac.in

S. Das et al. (eds.), *Proceedings of the 1st International Conference on Fluid, Thermal and Energy Systems*, Lecture Notes in Mechanical Engineering,
https://doi.org/10.1007/978-981-99-5990-7_3

Much research has investigated carbon capture and sequestration technology from coal-fired power plant set-ups. For example, an integrated gasification combined cycle (IGCC) with pre-combustion removal of carbon and carbon capture from a post-combustion coal-fired power plant [2–4]. One of the researchers studied the effect of parameters on CCS and the source of carbon. Filonchyk and Peterson carried out the preliminary analysis for toxic emissions from the coal-fired power plant, and analysis was performed using a life cycle assessment tool, which anticipates every relevant innovation regarding the emission analysis [5]. It was observed that coal-fired emission analysis was performed and observed that greenhouse gases were decreased by 69 wt% when using hard coal. It was confirmed by the analysis of methane discharge from coal-fired power plants when using hard coal to produce power and depended on the coal storage network. CO_2 emissions from coal-fired power plants were reduced due to the implementation of CCS technology and supply chain management in the transport of CO_2, which gained the most relevance in newly set-up coal-fired power plants. In any case, scientists are considering post-combustion innovation with carbon capture and sequestration in coal-fired power plants.

A few research have been conducted to study the ability of CFD modeling for hydrodynamics of liquid stream and the absorption capacity of gases/liquid in silica gel in packed-bed columns [6–8]. The absorption of gas/liquid in silica gel was tested in two designed sizes retentive. The complete CFD model was established for each of the experiments. There are a few existing innovations. For example, physical absorption, adsorption, and membrane separation have been proposed to capture and sequestrate CO_2 from flue gases during combustion. A few researchers proposed a complex computational mass transfer model (CMT) to demonstrate the synthetic absorption process with heat impact in packed column sections [9]. The element of the proposed model can anticipate the concentration, temperature, and velocity distributions on the column without expecting the turbulent Schmidt number or utilizing the tentatively estimated turbulent mass transfer diffusivity. In the scientific articulation accompanied by CFD and CHT, the ordinary strategies for k-ε and k-ω are utilized for closing the momentum and mass transfer equations [10].

The study's objective was to simulate a parallel plate absorption column using the software FLUENT v6. The simulations were performed for the different liquid hold-up volumes by fixing the mass transfer area. Furthermore, simulations continued with variable liquid flow rates with a fixed gas flow rate to ensure maximum CO_2 absorption with low-pressure drops. The fluctuation in CO_2 content at the outlet was investigated concerning the operating conditions.

2 Design and Operating Procedure

Figure 1 depicts the absorption column modeling, which included geometry and was created using GAMBIT software. Real-time reactor dimensions are considered to design the absorption column. The absorption column in this study was developed using a packed-bed reactor model with water and flue gas inlets at the top and

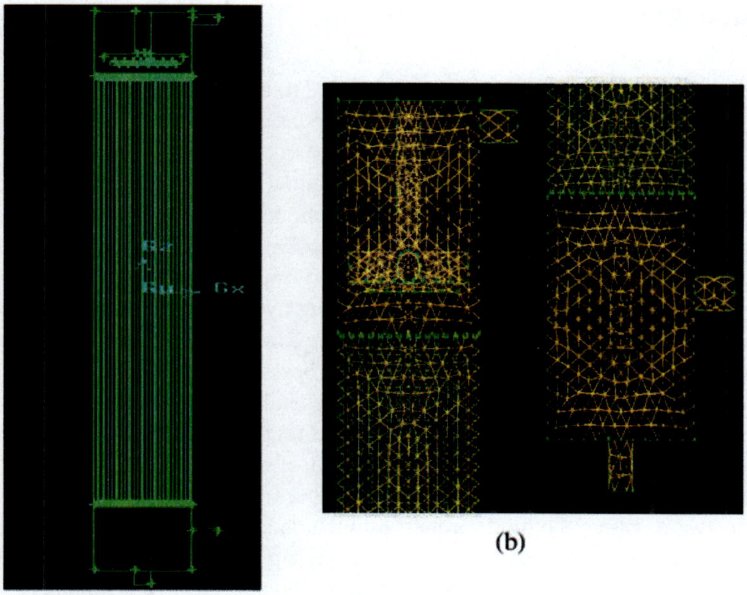

(a) (b)

Fig. 1 Complete modeling of absorption column in GAMBIT software **a** designed absorption column **b** meshed absorption column

bottom, respectively. To remove the liquid and gaseous products, outlet values are provided at the bottom and top of the absorption column. Instead of packing materials, parallel plates with definite spacing were introduced in the column to study the CO_2 absorption capability in water. Designed parallel plates are assumed to support the absorption column while operating at industrial scales.

Initially, the column was purged with water from the top with a distributor provided over parallel plates/packing material and ensured the column was uniformly wetted pyrolysis of biomass residue containing about 4 wt% of CO_2 in non-condensable gases. Total flue gas or fuel gas containing CO_2 was allowed to pass through the absorption column in a distributing space below the parallel plates. It was ensured that gas distribution in the column was uniform. The idea is to capture the CO_2 in water entering from the top and leaving as CO_2-rich water from the bottom. At the same time, non-condensable gases leave the top of the column as dilute or lean gas.

3 Modeling

Three laws of conservation concepts are studied to comprehend the fluid's physical properties. They are mass conservation, energy conservation, and momentum conservation, i.e., Newton's law was followed. CFD is defined as the art of determining the numerical solution for hydrodynamics fluid flow using the space vector and time condition to solve the numerical condition of the total stream divided into smaller entities of the control volume. CFD software, which organizes codes and pre-defines models, was used to solve hydrodynamic fluid flow problems. CFD has a sophisticated user interface for modeling and simulating hydrodynamic fluid flow streams. This complex user interface includes a pre-processor that includes basic design, a solver that simulates the models, and a post-processor that displays the solution of iterated values and fluid vectors to help comprehend the model.

CFD modeling can be used to solve a wide range of fluid dynamics problems. The quality of the model and the number of control volumes determine the difficulty of modeling in a CFD solver. Mass transfer processes are proposed to be solved in CFD software, where the model codes are included for packed beds and other unit operation networks. In chosen models that are subjected to smaller entities based on local conditions, CFD solves the spiral and pivotal fluid flow streams in mass transfer operations and provides systems efficiency. The CFD models, on the other hand, necessitate the selection of distinct conclusion models in order to capture data lost during the operation.

The conservation equations dictate the modeling and simulation in CFD software. In the present study, along with conservation equations, Reynolds-averaged Navier–Stokes (RANS) equations are synchronous to solve the hydrodynamic fluid flow problems. The equations and the k-ε turbulence display were accompanied to utilize the complete CFD modeling. Equations (1)–(11) shown below are model equations used to solve the chosen problem.

Mass conservation equation

$$\frac{\partial u}{\partial x} + \frac{\partial v}{\partial y} + \frac{\partial w}{\partial z} = 0 \tag{1}$$

3.1 Momentum Equation

$$\frac{\partial u}{\partial t} + div(uU) = -\frac{1}{\rho}\frac{\partial p}{\partial x} + v div(grad\, u) + \left[-\frac{\partial u^{2\prime}}{\partial x} - \frac{\partial u\, v'}{\partial y} - \frac{\partial u\, w''}{\partial z}\right] + S_{MX} \tag{2}$$

$$\frac{\partial v}{\partial t} + div(vU) = -\frac{1}{\rho}\frac{\partial p}{\partial y} + vdiv(grad\, v) + \left[-\frac{\partial v^{2\prime}}{\partial y} - \frac{\partial u\acute{v}}{\partial x} - \frac{\partial v\ddot{w}}{\partial z} \right] + S_{My} \tag{3}$$

$$\frac{\partial w}{\partial t} + div(wU) = -\frac{1}{\rho}\frac{\partial p}{\partial z} + vdiv(grad\, w) + \left[-\frac{\partial w^{2\prime}}{\partial z} - \frac{\partial u\ddot{w}}{\partial x} - \frac{\partial v\ddot{w}}{\partial y} \right] + S_{Mz} \tag{4}$$

3.2 *Scalar Quantity*

$$\frac{\partial \varnothing}{\partial t} + div(\varnothing U) = div\left(\Gamma_{\varnothing} grad\, \varnothing\right) + \left[-\frac{\partial u\acute{\varphi}}{\partial x} - \frac{\partial v\acute{\varphi}}{\partial y} - \frac{\partial w\acute{\varphi}}{\partial z} \right] + S_{\varphi} \tag{5}$$

General notation:

$$-\rho\acute{u}_i\acute{\varphi} = \Gamma_t\left[-\frac{\partial \varnothing}{\partial x_i} \right] \tag{6}$$

$$-\rho\acute{u}_i\acute{u}_j\tau_{ij} = \mu_t\left[\frac{\partial u_i}{\partial x_j} + \frac{\partial u_j}{\partial x_i} \right] \tag{7}$$

The terms $\acute{u}_i$ and $\acute{u}_j$ were characterized using a time-averaged velocity model, which was ruled by the Reynolds term. In turn, the turbulence model was demonstrated to change μ_t using computation methodologies. This model impacts turbulence, which anticipates the outcomes of models used in solving the absorption of CO_2 with water in a packed-bed column. Two standard models chosen from the k-ε family are RNG and the standard model. The k-ε solves the (k) turbulent kinetic vitality, and the other (ε) energy scattering rates are shown in the below Eqs. (8). Equations (8) and (9) are solved using the conditions (10) and (11) equations.

$$\frac{\partial(\rho k)}{\partial t} + \frac{\partial(\rho u_i k)}{\partial x_i} = \left[\frac{\partial}{\partial x_i}\left(\rho_{\sigma_k}^{v_{eff}} \frac{\partial k}{\partial x_i} \right) + \rho(P_k - \varepsilon) \right] \tag{8}$$

$$\frac{\partial(\rho k)}{\partial t} + \frac{\partial(\rho u_i \varepsilon)}{\partial x_i} = \left[\frac{\partial}{\partial x_i}\left(\rho_{\sigma_\varepsilon}^{v_{eff}} \frac{\partial \varepsilon}{\partial x_i} \right) + S_\varepsilon \right] \tag{9}$$

$$\mu_T = \rho C_\mu \frac{k^2}{\varepsilon} \tag{10}$$

$$v_{eff} = \frac{\mu_{eff}}{\rho} \text{ and } \mu_{eff} = \mu + \mu_T \tag{11}$$

Development of RNG and a Realizable model, the limitation of standard mass was exploited. The semi-experimental models described with the standard k-ε model depend on the transport condition of turbulent kinetic energy and energy dissipation rate, which are not adequate for packed-bed absorption columns. To overcome the problem, RNG models were used.

Methods for an irregular compelling capacity in the Navier–Stokes conditions in the RNG k-ε model depict the impact of small-scale turbulence. Turbulent viscosity and energy dissipation rate, including other transport conditions, demonstrate the k-ε model's feasibility along with improvement ongoing with these models. The mean-square turbulence variance was used to check the corrected condition of the transport equations. The streamlines, flow, vortices, and revolutions in the standard k-ε model and the RNG k-ε model were enhanced, and feasibility was tested. Using commercial CFD programming Fluent software, numerical analysis was performed to evaluate CO_2 absorption from flue gas in water.

4 Examine the Mesh

The rectangular cuboid design of the absorption column was selected for the CO_2 sequestration process. Column height was 0.92 m, and width was 0.08 m. Parallel plates were created as packing material having a length of 0.1 m and width of 0.15 m with spacing between two plates of 0.3 cm in the first configuration and 0.3 cm spacing increase in the next configuration till the total plate spacing distance increase to 0.9 cm. The meshing of plates was performed in the same Gambit software. All the plates were divided into small tetrahedral cells of around 300,000–12,000,000 quantities. The current study's modeling contains a standard pressure and simple pressure–velocity coupling method, in which the upwind discretization conspires for energy, force, turbulence kinetic, energy, and dispersal energy is the same type of solver utilized in three observations.

A tolerance limit of 10^{-7} was set for all parameters. Skewness can significantly influence the exactness and strength of the CFD arrangement. GAMBIT gives a few quality measures to evaluate the nature of work. On account of skewness estimates, Equi Angle Skew and Equisized Skew, for instance, more minor qualities are progressively alluring. It is additionally critical to confirm that most of the components in our work have positive zone/volumes.

5 Result and Discussion

The use of CO_2 in algae growth reduces CO_2 levels in the atmosphere, and this is known as algae-based CCS technology. The rate of CO_2 uptake by algae was prolonged. As a result, absorbed CO_2 in water escapes into the atmosphere during bio-sequestration. To lower the desorption rate, simulations are performed to establish

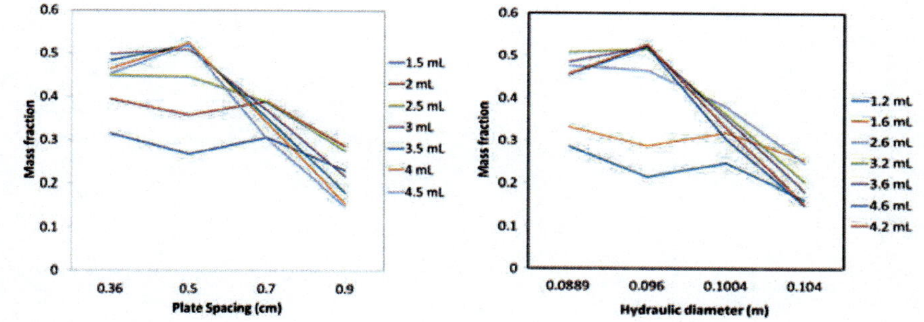

(a) (b)

Fig. 2 CO_2 mass fraction recovery from absorption column **a** plate spacing **b** hydraulic diameter

the water flow rate so that CO_2 absorption recovery is very low, which may then be used for microbial growth in the bio-sequestration process.

The mass fraction of CO_2 recovery in the absorption column with varying plate spacing for varied flow rates is shown in Fig. 2a. The mass fraction of CO_2 reduces as plate spacing increases in the absorption column. The reason could be that the surface area available for absorption between the fluids flowing was reduced. Figure 2b shows the mass fraction of CO_2 recovery in a parallel plate absorption column with various hydraulic diameters (D_h). The mass fraction of CO_2 drops as the hydraulic diameter of the absorption column increases. The reason could be that the mass transfer area available for absorption between the fluids flowing was reduced. The mass fraction of CO_2 recovery decreases as plate spacing increases, from $D_h = 0.0889$ m to $D_h = 0.1040$ m. The proportion of CO_2 recovery for $D_h = 0.096$ m is nearly the same as that of $D_h = 0.0889$ m until a liquid flow rate of 3 mL/min. As discussed, CO_2 recovery increases as the liquid flow rate increases. All the simulations with different configurations and various fluid flow rates were performed, and it was discovered that each iteration takes approximately 7–8 h to converge the fluid flow equations.

Figure 3 shows that CO_2 recovery varies with different liquid flow rates. As plate spacing increases, the mass fraction of CO_2 recovery decreases as the fluid flow rate increases beyond the optimum flow rate, and CO_2 recovery researches plateau. The percent of CO_2 recovery increases with an increase in liquid flow rate until it reaches optimum. After that, CO_2 recovery decreases with an increasing liquid flow rate. The parameter changed with liquid flow rates for a constant gas flow rate. The packed-bed model was solved using CFD modeling software, which yielded a numerical solution based on the governing equations and related boundary conditions. Fluent software was used to solve the models and determine the porosity and absorption capacity of the packing bed. Simulating the absorption column problem with the upwind discretization method was done using the SIMPLE algorithm technique.

Modeling and simulation convergence was achieved by normalizing both with a residue of 10^{-6}. Finally, a solution was found when the flow equation for each node of the object chosen in the field was less than 10^{-12} between iterations. Figure 4 shows

Fig. 3 CO_2 mass fraction recovery based on the liquid flow rate (mL/min)

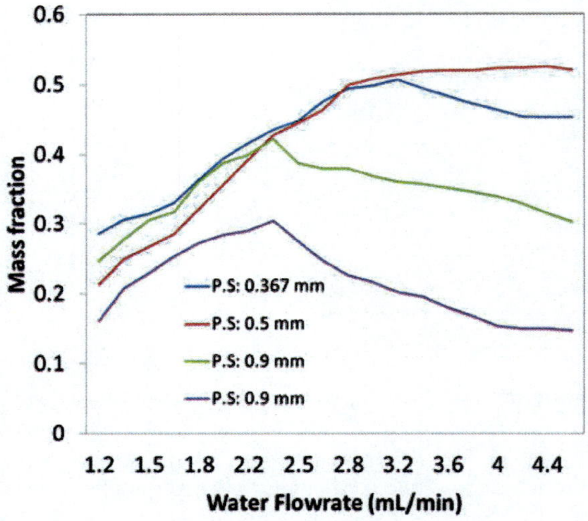

that successive iterations of residuals are 10^{-12}. The optimum working condition of the parallel plate CO_2 absorption column utilizing water as a solvent was determined using CFD simulations software. The simulation was performed for various liquid flow rates at a fixed gas flow rate to determine optimal liquid flow parameters that allow CO_2-solvated water to be used flawlessly without any adjustments.

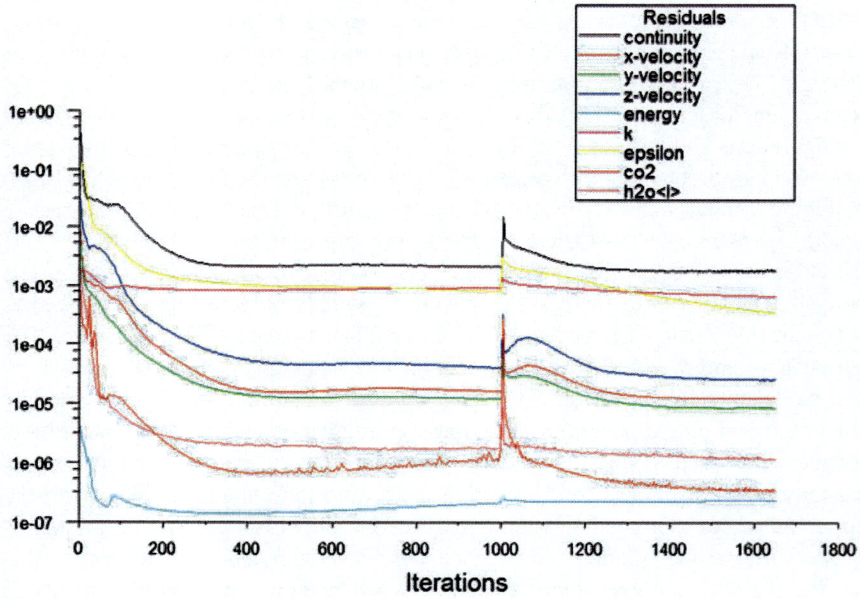

Fig. 4 Residuals of process in FLUENT solver

For a liquid flow rate of 2 $mLmin^{-1}$, the maximum recovery of CO_2 was calculated for all the different plate spacing. Thus, the configuration of the absorption column with hydraulic diameter $D_h = 0.096$ m and plate spacing of 0.5 cm found a maximum recovery of 10.5% CO_2. The optimum configuration of the absorption column was found to save a lot of energy and water requirement for carbon sequestration.

6 Conclusion

CFD modeling in Ansys Fluent software provides a broader knowledge of hydrodynamics fluid flow, where real-time experimental investigations are not economically feasible and impossible. Carbon capture from the thermochemical conversion of biomass was simulated in fluent software, and it was discovered that inorganic carbon could be captured and used for microalgae development. In the specially designed absorption column, CO_2 recovery was found to be high when plate spacing was very small. For hydrodynamics and mass transport processes, CFD simulations allow performing experiments from small-scale to large-scale plants. CFD modeling and simulations are quick and effective methods for calculating the risk of any commercial unit operations or unit processes.

References

1. Arnette AN (2017) Renewable energy and carbon capture and sequestration for a reduced carbon energy plan: an optimization model. Renew Sustain Energy Rev 70:254–265. https://doi.org/10.1016/j.rser.2016.11.218
2. Corsten M, Ramírez A, Shen L, Koornneef J, Faaij A (2013) Environmental impact assessment of CCS chains—lessons learned and limitations from LCA literature. Int J Greenh Gas Control 13:59–71. https://doi.org/10.1016/j.ijggc.2012.12.003
3. Shackley S, Verma P (2008) Tackling CO_2 reduction in India through use of CO_2 capture and storage (CCS): prospects and challenges. Energy Policy 36:3554–3561. https://doi.org/10.1016/j.enpol.2008.04.003
4. Cuéllar-Franca RM, Azapagic A (2015) Carbon capture, storage and utilisation technologies: a critical analysis and comparison of their life cycle environmental impacts. J CO2 Util 9:82–102. https://doi.org/10.1016/j.jcou.2014.12.001
5. Filonchyk M, Peterson MP (2023) An integrated analysis of air pollution from US coal-fired power plants. Geosci Front 14:101498. https://doi.org/10.1016/j.gsf.2022.101498
6. Gbadago DQ, Oh H-T, Oh D-H, Lee C-H, Oh M (2020) CFD simulation of a packed bed industrial absorber with interbed liquid distributors. Int J Greenh Gas Control 95:102983. https://doi.org/10.1016/j.ijggc.2020.102983
7. Xie P, Lu X, Yang X, Ingham D, Ma L, Pourkashanian M (2017) Characteristics of liquid flow in a rotating packed bed for CO_2 capture: a CFD analysis. Chem Eng Sci 172:216–229. https://doi.org/10.1016/j.ces.2017.06.040
8. White J (2016) CFD simulation and experimental analyses of a copper wire woven heat exchanger design to improve heat transfer and reduce the size of adsorption beds. Computation 4:8. https://doi.org/10.3390/computation4010008

9. Xu Y, Chang W, Chen X, Jin B (2023) CFD modeling of MEA-based CO_2 spray scrubbing with computational-effective interphase mass transfer description. Chem Eng Res Des 189:606–618. https://doi.org/10.1016/j.cherd.2022.11.018
10. Martín M, Montes FJ, Galán M (2009) Mass transfer from oscillating bubbles in bubble column reactors. Chem Eng J 151:79–88. https://doi.org/10.1016/j.cej.2009.01.046

Numerical Analysis on the Effect of Fuel Injection Location on Combustion Characteristics in a Trapped Vortex Combustor

S. Subramanian, M. Chindesh, and K. M. Parammasivam

Abstract The present work focuses on the numerical simulations of non-reactive and reactive cases in a trapped vortex combustor. Numerical simulations are carried out in ANSYS FLUENT to analyze the effects of fuel injection locations on mixing and combustion characteristics in a model trapped vortex combustor. Three different combinations of fuel injection viz (i) injection in cavity alone, (ii) injection in both main stream and cavity, and (iii) injection in mainstream alone were analyzed using steady, 2D RANS Non-premixed combustion model for mainstream air velocities of 10 and 30 m/s. From the simulations, it has been observed that fuel injection location plays a vital role in mixing and complete combustion both inside the cavity and mainstream. The species concentration contours for the 3 different cases indicate that fuel injection on top of the cavity enhances combustion and also improves the strength of the locked vortex by helping in shear layer entrainment. Static temperature plots obtained for the different cases show that fuel injection in both cavity and mainstream is the optimum one with minimum partial combustion and highest lean burn rates.

Keywords Trapped-vortex combustor · Complete combustion · Gas turbine emissions · Non-premixed combustion · Numerical simulations

1 Introduction

Emission reduction in Gas turbine combustors is a serious challenge that engages researchers across worldwide for decades. Turbines when operating under fuel lean conditions can have very low emissions and very high efficiencies. Thermal nitric oxide formation is lower in lean combustion because of reduced flame temperature. Hydrocarbon and carbon monoxide emissions are reduced as combustion is generally complete with excess air [1]. As a result, when compared to fuel rich conditions, lower

S. Subramanian (✉) · M. Chindesh · K. M. Parammasivam
Department of Aerospace Engineering, Anna University, MIT Campus, Chennai 600044, India
e-mail: subramanians4005@gmail.com

S. Das et al. (eds.), *Proceedings of the 1st International Conference on Fluid, Thermal and Energy Systems*, Lecture Notes in Mechanical Engineering,
https://doi.org/10.1007/978-981-99-5990-7_4

pollutant emissions are produced in lean combustion. Trapped-vortex combustion is an emerging combustion technology for achieving low emissions in gas turbine combustors. Trapped-vortex combustors can be operated on a very lean mixture ratio and hence it helps in achieving high combustion efficiency and low overall emission levels.

One of the main challenges of operating gas turbine combustors at lean conditions is maintaining a stable flame at all points of time in operation. As such research on flame stability gains traction when it comes to design of lean combustors. Conventionally flame stability is achieved using swirlers in gas turbine combustors. With the advent of lean combustors such as Lean Direct Injection (LDI), Lean premixed prevaporized (LPP) combustors, Trapped-vortex combustors (TVC), bluff bodies such as v-gutters, backward facing step, wedges were employed. A trapped vortex combustor achieves flame stabilization by means of holding a flame in a cavity. The flame in the cavity of a TVC is stable as the locked vortex enhances mixing and shielding the flame from high-velocity mainstream flow.

The present work focuses on the effects of injection location on complete combustion in a trapped vortex combustor. This section is followed by a literature review where past work on numerical simulations in TVC is analyzed critically and objectives are framed. The third section focuses on numerical methodology followed by results and discussions.

2 Literature Review and Objective

The utilization of trapping a stable vortex within a cavity for combustion was first demonstrated by Hsu et al. in their study on the performance of a trapped vortex combustor [2]. The experiment involved two separate airflow streams: the annular airflow and the primary airflow directly injected into the cavity. The main objective was to optimize the cavity length and depth to effectively trap a vortex while minimizing pressure drop along the combustor. The study determined that an optimum length-to-forebody diameter ratio of 0.59 was required to achieve vortex locking. Notably, this research conducted experiments with an unconfined flame inside the cavity, without an annular casing. Temperature measurements were taken using a CARS setup, revealing that increasing the primary airflow led to higher vortex temperatures. Consequently, the flame length anchored on the cavity decreased, resulting in reduced peak combustion efficiency. Additionally, the study found that maintaining an overall equivalence ratio of 0.05 in the main flow ensured a stable flame, which provided a broader range compared to bluff body or swirl stabilized combustors.

Subsequently, numerical studies were conducted to investigate stable combustion in a trapped vortex combustor using a forebody-spindle-disc geometry setup [3, 4]. This marked the first instance of numerically analyzing reacting flows within a trapped vortex combustor. The researchers obtained residence time and cavity flow entrainment data by observing the movements of seeder particles in the main flow and cavity flow. Further numerical simulations focused on rectangular/dump-type combustors with cavities [5–7]. Specifically, the simulations examined the arrangement of radial struts and their impact on combustion characteristics such as ignition and blowout [5]. However, the simulations did not investigate the problem of effectively maintaining the trapped vortex within the cavities through changes in cavity flow rate, thus providing an opportunity for further research. Non-reacting CFD simulations revealed that the cavity flow struggled to reach the tail of the bluff body due to insufficient protection from slenderer struts. An analysis of the flame structure within the cavity highlighted that a portion of the injected fuel was consumed inside the cavity, while the remaining portion burned in the main flow passage area [8]. Moreover, the location and angle of holes in the cavity were studied, revealing that increasing the distance between the primary injection and the bottom wall of the cavity, while keeping the inlet angle constant, disrupted the flow structure [9]. Subsequent investigations on the flow pattern of single and double vortex structures in the cavity contributed to an understanding of the differences in cavity-stream-mainstream mixing mechanisms, ultimately leading to improved mixing conditions [6]. Additionally, studies focused on flow structures in the quench zone of an RQL combustor, where the penetration distance of the flame into the mainstream was found to be a ratio of the cavity width [10].

Numerical simulations of the trapped vortex combustor model by Zhang et al. encompassed the use of RANS [11, 12]. These simulations explored the factors influencing the combustion characteristics of the trapped vortex combustor, including the influence of air injection mode and flow resistance [11]. The temperature distribution curve played a crucial role in examining the variation of combustion efficiency under different operating conditions. In the reactive case, where kerosene served as the fuel, the effects of air velocity and spray cone half-angle on fuel concentration were investigated [12]. The simulations revealed that, in the non-reacting case, the distribution range of gaseous fuels within the cavity decreased with increasing inlet air velocity. Conversely, increasing the spray cone half-angle of the nozzle expanded the distribution range of kerosene droplets and vapor within the cavity [11].

From the literature survey, it has been found that the plenty of numerical analysis on TVC focuses on geometry and performance optimization. The effects of fuel injection location on complete combustion have not been studied in detail. The present study aims at exploring the effects of fuel injection location in TVC on combustion characteristics. The next section explains the geometry and numerical methodology.

Fig. 1 Experimental test facility in the Combustion Lab of Aerospace Engineering Department, MIT, Anna University-Chennai

3 Materials and Methods

3.1 Experimental Setup

The experimental setup comprises a dump-type combustion chamber with a square cross section of side 8.5 cm. The setup is shown in Fig. 1. The setup can be run under atmospheric pressure condition or under elevated pressures of up to 4 bar which is supplied by a compressor storage tank facility.

The cavity in the combustion chamber has an L/D ratio of 1.2 which is chosen from literature. The cavity has a width of 10.2 cm and a depth of 8.5 cm.

The chamber length past the cavity is 50 cm. Flame is anchored in the cavity along with a bluff body.

Premixed combustion is carried out with methane as fuel.

The setup is provisioned in such a way to change the fuel flow rate and airflow rate together to experiment with various rich and lean burning conditions. The whole combustion chamber is machined with stainless steel 404 material. Provisions are made to conduct optical diagnostics and appropriate optical windows are stationed at the combustion chamber.

3.2 Geometry

Numerical analysis has been carried out in a simplified combustion chamber without compromising on the overall flow structure. Geometry for the numerical analysis is designed similarly to that of the experimental chamber. The combustion chamber is 80 cm in length and has a square cross section with a side of 8.5 cm. The cavity L/D ratio is fixed at 1.2 with a cavity width of 10.2 cm and a depth of 8.5 cm. A cavity guide wane with a length of 4 cm and depth of 6 cm is placed to channelize the airflow into the cavity. Two bluff body configurations with three different positioning have been tested numerically. The first two testing schemes are with a wedge bluff body

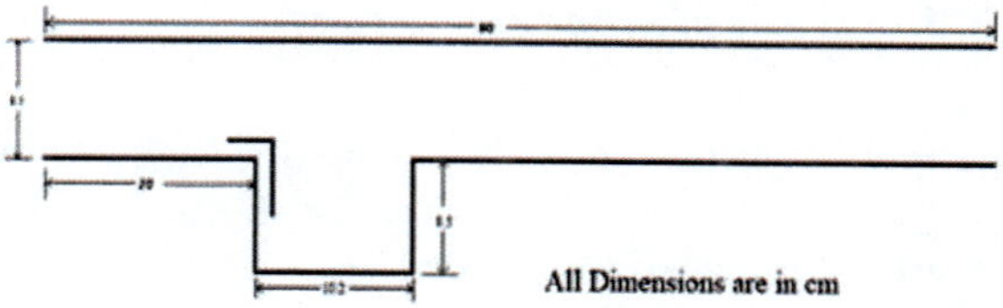

Fig. 2 Schematic of TVC

Table 1 Mesh details

No. of elements	112,752
Elements size	2 mm, 0.5 mm
No. of nodes	113,826
Y^+ value	22.84

with 2 cm side and 60° angle between the sides. The third testing scheme is that of an integrated bluff body and cavity guide vane. In this scheme, a triangular wedge is integrated throughout the length of the strut with a side of 1 cm.

The schematic of the combustion chamber for the two bluff body configurations is shown in Fig. 2. Details of mesh are explained in Table 1 and Fig. 3.

The reactive flow simulation conditions are listed in Table 2.

Fig. 3 Meshing

Table 2 Reactive flow: simulation conditions

S. No.	Solution setup	Input
1	General	Pressure-based solver, 2D planar
2	Model	Realizable k-ε, non-premixed combustion
3	Materials	Air, aluminum, pdf mixture
4	Boundary conditions	Inlet—air velocity inlet = 10, 30 m/s; fuel velocity = 5 m/s; gauge pressure = 101,325 Pa Outlet—pressure outlet = 101,325 Pa Wall—no slip, stationary
5	Solution scheme	SIMPLE

3.3 Grid Independence Study

A grid independence study has been carried out using four different grids with varying number of elements to ensure that the solution was independent of the grid. 4 grids, viz., G1, G2, G3, and G4 with 85,000, 103,642, 112,752, and 125,368 elements, respectively, were simulated and their normalized axial velocity profiles were compared. It has been observed from the grid independence study that the grids G3 and G4 perform very similarly to each other and thus grid G3 is utilized for the present work to have higher accuracy and optimum time resolution. The normalized axial velocity profile for the different grids is shown in Fig. 4.

The convergence criteria for the present simulations were set as 10-6. The simulation results were validated with the experimental results of Zhang et al. [12]. In the work [12], temperature rise past the cavity was measured when the fuel injection was done on the cavity alone and the variation is linear as in the case with present simulations. The simulation results of the present work are in line with the experimental results of Zhang et al. [12]

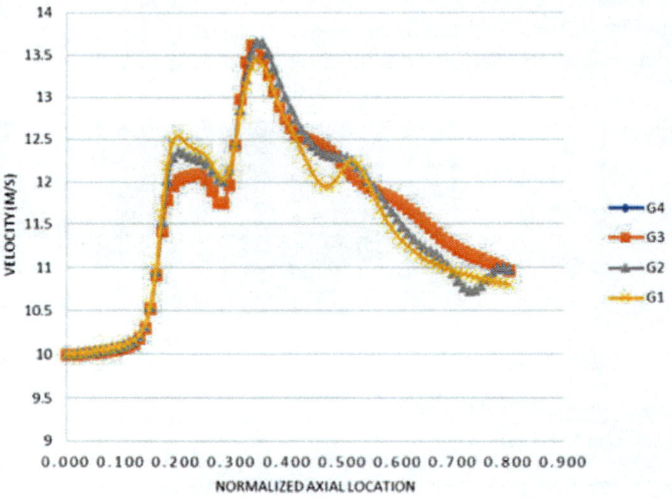

Fig. 4 Grid independence study

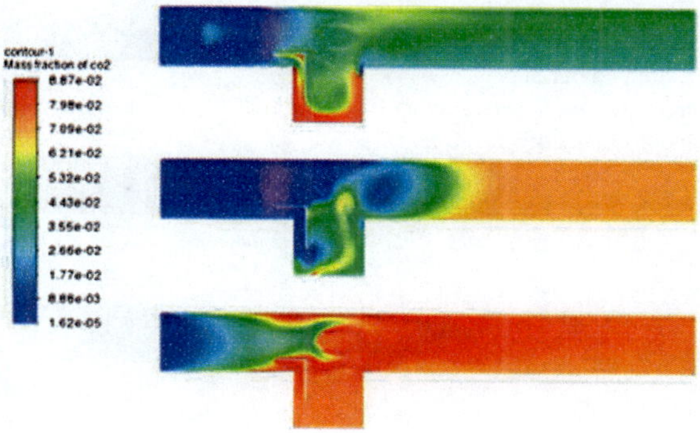

Fig. 5 Mass fraction of CO_2 in TVC for fuel injection **a** both cavity and mainstream, **b** only cavity, **c** only mainstream (10 m/s)

4 Results and Discussion

4.1 Complete Combustion Regimes: CO_2 Contours

CO_2 contours indicate areas of complete combustion inside TVC. CO_2 contours for three different fuel injection cases at two different velocities are compared and shown in Figs. 4 and 5. When fuel is injected from the cavity aftwall top, complete combustion mostly occurs in the mainstream flow as air supply is abundant there and some level of mixing occurs inside cavity as a result of prevailing locked vortex in cavity. The effect of bluff body in aiding complete combustion for the case is very low as evident from CO_2 contours of 10 and 30 m/s velocities.

The most effective case of complete combustion in mainstream occurs when the fuel is injected both from cavity top and mainstream locations. This can also be inferred from the static temperature rise plot which is explained in Sect. 4.3. In the case of fuel injection from the mainstream alone, most of the air in the mainstream upstream of the guide vane participates in the combustion process, and hence very little complete combustion takes part inside the cavity. This is indicated in Fig. 5.

4.2 Incomplete Combustion Regimes: CO and NO Contours

CO contours indicate incomplete and partial combustion that takes place inside TVC. Figures 6 and 7 represent CO contours for mainstream air velocities of 10 m/s and 30 m/s respectively. CO formation occurs when there is a deficiency in O_2 supply inside mixing and burning zones. CO formation is lowest when fuel is injected only in the mainstream. As all the air in this case takes part in combustion process, CO formation is reduced. Fuel injection in the mainstream and cavity leads to the higher rate of production of CO inside cavity due to a deficiency of O_2 in cavity. CO production is reduced when fuel injection happens only in cavity top. Some of the fuel injected

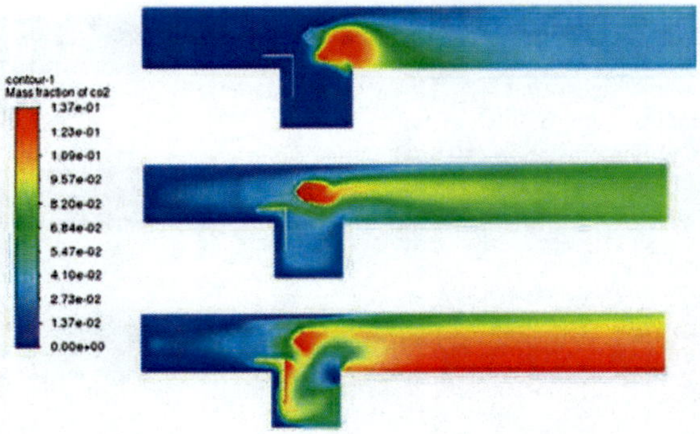

Fig. 6 Mass fraction of CO_2 in TVC for fuel injection **a** both cavity and mainstream, **b** only cavity, **c** only mainstream (30 m/s)

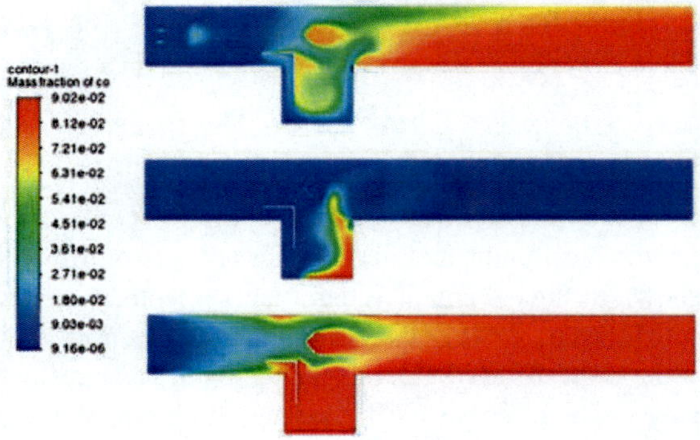

Fig. 7 Mass fraction of CO in TVC for fuel injection **a** both cavity and mainstream, **b** only cavity, **c** only mainstream (10 m/s)

will undergo combustion in mainstream and this balances the O_2 deficiency for CH_4 burning in cavity. This leads to a reduced CO production.

NO production is associated with regions of high static temperature and longer residence time. Accordingly, from Figs. 8 and 9, it can be inferred that NO production is maximum in the case of fuel injection from both cavity top and mainstream at 30 m/s (Fig. 10).

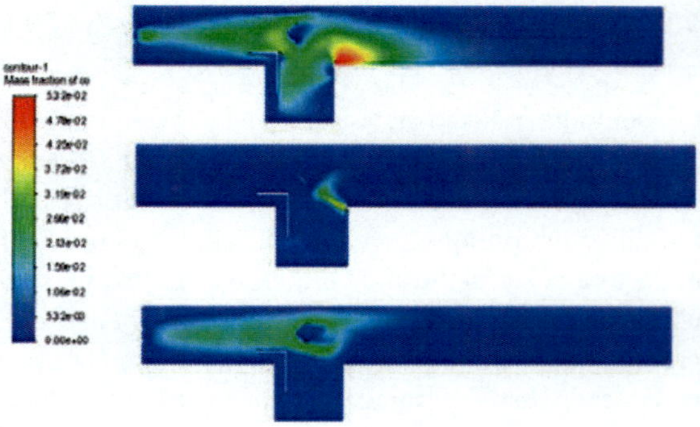

Fig. 8 Mass fraction of CO in TVC for fuel injection **a** both cavity and mainstream, **b** only cavity, **c** only mainstream (30 m/s)

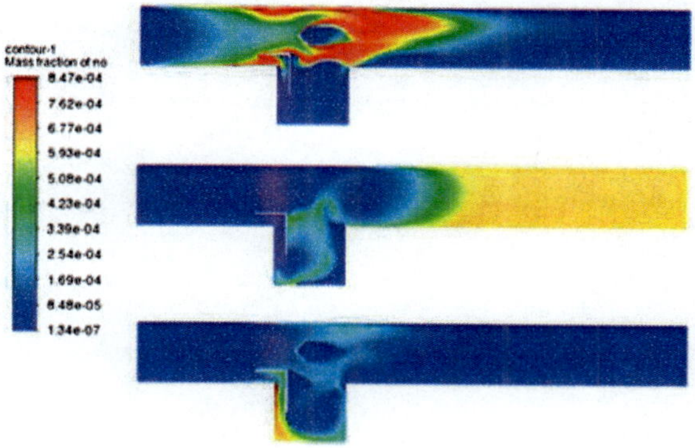

Fig. 9 Mass fraction of NO in TVC for fuel injection **a** both cavity and mainstream, **b** only cavity, **c** only mainstream (10 m/s)

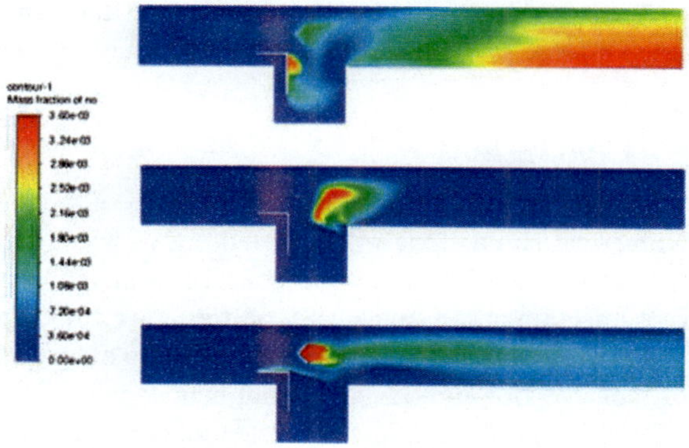

Fig. 10 Mass fraction of NO in TVC for fuel injection **a** both cavity and mainstream, **b** only cavity, **c** only mainstream (30 m/s)

Higher temperature in this case leads to higher NO rate. NO production is also affected by mainstream air velocity. Higher the velocities, higher the air availability for a fixed fuel flow rate which results in higher rate of complete combustion. This results in maximum NO production.

4.3 Static Temperature Plot

The completeness of combustion inside TVC and a clear indication of optimized fuel injection location can be inferred from static temperature plots for different mainstream air velocities (Fig. 11).

When the fuel injection is in the cavity only, the temperature rise which is an indicator of complete combustion is better along the axial line on top of the cavity. Formation of CO results in lesser static temperature rise as evident from fuel injection in both mainstream and cavity cases. A sudden drop in static temperature around 20% indicates a lack of oxygen for complete combustion inside the cavity.

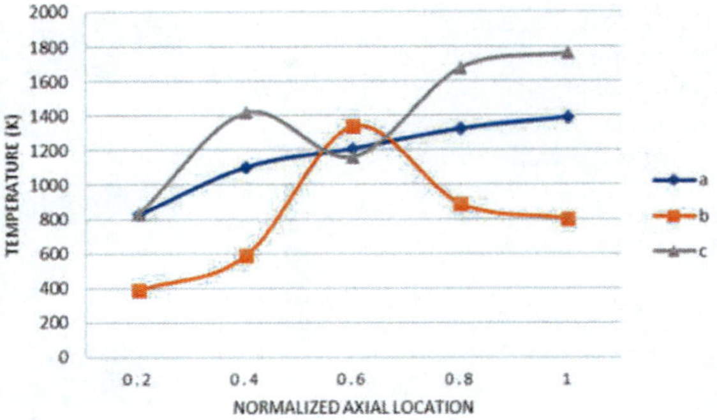

Fig. 11 Mean temperature variation across axial distance for case **a** fuel injection at mainstream only, **b** fuel injection at cavity only, **c** fuel injection both at mainstream and cavity

It can be concluded that for the given velocities, fuel injection at the cavity only performs well followed by the case of both cavity and mainstream.

5 Conclusions

Numerical simulations were carried out in TVC for three different fuel injection locations at two different velocities. The results indicate that for the given velocities, fuel injection at the cavity only performs well followed by the case of both cavity and mainstream. Also, a drop in temperature of around 20% is recorded for the case of fuel injection in both cavity and mainstream. This indicates a lack of O_2 inside the cavity in this case. It is suggested to carry out guide vane modification to enable a high airflow rate in the cavity at high velocities.

References

1. Derek D, Peter T (2016) Lean combustion: technology and control, 2nd edn. Academic Press
2. Hsu KY et al (1998) Characteristics of a trapped-vortex combustor. J Propul Power 14(1):57–65
3. Katta VR, Roquemore WM (1998) Study on trapped-vortex combustor- effect of injection on flow dynamics. J Propul Power 14(3):273–281
4. Hendricks RC et al (2001) Experimental and computational study of trapped vortex combustor sector rig with high-speed diffuser flow. Int J Rotat Mach 7(6):375–385
5. Jin Y et al (2012) Design and performance of an improved trapped vortex combustor. Chin J Aeronaut 25(6):864–870
6. Jin Y et al (2014) Effect of cavity-injector/radial-strut relative position on performance of a trapped vortex combustor. Aerosp Sci Technol 32(1):10–18
7. Jin et al (2014) Numerical investigation on flow structures of a laboratory-scale trapped vortex combustor. Appl Therm Eng 66(1–2):318–327
8. Li Q et al (2013) Experimental investigation of airflow distribution for a novel combustor mode. Appl Mech Mater 284–287:743–747
9. Xing F et al (2012) Experiment and simulation study on lean blow-out of trapped vortex combustor with various aspect ratios. Aerosp Sci Technol 18:48–55

10. Jiang B et al (2015) Computational study on flow structures of quench zone in a rich-quench-lean trapped-vortex combustor. Energy Procedia 66:321–324
11. Zhang RC et al (2018) Effect of swirl field on the fuel concentration distribution and combustion characteristics in gas turbine combustor with cavity. Energy 162:83–98
12. Zhang RC et al (2019) Influence of injection mode on the combustion characteristics of slight temperature rise combustion in gas turbine combustor with cavity. Energy 179:603–617

A Systematic Review on Solar-Driven Interfacial Evaporation for Desalination

Jackson Ondiko, Jaap Hoffmann, Ben Sebitosi, K. Ashish Chandran, and C. S. Sujith Kumar

Abstract Traditional literature reviews and state-of-art approaches have been a center of attention in assessing the progress of solar-driven interfacial evaporation for desalination (SDIED). Moreover, these approaches focused on the development of photothermal materials, structural design, and salt-resistant strategies. However, a quantitative analysis review on the progress of surface temperature, evaporation rate, and efficiency of the photothermal structure has scarcely been reported. Considering the quality, reliability, and transparency, a systematic review method was adopted in this study to quantitatively analyze the variables stated. The articles from 2016 to 2021 were considered for this review. The average surface temperature, evaporation rate, and efficiency determined were 44.68 °C, 1.53 kg m^{-2} h^{-1}, and 83.95%, respectively, over 5 years of study. Moreover the higher the surface temperature of the solar absorber, the lower the evaporation rate and efficiency was realized.

Keywords Systematic review · PRISMA · Interfacial · Surface temperature · Evaporation rate · Efficiency · Photothermal structure

1 Introduction

Although 75% of the earth's surface is covered by water, freshwater resources account for only 0.5%, thus resulting in a serious shortage of freshwater in the world [1]. About 844 million people in the world still do not have access to basic water supplies and 79% of them are rural residents [2]. Moreover, agricultural activities, population growth, and water pollution have largely contributed to water shortages. Reverse

J. Ondiko (✉) · J. Hoffmann · B. Sebitosi
Department of Mechatronics and Mechanical Engineering, Stellenbosch University, Matieland 7602, South Africa
e-mail: jacksonondiko3@gmail.com

K. Ashish Chandran · C. S. Sujith Kumar
Department of Mechanical Engineering, National Institute of Technology Calicut, Kozhikode, India

S. Das et al. (eds.), *Proceedings of the 1st International Conference on Fluid, Thermal and Energy Systems*, Lecture Notes in Mechanical Engineering,
https://doi.org/10.1007/978-981-99-5990-7_5

osmosis technology has been widely used due to its applicability in areas far away from large water bodies and high-quality freshwater supply. However, this technology is still inadequate to meet the increasing demand for freshwater due to its high energy consumption, large centralized installations, and advanced supporting infrastructure which is not feasible in rural areas [1, 3] naturally, water evaporation takes place at the interface between water and air. Therefore, absorbing solar radiation and effectively converting the energy into heat at the water surface increases the evaporation rate. However, the evaporation efficiency is too low to meet the actual demand for freshwater due to poor photothermal conversion of water and great heat loss in the bulk water. This has therefore triggered research concerns about SDIED [3]. In this technology, solar energy is absorbed by the photothermal material which converts to heat energy that is localized at the water–air interface to generate vapor that is then condensed to produce freshwater. SDIED is now becoming a promising solution for people in remote areas. It's affordable to get clean water without extra energy and negligible environmental impact because solar energy is free and naturally available.

Reviews in SDIED technology have tremendously increased in recent years. Most of them focus on type, preparation method, structural design, salt-resistance and solar-heat conversion of photothermal materials, and their evaporation performance. However, no review has been reported on the quantitative analysis of surface temperature, evaporation rate, and efficiency of the photothermal structure. The structure is a floating absorber that comprises a thin photothermal layer that absorbs and converts solar energy into heat energy, a thermal barrier that prevents the dissipation of localized heat at the layer into bulk water, and micro-channels that ensure a continuous flow of water to the surface for evaporation as illustrated in Fig. 1. In addition, most previous reviews used traditional methods [4–8] and state-of-the-art [9] approach, and therefore no systematic literature review approach was identified. Application of traditional or narrative literature methods provides a lack of scientific rigor, unreliability, research biases, and invalidity [10, 11] because the researcher collects data depending on his/her judgment which may result in the omission of essential articles during data extraction.

In this review, a systematic literature review approach was adopted to analyze the evaporation surface temperature, evaporation rate, and efficiency of the photothermal structure. Perhaps this will help to reduce biases in data extraction [12]. Moreover, this will provide areas of concern for exploration by researchers in this field. This review is organized into four sections. Section 1 is the introduction. Section 2 is the methodology using preferred reporting items for systematic reviews and Meta-analysis (PRISMA) [13]. Section 3, results and analysis are presented. Lastly, the summary and future agenda are discussed in Sect. 4.

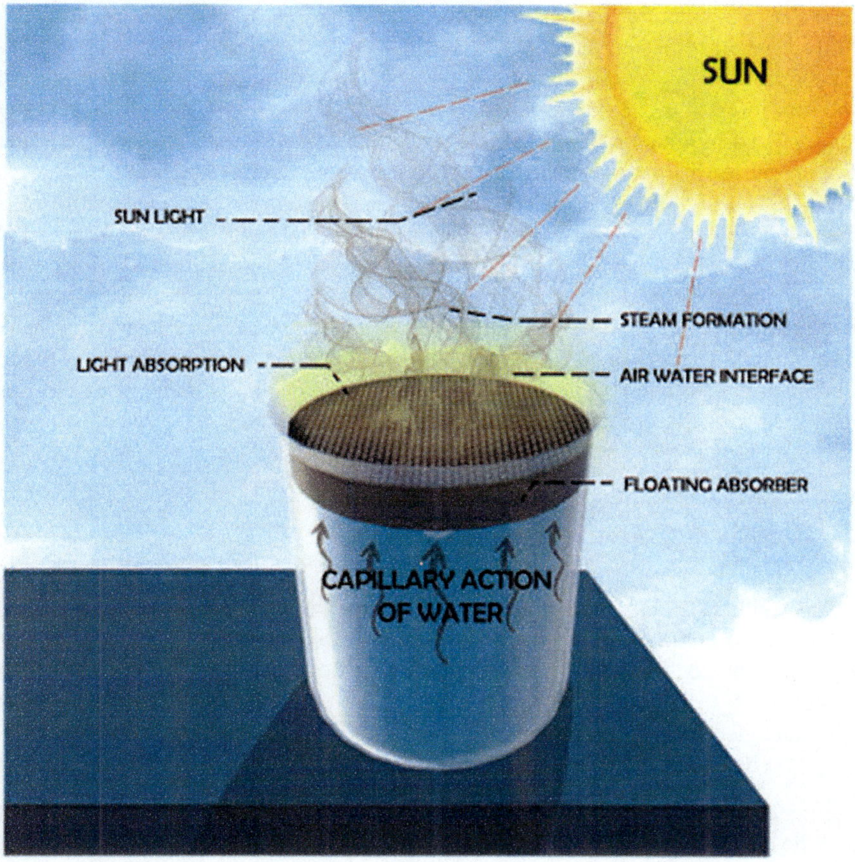

Fig. 1 Schematic representation of floating absorber-based solar-driven steam generation [3]

2 Methodology

This systematic review was conducted by following the PRISMA format. An online library was utilized to search for available literature. In advanced search, the search term used was "Any field contains 'solar-driven desalination/water purification, AND solar-driven distillation, AND interfacial evaporation.'" In the screening process, all articles and the 2016–2021 year of publication were selected. Science technology, materials science and multidisciplinary, nanoscience and nanotechnology, water purification, solar energy, evaporation, photothermal conversion, efficiency, and evaporation rate were preferred in the subject area. The articles written in English were chosen and the results were personalized. The screened articles were exported as a CSV file to Google spreadsheet on January 4, 2022. The titles and abstracts of the uploaded articles were reviewed and the existing literature reviews and non-related articles were removed. The full-text articles were downloaded and the content analysis was done. For uniformity of results, content analysis was based on scalability, type of feedwater, durability, and field-weighted citation impact in the Scopus database. All articles done indoor and on laboratory scale were selected because of minimal impact by external conditions, e.g., wind, varying humidity, and temperature experienced in outdoor set-ups. Explorations done using wastewater and seawater were preferred in this study because they are the preferred target types of feedwater in SDIED. The eligible articles selected had both surface temperature, evaporation rate, and efficiency of the evaporative structure under light intensity 1 kW/m^2 for a period of 1 h. The quality of the published articles was vetted by taking the articles with field-weighted citation impact above a value of 2 in the Scopus database. Descriptive analysis of the distribution of the articles and databases over the last 5 years was studied. Analysis of surface temperature, evaporation rate, and efficiency of the photothermal structures was conducted. Thereafter the summary and research gaps identified were discussed.

3 Results and Analysis

In this study, 28 articles were reviewed as shown in Table 1. The available data found were 182. After the screening process, 83 articles were recorded. The review articles noted and deleted from the spreadsheet data were 24. After thorough scrutiny of the content analysis regarding the study variables, 47 articles were discarded. No articles were included from other sources.

(a) Articles and database distribution per year

Figure 2 shows the distribution of articles over the last 5 years. It was observed that the distribution has been increasing. From this study, 2021 recorded the highest number of articles compared to the previous years. Moreover, it was noted that the articles recorded were sourced from various databases. Figure 3 illustrates the highest percentage (10.8%) of the database was not specified and the least percentage (1%) of the database used was ABI/INFORM collection. This indicates an increase in knowledge, study, and publications by the researchers in SDIED which agrees with the existing literature.[3]

(b) Thematic analysis

The main objective of SDIED is to meet the increasing demand of freshwater in the remote areas. Therefore it requires the SDIED to have a high evaporation rate to increase the productivity of freshwater and relatively sufficient photothermal efficiency to enhance the stability and durability of the photothermal material. Moreover, surface temperature should be relatively at an optimum value so as to reduce thermal losses on the glass cover as well as increase the rate of natural condensation of the vapor in case of a closed solar still system. In Fig. 4, whenever the average surface temperature is high, the evaporation rate and efficiency are relatively low. The high surface temperature of the absorber creates a temperature gradient with the

Table 1 Results for systematic process

Impact factor	Year	Solar absorber	Surface temperature (°C)	Evaporation rate (kg m^{-2} h^{-1})	Efficiency (%)	References
29.14	2020	Konjac gucomannan (KGM)	30.3	3.2	90	[14]
19.58	2019	Multilayer polypyrrole Nanosheets	40	1.38	92.12	[15]
18.72	2020	Biomimetic 3D evaporator	33.1	2.63	96	[16]
14.82	2018	Cellular carbon sponge	47	1.31	74	[17]
11.2	2019	A 3D organic bucky sponge	43.7	1.35	87.4	[18]
10.43	2020	Pure PVA, graphene, and hydrogel structured evaporator	45.5	1.77	92	[19]
9.81	2019	Hydrogen-based ultrathin membrane	37	2.4	80	[20]
9.31	2017	Black TiOx nanoparticles	36	0.8	50.3	[21]
8.69	2020	Printed air-laid paper	45	1.75	81.3	[22]
8.42	2020	Defect-rich semiconductor HNb3O8 (D-HNb3O8) Nanosheets and polyacrylamide	46	1.401	71	[23]
8.02	2018	3D silica gel	41.4	1.356	85	[24]
8.01	2019	Three-dimensional (3D) MXene architecture	39	1.41	88	[25]
7.61	2020	Photothermal catalytic (PTC) gel with a hydrophobic membrane	52	1.49	93.8	[26]
7.16	2021	Cross-linked nanocellulose/PVA composite aerogel	45	1.4	88	[27]
6.92	2019	Hanging, hydrophilic photothermal arc-shaped fabric (PANi-cotton fabric)	40	1.94	89.9	[28]
5.67	2021	3D-MWCNTs@PVDF-HFP/radish	36.9	2.35	121.6	[29]
5.54	2018	Contactless solar evaporation structure	133	0.0026	40	[30]
4.31	2021	PVDF/graphene	40.6	1.2	84	[31]
4.09	2019	rGO/nickel foam	63.3	1.33	83.4	[32]
3.98	2020	Ferric tannate coated-membrane	33.9	1.33	63.9	[33]
3.83	2020	Oleophobic-basalt fiber-based (O-BFT)	40.1	1.49	90	[34]
2.95	2021	CB-SA-coated ramie fabric evaporator	43.4	1.81	96.6	[35]

(continued)

Table 1 (continued)

Impact factor	Year	Solar absorber	Surface temperature (°C)	Evaporation rate (kg m^{-2} h^{-1})	Efficiency (%)	References
1.97	2021	Pencil-traced-graphite	47.6	1.28	86.48	[36]
1.86	2018	$CuCr_2O_4/SiO_2$ composite membrane	37	1.32	88	[37]
1.58	2021	C-CuO composite	44.8	1.34	92.6	[38]
1.41	2020	Polypyrrole (PPy)-coated filter paper	34.4	1.21	92.2	[39]
1.34	2020	Hydrophobic carbonized coffee grounds	37	1.05	71	[40]
1.19	2020	Dual functional film	38	1.64	82.1	[41]

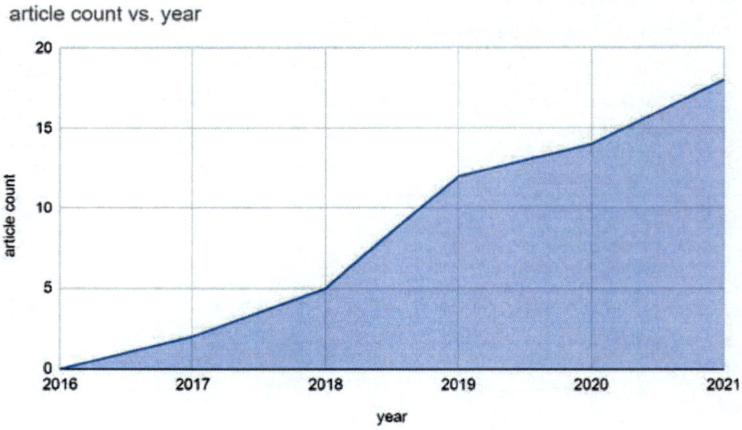

Fig. 2 Articles count versus year

surrounding environment. This results in heat transfer through conduction to the bulk water, convection, and radiation to the ambient surroundings. Therefore heat energy is reduced due to thermal losses leading to a low evaporation rate and efficiency.

In Table 2, the average surface temperature determined was 44.68 °C. From this review, it is clear that the vapor generated in most studies is relatively cool. Therefore the molecular kinetic energy is low hence low saturation vapor pressure. From 2018 to 2019, a significant drop in the average drop was recorded. The decrease in average temperature was interpreted by the evolution from metal-based [30, 37] and plasmonic-based [24] materials to polymer-based [15, 25, 32] and biomass-based [14] materials. From 2019 to 2021, a relatively constant trend was noticed due to great efforts in the research to study and fabricate materials with higher photothermal conversion efficiency with inbuilt wettability capabilities. For example fabrication of C-CuO composite [38].

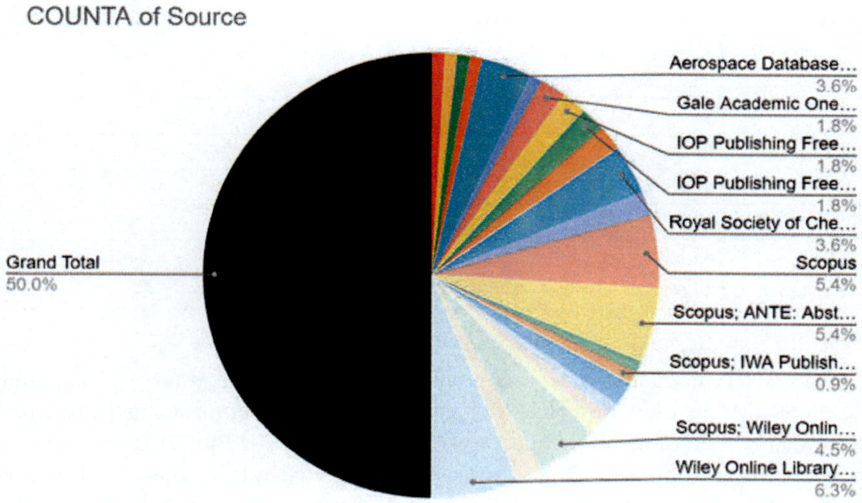

Fig. 3 Sources percentage

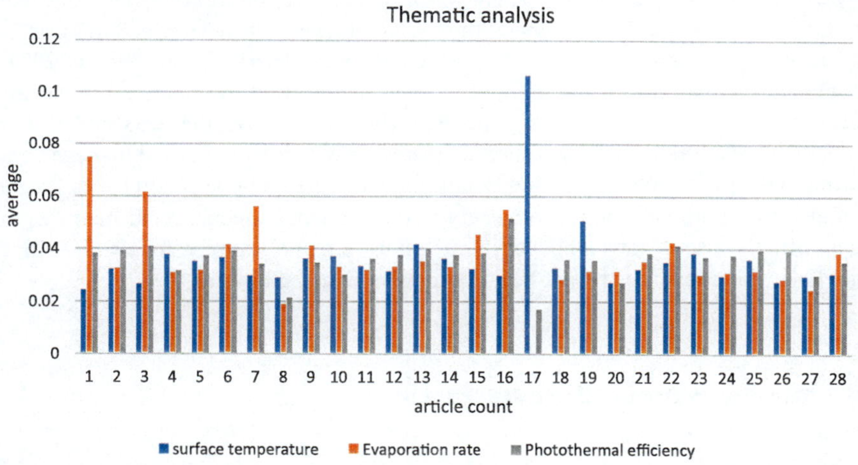

Fig. 4 Analysis

The average evaporation rate recorded was 1.53 kg m^{-2} h^{-1} which is approximately 1.5 L of water collected per hour if the water collection efficiency is assumed to be 100%. This study also assumed a constant 1 kW/m^2 solar influx for 8 h on a sunny day in a homestead of 5 people. The total liters collected in a day was calculated as

$$1.53 \times 8 = 12.24\,\mathrm{litresperday}$$

Table 2 Average of the variables versus year

Year	Average of surface temperature	Average of evaporation rate (kg m^{-2} h^{-1})	Average of efficiency (%)
2017	36	0.8	50.3
2018	64.6	1	71.75
2019	43.83	1.64	86.80
2020	39.57	1.72	83.94
2021	43.05	1.56	94.88
Grand total	44.68	1.53	83.95

The water production rate was relatively lower than the water consumption recommended by the World Health Organization (WHO) per person and also from the existing literature. An increment was noticed from 2017 to 2020 and slightly decreased in 2021 from the Table. The rise was interpreted by the tremendous increase in the utilization of materials with good wettability properties to provide a continuous flow of water to the evaporation surface. For example filter paper strip [39], cotton [28, 32], temperature gradient [16], air-laid paper [15, 22], and inbuilt super-hydrophilic properties [17, 21]. There was a slight reduction from 2020 to 2021 due to much focus by the researchers on improving efficiency as determined in Table 1.

A progressive elevation of average efficiency was observed from 2017 to 2021. Efficiency is the ratio of the amount of heat energy used in interfacial water–air evaporation to the total solar energy supplied. The elevation over the years studied was due to the development of photothermal materials with inbuilt thermal management properties [26] to confine heating within the structure, cone shaped [16] to create temperature gradient hence absorb energy from ambient temperature, the narrow gas gap [30] and use of low conductivity materials, e.g., polystyrene foams, 3D silica gel [24], underneath to prevent heat dissipation to the bulk water.

4 Summary and Future Research

SDIED is considered as one of the promising solutions to address freshwater shortages in rural areas. In this review, a systematic approach through PRISMA format to collect data from the Stellenbosch University online library. 50 articles were collected based on the content analysis of surface temperature, evaporation rate, and efficiency. Article and database distribution, surface temperature, evaporation rate, and efficiency were studied and discussed.

Remarkable progress was realized in the last five years of study in the field of SDIED, through the increase in the number of publications every year and the development of photothermal materials to increase efficiency. However, there are a few limitations that were realized and recommended to be researched on.

- Due to low saturation vapor pressure, there is a higher tendency of condensation on the evaporation surface. Moreover, low temperature and moist air near the surface in an enclosed system may create favorable conditions to enhance the growth of microorganisms. Therefore this provides a gap for research on mechanisms to curb the effect of the favorable conditions on the breeding of micro-organisms within a closed solar still system.
- The average evaporation rate for the five years reviewed was 1.53 kg m^{-2} h^{-1} which is still close to the theoretical evaporation rate limit which is 1.47 kg m^{-2} h^{-1}. The evaporation rate determined is still low to meet the practical demand for water production. This triggers the need for more exploration to increase water supply far above the theoretical value.
- There is a need to find the optimum temperature of the absorber because it determines the heat loss through convection and radiation to the ambient surrounding. This would help to increase evaporation rate and efficiency by reducing heat loss.

Acknowledgements The authors would like to thank the Mobility of African scholars for transformative engineering training, scholarship, and study mobility program for facilitation and funding of studies in the university.

References

1. Gan Q, Zhang T, Chen R, Wang X, Ye M (2019) Simple, low-dose, durable, and carbon-nanotube-based floating solar still for efficient desalination and purification. ACS Sustain Chem Eng 7:3925–3932. https://doi.org/10.1021/acssuschemeng.8b05036
2. Omarova A, Tussupova K, Hjorth P, Kalishev M, Dosmagambetova R (2019) Water supply challenges in rural areas: a case study from central Kazakhstan. Int J Environ Res Public Health 16(5). https://doi.org/10.3390/IJERPH16050688
3. Liu X, Mishra D, Wang X, Peng H (2020) Towards highly efficient solar-driven interfacial evaporation for desalination. pubs.rsc.org. Accessed 08 Jan 2022 [Online]. https://pubs.rsc.org/en/content/articlehtml/2020/ta/c9ta12612k
4. Gao M, Peh CK, Meng FL, Ho GW (2021) Photothermal membrane distillation toward solar water production. Small Methods 5(5):e2001200. https://doi.org/10.1002/smtd.202001200
5. Tao P et al (2018) Solar-driven interfacial evaporation. Nat Energy 3(12):1031–1041. https://doi.org/10.1038/s41560-018-0260-7
6. Gao M, Zhu L, Peh CK, Ho GW (2019) Solar absorber material and system designs for photothermal water vaporization towards clean water and energy production. Energy Environ Sci 12(3):841–864. https://doi.org/10.1039/c8ee01146j
7. Wang P (2018) Emerging investigator series: the rise of nano-enabled photothermal materials for water evaporation and clean water production by sunlight. Environ Sci Nano 5(5):178–189. https://doi.org/10.1039/c8en00156a
8. Deng Z et al (2017) The emergence of solar thermal utilization: solar-driven steam generation. J Mater Chem A Mater Energy Sustain 5(17):7691–7779. https://doi.org/10.1039/c7ta01361b
9. Xu K, Wang C, Li Z, Wu S, Wang J (2021) Salt mitigation strategies of solar-driven interfacial desalination. Adv Funct Mater 31(8):2007855. https://doi.org/10.1002/adfm.202007855
10. Noblit GW, Dwight Hare R (1988) Meta-ethnography: synthesizing qualitative studies, p 88. Accessed 12 Mar 2022 [Online]. https://books.google.com/books/about/Meta_Ethnography.html?id=fQQb4FP4NSgC

11. Grant MJ, Booth A (2009) A typology of reviews: an analysis of 14 review types and associated methodologies. Health Info Libr J 26(2):91–108. https://doi.org/10.1111/J.1471-1842.2009.00848.X
12. Tranfield D (2013) Towards a methodology for developing evidence-informed management knowledge by means of systematic review by David R. Tranfield, David Denyer, Palminder Smart :: SSRN. Accessed 12 Mar 2022 [Online]. https://papers.ssrn.com/sol3/papers.cfm?abstract_id=447301
13. Liberati A et al (2009) The PRISMA statement for reporting systematic reviews and meta-analyses of studies that evaluate health care interventions: explanation and elaboration. J Clin Epidemiol 62(10):e1–e34. https://doi.org/10.1016/J.JCLINEPI.2009.06.006
14. Guo Y, Lu H, Zhao F, Zhou X, Shi W, Yu G (2020) Biomass-derived hybrid hydrogel evaporators for cost-effective solar water purification. Adv Mater (Weinheim) 32(11):e1907061. https://doi.org/10.1002/adma.201907061
15. Wang X, Liu Q, Wu S, Xu B, Xu H (2019) Multilayer polypyrrole nanosheets with self-organized surface structures for flexible and efficient solar–thermal energy conversion. Adv Mater (Weinheim) 31(19):e1807716. https://doi.org/10.1002/adma.201807716
16. Wu L et al (2020) Highly efficient three-dimensional solar evaporator for high salinity desalination by localized crystallization. Nat Commun 11(1):521–521. https://doi.org/10.1038/s41467-020-14366-1
17. Zhu L, Gao M, Peh CKN, Wang X, Ho GW (2018) Self-contained monolithic carbon sponges for solar-driven interfacial water evaporation distillation and electricity generation. Adv Energy Mater 8(16):1702149. https://doi.org/10.1002/aenm.201702149
18. Zhu L, Ding T, Gao M, Peh CKN, Ho GW (2019) Shape conformal and thermal insulative organic solar absorber sponge for photothermal water evaporation and thermoelectric power generation. Adv Energy Mater 9(22):1900250. https://doi.org/10.1002/aenm.201900250
19. Lei W et al (2020) Hierarchical structures hydrogel evaporator and super hydrophilic water collect device for efficient solar steam evaporation. Nano Res 14(4):1135–1140. https://doi.org/10.1007/s12274-020-3162-5
20. Lu H et al (2021) High-yield and low-cost solar water purification via hydrogel-based membrane distillation. Adv Funct Mater 31(19):2101036. https://doi.org/10.1002/adfm.202101036
21. Ye M et al (2017) Synthesis of black TiO_x nanoparticles by Mg reduction of TiO_2 nanocrystals and their application for solar water evaporation. Adv Energy Mater 7(4). https://doi.org/10.1002/aenm.201601811
22. Xu J et al (2020) Solar-driven interfacial desalination for simultaneous freshwater and salt generation. Desalination 484:114423. https://doi.org/10.1016/j.desal.2020.114423
23. Yang M, Tan CF, Lu W, Zeng K, Ho GW (2020) Spectrum tailored defective 2D semiconductor nanosheets aerogel for full-spectrum-driven photothermal water evaporation and photochemical degradation. Adv Funct Mater 30(43). https://doi.org/10.1002/adfm.202004460
24. Gao M, Peh CK, Phan HT, Zhu L, Ho GW (2018) Solar absorber gel: localized macro-nano heat channeling for efficient plasmonic Au nanoflowers photothermic vaporization and triboelectric generation. Adv Energy Mater 8(25):1800711. https://doi.org/10.1002/aenm.201800711
25. Zhao X et al (2019) Macroporous three-dimensional MXene architectures for highly efficient solar steam generation. J Mater Chem A Mater Energy Sustain 7(17):1446–1455. https://doi.org/10.1039/c9ta00176j
26. Gao M, Peh CK, Zhu L, Yilmaz G, Ho GW (2020) Photothermal catalytic gel featuring spectral and thermal management for parallel freshwater and hydrogen production. Adv Energy Mater 10(23):2000925. https://doi.org/10.1002/aenm.202000925
27. Yang L et al (2021) Marine biomass-derived composite aerogels for efficient and durable solar-driven interfacial evaporation and desalination. Chem Eng J 417:128051. https://doi.org/10.1016/j.cej.2020.128051
28. Liu Z, Wu B, Zhu B, Chen Z, Zhu M, Liu X (2019) Continuously producing water steam and concentrated brine from seawater by hanging photothermal fabrics under sunlight. Adv Funct Mater 29(43):1905485. https://doi.org/10.1002/adfm.201905485

29. Lv B et al (2021) A self-floating, salt-resistant 3D Janus radish-based evaporator for highly efficient solar desalination. Desalination 510:115093. https://doi.org/10.1016/j.desal.2021.115093
30. Cooper TA et al (2018) Contactless steam generation and superheating under one sun illumination. Nat Commun 9(1):5086–5086. https://doi.org/10.1038/s41467-018-07494-2
31. Huang C et al (2021) Tailoring of a piezo-photo-thermal solar evaporator for simultaneous steam and power generation. Adv Funct Mater 31(17):2010422. https://doi.org/10.1002/adfm.202010422
32. Shan X et al (2019) Porous reduced graphene oxide/nickel foam for highly efficient solar steam generation. Nanotechnology 30(42):425403–425403. https://doi.org/10.1088/1361-6528/ab3127
33. Zhang C et al (2020) Ferric tannate photothermal material for efficient water distillation. Environ Sci: Water Res Technol 6(4):911–915. https://doi.org/10.1039/c9ew01016e
34. Chen L, Xia M, Du J, Luo X, Zhang L, Li A (2020) Super hydrophilic and oleophobic porous architectures based on basalt fibers as oil-repellent photothermal materials for solar steam generation. Chemsuschem 13(3):493–500. https://doi.org/10.1002/cssc.201902775
35. Gao C, Zhu J, Bai Z, Lin Z, Guo J (2021) Novel Ramie fabric-based draping evaporator for tunable water supply and highly efficient solar desalination. ACS Apple Mater Interfaces 13(6):7200–7207. https://doi.org/10.1021/acsami.0c20503
36. Tariq MZ, Hani Z, La M, Choi D, Park SJ (2021) Pencil-traced-graphite on cellulose: a rapid and solvent-less approach for solar steam generation. Int J Energy Res 45(4):6395–6404. https://doi.org/10.1002/er.6219
37. Shi Y, Li R, Shi L, Ahmed E, Jin Y, Wang P (2018) A robust $CuCr_2O_4/SiO_2$ composite photothermal material with underwater black property and extremely high thermal stability for solar-driven water evaporation. Adv Sustain Syst 2(3). https://doi.org/10.1002/adsu.201700145
38. Cao H, Jiao S, Zhang S, Xu J, Wang H, Guo C (2021) In situ synthesizing C-CuO composite for efficient photo-thermal conversion and its application in solar-driven interfacial evaporation. Int J Energy Res 45(5):7829–7839. https://doi.org/10.1002/er.6367
39. Zhang Y et al (2020) In situ chemo-polymerized polypyrrole-coated filter paper for high-efficient solar vapor generation. Int J Energy Res 44(2):1191–1204. https://doi.org/10.1002/er.5012
40. Wang C-F et al (2020) Preparation of efficient photothermal materials from waste coffee grounds for solar evaporation and water purification. Sci Rep 10(1):12769–12769. https://doi.org/10.1038/s41598-020-69778-2
41. Xu J et al (2020) All-day freshwater harvesting through combined solar-driven interfacial desalination and passive radiative cooling. ACS Apple Mater Interfaces 12(42):47612–47622. https://doi.org/10.1021/acsami.0c14773

Numerical Investigation of Cavitation Behavior for Dodecane and OME_3 Fuel in ECN Spray C Injector Nozzle

Srijna Singh, Akhil Ailaboina, Michele Battistoni, Mohammad Danish, and Kaushik Saha

Abstract This work numerically investigates the cavitation characteristics in the Engine Combustion Network (ECN) Spray C037 nozzle using two different fuels (n-Dodecane and OME_3). The cavitation modeling is performed using the CONVERGE code. RNG $k - \epsilon$ turbulence model is used to determine the effect of turbulence, and phase change inside the nozzle is predicted using the Homogeneous Relaxation Model (HRM) coupled with a multiphase mixture model. For model validation, the simulated mass flow rate and cavitation contour shape of n-dodecane fuel are compared with the experimental result provided in the literature. Finally, the comparison of results for n-dodecane and OME_3 fuel is presented for the Spray C nozzle.

Keywords Spray C nozzle · n-Dodecane · OME_3

1 Introduction

Cavitation in nozzle has been an extensive research topic for a few decades [1–4]. In certain regions, whenever the local pressure of flowing liquid falls below the vapor pressure of the fuel concerned, the phenomenon of cavitation occurs [5]. The cavitation may affect the fuel spray characteristics, discharge coefficient, atomization process, and fluid flow field boundary conditions. Ranz [6] carried out the primary experimental work on cavitating nozzles with the liquid jet spray system. The focus of this study was on the breaking of jet due to local inertial stresses and their effect

S. Singh · M. Danish
Department of Mechanical Engineering, Bennett University, Greater Noida, Uttar Pradesh 201310, India

A. Ailaboina · K. Saha (✉)
Department of Energy Science and Engineering, IIT Delhi, New Delhi 110016, India
e-mail: kaushiksaha@dese.iitd.ac.in

M. Battistoni
Department of Engineering, University of Perugia, 06125 Perugia, Italy

S. Das et al. (eds.), *Proceedings of the 1st International Conference on Fluid, Thermal and Energy Systems*, Lecture Notes in Mechanical Engineering,
https://doi.org/10.1007/978-981-99-5990-7_6

on spray characteristics in front of an orifice. Dabiri et al. [7] performed a numerical simulation of two-phase incompressible flow to discover the potential location of cavitation by comparing the conditions based on the total-stress criterion and pressure criterion in an axis-symmetric orifice using the finite volume method. They observed the substantial cavitation region in the flow field for the total-stress criterion as compared to the pressure criterion. The accuracy of numerical results for multiphase flow in nozzles also depends on the type of cavitation model used. Saha et al. [8] developed a cavitation model and tested its performance compared to two already existing cavitating models: Schnerr-Sauer and Zwart-Gerber-Belamri in Winklhofer nozzle [9]. The model proposed by Saha et al. [8] worked better than the existing models in ANSYS Fluent when a single fluid or mixture multiphase approach was used. Battistoni et al. [10] compared the prediction capabilities of two different cavitation models in a simple nozzle, one is a homogeneous mixture model, and another is the multi-fluid non-homogeneous model. They found that the quantitative mixture model overpredicts the void fraction as compared to the multi-fluid model when both results were compared with the experimental work.

The cavitation in injector nozzles is sensitive to geometric features such as inlet corner radius, degree of taper, and surface roughness. Schmidt et al. [3, 11, 12] performed the experimental and numerical study on different cavitating and non-cavitating injector nozzles like sharp-edged nozzle, round tip nozzle, and tapered nozzle hole. They found that sharp inlet nozzles initiate the cavitation, whereas rounded nozzles delay the cavitation due to reduced flow separation. Tapered nozzles were found effective for suppressing the cavitation. Örley et al. [13] numerically simulated the fully compressible two-phase homogeneous mixture model to examine the cavitation and turbulent scale effects in a nine-hole common rail-diesel injector using an implicit LES turbulence model. Small-scale turbulence was detected during opening and closing, and during the injection phase, a large vortical structure was formed. Moreover, vortex cavitation was observed at the outlet due to the tapered shape of nozzle holes.

Chaves et al. [14] conducted an experiment on transparent nozzles of different sizes to analyze the cavitation for sharp, tapered, and rough surface geometries. A modified laser-two-focus velocimeter was used to measure flow velocity and discharge. They observed that sharp geometries are more prone to cavitation than tapered or rounded nozzles due to sudden flow separation. Additionally, they concluded that rough surfaces are more susceptible to cavitation wear due to the availability of more sites for bubble formation. These bubbles rapidly collapse, producing shock waves that are strong enough to generate consequential damage to the surface like pitting, erosion, etc. Andriotis et al. [15] studied the origin, formation, development, lifetime, and influence of string cavitation on a three-dimensional transparent acrylic replica of the five-hole nozzle of vortex type. Their crucial observation was that there is an uneven dispersion of liquid due to string cavitation, and to prevent the sudden pressure drop tapered holes acts as a better alternative.

Brandon et al. [16] performed an experimental study on Engine Combustion Network (ECN) Spray C and Spray D diesel injectors using the optical and X-ray diagnostic techniques to study the effect of geometrical features on cavitation formation. They observed the asymmetric sheath of fuel vapor in Spray C, which affects the fuel spray structure near the nozzle exit. In contrast, no cavitation was observed in Spray D. Similarly, Maes et al. [17] performed experiments on Spray C and Spray D to study the effect of cavitation on spray penetration, combustion characteristics, and soot formation of n-dodecane fuel in the constant pressure vessel of multiple combustion facilities. They concluded that soot formation was sensitive to the main species of the ambient gas, i.e., carbon dioxide, water, and nitrogen, in the presence of constant oxygen. Tekawade et al. [18] performed an experiment on a Spray C nozzle to study the 3D flow field inside the diesel injector using a synchrotron X-ray source. They found strong flow separation and vapor-filled cavities related to the nozzle's asymmetric inlet corner.

It has been observed from the literature that people have studied cavitation in Spray C using n-dodecane and commercial diesel fuel only. To the best of our knowledge, no one has reported using alternative fuels like oxymethylene ether (OME_n). OME_n is a synthetic fuel, promising absorbents for carbon dioxide, and can be used as a drop-in replacement for conventional diesel fuel [19]. This paper compares cavitation characteristics in spray C for n-dodecane and OME_3 fuel using the CONVERGE v3.0 code. The fuel properties for n-dodecane and OME_3 are mentioned in the literature [19, 20].

2 Model Formulation

The Engine Combustion Network (ECN) injector nozzle Spray C037 is considered in this study. Spray C is a single orifice diverging nozzle with a nominal hole diameter of 200 μm. It is manufactured with a sharp inlet corner to produce cavitation and has 5% hydro erosion to maintain flow rate consistency in the injector [20]. The three-dimensional part image of Spray C and the section of the injector nozzle hole using X-ray tomography is shown in Fig. 1 [18].

The simulation is performed using the CONVERGE code, in which mesh is generated automatically during run time. A vertical cut plane showing the mesh with a maximum cell size of 500 μm and a minimum cell size of 15.625 μm is shown in Fig. 2. A truncated cone-shaped fixed embedding is used near the hole to improve the accuracy around the zone of interest. Five level of embedding is used, according to which the smallest cell has a dimension $2^5 = 32$ times smaller than the base/maximum grid size in the domain [21]. In our simulation, the base grid size is 500 μm, and 15.625 μm is the minimum cell size. There were around 1.6 million cells in the domain. The simulation was performed on 40 cores, and it took approximately 9 days to reach the time of 1 ms in Spray C for both n-dodecane and OME_3 fuel.

The inlet of the nozzle is specified with the x-component velocity of 5.4 m/s and outlet as a pressure outlet, and the walls are kept in the no-slip condition with a needle as a moving wall. The standard law-of-wall function is used to model the boundary layer. The transient simulation is run until both the inflow and outflow mass

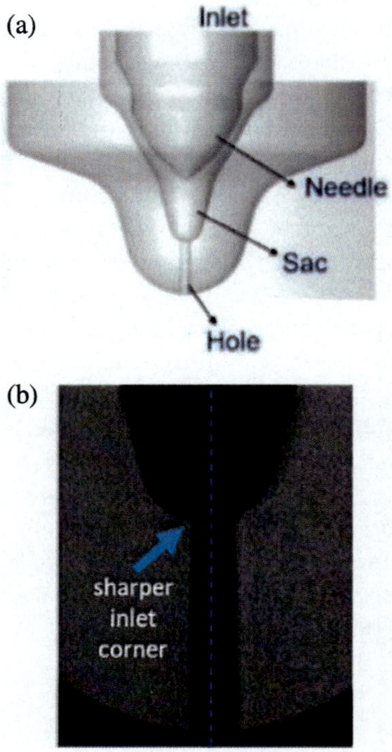

Fig. 1 Surface rendering of X-ray tomography of ECN **(a)** Spray C injector nozzle and **(b)** asymmetrical sharp corner hole of Spray C [18]

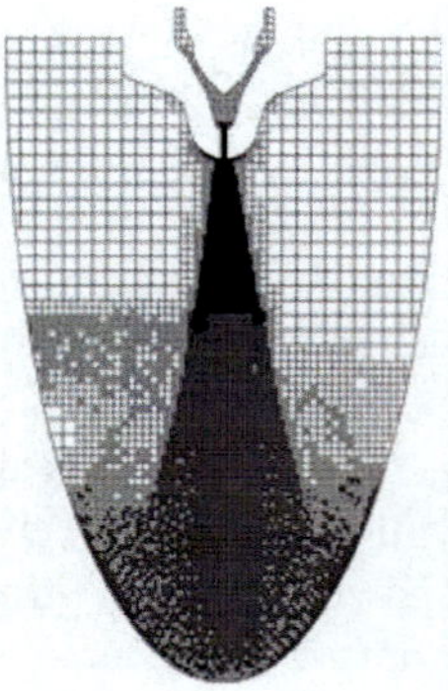

Fig. 2 Vertical cut-plane showing the mesh with 15.625 μm as minimum grid size and 500 μm base grid size

flow rates are stabilized, and results are presented at 1 ms. In the liquid fuel, very small mass fraction of non-condensable gas ($N_2 = 10^{-6}$ mass fraction) is used as a nucleation site for cavitation initiation. Pressure implicit with the splitting operator (PISO) algorithm is used for the pressure–velocity coupling. Reynolds number based on the fuel properties exit velocity, and hole diameter for n-dodecane and OME_3 is 32691 and 129,223, respectively. The cavitation number based on the inlet, back,

Table 1 Operating conditions used in the simulation

Parameters	Spray C
Inlet velocity (m/s)	5.4
Chamber pressure (MPa)	2
Chamber temperature (K)	303
Fuel temperature (K)	338
Chamber fluid	N_2

and saturation pressure for n-dodecane and OME_3, are 84 and 94, respectively [5]. The operating condition used in the simulation is given in Table 1.

3 Governing Equations

The single fluid, i.e., multiphase mixture model, is used in this study. The mass, momentum, and energy conservation equations are as follows [21].

$$\frac{\partial \rho}{\partial t} + \frac{\partial u_j \rho}{\partial x_j} = 0 \tag{1}$$

$$\frac{\partial \rho u_i}{\partial t} + \frac{\partial \rho u_i u_j}{\partial x_j} = -\frac{\partial p}{\partial x_i} + \frac{\tau_{ij}}{\partial x_j} \tag{2}$$

$$\frac{\partial \rho e}{\partial t} + \frac{\partial \rho e u_j}{\partial x_j} = -p\frac{\partial u_j}{\partial x_j} + \tau_{ij}\frac{\partial u_i}{\partial x_j} + \frac{\partial}{\partial \varkappa_j}\left(k_{eff}\frac{\partial T}{\partial x_j}\right) + \frac{\partial}{\partial x_j}\left(\rho \sum_m D_{eff} h_m \frac{\partial Y_m}{\partial x_j}\right) \tag{3}$$

where the effective (viscous + turbulent) stress tensor is given by

$$\tau_{ij} = \frac{\partial}{\partial x_j}\left[(\mu + \mu_t)\left(\frac{\partial u_i}{\partial x_j} + \frac{\partial u_j}{d x_i} - \frac{2}{3}\frac{\partial u_k}{\partial x_k}\right)\right] \tag{4}$$

Awhere μ is the molecular viscosity and μ_t is the turbulent viscosity which is modeled as

$$\mu_t = C_\mu \rho \frac{k^2}{\epsilon} \tag{5}$$

using RNG $k - \epsilon$ turbulence model. The mixture density is calculated from.

$$\rho = \sum \rho_m \alpha_m \tag{6}$$

$$\alpha_m = Y_m \frac{\rho}{\rho_m} \tag{7}$$

where ρ is the mixture density and ρ_m is the species density in the mixture. The species used are liquid fuel, vapor fuel, and N_2. The Ideal Gas Equation is used to calculate the densities of all the gaseous species. α_m and Y_m represent the volume fraction and mass fraction of the individual species, respectively. The species conservation equation is used to determine the individual mass fractions:

$$\frac{\partial \rho Y_m}{\partial t} + \frac{\partial u_j \rho Y_m}{\partial x_j} = \frac{\partial}{\partial x_j}(\rho D \frac{\partial Y_m}{\partial x_j}) + S_m \tag{8}$$

where S_m is the source or sink term. The non-condensable gases do not have the source or sink term. The S_m is estimated using Homogeneous Relaxation Model which is described in the Sect. 3.2.

3.1 Turbulence Model

The Re-Normalization (RNG) $k - \epsilon$ turbulence model is a two-equation model that solves the turbulent kinetic energy (κ) and turbulent dissipation rate (ϵ). This model was developed to consider the effect of smaller length scales with the help of production term in the dissipation transport equation. The RNG $k - \epsilon$ model equations are [21]:

$$\frac{\partial \rho k}{\partial t} + \frac{\partial \rho u_j k}{\partial x_j} = \tau_{ij} \frac{\partial u_i}{\partial x_j} + \frac{\partial}{\partial x_j} \frac{\mu}{Pr_k} \frac{\partial k}{\partial x_j} - \rho \epsilon \tag{9}$$

$$\frac{\partial \rho \epsilon}{\partial t} + \frac{\partial \rho u_j \epsilon}{\partial x_j} = \frac{\partial}{\partial x_j}\left(\frac{\mu}{P_{r_\epsilon}} \frac{\partial \epsilon}{\partial x_j}\right) + C_{\epsilon 3} \rho \epsilon \frac{\partial u_j}{\partial x_j} + \left(C_{\epsilon 1} \frac{\partial u_i}{\partial x_j} \tau_{i_j} - C_{\epsilon 2} \rho \epsilon\right) \frac{\epsilon}{k} - \rho R \tag{10}$$

where $R = \{[C\mu\eta^3(1 - \eta/\eta_0)]/(1 + \beta\eta^3)\}[\varepsilon^2/k]$ and $\eta = (k/\epsilon)|S_{ij}|$. The model constants for the RNG $\kappa - \epsilon$ model are as $C_\mu = 0.0845$; $C_{\epsilon 1} = 1.42$; $C_{\epsilon 2} = 1.68$; $C_{\epsilon 3} = -1.0$; $\beta = 0.012$.

3.2 Homogeneous Relaxation Model (HRM)

The HRM model evaluates the source term S_m in the fuel species conservation equation. It determines the rate at which local vapor quality proceeds toward the equilibrium vapor quality. The quality of vapor (x) is calculated as [21]

$$x = \frac{m_{vap}}{m_{vap} + m_{liq}} \tag{11}$$

The S_m is estimated using rate of change of local vapor quality (Dx/Dt). The term Dx/Dt is expressed in the HRM model with the help of time scale θ as

$$\frac{Dx}{Dt} = \frac{\overline{x} - x}{\theta} \tag{12}$$

$$\theta = \theta_0 \alpha^{-0.54} \psi^{-1.76} \tag{13}$$

where $\theta_0 = 3.84 \times 10^{-7}$; $\psi = \frac{|p_{sat} - p|}{p_{crit} - p_{sat}}$; $\alpha = \alpha_v + \alpha_{N_2}$.

4 Results and Discussion

In this section, the cavitation characteristics for n-dodecane fuel and OME_3 have been studied for the cavitating Spray C nozzle in the CONVERGE v3.0 code. All results are presented at 1 ms of physical time. In Fig. 3, the model validation is illustrated by comparing the void fraction contour on six different slices perpendicular to the nozzle flow axis, obtained from simulation for n-dodecane fuel with the experimental results in the literature [14]. The first slice is taken at the nozzle entrance, i.e., $x = -974$ μm, and the sixth slice is taken on the downstream side, i.e., $x = -134$ μm. The void fraction contour shows that cavitation begins at the sharp inlet corner of the nozzle and extends to the nozzle's exit. Most of the nozzle cross-section is filled with a liquid phase enclosed with a thin layer of vapor near the wall. The presence of a vapor cavity shows the flow separation at the sharp corner of Spray C, which is predicted well using simulation. Next, the comparison of results obtained through simulation for n-dodecane and OME_3 is presented.

Figure 4 compares the mass flow rate of n-dodecane and OME_3 fuels from the nozzle exit to the chamber. Initially, there is fluctuation in the mass flow rate value till 0.2 ms due to pressure oscillations in the sac region, and later the mass flow rates become almost constant. The stabilized mass flow rate of n-dodecane fuel is 10.42 mg/ms, which is consistent with the literature [17, 20, 22], whereas the mass flow rate for OME_3 is 12 mg/ms. This difference in mass flow rate is mainly due to difference in density and viscosity of both fuels. At 303 K, the density of n-dodecane is 718 kg/m^3, and for OME_3 it is 1032 kg/m^3, whereas the viscosity of n-dodecane is 0.0028 Pa s and OME_3, it is 0.001 Pa s. The higher density and lower viscosity of fuel results in higher mass flow rate of OME_3 as compared to n-dodecane [22].

It will be more meaningful to compare the mass fraction profiles of n-dodecane and OME_3 fuels inside the nozzle to understand how the cavity is forming using these two fuels. Figure 5 compares the mass fraction contour of vapor for n-dodecane and OME_3 fuel. It is evident from the contour that the mass fraction of OME_3 is higher than the n-dodecane at the asymmetric wall of the hole resulting in more cavitation

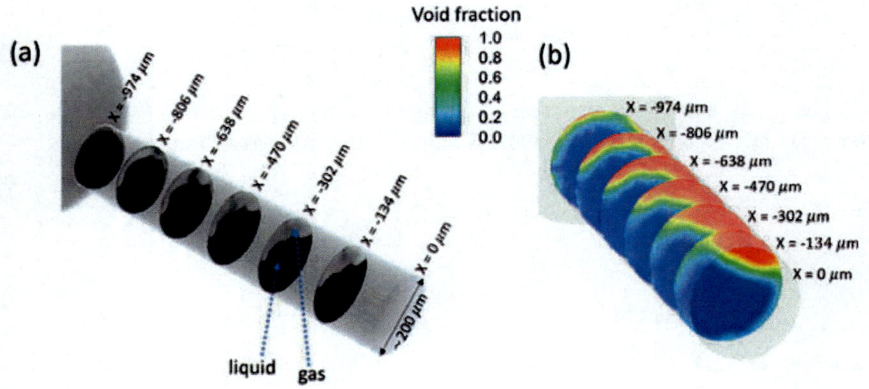

Fig. 3 Comparison between **(a)** intensity map from X-ray tomography experiments [18] with **(b)** void fraction contours from simulation for n-dodecane

Fig. 4 Mass flow rate versus time in Spray C nozzle for n-dodecane and OME_3

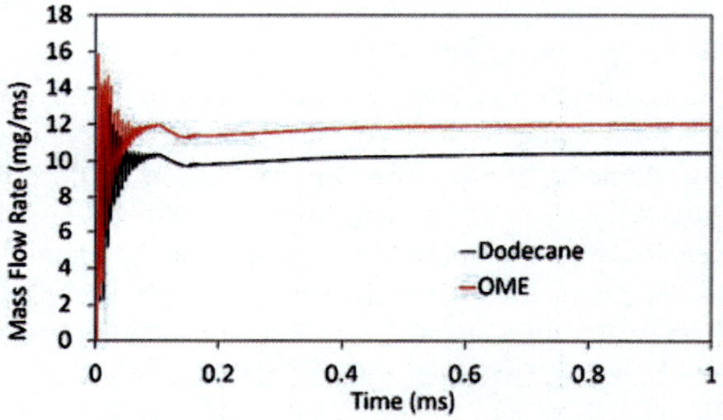

than n-dodecane. Reynolds number of OME_3 is almost four times higher and viscosity being nearly three times lower for OME_3 as compared to n-dodecane, the extent of flow separation for OME_3 will be higher, and such phenomenon has been reported for different fuel property effects in the literature [5, 22].

Next, in Fig. 6, the density contour of both fuels is presented. The center of the nozzle is filled with the higher density liquid fuel, whereas the wall is covered with the low-density gas, which can be seen in the contour. The density at the inlet of the nozzle for OME_3 is around 1000 kg/m^3, while for dodecane, it is around 800 kg/m^3.

5 Conclusions

This paper compares the numerical results obtained for n-dodecane and OME_3 fuel using commercial CONVERGE code in Engine Combustion Network (ECN) Spray C nozzle. The objective was to study the cavitation behavior of these fuels inside

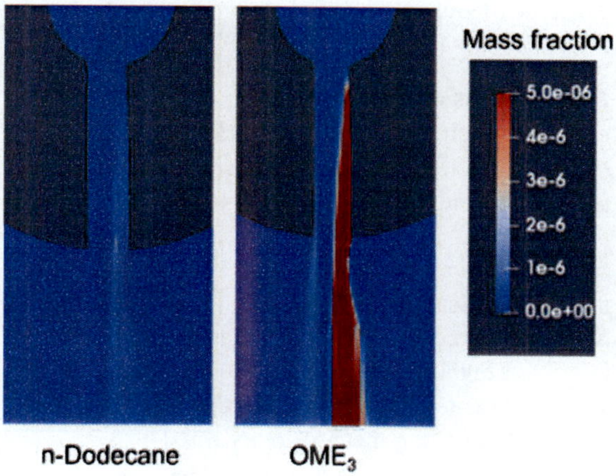

Fig. 5 Mass fraction contours for n-dodecane and OME_3

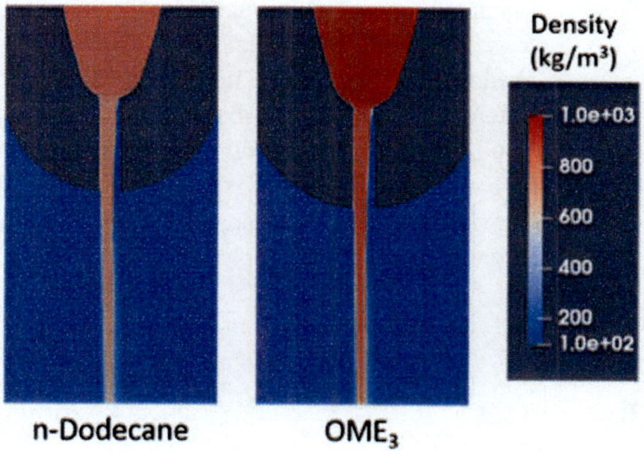

Fig. 6 Density variation contours for n-dodecane and OME_3

the injector nozzle. The Homogeneous (HRM) model coupled with the multiphase mixture model is used for cavitation modeling, and RNG $k - \epsilon$ is used to model the turbulence effect. The void fraction contour and the mass flow rate of n-dodecane fuel are compared with the experimental results provided in the literature. The mass flow rate and contours of mass fraction and density are presented for OME_3 and n-dodecane fuels comparison.

The simulations predict the flow separation at the sharp asymmetric corner of the nozzle. The flow separation at the corner initiates the cavitation along the wall to the nozzle exit. The mass flow rate obtained for n-dodecane is 10.42 mg/ms and for OME_3 is 12 mg/ms. The mass flow rate difference is mainly attributed to the higher density of OME_3 fuel. The OME_3 fuel produced more cavities than n-dodecane fuel in the Spray C nozzle.

Nomenclature

D	Mass diffusivity [m^2/s]
e	Specific internal energy [J/kg]
h_m	Specific enthalpy of mth species [kJ/kg]
k	Turbulent kinetic energy [m^2/s^2]
p	Local pressure [Pa]
p_{sat}	Saturation pressure [Pa]
p_{crit}	Critical pressure [Pa]
S_m	Source term in species transport equation.
t	Time [s]
T	Local cell temperature [K]
u_i	Advecting mean velocity [m/s]
u_j	Advected mean velocity [m/s]
x	Local cell vapor quality
$\overline{x}$	Local cell equilibrium quality
Y_m	Mass fraction of mth species
α	Void fraction (vapor and non-condensable gases)
ϵ	Rate of dissipation of turbulent kinetic energy [m^2/s^3]
θ, θ_0	Equilibrium timescale and empirical time constants [s]
μ, μ_t	Dynamic and turbulent viscosity coefficient [kg/m s]
ρ, ρ_m	Density of mixture and density of species in mixture [kg/m^3]
τ_{ij}	Reynolds stress [Pa]

References

1. Chen Y, Heister S (1992) Computational modeling of cavitated injector flows. In: 28th International JP conference
2. Arcoumanis AC, Flora H, Gavaises M, Kampanis N, Horrocks R, Arcoumanis C, Flora H, Gavaises M, Kampanis N, Co FM (1999) Investigation of cavitation in a vertical multi-hole injector. SAE Trans 661–678
3. Schmidt DP, Rutland CJ, Corradini ML (1997) A numerical study of cavitating flow through various nozzle shapes. SAE technical paper series 1
4. Arcoumanis C, Flora H, Gavaises M, Badami M (2010) Cavitation in real-size multi-hole diesel injector nozzles. SAE technical paper series 1
5. Saha K (2014) Modelling of cavitation in nozzles for diesel injection applications, p 144
6. Ranz WE (1958) Some experiments on orifice sprays. Can J Chem Eng 36:175–181
7. Dabiri S, Sirignano WA, Joseph DD (2007) Cavitation in an orifice flow. Phys Fluids 19:1–9
8. Saha K, Li X (2015) Assessment of cavitation models for flows in diesel injectors with single- and two-fluid approaches. J Eng Gas Turbines Power 138:011504
9. Winklhofer E, Kull E, Kelz E, Morozov A (2001) Comprehensive hydraulic and flow field documentation in model throttle experiments under cavitation conditions. In: Proceedings of the ILASS-Europe conference, Zurich, pp 574–579
10. Battistoni M, Som S, Longman DE (2014) Comparison of mixture and multifluid models for in-nozzle cavitation prediction. J Eng Gas Turbines Power 136

11. Schmidt DP, Rutland CJ, Corradini ML (2014) SAE technical cavitation in two-dimensional asymmetric nozzles
12. Duke D, Swantek A, Tilocco Z, Kastengren A, Fezzaa K, Neroorkar K, Moulai M, Powell C, Schmidt D (2014) X-ray imaging of cavitation in diesel injectors. SAE Int J Engines 7:1003–1016
13. Örley F, Hickel S, Schmidt SJ, Adams NA (2017) Large-Eddy simulation of turbulent, cavitating fuel flow inside a 9-hole diesel injector including needle movement. Int J Engine Res 18:195–211
14. Chaves H, Knapp M, Kubitzek A, Obermeier F, Schneider T (2010) Experimental study of cavitation in the nozzle hole of diesel injectors using transparent nozzles. SAE technical paper series 1
15. Andriotis A, Gavaises M, Arcoumanis C (2008) Vortex flow and cavitation in diesel injector nozzles. J Fluid Mech 610:195–215
16. Sforzo BA, Matusik KE, Powell CF, Kastengren AL, Daly S, Cenker E, Pickett LM, Crua C, Manin J (2019) Fuel nozzle geometry effects on cavitation and spray behavior at diesel engine conditions. In: Proceedings of the 10th international symposium on cavitation, pp 474–480
17. Maes N, Skeen SA, Bardi M, Fitzgerald RP, Malbec LM, Bruneaux G, Pickett LM, Yasutomi K, Martin G (2020) Spray penetration, combustion, and soot formation characteristics of the ECN Spray C and Spray D injectors in multiple combustion facilities. Appl Therm Eng 172:115136
18. Tekawade A, Sforzo BA, Matusik KE, Kastengren AL, Powell CF (2020) Time-resolved 3D imaging of two-phase fluid flow inside a steel fuel injector using synchrotron X-ray tomography. 1–10
19. Kulkarni A, García EJ, Damone A, Schappals M, Stephan S, Kohns M, Hasse H (2020) A force field for poly(oxymethylene) dimethyl ethers (OME n). J Chem Theory Comput 16:2517–2528
20. Westlye FR, Battistoni M, Skeen SA, Manin J, Pickett LM, Ivarsson A (2016) Penetration and combustion characterization of cavitating and non-cavitating fuel injectors under diesel engine conditions. SAE technical paper 2016–April
21. Saha K, Som S, Battistoni M (2017) Investigation of homogeneous relaxation model parameters and their implications for gasoline injectors. At Sprays 27:345–365
22. Payri R, Gimeno J, Cuisano J, Arco J (2016) Hydraulic characterization of diesel engine single-hole injectors. Fuel 180:357–366

Experimental Investigation and the CFD Modeling of a Invelox Type Wind Turbine

Hiranmoy Samanta, Debajit Saha, Kamal Golui, and Rajesh Dey

Abstract The peculiar design confronts the obstacles faced by conventional turbines and removes the barrier of tower-loading turbines. Compared to the standardization of wind energy generation by massive blades attached at the top of the wind tower this model is different by keeping in mind the economic perspective as well as the people and wildlife. The Invelox design perspective captures the wind Omni-directionally and is directed toward the funnel-shaped design leading to the ground level, the impact of the high velocity wind energy generation by striking the blades of the turbine. The objective of this work is to design and modeling of an Invelox turbine and to understand l fluid flow pattern and analysis inside the Invelox where the wind turbines are placed with the standardization. Experimental and Numerical Modeling is done and validated against literature data. Multiple variations are implied to check and optimize the computations and the relation between wind flow direction and inner geometry of the system. The study shows it is possible to orient the flow of wind by capturing, accelerating, and concentrating at a specified optimization level to obtain maximum power output. The results present Invelox as a better way of harnessing wind energy over the other available wind-powered mechanisms used for energy harnessing.

Keywords Wind energy · Nonconventional energy · Ducted turbines · Wind velocity · FVM · CFD

H. Samanta (✉) · K. Golui · R. Dey
Department of Mechanical Engineering, Gargi Memorial Institute of Technology, Kolkata 700144, India
e-mail: hiranmoy.me_gmit@jisgroup.org

D. Saha
Department of Mechanical Engineering, NIT Sikkim, Sikkim 737139, India

S. Das et al. (eds.), *Proceedings of the 1st International Conference on Fluid, Thermal and Energy Systems*, Lecture Notes in Mechanical Engineering,
https://doi.org/10.1007/978-981-99-5990-7_7

1 Introduction

The Invelox design perspective captures the wind Omni-directionally and is directed toward the funnel-shaped design leading to the ground level; the impact of the high velocity wind energy generation by striking the blades of the turbine. An experimental setup is developed by keeping the aspect ratio and experiments were done to obtain the flow parameters. A FVM-based CFD model is used and numerical results were presented. Variations were implemented to optimize the computations and to check the relation between wind flow direction and inner geometry of the system. The study shows it is possible to orient the flow of wind by capturing, accelerating, and concentrating at a specified optimization level to obtain maximum power output. The FVM method is used. ANSYS platform is used to represent the modeling.

2 Literature Review and Objective

Farokhzade and Maghrebi [1] stated that the performance of Invelox turbine depends on the direction of the angle of wind. They studied and improved the performance of Invelox for different wind directions and also by changing the number of blades. They also studied that the escaping of wind can be reduced by increasing the length of the upper funnel. Ahmed et al. [2] proposed the designing and modeling of an Invelox and studied the flow field inside the turbine. It is found that higher wind velocity improves the output power. Shalar et al. [3] studied the design and modeling as well as the relevance of replacing traditional wind turbines with Invelox turbines. Golozar et al. [4] introduced the concept of an aerodynamic controllable roof structure to improve the effectiveness of the Invelox turbine. This helps in improving the efficiency as well as taking care that no wind is escaping the system. Simulations of computational fluid dynamics had been done to validate the capability of the mechanism and fuzzy controls had been used to control roof orientations. Ratna Reddy et al. [5] have simulated using CFD to study the relation of the geometry of the system and the wind direction in Invelox turbine by varying the vane heights and throat diameters. The outcome has been stated that capturing, acceleration and concentrating is possible to increase the power output. Ding and Guo [6] proposed a straight-through layout with a windshield to improve the aerodynamic performance of Invelox.

The goals of the current work are:

- Experimental investigation of the Invelox model prototype.
- Check the adequacy of the framework under different free-stream wind speeds.
- Really take a look at the omnidirectional property of the framework.
- Compare and validate the computational outcomes with the previous work outcomes.
- To further attempt and change and develop the system toward more efficient and useful one.

- The different configuration setup and experimentation.

3 Features of Invelox Turbine

(1) There is a decoupling between the inlet and the turbine. This shows that, contingent upon the expected speed proportion and ecological circumstances, the inlet size may be fluctuated while keeping the turbine as little as could really be expected.
(2) The Wind Turbine Generator (WTG) will be located at ground level to reduce operation and maintenance costs by separating the intake and turbine locations [7, 8].
(3) Invelox can be planned with gear proportions of at least 6 because of the partition of the admission and vents where the turbine is found. This permits the turbine to work at higher breeze speeds, producing greater power. More modest edges with productive fast generators are frequently utilized.
(4) By working at higher breeze speeds, the edge size is diminished by up to 85%, bringing about material, assembling, transportation, and establishment cost investment funds [8].
(5) Since the admission is omnidirectional, there is no requirement for enormous courses or engines to turn it toward the breeze [9, 10].
(6) Invelox can be built for power appraisals going from 500 W to 25 MW. Contingent upon how much air caught.
(7) Since the inlet and turbine are decoupled, a few admissions can be connected to working on mass stream and subsequently power yield [7–9].

4 Methodology

The basic steps involved in the designing, as well as the simulation of Invelox type wind turbine is being described with the effect of (Figs. 1 and 2; Table 1)

i. Intake design of the hopper.
ii. The channel carrying and accelerating the inlet wind.
iii. Boosting of the wind speed by a Venturi flume.
iv. Conversions of wind energy.
v. Diffuser.

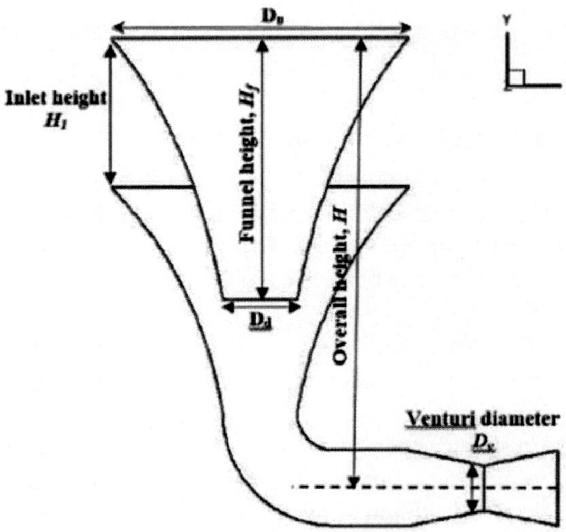

Fig. 1 The original design of Invelox Allalei and Andreopoulos [11]. The drawing is not to scale

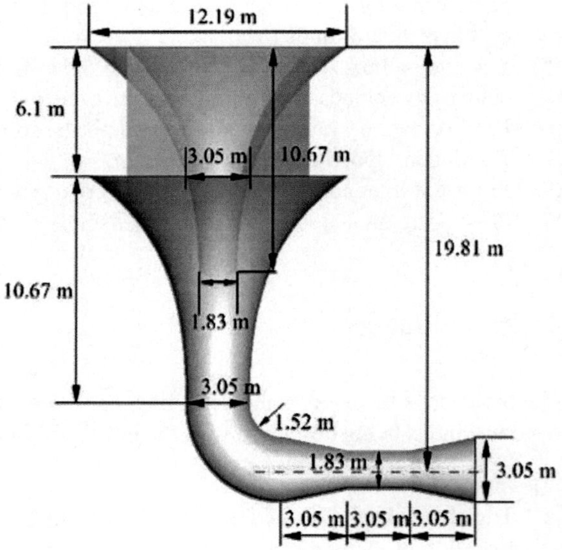

Fig. 2 The Invelox design of Allaei and Andreopoulos [11] with the proportional dimensions

Table 1 Dimensional parameter of the Invelox with the scale factor for the present work

Parameter	Symbol	Dimension [m]	Scale factor 1:28
Upper funnel diameter	D_u	12.19	0.435
Overall height	H	19.81	0.708
Funnel height	H_f	10.67	0.381
Lower funnel diameter	D_d	1.86	0.066
Inlet height	H_I	6.1	0.218
Venturi diameter	D_v	1.86	0.066

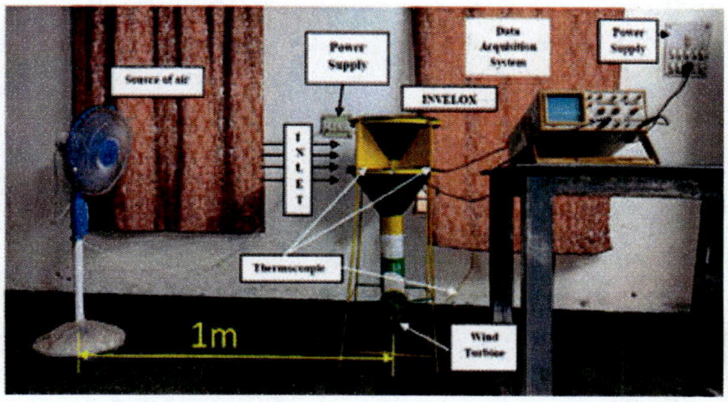

Fig. 3 Experimental set up of Invelox turbine

5 Materials and Methods

5.1 Experimental Set Up

The experimental setup consists of the Invelox turbine with the proper Model to prototype ratio. The numerical results were validated with the experimental data after a successful grid independency test. The experimental data validated with Abe and Ohya [12]. The numerical data shows a good agreement with the experimental data. Probes are used to measure the wind velocity and wind pressure at different locations of the setup. The thermocouples are connected with a data logger followed by a computer (Figs. 3 and 4).

6 Mathematical Model

6.1 Numerical Modeling

CFD utilizes various calculations and mathematical procedures to acquire legitimate arrangements and to assess the issue that contains a liquid stream. The present CFD displaying depends on the rudimentary conditions of liquid elements. Those are

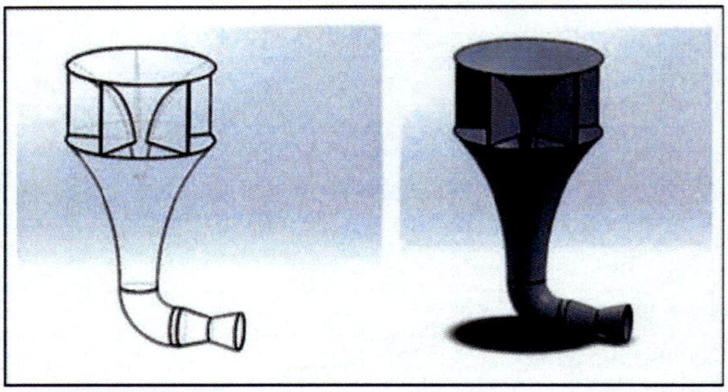

Fig. 4 Complete design model

1. Mass conservation.
2. Momentum conservation.
3. Energy conservation.

It is found that the $k - \varepsilon$ turbulence model affects the average velocity of the venturi. The data is compared with the numerical data of Allaei and Andrianopoulos [11], there is a relative error of 0.8% and the velocity is found to be 7 m/s.

6.2 Reynolds-Averaged Navier–Stokes Equations

The conservation form of Navier–Stokes equations are:

$$\frac{\partial u_i}{\partial x_i} = 0 \tag{1}$$

$$\rho \frac{\partial (u_j u_i)}{\partial x_j} = -\frac{\partial p}{\partial x_i} + \frac{\partial (2\mu s_{ij})}{\partial x_j} \tag{2}$$

And, $s_{ij} = \frac{1}{2}(\frac{\partial u_i}{\partial x_j} + \frac{\partial u_j}{\partial x_i})$ is the strain-rate tensor. Therefore, Eq. (2) can be written as

$$\rho u_j \frac{\partial (u_i)}{\partial x_j} = -\frac{\partial p}{\partial x_i} + \mu \frac{\partial^2 u_i}{\partial x_i \partial x_j} \tag{3}$$

The field variables u_i and p must be expressed as the sum of mean and fluctuating components as a result of turbulent flow characteristics.

$$u_i = U_i + \acute{u}_i \text{ and } p_i = P_i + \acute{p}_i \tag{4}$$

The bar represents the time average. Equations (4 and 5) are obtained by inserting these definitions into Eqs. (1 and 2).

$$\frac{\partial U_i}{\partial x_i} = 0 \tag{5}$$

$$\rho \frac{\partial (U_j U_i)}{\partial x_j} = \frac{\partial P}{\partial x_i} + \frac{\partial}{\partial x_j}\left(2\mu S_{ij} - \rho \overline{u_i' u_j'}\right) \tag{6}$$

In Eq. (6), $S_{ij} = \frac{1}{2}(\frac{\partial U_i}{\partial x_j} + \frac{\partial U_j}{\partial x_i})$ and $\tau_{ij} = -\overline{u_i' u_j'}$ are the mean strain-rate tensor and the Reynolds stress tensor, respectively. So, Eq. 6 can be rewritten as follows:

$$U_j \frac{\partial (U_i)}{\partial x_j} = -\frac{\partial P}{\partial x_i} + \nu \frac{\partial^2 U_i}{\partial x_i \partial x_j} - \frac{\partial \overline{u_i' u_i'}}{\partial x_j} \tag{7}$$

There are also various turbulence models that can be used to define unknown quantities. The $k - \varepsilon$ model is employed in this paper to achieve this purpose.

6.3 Turbulence Model

In a $k - \varepsilon$ model two transport equations are used which signifies the turbulent traits of a turbulent flow. One of these traits is turbulent kinetic energy denoted by k, which signifies the energy in turbulence, and the other trait is called specific dissipation which is denoted by w. Specific dissipation (w) is used to forecast the turbulence scale. $k - \varepsilon$ turbulence model for shear stress transport is interpreted as a model of eddy-viscosity that comprises two equations. For the inner region of a boundary layer, this model uses a k-formulation. According to the properties of the shear stress transport $k - \varepsilon$ turbulence model, it can be used all the way down to the wall through the viscous sub-layer. So, this model can be used for flows that are turbulent and have low Reynold's number without any supplementary damping functions. In addition, in the free-flow, the formula can be combined with the k-model, reducing the sensitivity of the model to the turbulent characteristics of the input free-flow. Another advantage of the $k - \varepsilon$ model is that it can accurately predict adverse pressure gradients and separate flow. In places where normal distortion is significant, such as stagnation and severe acceleration, the $k - \varepsilon$ model produces slightly too high levels of perturbation. The kinetic energy and the specific dissipation rate are calculated as follows in this method:

$$\frac{\partial k}{\partial t} + U_j \frac{\partial k}{\partial x_j} = P_k - \beta^* k\omega + \frac{\partial}{\partial x_j}\left[(\nu + \sigma_k \nu_T)\frac{\partial k}{\partial x_j}\right] \tag{8}$$

7 Results and Discussion

Quantitative results were obtained for the present study and compared with the experimental results of Abe and Ohya [12]. Pressure and velocity contours obtained using turbulence models have been presented (Figs. 5 and 6).

The dimensionless speed (speed partitioned by the breeze speed) shapes on a middle cross segment of Invelox, utilizing different choppiness models, are introduced in Fig. 7.

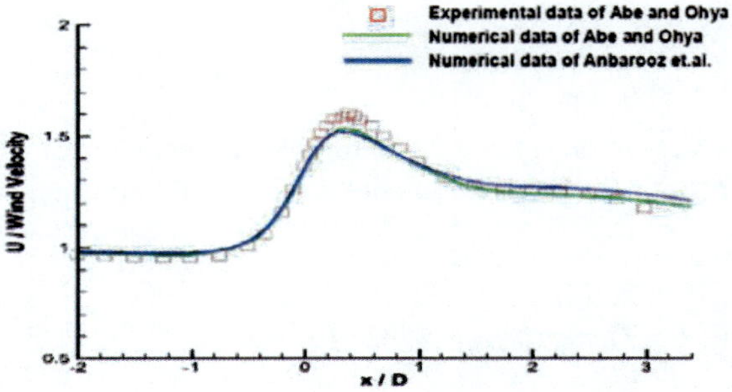

Fig. 5 Reference graph of Abe and Ohya [12] and Anbarsooz et al. [13]

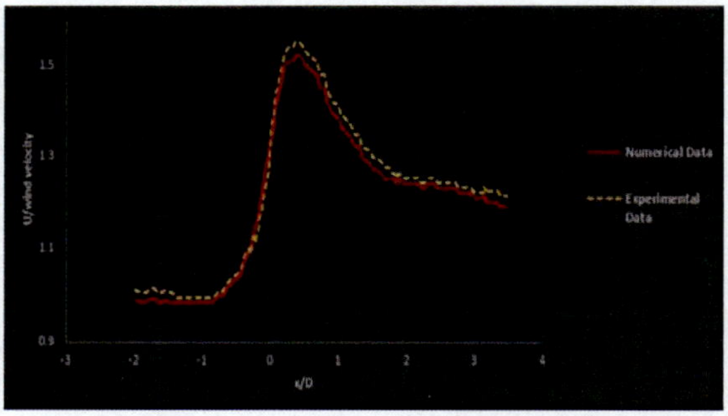

Fig. 6 Comparison of numerical and experimental data with Abe and Ohya [12] and Anbarsooz et al. [13]

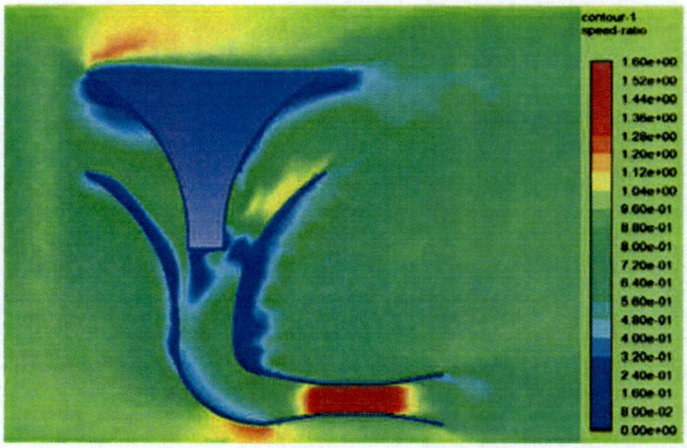

Fig. 7 Standard $k - \varepsilon$ velocity contour on median cross section

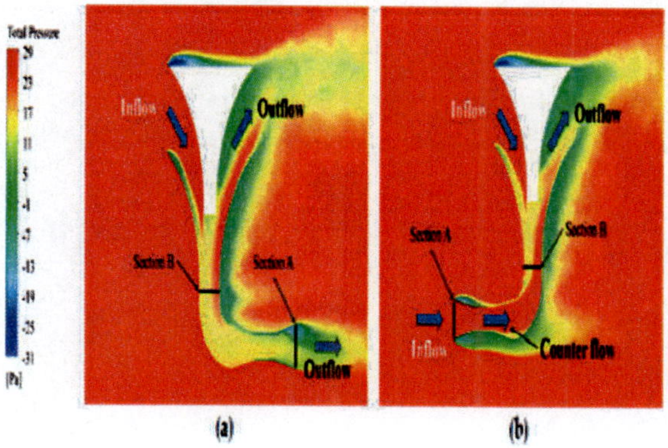

Fig. 8 Pressure contours at **a** $\theta = 0°$ **b** $\theta = 180°$

The above image shows the pressure contours in the vertical plane of the Invelox at 0° and 180°. At 0° the average total pressure of section B is greater than that of section A. At 180° the total pressure of section B is lesser than section A (Fig. 8).

Figure 9 shows the velocity vector distribution in the numerical domain with the specified wind velocity as well as the whole inlet section is open.

The admission catches the free-stream flow in rather complex manner. As displayed in the plot of speed vectors in Fig. 10, part of the approaching flow encroaches on the mass of the front quadrant of the admission shaped by the four parceling baffles and is redirected downwards inside the conveyance framework. One more piece of the approaching free stream is deflected to the sides of the admission and it isolates at the tip of the two fins. The flow inside the channels seems, by all accounts, to be non-uniform and there is a division zone in the rearward segment. Generally, the admission catches a significant measure of free-stream flow notwithstanding the flow division which is additionally connected with a little piece of flow leaving the framework at its toward the back.

An impressive decrease in how much the getting away from the air has happened that is displayed in Fig. 11. In any case, the way of the air has been limited and thus,

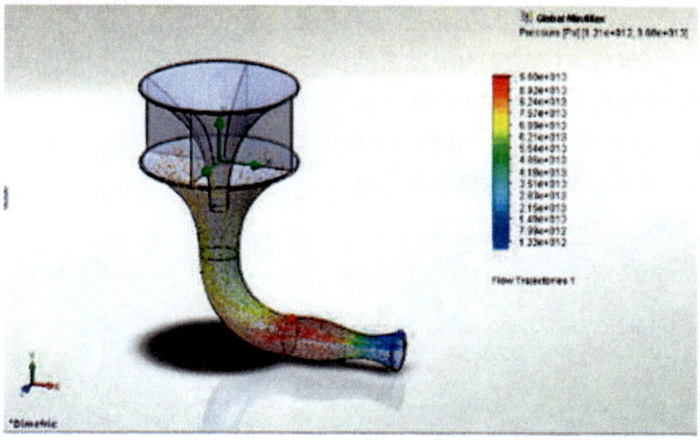

Fig. 9 Velocity vector in the numerical domain with 100% inlet condition

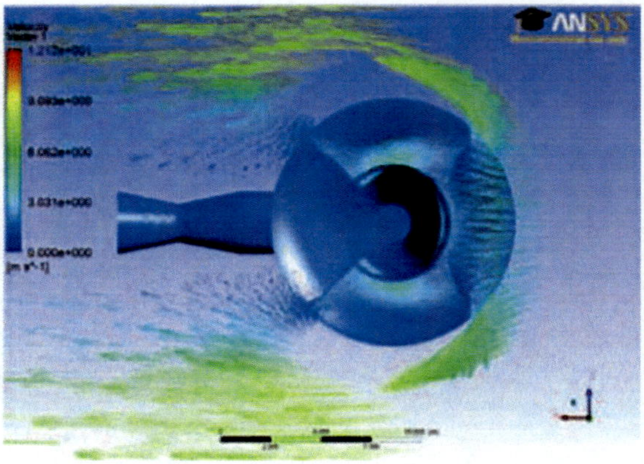

Fig. 10 Velocity vectors predicted at the top view on a horizontal plane which is perpendicular to the intake symmetric axis

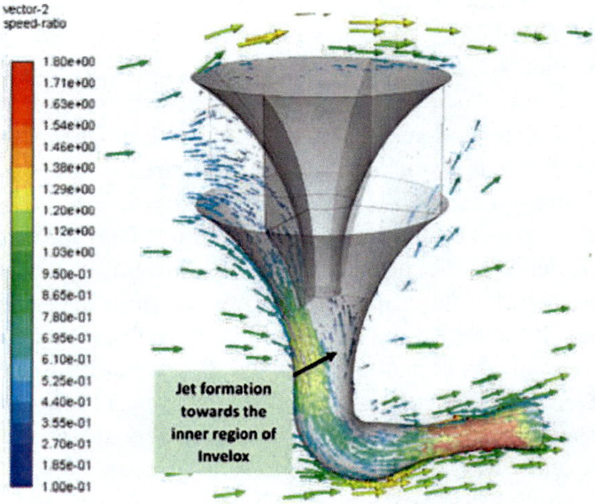

Fig. 11 Velocity vector distribution of Invelox at the median cross section. wind velocity of 7 m/s

the strain drop of the way is raised. This prompts a decrease in the volumetric flow pace of the air entering Invelox.

Recreation results showed that these cases have mediocre execution contrasted with the situation where 1/3 of the admission was open. The explanation is that some part of the air escapes from these openings as displayed in Figs. 12 and 13 for the episode wind speed of 7 m/s.

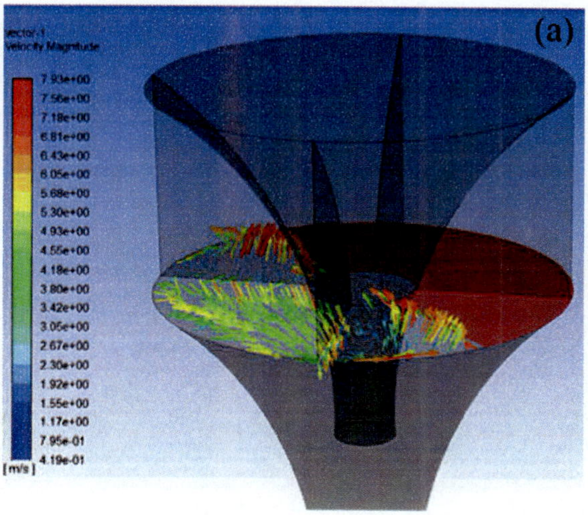

Fig. 12 Velocity vectors on the open faces of the Invelox entry at inlet, wind velocity of 7 m/s. **a** Half of the intake is open, and **b** 1/3 of the intake is open

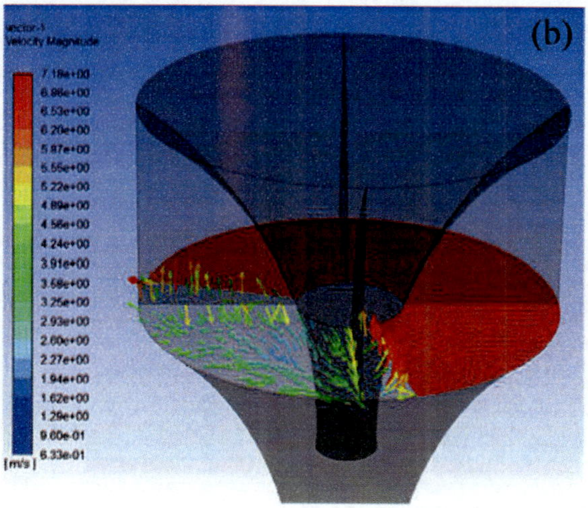

Fig. 13 Velocity vectors on the open faces of the Invelox entry at inlet, wind, 1/3 of the intake is open

8 Conclusions

- Invelox wind turbine framework is an Innovative ducted kind of wind turbine. The equipment accumulates the wind stream from every direction horizontal mainly and pushes the speeding up wind stream through the venturi tube.
- The accelerated proportion of air current in various wind directions is concentrated. This model has been formulated through modeling by mathematical and Numerical means.

- It has been found that the acceleration ratio will disintegrate quickly at the point when the point between the wind stream course and the center line of the venturi passes beyond 90°.
- The other turbulence model line RANS and other models can be checked. This paper considered only the $k - \varepsilon$ model.
- The opening section of the Invelox with the different percentages of opening shows the mass flow rate in the section.

References

1. Farokhzade A, Maghrebi MJ (2021) Inlet parameters effects of Invelox on the aerodynamic performance using numerical simulation. J Appl Fluid Mech 14(5):1511–1520. ISSN: 1735-3572
2. Ahmed AA, Leevans AL, Astle M, Sanju Edwards M (2021) Design and fabrication of Invelox wind turbine. IRJET 08(Issue: 04). e-ISSN: 2395-0056; p-ISSN: 2395-0072
3. Shalar S, Yadav P, Chavan N, Birajdar S (2021) Design and fabrication of wind operated "Invelox". Int Res J Eng Technol (IRJET) 08(03). e-ISSN: 2395-0056
4. Golozar A, Shirazi FA, Siahpour S, Khakiani FN, Osguei KG (2020) A novel aerodynamic controllable roof for improving performance of INVELOX wind delivery system. Wind Eng 1–14
5. Ratna Reddy T, Indira Priyadarsini Ch, Hareesh Kumar P (2020) Performance evaluation of invelox wind turbine using CFD technique. J Inf Comput Sci 13(12). ISSN: 1548-7741
6. Ding L, Guo T (2020) Numerical study on the power efficiency and flow characteristics of a new type of wind energy collection device. Appl Sci 10:7438
7. Ding L (2020) Study of Invelox wind turbine considering atmospheric boundary layer: based on numerical simulation. J Phys Conf Ser 1600:012063. FMIA 2020
8. Venkatramakrishnan R, Pandey JK, Mondal AK, Karn A (2020) Low speed wind turbines for power generation: a review. J Adv Res Fluid Mech Therm Sci 67(1):146–169
9. Gohar GA, Manzoor T, Ahmad A, Hameed Z, Saleem F, Ahmad I, Sattar A, Arshad A (2019) Design and comparative analysis of an Invelox wind power generation system for multiple wind turbines through computational fluid dynamics. Adv Mech Eng 11(4):1–10
10. Solanki AL, Prof. Kayasth BD, Prof. Bhatt H (2017) Design modification & analysis for venturi section of INVELOX system to maximize power using multiple wind turbine. IJIRST 3(Issue 11):2349–6010. ISSN (online)
11. Allaei D, Andreopoulos Y (2014) INVELOX: description of a new concept in wind power and its performance evaluation, Elsevier. Energy 1–9
12. Abe KI, Ohya Y (2004) An investigation of flow fields around flanged diffusers using CFD. J Wind Eng Ind Aerod 92(3e4):315e30
13. Anbarsooz M, Amiri M, Rashidi I (2019) A novel curtain design to enhance the aerodynamic performance of Invelox: a steady-RANS numerical simulation, Elsevier. Energy 168:207–221
14. Gavade AA, Mulla AS, Ransing AM, Sane NM (2018) Design and manufacturing of INVELOX to generate Wind power using nonconventional energy sources. IJARSE 7(Special Issue No. 3). ISSN: 2319-8354

Nanofluids for Heat Transfer Augmentation in Heat Exchangers—An Overview of Current Research

P. Adarsh Varma, Ch. SatyaPriya, M. Prashanth, P. Mukesh, B. Sai Sri Nandan, G. Srinivas, M. Sandeep Kumar, and T. Srinivas

Abstract In the heat exchanger system, nanofluids are used to improve the effectiveness of passive heat transfer methods. Many researchers have demonstrated the wide range of engineering applications for nanofluids. In this study, heat transfer in heat exchangers is enhanced by the use of nanofluids. Nanofluids have a better thermal conductivity in heat exchangers than conventional fluids. When compared to pure liquids, the mixing stream had superior heat transfer properties because of the nano-sized particles in nanofluids. Despite the fact that many heat exchangers outperform others, nanofluids are the only ones that do so due to their thermal conductivity and faster heat transfer rates. This review paper looks at and discusses the findings and observations of various research groups on nanofluids as an employed way. Furthermore, the conclusions of various authors are evaluated, and points of argument are discussed. As a result, using nanofluid in a heat exchange system is likely to improve thermal transmission performance.

Keywords Base fluid · Nanoparticles · Nanofluid · Thermal conductivity · Heat transfer rate

---

P. Adarsh Varma · Ch. SatyaPriya · M. Prashanth · P. Mukesh · B. Sai Sri Nandan · G. Srinivas · T. Srinivas (✉)
Chemical Engineering Department, B V Raju Institute of Technology, Narsapur, Medak, Telangana State 502313, India
e-mail: dr.t.srinivas85@gmail.com

M. Sandeep Kumar
Mechanical Engineering Department, B V Raju Institute of Technology, Narsapur, Medak, Telangana State 502313, India

S. Das et al. (eds.), *Proceedings of the 1st International Conference on Fluid, Thermal and Energy Systems*, Lecture Notes in Mechanical Engineering,
https://doi.org/10.1007/978-981-99-5990-7_8

1 Introduction

Heat exchangers, also known as heat transferring equipment, allow thermal energy to flow among two or additional liquids that are at various temperatures. There are many industries that use heat transfer devices, including power generation, food, process, and chemical, as well as manufacturing, electronics, air-conditioning, space, and refrigeration applications. In terms of global economics, energy, materials, and space, saving energy prompted an increase in efforts to develop more cost-effective heat exchange equipment. As a result of these efforts, the fundamental dimensions of heat exchange device for a certain thermal capability has been reduced [1]. Therefore, the more effective heat purposes in a heat exchanger are to reduce the heat exchanger's size, which is necessary for thermal efficiency (capacity), to improve the capability and function of an current heat exchanger, or to diminish pumping power. In addition to improving heat transfer in heat transfer fluids and heat exchangers, many investigations have been carried out [2].

A device that enables the transfer of heat from one liquid to another is a heat exchanger. Heat exchangers can also be utilized for heating as well as cooling. To prevent cross-contamination, the liquids could be detached by a solid object, or maybe they're in touch directly. In chemical plants, petrochemical plants, electricity generation and distribution, petroleum refineries and petroleum plants, natural gas processing, and sewage treatment, they are widely used [3].

A variety of heat exchanger issues can lead to poor productivity or, in certain situations, the complete shutdown of the industrial heat exchanger. The researchers use a variety of approaches to elucidate them. Because the most recent research supports utilizing nanofluids in heat exchangers, the work in question is taken into account. Other methods to avoid vibration problems, heat exchanger secretion, enhanced heat exchanger power usage, aisle separation (thermal leakage), and contamination should be considered in other articles [4].

But there is a quick brief of the difficulties that can happen in heat exchangers and why nano is a good idea. Contemplating the foregoing, the accessibility of superior-effectiveness heat exchangers is one of the most basic requirements in many businesses and research. As a result, the researchers suggested new approaches. The purpose of passive heat transfer mechanisms and variations in the rheological properties of fluids are being researched to find solutions to these problems. These techniques work by reducing the formation of a laminar sublayer, improving the number of interruptions, improving the efficient heat transfer surface, creating secondary flows or vortices, and improving fluid blending [5]. Many methods for increasing the heat transfer level in these methods have been proposed because heat exchangers show such a significant function in various electronics, transportation, manufacturing processes, including heat sources, and industrial processes. Most of these procedures rely on structural changes, such as increased thermal exteriors (fins), thermal exterior pulsation, and fluorescence. Heat transmission and compression in high-energy process equipment will be a challenge for these techniques to meet [6].

2 Improved Heat Transfer in Nanofluids Heat Exchangers

Plate type heat exchangers, double pipe heat exchangers, shell and tube heat exchangers, shell and helical coil heat exchangers, and fin type heat exchangers are just a few examples of the many types of heat exchangers available. In the tables below, we summarized heat transfer development using nanomaterials in different heat exchangers (Tables 1, 2, 3, 4 and 5).

Table 1 An overview of experimental work on a double pipe heat exchanger with nanofluids

Refs.	Investigator	Nanomaterial—fluid	Remarks
[7]	Bahiraei and Hangi (2013)	Mn–Zn ferrite magnetic—water-based nanofluid	They studied how pressure drop and heat transfer were affected by magnetic fields and nanofluids. For maximum heat transfer and little pressure drop, they reported the ideal conditions
[8]	C.S. Reddy and V.V. Rao (2014)	TiO_2—ethylene glycol water based	They discovered that when 0.02% nanofluid is added to reference fluid, the heat transfer coefficient and friction factor increase with 10.69 and 13.85%, respectively
[9]	Marjan et al. (2016)	Graphene with nitrogen-doped water-based nanofluids	They reported that for a given pumping energy, heat withdrawal, heat transfer rate and the power usage for nanofluids were significantly higher than that for liquid throughout all studies reported. They also demonstrated a nearly 16.2% increase in heat transfer coefficient on average
[10]	Ravi Kumar et al. (2017)	Fe_3O_4/water	They found that, in comparison to water data, the *Nu* improvement for the nanofluid's 0.06% volume fraction is 14.7%, and it increases to 41.29% when combined with the longitudinal tape spacer

(continued)

Table 1 (continued)

Refs.	Investigator	Nanomaterial—fluid	Remarks
[11]	Hussein (2017)	Aluminum nitride—EG	In comparison to conventional fluids, he claimed that the use of hybrid nanofluids with low volume parts increased the heat transfer efficiency by 160%
[12]	Milani Shirvan et al. (2017)	Al_2O_3—water	They observed that the mean Nusselt number increased by 57.70% when combined with the nanoparticle concentration and Reynolds number
[13]	Maddah et al. (2018)	Al_2O_3–TiO_2/water	If used as a fluid for heat transfer instead of ordinary water, they made the discovery that the utilization of twisted tapes and nanostructured materials results in an increase in exergy competence
[14]	Bahmanu et al. (2018)	Alumina/water	They discovered that when the Reynolds number and the volume fraction of nanoparticles were combined, the Nusselt number and the convection heat transfer coefficient increased by 32.7 and 30%, respectively
[15]	V Nageswarw et al./2019	CuO/water	When related to base fluid, augmentation in the *Nu* was about 18.6% at 0.06% volume concentration
[16]	Sumit Kr. Singh et al./2020	Al_2O_3_MgO hybrid nanofluid	In this study, it was determined that a hybrid nanofluid and tapered wire coil arrangement was a viable choice for enhancing a double pipe heat exchanger's hydrothermal properties

Table 2 Experimental studies on the efficiency of a Shell and tube heat exchanger employing various nanofluids are summarized

Refs.	Author/year	Nanoparticle/base fluid	Observations
[17]	Kanjirakat Anoop et al./2013	Water—SiO_2	Heat transfer coefficient of nanofluids enhanced with nanoparticles intensity in heat exchangers, according to experimental results
[18]	K.V. Sharma et al./ 2014	TiO_2 and SiO_2/ water	Heat transfer coefficients increased by 26% with 1.0% TiO_2 nanofluid concentration and 33% with 3.0% SiO_2 nanofluid concentration

(continued)

Table 2 (continued)

Refs.	Author/year	Nanoparticle/base fluid	Observations
[19]	Nishant Kumar/ 2016	Fe_2O_3/water Fe_2O_3/ ethylene glycol	They discovered that as nanoparticle content rises, thermal conductivity and overall heat transfer improved due to particle interaction. The thermal conductivity of the conventional fluids improved with temperature
[20]	Milad Rabbani Esfahani, et al./2017	Graphene oxide/DI water	With both turbulent and laminar flow conditions, graphene oxide nanofluids were found to be more effective as the hot fluid
[21]	Ramtin Barzegarian, et al./2017	Al_2O_3/water	Because of this, nanofluids outperformed base fluids in heat transfer, increasing the Nusselt number from 9.7 to 20.9 to 29.8% while also improving overall heat transfer by 5.4 to 11.0% (and the Nusselt number from 9.7 to 19.9 to 29.8%)
[22]	V. Rambabu et al./ 2017	TiO_2/water	The introduction of Nano particles at different volume concentrations (0.05, 0.1, 0.15, and 0.2%) and flow rates results in a significant improvement in heat transfer characteristics
[23]	S. Anitha et al./ 2019	Al_2O_3-Cu /water	They found that hybrid nanofluid heat transfer coefficient has a 139% greater than pure water and a 25% greater heat transfer coefficient than water-Cu nanofluid. Furthermore, the significant increase in Nusselt number exceeds 90% for hybrid nanofluids
[24]	Z. Said et al./2019	CuO/water	They discovered using the Al_2O_3 at a volume concentration of 0.3% improved thermodynamic efficiency by 24.21%
[25]	Mohammad Fares, et al./2020	Graphene /water	They discovered that at 0.2 wt%, the heat transfer coefficient was 29%, and the mean thermal efficiency was 13.7%
[26]	Aysan Shahsavar G et al./2021	Fe_3O_4-CNT/water	They discovered that increasing hybrid nanofluid concentration and Re improved heat transfer in heat exchangers

Table 3 On the effectiveness of a Shell and Helical heat exchanger with various nanofluids, experimental studies are summarized here

Refs.	Author/year	Nanoparticles/base fluid	Observations
[27]	Vinod Kumar /2015	Al_2O_3/water	He indicated that higher the mass flow rate and the Reynolds number led to an increase in the amount of heat that could be transferred by convection. He used 0.2 vol%
[28]	Khoshvaght, et al./2016	Cu/water	They showed that changing the pressure drop increased the heat transfer enhancement caused by copper nanoparticle suspension in water and that the enhancement was amplified by boosting the nanoparticle intensity
[29]	T. Srinivas, et al./2016	TiO_2	They found that when TiO_2/water nanofluids were compared to water, the heat exchanger's effectiveness increased by 26.8%, indicating that heat transfer was intensified
[30]	P.J. Fule, et al./2017	CuO/water	At 0.1 and 0.5 vol%, respectively, the heat transfer coefficient is found to be 37.3 and 77.74% greater than the base fluid
[31]	Amol F. Niwalker et al./2018	SiO_2/water	This fluid has a 29% higher heat transfer coefficient than the base fluid. Furthermore, friction and pressure drop are improved by 52.61 and 62.60%, respectively, when related to the base fluid

(continued)

Table 3 (continued)

Refs.	Author/year	Nanoparticles/base fluid	Observations
[32]	K. Palanisamy, et al./2019	Water/MWCNT	At 0.1, 0.3, and 0.5% volume intensity, the Nusselt number is 28, 52%, and the pressure drop is 16, 30, and 42% higher than the base fluid
[33]	Neeraj R. Koshta, et al./2019	Graphene oxide-TiO_2/water	The percent heat transfer coefficient improvement for a 0.25% volumetric intensity of the rGO/TiO_2 nanocomposite in the base liquid was 35.7%
[34]	Kriti Singh, et al./2020	CNT/water	At Reynolds number 5000, CNT nanofluid has a 62.62% higher overall heat transfer coefficient than water
[35]	Abhishek Lanjewar, et al./ 2020	CuO-Polyaniline/water	When 0.2% vol. 1 wt% CuO—PANI nanocomposite was added to a base fluid, the heat transfer coefficient increased 37%
[36]	D. Saratha Chandra, et al./ 2021	Cu–Ni/water	Because of its consistency in maintaining a constant temperature, 0.04% volumetric Cu–Ni/H_2O with 12 rings appears to be more dominant in food processing applications

Table 4 Experimental investigations on the efficiency of plate type heat exchangers using various nanofluids are summarized

Refs.	Author/year	Nanoparticle/base fluid	Observations
[37]	Pantzali et al./2017	CuO	They discovered that the heat exchanger operates in laminar conditions, making nanofluids an effective option
[38]	Shalkevich et al./ 2018	Gold nanoparticle	All of the experimental tests were repeated twice, and the thermal conductivity of gold nanoparticles was improved using a plate type heat exchanger
[39]	Sözen A et al./2019	Kaolin—deionized water	The use of kaolin—deionized water nanofluid as a working fluid improved the mean heat transfer coefficient by 9.3%, according to the authors
[40]	Bhattad A et al./ 2019	Al_2O_3—MWCNT hybrid nanofluid	They observed a negligible increase of 0.08% in pump work and a 12.46% increase in the performance index for Al_2O_3—MWCNT (4:1) hybrid nanofluid heat transfer coefficient
[41]	Talari VK et al./ 2019	Al_2O_3/water	They claimed that using nanofluids in heat exchangers could increase heat transfer efficiency
[42]	Behrangzade et al. / 2019	Ag/Water	When water contains 100 parts per million of silver nanoparticles, the overall heat transfer coefficient of the plate heat exchanger increases by 17%
[43]	Khanlari A et al. / 2019	TiO_2/DI water and Kaolin/DI water	The heat transfer rate of TiO_2/ deionized water was 12%, while kaolin/ deionized water was 18%
[44]	Kumar SD, et al./ 2019	CNT, Al_2O_3, surfactant with deionized water	They showed that nanofluids can replace conventional fluids in heat transfer applications where large volumes of conventional fluids are required
[45]	Mehrali et al./2020	Graphene nanoplatelet (GNP)	GNP nanofluids with 300, 500, and 750 m^2/g particular surface areas have thermal efficiency considerations of 1.51, 1.74, and 1.89, respectively
[46]	Hussein M Maghrabiea, et al./ 2021	CeO_2/Water	With a CeO_2/water concentration of 0.75 vol.%, the best performance was achieved, with a 16% increase in the performance index

Table 5 Experimental investigations on the effectiveness of a Finned tube heat exchanger using numerous nanofluids are summarized

Refs.	Author/year	Nanoparticle/base fluid	Observations
[47]	Y. Vermahmoudi, et al./2014	Fe_2O_3/water	The heat transfer rate and total heat transfer coefficient increase by 13 and 11.5%, respectively, when associated with the base fluid (DI water), at an intensity of 0.65 vol.%
[48]	Dr. Qasimb S. Mahdi, et al./ 2016	Al_2O_3/water	It was discovered that as the concentration of nanoparticles increased, H.T. increased as well, reaching a maximum of 12.4% at 0.8 vol%. In addition, improving the intensity of nanoparticles to a maximum of 19.22% at 0.8 vol% increased inner-side heat transfer
[49]	DR. Zena K. Kadhim, et al./ 2016	MGO/water	The airside heat transfer coefficient increases with nanoparticle concentration. At 0.75% concentration, the increase over the base fluid was 15.85%
[50]	Mohammad. SikindarBaba, et al./2017	Fe_3O_4 /water	They noticed that the heat transfer rate in three tubes is 70–80% higher than any other type when particle Reynolds number and concentration are considered
[51]	Behrouz Raei, et al./2018	Gamma alumina/water	When 0.2 wt% alumina nanoparticles are added to water, the heat transfer coefficient and friction factor increased by 5 and 20% correspondingly
[52]	Hassan Hajabdollahi, et al./ 2020	Al_2O_3, CuO—water	They discovered that at 0.025, 0.05, and 0.075 vol concentrations, a Reynolds number of 3000 increased 1.92, 15.08, and 22.46% for Al_2O_3 and 10.98, 35.30, and 46.11% for CuO
[53]	Budi Kristiawan, et al./2020	TiO_2/water	Their findings add to our knowledge of how helical micro fins inside tubes and nanofluids can improve heat transfer

3 Conclusions

This paper reviews all the important papers on heat exchangers and nanofluids that have been published in the last few years (HEs). This comprehensive review looked at the outcomes of nanofluid on heat transfer in five various kinds of heat exchangers. The key findings are summarized below:

- The use of nanofluid in all five heat exchangers has increased in the last decade as a result of its promise and a significant growth in thermal conductivity over the pure fluid.
- The optimum volumetric concentration of nanofluid is a point at which the heat transfer rate increases as the volume concentration of nanoparticles rises.
- It also increases viscosity and friction, lowering pumping energy. Almost all studies found that the desired thermal performance, heat transfer augmentation, entropy generation decline, and exergy destruction reduction were all better than the base fluids. Nanofluids for industrial use, on the other hand, necessitate high concentrations and kilos of nanoparticles, which have yet to be proven cost-effective.
- It was also discovered that the working nanofluids temperature has a significant impact on heat exchanger efficiency improvement.

4 Recommendations for Future Work

- The most research is required to ascertain how a nanofluid mixture affects a heat exchanger's convective heat transfer coefficients, containing nanoparticle size, shape, pH variation, surfactant addition, sonication time, and agglomeration.
- The heat transfer performance of a variety of hybrid nanofluids can be improved by changing the statistical constraints, such as plate thickness, plate pitch, corrugation pattern, and corrugation angle.
- The effects of different surfactants on the thermal stability and properties of nanofluid mixtures.

References

1. Pordanjani AH, Aghakhani S, Afrand M, Mahmoudi B, Mahian O, Wongwises S (2019) An updated review on application of nanofluids in heat exchangers for saving energy. Energy Convers Manag 198
2. Choi SUS, Eastman JA Enhancing thermal conductivity of fluids with nanoparticles. DOE Contracm Number: W-31109-ENG-38
3. Xuan Y, Roetzel W (2000) Conceptions for heat transfer correlation of nanofluids. Int J Mass Heat 43(19):3701–3703
4. Philip J, Shima PD, Raj B (2007) Enhancement of thermal conductivity in magnetite based Nanofluid due to chainlike structures. 91
5. Buzea C, Pacheco II, Robbie K (2007) Nanomaterials and nanoparticles: sources and toxicity. Biointerphases 2
6. Hong KS (2004) Thermal conductivity of Fe nanofluids depending on the cluster size of nanoparticles. Appl Phys Lett 88(3)
7. Bahiraei M, Hangi M (2013) Investigating the efficacy of magnetic nanofluid as a coolant in a double-pipe heat exchanger in the presence of magnetic field. Energy Convers Manage 76:1125–1133
8. Reddy MCS, Rao VV (2014) Experimental investigation of heat transfer coefficient and friction factor of ethylene glycol water-based TiO_2 nanofluid in double pipe heat exchanger with and without helical coil inserts. Int Commun Heat Mass Transf 50:68–76
9. Goodarzi M, Kherbeet ASh, Afrand M, Sadeghinezhad E, Mehrali M (2016) Investigation of heat transfer performance and friction factor of a counter-flow double-pipe heat exchanger using nitrogen-doped, graphene-based nanofluids. Int Commun Heat Mass Transf 76:16–23
10. Ravi Kumar NT, Bhramara P, Addi BM, Syam Sundar L, Singh MK, Sousa ACM (2017) Heat transfer, friction factor and effectiveness analysis of Fe3O4/water nanofluid flow in a double pipe heat exchanger with return bend. Int Commun Heat Mass Transf 82:155–163
11. Hussein AM (2017) Thermal performance and thermal properties of hybrid nanofluid laminar flow in a double pipe heat exchanger. Exp Therm Fluid Sci 88:37–45
12. Shirvan KM, Mamourian M, Mirzakhanlari S, Ellahi R (2017) Numerical investigation of heat exchanger effectiveness in a double pipe heat exchanger filled with nanofluid: a sensitivity analysis by response surface methodology. Powder Technol 313:99–111
13. Maddah H, Aghayari R, Mirzaee M, Ahmadi MH, Sadeghzadeh M, Chamkha AJ (2018) Factorial experimental design for the thermal performance of a double pipe heat exchanger using Al2O_3-TiO_2 hybrid nanofluid. Int Commun Heat Mass Transf 97:92–102
14. Bahmani MH, Sheikhzadeh G, Zarringhalam M, Akbari OA (2018) Investigation of turbulent heat transfer and nanofluid flow in a double pipe heat exchanger. Adv Powder Technol 29(2):273–282
15. Nageswara Rao V, Ravi Sankar B (2019) Heat transfer and friction factor investigations of CuO nanofluid flow in a double pipe U-bend heat exchanger. Mater Today Proc 18:207–218
16. Singh SK, Sarkar J (2020) Improving hydrothermal performance of hybrid nanofluid in double tube heat exchanger using tapered wire coil turbulator. Adv Powder Technol
17. Anoop K, Cox J, Sadr R Thermal evaluation of nanofluids in heat exchangers
18. Azmi WH, Sharma KV, Saram PK, Mamat R, Najafi (2014) Heat transfer and friction factor of water based TiO_2 and SiO_2 nanofluids under turbulent flow in a tube. Int Commun Heat Mass Transf 59:30–38
19. Kumar N, Sonawane SS () Experimental study of Fe_2O_3/water and Fe_2O_3/ethylene glycol nanofluid heat transfer enhancement in a shell and tube heat exchanger Int Commun Heat Mass Transf
20. Esfahani MR, Languri EM (2017) Exergy analysis of a shell-and-tube heat exchanger using graphene oxide nanofluids Therm Fluid Sci 83:100–106
21. Barzegarian R, Aloueyan A, Yousefi T (2017) Thermal performance augmentation using water based Al_2O_3-gamma nanofluid in a horizontal shell and tube heat exchanger under forced circulation. Int Commun Heat Mass Transfer 86:52–59

22. Kunwar A, Gautam AK, Rambabu K (2016) Design of a compact U-shaped slot triple band antenna for WLAN/WiMAX applications. Int J Electron Commun
23. Anitha S, Thomas T, Parthiban V, Pichumani M (2019) What dominates heat transfer performance of hybrid nanofluid in single pass shell and tube heat exchanger? Adv Powder Technol
24. Said Z, Assad MEH, Hachicha AA, Bellos E, Abdelkareem MA, Alazaizeh DZ, Yousef BA (2019) Enhancing the performance of automotive radiators using nanofluids. Renew Sustain Energy Rev 112:183–194
25. Fares M, Al-Mayyahi M, Al-Saad M (2020) Heat transfer analysis of a shell and tube heat exchanger operated with graphene nanofluids. Case Stud Therm Eng 18
26. Goldanlou AS, Sepehrirad M, Papi M, Hussein AK, Afrand M, Rostami S (2020) Heat transfer of hybrid nanofluid in a shell and tube heat exchanger equipped with blade-shape turbulators. J Thermal Anal Calorim
27. Vinodkumar KV, Tharakeshwar TK (2015) Improvement of heat transfer coefficients in a shell and helical tube heat exchanger using water/Al2O3 Nanofluid. IRJET J 2(3)
28. Khoshvaght-Aliabadi M, Khaligh SF, Tavassoli Z (2018) An investigation of heat transfer in heat exchange devices with spirally-coiled twisted-ducts using nanofluid. Appl Therm Eng 143:358–375
29. Srinivas T, Vinod V (2016) Heat transfer intensification in a shell and helical coil heat exchanger using water-based nanofluids. Chem Eng Process Process Intensif
30. Fule PJ, Bhanvase BA, Sonawane SH (2017) Experimental investigation of heat transfer enhancement in helical coil heat exchangers using water based CuO nanofluid. Adv Powder Technol 28(9):2288–2294
31. Niwalkar AF, Kshirsagar JM, Kulkarni K (2019) Experimental investigation of heat transfer enhancement in shell and helically coiled tube heat exchanger using SiO2/water nanofluids. Mater Today Proc 18(3):947–962
32. Palanisamy K, Mukesh Kumar PC (2019) Experimental investigation on convective heat transfer and pressure drop of cone helically coiled tube heat exchanger using carbon nanotubes/water nanofluids. Heliyon 5(5)
33. Koshta NR, Bhanvase BA, Chawhan SS, Barai DP, Sonawane SH (2020) Investigation on the thermal conductivity and convective heat transfer enhancement in helical coiled heat exchanger using ultrasonically prepared rGO–TiO2 nanocomposite-based nanofluids. Indian Chem Eng 62(2):202–215
34. Singha K, Sharma SK, Gupta SM (2021) An experimental investigation of hydrodynamic and heat transfer characteristics of surfactant-water solution and CNT nanofluid in a helical coil-based heat exchanger. Mater Today Proc 43(6):3896–3903
35. Lanjewar A, Bhanvase B, Barai D, Chawhan S, Sonawane S (2020) Intensified thermal conductivity and convective heat transfer of ultrasonically prepared CuO–polyaniline nanocomposite based nanofluids in helical coil heat exchanger. Periodica Polytech Chem Eng 64(2):217–282
36. Chandra DS (2021) Experimental analysis of heat transfer coefficient in counter flow shell and helical coil tube heat exchanger with hybrid nanofluids to enhance heat transfer rate using in food processing industries. Turk J Comput Math Educ (TURCOMAT) 12(2):2868–2875
37. Pantzali MN, Mouza AA, Paras SV (2009) Investigating the efficacy of nanofluids as coolants in plate heat exchangers (PHE). Chem Eng Sci 64(14):3290–3300
38. Shalkevich N, Escher W, Bürgi T, Michel B, Si-Ahmed L, Poulikakos D (2009) On the thermal conductivity of gold nanoparticle colloids. Langmuir 26(2):663–670
39. Sözen A, Khanları A, Çiftçi E (2019) Heat transfer enhancement of plate heat exchanger utilizing kaolin-including working fluid. Proc Inst Mech Eng Part A J Power Energy 1–9
40. Bhattad A, Sarkar J, Ghosh P (2020) Hydrothermal performance of different alumina hybrid nanofluid types in the plate heat exchanger. J Therm Anal Calorim 139:3777–3787
41. Talari VK, Thamida SK, Sastry RC (2018) Determination of optimum concentration of nanofluid for process intensification of heat transfer using corrugated plate type heat exchanger. Chem Prod Process Model 14(1):00–02

42. Behrangzade A, Mahdi M (2016) The effect of using nano-silver dispersed water-based nanofluid as a passive method for energy efficiency enhancement in a plate heat exchanger. Appl Therm Eng 102:311–317
43. Kumar SD, Purushothaman K (2018) Enhancement of thermal conductivity in a plate heat exchanger by using nanoparticles CNT, Al2O3, surfactant with de-ionised water as a coolant. Int J Ambient Energy 42(6):648–651
44. Mehrali M, Sadeghinezhad E, Rosen MA, Latibari ST, Mehrali M, Metselaar HS, Kazi SN (2015) Effect of specific surface area on convective heat transfer of graphene nanoplatelet aqueous nanofluids. Exp Thermal Fluid Sci 68:100–108
45. Tiwari AK, Ghosh P, Sarkar J (2013) Performance comparison of the plate heat exchanger using different nanofluids. Exp Thermal Fluid Sci 49:141–151
46. Vermahmoudi Y, Peyghambarzadeh SM, Hashemabadi SH, Naraki M (2014) Experimental investigation on heat transfer performance of /water nanofluid in an air-finned heat exchanger. Eur J Mech B/Fluids 44:32–41
47. Mahdi QS, Hussein KA, Isfasy AM (2016) Effect of Al2O3 nanofluid on heat transfer characteristics for circular finned tube exchanger. Int J Mech Eng Technol (IJMET) 7(3):86–101
48. Kadhim ZK, Kassim MS, Hassan AYA (2016) Effect of MGO nanofluid on heat transfer characteristics for integral finned tube heat exchanger. Int J Mech Eng Technol (IJMET) 7(2):11–24
49. Baba MS, Raju AVSR, Rao MB (2018) Heat transfer enhancement and pressure drop of Fe3O4-water nanofluid in a double tube counter-flow heat exchanger with internal longitudinal fins. Case Stud Thermal Eng 12:600–607
50. Raei B, Peyghambarzadeh SM, Asl RS (2018) Experimental investigation on heat transfer and flow resistance of drag-reducing alumina nanofluid in a fin-and-tube heat exchanger. Appl Therm Eng 144:926–936
51. Hajabdollahi H, Dehaj MS (2021) Experimental study and optimization of friction factor and heat transfer in the fin and tube heat exchanger using nanofluid. Appl Nanosci 11:657–668
52. Kristiawan B, Rifai AI, Enoki K, Wijayanta AT, Miyazaki T (2020) Enhancing the thermal performance of TiO_2/water nanofluids flowing in a helical microfin tube. Powder Technol 376:254–262

Prediction of Heat Transfer Performance of Heat Pipe Using Machine Learning Approach

Sudev Das, Ashutosh K. Jha, R. Johnsan, Chandra S. Bestha, and V. Teja Reddy

Abstract The heat pipe is widely used for electronic cooling. In the simplest configuration of heat pipe, it consists of a cylindrical vessel having three sections (Evaporator, Adiabatic, Condenser). It has a wide range of applications, from smartphones to satellites. Different working fluids are used in the heat pipe based on the operating conditions. This paper discusses the machine learning approach along with some analytical methods to predict the performance of Heat Pipes. Regression, Random Forest, Support Vector Machine (SVM), and Gaussian Matrix Model (GMM) have been tested to find an optimal model based on 2500 experimental data points collected from different research papers and to show the relationship between different parameters. Our objective is to predict the heat transfer performance of a heat pipe by using different machine learning algorithms.

Keywords Heat pipe · Heat transfer performance · Random forest · SVM · Regression analysis · GMM

1 Introduction

A heat pipe is a device used for heat transfer and has a very high heat transfer coefficient [1]. It works on the principles of a combination of thermal conductivity and phase transition. In its simplest configuration, it consists of a cylindrical vessel that is filled with working fluid [2]. Working fluids are selected based on the temperature ranges at which one wants to operate the heat pipe, for example, liquid helium (−271 to −269 °C), Acetone (−48–125 °C), Water (1–325 °C), Lithium (925–1,825 °C) [3]. It consists of three sections named, Condenser section, Adiabatic section, and Evaporator section. The adiabatic part is directly attached to the heat source. Working fluid inside it gets vaporized due to heating of the source due to which pressure in this section increases. This increase in pressure creates a driving force under the influence

S. Das (✉) · A. K. Jha · R. Johnsan · C. S. Bestha · V. Teja Reddy
Department of Chemical Engineering, NIT, Calicut 673601, Kerala, India
e-mail: sudev@nitc.ac.in

S. Das et al. (eds.), *Proceedings of the 1st International Conference on Fluid, Thermal and Energy Systems*, Lecture Notes in Mechanical Engineering,
https://doi.org/10.1007/978-981-99-5990-7_9

of this pressure driving force vapor is transported to the condenser. At the condenser part, this vapor loses heat and is condensed back to liquid. The liquid at the condenser section is pumped again to the evaporator portion due to the action of the capillary of the wick structure. The adiabatic section that is in the middle of the condenser and evaporator section has a very small temperature difference [4–8]. Due to its efficient and effective heat transfer characteristics, it is widely used for electronic cooling in smartphones, laptops, and also for space applications, and satellites [9–11].

Machine learning approaches are slightly different from deep learning approaches, first, one is used when data points are not sufficient, and the second one is used when data points are sufficient. To predict the performance of heat pipe, we are using three machine learning (Supervised learning) models for the project, these are Random Forest, Regression, Gaussian Matrix Model (GMM), and Support Vector Machine (SVM). The model validation is decided by the value of the coefficient of determination that is *R* square and the best model is chosen by the maximum value of *R* square.

2 Scope and Objective

A heat pipe is an effective thermal solution, particularly for the application of high heat flux. This technology can be applied on any scale, and it is viable commercially and can be implemented easily [4]. Despite a lot of research and studies, it is still challenging to predict the optimal parameters for heat pipes. So, the approach used in this project tried to cover a wide range of operating conditions and parameters and tried to predict the optimal parameters by combining the data collected over a period of the last two decades.

Its market is estimated to register a CarGurus of over 4% during the forecast period 2020–2027. Heat pipes are the best choice for cooling electronic devices because their effective thermal conductivity can be up to several thousand times higher than that of a copper rod when the same length is compared. Many of today's electronic devices require cooling beyond the capability of standard metallic heat sinks. Nowadays, the focus of the automobile sector is on electric vehicle. As the demand of electric vehicles rises, there will be the demand for effective heat transfer devices.

The main aim of this project is to predict the optimal value of parameters for the development of heat pipes for effective and efficient heat transfer with low hindrance. The specific objectives of this work are divided into two parts: first to develop and check the fit of a machine learning algorithm that can predict the performance of a heat pipe based on different input parameters; such as geometry, working condition, etc. Second is to predict the optimum performance of a heat pipe using four different machine learning approaches.

Table 1 Parameters and their range

Parameters	Data range
Heat load Q (W)	140–420
Thermal resistance R_{th} (K/W)	0.2–1.4
Length of heat pipe L (mm)	100–1000
Fill ratio (%)	0–90
Coefficient of heat transfer h (W/m^{2}K)	11–6000

3 Methodology

The proposed method consists of four sub-processes. Firstly, data collections are made, followed by an analysis of data distribution through visualization technique. The third step is to develop and train the different proposed models and finally, to optimize the data analysis. Detailed information regarding each steps is given below.

3.1 *Data Collection*

Around 2500 sets of experimental data were extracted from different research papers [9, 12, 14–22] to predict the optimal parameter of heat pipes. The data covers a wide range of working fluid and operating conditions for heat pipes made up of Cu that are generally used for electronics cooling were collected. The range of parameters is mentioned in Table 1. The table shows a wide range of data for different types of working fluids such as water, Methanol, N-Pentane, Acetone, Ethanol, Kerosene, and DI Water. After the collection of data, different machine learning algorithms were trained using these data.

3.2 *Regression Analysis*

This method of machine learning is based on statistical modeling, and it is used to get a relationship between the output variable and input variable. It can be expressed as Eq. (1).

$$y = b_0 + bx + e \quad (1)$$

where: y—dependent variable, x- intercept, b—slope and e—residual error.

In this, the outcome variable, that is predicted data, is calculated by assuming a linear or no-linear function of predictor variables that are collected data points. For checking the acceptance of a model, one needs to calculate the R square value, which is the coefficient of determination.

3.3 Gaussian Matrix Modeling (GMM)

This method of machine learning is a probabilistic method. The data have been collected from [13–19] and are classified into different classes or into clusters. The classification of data points is achieved by assigning every data point to most likely classes or clusters and probability is calculated for each class. The distribution of the input data points follows a normal distribution and it is a mixture of a set of Gaussian distributions, which can be expressed as Eq. (2).

$$p(x) = \sum \pi_k N(x|\mu k, \sum_k) \tag{2}$$

where, x represents data points; K represents the number of components; μ and $\sum k$ represents the mean and covariance parameter of Gaussian function respectively; N represents Normal Distribution and $\prod k$: Mixing Coefficient ($0 \leq \prod k \leq 1$)

For modeling of GMM, the collected data were trained and classified into different classes in such a way that the variance of each class would be the same. Class with higher proportion and coefficient of determination is chosen for validation of model.

3.4 Random Forest

Random Forest is a model that uses ensemble learning methods for regression. In the Ensemble learning method, predictions from multiple algorithms combine together to give a more accurate model. Random forest builds multiple decision trees and decisions from each tree merge together to give a stable and more accurate result than comparison to a single decision model.

3.5 Support Vector Machine Regression

SVR is basically one of the regression algorithms that is used in choosing best fit line, which is of minimum errors, it creates a hyperplane and two to five boundary layers with a tolerable error margin (ϵ).

4 Results and Discussion

Results of proposed models for different parameters have been discussed briefly in the below section.

4.1 Gaussian Matrix Modeling (GMM)

For the heat pipe, thermal resistance increases with the length of the heat pipe, when other parameters are constant because if the length increases, the temperature drop is increasing and it leads to higher thermal resistance. In Fig. 1, the dotted line represents the predicted data by the GMM whereas the blue dot point represents the actual data. The value of R square is 0.931, which means more than 93% of the data show good fitness as shown in Fig. 1. The R square is also known as the coefficient of determination.

From Fig. 2, it was observed that for a heat pipe heat transfer coefficient increases with the heat load where heat transfer coefficient is heat transfer or heat load per unit area. The dotted line that is predicted data by GMM coincided with the blue point that is actual data. The Value of R square is 0.997, which means more than 99% of the data show good fitness. Figures 1 and 2 shows 0.931 and 0.997 R square value, respectively, in other words, the GMM algorithm is showing the performance of around 93% for the thermal resistance of the heat pipe and approximately 99% for the heat transfer coefficient of the heat pipe.

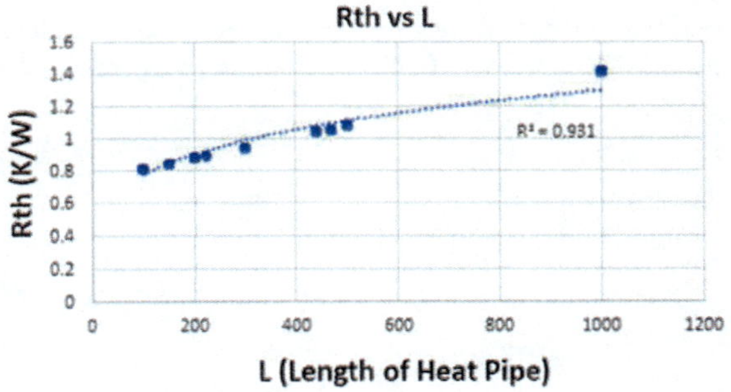

Fig. 1 R_{th} versus L by Gaussian matrix modeling (GMM)

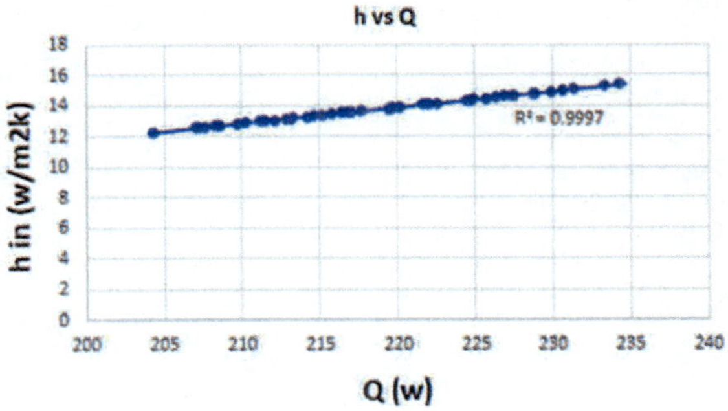

Fig. 2 h versus Q by Gaussian matrix modeling (GMM)

4.2 Regression Analysis

From Fig. 3, it was observed that thermal resistance increases with the length of the heat pipe. The R square value is 0.9282, which means approximately 93% of the data show good fitness.

From Fig. 4, by regression analysis, the value of the coefficient of heat transfer is increasing with the load, and the coefficient of determination is 0.9005, which means 90% of the data show good fitness.

Figures 3 and 4 show approximately 0.9282 and 0.9005 R square values, respectively, in other words, the Regression algorithm is showing 92% performance for the thermal resistance of the heat pipe and 90% performance for the heat transfer coefficient of the heat pipe.

For Fig. 5 the value of the coefficient of determination is 0.934 which means it shows more than 93% fitness, in other words, more than 93% of predicted data coincide with actual data of thermal resistance.

The value of the coefficient of heat transfer and thermal resistance is increasing with the length of the heat pipe and heat load, respectively, by both GMM and regression methods. The value of the coefficient of determination for thermal resistance versus length of heat pipe by GMM method is 0.931, whereas by regression method is 0.929 and for heat transfer coefficient versus heat load coefficient of determination by GMM method is 0.9997, whereas by regression method 0.9005. However, the

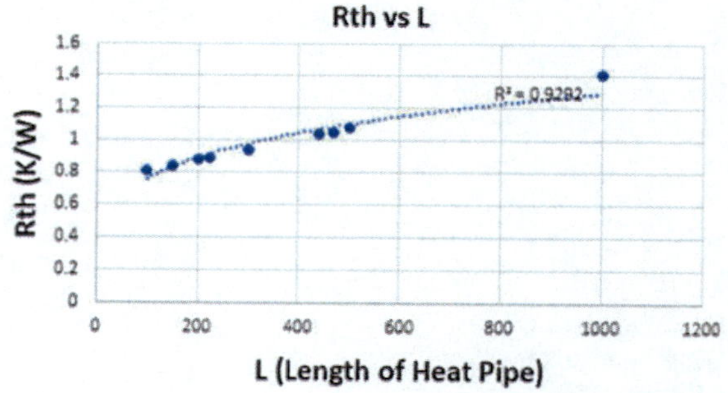

Fig. 3 R_{th} versus L by regression analysis

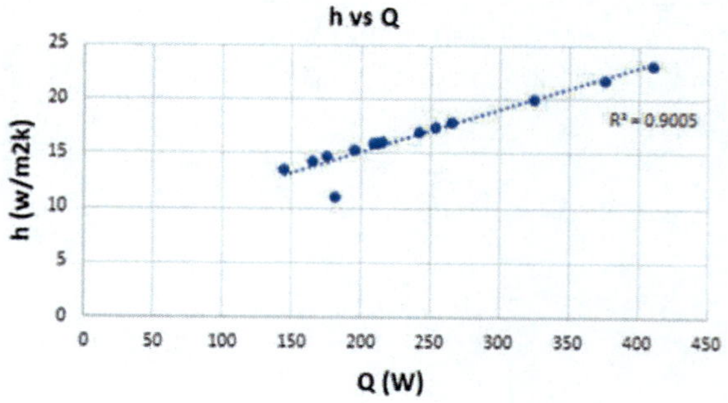

Fig. 4 h versus Q by regression analysis

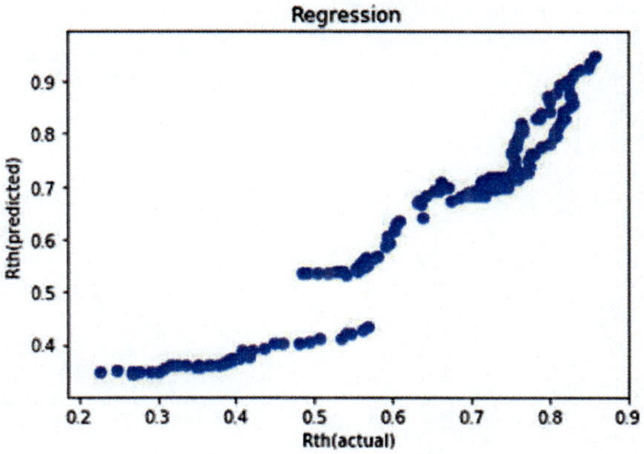

Fig. 5 Comparison between R_{th} predicted and R_{th} actual

value of the coefficient of determination is greater than 0.9 for both methods which means both methods of machine learning are acceptable. The performance of heat pipes for thermal resistance and heat transfer coefficient is higher by GMM, which means the GMM algorithm is better than the regression method.

4.3 *Random Forest Analysis*

For Fig. 6, the value of the coefficient of determination is 0.9979, which means it shows more than 99% fitness, in other words, more than 99% of predicted data coincide with actual data of thermal resistance.

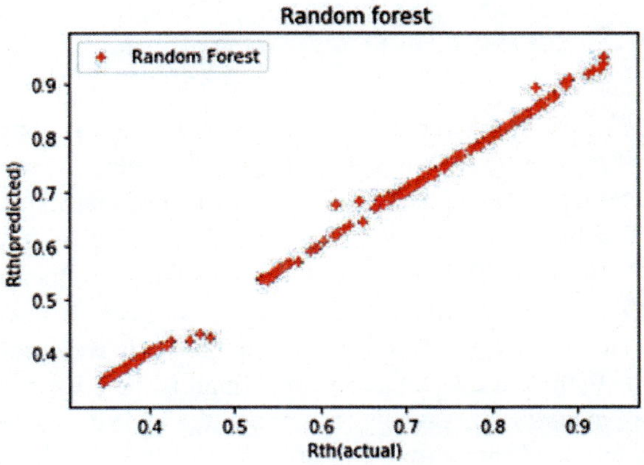

Fig. 6 Comparison of R_{th} predicted versus R_{th} actual

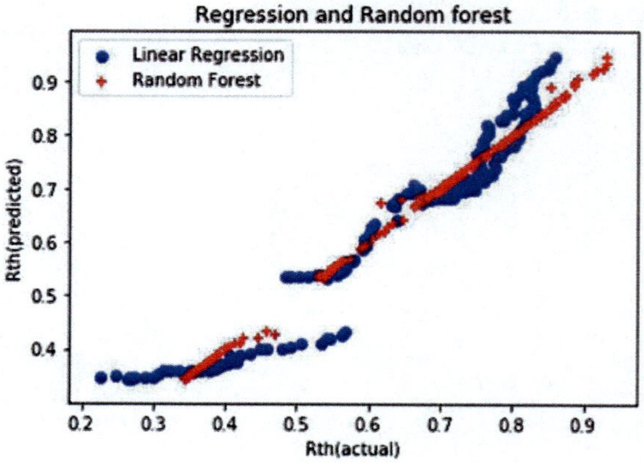

Fig. 7 Comparison of R_{th} predicted and actual using regression and random forest

4.4 Comparison Between Regression and Random Forest

The values of the coefficient of determination are 0.9282 and 0.9979 by the regression method and Random Forest method, respectively. The Random Forest algorithm shows a higher value of R square than Regression, in other words, the Regression method has coefficient of determination of 0.9282 for the thermal resistance of heat pipe whereas the Random Forest method has coefficient of determination of 0.9979 for the thermal resistance of heat pipe. So, among these two Random Forest algorithms are showing higher performance in predicting the performance of heat pipe. Figure 7 clearly shows Random Forest that shows better performance than Linear Regression.

4.5 Support Vector Machine

For Fig. 8, the value of the coefficient of determination is 0.979 for SVM model which means it shows more than 97% fitness, in other words, more than 97% of predicted data coincide with actual data of thermal resistance or the model shows 97% performance.

Note: As SVM does not take real values into the account so it is modified to get the required results, like 0 on the graph means not zero values but it is the span of values of R_{th} (predicted) from 3 to 3.5, similarly for other value ranges.

Table 2 shows the R square, which is the coefficient of determination value for all machine learning algorithms and the RMSE (Root Means Square Error) value as well, in a comparative format.

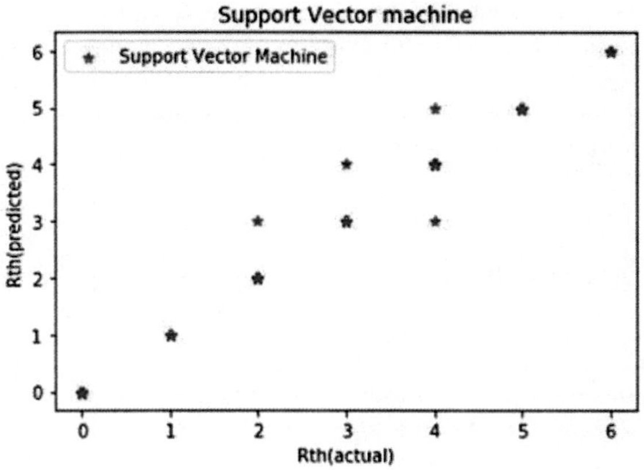

Fig. 8 Comparison of R_{th} predicted versus R_{th} actual using SVM

Table 2 Consolidated result for all ML algorithms

Machine learning approach	*R* square	RMSE
GMM		
• *h* versus *Q*	0.9997	9.44E-15
• R_{th} versus *L*	0.931	9.04E-16
Regression		
• *h* versus *Q*	0.9005	8.84E-14
• R_{th} versus *L*	0.9282	9.57E-17
• R_{th} (actual) versus R_{th} (predicted)	0.934	1.89E-03
Random forest	0.9979	5.794E-05
SVM	0.979	7.381E-04

5 Conclusions

For heat pipe, amongst all four-machine learning approaches, the Random Forest algorithm shows the highest coefficient of determination for thermal resistance, which means it has the highest performance, which is 99.79%, in other words, the Random Forest model is the best model. The length of the Heat pipe should be 140–150 mm, we have verified it from the Random Forest model, which fits similar to the result obtained from the work done by others. Also, the diameter should be in the range of 2–4 mm. Each working fluid has different thermophysical properties and operating temperature range and based on it they show different variations with a bond number.

At last, it was concluded that an 85% fill ratio gives optimum performance. Also, it was seen that when the diameter increases, the coefficient of heat transfer decreases, this is due to a decrease in pressure that causes a capillary rise in the wick. For low heat load, thermal resistance increases with an increase in the angle of inclination whereas for high heat load thermal resistance is maximum at around 10 inclinations.

For effective transfer of heat, it is necessary that there is a phase change of working fluid.

Acknowledgment The author (Sudev Das) thanks SERB/DST, India (EEQ/2021/000283), for the necessary financial support to compete in this work.

Nomenclature

SVM	Support Vector Machine
GMM	Gaussian Matrix Modeling
RMSE	Root Mean Square Error

References

1. Cengel YA (2003) Heat transfer, 2nd edn. McGraw-Hill
2. Garner SD, PE (1996) Heat pipes for electronics cooling, vol 2
3. Schwiegel M (2005) European patent No. 1812527A1
4. Zohuri B (2020) Heat pipe—an overview. Science Direct. https://doi.org/10.1016/B978-0-12-819819-3.00001-8
5. Thermopedia (2022) ISBN: 978-1-56700-456-4
6. Saied W (1995) Fundamentals of heat pipes. Tech. Rep.
7. Advanced Thermal Solution, LLC
8. Byon C (2016) Heat pipe and phase change heat transfer technologies for electronics cooling. https://doi.org/10.5772/62328
9. Mozumder AK, Shaiful M, Chowdhury H, Akon AF (2011) Characteristics of heat transfer for heat pipe and its correlation. ResearchGate. https://doi.org/10.5402/2011/825073
10. Elnager M, Edwan E (2016) Heat pipes for computer cooling applications. https://doi.org/10.5772/62279
11. Shioga T, Mizuno Y (2015) Micro loop heat pipe for mobile electronics applications. 50–55. https://doi.org/10.1109/SEMI-THERM.2015.7100139
12. Khandekar S, Groll M (2003) On the definition of PHP: an overview. In: Minsk international seminar 2003
13. Gangacharyulu D, Tathgir RG (2007) Heat transfer studies of a heat pipe. https://doi.org/10.1080/01457630701421810
14. Pachghare PR, Mahalle AM (2012) Thermal performance of closed loop pulsating heat pipe using pure and binary working fluid. https://doi.org/10.5098/fhp.v3.3.3002
15. Naruka DS, Dwivedi R, Singh PK (2020) Experimental inquisition of heat pipe: performance evaluation for different fluids. https://doi.org/10.1080/08916152.2020.1713254
16. Jafari D, Di Marco P, Filippeschi S, Franco A (2017) An experimental investigation on the evaporation and condensation heat transfer of two-phase closed thermosyphons. https://doi.org/10.1016/j.expthermflusci.2017.05.019
17. Raghuram J, Phani Kumar KVNK, Khiran GV, Snehith K, Bhanu Prakash S (2017) Thermal performance of a selected heat pipe at different tilt angles. https://doi.org/10.1088/1757-899X/225/1/012043
18. Senthilkumar R, Vaidyanathan S, Sivaraman B (2011) Performance investigation of heat pipe using aqueous solution of n-Pentanol with different inclinations. https://doi.org/10.1007/s12206-011-0207-4

19. Abdulshaheed AA, Wang P, Huang G, Chen L (2018) High performance copper-water heat pipes with nanoengineered evaporator sections. https://doi.org/10.1016/.12.1140017-9310/2018
20. Mozumder AK, Chowdhury MSH, Akon AF (2011) Characteristics of heat transfer for heat pipe and its correlation. https://doi.org/10.5402/2011/825073
21. Goshayeshi HR, Goodarzi M, Dahari M (2015) Effect of magnetic field on the heat transfer rate of kerosene/Fe2O3 nanofluid in a copper oscillating heat pipe. Exp Therm Fluid Sci 68:663–668. https://doi.org/10.1016/j.expthermflusci.2015.07.014. ISSN 08941777
22. Gou X, Zhang R, Xu G, Zhao D (2020) Comparative study of low temperature denitration performance of Mn-Ce/ACFA-TiO2 catalysts under oxy-fuel and air-fuel combustion flue gases. Energy Reports 6:15451552. https://doi.org/10.1016/j.egyr.2020.06.005. ISSN23524847

Numerical Analysis to Investigate the Effect of Stenosis Shape on the Hemodynamics of Flow Through a Straight-Cylindrical Artery

A. Equbal and P. Kalita

Abstract Atherosclerosis is a vascular disorder caused by plaque buildup. It decreases the arterial diameter through plaque deposition and arterial wall thickening. This condition causes serious problems in the blood-circulatory system. The present work investigates the effect of stenosis shape on the hemodynamics through a straight-cylindrical artery using the Carreau viscosity model. Here, the different stenosis shapes, namely, cosine, circular, and elliptic shapes are studied at the same degree of stenosis. The numerical model is validated against Zhou et al.'s investigation. Our result shows that the cosine shape stenosis has the highest maximum wall shear stress among the different shapes of stenosis. This type of research has the potential to understand the pathogenesis of vascular diseases.

Keywords Atherosclerosis · Stenosis · Carreau model · Hemodynamics · Numerical simulation

1 Introduction

Large number of deaths due to cardiovascular diseases (CVDs) across the world is a matter of major concern. In 2019, 179 lakh individuals passed away owing to from CVDs, which accounted for 32% of all fatalities worldwide. Out of such deaths, 85% of cases were due to heart attacks and strokes only. Heart and blood vessel issues are grouped together as cardiovascular diseases. A few examples include coronary heart disease, cerebrovascular disease, and peripheral arterial disease [1]. The most prevalent cause of coronary heart disease is plaque deposition in the inner wall of a coronary artery that causes it to narrow or get clogged. This process disrupts the normal blood flow and creates areas of severe fluid stress. This flow behavior can lead to major pathological issues such as endothelial damage, hemolysis, thrombosis, and other arterial injuries. As a result, many experts believe that knowing the dynamics

A. Equbal (✉) · P. Kalita
Department of Mechanical Engineering, Tezpur University, Nappam 784028, India
e-mail: asifraj321@gmail.com

S. Das et al. (eds.), *Proceedings of the 1st International Conference on Fluid, Thermal and Energy Systems*, Lecture Notes in Mechanical Engineering,
https://doi.org/10.1007/978-981-99-5990-7_10

of blood flow in the artery will help them better understand the disease and how it progresses [2]. To assess the efficiency of blood flow in such situations, it is necessary to know the velocity, pressure, and wall shear stress distribution. In such cases, computational hemodynamics act as an appropriate measurable tool in the study of diseased arteries [3].

2 Literature Review and Objective

Blood vessel disorders are caused by plaque build-up called atherosclerosis. Different fibrofatty deposits that might trigger plaque development include cholesterol, lipid, calcium, as well as a particular type of white blood cell [4, 5]. It is a leading cause of death, affecting large and medium-diameter blood vessels. Over the past few decades, several investigations have been performed to model flow through blood vessel. Young et al. [6] performed a set of steady flow in vitro tests for locally constricted tubes in which significant hydrodynamic parameters such as pressure drop, separation, and turbulence were taken into account. The findings demonstrated that the size and geometry of the stenosis had a significant effect on the hemodynamics of blood flow. Ang et al. [7] analyzed mathematical modeling of blood flow through an artery with an asymmetric stenosis. They concluded that the pressure drops were directly proportional to the degree of stenosis, and the maximum shear stresses for both stenosed and unstenosed side were reached just prior to the center of the stenosis. Zhou et al. [8] performed computational fluid dynamics (CFD) analysis on a standard axisymmetric tube model. They found that the wall shear stress (WSS) computed with the Carreau model was somewhat higher than that obtained with the Newtonian model, but both had similar patterns. Yao et al. [9] examined a three-dimensional numerical model of elliptic stenosis in a bent blood vessel. They came to the conclusion that even mild stenosis at high curved parts could have worse consequences than larger percentage stenosis on a straight artery. Nandakumar et al. [10] performed CFD analysis for pulsatile flows through a two-dimensional channel. The result showed that a recirculation zone develops immediately after the occlusion and the fractional change between the maximum wall shear stresses diminishes as the pulsatile character of the flow increases. Stroud et al. [11] investigated the influence of stenosis shape and surface irregularity on blood flow via a stenosed channel. They concluded that the maximal shear stress for axisymmetric shape occurs at the neck, while the shear stress along the stenosis was substantially reduced for asymmetric geometry. Ahamad et al. [12] investigated the effect of bending and blockage on the hemodynamics in the coronary blood vessel by using the Carreau model. They concluded that the cumulative impact of blockage and vascular bend had a considerable influence on blood behavior. Razai et al. [13] examined the pulsatile blood flow in an occluded carotid vessel by employing distinct non-Newtonian rheological theories. They discovered that all the blood viscosity models had the same variation of WSS and radial velocity changes. The magnitudes of these parameters, however, varied from one model to another. Freidoonimehr et al. [14] investigated the effects of stenosis shape on the hemodynamics of a coronary artery. They discovered that judging the seriousness of a coronary artery obstruction simply on stenosis degree without taking into account the cross-sectional shape resulted in an underestimation of hemodynamic parameters. Abugattas et al. [15] performed numerical analysis of blood flow through branched blood vessels by employing distinct non-Newtonian constitutive models. They concluded that the minimal WSS were common in areas where blockage developed. Although Computational Fluid Dynamics (CFD) techniques have been used to analyze various arterial stenosis cases, but the results may lack general applicability. To the best of our knowledge, the impacts of blockage shape on the hemodynamics via straight arteries have not been reported yet. Consequently, the present study investigates the effects of stenosis in a straight cylindrical artery with different stenosis shapes, namely, cosine, elliptic, and circular.

3 The Models of Artery

The cylindrical tube models of different shapes of stenosis are shown in Fig. 1.

All the models possess the same degree of stenosis and same stenosis length i.e., 75% and 10 mm, respectively. The shape of stenosis is considered as cosine, circular and elliptic so that its effect on hemodynamics of blood flow is investigated. Based on Zhou et al.'s literature the diameter of artery model is considered as 5 mm and

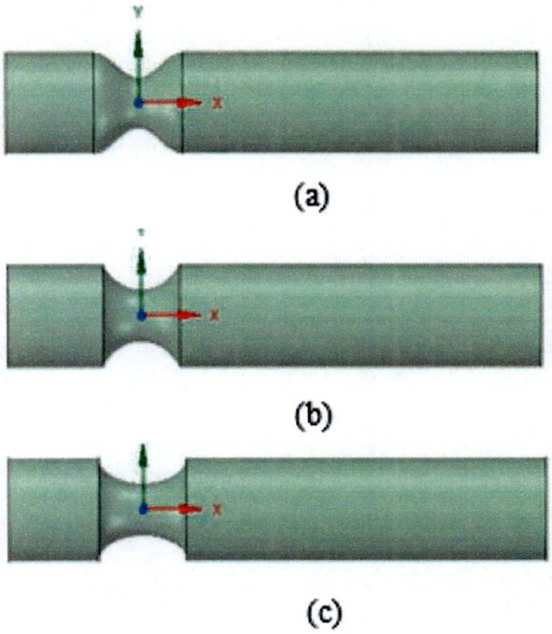

Fig. 1 Stenosed cylindrical artery model **a** cosine shape; **b** circular shape; **c** elliptic shape

the position of inlet and outlet boundaries are kept 15 mm left and 45 mm right from the center of stenosis, respectively.

3.1 Model Validation

For validating the present model, we built another model with the same shape and dimensions as Zhou et al. [8]. Figure 2 shows the comparison of wall pressure at 75% stenosis and Fig. 3 shows the comparison of wall shear stress at 75% stenosis. Both the graphs indicate that the current results are quite consistent with Zhou et al.'s investigation. Thus, it clearly demonstrates that the current model can accurately calculate the blood flow in a tube.

4 Governing Equations and Numerical Schemes

In this investigation, the blood is considered to be incompressible, homogenous, and non-Newtonian. The blood vessel wall is assumed to be rigid and influence of body force is neglected. Under these assumptions, the blood flow through vessel can be represented by continuity and momentum equation as follows:

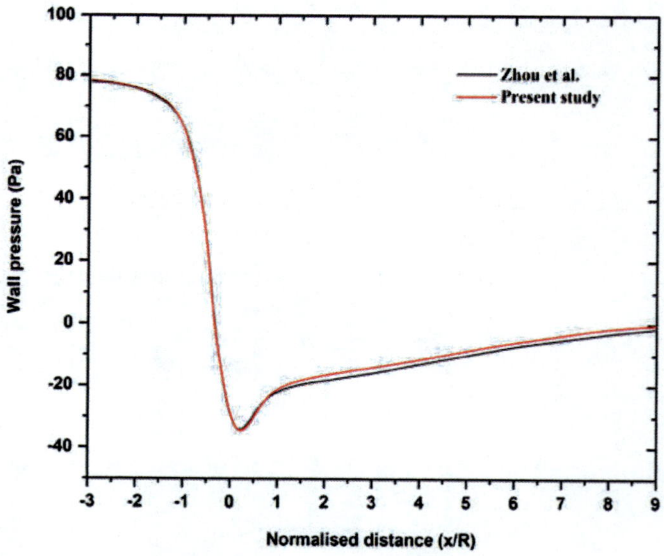

Fig. 2 Comparison of wall pressure between the result of present study and Zhou et al. [8] at 75% stenosis degree

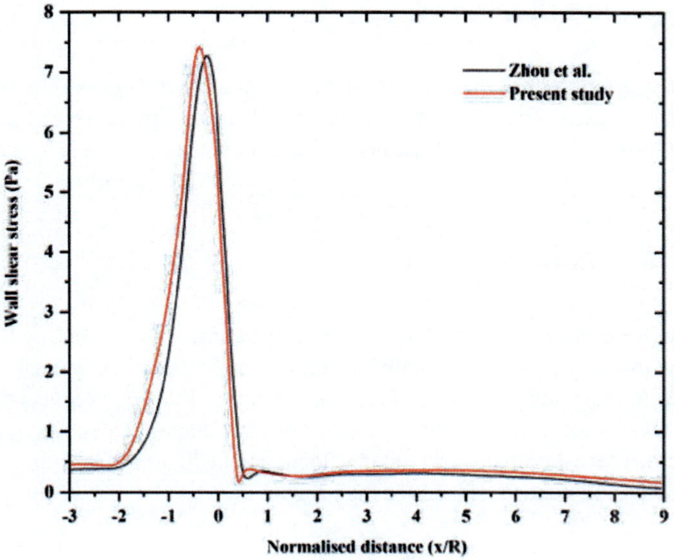

Fig. 3 Comparison of wall shear stress between the result of present study and Zhou et al. [8] at 75% stenosis degree

$$\nabla \cdot \vec{U} = 0$$

$$\frac{\partial \rho \vec{U}}{\partial t} + \nabla \cdot \left(\rho \vec{U} \vec{U} \right) = -\nabla p + \nabla \cdot \vec{\vec{\tau}}$$

where $vecU$, ρ, p and $\vec{\vec{\tau}}$ are the velocity vector; blood density with a value of 1060 kg/m^3 [16], the pressure and the viscous stress tensor, respectively. The viscosity is computed by using the Carreau viscosity model [9, 14]

$$\mu = \mu_\infty + (\mu_0 - \mu_\infty)[1 + (\lambda \dot{\gamma})^2]^{(n-1)/2}$$

where $\dot{\gamma}$ is local shear rate; λ is relaxation time; μ_0 and μ_∞ are the viscosity at zero and high shear rate, and n is the power index. The values of all the four coefficients for human blood are given in [17]. In the cases, the blood flow rate at inlet is 0.0456 L/min [8]. At the outlet, fully developed flow condition is adopted. No slip condition is employed at arterial wall. The finite volume-based CFD package fluent 19.0 is used to solve the laminar flow models. The SIMPLE algorithm is used to solve the governing equations of blood flow. The solution is deemed to have reached convergence when the residual of all the variables fall below the thresold of 10^{-6}.

5 Results and Discussion

The effects of different shapes of stenosis on the hemodynamics are investigated in this study. It also examines the numerical results of flow behavior, which have a major impact in the development and treatment of arterial disease in medical settings.

5.1 *Flow Pattern*

Figure 4 represents flow patterns of different shapes of stenosis including the normal case. Flow patterns change drastically as the flow approaches the stenosis region. It is obvious because pressure energy gets converted into kinetic energy as the area of cross-section is reduced by plaque deposition. All the shapes of stenosis have similar flow pattern. There is a small variation in velocity about 20 mm/s from circular shapes to other shapes of stenosis at the blockage region.

5.2 *Axial Velocity*

Figure 5 shows the variation of axial velocity versus normalized distance for all shapes of stenosis including the normal case. From the plot, it is clear that the variation in axial velocity profile is not the same for different shapes of stenosis. However, the elliptic shape has the highest maximum velocity 0.45 m/s and circular shape has the lowest maximum velocity 0.41 m/s at the center of stenosis. The cosine shape has velocity between the two.

5.3 *Wall Pressure*

Figure 6 shows wall pressure distributions for all the shapes of stenosis including the normal case. Pressure distribution changes drastically as it approaches the stenosis region. It is obvious because kinetic energy gets converted into pressure energy as the area of cross-section is reduced by plaque deposition. The pressure variations do not vary much among the different shapes of stenosis. However, the cosine shape stenosis has the maximum pressure compared with the other stenosis shapes prior to the stenosis center. Under normal circumstance, the blood flow is completely developed. As a result, the axial pressure decreases in a linear fashion.

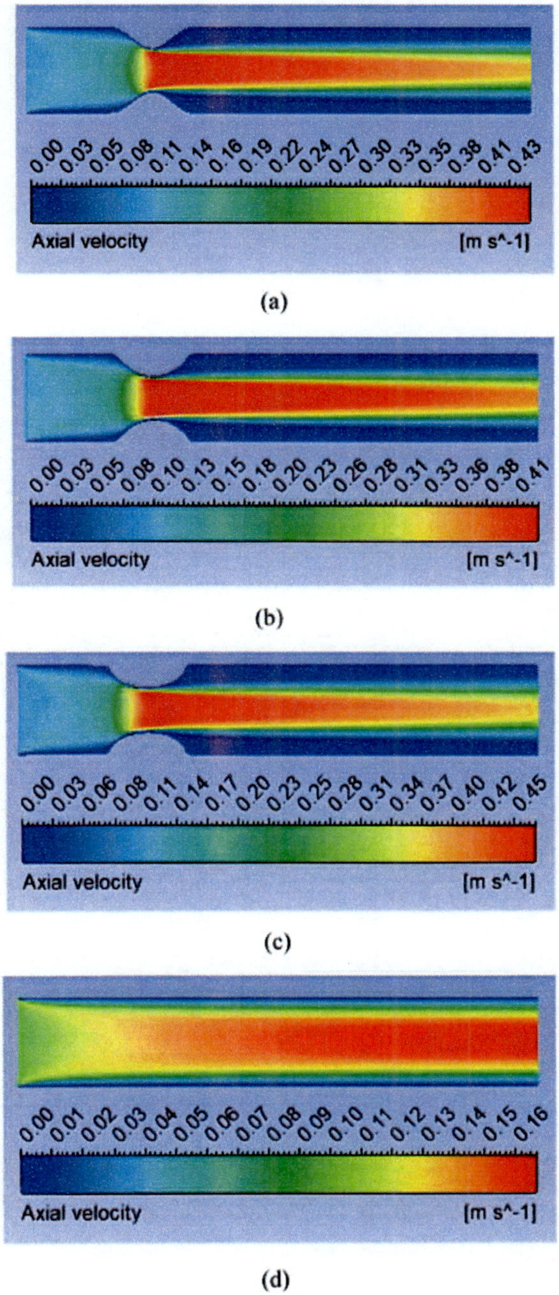

Fig. 4 Flow patterns: **a** cosine shape; **b** circular shape; **c** elliptic shape; **d** normal case

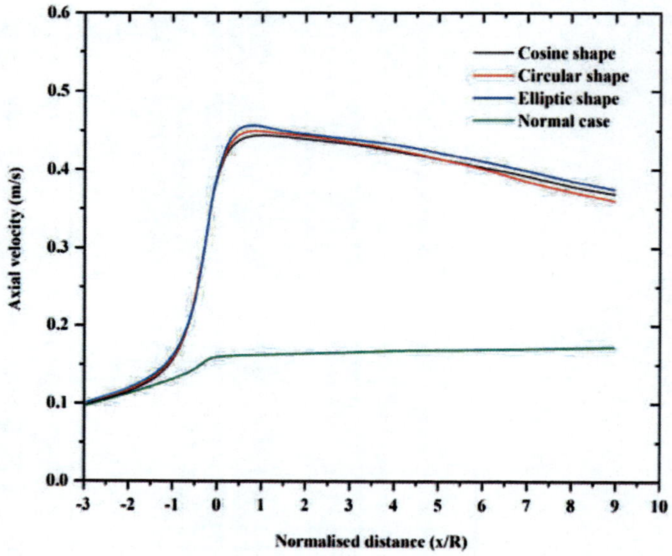

Fig. 5 Axial velocity distributions for different stenosis shapes and the normal case

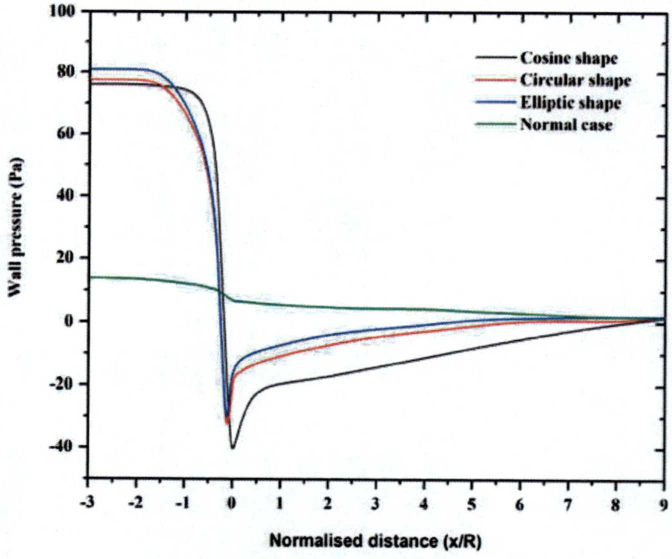

Fig. 6 Axial wall pressure distribution for different stenosis shapes and the normal case

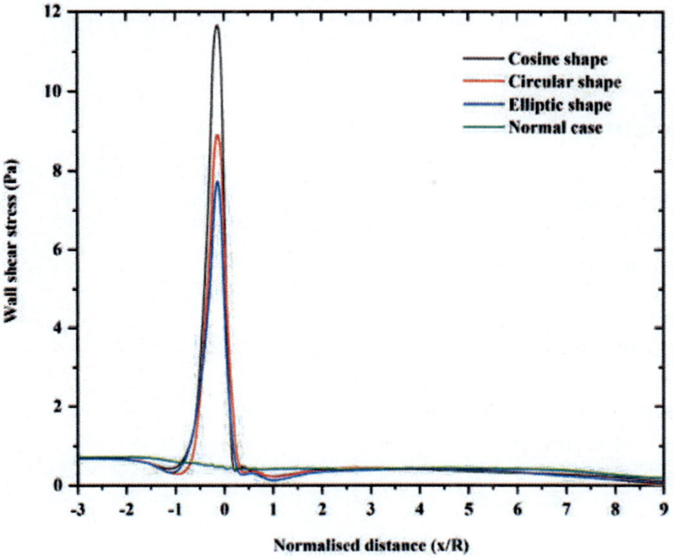

Fig. 7 Wall shear stress distribution for different stenosis shapes and the normal case

5.4 Wall Shear Stress

Wall shear stress is a very important hemodynamic parameter in blood flow analysis through arteries [18]. Figure 7 shows the wall shear stress distribution versus normalized distance. From the graph, it is evident that the cosine shape stenosis has got the highest wall shear stress at the blockage region. This is true because WSS increases with decreasing stenosis length [19]. According to Zhou et al.'s literature, a stenosis of the same shape, with a length of 20 mm, leads to a wall shear stress (WSS) value of 7.8 Pa.

6 Conclusions

In this study, the influence of stenosis shape on hemodynamics is numerically examined by using the Carreau viscosity model. The results show that the cosine shape stenosis has the highest wall shear stress at constriction region compared with the other shapes of stenosis. However, the variation of WSS has the same trend for all shapes. This type of research has a potential to understand the pathogenesis of vascular diseases. This piece of work might be useful to resolve pathological issues such endothelial damage, hemolysis, thrombosis, and other arterial injury. Studies on the influence of other shapes of stenosis and pulsating flow boundary condition should be explored in the future research.

Acknowledgements The authors declare that they have no conflicts of interest.

Nomenclature

$\overrightarrow{U}$	Velocity vector [m^2]
ρ	Blood density [kg/m^3]
p	Pressure [Pa]
$\overrightarrow{\overrightarrow{\tau}}$	Viscous stress tensor [Pa]
μ	Viscosity [Pa-s]
μ_0	Viscosity at zero shear rate [Pa-s]
μ_∞	Viscosity at high shear rate [Pa-s]
λ	Relaxation time [s^{-1}]
$\dot{\gamma}$	Shear rate [s]

References

1. World Health Organization (2022) Cardiovascular diseases (CVDs). Retrieved March 2, 2022 from https://www.who.int/news-room/fact-sheets/detail/cardiovascular-diseases-(cvds), 11 June 2021.
2. Giddens DP, Zarins CK, Glagov S (1993) The role of fluid mechanics in the localization and detection of atherosclerosis. Am Soc Mech Eng 115:588–594
3. Zhang B, Jin Y, Wang X, Zeng T, Wang L (2017) Numerical simulation of transient blood flow through the left coronary artery with varying degrees of bifurcation angles. J Mech Med Biol 17(01):1750005
4. Freidoonimehr N, Arjomandi M, Sedaghatizadeh N, Chin R, Zander A (2020) Transitional turbulent flow in a stenosed coronary artery with a physiological pulsatile flow. Int J Numer Methods Biomed Eng 36(7):e3347
5. Wang T, Butany J (2017) Pathogenesis of atherosclerosis. Diagn Histopathol 23(11):473–478
6. Young DF, Tsai FY (1973) Flow characteristics in models of arterial stenoses—I. Steady flow. J Biomech 6(4):395–410
7. Ang KC, Mazumdar JN (1997) Mathematical modelling of three-dimensional flow through an asymmetric arterial stenosis. Math Comput Model 25(1):19–29
8. Zhou Y, Lee C, Wang J (2018) The computational fluid dynamics analyses on hemodynamic characteristics in stenosed arterial models. J Healthc Eng 2018(4312415):6
9. Yao H, Ang KC, Yeo JH, Sim EKW (2000) Computational modelling of blood flow through curved stenosed arteries. J Med Eng Technol 24(4):163–168
10. Nandakumar N, Sahu KC, Anand M (2015) Pulsatile flow of a shear-thinning modefor blood through a two-dimensional stenosed channel. Eur J Mech B Fluids 49:29–35
11. Stroud JS, Berger SA, Saloner D (2000) Influence of stenosis morphology on flow through severely stenotic vessels: implications for plaque rupture. J Biomech 33(4):443–455
12. Ahamad NA, Kamangar S, Badruddin IA (2018) The influence of curvature wall on the blood flow in stenosed artery: a computational study. Bio-Med Mater Eng 29(3):319–332
13. Razavi A, Shirani E, Sadeghi MR (2011) Numerical simulation of blood pulsatile flow in a stenosed carotid artery using different rheological models. J Biomech 44(11):2021–2030

14. Freidoonimehr N, Chin R, Zander A, Arjomandi M (2021) Effect of shape of the stenosis on the hemodynamics of a stenosed coronary artery. Phys Fluids 33(8):081914
15. Abugattas C, Aguirre A, Castillo E, Cruchaga M (2020) Numerical study of bifurcation blood flows using three different non-Newtonian constitutive models. Appl Math Model 88:529–549
16. Amiri MH, Keshavarzi A, Karimipour A, Bahiraei M, Goodarzi M, Esfahani JA (2019) A 3-D numerical simulation of non-Newtonian blood flow through femoral artery bifurcation with a moderate arteriosclerosis: investigating Newtonian/non-Newtonian flow and its effects on elastic vessel walls. Heat Mass Transf 55(7):2037–2047
17. Tabakova S, Nikolova E, Radev (2014) Carreau model for oscillatory blood flow in a tube. In: AIP conference proceedings, vol 1629, no 1. American Institute of Physics, pp 336–343
18. Bahrami S, Norouzi M (2018) A numerical study on hemodynamics in the left coronary bifurcation with normal and hypertension conditions. Biomech Model Mechanobiol 17(6):1785–1796
19. Song J, Kouidri S, Bakir F (2020) Numerical study of hemodynamic and diagnostic parameters affected by stenosis in bifurcated artery. Comput Methods Biomech Biomed Eng 23(12):894–905

Numerical Analysis of Cross-Flow Hydro Turbine with Different Number of Blades

Kundan Kumar Sahu, Charu Parashar, and Priyanka Dhurvey

Abstract Hydropower is one of the most desirable energy sources to be used if it is as competitive and cost-effective as any other energy source. One cost-conscious measure in hydropower is the efficiency of turbines, which is the subject of the study reported here. Despite its reduced efficiency, the cross-flow hydro turbine has proven to be effective at low discharge and low head. This study aims to make a cross-flow hydro turbine more efficient. This study conducted a numerical study of a cross-flow hydro turbine (CFHT) using ANSYS FLUENT and a multi-physics finite volume solver (FVM). The shear stress transport (SST) k-ω turbulence model was used to perform 3D unsteady simulations. A CFHT's performance was evaluated by increasing the blade numbers from 18 to 21. When the θ_s (nozzle entrance arc angle) was 65°, the rotational speed was 600 rpm, and the β_1 was 10°, with 20 blades, the optimal efficiency was found to be 95.75%.

Keywords Cross-flow hydro turbine · Number of blades · Turbine performance · Numerical analysis · Efficiency

1 Introduction

Environmental concerns, rising global power demand, fixed fossil fuels, improved efficiency, and other factors have all contributed to the growth of renewable energy generation [1]. Compared to other renewable energy sources like solar, wind, biomass, and others, hydropower systems have grown in popularity because of the low generation costs, ease of preservation, consistency, availability of water power potential, superior efficiency, and other factors [2]. The CFHT is an impulse turbine working under a low head and low flow [3]. Various researchers have modified some geometrical parameters to improve CFHT efficiency [4].

K. K. Sahu (✉) · C. Parashar · P. Dhurvey
Department of Civil Engineering, MANIT, Bhopal, Madhya Pradesh 462003, India
e-mail: kundansahu234@gmail.com

S. Das et al. (eds.), *Proceedings of the 1st International Conference on Fluid, Thermal and Energy Systems*, Lecture Notes in Mechanical Engineering,
https://doi.org/10.1007/978-981-99-5990-7_11

2 Literature Review

In recent years, it has been discovered that an effort has been made to improve the CFHT's performance through the use of advanced sensors, monitoring, optimization, software tools, and other methods [5]. According to their power rating, Yahya et al. [6] categorized micro-hydro turbines. Small-scale applications can make use of Pelton, Turgo, and cross-flow turbines, among other types [6]. By modifying the nozzle form, Acharya et al. [7], Adhikari and Wood [8] performed a CFD simulation of CFHT and got 76.6 and 87% efficiency, respectively. Patel et al. [9] used a computational model to investigate the effect of mass flow rates at 354 rpm on rotational speed. According to their findings, the CFHT's maximum efficiency climbed from 67.1 to 95.08% as the mass flow rate increased. Patel et al. [10] calculated the maximum efficiency of CFHT using a mass flow rate of 0.315 m^3/s and 18 blades, resulting in an efficiency of 84% at 300 rpm. The CFHT has also been the subject of experimental studies to examine the effects of different runner speeds, blade counts, and flow rates [11]. When the flow rate was 0.0091 m^3/s and the number of blades 20, Saini and Singal [3] and Kim et al. [12] completed CFHT's numerical analysis, and they stated that at 650 rpm, the maximum efficiency is 75%. However, numerical research by Andrade et al. [13] and Kunda et al. [14] indicated at 900 rpm, the highest efficiency is 72.8%. using 25 numbers of blades. Oliy and Ramayya [15] used ANSYS CFX to do a numerical investigation of CFHT with a flow rate of 13.7 m^3/s and several blades of 20 with a blade inlet angle of 18°, reporting efficiency of 79% at 500 rpm. According to Chichkhede et al. [16], the ideal blade angle is between 24 and 30°. However, because the amount of momentum transferred to a curved surface is linked to its curvature when the flow strikes it, this angle may exceed the specified range. It has been discovered that as the curvature of the blade increases, so does the momentum transmission. The hydropower system provides the majority of India's electricity generation in the northeast. Because the CFHT is suited for low flow rate and low head settings, it can be used in a variety of situations [17, 18]. As a result, this turbine was chosen for the current research. The inlet blade angle (β_1) was considered by the majority of the investigators to be in the range of 30–40°. However, in this study, a low value of β_1 was used with a total of 20 blades, which resulted in higher efficiency. As a result, the major goal of this study is to increase the efficiency of CFHT by altering the blade numbers in a runner from 18 to 21. The highest efficiency has been established with several blades of 20 according to the three-dimensional unsteady Reynolds Averaged Naiver-Stokes (RANS) equation data.

3 System Description

Figure 1 is a two-dimensional illustration of a CFHT. The water flow's kinetic energy is increased incrementally, and the water flow to the turbine blades is controlled by the nozzle. Water hits the top half of the blades in the first stage of the CFHT runner,

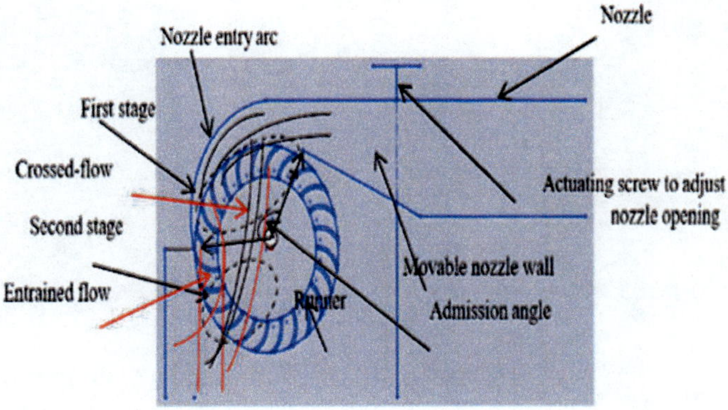

Fig. 1 The design of a cross-flow turbine [14]

then flows to the rotor's air space area and hits the lower section of the blades in the second stage.

Several researchers have identified the CFHT's numerous geometric parameters to increase its performance. In this study, 3D unsteady simulations were done using ANSYS FLUENT and a multi-physics finite volume method (FVM) solver to evaluate the effect of the blade numbers.

3.1 *Model Geometry*

Figures 2 depict the CFHT computational model created in the ANSYS Workbench. As presented in Fig. 2a–c, the three-dimensional model includes fluid domains of the non-rotating domain, rotating domain, and casing with nozzle. Table 1 [8] shows the dimensions of the geometric characteristics of CFHT.

3.2 *Meshing*

ANSYS Meshing was used to discretize the computational domain. As shown in Figs. 3a–c, unstructured tetrahedron grids were used to discretize the domain. However, unstructured grids also have been recommended by researchers [16, 19]. The grids are created using the face and body sizing processes. With a 100 mm element size, the body sizing was used to create a very fine meshing in the domain's rotating area, accurately depicting the flow mechanics. The non-rotating structured grids portion and nozzle, on the other hand, were discretized using a body sizing procedure with a 300 mm element size (Table 2).

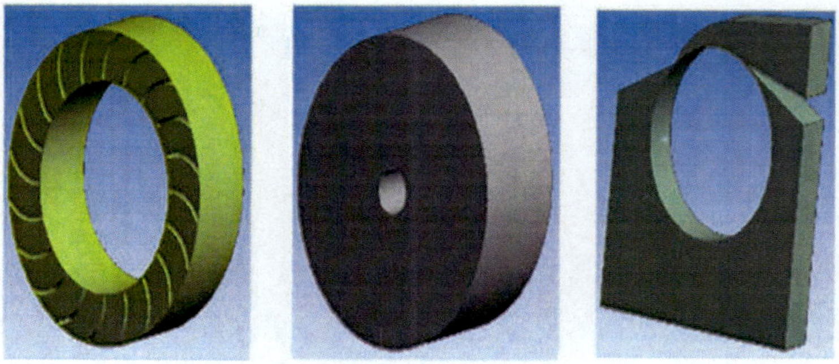

(a) Rotating domain (b) Non-rotating domain (c) Casing with nozzle

Fig. 2 The CFHT's computational domain at different parts

Table 1 Details of the numerical study's cross-flow hydro turbine design parameters

Parameters	Specifications
Outer diameter of runner (D_1)	31.6 cm
Inner diameter of runner (D_2)	21.172 cm
Number of blades (N_s)	20
Blade angles at the outer and inner periphery ($\beta 1$ and β_2)	30° and 90°
Nozzle entry arc angle (θ_s)	69°
Runner and nozzle width (W)	15 cm
The thickness of blade (t)	0.3 cm
Height of nozzle throat (h_0)	6.5 cm

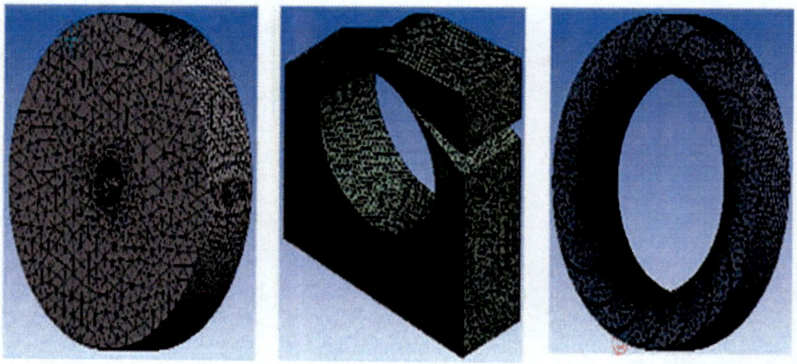

(a) Non-rotating domain (b) Nozzle with casing domain (c) Rotating domain

Fig. 3 The meshing of the CFHT: **a** non-rotating domain, **b** Nozzle with casing domain, and **c** Rotating domain

Table 2 Mesh information for the three flow computational domains

Computational domains	Mesh statistics	
	Nodes	Elements
Casing with nozzle	19,810	96,696
Rotating domain	795,656	4,058,024
Non-rotating domain	9179	47,080

Table 3 Three-dimensional model's boundary conditions [9]

Boundary conditions	Boundary condition details
Inlet	5.833 m/s
Outlet	Atmospheric pressure
Blade walls	Moving wall
Rotating domain walls	Moving wall
Nozzle with the casing walls	Wall
Non-rotating domain walls	Wall

3.3 Boundary Conditions

Table 3 lists the boundary conditions for the computing domain.

Compared to other values, the 5.833 m/s constant inlet flow velocity was shown to be more efficient [8, 11]. Three sub-domains make up the computational domain: non-rotating domain, nozzle with the casing, and rotating component rotating domain.

3.4 Details of the Solver Setup

The CFHT is surrounded by a turbulent flow field. As a result, 3D unsteady simulation using the SST k-ω turbulence model was used in this study. The SST k-ω model combines the best qualities of the k-ω and k-ε models, as well as improved prediction skills, as reported by several researchers [20–23]. The non-dimensional wall distance (y+) has a value that is less than one. As a result, the flow is resolved at the blade's wall borders. The simulation will be less accurate as y+ increases. The simulations are run with a time step size of 0.001 s and 200-time steps. The time step size is derived using the rotor's 1° rotation. Per time step, the number of iterations is kept constant at 20. The solution's convergence criterion has been established at 10^{-6}. The second-order upwind technique discretizes the momentum conservation equations' conservative and temporal aspects.

3.5 Performance Parameter Estimation

The equation used for the calculation of efficiency [24].

$$\eta = \left[\frac{P_{out}}{P_{in}}\right] \tag{1}$$

P_{in} and P_{out} are the input and output power, respectively, and have been calculated as follows:

$$P_{out} = \theta \times \omega \tag{2}$$

$$P_{in} = \rho \times g \times Q \times H \tag{3}$$

The output torque and the angular velocity are represented by θ and ω, respectively.

On the other hand, the density of the water, water flow rate, gravitational acceleration, and head are represented by ρ, Q, g, and H, respectively.

4 Validation of the Model

The numerical work of Ranjan et al. [25] and Dakers and Martin's experimental work [11] were used to validate the current 3D unsteady simulations. Because of numerous losses encountered throughout the tests, the current computational results are thought to have over-predicted the experimental work. The current numerical results, however, indicate a similar pattern to that of Ranjan et al. [25]. However, when compared to the observation made by Ranjan et al. [25] at 600 rpm, the current numerical simulations showed a 2.1% inaccuracy shown in Fig. 4.

5 Results and Discussion

The functioning of a CFHT was evaluated in this study by enhancing the blade numbers from 18 to 21, using a 10° β_1, a 65° θ_s, and a turbine rotational speed of 600 rpm. The flow study was performed with the use of pressure and velocity contours in the ANSYS FLUENT environment using the FVM solver.

When water enters the blade channel of the first stage, the runner emerges from between the turbine blades and transfers kinetic energy to them. The water then proceeds to the second stage bypassing the open space within the runner and impacting the lower turbine blades. As a result, the flow's velocity has decreased slightly, and it has changed direction, guided by blade curvature, to exit the first stage.

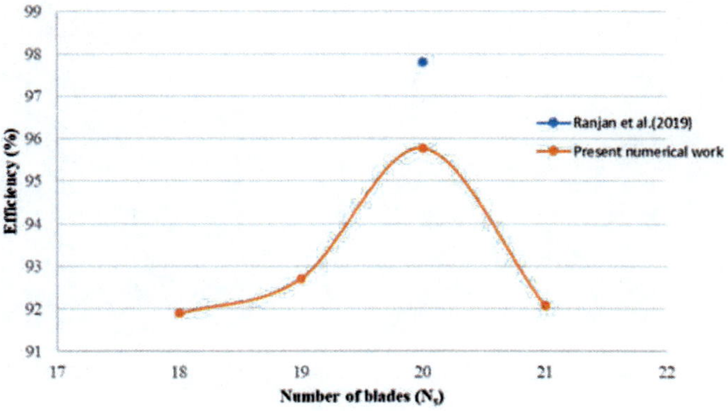

Fig. 4 The current study's results are compared to numerical results

Then it moves on to the second stage, where it lessens velocity and alters the direction, transferring some of the blades' momentum. The flow then exits the first stage and enters the second stage, Because of the higher runner speed, it is spliced over in the casing. The pressure was gradually reduced as the flow entered the blade channel from the intake to the exit of the first stage. Finally, the flow was discharged via the casing exit at the same pressure. On the other side of the flow, the nozzle has the maximum pressure, while the turbine shaft has the lowest. At different regions of the turbine, the pressure does not remain constant. The highest pressure values at the nozzle periphery are presented in Fig. 5.

The highest efficiency was found to be 95.75% at 600 rpm using the current three-dimensional unsteady analysis with a 65°, β_1. It has been discovered that as the number of blades is reduced, the efficiency decreases. This could be explained by the fact that as the number of turbine blades grows, more energy is transferred from the water to the turbine blades. The efficiency of the CFHT has been investigated with the number of blades used, as shown in Figs. 6, 7, and 8, and Table 4.

6 Conclusions

The efficient use of existing water power potential, particularly in India's northeast, drives the development of micro and tiny hydropower systems. three-dimensional unsteady simulations were performed with the use of the ANSYS FLUENT finite volume technique solver, utilizing the SST k-ω turbulence model. In the current analysis, the highest efficiency was found to be 95.75% at, 65° at 600 rpm when β_1, 10°.

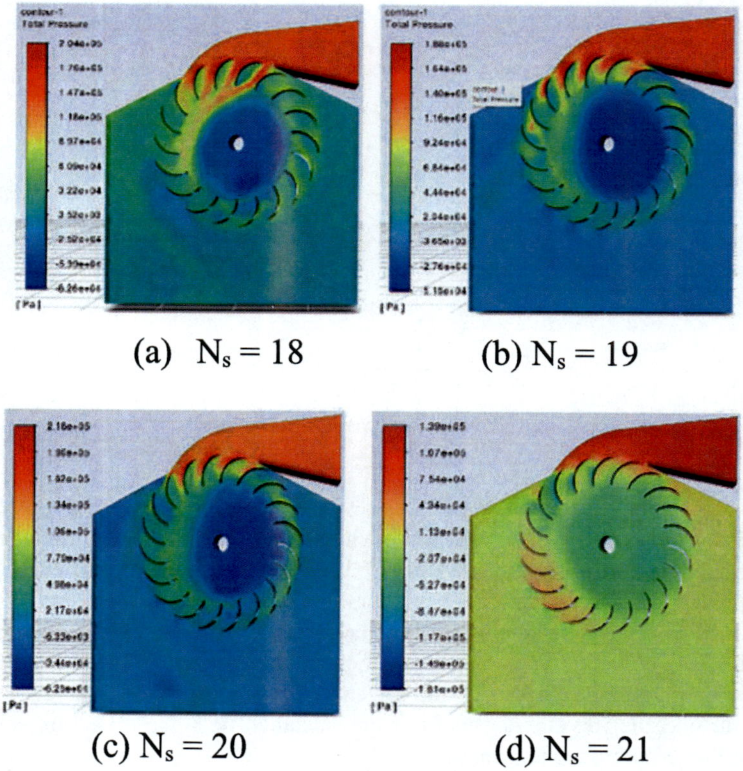

Fig. 5 Total Pressure contours of the model for different number of blades

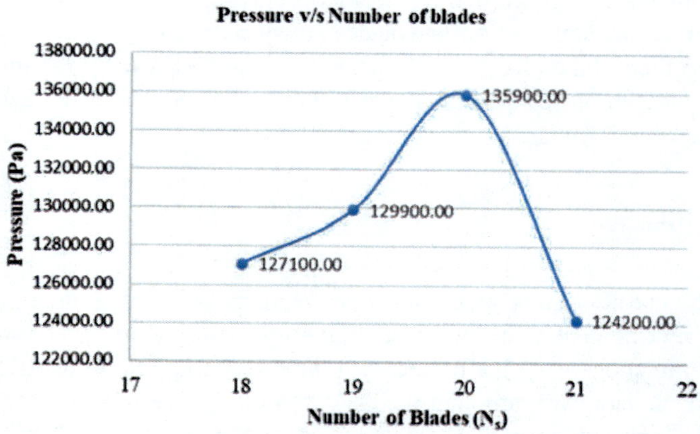

Fig. 6 Variation of inlet pressure with a number of blades

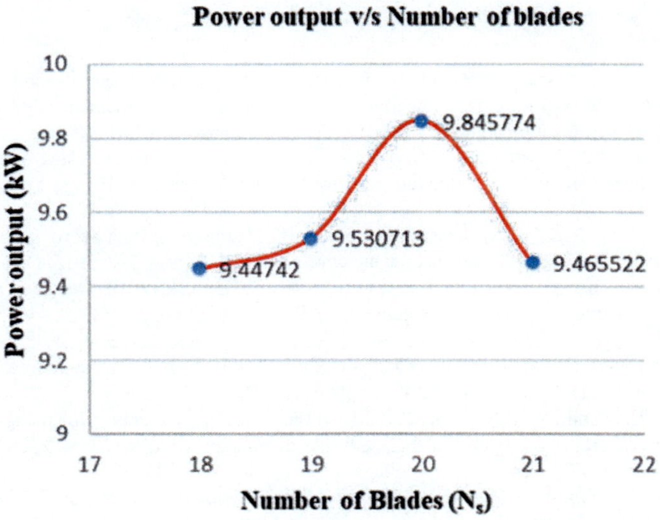

Fig. 7 Variation of power output with a number of blades

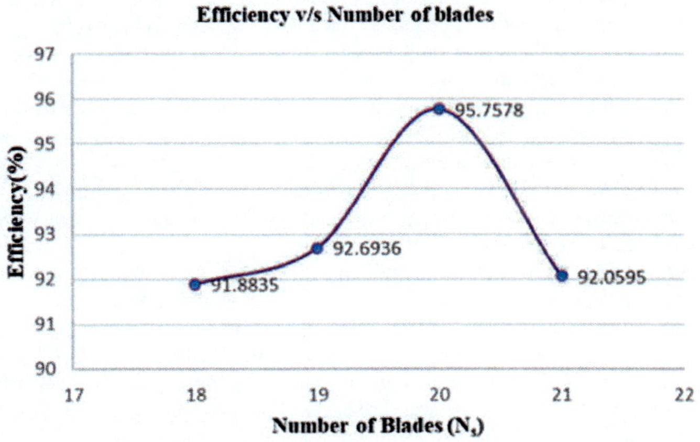

Fig. 8 Variation of efficiency with the number of blades

Table 4 Cross-flow hydro-turbine efficiencies at different blade numbers

Number of blades, N_s	Rotational speed, N (rpm)	Efficiency, η (%)
18	600	91.23
19	600	92.41
20	600	95.75
21	600	93.24

References

1. Zhou J, Lu P, Li Y, Wang C, Yuan L, Mo L (2016) Short-term hydro-thermal-wind complementary scheduling considering the uncertainty of wind power using an enhanced multi-objective bee colony optimization algorithm. Energy Convers Manage 123:116–129
2. Date A, Date A, Akbarzadeh A (2013) Investigating the potential for using a simple water reaction turbine for power production from low head hydro resources. Energy Convers Manage 66:257–270
3. Saini RP, Singal SK (2015) Development of crossflow turbine for Pico hydro. In: International conference on hydropower for sustainable development, February 5–7, 2015, Dehradun, India
4. Chattha JA, Khan MS, Wasif ST, Ghani OA, Aunt MO, Hamid Z (2010) Design of a cross-flow turbine for a micro-hydro power application. ASME Power Conf 49354:637–644
5. Ahmad KY, Wan N, Wan AM, Zulkifli O (2014) Pico-hydro power generation using dual Pelton turbines and single generator, Paper No. PEOCO2014. In: IEEE 8th international power engineering and optimization conference, pp 579- 584
6. Yahya AK, Munim WNWA, Othman Z (2014) Pico-hydro power generation using dual Pelton turbines and single generator. In: IEEE 8th international power engineering and optimization conference, pp 579–584
7. Acharya N, Kim CG, Thapa B, Lee YH (2015) Numerical analysis and performance enhancement of a cross-flow hydro turbine. Renew Energy 80:819–826
8. Adhikari RC, Wood DH (2017) A new nozzle design methodology for high-efficiency crossflow hydro turbines. Energy Sustain Dev 41:139–148
9. Patel JD, Patel KD, Patel DA (2015) To examine the effect of mass flow rate on crossflow turbine using computational fluid dynamics. Int J Eng Res Technol 4(5):1094–1096
10. Patel M, Oza N, Patel K (2016) Computational fluid dynamic analysis of cross-flow turbine. Int J Innov Res Sci Eng Technol 5(9):18936–18944
11. A. J. Dakers, and G. Martin, Development of a simple cross-flow water turbine for rural use, Agricultural Engineering Conference, Resources efficient use and conservation preprints of papers, Australia Institution of Engineers, 35, 1982.
12. Kim IC, Wata J, Ahmed MR, Lee YH (2012) CFD study of a ducted cross-flow turbine concept for high efficiency tidal current energy extraction. In: Proceedings of Asian wave and tidal energy conference, pp 400–405
13. Andrade JD, Curiel C, Kenyery F, Aguillon O, Vasquez A, Asuaje M (2011) Numerical investigation of the internal flow in a Banki turbine. Int J Rotating Mach
14. Kaunda CS, Kimambo CZ, Nielsen TK (2014) A numerical investigation of flow profile and performance of a low-cost crossflow turbine. Int J Energy Environ 5(3):275–296
15. High GB, Ramayya AV (2017) Design and computational fluid dynamic simulation study of high-efficiency cross flow hydro-power turbine. Int J Sci Technol Soc 5(4):120–125
16. Chichkhede S, Verma V, Gaba VK, Bhowmick S (2016) A simulation-based study of flow velocities across cross-flow turbine at different nozzle openings. Procedia Technol 25:974–981
17. Shahidul MI, Tarmizi SSS, Yassin A, Zen H, Hung TC, Djun LM (2015) Modeling the energy extraction from in-stream water by multi-stage blade system of cross-flow micro-hydro turbine. Procedia Eng 105:488–494
18. Raman N, Hussein I, Palanisamy K (2009) Micro-hydro potential in West Malaysia. Int Conf Energy Environ 3:348–359
19. Kumar A, Saini RP (2017) Performance analysis of a Savonius hydrokinetic turbine having twisted blades. Renew Energy 108:502–522
20. Silva PASF, Vaz DATDR, Britto V, Oliveira TFD, Vaz JR, Junior ACB (2018) A new approach for the design of diffuser-augmented hydro turbines using the blade element momentum. Energy Convers Manage 165:801–814
21. Talukdar PK, Sardar A, Kulkarni V, Saha UK (2018) Parametric analysis of model Savonius hydrokinetic turbines through experimental and computational investigations. Energy Convers Manage 158:36–49

22. Jiyun D, Hongxing Y, Zhicheng S, Xiaodong G (2018) Development of an inline vertical cross-flow turbine for hydropower harvesting in urban water supply pipes. Renew Energy 127:386–397
23. Alom N, Borah B, Saha UK (2018) An insight into the drag and lift characteristics of modified Bach and Benesh profiles of Savonius rotor. Energy Procedia 144:50–56
24. Huang SR, Ma YH, Chen CF, Seki K, Aso T (2014) Theoretical and conditional monitoring of a small three-bladed vertical-axis micro-hydro turbine. Energy Conv Manage 86:727–734
25. Ranjan RK, Alom N, Singh J, Sarkar BK (2019) Performance investigations of the cross-flow hydro turbine with the variation of blade and nozzle entry arc angle. Energy Convers Manage 182:41–50

Thermal Characteristics of Synthetic Jet Impingement at Different Strouhal Numbers for Electronics Cooling

Pawan Sharma, Santosh K. Sahu, and Harekrishna Yadav

Abstract In the present experimental investigation, the effect of Strouhal number on the thermal performance of synthetic jets is studied. Experiments are carried out with an 8 mm orifice diameter plate for two different Strouhal numbers (Sr = 0.04 and 0.66) at constant Re = 2280. The hot wire anemometry is used to gauge the flow velocity, while an Infrared thermal imaging camera is used to study the thermal characteristics. The results showed that Sr plays a significant role in the variation of the heat transfer rate for a given surface distance at constant jet Reynolds number. At a larger surface distance (z/d = 16), a low Sr jet produces 19% better heat transfer than a high Sr jet. However, for a lower surface distance (z/d = 6), a jet with a high Sr exhibits a 9% higher heat transfer rate. Further, by reducing the surface distance up to z/d = 2, unforeseen enhancement in heat transfer is observed with a low Sr. This occurs because of the flow circulation at low surface distance, which would have a more substantial effect at the higher Sr.

Keywords Synthetic Jet · Strouhal number · Flow recirculation · Power spectral density

Nomenclature

D	Orifice diameter (mm)
Sr	Strouhal number
T	Orifice plate thickness (mm)
Nu_0	Stagnation point Nusselt number
Re	Reynolds number
T_s	Surface temperature of test foil (°C)
T_a	Ambient temperature (°C)
V	Voltage drop across foil

P. Sharma (✉) · S. K. Sahu · H. Yadav
Department of Mechanical Engineering, Indian Institute of Technology, Indore, India
e-mail: phd1901203008@iiti.ac.in

S. Das et al. (eds.), *Proceedings of the 1st International Conference on Fluid, Thermal and Energy Systems*, Lecture Notes in Mechanical Engineering,
https://doi.org/10.1007/978-981-99-5990-7_12

V_{rms} Root mean square amplitude (V)
z/d Dimensionless orifice to surface distance
SJ Synthetic jet
PSD Power spectral density

1 Introduction

With the ever-increasing demand for high-end electronic devices with compact size, the density of transistors on ICs rises exponentially [1]. Such dense electronic systems will cause the rise of system temperature and eventually its overheating. This overheating of electronic devices accounts for their 55% failure and, consequently, less reliability in operations. So, an efficient thermal dissipation system is required to cool the electronic devices for their best functioning. In this regard, synthetic jet (SJ) is a promising candidate for thermal management of future electronics [2, 3]. A SJ is a novel zero-net-mass flux device that synthesizes its jet with the surrounding fluid. A typical SJ works like a Helmholtz resonator and its assembly consists of an actuator, cavity, and an orifice plate [4]. The SJ's advantages include its light weight, compact design, lower power consumption, and high reliability.

In SJ actuation, actuating frequency with corresponding Strouhal number (Sr) plays an important role in the emanating vortices from the orifice exit [5]. Also, the frequency of vortex shedding varies with surface distance due to the response of jet shear layer. In the present investigation, the effect of actuating frequency on the vortice formation and their impact on the thermal performance are studied.

2 Literature Review and Objective

It is found that excitation frequency and corresponding Strouhal number play a critical role in the thermal characteristics of impinging jets. Pavlova and Amitay [6] studied the excitation frequency (f = 420 and 1200 Hz) effect on SJ cooling performance. They argued that at lower orifice to surface distance (z/d < 10), high frequency jet (f = 1200 Hz) outperformed the low frequency jet. However, at higher surface distance (z/d > 15), low-frequency jet (f = 420 Hz) dissipates more heat w.r.t. to high-frequency jet. They argued that for a low-frequency jet, large wavelength between coherent structure causes vortex ring impingement on the surface separately, while a high-frequency jet is associated with disintegration and fusion of vortices. At low surface distance (z/d = 2), Vejrazka et al. [7] explored the effect of excitation for excitation frequency corresponding to Sr = 0.2–4 and Re = 10,000. When excited with frequency near to Sr = 0.017, by contrast, small vortices are produced and consequently flow fluctuations are attenuated.

The small-size vortices are incapable to cause the flow separation and thus the stream of fluids remain attached to the wall. However, the low-frequency excitation causes large-size vortice formation than those with unexcited jet. Tianshu and Sullivan [8] studied the excited impinging jet at a small distance ($z/d \leq 2$). They also observed that at a frequency excitation of $Sr = 0.85$ forms bigger and structured vortices which causes local reduction in heat transfer. However, heat transfer is increased by exciting at $Sr = 1.6$, which induces arbitrary and irregular small-size vortices.

3 Materials and Methods

3.1 Test Facility and Procedure

The experimental facility in present investigations consists of different components as shown in Fig. 1. It comprises a SJ actuator assembly which uses a loud-speaker of 73 mm diaphragm size and 8 mm diameter orifice plate ($t/d = 0.31$). The input sinusoidal signal to the SJ actuator is supplied by using a power oscillator (Syscon instruments). The amplitude of input signal is maintained constant (4 V_{rms}), and frequency is varied as per the experimental requirement. The test foil (AISI-304) having dimensions of 90 mm × 90 mm is heated by a direct current power source for a constant heat flux of 2 kw/m^2. The back side of the test surface is painted with flat black paint to achieve uniform emissivity ($\varepsilon = 0.95$). Further, to measure the flow velocity at orifice exit, a constant temperature hot wire anemometer (Mini CTA 54 T42) with a single wire probe is employed. To study the heat transfer characteristics, the thermal imaging technique or IR camera is employed (FLIR: A655sc) which can gauge the temperature of the test surface in the range of 0–2000 °C with an uncertainty of ±1.5%. Further, a stepper motor-controlled traversing stand is used to maintain the surface distance between the orifice plate and heated foil. All the experiments are conducted inside a closed environment to abstain from surrounding effects.

3.2 Data Reduction

To measure the centerline mean velocity, instantaneous velocity during ejection stroke is considered (half of one complete cycle), and it is calculated as

$$U_0 = \frac{1}{T}\int_0^{T/2} u(t)\,dt \tag{1}$$

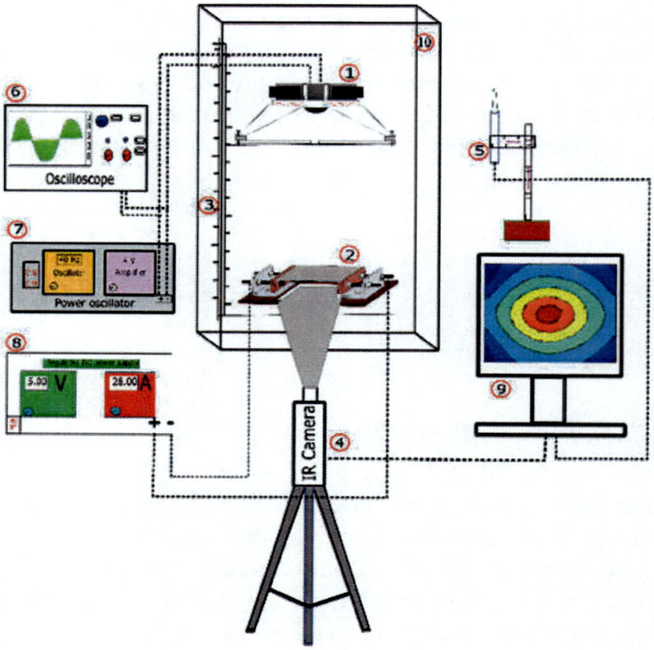

1	SJ actuator	6	Oscilloscope
2	Heating plate	7	Power oscillator
3	Traversing stand	8	Dc power supply
4	IR camera	9	computer
5	HWA probe	10	Cubical enclosure

Fig. 1 Schematic of the experimental setup used in present investigation

The Reynolds number (Re) based on mean centerline velocity can be calculated as

$$\mathrm{Re} = \frac{U_0 \times d}{\nu} \tag{2}$$

where d is the orifice diameter and ν is the kinematic viscosity of the fluid.

The Strouhal number is a measure of the ratio of the inertial forces due to local acceleration to the convective acceleration. It is the inverse of dimensionless stroke length

$$Sr = \frac{f \times d}{U_0} \tag{3}$$

where f is vortex shedding frequency (or Strouhal frequency).

The Nu_{avg} is computed as follows:

$$Nu_{avg} = \frac{h_{avg} \times d}{k_f} \tag{4}$$

The average heat transfer coefficient (h_{avg}) can be computed as below

$$h_{avg} = \frac{q_{conv}}{(T_s - T_a)} \tag{5}$$

Here, T_s and T_a are the temperature of test foil and its ambience respectively, while q_{conv} is the net heat supplied to the test specimen

$$q_{conv} = q_j - q_{loss} \tag{6}$$

where q_j denotes the net heat flux supplied to the test specimen, while q_{loss} signifies the total heat loss (in the form of convection and radiation) from the target surface. The net heat flux across the test specimen is measured in terms of input power (V = voltage drop across test surface; I = flowing current) and it is computed as below

$$q_j = \frac{V \times I}{A} \tag{7}$$

The heat loss from the test surface is computed by no-flow experiment [9], and the heat loss equation for our test surface is stated w.r.t. $(T_s - T_a)$ as below

$$Q_{loss} = 0.23143 \times (T_s - T_a) - 0.31364 \tag{8}$$

A maximum of 11% heat loss is computed which combines the heat loss due to natural convection and radiation from the foil surface (from both top and bottom sides).

4 Results and Discussion

4.1 Validation of Present Results

To validate the heat transfer results, reduced stagnation point Nusselt number ($Nu_{0,r}$) is considered for comparison with the existing literature. Gil and Wilk [10] used a 150 mm diaphragm size actuator with other operating parameters are input V_{rms} =

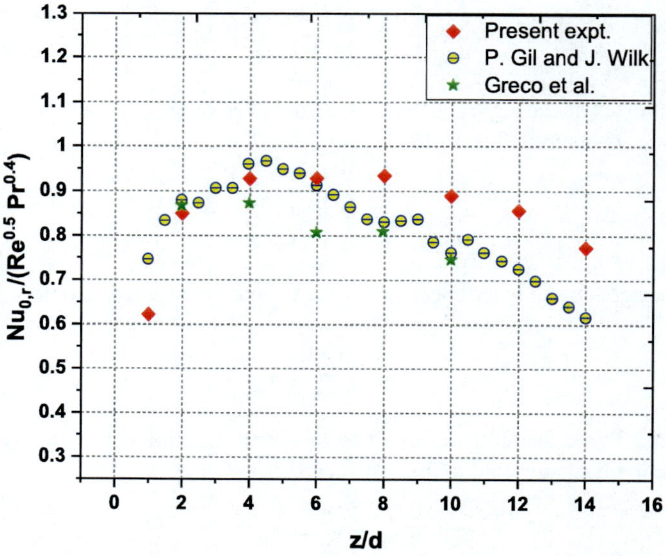

Fig. 2 Results of reduced stagnation point Nusselt number as compared to the literature data

1–6 V, Re = 3600–22,950, and L_0 = 0.84–170.5. While Greco et al. [11] did their study by using a SJ actuator with 208 mm diaphragm size which operated at Re = 5250 and L_0/d = 5–20. In our study, we employed a 73 mm diaphragm size actuator operating at 4 V_{rms}. As can be seen, the heat transfer results in our case match close to [10, 11] (Fig. 2).

5 Results and Discussion

The experiments are conducted by using a constant orifice diameter of 8 mm. The frequency of SJ actuation is varied for corresponding Strouhal number of 0.04–0.8 and at constant Re = 2280.

In the case of SJ, the flow velocity from the orifice exit may be varied by actuating the SJ actuator (loudspeaker) at different frequencies. Figure 3 shows the variation of the Strouhal number and their corresponding Reynolds number (Re) with different input frequencies. Here, it can be seen that initially with the rise in the actuating frequency from 10 to 80 Hz, the mean centerline velocity rises. It occurs due to the fact that resonance frequency occurs at 80–90 Hz for a 73 mm diaphragm size actuator. With further rise in input frequency beyond 80 Hz, the mean centerline velocity shows a downward trend. The Strouhal number (based on Eq. 3) shows an increasing trend with the rise in actuating frequency which is also shown in Fig. 3. In our study, SJs corresponding to a constant Re = 2280 are considered. Here, SJ

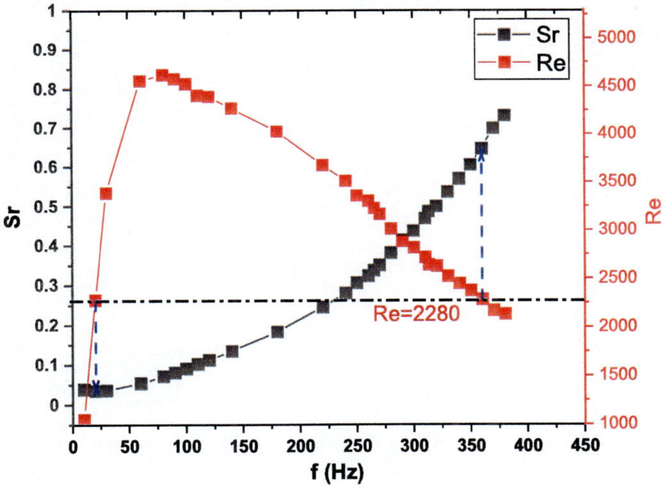

Fig. 3 Variation of Strouhal number and their corresponding Reynolds number (Re) with different SJ actuation frequencies

operating at two different frequencies (f = 20 Hz; Sr = 0.04) and (f = 365 Hz; Sr = 0.66) are considered for heat transfer comparison which are having the same Reynolds number of 2280 (Table 1).

Figure 4. shows the stagnation point Nusselt number variation with axial surface distance at two Strouhal numbers (Sr = 0.04 and 0.66). In the near field (z/d = 2), low Sr SJ (Sr = 0.04) has higher cooling effect as compared to Sr = 0.66. This behavior of better cooling performance of low Sr in near-surface distance (z/d = 2) is strongly influenced by the negative impact of flow recirculation, since a low Sr jet with f = 20 Hz has quite less recirculation compared to high Sr SJ (f = 365 Hz). On increasing the distance between the orifice to test surface (z/d = 6), high Sr jet overcomes the recirculation effect and exhibits 9% better heat transfer compared to low Sr jet (Fig. 5). According to Liu and Sullivan [8], heat transfer performance with low Sr excitation reduces at low distance (z/d ≤ 2). It occurs because low Sr jet promotes large-scale vortex pairing due to which the separation of the wall boundary layer degrades [12]. However, on further increasing the surface distance to z/d = 16, low Sr jet produces 19% better heat transfer compared to high Sr jet (Fig. 4).

Further, the average Nu variation for a range of Sr at a constant amplitude of 4 V_{rms} is shown in Fig. 6. Here, it can be observed that, initially, the heat transfer performance increases with the increase in Sr (Sr = 0.04–0.1). This corresponds due to the rise of mean ejection velocity (or Reynolds number) from 20 Hz (Sr = 0.04) to 110 Hz (Sr = 0.1). As mentioned in Fig. 3, the mean ejection velocity initially increases with excitation frequency, reaches maximum at 80 Hz, and then shows a downward trend. Further following a similar pattern of Re, the average Nu decreases with further rise in Sr from 0.1 to 0.8.

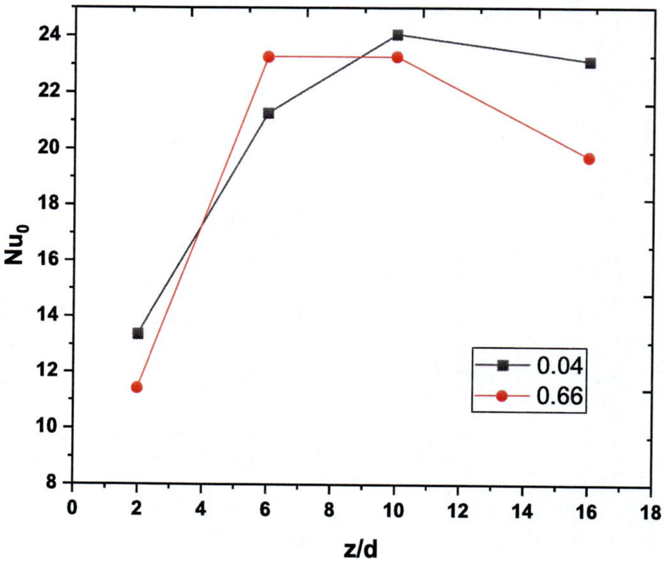

Fig. 4 The stagnation point Nusselt number variation with axial surface distance at two Strouhal numbers (Sr = 0.04 and 0.66)

Figure 7 shows the power spectrum of SJ velocity with different Strouhal numbers (Sr = 0.04 and 0.66) at the orifice exit. The energy associated with the emanating vortices in the SJ flow field is calculated using Fast Fourier transformation (FFT) and a MATLAB code. PSD gives information about the energy distribution of the velocity flow field in the entire frequency range [13]. The peaks in PSD plots exhibit the energy distribution of velocity signal at excitation frequency/first harmonics (f) and after that at subsequent harmonics (2f, 3f, etc.). At orifice exit, SJ with low Sr = 0.04 is dominated by distinct peaks at f (20 Hz), 2f (40 Hz), 3f (60 Hz), etc. The higher peak of the second harmonics w.r.t. the first is due to the suction effect [14]. These distinct peaks indicate the presence of larger vortex rings with less turbulence in SJ flow field for low Sr SJ. However, the PSD plot of high Sr SJ (Sr = 0.66) shows that the first harmonic spectral value is barely higher compared to the low Sr (Sr = 0.04) jet. Also, fewer peaks with relatively less energy content occur due to transition of energy into smaller vortices which increases turbulence and hence more cooling effect is anticipated at low surface distance for high Sr jet.

Fig. 5 Contour map of local Nusselt number for SJ impingement at z/d = 6 for a) Sr = 0.04 and b) 0.66

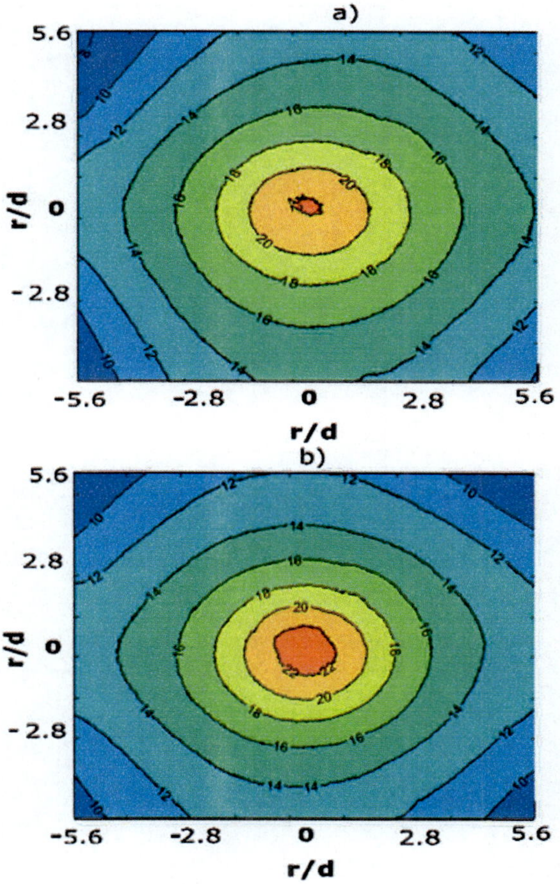

6 Conclusions

In the present experimental study, the effect of Strouhal number (Sr = 0.04 and 0.66) on thermal performance of a synthetic jet is studied at constant Re = 2280. The heat transfer performance for an SJ excitation is studied at a normalized surface distance of z/d = 2, 6, 10, 16. Heat transfer results reveal that at low surface distance (z/d = 2), low Sr jet exhibits better heat transfer because of the flow recirculation. At z/d = 6, the high Sr (0.66) jet overcomes the recirculation effect and exhibits 9% better heat transfer compared to low Sr (0.04) jet. It occurs because low Sr jet promotes large-scale vortex pairing due to which separation of the wall boundary layer becomes difficult. However, on further increasing the surface distance to z/d = 16, low Sr jet produces 19% better heat transfer compared to high Sr jet. PSD data shows that SJ with low Sr = 0.04 is dominated by distinct peaks which indicate the presence of larger vortex rings in SJ flow field. However, high Sr SJ (Sr = 0.66) shows fewer

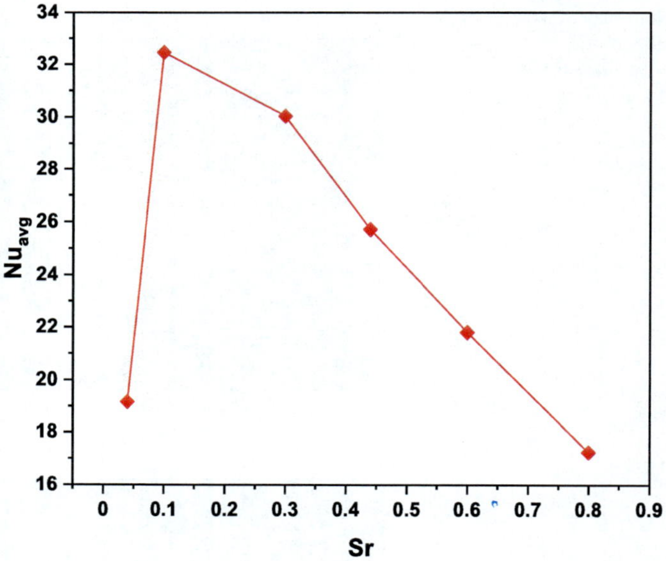

Fig. 6 The average Nusselt number variation for a range of Sr at constant amplitude of 4 V_{rms}

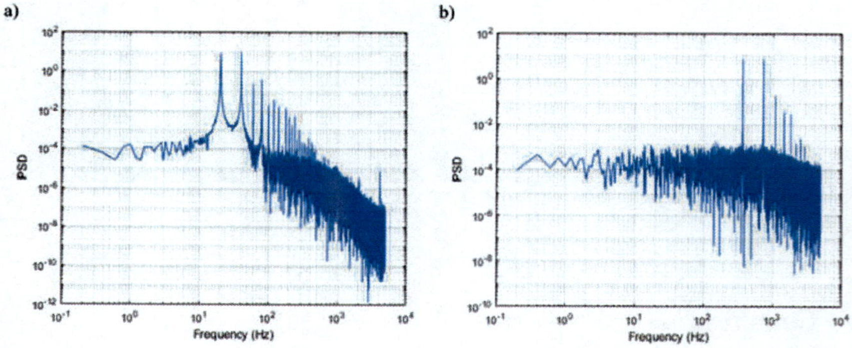

Fig. 7 Distribution of power spectral density of SJ velocity at orifice exit at different Strouhal numbers, **a** Sr = 0.04 and **b** Sr = 0.66

Table 1 Actuating parameters of the input signal for different Strouhal numbers

Actuating frequency, f (Hz)	Mean ejection velocity, U (m/s)	Reynolds number (Re)	Strouhal number (Sr)
20	8.8	2280	0.04
365	8.8	2280	0.66

peaks with relatively less energy content which occurs due to the transition of energy into smaller vortices.

References

1. Moore GE (1998) Cramming more components onto integrated circuits. Proc IEEE 86(1):82–85. https://doi.org/10.1109/JPROC.1998.658762
2. Garimella SV, Singhal V, Liu D (2006) On-chip thermal management with microchannel heat sinks and integrated micropumps. Proc IEEE 94(8):1534–1548. https://doi.org/10.1109/JPROC.2006.879801
3. Sharma P, Joshi J, Singh PK, Sahu SK, and Yadav H (2021) Heat transfer enhancement of an impinging synthetic air jet using sharp-edged orifice. In: Proceedings of 26th national 4th international ISHMT-ASTFE heat mass transfer conference, December 17–20, 2021, IIT Madras, Chennai-600036, Tamil Nadu, India, pp 1835–1840. https://doi.org/10.1615/IHMTC-2021.2770
4. Sharma P, Singh PK, Sahu SK, Yadav H (2021) A critical review on flow and heat transfer characteristics of synthetic jet. Trans Ind Natl Acad Eng 0123456789:1–32. https://doi.org/10.1007/s41403-021-00264-5
5. Yadav H, Agrawal A, Srivastava A (2016) Mixing and entrainment characteristics of a pulse jet. Int J Heat Fluid Flow 61:749–761. https://doi.org/10.1016/j.ijheatfluidflow.2016.08.006
6. Pavlova A, Amitay M (2006) Electronic cooling using synthetic jet impingement. J Heat Transfer 128(9):897–907. https://doi.org/10.1115/1.2241889
7. Vejrazka J, Tihon J, Mart P, Sobolík V (2005) Effect of an external excitation on the flow structure in a circular impinging jet. Phys Fluids 17(10). https://doi.org/10.1063/1.2084207
8. Tianshu L, Sullivan JP (1996) Heat transfer and flow structures in an excited circular impinging jet. Int J Heat Mass Transf 39(17):3695–3706. https://doi.org/10.1016/0017-9310(96)00027-0
9. Yogi K, Krishnan S, Prabhu SV (2021) Experimental investigation on the local heat transfer with an unconfined slot jet impinging on a metal foamed flat plate. Int J Therm Sci 169. https://doi.org/10.1016/j.ijthermalsci.2021.107065
10. Gil P, Wilk J (2020) Heat transfer coefficients during the impingement cooling with the use of synthetic jet. Int J Therm Sci 147:106132. https://doi.org/10.1016/j.ijthermalsci.2019.106132
11. Greco CS, Paolillo G, Ianiro A, Cardone G, de Luca L (2018) Effects of the stroke length and nozzle-to-plate distance on synthetic jet impingement heat transfer. Int J Heat Mass Transf 117:1019–1031. https://doi.org/10.1016/j.ijheatmasstransfer.2017.09.118
12. Choudhary K, Lote C, Sharma P, Sahu SK, Yadav H (2021) Investigation of fluid flow and heat transfer characteristics of synthetic jet impingement at different Strouhal number. In: Proceedings of 26th national 4th international ISHMT-ASTFE heat mass transfer conference, December 17–20, IIT Madras, Chennai-600036, Tamil Nadu, India, pp 1357–1362, Feb. 2021. https://doi.org/10.1615/IHMTC-2021.2050
13. Tesař V, Kordík J (2011) Spectral analysis of a synthetic jet. Sens Actuators, A Phys 167(2):213–225. https://doi.org/10.1016/j.sna.2011.02.011
14. Kumar A, Saha AK, Panigrahi PK, Karn A (2019) On the flow physics and vortex behavior of rectangular orifice synthetic jets. Exp Therm Fluid Sci 103:163–181. https://doi.org/10.1016/j.expthermflusci.2019.01.020

A Short Review on Wellbore Heat Exchangers

Srinivas Chappidi, Ankesh Kumar, and Jogender Singh

Abstract Wellbore geothermal heat exchanger (GHE) with a single well system is a promising technology for geothermal energy extraction from hot dry rock resources and abandoned wells. In comparison to the conventional two-well geothermal energy extraction systems, a single-well geothermal heat exchanger may reduce the cost and environmental impact. However, the optimization of the wellbore GHE is necessary to achieve higher thermal power due to higher outlet fluid temperature. Thus, the paper briefly reviews various wellbore GHEs and their thermal performance. The effect of the various parameters viz. geothermal gradient, inlet fluid temperature, mass flow rate, and operating time on the system's thermal performance is also reviewed. Thus, the present study may aid the development of novel wellbore GHEs with higher heat extraction efficiencies.

Keywords Geothermal energy · Renewable energy · Wellbore geothermal heat exchanger · Coaxial · U-Tube

1 Introduction

Rapidly growing population, energy deficit, and environmental pollution are three major problems for global sustainable development [1]. Therefore, to overcome energy and environmental challenges, many countries are looking for alternate renewable resources of energy to conventional fossil fuels. Geothermal energy gained huge recognition as a promising alternative to fossil fuel-based energy due to its renewability, long-term sustainability and low impact on the environment [2].

Hydrothermal and Hot Dry Rock (HDR) resources are widely available throughout continents, and may help in reducing fossil fuel consumption as well as import of the

S. Chappidi (✉) · A. Kumar
Department of Civil Engineering, SVNIT Surat, Surat, Gujarat 395007, India
e-mail: Srinivaschappidi954@gmail.com

J. Singh
Department of Chemical Engineering, SVNIT Surat, Surat, Gujarat 395007, India

S. Das et al. (eds.), *Proceedings of the 1st International Conference on Fluid, Thermal and Energy Systems*, Lecture Notes in Mechanical Engineering,
https://doi.org/10.1007/978-981-99-5990-7_13

energy. The conventional heat extraction from the hydrothermal reservoirs requires a well or multiple wells to extract the geothermal fluid. These geothermal fluids may be ecotoxic in nature. Hence, geothermal fluids must be treated properly before reinjection. This process involves several technical and environmental challenges such as corrosion of pipes, water pollution, and land subsidence [3]. HDR geothermal energy extraction also requires both injection and production wells to circulate a working fluid through an artificial fractured reservoir. This process also involves high initial costs for wellbore drilling and hydraulic fracturing of the reservoir.

Abandoned wells in oil and gas fields are emitting methane and other gases. In the US, methane is released at an average of 6.0 g/h per well [4]. So, there is a distinct need to handle these abandoned wells. On the other hand, these abandoned wells have higher temperatures while going deep. The effective solution to these hurdles is to utilize the abandoned wells for geothermal energy extraction. This will reduce the new geothermal well drilling cost and also scale up the viability of abandoned wells [5]. Hence, several studies have been carried out to extract geothermal energy through a single well without producing binary fluids using wellbore heat exchangers.

Currently, around 88 countries are utilizing 10,20,887 TJ/yr of geothermal energy for direct heat applications. This results in the reduction of 596 million barrels of oil consumption as well as 252.6 million tons of CO_2 emissions [6]. Approximately, 28 countries are producing electricity from geothermal heat with 15,950.46 MWe installed geothermal power capacity till 2020 [7]. India is not on the list of geothermal power-producing countries. Only a pilot plant of 5 kW was installed in Manikaran. India is utilizing around 357.644 MWt of geothermal energy for direct heat applications [8]. A U-shaped pipe network with a heat pump system is installed at Manali, Himachal Pradesh, for heating buildings adjacent to the SASE plant, which reported 67% energy saving [9]. According to Government of India, ONGC is going to establish first-ever geothermal power plant in Ladakh. The Ladakh Autonomous Hill Development Council (Leh) and the Union Territory of Ladakh signed an MOU with ONGC on 6 February 2021 [10].

2 Background

In 1970, Los Alamos National Laboratory developed a prototype of Enhanced Geothermal System (EGS) with injection, production wells, and an artificial fractured reservoir to study the heat extraction from HDR [11]. The researchers have proposed alternative methods to extract deep geothermal energy particularly using a single well with GHE [11–16]. Nalla et al. (2005) performed a numerical analysis to assess the ability of wellbore GHE to extract heat from a 5563 m deep reservoir at 350 °C temperature [12]. The results show that the wellbore GHE system is producing an outlet temperature of 253 °C at wellhead. Cheng et al. (2013) carried out a numerical analysis for heat extraction from abandoned wells using wellbore GHE at a 6000 m deep reservoir at 255 °C temperature [13]. They reported that the system is capable of producing 137 °C outlet temperature at wellhead. The wellbore GHE system was also employed for heat extraction from horizontal as well as multi-directional wells without producing binaries [5, 14].

3 Wellbore Heat Exchanger

The wellbore heat exchanger works by injecting cold fluid into the well up to the required depth and extracting the hot fluid at the wellhead from the production tubing [12]. The wellbore GHE extracts heat without circulating the working fluid into the surrounding formation. Heat transfer is mainly due to the conduction between formation and working fluid, and convection within working fluid. Coaxial wellbore heat exchanger (CGHE) and U-Tube wellbore heat exchanger are the two major GHEs that are utilized in wellbores. Quaggiotto et al. (2019) carried out a comparative analysis between CGHE and U-Tube GHE under the same working condition [15]. They reported that CGHE augments heat transfer by 2.9–5.3% with a 50% reduction in borehole thermal resistance due to large probe volume.

3.1 U-Tube Geothermal Heat Exchanger

A U-shaped tube is placed inside the well and the gap between the well casing, and the U-tube is filled with a grout material of relatively higher thermal conductivity. The working fluid is typically pumped through the U-tube to extract heat from the reservoir. U-tube GHE has the advantage of being able to operate reliably as a long-term geothermal heat extraction system [16]. Gharibi et al. (2018) performed a numerical analysis to assess the ability of U-tube GHE to extract heat from abandoned wells [16]. They extracted a maximum outlet temperature of 388 K at wellhead from 3861 m depth.

3.2 Coaxial Double Pipe Wellbore Heat Exchanger (CGHE)

The CGHE contains two pipes, one is an outer pipe and another one is an inner pipe which is known as tubing. The working fluid is pumped into the annulus between the outer and inner pipes which extracts heat from the rock formation. The tubing needs to be thermally insulated to minimize the heat losses between the hot producing fluid and the cold injecting fluid. Finally, the hot producing fluid will be utilized for various purposes [17].

CGHE is a more promising technology than U-tube in terms of heat exchange efficiency [14, 18]. Wang et al. (2018) and Cvetkovski et al. (2014) have reported that as compared to U-tube, CGHEs have a larger surface area to exchange heat [14, 19]. Furthermore, in the case of abandoned wells, constructing a CGHE may be a significant choice compared to the U-tube heat exchanger, because the outer pipe (casing) that already exists will save money and time. Also, a double pipe heat exchanger has less thermal resistance and thus results in better thermal efficiency. The summary of various heat exchangers is presented in Table 1.

4 Case Study

Macenic and Kurevija (2017) carried out the research to use dry and abandoned oil wells in the Pannonian basin of Croatia for geothermal energy extraction. The heat extraction study was conducted on the Pcelic-1 well in that basin and reported a maximum heat extraction of 1750 MWh per year by using CGHE [25]. A summary of various geothermal projects that are using wellbore heat exchangers is presented in Table 2.

5 Thermal Performance

The thermal performance of the wellbore GHE is highly heterogeneous due to various thermal properties of in situ rock or soil, type of heat exchanger, depth of the wellbore GHE, working fluid properties, flow rate, insulation of tubing, the thermal conductivity of grout and casing, etc.

Kujawa et al. (2006) proposed a CGHE with water as a heat-carrying fluid for the extraction of geothermal energy [29]. Davis and Michaelides 2009 studied the power-producing ability of CGHE from abandoned wells using local wells [30]. The study reported that the abandoned wells are capable of generating electricity. Later, many researchers studied the thermal performance of wellbore GHE and investigated the influence of different parameters such as well depth, formation properties, working fluid properties, circulation rate, injection temperature, insulation length, and properties on the heat extraction [13, 16, 31–34].

The working fluid injection temperature has a substantial effect on the heat extraction capacity of the wellbore GHE. The variation of thermal power of the system and outlet temperature of the producing fluid concerning inlet temperature is presented in Fig. 1. The increase in inlet fluid temperature will increase the outlet fluid temperature linearly but has a negative impact on the thermal power of the system. Hence, optimum inlet fluid temperatures are more feasible to direct heat applications as well as power generation.

Mass flow rate is a crucial and controllable parameter which influences the heat extraction capacity of the wellbore GHE. Many researchers investigated the effect of the mass flow rate of the injection fluid on the outlet temperature and thermal power [20, 32, 33, 35, 37]. From Fig. 2, it is observed that thermal power increases as the mass flow rate increases, while outlet temperature decreases with an increase in mass flow rate. The larger mass flow rate of the fluid injection involves more fluid

Table 1 Summary of various wellbore geothermal heat exchangers

References	Geometry	Details of the study	Main findings
Gharibi et al. [16]		Studied the heat extraction from the abandoned oil wells of 3861 m depth with 412 K bottom hole temperature	The U-tube system produces 388 K outlet temperature for an inlet temperature of 303 K, velocity 0.5 m/s
Bouhacina et al. [20]	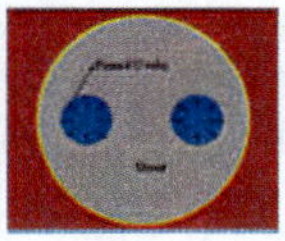	Proposed a novel approach of U-tube heat exchanger with 2 mm thick 4 mm length fins to extract heat from 35 m deep borehole	Finned U-tube BHE enhances the heat extraction by 7% compared to smooth pipe U-tube BHE
Jahanbin [21]		Developed a novel approach of elliptical-shaped U-tube GHE to enhance the heat extraction. The analysis was conducted for 100 m and 200 m depths	The elliptical-shaped U-tube GHE reduced the wellbore thermal resistance by over 17% and showed better thermal performance compared to circular U-tubes
Song et al. [22]		Numerical analysis of geothermal heat extraction from 50 m depth using spiral tube BHE is studied for different spiral lengths and helix diameters	Spiral tube BHE system with the centre vertical tube as inlet produced 25% more thermal power compared to single U-tube GHE
Song et al. [22]		Multiple U-tube heat exchangers were connected in parallel in a 50 m deep borehole. The author carried out analysis for two, three, six, and twelve U-tube configurations connected in parallel	12 U-tube GHE produced 75 and 22% more thermal power than single and double U-tube GHE, respectively
Wight and Bennet [23]	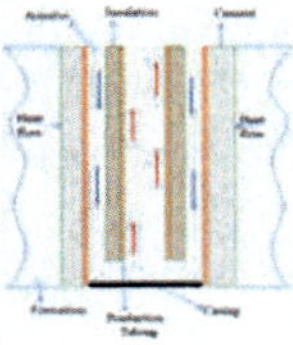	A CGHE is used to extract heat from orphaned oil wells with water as a circulating fluid. A wellbore of 4200 m depth with 31 °C/km thermal gradient produces 130 °C temperature at the wellhead with 2.5 kg/s mass flow rate	Approximately 109–630 kW net power was obtained at various mass flow rates from a binary cycle geothermal power plant. It was reported that the abandoned wells are suitable for producing economic geothermal electricity

(continued)

Table 1 (continued)

References	Geometry	Details of the study	Main findings
Cheng et al. [24]	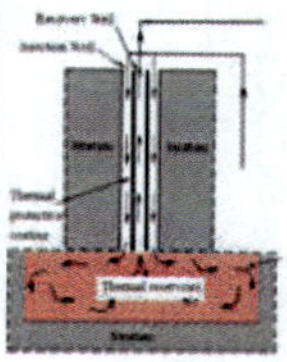	Proposed a method of heat extraction from decommissioned wells using coaxial double pipe heat exchanger with artificially fractured reservoir. Working fluid circulates into the reservoir and is collected from the centre tube. The analysis was carried out for different depths up to 4500 m	It is reported that the heat extraction efficiency of the system increases up to 13% with the reservoir depth. Heat extraction also increases with reservoir length up to certain limit, after that it remains constant
Wang et al. [17]	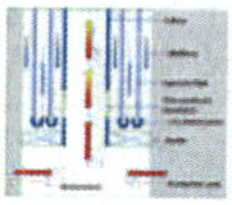	Proposed a novel approach to generate electricity inside the abandoned oil well of high water cut with a thermo-electric generator (TEG). The study is conducted up to 3000 m depth borehole	4.1% efficiency was achieved by the TEG system. The system can also produce binaries at the wellhead which are suitable for power generation by ORC or direct heat applications

in heat extraction per unit time, which increases the system's thermal power [38]. Also, a higher circulation rate leads to the reduction in circulation time of working fluid inside the GHE, which leads to a reduction in the producing fluid temperature.

The formation geothermal gradient has a significant effect on heat extraction in the vicinity of the geothermal well [23]. It indicates the quality of the thermal resources in a region. The variation of thermal power and fluid outlet temperature concerning the geothermal gradient is presented in Fig. 3. An increasing geothermal gradient increases the system's thermal power and outlet fluid temperature. This is due to the region with a high geothermal gradient, which will have higher temperatures along the well facing and bottom hole leading to higher heat transfer.

Figure 4 presents the variation in well thermal power and outlet fluid temperature of the system during its operating time. The well thermal power and outlet fluid temperature of the system reduce drastically within the short period of the operating time due to sudden changes in temperature near the wellbore. After a few days of operation, the reduction in the outlet temperature will be very low and the system attains a stable state.

6 Conclusions

The wellbore GHE is the key technology to extract heat energy without binaries from HDR resources and abandoned wells. The present paper reviews on various types of wellbores GHEs available and the various key performance parameters

Table 2 Geothermal projects with deep wellbore heat exchangers

Location	Geometry/image	Reservoir temp., °C	Description
Weggis, Switzerland [26]		78 °C	A single-well GHE is used to extract heat for heat pump application. The inner pipe of the GHE is vacuum insulated up to 1780 m depth. The single well delivered 220 MWh/yr heat energy with a 40.5 °C production temperature
Cornwall, UK [27]	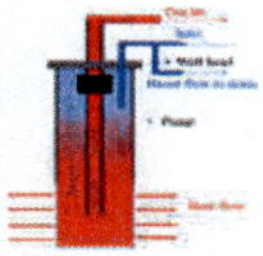		A deep single-well system with coaxial BHE is employed to extract heat from 2000 m depth. The system achieved a 69 °C delivery temperature and estimated output of 363 kW thermal power with only a 7 kW input. This system achieved coefficient of performance of 52
Hawaii, USA [28]		110 °C	A coaxial BHE is installed up to 876 m in a single wellbore of 1962 m depth and 0.18 m diameter. Double vacuum insulated tube is used as an inner tube. The system achieved 98 °C output temperature and maximum heat power of about 370 kW for 30 °C input temperature and 80 m^3/min flow rate
Aachen, Germany [28]		85 °C	A coaxial BHE is installed in a single wellbore of 2500 m depth to supply heat for space cooling and heating. The system supplies heat with temperature ranging from 25 to 55 °C with circulation rate of 10 m^3/hr

along with their effects on the heat extraction efficiency of the system. The enclosed plots represent the importance of numerous aspects of heat exchangers and their applications.

Coaxial double pipe GHE is a promising technology compared to the U-tube heat exchangers to extract geothermal energy due to less thermal resistance, higher heat extraction area, and volume of fluid flows through the pipe.

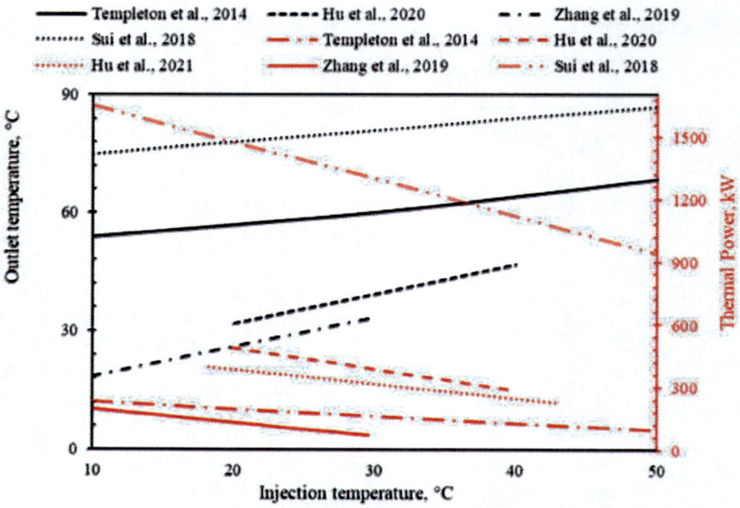

Fig. 1 Influence of inlet fluid temperature on outlet temperature

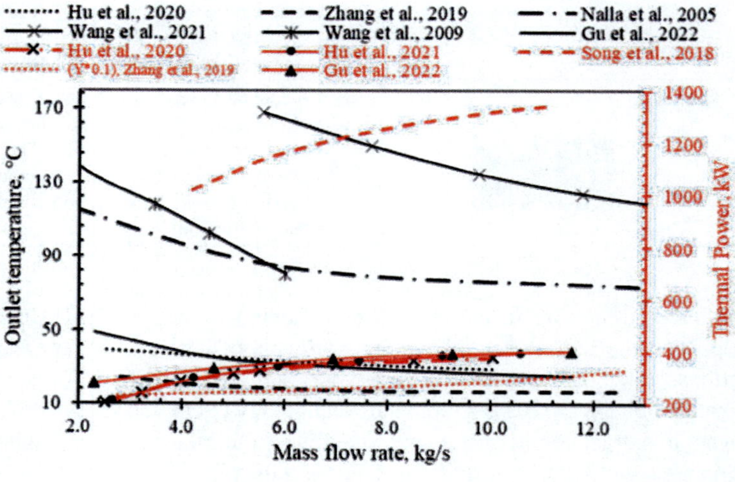

Fig. 2 Variation of thermal power and outlet temperature with the mass flow rate

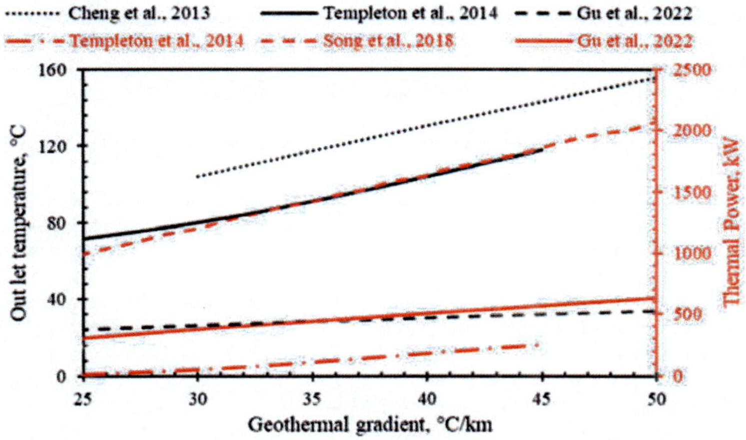

Fig. 3 Effect of geothermal gradient on outlet fluid temperature and thermal power of the system

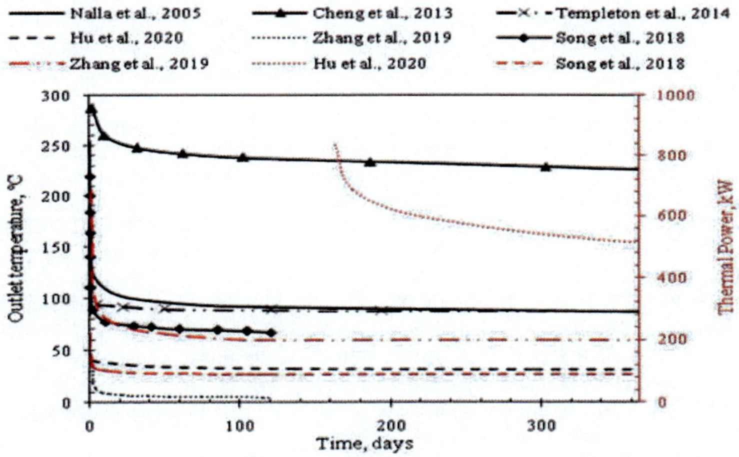

Fig. 4 Variation of outlet fluid temperature and thermal power with operation time

References

1. Zhu J, Hu K, Lu X, Huang X, Liu K, Wu X (2015) A review of geothermal energy resources, development, and applications in China: Current status and prospects. Energy 93:466–483
2. Singh HK, Chandrasekharam D, Trupti G, Mohite P, Singh B, Varun C, Sinha SK (2016) Potential geothermal energy resources of India: a review. Curr Sustain/Renew Energy Rep 3:80–91
3. Alimonti C, Soldo E, Bocchetti D, Berardi D (2018) The wellbore heat exchangers: a technical review. Renew Energy 123:353–381
4. Williams JP, Regehr A, Kang M (2021) Methane emissions from abandoned oil and gas wells in Canada and the United States. Environ Sci Technol 55:563–570

5. Harris BE, Lightstone MF, Reitsma S (2021) A numerical investigation into the use of directionally drilled wells for the extraction of geothermal energy from abandoned oil and gas wells. Geothermics 90:1019–1094
6. Lund JW, Boyd TL (2016) Direct utilization of geothermal energy 2015 worldwide review. Geothermics 60:66–93
7. Huttrer GW (2020) Geothermal power generation in the world 2015–2020 update report. In: Proceedings world geothermal congress, pp 1–17
8. Chandrasekharam D, Chandrasekhar V (2020) Geothermal energy resources of India: country update. In: Proceedings, world geothermal congress 2020. Reykjavik, Iceland, 13 p.
9. Pandey P, Trivedia S, Prasad S (2017) Comparative study of the utilization of geothermal energy in various ways in India and Rest of the World. Int Res J Eng Technol (IRJET) 4(6):1757–1763
10. https://www.ongcindia.com/wps/wcm/connect/en/media/press-release/geothermal-energy-ladakh#:~:text=Energy%20major%20ONGC%20will%20implement,Leh%20on%206%20February%202021.
11. Huang W, Cao W, Jiang F (2018) A novel single-well geothermal system for hot dry rock geothermal energy exploitation. Energy 162:630–644
12. Nalla G, Shook GM, Mines GL, Bloomfield KK (2015) Parametric sensitivity study of operating and design variables in wellbore heat exchangers. Geothermics 34(3):330–346
13. Cheng WL, Li TT, Nian YL, Wang CL (2013) Studies on geothermal power generation using abandoned oil wells. Energy 59:248–254
14. Wang G, Song X, Shi Y, Yang R, Yulong F, Zheng R, Li J (2021) Heat extraction analysis of a novel multilateral-well coaxial closed-loop geothermal system. Renew Energy 163:974–986
15. Quaggiotto D, Zarrella A, Emmi G, De Carli M, Pockelé L, Vercruysse J, Bernardi A (2019) Simulation-based comparison between the thermal behavior of coaxial and double U-tube borehole heat exchangers. Energies 12(12):2321
16. Gharibi S, Mortezazadeh E, Bodi SJHA, Vatani A (2018) Feasibility study of geothermal heat extraction from abandoned oil wells using a U-tube heat exchanger. Energy 153:554–567
17. Wang K, Yuan B, Ji G, Wu X (2018) A comprehensive review of geothermal energy extraction and utilization in oilfields. J Petrol Sci Eng 168:465–477
18. Kurnia JC, Shatri MS, Putra ZA, Zaini J, Caesarendra W, Sasmito AP (2022) Geothermal energy extraction using abandoned oil and gas wells: techno-economic and policy review. Int J Energy Res 46(1):28–60
19. Cvetkovski CG, Reitsma S, Bolisetti T, Ting DSK (2014) Ground source heat pumps: should we use single U-bend or coaxial ground exchanger loops? Int J Environ Stud 71(6):828–839
20. Bouhacina B, Saim R, Oztop HF (2015) Numerical investigation of a novel tube design for the geothermal borehole heat exchanger. Appl Therm Eng 79:153–162
21. Jahanbin A (2020) Thermal performance of the vertical ground heat exchanger with a novel elliptical single U-tube. Geothermics 86
22. Song X, Wang G, Shi Y, Li R, Xu Z, Zheng R, Li J (2018) Numerical analysis of heat extraction performance of a deep coaxial borehole heat exchanger geothermal system. Energy 164:1298–1310
23. Wight NM, Bennett NS (2015) Geothermal energy from abandoned oil and gas wells using water in combination with a closed wellbore. Appl Therm Eng 89:908–915
24. Cheng WL, Liu J, Nian YL, Wang CL (2016) Enhancing geothermal power generation from abandoned oil wells with thermal reservoirs. Energy 109:537–545
25. Macenić M, Kurevija T (2018) Revitalization of abandoned oil and gas wells for a geothermal heat exploitation by means of closed circulation: case study of the deep dry well Pčelić-1. Interpretation 6(1):SB1–SB9
26. Falcone G, Liu X, Okech RR, Seyidov F, Teodoriu C (2018) Assessment of deep geothermal energy exploitation methods: the need for novel single-well solutions. Energy 160:54–63
27. Collins MA, Law R (2017) The development and deployment of deep geothermal single well (DGSW) technology in the United Kingdom. Eur Geol J 43
28. Sapinska-Sliwa A, Rosen MA, Gonet A, Sliwa T (2016) Deep borehole heat exchangers—a conceptual and comparative review. Int J Air-Cond Refrig 24(01):1630001

29. Kujawa T, Nowak W, Stachel AA (2006) Utilization of existing deep geological wells for acquisitions of geothermal energy. Energy 31:650–664
30. Davis AP, Michaelides EE (2009) Geothermal power production from abandoned oil wells. Energy 34(7):866–872
31. Templeton JD, Ghoreishi-Madiseh SA, Hassani F, Al-Khawaja MJ (2014) Abandoned petroleum wells as sustainable sources of geothermal energy. Energy 70:366–373
32. Hu X, Banks J, Wu L, Liu WV (2020) Numerical modeling of a coaxial borehole heat exchanger to exploit geothermal energy from abandoned petroleum wells in Hinton, Alberta. Renew Energy 148:1110–1123
33. Hu X, Banks J, Guo Y, Huang G, Liu WV (2021) Effects of temperature-dependent property variations on the output capacity prediction of a deep coaxial borehole heat exchanger. Renew Energy 165:334–349
34. Sui D, Wiktorski E, Røksland M et al (2019) Review and investigations on geothermal energy extraction from abandoned petroleum wells. J Petrol Explor Prod Technol 9:1135–1147. https://doi.org/10.1007/s13202-018-0535-3
35. Zhang Y, Yu C, Li G, Guo X, Wang G, Shi Y, Tan Y (2019) Performance analysis of a down hole coaxial heat exchanger geothermal system with various working fluids. Appl Therm Eng 163
36. Gu F, Li Y, Tang D, Gao Y, Zhang Y, Yang P, Ye H (2022) Heat extraction performance of horizontal-well deep borehole heat exchanger and comprehensive comparison with the vertical well. Appl Therm Eng 211:118426
37. Wang Z, McClure MW, Horne RN (2019) A single-well EGS configuration using a thermosiphon. Stanford University, Stanford, California, In Workshop on Geothermal Reservoir Engineering
38. Li Z, Zheng M (2009) Development of a numerical model for the simulation of vertical U-tube ground heat exchangers. Appl Therm Eng 29(5–6):920–924

Parametric Investigation of Droplet Generation Inside T-Junction Microchannel

Santosh Kumar Jena, Tushar Srivastava, and Sasidhar Kondaraju

Abstract Droplet generation inside the microchannel and its manipulation are one of essential droplet microfluidics processes. Many researchers study the physics behind droplet generation inside the T-junction microchannel; still, a thorough understanding of the concepts behind droplet formation is missing. The current investigation focuses on the parametric study of aqueous droplets in oil for a T-channel. This study examines the impact of numerous parameters, including the continuous phase capillary number (Ca_c), flow rate ratio (ϕ), and various fluid combinations, as well as the dispersed phase Weber number (We_d), on the droplet size and frequency of droplet formation within a T-channel. Different flow patterns like squeezing, dripping, and jetting regimes are observed from flow visualization. The variation of droplet length due to We_d is studied for a constant Ca_c. In-house experiments are conducted for Ca_c = 0.0004–0.2, flow rate ratios ($\phi = 1, 0.5, 0.25$), and for varying We_d.

Keywords T-junction · Capillary number · Weber number · Flow rate ratio · Frequency

Nomenclature

ρ	Density of liquid [kg/m^3]
μ	Viscosity of liquid [mPa s]
σ	Oil-water interfacial tension [mN/m]
Q	The flow rate of liquid [μl/min]
w	Channel width [μm]
L	Droplet length [μm]
Ca_c	Continuous phase capillary number
We_d	Dispersed medium Weber number

S. K. Jena (✉) · T. Srivastava · S. Kondaraju
School of Mechanical Sciences, Indian Institute of Technology Bhubaneswar, Argul, Odisha 752050, India
e-mail: skj10@iitbbs.ac.in

S. Das et al. (eds.), *Proceedings of the 1st International Conference on Fluid, Thermal and Energy Systems*, Lecture Notes in Mechanical Engineering,
https://doi.org/10.1007/978-981-99-5990-7_15

ϕ Flow rate ratio
f Frequency of droplet generation [Hz]

1 Introduction

Microfluidics has been continuously growing in the past three decades due to its vast possibility in scientific research and technical and medical applications. It has a broad spectrum of use, from fundamental concepts to application or benefit to society. It has become an interdisciplinary branch of engineering, physics, chemistry, biology, medicine, and biotechnology [1, 2]. It has unique properties over macro-fluidics, such as laminar flow, negligible gravity, capillary phenomenon, and dominance of surface forces. Amid various classifications of microfluidics, droplet microfluidics emerged as a new versatile field of research because of its widespread application and several intriguing advantages [3–5]. The immiscible nature of two liquids, smaller sample volume, and spacing requirements extend droplet microfluidics among researchers to explore the physics in detail. Along with high throughput, the benefits of droplet microfluidics can be extended to improve analytical efficiency, and process control having smaller reagent usage. The vital points to be considered in droplet microfluidics are droplet formation and breakup, controlling droplets, droplet mixing, droplet sorting, etc. Droplet formation inside the microchannel and control are most important as they affect other processes.

2 Literature Review and Objective

Generally, active and passive methods were used for droplet generation inside the microchannel. Zhu et al. [6] carried out a thorough study of these methods. Among the two approaches, passive strategies are mainly used due to fabrication simplicity in the passive form. The necking and breakup of the interface take place because of channel geometry and flow rate. Three commonly used methods are co-flow, cross-flow (T-junction), and flow focusing, among different passive techniques. Due to monodispersed droplet production and easy fabrication technique, the T-junction is an extensively used channel in droplet microfluidics system. In T-junction, the two fluids, i.e., continuous and dispersed phase flows, are orthogonal. Thorsen et al. [7] first produced aqueous droplets suspended in oil using a T-junction. They showed that shear stress along with interfacial tension were the main reason behind the droplet dynamics inside the microchannel. Later, Garstecki et al. [8] showed that droplet breakups were affected by the balance of dispersed and continuous phase pressures at the junction. They explained the physics at low capillary numbers. For the squeezing regime, they developed a scaling rule to forecast droplet size, showing the flow rate ratio dependency. Using numerical investigation, De Menech and colleagues

[9] identify three unique droplet formation regimes at the T-junction: squeezing, dripping, and jetting. Christopher and co-workers [10] reported that the break-up of droplets in T- shaped microchannel from squeezing to dripping regime depends upon both flow rate ratio and capillary number. Further, Nisisako et al. [11] reported nanoliter-sized water droplet formation in an oil-based continuous phase T-junction.

The study shows that all regimes are classified based on the continuous phase's capillary number (Ca_c). Capillary number (Ca) is a dimensionless ratio of a system's viscous force to its surface tension: $Ca = \frac{Viscous force}{surface tension} = \frac{\mu u}{\sigma}$, where μ is the fluid's viscosity, u is the velocity of the liquid, and σ is surface tension. It is used when one immiscible fluid moves in another immiscible fluid. This number shows the dominance of viscous force or surface tension depending upon whether Ca < 1 or Ca > 1. It was studied that the major forces responsible for droplet generation were a shear force, interfacial tension, and pressure drop across the droplet. The effect of inertial force and gravity was neglected in all cases. Many correlations and scaling models were proposed by several authors [12–14] for different regimes of droplet production. Parameters like geometry and dimension of the channel, wetting properties of solid surfaces, the fluid viscosity, interfacial tension, and flow rate of immiscible liquids add complexity to investigating the dynamics of droplet formation activity [15–17]. Recently, Jena et al. [18] explained how dispersed phase inertial force affects the prediction of droplet size. They developed an analytical model for different fluid combinations having a broad spectrum of flow rate ratios. Still, many concealed concepts need to be adequately understood and addressed. The purpose of this work is to gain a deeper comprehension of droplet production in microchannels with a T-junction. The current work focuses on the effect of flow rate ratio and capillary number on droplet size for different fluids by conducting a large set of systematic experiments. However, in this work, we conduct investigations for various fluids over a wide range of Ca_c (0.0005–0.2374), flow rate ratio (ϕ = 0.25, 0.5, 1), and Weber number (We_d). Weber number is defined as the competition between inertial force and interfacial tension. $We = \frac{Inertia force}{Interfacial tension} = \frac{\rho u^2 l_c}{\sigma}$, where ρ is the density of the fluid, u is the characteristic velocity, l_c is the typical length, and interfacial tension is σ. This number is helpful in droplet and bubble formation and thin-film flows. Weber number is vital in the droplet to break up when viscous forces are negligible compared to inertia and interfacial tension. For high throughput, high-speed flows' Weber number is essential as inertial force comes into play.

This study demonstrates several regimes, such as squeezing, dripping, and jetting, for diverse fluids throughout a wide range of the capillary number of the continuous phase. In addition, the frequency of droplet production is investigated. The work also examines how droplet size and w_{gap} are affected by the dispersed phase's Weber number. The novelty of the work is that the study reveals the impact of the Weber number of the dispersed phase (We_d) on droplet size and the distance between the droplet interface and channel's top wall. The investigation of Weber's number of dispersed phase can explore the effect of the inertial force exerted by the dispersed phase on the production of droplets. It can be found that droplet size increases and w_{gap} decreases with an increase on We_d. An extension of the work can classify different regimes of droplet generation based on Weber's number of the dispersed phase.

3 Materials and Methods

An in-house experimental setup is developed for performing the experiments. Readers should refer to Jena et al. [18] for a more detailed explanation of the schematic diagram of the setup. Current experiments are conducted using silicone oil, mineral oil, and hexadecane oil (Sigma-Aldrich) as the continuous fluid, and distilled water (DIW) serves as the dispersed medium. Table 1 lists the physical characteristics of continuous fluids. DIW is being employed as a dispersed phase medium in the current research with a density of 1000 kg/m^3 and dynamic viscosity of 0.99 mPa s.

The microchannel fabrication is carried out using CNC micro milling and thermal bonding method. A polymethylmethacrylate (PMMA) sheet of dimension 50 mm x 30 mm × 5 mm is used for making the T-channel. An end mill of 200 μm diameter is used throughout the channel preparation. The channel cross-section is rectangular. A constant depth of 200 μm is specified for all channels. Uniform widths are taken for both continuous and dispersed phase liquids. The total channel length is 32 mm. The dispersed phase channel length is 9 mm, and the distance between the continuous phase inlet to the junction is 9 mm. To make the channel hydrophobic, aquapel (Pittsburg Glass Works LLC. U.S.) treatment is carried out. An optical profilometer (Wyko NT9100, U.S.) is used to measure the roughness, and found to be less than 3.0 μm. The fluids are pumped into the channel using a syringe pump (Holmarc Inc.) that has two separate channels. The channel's outflow is open to the atmosphere. The droplet production process is observed using an inverted microscope (made by Olympus, Japan). To capture and examine the droplet formation process, a high-speed camera (Photron FASTCAM Mini UX100), having 4000 fps at full resolution, is used. All the experiments are conducted at room temperature.

The shape of the droplet and different geometrical parameters that affect the droplet generation inside the T-junction are shown below (Fig. 1).

Q_c, Q_d are flow rates, channel widths are represented as w_c, and w_d, and μ_c and μ_d are the viscosities of the fluids. Continuous and dispersed phase fluids are denoted by the subscripts c and d, respectively [18]. L is the length of the droplet. The dimensionless terms used are flow rate ratio (ϕ), which is the ratio of the flow

Table 1 The following table lists the physical features of continuous medium fluids

Continuous phase liquid	Density in kg/m^3	Viscosity in mPa s	Interfacial tension in mN/m
Hexadecane oil	773	2.38 ± 0.02	43.60 ± 0.10
Mineral oil	840	23.65 ± 0.02	42.27 ± 0.03
Silicone oil	960	89.72 ± 0.04	49.48 ± 0.02

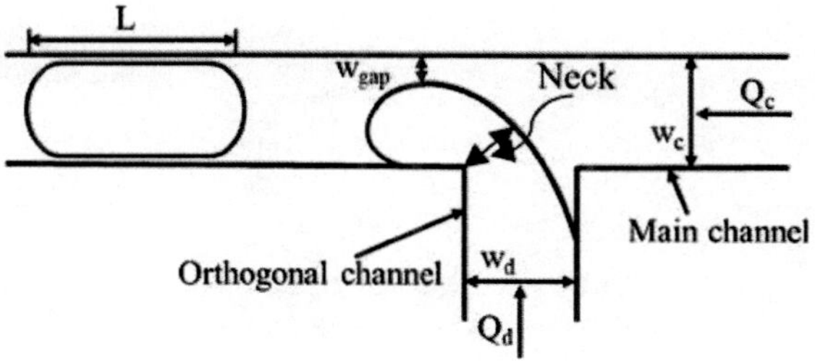

Fig. 1 Shape of the droplet inside T-junction. Dispersed phase liquid moves via the orthogonal channel while continuous phase liquid moves through the main channel

rate of the dispersed phase to the continuous phase, viscosity ratio (λ) is the viscosity of the dispersed to the continuous phase, and Capillary number of the continuous phase is Ca_c and Weber number of the dispersed phase is We_d.

4 Results and Discussion

Extensive studies are undertaken with $Ca_c = 0.0004–0.2$ at flow rate ratios of 1, 0.5, and 0.25 for mineral oil, hexadecane, and silicone oil, which are considered continuous phase fluids. In the current work, the flow rate ratio ϕ is varied by fixing Q_c and varying Q_d. Photron PFV software is used for measuring the droplet length.

4.1 Effect of Ca_c on Different Regime Transitions

For diverse fluid combinations, flow visualization reveals squeezing, dripping, and jetting regimes of droplet production inside a T-junction. The experimental results show that the regimes vary for the same capillary number of the continuous phase for different flow rate ratios. Besides, the regimes differ for other liquids and for different Cac and ϕ due to dissimilarity in interfacial tension and viscosity. For a 0.5 flow rate ratio, taking mineral oil and hexadecane oil as the continuous phase liquids, three different regimes are shown in Fig. 2. For the two fluids, the regimes occurred at different Ca_c.

However, as a continuous phase of hexadecane oil, the jetting regimes occur at a low capillary number. The reason is for the same capillary number; the velocity of the continuous phase is higher for hexadecane oil compared to other liquids due to low viscosity. This results in a high shear force on the interface, and its length increases rapidly downstream of the channel. This creates instability in the dispersed phase, which causes the jetting regime to occur.

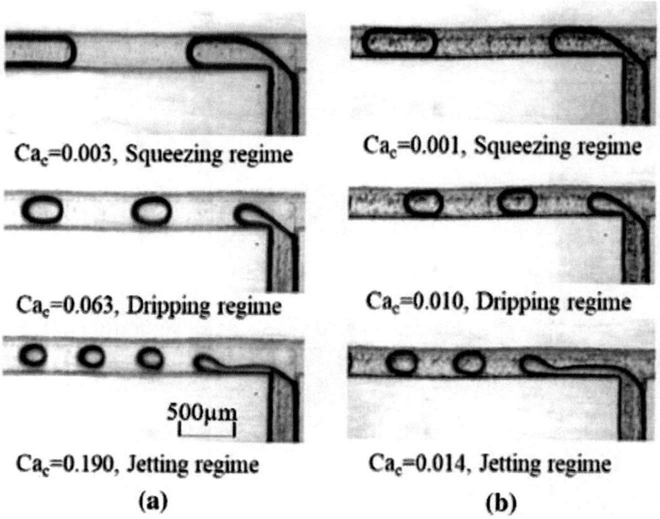

Fig. 2 Different regimes of droplet generation for **a** Mineral oil and **b** Hexadecane oil. For $\phi = 0.5$, the results are shown

4.2 *Variation of Droplet Size with Weber Number of the Dispersed Phase (We_d)*

In another study, we have conducted experiments by fixing the Ca_c and varying the We_d.

As can be seen in Fig. 3, a rise in Ca_c results in a decrease in the non-dimensional droplet length (L/w_c). It can be seen from the figure that there is a correlation between a rise in We_d and an increase in the size of the droplet for a given Ca_c. In each of the four scenarios, the result demonstrates behaviour that is comparable. When We_d increases, the dispersed phase's flow rate increases, decreasing the penetration time between the two fluids. As a consequence of this, the interface drags to a more considerable extent before breaking up, which eventually increases droplet length.

4.3 *Effect of We_d on w_{gap}*

For a constant $Ca_{c.}$ results are shown below, using mineral oil as a continuous phase (Fig. 4).

From our experimental observation, it is found that w_{gap} decreases as We_d increases. The change in We_d has an inverse effect on $w_{g,}$ as seen from the figure. As We_d increases, the resistance from the dispersed phase increases, increasing the

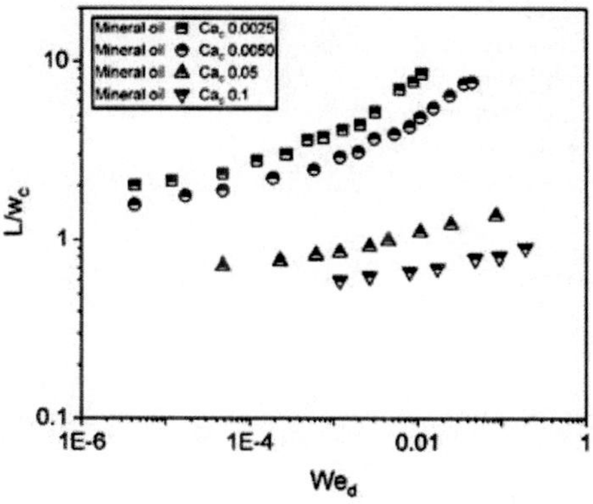

Fig. 3 Non-dimensional droplet length (L/w_c) is plotted against We_d for varying Ca_c

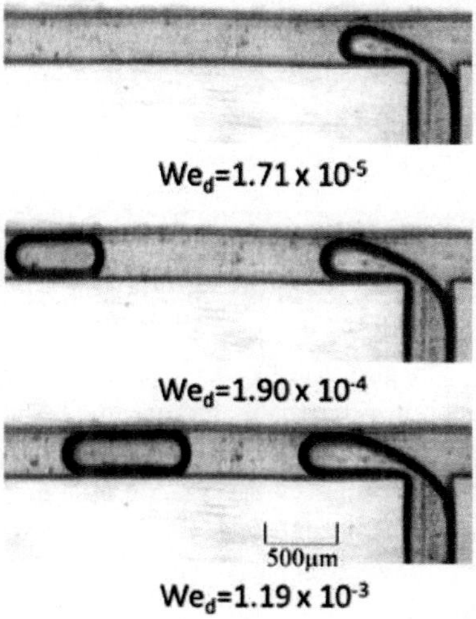

Fig. 4 The effect of We_d on w_g when mineral oil is used as continuous phase fluid. The capillary number is maintained constant ($Ca_c = 5 \times 10^{-3}$)

droplet's length and reducing w_g. For a large set of experiments, the findings are validated qualitatively and quantitatively.

4.4 *Effect of Flow Rate Ratio (Φ) and Ca_c on the Frequency of Droplet Production*

The frequency of droplet formation is measured as the number of droplets produced per unit time. The results in Fig. 5 show the droplet generation frequency as a function of Ca_c. The droplet frequency is calculated from the videos captured during the experiments. It was previously found that while the capillary number of continuous

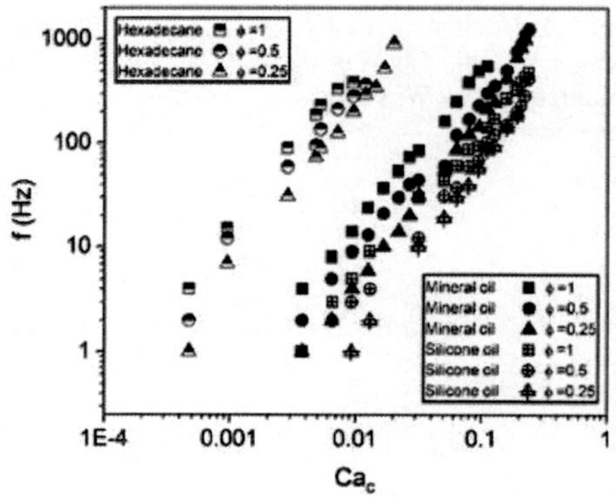

Fig. 5 The effect of Capillary number of the continuous phase (Ca_c) on frequency of droplet generation for three flow rate ratios ($\phi = 1$, 0.5, and 0.25)

Table 2 Key results for different Ca_c, We_d, and ϕ obtained from the experiments are shown below

Ca_c range	We_d range	Flow rate ratio (ϕ) = 1, 0.5, and 0.25	
		L/w_c range	Frequency (Hz) range
3×10^{-4}–2×10^{-1}	1×10^{-6}–3×10^{-1}	0.5–8.0	1–850

phases increases for a fixed dispersed phase fluid, the frequency of droplet formation increases. The current work shows a similar trend for different ratios of flow rates, i.e., the pace at which droplets are formed speeds up as the relative flow rate goes up.

In addition to that, for a particular flow rate ratio, the droplet production rate increases when Ca_c increases. The present work shows that for a fixed Ca_c, the droplet generation frequency is high at a higher ratio of flows. In current experiments, the flow rate ratio increases with the dispersed phase flow rate, which decreases the time of penetration between the two phases. A greater relative flow rate requires less time for the interface to move, which increases the frequency of droplet production within the channel. From the plot, it can be shown that the droplet formation rate increases almost linearly for different flow rate ratios.

5 Conclusions

In this paper, we conducted a comprehensive study of a wide variety of variable parameters inside the T-junction. Along with the capillary number of the continuous phase, the flow rate ratio and Weber number of the dispersed phase are examined for exploring the physics of droplet generation inside the T-channel. It is observed that droplet size decreases with an increase in Ca_c but increases with an increase on We_d for a fixed Ca_c. The regimes like squeezing, dripping, and jetting are investigated for distinct fluid combinations and flow rate ratios. An analytical model needs to be developed to understand physics properly. Further study can explore the effect of inertial force on droplet generation. We also found that the droplet production frequency increases with an increase in the flow rate ratio. Although our experiments show good agreement with others' work, more testing is necessary to understand the physics better. Our work gives new insight into droplet generation and forecasts droplet's length and frequency within the T-junction.

Acknowledgements We acknowledge the contribution of the IIT Bhubaneswar Robotics Laboratory for assisting in the fabrication of the microchannel.

References

1. Stone HA, Kim S (2001) Microfluidics: basic issues, applications, and challenges. AIChE J 47(6):1250
2. Ong SE, Zhang S, Du H, Fu Y (2008) Fundamental principles and applications of microfluidic systems. Front Biosci 13(7):2757–2773
3. Seemann R, Brinkmann M, Pfohl T, Herminghaus S (2011) Droplet based microfluidics. Rep Prog Phys 75(1):016601
4. Shang L, Cheng Y, Zhao Y (2017) Emerging droplet microfluidics. Chem Rev 117(12):7964–8040
5. Teh SY, Lin R, Hung LH, Lee AP (2008) Droplet microfluidics. Lab Chip 8(2):198–220
6. Zhu P, Wang L (2017) Passive and active droplet generation with microfluidics: a review. Lab Chip 17(1):34–75
7. Thorsen T, Roberts RW, Arnold FH, Quake SR (2001) Dynamic pattern formation in a vesicle-generating microfluidic device. Phys Rev Lett 86(18):4163
8. Garstecki P, Fuerstman MJ, Stone HA, Whitesides GM (2006) Formation of droplets and bubbles in a microfluidic T-junction—scaling, and mechanism of break-up. Lab Chip 6(3):437–446
9. De Menech M, Garstecki P, Jousse F, Stone HA (2008) Transition from squeezing to dripping in a microfluidic T-shaped junction. J Fluid Mech 595:141–161
10. Christopher GF, Noharuddin NN, Taylor JA, Anna SL (2008) Experimental observations of the squeezing-to-dripping transition in T-shaped microfluidic junctions. Phys Rev E 78(3):036317
11. Nisisako T, Torii T, Higuchi T (2002) Droplet formation in a microchannel network. Lab on a Chip 2(1):24–26; Garstecki P, Fuerstman MJ, Stone HA, Whitesides GM (2006) Lab on a Chip 6:437
12. Xu JH, Li SW, Tan J, Wang YJ, Luo GS (2006) Preparation of highly monodisperse droplet in a T-junction microfluidic device. AIChE J 52(9):3005–3010
13. Gupta A, Kumar R (2010) Effect of geometry on droplet formation in the squeezing regime in a microfluidic T-junction. Microfluid Nanofluid 8(6):799–812
14. Qin N, Wen JZ, Ren CL (2017) Highly pressurized partially miscible liquid-liquid flow in a micro-T-junction. I Exp Observ Phys Rev E 95(4):043110
15. Liu H, Zhang Y (2009) Droplet formation in a T-shaped microfluidic junction. J Appl Phys 106(3):034906
16. Tice JD, Lyon AD, Ismagilov RF (2004) Effects of viscosity on droplet formation and mixing in microfluidic channels. Anal Chim Acta 507(1):73–77
17. Sivasamy J, Wong TN, Nguyen NT, Kao LTH (2011) An investigation on the mechanism of droplet formation in a microfluidic T-junction. Microfluid Nanofluid 11(1):1–10
18. Jena SK, Bahga SS, Kondaraju S (2021) Prediction of droplet sizes in a T-junction microchannel: effect of dispersed phase inertial forces. Phys Fluids 33(3):032120

Hydraulic Design Optimization of Flow Distribution Device in Bottom Header of IHX for SFR

Amit K. Chauhan, M. Rajendrakumar, and K. Natesan

Abstract An intermediate heat exchanger (IHX) of a sodium-cooled fast reactor is a typical shell and tube-type counterflow heat exchanger. Primary sodium flows on the shell side, and secondary sodium flows on the tube side. There is a central downcomer and 25 rows of tubes. Tube rows in the periphery of the IHX see primary sodium at high temperatures compared to the inner rows of a tube. Non-uniform heat exchange increases the temperature difference between sodium from different rows, thereby increasing thermal damage to the tube sheet. A flow distribution device in the bottom header would regulate higher flow in the outer rows, whose design is optimized using a 3D CFD hydraulic analysis. The effect of the conical diffuser and vertical baffle on the flow distribution inside the bottom header is quantified. A vertical baffle of 225 mm in height (located 12 mm below the tube sheet) provided after the 18th row rendered a flow distribution very close to the desired one. The deviation between expected and estimated flow reduces to ~10%.

Keywords Hydraulic design · Flow distribution device · CFD · Intermediate heat exchanger · Sodium-cooled fast reactor

1 Introduction

Three thermal loops are operating in equilibrium in a typical pool-type sodium-cooled fast reactor, including the conventional steam-water system [1, 2]. The primary sodium loop confines within the boundary of the main vessel. Primary sodium extracts heat from the core and transfers heat to secondary sodium in Intermediate Heat Exchanger (IHX). Subsequently, secondary sodium transfers heat to water in a steam generator to convert it into steam [2]. Intermediate Heat Exchanger (IHX) is a typical

A. K. Chauhan (✉) · M. Rajendrakumar · K. Natesan
Thermal-Hydraulics Section, Thermal Hydraulic Division, Reactor Design and Technology Group, Indira Gandhi Centre for Atomic Research, Kalpakkam, Kancheepuram, Tamil Nadu 603102, India
e-mail: amitchauhan@igcar.gov.in

S. Das et al. (eds.), *Proceedings of the 1st International Conference on Fluid, Thermal and Energy Systems*, Lecture Notes in Mechanical Engineering,
https://doi.org/10.1007/978-981-99-5990-7_16

shell and tube counterflow heat exchanger with primary sodium on the shell side and secondary sodium on the tube side [1]. Primary sodium enters the shell side of IHX through an inlet window at the top and exits through an outlet window located at the bottom of IHX. The secondary sodium enters IHX from the top into a downcomer at the center of IHX. It flows down through the downcomer to an inlet plenum, also called the bottom header. It then flows upwards through the tubes, as shown in Fig. 1. There are 3900 tubes arranged in a circular pitch surrounding the central downcomer in 25 rows.

The heat exchanged between primary and secondary sodium across the tube wall is not the same due to the following reasons: (a) cross-flow of primary sodium at the inlet and outlet windows due to which secondary sodium in inner rows are at a lower temperature, (b) primary sodium flow in inner rows is lower compared to outer rows, and (c) secondary sodium flow inside the tubes is not identical. Consequently, the temperature of secondary sodium at the outlet of various tubes is not the same resulting in thermal loading of tube sheet and other structures of IHX [3]. Since the secondary sodium flowing in the outer rows receives significantly larger heat than the inner rows, the temperature of secondary sodium at the outlet of various tubes can be made more uniform by admitting more flow through the outer rows. Such flow distribution is possible by increasing the hydraulic resistance of the inner rows compared to the outer rows. The desirable option is to introduce a flow distribution device (FDD) [4] (see Fig. 2) in the bottom header to accomplish the required flow zoning, which is an integrated device having a circular orifice plate and a vertical baffle, as shown in Fig. 2. Various flow zoning options have been investigated based on a study carried out using a simplified 1D network model. The outer seven rows of tubes in IHX must have a 30% more flow rate than the 18 inner rows. The porous body approach was used to simulate the IHX of PFBR [4]. Gajapathy et al. [4] investigated the flow distribution in the bottom header of IHX due to FDD and mixing devices. Primary flow in the shell side of the shell and tube-type heat exchanger is typically solved using 3D governing equations. To incorporate the heat transfer across the heat exchanger's shell and tubes, Patankar and Spalding [5] solved 3D governing equations for primary fluid along with the 1D governing equations for the secondary fluid and 1D governing equations for tube wall temperature. However, distributed resistance coefficients in the momentum equations of the primary fluid are incorporated as tube effects.

Mochizuki and Takano [6] used test data of IHX from Monju and Joyo [7] reactors and proposed an empirical correlation and concluded that at Peclet number (Pe) more significant than 30, the correlations proposed by Seban and Shimazaki [7] could be used for calculating Nusselt number. In the available literature, it is observed that most of these studies are based on the porous body approach. Also, the numerous tubes of the IHX are not modeled because of computational difficulty. In the present study, 3D CFD hydraulic analyses have been carried out by modeling individual tubes to account for pressure drops on the flow distribution among tubes (Fig. 3).

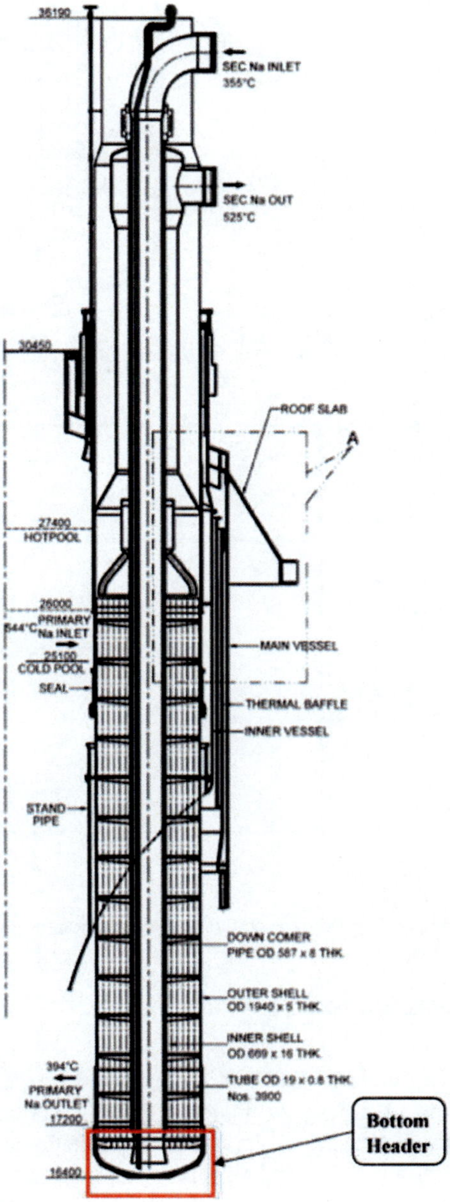

Fig. 1 Schematic of IHX

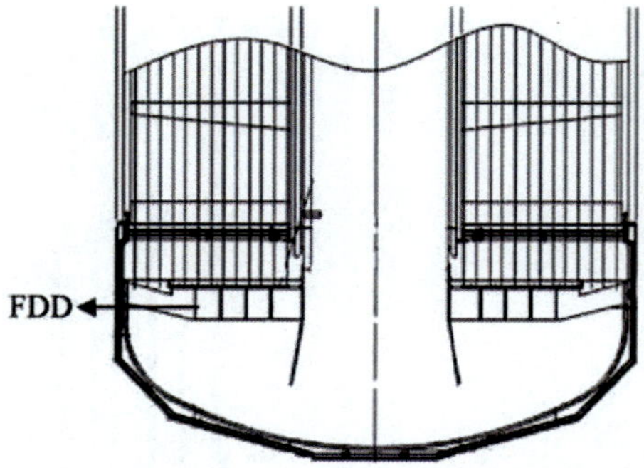

Fig. 2 Schematic of secondary sodium bottom header with FDD in a typical SFR

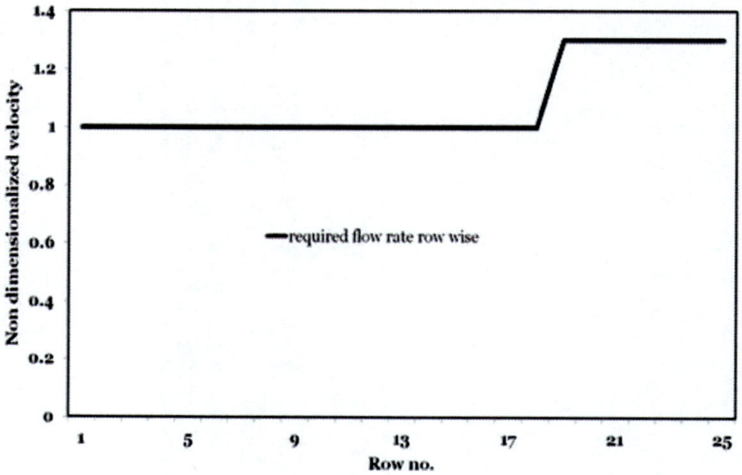

Fig. 3 The desired sodium flow distribution (normalized) among various rows of tubes

2 Numerical Modeling

The tubes of the IHX are arranged in a circular pitch, with 84 tubes in the first row. A 30° sector model of the IHX bottom header is considered for the analysis, as shown in Fig. 4a, b. The computational domain comprises 325 tubes, i.e., 306 full tubes and 38 half tubes. There are 25 rows numbered rows 1 to 25 in a radial outward direction. The complete axial length of IHX tubes is modeled, and the outlet boundary condition is specified at the top of the tubes, as shown in Fig. 4a. The choice of modeling complete tube length is made to account for the effect of tube pressure drop on the flow distribution. Figure 4c shows the 3D mesh generated. Finer mesh with the hexahedral cell is generated near most walls to capture the near-wall effect accurately. Also, mesh quality and y + value (considered ~30–300) are respected near the tube region. The total mesh count is approximately 1.2 million.

The mass flow rate of secondary sodium entering the domain under consideration is 1728 kg/s. Thus, a uniform velocity of ~8 m/s is imposed as an inlet boundary condition at the top of the downcomer model. Zero-gauge pressure is specified at the exit of all the tubes modeled. No-slip boundary condition is imposed over all the walls. Symmetric planes of the 30° sector are specified as symmetry boundary conditions as shown in Fig. 4a, b. Steady-state incompressible flow of sodium is considered in the present study. Density and dynamic viscosity of sodium are considered as constant at 848 kg/m^3 and 2.64×10^{-4} kg/ms, respectively. The high Re $k - \varepsilon$ turbulence

Fig. 4 Computational model and mesh with boundary conditions

model is used to consider the effect of turbulence on fluid flow. ANSYS FLUENT 19.2 [8] is used to carry out the simulation. The 3D steady-state incompressible flow of Newtonian fluid in the shell and tube-type heat exchanger is governed by fundamental equations of conservation of mass and momentum as shown below [9]:

Continuity

$$\frac{\partial}{\partial x_j}(\rho u_j) = 0 \tag{1}$$

Momentum

$$\frac{\partial}{\partial x_j}(\rho u_j u_i - \tau_{ij}) = -\frac{\partial p}{\partial x_i} + s_i \tag{2}$$

where s_i = source $\tau_{ij} = 2\mu s_{ij} - \frac{2}{3}\mu\frac{\partial u_k}{\partial x_k}\delta_{ij} - \rho\overline{u_i{}'u_j{}'}$ and $s_{ij} = \frac{1}{2}\left(\frac{\partial u_i}{\partial x_j} + \frac{\partial u_j}{\partial x_i}\right)$.

Turbulence characteristic of the fluid flow in the present computational domain is solved using the standard $k - \varepsilon$ turbulence model as shown in Eqs. 3 and 4 [9].

Turbulent kinetic energy:

$$\frac{\partial}{\partial x_j}\left[\rho u_j k - \left(\mu + \frac{\mu_t}{\sigma_k}\right)\frac{\partial k}{\partial x_j}\right] = -\rho\varepsilon - \frac{2}{3}\left(\mu_t\frac{\partial u_i}{\partial x_i} + \rho k\right)\frac{\partial u_i}{\partial x_i} \tag{3}$$

Turbulent dissipation rate:

$$\frac{\partial}{\partial x_j}\left[\rho u_j \varepsilon - \left(\mu + \frac{\mu_t}{\sigma_\varepsilon}\right)\frac{\partial \varepsilon}{\partial x_j}\right] = - c_{\varepsilon 2}\rho\frac{\varepsilon^2}{k} + c_{\varepsilon 1}\frac{\varepsilon}{k}\left[\mu_t P - \frac{2}{3}\left(\mu_t\frac{\partial u_i}{\partial x_i} + \rho k\right)\frac{\partial u_i}{\partial x_i}\right] \tag{4}$$

where $\mu_t = \frac{C_\mu \rho k^2}{\varepsilon}$ and model constants are $C_\mu = 0.09$, $\sigma_k = 1.0$, $\sigma_\varepsilon = 1.22$, $C_{\varepsilon 1} = 1.44$, $C_{\varepsilon 2} = 1.92$.

3 Results and Discussion

Different geometrical configurations for the flow distribution device were analyzed to arrive at a suitable arrangement. Details of these studies are discussed in the following subsections.

3.1 Bottom Header Without a Conical Diffuser

Numerical simulations of the IHX bottom header without a conical diffuser were conducted to analyze its effect. The downcomer outlet is terminated at the bottom of the tube sheet without the conical diffuser attached to it. The predicted pressure distribution in the header is shown in Fig. 5a. High pressure is observed below the downcomer at the header wall attributed to stagnate sodium jet from the downcomer. Thus, as expected, the bottom region below the downcomer nearly becomes a dead zone, as evident in Fig. 5b. Velocity distribution through 25 rows of tubes is shown in Fig. 5c. The velocity on the y axis is presented as non-dimensionalized form, i.e., a ratio of tube velocity (v_t) and downcomer velocity (v_{dc}). Because of the dome-shaped bottom header, the flow naturally takes a smooth 180° turn and goes preferentially in the outer rows, evident from Fig. 5c. It is observed that the conical diffuser helps in the smooth expansion of the jet thereby reducing the regions of sharp velocity interfaces. This helps in the minimization of eddy production and hence the pressure drops. The velocity distribution predicted from these studies is helpful for detailed flow-induced vibration analysis of other immersed structures inside the header.

3.2 Bottom Header with Conical Diffuser

Numerical simulations are carried out with the conical diffuser considered at the bottom of the downcomer [see Fig. 4]. Figure 6a, b shows the predicted static pressure distribution and velocity distribution in the bottom header. It can be observed that the diffuser guides sodium flow, and it expands before impinging at the bottom of the header, thus lowering the pressure value compared to an earlier case of the bottom header with no diffuser. However, there is no significant difference in the flow distribution through the tubes compared to the earlier case of the bottom header without a diffuser. The distribution of average velocity through various tube rows is shown in Fig. 5c. Thus, it is concluded from the study that by varying the geometrical configuration of the diffuser, significant changes in the flow distribution among the tubes cannot be achieved.

3.3 Bottom Header with Conical Diffuser and Vertical Baffle

In this case, a vertical baffle shell is provided right below the tube sheet located radially after the 18th row of tubes, as shown in Fig. 7. The conical diffuser below the downcomer is also considered.

Parametric studies were carried out by considering different lengths for the vertical baffle, viz., 22.5 mm, 45.0 mm, 90.0 mm, 135.0 mm, 185.0 mm, 225.0 mm, and

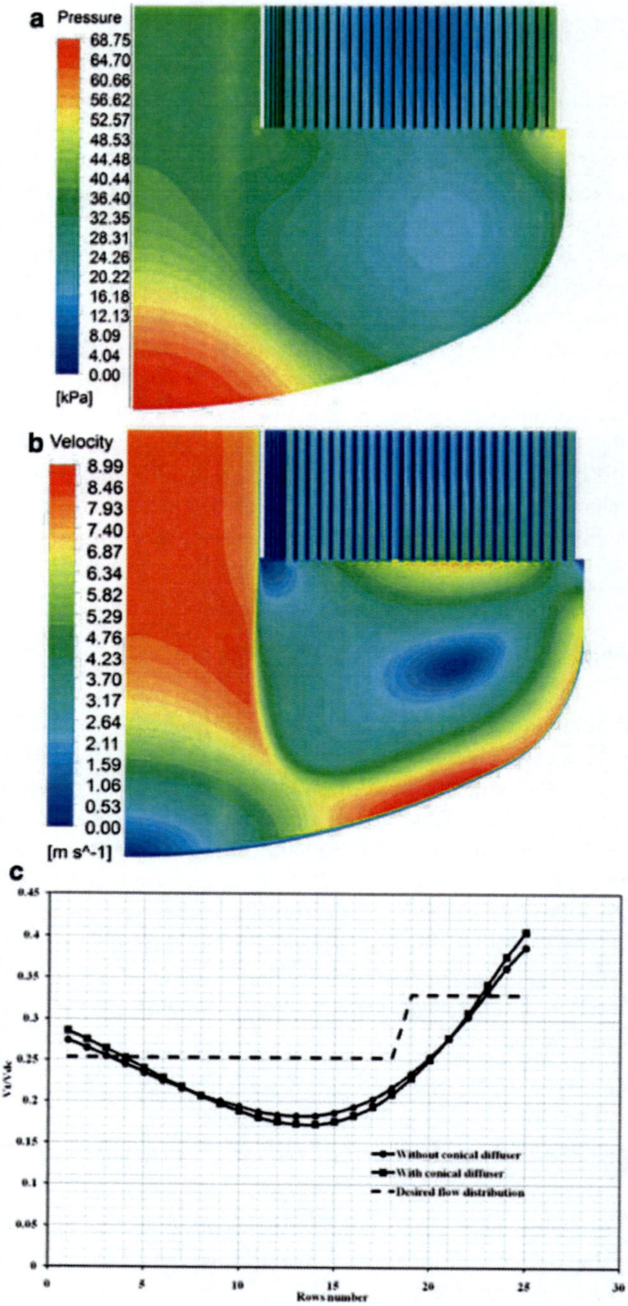

Fig. 5 **a** Static pressure distribution and **b** velocity distribution in IHX header without conical diffuser. **c** Velocity distribution in the case with and without a conical diffuser

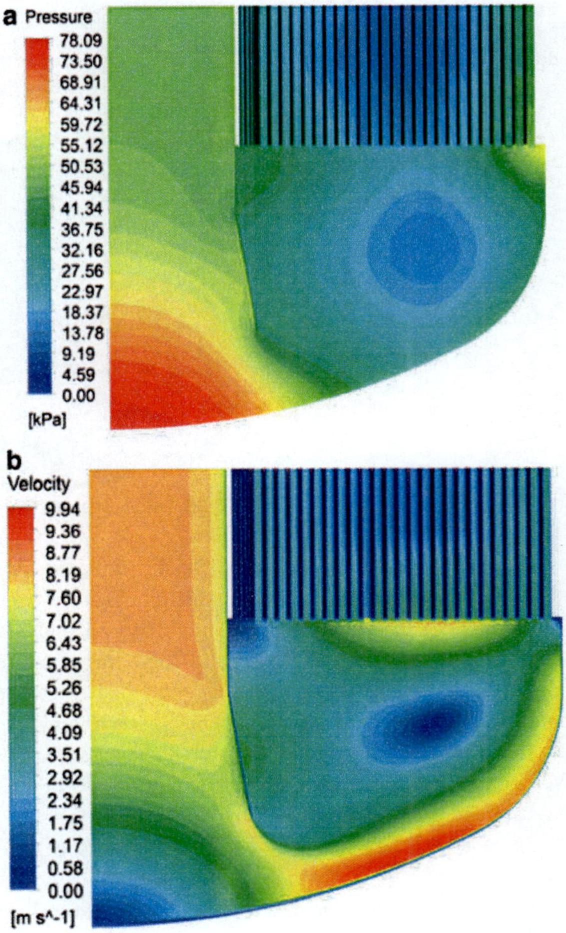

Fig. 6 **a** Static pressure distribution and **b** Velocity contour in IHX bottom header with conical diffuser

265.0 mm. Figure 8 shows the predicted distribution of non-dimensionalized velocity, V_t/V_{dc}, through various rows of tubes for different cases.

As the baffle length increases, flow distribution among the tubes approaches closer to the desired one. In the 17th row, the average velocity takes a dip and increases until the 25th row, as evident from Fig. 8a. This dip in the velocity is attributed to the positioning of the baffle near this location. The sodium flow gets restricted physically due to the vertical baffle, leading to flow recirculation at the baffle location and behind it. Consequently, the flow's kinetic energy in the header decreases at the recirculation zone, which lowers the velocity. As the height of the baffle plate increases, the strength of circulation increases, resulting in a lowering of average velocity through tubes from 1st to 17th row, as evident from Fig. 8a. Flow distribution predicted for the cases with baffle lengths 185.0 mm, 225.0 mm, and 265.0 mm are very close to the desired flow distribution.

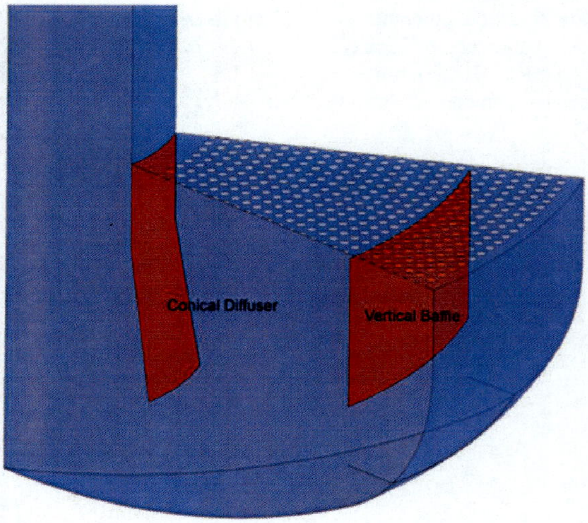

Fig. 7 Schematic of IHX bottom header with conical diffuser and vertical baffle

As welding baffle shell to tube sheet is not advisable due to mechanical strength, it should be welded to the header wall. Thus, there would be a clearance between the tube sheet and the vertical baffle. Therefore, an analysis of baffle configurations with heights of 185.0 mm, 225.0 mm, and 265.0 mm is carried out by considering an axial gap of 12 mm between the tube sheet and the baffle. The opening of 12 mm is chosen from design and welding considerations. This minimum axial space will provide accessibility to an operator. Figure 8b shows the predicted non-dimensionalized velocity distributions through various tube rows in these cases.

Interestingly, the dip in the flow rate at the 17th row significantly reduces compared to the case with a baffle connected to the tube sheet. The minimum velocity point has moved from the 17th row to the 15th or 16th row. The axial gap between the tube sheet and baffle plate is responsible for the shift in the minimum velocity from the 17th row to the 15–16th row. The fluid escapes to inner rows through this gap, lowering the strength of circulation. Hence, the minimum velocity location shifts inward. Deviation [(desired flow rate-actual flow rate)/desired flow rate] of the velocity distribution through various tube rows concerning the desired flow distribution is shown in Fig. 8c. The deviation is less than 20% in all the tube rows except the 18th row. The average absolute variation concerning the height of the baffle is shown in Fig. 8d. It is recommended to consider the baffle height of 225 mm. Figure 9 shows the velocity contour and vector plot with streamlines in the bottom header for the recommended baffle configuration. The temperature distribution of secondary sodium at the tube outlet must be re-assessed considering the above flow distribution achieved to confirm its acceptability.

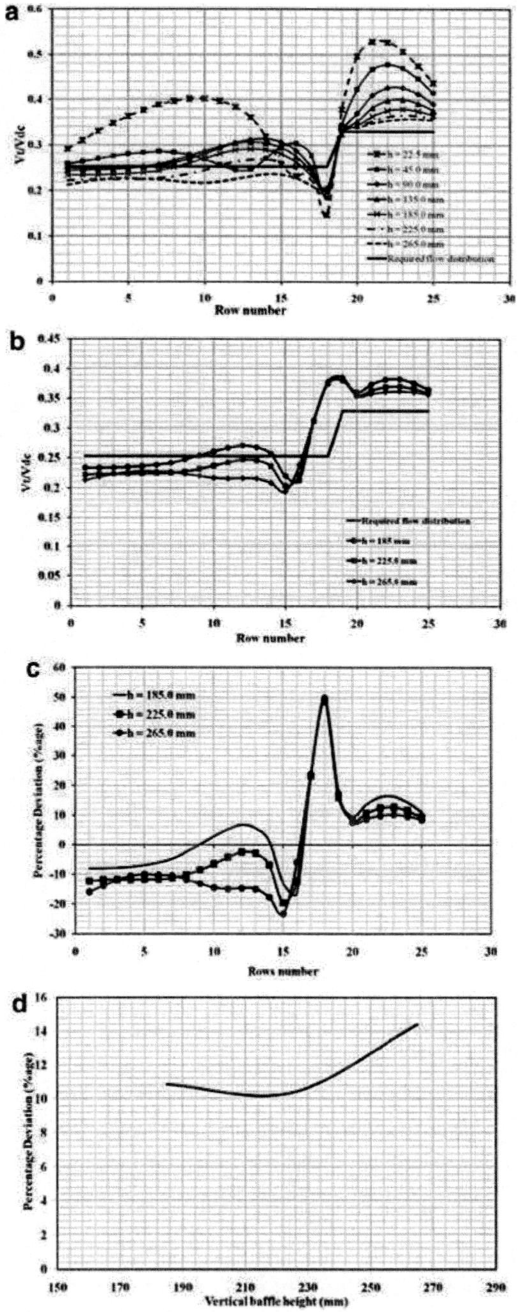

Fig. 8 Velocity distribution through **a** various tube rows for different baffle heights, **b** tube rows for different baffle heights (baffle located 12 mm below the tube sheet), **c** Deviation of estimated flow from desired flow, and **d** Absolute deviation of flow for the desired one for different baffle heights

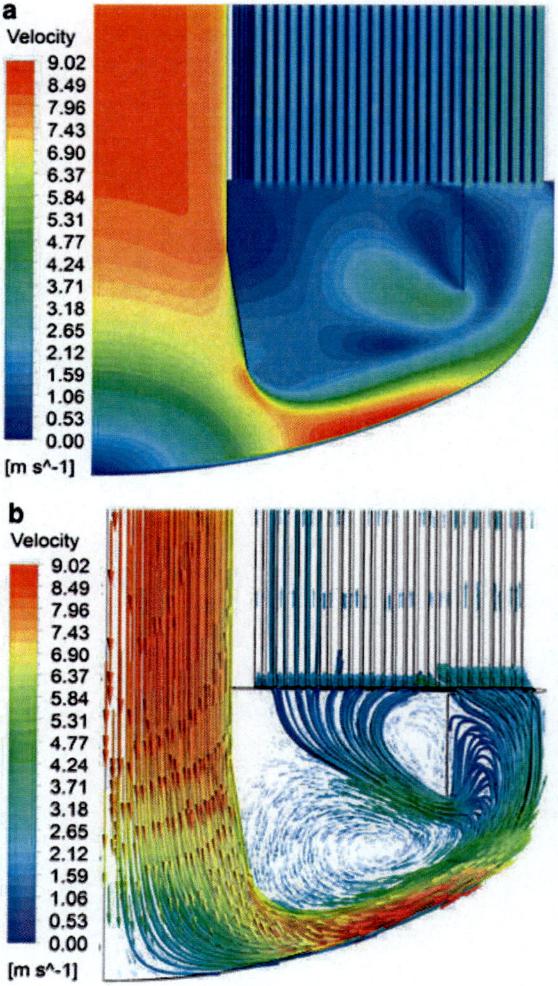

Fig. 9 **a** Velocity contour plot on the symmetric plane; **b** streamlines with velocity vector for vertical baffle height of 225 mm

4 Conclusions

3D CFD study of the bottom header of IHX for typical sodium-cooled fast reactors is carried out using ANSYS FLUENT, to explore the ways to simplify the design of the flow distribution device provided in the secondary sodium bottom header.

Parametric studies have been carried out to achieve the desired flow distribution. The effect of the conical diffuser and vertical baffle on the flow distribution inside the bottom header is quantified. Parametric studies by varying the height of the baffle are carried out. A vertical baffle of 225 mm height (located 12 mm below the tube sheet) provided after the 18th row rendered a flow distribution very close to the desired one.

With this arrangement, the average absolute deviation between the flow distribution achieved and the desired one is ~10%.

Nomenclature

C_p	Heat capacity J/kgK
$C_{\varepsilon 1}$, $C_{\varepsilon 2}$	Turbulent model constant --
K	Turbulent kinetic energy m^2/s^2
Nu	Nusselt Number
P	Piezometric pressure Pascal
Pr	Prandtl number
s_{ij}	Rate of strain tensor
s_i	Momentum source
u_i	Velocity component in i direction m/s
u'	Fluctuation about the ensemble average velocity m/s
x_i	Cartesian coordinatem
δ_{ij}	Kronecker delta
ε	Dissipation rate
ρ	Density Kg/m^3
σ	Turbulent Prandtl number
τ_{ij}	Viscous stress tensor components
v	Velocity m/s
μ	Dynamic viscosity Kg/ms
μ_t	Turbulent dynamic viscosity Kg/ms

References

1. Chetal SC, Balasubramaniyan V, Chellapandi P, Puthiyavinayagam P, Raghupathy S (2006) The design of the prototype fast breeder reactor. Nucl Eng Design 236:852–860
2. PuthiyaVinayagam P, Chellapandi P (2014) Sustainable energy security from fast breeder reactors. 6th Nuclear Energy Conclave, New Delhi
3. Chellapandi P, Chetal SC, Raj B (2009) Thermal striping limits for components of sodium cooled fast spectrum reactors. Nucl Eng Des 239:2754–2765
4. Gajapathy R, Velusamy K, Sundararajan T (2008) Thermal hydraulic investigations of intermediate heat exchanger in a pool-type fast breeder reactor. Nucl Eng Des 238(7):1577–1591
5. Patankar SV, Spalding DB (1975) Calculation procedures for the transient and steady-state behavior of shell-and-tube heat exchangers. In: Heat exchanger design and theory sourcebook, International Centre for Heat and Mass Transfer, 5thk Seminar, Proceedings, pp 155–176
6. Mochizuki H, Takano M (2009) Heat transfer in heat exchangers of sodium cooled fast reactor systems. Nucl Eng Des 239(2):295–307

7. Seban RA, Shimazaki TT (1951) Heat transfer to a fluid flowing turbulently in a smooth pipe with walls at constant temperature. Trans ASME 73:803–809
8. ANSYS Inc (2015) ANSYS fluent theory guide 12.0
9. White FM (2011) Fluid mechanics. McGraw-Hills, New York, USA

Graphical Approach for Image Improvement on a Cylindrical Receiver by the Center-Oriented Aiming Strategy of a Heliostat

Sainath A. Waghmare

Abstract The size and shape of the sun's image on the receiver surface is an essential aspect since it directly affects the optical performance and power generation in the central power tower. The combined graphical and Monte Carlo ray tracing (MCRT) is performed to evaluate optical characteristics on the central cylindrical receiver by a center-oriented aiming strategy for heliostats. A graphical ray tracing (GRT) model is established to determine target vectors and the exact orientation of the heliostat with the receiver. It is followed by MCRT analysis which predicts image size and shape of the sun on the receiver surface. The recently developed Center-Oriented Spinning-Elevation (COSE) tracking method produces a compact and better profile on the receiver surface than Spinning-Elevation (SE) and traditional Azimuth-Elevation (AE) methods.

Keywords Azimuth-elevation · Graphical ray tracing · Heliostat · Monte Carlo ray tracing · Spinning-elevation · Tracking angle

Nomenclature

A_s	Solar azimuth angle (deg.)
$\vec{B}_i$	Bisector vector (–)
$\vec{B}_g$	Positional vector of heliostat (–)
$\vec{B}_{g'}$	Perpendicular vector to $\vec{B}_g$ (–)
B_o	Bisector vector on ground (–)
D	Distance from tower to heliostat (unit)
D_R	Cylindrical receiver diameter (unit)
d_l	Heliostat diagonal (unit)

S. A. Waghmare (✉)
Department of Mechanical Engineering, FCRIT, Vashi, Navi Mumbai 400703, Maharashtra, India
e-mail: sainathwaghmare1990@gmail.com

Department of Mechanical Engineering, MGMCET, Navi Mumbai 410209, Maharashtra, India

S. Das et al. (eds.), *Proceedings of the 1st International Conference on Fluid, Thermal and Energy Systems*, Lecture Notes in Mechanical Engineering,
https://doi.org/10.1007/978-981-99-5990-7_17

H	Height of the tower (unit)
H_m	Height of the mirror (unit)
L_i	Length of incidence ray (unit)
L_g	Length of bisector on ground (unit)
L_r	Length of reflection ray (unit)
$\vec{N}$	Heliostat normal (–)
$\vec{N}_P$	Normal to bisector's plane (–)
$\vec{P}_B$	Bisector plane (–)
$\vec{P}_I$	Incidence plane (–)
$\vec{P}_S$	Vertical sagittal plane (–)
$\vec{t}_i$	Incidence vector (–)
$\vec{t}_r$	Reflection vector (–)
$X_{R\max}$	Maximum image-width on cylindrical receiver (unit)
$X_{rp\max}$	Maximum image-width on receiver plane (unit)
$Y_{rp\max}$	Maximum image-height (unit)
W_m	Width of the mirror (unit)

Greek symbols

α_s	Solar elevation angle (deg.)
Φ	Latitude (deg.)
Ω_1	Motor-1 rotation (deg.)
Ω_2	Motor-2 rotation (deg.)
ω	Hour angle (deg.)

Subscripts

$\hat{i}, \hat{j}, \hat{k}$	Unit vectors in X, Y, and Z dir. (–)

1 Introduction

Heliostat tracking is one of the crucial tasks in power generation from a solar power tower. The two-axis tracking procedure has a great tendency to concentrate rays and generate maximum power at the central receiver [1, 2]. The Azimuth-Elevation (AE) tracking is a widespread method in which the movement of the heliostat in azimuth and elevation direction provides an easy and effective way of tracking and concentrating the incidence rays onto the receiver [3, 4]. The Spinning-Elevation (SE) tracking method is an advanced one in which a heliostat is continuously focusing the target [5–7]. It is based on the target-oriented tracking technique. Chen et al. [8] have developed a general sun-tracking formula for the AE and SE with special cases like target-oriented, latitude-oriented, polar-oriented, etc. These tracking algorithms give accurate orientation angles for the heliostat; however, they provide a bit of complexity in the calculation as large transformation matrices and expressions are involved. Waghmare and Puranik [9] developed a center-oriented spinning-elevation (COSE) technique in which a heliostat is focusing the center of the tower instead of a receiver. This aiming strategy significantly reduces the tracking effort compared to AE and SE techniques [9].

The reflected image of the concentrator is an essential factor in assessing the central receiver thermal performance as it evaluates the optical losses [10, 11] and helps finalize the receiver dimensions [12, 13]. The effect of the aiming strategy and heliostat tracking methods on spot size and receiver size is not reported in the literature. The relationship between a heliostat and target is maintained in the SE method, significantly reducing the optical aberration on the receiver surface. Although the SE method provides a smaller spot size on the receiver plane than AE, the image shape is dominant in calculating spillage losses and intercept efficiency [14].

This research aims to implement a ray tracing procedure to determine the size and shape of an image produced on the receiver along with flux concentration for various tracking modes. A novel combined approach of Graphical Ray Tracing (GRT) [15] and Monte Carlo Ray Tracing (MCRT) [11] is applied for evaluation. Based on it, a comparative study is conducted to optimize the cylindrical receiver diameter. The combined approach of GRT and MCRT not only helps in deciding the aiming strategy for heliostats, but it also obtains image spread and flux profiles. The results highlight the influence of tracking modes on the image spread and receiver size.

2 Geometrical Formulation for Tracking Methods

A GRT algorithm with solar geometry is developed in the Cartesian coordinate system. The sun location is expressed in Azimuth angle A_s and Elevation angle α_s. The center point of the tower is at $O(x_o, y_o, z_o) = (0, 0, 0)$. A heliostat and a receiver-target are expressed in terms of points.

Let, Target point, $T = [x_t, y_t, z_t]$, Incident point, $I = [x_i, y_i, z_i]$, and Heliostat intercept point $P = \left[x_p, y_p, z_p\right]$. The incidence point $[x_i, y_i, z_i]$ represents the sun. The point is instantaneous and is composed of solar azimuth and elevation angle.

$$x_i = x_p - (\cos\alpha_s \cos A_s) \tag{1}$$

$$y_i = y_p + (\cos\alpha_s \sin A_s) \tag{2}$$

$$z_i = z_p + \sin\alpha_s \tag{3}$$

$$\alpha_s = \sin^{-1}(\sin\delta\sin\Phi + \cos\delta\cos\omega\cos\Phi) \tag{4}$$

$$A_s = \cos^{-1}\left(\frac{\sin\delta\cos\Phi - \cos\delta\cos\omega\sin\Phi}{\cos\alpha_s}\right) \tag{5}$$

A vector joins the points I and P is an incidence vector $\vec{t_i}$ which is reflected toward a target T via reflected vector $\vec{t_r}$:

$$\vec{t_i} = \left(\frac{x_i - x_p}{L_i}\right)\hat{i} + \left(\frac{y_i - y_p}{L_i}\right)\hat{j} + \left(\frac{z_i - z_p}{L_i}\right)\hat{k} \tag{6}$$

$$\vec{t_r} = \left(\frac{x_t - x_p}{L_r}\right)\hat{i} + \left(\frac{y_t - y_p}{L_r}\right)\hat{j} + \left(\frac{z_t - z_p}{L_r}\right)\hat{k} \tag{7}$$

$$L_i = \sqrt{(x_i - x_p)^2 + (y_i - y_p)^2 + (z_i - z_p)^2} \tag{8}$$

$$L_r = \sqrt{(x_t - x_p)^2 + (y_t - y_p)^2 + (z_t - z_p)^2} \tag{9}$$

$\vec{N}$ is normal to the heliostat surface and the point I is continuously changing its position with time. Points P and T are fixed points on the heliostat field. Therefore, the heliostat has to orient by two tracking motors to concentrate incoming rays on the target. It is done by aligning $\vec{N}$ with the angle bisector $\vec{B_i}$ of the incident and reflected vectors which is given as

$$\begin{aligned}\vec{B_i} = &\left[\left(\frac{x_i - x_p}{L_i}\right) + \left(\frac{x_t - x_p}{L_r}\right)\right] \\ &+ \left[\left(\frac{y_i - y_p}{L_i}\right) + \left(\frac{y_t - y_p}{L_r}\right)\right] \\ &+ \left[\left(\frac{z_i - z_p}{L_i}\right) + \left(\frac{z_t - z_p}{L_r}\right)\right]\hat{k}\end{aligned} \tag{10}$$

The angular adjustment of the two-axis tracking method Ω_1 and Ω_2 depends on the tracking mode and rotating coordinates for heliostat orientation.

2.1 Azimuth-Elevation Tracking Method

In this technique, a heliostat is tracked by two motors, i.e., Ω_1 in the azimuthal direction and Ω_2 in the elevation direction. A GRT model for the AE tracking system is shown in Fig. 1. The aim is to orient a heliostat to align $\vec{N}$ with $\vec{B_i}$ from time to time by adjusting Ω_1 and Ω_2. Initially, the heliostat is placed by facing the East so

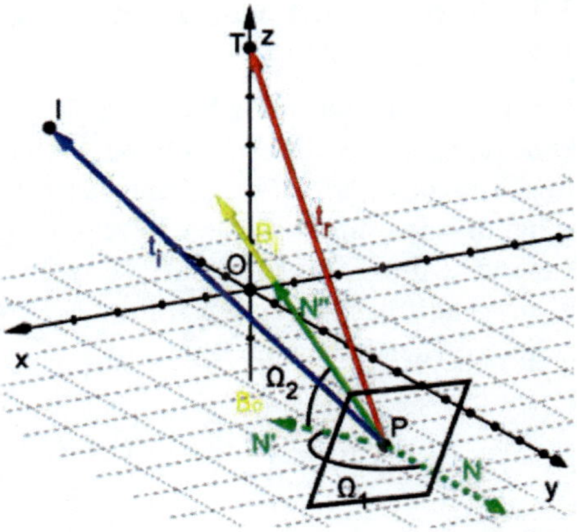

Fig. 1 GRT model for AE tracking method

that the normal of heliostat $\vec{N}$ is parallel to Y-axis. The angle between bisector and ground is nothing but an elevation-angle input Ω_2, therefore $\vec{B}_i$ is projected on the XY-plane to obtained the ground-bisector vector $\vec{B}_o$ which is given by

$$\vec{B}_o = \left[\left(\frac{x_i - x_p}{L_i}\right) + \left(\frac{x_t - x_p}{L_r}\right)\right]\hat{i} + \left[\left(\frac{y_i - y_p}{L_i}\right) + \left(\frac{y_t - y_p}{L_r}\right)\right]\hat{j} + 0\hat{k} \tag{11}$$

And,

$$\Omega_2 = \cos^{-1}\left[\frac{\vec{B}_i \cdot \vec{B}_o}{\left|\vec{B}_i\right|\left|\vec{B}_o\right|}\right]. \tag{12}$$

$$\Omega_1 = \cos^{-1}\left[\frac{\vec{B}_o \cdot \vec{Y}_o}{\left|\vec{B}_o\right|\left|\vec{Y}_o\right|}\right] \tag{13}$$

$$\vec{Y}_o = 0\hat{i} + 1\hat{j} + 0\hat{k} \tag{14}$$

Therefore, the heliostat has to rotate through the tracking angles of Ω_1 to align $\vec{N}$ with $\vec{B}_o$ becomes $\vec{N}'$ and then through angle Ω_2 to align rotated normal $\vec{N}'$ with $\vec{B}_i$ to obtain the correct tracking position.

2.2 Spinning-Elevation Tracking Method

In this case, a heliostat is facing the target. Therefore, a line joining T and P acts as a spinning-axis which is nothing but the direction of $\vec{t}_r$ as well as $\vec{N}$ at the initial step. The idea is same, i.e., alignment of $\vec{N}$ with $\vec{B}_i$ which is produced by adjusting spinning-angle input Ω_1 and elevation-angle input Ω_2. A GRT model for the SE method is illustrated in Fig. 2. $\vec{B}_g$ is the position vector from P to O and $\vec{B}_{g\prime}$ is its perpendicular vector which lies on the ground, and at the right-hand side of the heliostat when facing the target:

$$\vec{B}_g = \left(\frac{x_o - x_p}{L_g}\right)\hat{i} + \left(\frac{y_o - y_p}{L_g}\right)\hat{j} + \left(\frac{z_o - z_p}{L_g}\right)\hat{k} \tag{15}$$

$$\vec{B}_{g'} = \left(\frac{y_o - y_p}{L_g}\right)\hat{i} - \left(\frac{x_o - x_p}{L_g}\right)\hat{j} + \left(\frac{z_o - z_p}{L_g}\right)\hat{k} \tag{16}$$

$$L_g = \sqrt{(x_p)^2 + (y_p)^2 + (z_p)^2} \tag{17}$$

Here, the positive value of the term $\left(\frac{z_o - z_p}{L_g}\right)\hat{k}$ in Eq. (16) gives the clockwise movement of motor-1 rotation and vice versa. Initially, an elevation-motor axis which is perpendicular to the spinning-axis is aligned with $\vec{B}_{g\prime}$. Spinning-motor rotates the heliostat until $\vec{B}_i$, $\vec{t}_i$, and $\vec{t}_r$ become co-planar. This plane is normal to the mirror surface and is termed as the plane of reflection $\vec{P}_B$. It is given by

$$\vec{P}_B = \vec{t}_i \vec{B}_i \tag{18}$$

Elevation-axis is normal to $\vec{P}_B$ and is given by

$$\vec{N}_P = \vec{t}_i \vec{B}_i \tag{19}$$

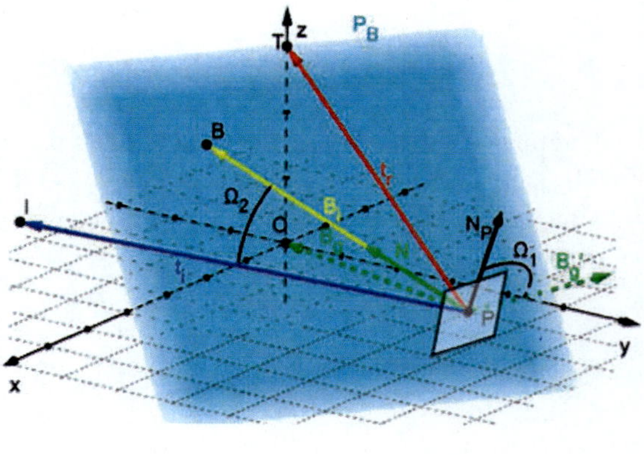

Fig. 2 GRT model for SE tracking method

Therefore, the angle between $\vec{N}_P$ and $\vec{B}_{g'}$ is an angular input Ω_1 by spinning-motor to make $\vec{B}_i$, $\vec{t}_i$, and $\vec{t}_r$ co-planar and $\vec{N}_P$ to be normal to the spinning-axis. It is given by

$$\Omega_1 = \cos^{-1}\left[\frac{\vec{B}_{g'} \cdot \vec{N}_P}{\left|\vec{B}_{g'}\right|\left|\vec{N}_P\right|}\right] \tag{20}$$

Here, $\vec{N}$ is still aligned with $\vec{t}_r$ unless it is rotated through Ω_2 by the elevation-motor to align $\vec{N}$ with $\vec{B}_i$. Ω_2 is nothing but an incidence angle, and it is given by

$$\Omega_2 = \cos^{-1}\left[\frac{\vec{t}_i \cdot \vec{B}_i}{\left|\vec{t}_i\right|\left|\vec{B}_i\right|}\right] \tag{21}$$

Here, point P is lying on the ground. If a heliostat height $h = z_p$ is considered, then the origin shifts to $O'(x_o, y_o, z_p)$ for that particular heliostat.

2.3 Center-Oriented Spinning-Elevation Tracking Method

Figure 3 shows the GRT model for the COSE tracking method. The working of this method is similar to general SE tracking except for the spinning-axis. It is along the line OP or $\vec{B}_g$, and the elevation-axis is perpendicular to the spinning-axis. $\vec{P}_S$ is the vertical sagittal plane of the heliostat, which is composed by $\vec{t}_r$ and $\vec{B}_g$. Spinning-motor rotates the heliostat with respect to $\vec{B}_g$ until $\vec{B}_i$ and $\vec{B}_g$ become co-planar. This plane is termed as incidence plane $\vec{P}_I$ as shown in Fig. 3. The expressions for incidence and sagittal plane are

$$\vec{P}_S = \vec{t}_r \vec{B}_g \tag{22}$$

$$\vec{P}_I = \vec{B}_i \vec{B}_g \tag{23}$$

The angle between $\vec{P}_S$ and $\vec{P}_I$ is a net rotation input for spinning-motor Ω_1 expressed as

$$\Omega_1 = \cos^{-1}\left[\frac{\vec{B}_{g'} \cdot \vec{N}_P}{\left|\vec{B}_{g'}\right|\left|\vec{N}_P\right|}\right] \tag{24}$$

This rotation makes the elevation-axis perpendicular to $\vec{P}_I$; thus, the angle between $\vec{B}_i$ and $\vec{B}_g$ is the elevation-motor input Ω_2:

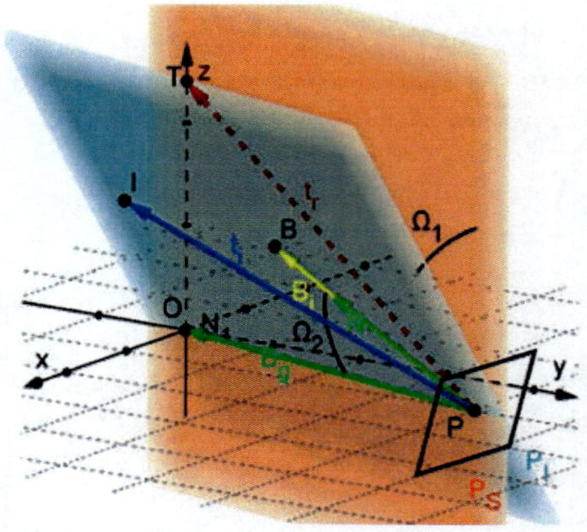

Fig. 3 GRT model for COSE tracking method

$$\Omega_1 = \cos^{-1}\left[\frac{\vec{B}_{g'} \cdot \vec{N}_P}{\left|\vec{B}_{g'}\right|\left|\vec{N}_P\right|}\right] \tag{25}$$

Here, a variation in heliostat height h is compensated by shifting the center point at $O'(x_o, y_o, z_p)$ for that particular heliostat.

3 Ray Tracing Simulation

Both GRT and MCRT perform the ray tracing analysis. The tracking method influences a reflected mirror image on the receiver. In GRT, the reflected image on the heliostat is formed by tracing rays (vectors) parallel to $\vec{t}_r$ and from the mirror surface to the receiver surface. The tracing points are the grid points formed as an array on the reflector surface lined along the mirror's width (W_m) and height (H_m). Note that the mirror's orientation by all the tracking methods remains the same. However, the rotation of the mirror with respect to $\vec{N}$ varies, which changes the shape of the image on the receiver. MCRT analysis is performed using SolTrace with solar disk size of 9.3 mrad Gaussian sun shape. The number of rays for analysis are taken as 5,000,000. The reflectivity of mirror 0.85, slope and specularity error as 0.95 mrad and 0.2 mrad are taken, respectively. Certainly, image shapes affect the receiver's optical efficiency and flux concentration [16]. The image spread [13] is measured in terms of maximum image-width $X_{rp\max}$ and maximum image-height $Y_{rp\max}$ on the receiver plane. It can be one of the deciding parameters for receiver diameter, associated with image-width and evaluated as image-width ratio obtained by $X_{rp\max}/d_l$ whereas d_l is the heliostat diagonal [13]. Furthermore, when the reflected image is projected on the cylindrical receiver surface, the maximum image width $X_{R\max}$ is presented by Eq. (26). Here, R is the receiver diameter:

$$X_{R\max} = \frac{2\pi R}{180}\left[\sin^{-1}\left(\frac{X_{rp\max}}{R}\right)\right] \tag{26}$$

4 Results and Discussion

Using a computer code, the tracking angles Ω_1 and Ω_2 for AE, SE, and COSE are calculated and geometries of the heliostat and receiver are developed in the GeoGebra mathematical tool accordingly. The GRT is performed by grid point tracing with double loop along the length and width of heliostat. These grid points are projected parallel to the reflected vector and plotted on the receiver surface to form an image in Fig. 4. Figure 5 shows the reflected image by the MCRT method. These figures can also highlight the relation between heliostat dimension and receiver diameter. The average of $X_{R\max}/d_l$ is considered for prescribed daylight hours and plotted with the non-dimensional term D_R/W_m. Here, W_m is kept constant of 20 units, and D_R varies from 20 to 60 units. Though $X_{rp\max}/d_l$ or $X_{R\max}/d_l$ is independent of heliostat aspect ratio (W_m/H_m) and gives consistent results for varying D [13], it is influenced by heliostat orientation and on-field position. It is well presented in Fig. 6. Image-width decreases as receiver diameter increases, irrespective of tracking techniques, heliostat position, and reflector area. Lower values of $X_{R\max}/d_l$ are desirable for improved optical performances, whereas $X_{R\max}/d_l > 1$ is treated as optical losses at the receiver. The minimum D_R for all types of tracking heliostats is 1.25 D_R/W_m for a standalone heliostat. Heliostats show a significantly compact image when placed at the North with the SE tracking method. Therefore, it is preferably used in the solar tower with a cavity receiver facing the North. Except for North heliostat, the COSE method produces better and compact images. Therefore, it can be helpful in a cylindrical receiver surrounded by heliostats spread 360° around the tower. The heliostats N-E and W-N, E and W, and E-S and S-W show the same results due to the symmetricity of the line joining North and South heliostats.

5 Conclusion

The GRT technique presented in this paper provides easy and better visualization of tracking and ray tracing analysis. Using center-orientation coordinates instead of target orientation, the size and shape of the sun image on the receiver surface have improved considerably. COSE in ray tracing analysis produces compact reflected images on the receiver surface; therefore, a receiver diameter can be reduced with a high concentration ratio.

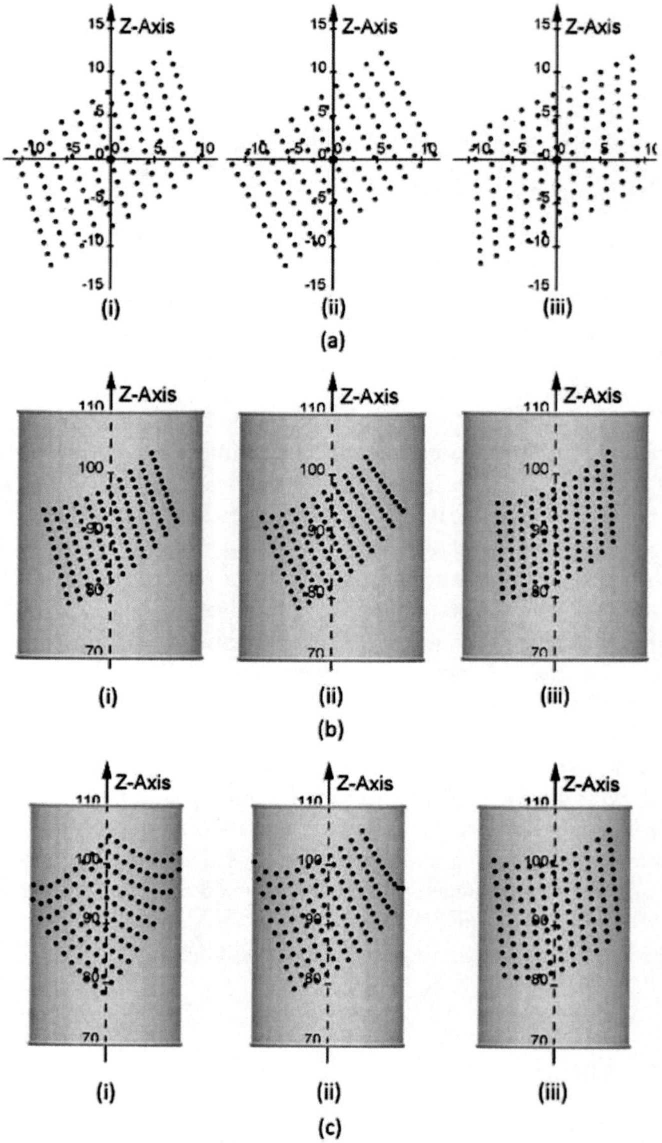

Fig. 4 The reflected image of (20 × 20) unit mirror with uniform grid points on the **a** reflector plane, **b** cylindrical receiver of radius 15 unit, and **c** 12.5 unit, with tracking methods (i) AE, (ii) SE, and (iii) COSE, on the South heliostat and traced on equinox days at 14:00 h

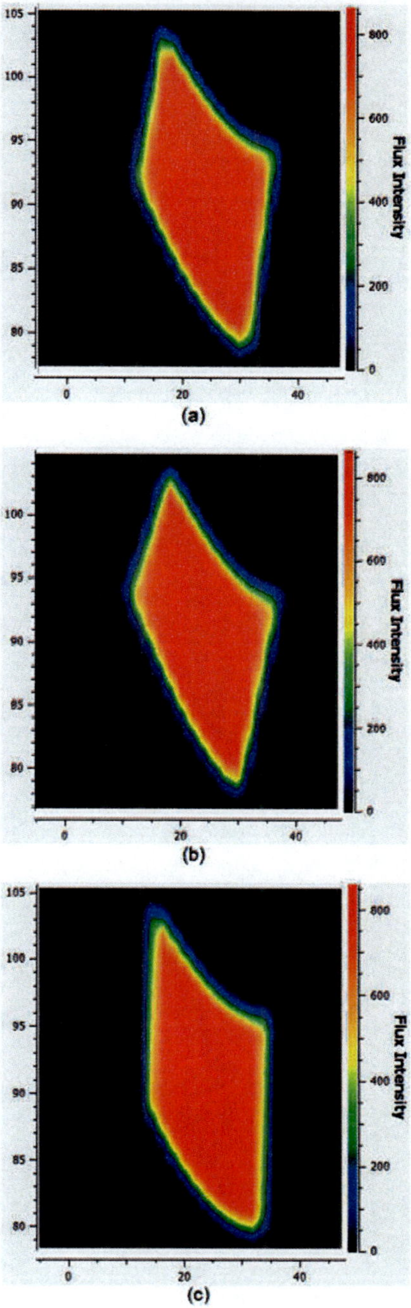

Fig. 5 The reflected image with flux intensity of (20 × 20) unit mirror of South heliostat and traced on equinox days at 14:00 h by MCRT on the receiver plane with tracking method: **a** AE, **b** SE, and **c** COSE

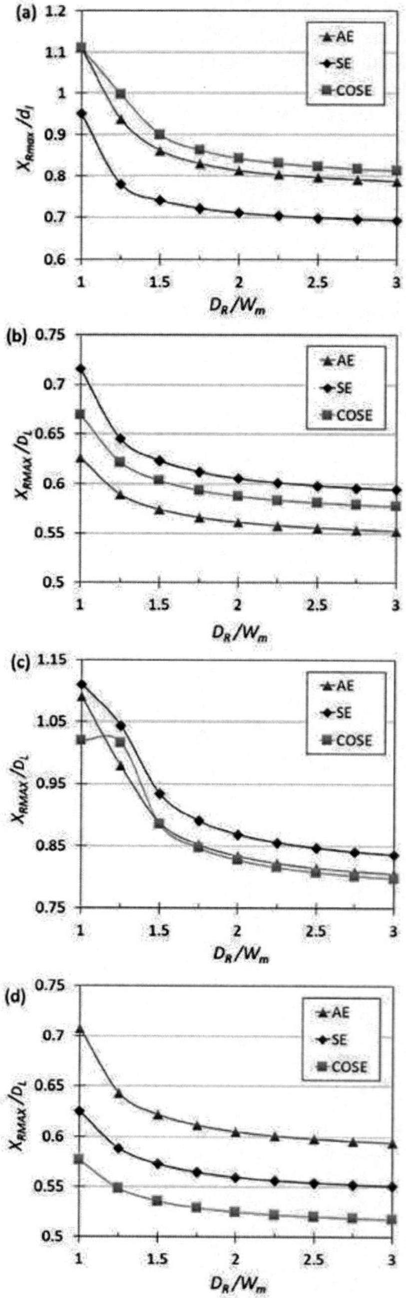

Fig. 6 Effect of image width on the various receiver diameter when the heliostat at **a** North and South, **b** North-East and North-West, **c** East and West, and **d** South-East and South-West

Acknowledgements We are very thankful to the Indian Institute of Technology Bombay (IIT Bombay), Mumbai, India, for providing the infrastructure and computation facilities for the research.

References

1. Saymbetov AK, Nurgaliyev MK, Tulkibaiuly Y et al (2018) Method for increasing the efficiency of a biaxial solar tracker with exact solar orientation. Appl Solar Energy 54(2):126–130
2. Samanta A, Dutta A, Neogi S (2012) A simple and efficient sun tracking mechanism using programmable logic controller. Appl Solar Energy 48(3):218–227
3. Riley D, Hansen C (2015) Sun-relative pointing for dual-axis solar trackers employing azimuth and elevation rotations. ASME J Sol Energy Eng 137(3):031108–031111
4. Grigoriev V, Milidonis K, Blanco M (2020) Sun tracking by heliostats with arbitrary orientation of primary and secondary axes. Sol Energy 207:1384–1389
5. Zaibel R, Dagan E, Karni J, Ries H (1995) An astigmatic corrected target-aligned heliostat for high concentration. Sol Energy Mater Sol Cells 37:191–202
6. Lim BH, Chong KK, Lim CS, Lai AC (2016) Latitude-orientated mode of non-imaging focusing heliostat using spinning-elevation tracking method. Sol Energy 135:253–264
7. Smit WJ (2020) Improvement of heliostat pointing accuracy by calibration at optimized dissimilar source vectors. Sol Energy 204:238–245
8. Chen YT, Lim BH, Lim CS (2006) General sun tracking formula for heliostats with arbitrarily oriented axes. ASME J Sol Energy Eng 128:245–250
9. Waghmare SA, Puranik BP (2022) Center-oriented aiming strategy for heliostat with spinning elevation tracking method. ASME J Sol Energy Eng 144:024503–024511
10. Collado FJ, Guallar J (2018) Fast and reliable flux map on cylindrical receivers. Sol Energy 169:556–564
11. Huang W, Yu L, Hu P (2019) An analytical solution for the solar flux density produced by a round focusing heliostat. Renew Energy 134:306–320
12. Sasidharan S, Dutta P (2021) Optical characterization of a fixed focus Scheffler reflector for pressurized solar receiver testing. Sol Energy 227:89–100
13. Venkatesh V, Rao BS, Srilakshmi G et al (2017) Correlation between central receiver size and solar field using flat heliostats. Appl Solar Energy 53(3):258–266
14. Chen YT, Kribus A, Lim BH, Lim CS, Chong KK, Karni J, Buck R, Pfahl A, Bligh TP (2004) Comparison of two sun tracking methods in the application of a heliostat field. ASME J Sol Energy Eng 126(1):638–644
15. Waghmare SA, Gulhane NP (2019) Evaluation and optimisation of receiver height in a compound parabolic collector with low acceptance angle. Int J Ambient Energy 40(7):675–686
16. García L, Burisch M, Sanchez M (2015) Spillage estimation in a heliostats field for solar field optimization. Energy Procedia 69:1269–1276

Sustainability Assessment of Hybrid Renewable Energy System: A Case Study from Indian Perspectives

Shivam Prajapati, Sanket Palchaudhary, Nishant Narvekar, and Ashis Acharjee

Abstract Hybrid power plants have proven to be a profitable energy system for supplying significant value of electricity or power from generated plants to the electrical system of grids, mainly as renewable energy shares in systems rise from 10 to 20% or more, and the prices of solar photovoltaic instruments, wind energy instruments, and battery storage all keep on falling. The purpose of this investigation is to describe various sustainability tools that may be used to identify renewable energy technologies for distant region power delivery. To comprehend the environmental, economic, and social ramifications of these technologies, various sustainability tools such as the life cycle assessment (LCA) for environment, life cycle costing, along with social assessment approach, and eco-efficiency analysis can be employed. While the different techniques may be useful in their own right, the study's objective and goal will determine their relevance. The eco-efficiency strategy was proven to be supportive in selecting environmentally friendly technologies with the least environmental cost for distant area power supply. All these pieces of work are substantial research endeavors, and more work is required to prioritize them.

Keywords Hybrid power plant · Renewable energy · Sustainability tools · Life cycle assessment

---

S. Prajapati · A. Acharjee
Department of Mechanical Engineering, National Institute of Technology, Agartala, India

S. Palchaudhary (✉)
Department of Mechanical Engineering, Sardar Vallabhbhai Patel Institute of Technology, Surat, India
e-mail: sanket.palchaudhary@gmail.com

N. Narvekar
Department of Mechanical Engineering, PIT, Vadodara, India

S. Das et al. (eds.), *Proceedings of the 1st International Conference on Fluid, Thermal and Energy Systems*, Lecture Notes in Mechanical Engineering,
https://doi.org/10.1007/978-981-99-5990-7_18

1 Introduction

To power the modern economy, present energy systems are primarily reliant on fossil fuels. Continuous usage of fossil fuels results in global resource scarcity, as well as negative environmental and health consequences. The current economy, which is dependent on fossil fuels, will lead to an unsustainable future. Renewable energy could handle future energy scarcity while also avoiding environmental damage on a global scale. Due to the rapid depletion of fossil fuels, an urgent need is being risen for alternative sources of energy to meet the ever-increasing energy demand. Another compelling reason to minimize our dependence on fossil fuels is the mounting threat of global warming. Power generation technologies that are environmentally friendly will play a significant part for power supply in the future [1].

Renewable energies are unlimited, environmentally friendly, and may be employed to be decentralized. They also have the advantage of being complementary, with a favorable integration between them. Solar photovoltaic energy, for example, provides electricity on sunny days, while wind generators provide more electricity on cool and breezy days, which are typically gloomy. Due to advancements in renewable energy technologies, HRES are becoming more popular for distant region power generation applications. A hybrid system for energy is made up of more than one renewable sources of energy that are combined to boost system efficiency and provide greater energy supply balance. Even then, due to variations in solar energy and wind energy generation, there will be periodic power outages. As a result, in addition to HRES, a battery can be given to meet the demands for load at that time. In addition, using a hybrid system with a (PV/WG) battery minimizes the amount of battery storage required. According to studies completed around the world, a hybrid PV/Wind/Battery system is a reliable source of electricity. To raise the output of HRES, a variety of computer software has been developed. Many of these studies have used the National Renewable Energy Laboratory's (NREL) Hybrid Optimization Model for Electric Renewable (HOMER) software. HOMER conducts a cost-benefit analysis of hybrid power systems. Inputs to HOMER will run an hourly simulation of every possible permutation of items entered, ranking the systems based on user-specified parameters like energy cost or capital costs [2].

2 Review of Literature

Highlighting HRES technologies for generation of power, and major concerns and obstacles in their design stage, generating technology selection, and unit sizing are analyzed by Abul Khair, Md. Ibrahim, and others. Implementations and benefits of hybrid energy systems, challenges and problems with hybrid energy systems, and a summary of energy storage methods for renewable energy sources are studied Ranjay Singh et al. [3]. The scale, cost, and reliability of HRES power output are all essential considerations in HRES planning. Review of optimization tools and limitations for which the HRES system is optimized, as well as the many types of storage and backup systems available for HRES Ajai Gupta et al. [4] shows modeling of a hybrid energy system with a focus on small hydro/micro-hydro power, biogas power, biomass (fuelwood) power, solar photovoltaic power as a renewable power source, and a battery bank, diesel power as a conventional power source Md. Ibrahim et al. [5]. Employment of the hybrid energy systems, along with benefits of hybrid energy systems, concerns and problems with hybrid energy systems, and a summary of technologies to store energy for systems of renewable energy are all discussed Ranjay Singh et al. [6]. Review of optimization tools and limitations for which the HRES system is optimized, as well as the many types of storage and backup systems available for HRES. Babatunde et al. [7] is based on an examination of two energy demand scenarios: user demand for energy efficient equipment (EET) and user demand for energy without efficiency (CED). Cuesta et al. [8] present a comprehensive evaluation of the most common tools for creating HRESs, with the major finding being that existing tools do not incorporate social variables, which represents a significant opportunity to improve the capabilities of such software packages.

3 Mathematical Modeling

A system for hybrid energy may include renewable energy conversion components such as wind turbines, solar panels, and hydro turbines, as well as non-renewable generators such as diesel generators, micro turbines, and devices to store energy such as batteries. All or a portion of it could be found in a hybrid energy system. To properly choose components and subsystems for optimal system sizing, the modeling approach aids in identifying and understanding the features of the components, as well as decision-making.

3.1 Hybrid Photovoltaic System

Solar radiation is the source of energy for PV systems, and total radiation on an inclined surface is calculated as

$$I_T = I_b R_b + (I_b + I_d) Rr \tag{1}$$

I_b is direct normal solar radiation; I_{d} is diffuse solar radiation; R_{d} and R_{r} are tilt factors for the diffuse and reflected parts of the solar radiations. Hourly output for power from the PV approach with an area A_{pv} (m^2), on an average day of *j*th month, when total radiation of I_{T} ($\mathrm{kW/m}^2$) is incident on PV surface, is given by

$$P_{sj} = I_{Tj} h A_{PV} \tag{2}$$

where h is the system efficiency. Also, the power output can be acquired by calculating the total current available from the PV cell (I_{pv}) and actual voltage of the PV module (Vpv). So, the power gained from the PV module can be expressed as

$$P_{pv} = I_{pv} V_{PV} \quad (3)$$

3.2 Hybrid Wind Energy System

Wind Turbine Power (P_{wt}): The available wind power is calculated by calculating the average wind speed at a specific point in time and the type of wind turbine employed. The power coefficient C_{Pwt} is used to estimate wind turbine performance. The turbine efficiency in converting aero kinetic power in the wind to mechanical power is expressed by C_{Pwt}, a non-dimensional quantity.

The use of an overall energy balance equation is the foundation of the wind-diesel system concept. The performance of a wind system is determined by the area's geography and environmental circumstances (wind speed, direction, and other factors).

4 Analysis of PV-WIND Hybrid Renewable Energy System

4.1 Environmental Assessment

Emission: Carbon dioxide, Sulfur dioxide, and Oxides of Nitrogen are among the gases emitted by an HRES. CO_2 and NOx are two of the six principal greenhouse gases. Emission is the most significant source of pollution in the environment, and it is linked to major environmental issues such as the greenhouse effect and acid rain. As a result, emission is used as a viable environmental headline evaluation.

Renewable fraction: The proportion of renewable energy generated divided by total energy generated is known as the renewable fraction of an HRES. It refers to the portion of renewable energy available in an HRES. More renewable energy resources are utilized when this proportion has a higher value. The renewable section has representativeness for environmental evaluation. It tends to be isolated into divisions, for example, PV portion (f_{PV}) and wind portion (f_{WG}). The two divisions are computed by the following equations:

$$F_{PV} = E_{PV}/E_{TOT} \quad (4)$$

$$F_{WG} = E_{WG}/E_{TOT} \quad (5)$$

energy generated by the system E_{total}, as expressed by the expression:

$$LCE = C_{A-TOTAL}/E_{total} \tag{6}$$

Another important factor is the Annual Life Cycle Cost—*ALCC*, which includes the Cost of Life Cycle—*LCC,* given by the equation

$$ALCC = CRF * LCC \tag{7}$$

Net Present Value: NPV, the Net present value, which includes the system's initial cost as well as repair and maintenance costs, is known as the net present value. The NPV compares the current value to the future value, accounting for inflation.

Payback period: The Payback Period (PBP) is computed by dividing the total investment cost by the first-year revenue by the energy saved, displaced, or produced. The number of years it takes to recover an investment from the overall cost of the system is the unit of measure in period analysis of investment recovery.

Ratio of Savings to Investment: The where E_{PV}, E_{WG} and E_{TOT} are the energy savings to investment ratio can be utilized to compare the related energy generation by solar photovoltaic, by wind generator and total generated energy, respectively.

4.2 *Economic Assessment*

Levelized cost of energy: In the examination of a hybrid system, the system cost analysis is equally crucial. A cost analysis for the system can be done in numerous ways. One of them is the *LCE* or is also called levelized cost of energy, which is interpreted as the relationship between the cost of the total annual system $C_{A\text{-}TOTAL}$ and the system's savings to the alternative energy system's costs. The SIR must be bigger than one in order to create positive net savings. High SIR values indicate greater savings compared to investment.

4.3 *Social Impacts*

Access to electricity: In underdeveloped countries, around a third of the overall population have no direct access to grid electricity. The electrification problem may be overcome in a few decades if green energy-based power plants were implemented, especially in rural locations. People can acquire green electricity at a lower cost than the grid by establishing several planned hybrid renewable-based power plants.

Job creation: Because fossil fuel technologies are often mechanized and capital-dependent, technology of renewable energy would be a more labor-oriented system than fossil fuel technologies. This means that in the renewable energy system, more occupation opportunities are generated per unit of energy obtained than in the fossil fuel system.

Fiscal impacts: Individual landowners can benefit from annual payments for lease when the solar PV cells or when the wind farms are installed on private land. Because hybrid renewable energy plants have tiny footprints, they may not have a significant impact on agricultural fields. As a result, landowners may gain from both leasing payments and agricultural work. In the case of community-owned water, on the other hand, in case of community-owned wind/solar farms, the facility's owners would be benefitted by selling electricity to the local utility.

5 Sustainability Assessment of the Hybrid Renewable Energy System (HRES)

Sustainability indicators have been proposed for qualitative and quantitative assessmentsof sustainability in the setting of rural and renewable energy. The triangle approach to Sustainable Development, which considers the three components of economic, social, and environmental sustainability, continues to be prominent. The framework of the indices of Sustainable Development is based on this approach. Various scholars have attempted to determine the elements of sustainability for rural and renewable energy initiatives. The followingarethefive dimensions/indicators proposed for rural electrification:

- Technical sustainability focused on maintaining the energy services during the economic lifetime of the initial investment.
- Economic sustainability focused on survival of the service beyond the economic lifetime of the initial investment.
- Social maintainability zeroed in on fair dispersion of the advantages presented by electrification.
- Environmental sustainability concentrated on the conservation of natural resources, avoiding degradation of the environment and preventing in-door air pollution.
- Institutional sustainability focused on survival of the organization and its capability to maintain sufficient performance with regards to the other prospects of sustainability.

Because of the intricate interaction of numerous variables such as technical, economic, environmental, social, and institutional factors, ensuring the sustainability of rural and systems of renewable energy can be a difficult undertaking. As a result, it is possible to utilize Multi-Criteria Decision Analysis to make such decisions, and Analytical Hierarchical Process is one of the MCDA tools that has been employed in this case of energy systems.

6 Results and Discussion

6.1 Case Study 1—Environmental Analysis

Process-based and input-output based LCA are the two most popular techniques to doing an LCA. The process-based LCA would be more appropriate for accurately describing the life cycle environmental implications of a given product, such as solar PV or wind power. The majority of renewable energy inputs were water use, ground resource consumption, and fossil fuel consumption. Land use impact, water impact, radon-222, Over the course of its life cycle, the hydrogen-3, radioactive chemicals, and waste heat emissions were among the model's manufacturing and operation outputs. Similarly, solar PV production methods released radon-222 and other radioactive chemicals.

Solar modules produce the majority of renewable electricity emissions, accounting for 85 percent of total emissions. Wind power module emissions are relatively low, accounting for about 15-16 percent of overall emissions.

integrated model's renewable power has a minor impact on resource use. Furthermore, the model's GHG emissions are negligible. The effects on human health and the ecology are essentially non-existent among the environmental consequences.

6.2 Case Study 2—Socio-Economic Analysis

Such an analysis is necessary to emphasize the significance of HRESs as a better way to raise rural electrification rates and, as a result, to draw decision-makers' attention to the importance of establishing appropriate policies to strengthen their promotion and implementation in the region.

The following parameters are used in the economic analysis: TAC is the total annualized price of the stand-alone system in dollars; N is the project's lifetime in years; C_{power} is the price of grid power in USD/kWh; E_{demand} is the total annual energy demand in kWh/yr. C_{inv} is the price of grid extension investment in USD/km; Com is the cost of grid extension operation and maintenance in USD/yr./km. The value added by goods and services created by the hybrid system, numerous welfare-related metrics designed to quantify economic performance, employment, and other aspects are all taken into consideration in social analysis.

6.3 Case Study 3—Technical Analysis

The portion of solar energy accessible on the surface area is crucial information for planning and analysis of a conversion (transformation) system of solar energy. The total quantity of solar energy that extends up to a certain unit surface area

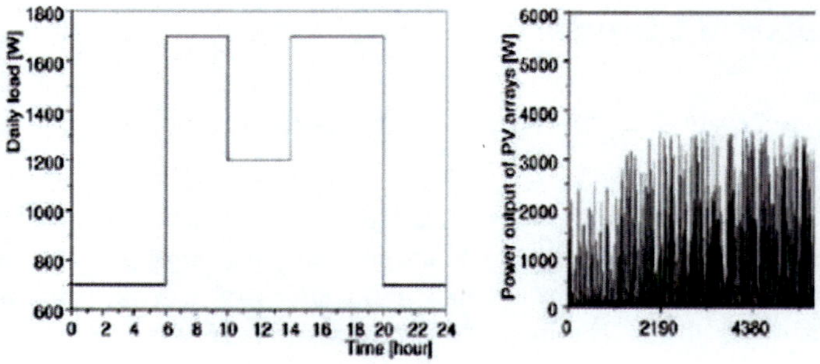

Fig. 1 The Daily Load Profile and Power output of PV array [10]

(irradiation, radiation energy) in a given time interval is an essential aspect from a practical standpoint (hour, day, month, and year).

Geographic latitude, season, direction, surface inclination, and weather conditions all influence the quantity of energy released. For wind data, average velocities of wind at different heights were considered for various locales.

The system in issue uses a variety of PV module configurations, as well as turbines with varying power outputs. Dynamical simulations of a solar PV or wind energy system with 20 solar PV modules (about 28m^2) and wind turbines of 1 kW, 3 kW, and 5 kW were performed. The daily load profile with respect to the PV array power output is given by Fig. 1.

It may be stated that the PV/wind system is appealing for usage in residential settings. Hydrogen storage is a strong option in terms of ease of use and cost effectiveness, and it can be utilized to generate steady electricity regardless of the weather. It is possible to achieve consistent electricity across a large area by using special equipment for electrical storage, such as batteries or hydrogen-based systems.

6.4 *Sustainability Assessment Result*

According to the findings, factors such as "energy availability," "reliability," "capital investment cost," "operational and maintenance cost," "technology compatibility for various end uses," and "operational and management capacity required" all play a significant role in the long-term viability of energy technologies.

According to many publications examining qualitative aspects that contribute to the long-term viability of rural and renewable energy projects, the following are the characteristics of successful and long-term renewable energy projects:

- Selecting appropriate technology through feasibility studies and surveys.
- Applying renewable energy with income-generating activities and partnering with groups of livelihoods.
- Giving availability to financing and micro credit.

Table 1 Life cycle impacts of the integrated solar PV and wind power models [9]

Impacts	Results
Human health (DALYs)	0.00086
Ecosystem (species/year)	0.0000020 9
Resource consumption (USD2013)	0.0021

- Providing the capacity and investments for local institutions.
- Motivating active involvement (and feedback) from groups.
- Enforcing technical standards and certifications.

Solar and other emerging technologies have a low sustainability score due to their high capital costs and low-capacity factors. Although the sustainability performance of standalone renewable energy systems has improved as a consequence of considerable capital cost reductions and increases in conversion efficiency and energy availability, it is likely to move forward to improve in the near future. However, hybrid systems that combine many sources of renewable energy have a considerably higher sustainability index and are a promising alternativetotraditional fossil-fuel-powered energy systems.

7 Conclusion

The scientific community has discovered untapped potential for hybrid renewable energy systems (HRES) through an analysis of advancements in system design. Currently, system design is mostly driven by technical considerations, with some economic analysis of return on investment thrown in for good measure. The need for cheap decision-making tools that integrate technical, economic, and environmental factors would help to alleviate the stress. Those identified future innovation opportunity areas would have a number of advantages, including a multidisciplinary approach to developing innovative business models and other enabling areas, greater integration of real-world cases in research,andincorporating environmentalbenefitsinto decision-making to encourage more investments.

Because hybrid system performance is dependent on weather conditions, transient study of the system for varying constraints such as solar radiation, wind velocity, and load demand is required to reduce issues associated to system dependability and operation. Because the price of energy (COE) sources employed in hybrid systems is so expensive, subsidies from the central and state governments are required.

Table 2 Emission of the studied integrated model, compared to on-grid electricity power [9]

	Electri city genera tion (kWh/ year)	Life cycle GHG emission factor ($kgCO_2$/ kWh)	Lifet ime (year)	Total GHG emiss ion (kgC O_2)
Solar	75,579	0.02	25	37,78
PV			years	9
modul				
e				
Wind	5,256	0.00377	25	495
power		2	years	
modul				
e				
On-gri	80,835	0.913	25	1,845,
d			years	058
electri				
city*				

References

1. Raghuwanshi SS, Arya R (2019) Economic and reliability evaluation of hybrid photovoltaic energy systems for rural electrification. Int J Renew Energy Res-IJRER
2. Murali R, et al (2015) Socio-technical assessment of solar photovoltaic systems implemented for rural electrification in selected villages of Sundarbans region of India. AIMS Energy
3. Quyen Le LUU et al (2020) Life Cycle Assessment (LCA) of an integrated solar PV and wind power system in Vietnam. J Asian Energy Stud
4. Yimen N et al, Optimal sizing and techno-economic analysis of hybrid renewable energy
5. Systems—a case study of a photovoltaic/Wind/Battery/Diesel System in Fanisau, Northern Nigeria. Processes (2020)
6. Bakić VV et al (2012) Techical analysis of photovoltaic/wind systems with hydrogen storage. Thermal Sci 865–875
7. Saheb-Koussa D et al (2011) Economic and Environmental analysis for grid-connected hybrid photovoltaic-wind power system in the arid region. Energy Procedia
8. Ghonima F, Ezzat M, Abdel-Salam TS (2020) Economical and technical assessment of hybrid renewable energy systems for optimal performance applied in Hurghada, Egypt. Int J Recent Technol Eng (IJRTE) 9:1
9. Mainalin B, Semida S (2013) Using a sustainability index to assess energy technologies for rural electrification. Elsevier
10. Torres A et al (2018) Technical and economic feasibility survey for wind and photovoltaic hybrid renewable energy system. A case study in Neiva-Huila, Colombia. Contemp Eng Sci 11:42
11. Dhital RP et al (2014) Framework for sustainability assessment of renewable energy projects in Nepal. Int J Anal Hierarchy Process

Assessment of Turbulence Models in Predicting the Mixing Characteristics of Circular Twin Jets

Chitimada Narendrakumar and K. P. Sinhamahapatra

Abstract In this paper, a computational fluid dynamic investigation of twin circular orifice jets was conducted for an orifice-to-orifice distance of 3.75 mm and a Mach number of 0.8. The RANS-based turbulence models are evaluated based on examining the jet properties like streamwise velocity, rate of decay, jet half-width and rate of spread. The results show that the *K-ω SST* turbulence model was effective in predicting the mixing properties like potential core, decay rate, jet half-width and spread rate. Numerical data obtained from the Reynolds stress model and K-ε turbulence models agree well with experimental data in the core region of the jet. In contrast, the transition and self-similar regions are better predicted by the shear stress transport *K–ω* turbulence model. Overall, the *K-ω SST* turbulence model was close in agreement with the experimental measurements in the near field and far field of the jet flow field.

Keywords Twin jets · Mixing · Turbulence models · Decay rate · Half-width

1 Introduction

The mixing behaviour of twin jets has been investigated experimentally and numerically for several years. Twin jets have various industrial uses including in boilers, gas turbines, combustors, thrust vectoring and stealth apart from applications such as mixing enhancement and noise suppression. A better imagination and understanding of the flow field of a dual jet can consequently help to control the effects of exhaust pollutants. In recent years, many researchers have applied experimental and numerical techniques to understand the mixing phenomena of twin jet flows. From previous research, it was identified that the mixing properties and the turbulent intensities of the flow depend on the exit Reynolds number (*Re*), the change in nozzle topology and the *S/D* between the nozzles [1, 2].

C. Narendrakumar (✉) · K. P. Sinhamahapatra
Department of Aerospace Engineering, IIT Kharagpur, Kharagpur, West Bengal 721302, India
e-mail: chnarendrakumar06@gmail.com

S. Das et al. (eds.), *Proceedings of the 1st International Conference on Fluid, Thermal and Energy Systems*, Lecture Notes in Mechanical Engineering,
https://doi.org/10.1007/978-981-99-5990-7_19

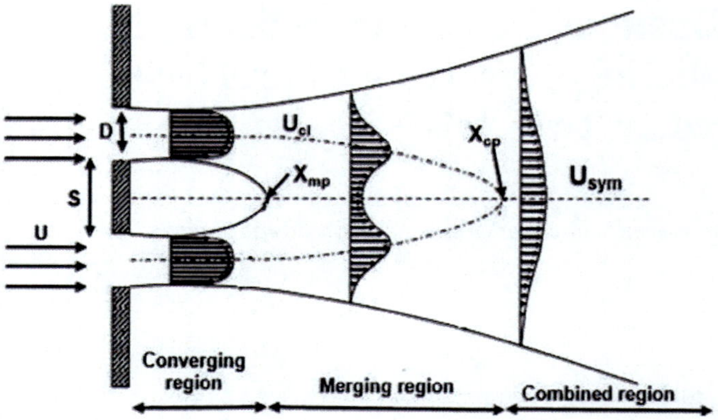

Fig. 1 Flow features of circular twin jet

The typical flow structures of the twin jet flow field are shown in Fig. 1. Understanding the twin jet flow characteristics is complicated compared to single jet, and the complexity is arising because of influence of each other on other negative static pressures are observed in the near field of a twin jet, causing the two jets to combine downstream [3]. This leads to three basic region formations in twin jets, which are named as converging, merging, and combined regions. The converging area is defined as the area between the nozzle exits to the location of the convergence point. The direction of the streamwise velocity changes from positive to negative at some point in the flow field, and it is termed as merging point [3, 4]. However, the definition does not apply to small distances between the nozzles, since they are ineffective for the formation of the recirculation area at the nozzle exit. The merging region lies in between the merging point and the combined point. After the merge region, the two jets behave as a single jet, and the velocity profiles look similar. The location where twin jets have completely merged into a single jet is called the combined point. The region after the combined point is called the combined region.

2 Literature Review and Objective

Many researchers have been investigating the effects of nozzle geometry, Reynolds number and spacing-to-diameter (*S/D*) ratio on the mixing properties of dual nozzles in recent years [2, 3]. Miller and Comings [3, 5] studied the mixing properties of plane twin jets at a specific distance and *Re*. They reported that a negative gauge pressure develops at the nozzle exit due to the pumping effect between the nozzles.

Anderson et al. [6] studied *S/D* variation at combined and merging points of twin jets. It has been found that the turbulent intensities of the combined jet increase progressively as the *S/D* ratio increases. Anderson et al. [6] evaluated the *K*-ε, *SST K*-ω and *RSM* turbulence models in predicting the mixing properties of twin jets.

They suggested that *K-ω* and *RSM* models are adequate for simulating twin jets. Nasr and Lai [7] also correlated the ability of standard *K-ε*, *RNG* and *RSM* models for 2*D* parallel twin jets. It was observed that the *RSM* model comes closest to the experimental results than the other two models and that the *K-ε* method slightly over-predicted the experimental results. Laban et al. [8] studied the *S/D* effects on circular twin jets at fixed *Re*. It was found that increasing the spacing between individual jets reduced the jet propagation rate and decreased the turbulent flow properties.

In recent years, several researchers have suggested the numerical technique of RANS modelling is good and accurate enough to understand the mixing properties of jets [6, 7]. Therefore, the present study uses the RANS approach over computationally intensive methods such as DNS and LES. So based on the availability of CFD tools in the present research fields, the present work aims to evaluate the three turbulence models for predicting the mixing properties of twin circular jets at a Mach number (*M*) of 0.8 and a Reynolds number of 2.74×10^5. The present discussed numerical simulations are performed in ANSYS Fluent 19.2 commercial software.

3 Numerical Methodology

Numerical simulations performed in the present investigation for circular nozzle are the same as those by Anderson et al. [6] investigated experimentally. The diameter (*D*) of the nozzle is 15 mm and the spacing distance (*S*) between the twin circular orifices is 3.75 mm. The circular jet models have an exit area of 176.625 mm^2 and mass flow rate of 0.06 kg/sec. Figure 2 shows the present studied twin circular jet geometry with relevant notations. In the present investigation, the twin jet numerical simulations are performed in the rectangular computational domain of size $30D \times 5.3D \times 5.3D$ with a mesh structure, as shown in Fig. 3. Jet *M* of 0.8 is considered as it is close to the actual exhaust jet velocity of a commercial jet airliner while taking off, and the corresponding jet exhaust velocity is 277.60 m/sec. The pressure-based coupled solver is used for steady-state RANS simulations. Air, as an ideal gas, is used as the jet fluid and viscosity calculation based on Sutherland's three coefficient method. The pressure and temperature for the domain are considered as ambient conditions, i.e., 101,325 Pa and 300 K. At the inlet of the domain, the pressure inlet boundary condition is given with a stagnation pressure of 154,453.80 Pa and a stagnation temperature of 300 K to simulate an adiabatic jet of M 0.8. At the outlet atmospheric conditions, i.e., the static pressure of 101,325 Pa and a temperature of 300 K are set. Far-field conditions are specified with ambient pressure and temperature of 101,325 Pa and 300 K.

Fig. 2 Twin circular jet

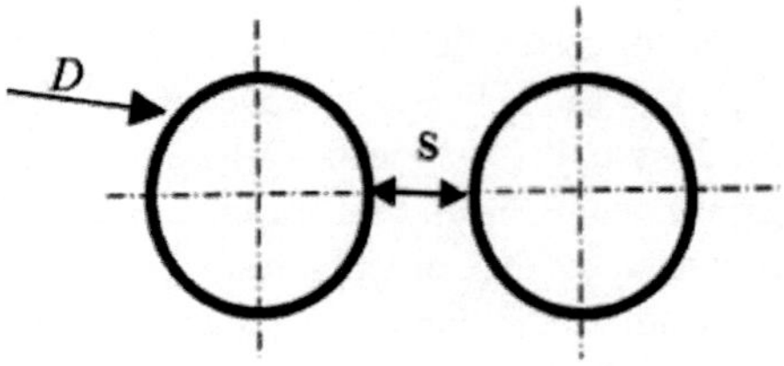

Fig. 3 Computational domain with boundary conditions

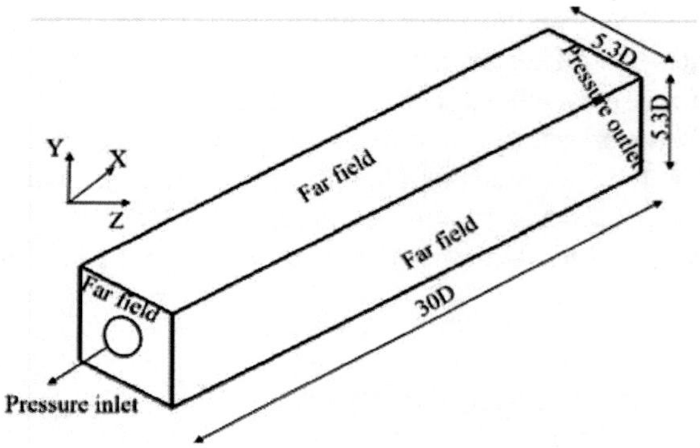

3.1 Turbulence Modelling

The *K-ε, K-ω SST* and *RSM* turbulence models are used in the present investigation to evaluate the mixing properties of circular twin jets. Later, the results obtained are validated with Anderson et al.'s experimental work to identify an appropriate turbulence model for further investigation. The present studied turbulence models are chosen from previous literature studies. The *K-ε* (rate of dissipation) turbulence model is commonly employed to simulate the mean properties in turbulent flows, and it is suitable for flows with relatively small pressure gradients and laminar flow fields. For the flow field with strong adverse pressure gradients or flow separations, this model is not preferable. The main advantage of the *SST K-ω*, according to Menter [9], is that it uses the *K-ω* model in the jet core field and shifts to the *K-ε* model in the far fields and mixing regions. In particular, the *SST K-ω* model is good and accurate for flows involving adverse pressure gradients, shock waves and transonic flows. Turbulent kinetic energy is denoted by *K*, and a specific dissipation rate is denoted by ω in the *K-ω SST* turbulence model. For complex turbulent flows, which includes flow separation and high degree of anisotropy, recirculation zones and the influence of mean rotational effects, the Reynolds stress model is much preferable.

4 Results and Discussions

The closely spaced twin jets are practically used in military and fighter aircraft for stealth applications. Whenever two jets enter the atmosphere from the nozzle exit, the interaction between individual jets rapidly occurs. Numerical simulations are done for twin circular free orifice jets to evaluate the three turbulence models by comparing them with the experimental work of Anderson et al. [6]. The centreline velocity (U_{cl}) and symmetry line velocity (U_{sym}) of the circular twin jet are normalized with the exit velocity (U_e) of the orifice and plotted against the axial (X) direction to comprehend the mixing characteristics. Statistical similarities between the two (top and bottom) jets around the line of symmetry allow the top jet to be used to estimate the potential core (*PC*), decay rate, half-width, and propagation rate.

4.1 Contours of Mean Velocity, Decay Rate, Half-Width

Contour plots of streamwise mean velocity are shown in Fig. 4 for three turbulence models. For all turbulence models, vena contracta effect is observed (Fig. 4) as there is acceleration of jet. Hence, the maximum value of velocity is attained just after the jet exit. The jet expands transversely due to the entrainment of ambient fluid into the shear layers, hence the spacing between contour planes increases with distance downstream.

To capture the flow regimes of a twin jet flow, *K-ε, K-ω SST* and *RSM* turbulence models are compared with the experimental work due to Anderson et al. [6]. The

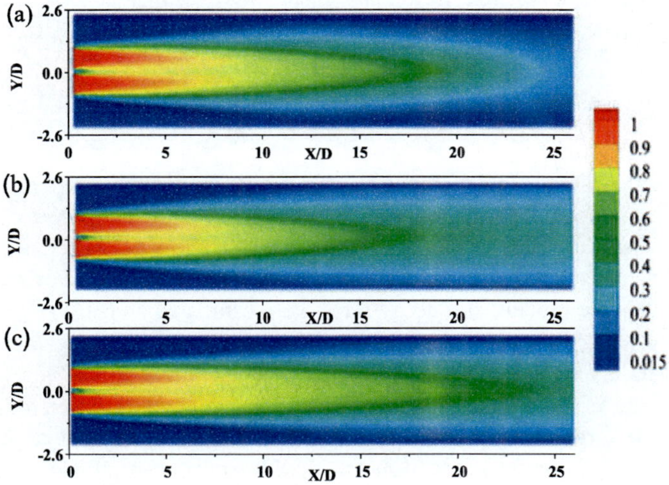

Fig. 4 Mean velocity contours (**a**) K-ε, (**b**) K-ω SST, (**c**) RSM

Fig. 5 Centreline velocity of circular twin jets

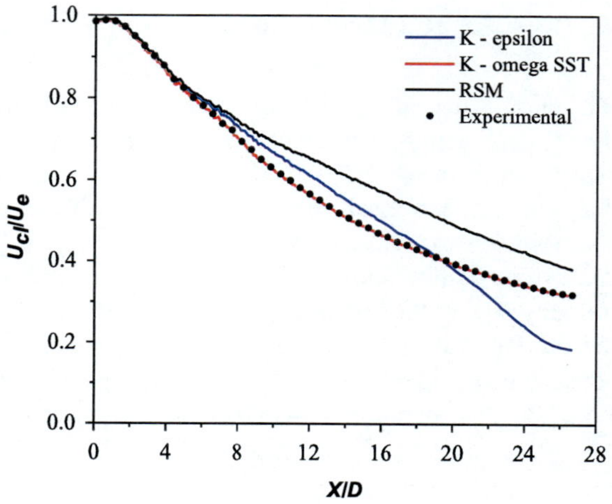

expansion of turbulent jets starts from the atmospheric fluid entrainment into the jet core. The enhanced mixing of jet with the quiescent fluid can be indicated by decay of centreline velocity and spreading of jet. Figure 5 shows the upper jet centreline normalized velocity. The distance between the jet exit and the point at which the centreline velocity is 0.98 U_e is defined as *PC* length [6], which is normally used to describe the near-field jet mixing capabilities. The *PC* region exists up to a length of *X/D* = 4 for all investigated turbulence models. Still, from the transition region onwards, the *K-ε* and *RSM* models are shown over the prediction of jet velocity compared to the experimental work of Anderson et al. [6]. However, the *K-ω SST* turbulence model agreed with the experimental results in all jet regions.

The decay rates of the present circular twin jets were estimated in a similar fashion of free single jets decay rate estimation that is fitting Eq. (1) to the U_{cl} decay profile:

$$\frac{U_e}{U_{cl}} = K_u\left(\frac{x - x_k}{D}\right) \quad (1)$$

where K_u and x_k are the rate of decay and the kinematic virtual origin, respectively. Figure 6 shows an initially flat area near the jet exit that characterizes the *PC*. Because of the atmospheric fluid entrainment into the core of the jet, a faster decay in the centreline velocity can be observed immediately next to the *PC*. For the present studied turbulence models, the decay rates were estimated as 0.16, 0.08 and 0.06 respectively with in the *X/D* range of 5 ≤ *X/D* ≤ 28. *K-ω SST* and *RSM* models showed better agreement with the experimental decay rate of 0.09.

Figure 7 shows the circular twin jet symmetry line velocity variation for *K-ε*, *K-ω SST* and *RSM* turbulence models. From Fig. 7, it is evident that out of the chosen models, the *SST K-ω* model is well-matched with Anderson et al. [6]. At the same time, *K-ε* and *RSM* turbulence models are good in predicting the mixing properties of a twin jet only in the jet core field. From *X/D* = 10 onwards, these two models are

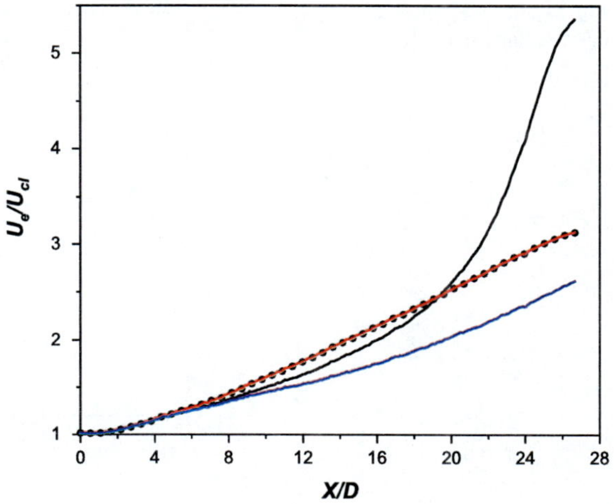

Fig. 6 Velocity decay for the upper jet

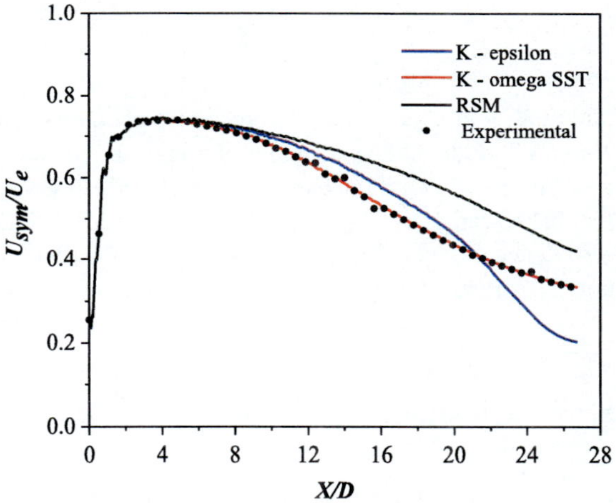

Fig. 7 Symmetry line velocity of circular twin jets

not matched with Anderson et al. [6] and show over-prediction of mixing properties in the downstream jet locations. Therefore, the *SST K-ω* turbulence model was used to perform the CFD simulations for the present study.

For quantifying the mixing properties of jet, the spreading rate is also used in addition to velocity decay along the symmetry line and centreline. The turbulent jet spread rate can be calculated by plotting the jet's half-velocity width along the non-dimensional length (*X/D*). The normalized half-velocity width profiles for the three-turbulence model, along with Anderson et al. [6], are shown in Fig. 8. The distance from the centreline of the jet to the point where the mean axial velocity drops to half is called the half-velocity width [5, 6]. For the present study, the quantification of spread rate is achieved by fitting Eq. (2) to the half velocity width profile:

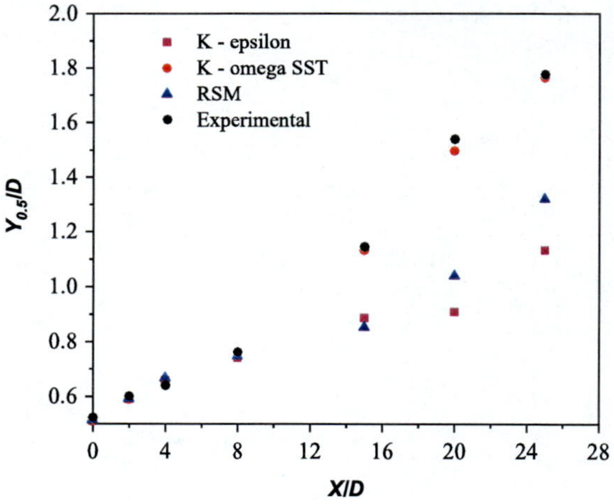

Fig. 8 Half-width of a twin circular upper jet

$$\frac{Y_{0.5}}{D} = K_s\left(\frac{x - x_{0s}}{D}\right) \tag{2}$$

where K_s and x_{0s} *are* rate of spread and geometric virtual origin, respectively. The *K-ω SST* model shows a sudden increase in spreading (Fig. 8) after exiting the jet and agrees well with the experimental work. However, compared to Anderson et al. [6], in the range of $3 \leq X/D \leq 26$, the spread of Anderson et al. [6] and *K-ω SST* turbulence model is 0.056, whereas that for *K-ε* and *RSM* turbulence models are 0.02 and 0.03, respectively.

5 Conclusions

A computational fluid dynamics study of jets with two circular orifices was performed using three *RANS*-based turbulence models *K-ε*, *K-ω SST* and *RSM*. The results show that from the selected turbulence models, *K-ω SST* turbulence has been shown to be effective in predicting jet properties in both the near field and far field. At the same time, turbulence models *K-ε* and *RSM* are fairly good at predicting the mixing properties of a twin jet only in the near field. From $X/D = 10$ onwards, these two models are not matched with Anderson et al. [6] and show over-prediction of mixing properties in the downstream jet locations.

Acknowledgements The first author is very much thankful to PARAM Shakti, IIT Kharagpur, and MHRD for computational support and funding.

Nomenclature

D	Exit diameter of the circular nozzle [mm]
K	Turbulent kinetic energy J/kg
M	Mach number –
Re	Reynolds number –
S	Spacing distance [mm]
X	Axial distance from the orifice exit [mm]
E	Rate of dissipation [m^2/s^3]
ω	Specific rate of dissipation [1/s]

References

1. Deo RC, Mi J, Nathan GJ (2008) The influence of reynolds number on a plane jet. Phys Fluids 20(7)
2. Harima T, Fujita S, Osaka H (2005) Turbulent properties of twin circular free jets with various nozzle MACH. In: ERCOFTAC international symposium on engineering turbulence modelling and measurements, pp 501–510
3. Miller D, Comings E (1960) Force-momentum fields in a dual-jet flow. J Fluid Mech 7(2):237–256
4. Mi J, Nathan G (2010) Statistical properties of turbulent free jets issuing from nine differently shaped nozzles. Flow Turbul Combust 2010:583–614
5. Quinn WR (2007) Experimental study of the near field and transition region of a free jet issuing from a sharp edged elliptic orifice plate. Eur J Mech B/Fluids 2007:583–614
6. Anderson E, Spall ER (2001) Experimental and numerical investigation of two dimensional parallel jets. J Fluids Eng 123:401–406
7. Nasr A, Lai JCS (1997) Two parallel plane jets: mean flow and effects of acoustic excitation. Exp Fluids 22(3):251–260
8. Laban A, Aleyasin SS, Tachie MF, Koupriyanov M (2019) Experimental investigation of nozzle spacing effects on characteristics of round twin free jets. J Fluids Eng 141(7):281–294
9. Menter FR (1994) Two-equation eddy-viscosity turbulence models for engineering applications. AIAA J 32(8):1598–1605

Investigation of Spray Behaviour Simulation to Predict Spray Tip Penetration Under Ultra-High Injection Pressure

Mehul P. Bambhania and Nikul K. Patel

Abstract The spray characteristics, in particular, the spray tip penetration and atomization of the droplet are known to affect the fuel combustion in the diesel engine. Researchers have enough understanding of the diesel spray characteristics, but less frequent literature is available for biodiesel with ultra-high injection pressure. The non-evaporating spray simulation has been carried out for a single-hole fuel injector nozzle with a 0.16 mm diameter. The diesel fuel has been considered to be injected with 300 MPa injection pressure. The discrete phase model (DPM), which is based on the Lagrangian drop Eulerian fluid method, is used to simulate spray characteristics. The discrete random walk (DRW) model has taken into account the turbulent diffusion of the spray droplet. The spray tip penetration (STP) length has been calculated with the Kelvin–Helmholtz Rayleigh–Taylor (KHRT), Wave and Stochastic Secondary Droplet (SSD) spray breakup models and compared with the existing experimental data. The Wave model and experimental findings show good agreement. Three different nozzle diameters (0.16 mm, 0.25 mm and 0.3 mm) have been used to compare the L/d ratio with STP length. Finally, the spray formation with biodiesel exhibits slightly better spray penetration but suppressed atomization as compared to diesel fuel.

Keywords Ultra-high pressure · Diesel spray · Spray breakup model · Biodiesel spray · Discrete phase model

1 Introduction

The rapid growth in automobiles increases energy demand and carbon emissions, which leads to the risk of climate change. It is important to understand the spray characteristics of new fuel for effective power output and fulfil the new emission

M. P. Bambhania (✉) · N. K. Patel
Department of Mechanical Engineering, Faculty of Technology & Engineering, The Maharaja Sayajirao University of Baroda, Vadodara, Gujarat, India
e-mail: mehul.bambania-med@msubaroda.ac.in

S. Das et al. (eds.), *Proceedings of the 1st International Conference on Fluid, Thermal and Energy Systems*, Lecture Notes in Mechanical Engineering,
https://doi.org/10.1007/978-981-99-5990-7_20

regulations. Researchers have enough understanding of the diesel spray characteristics, but less frequent literature is available for biodiesel with high injection pressure. The spray characteristics can be understood by two approaches: experimental and numerical. The macroscopic parameter like the spray tip penetration (STP) and the spray cone angle (SCA) concerning injection time can be measured with the use of a high-speed camera with image processing software. The microscopic parameter like velocity distribution and droplet size distribution can be obtained with the help of the Phase Doppler particle analyzer. For example, Payri et al. [1] reported macroscopic parameters, i.e. spray tip penetration and spray cone angle for biodiesel. In comparison to diesel, biodiesel exhibits longer spray penetration and a smaller spray cone angle. Wang et al. [2] carried out a detailed experimental study for the macroscopic parameter of spray for palm oil and used cooking oil at a high injection pressure of 300 MPa. A light scattering method was used to visualize the fuel spray. He observed that biodiesel gives a longer injection delay and penetration length. Compared to diesel, biodiesel has a narrower spray cone angle. Mohan et al. [3] investigated spray tip penetration, spray angle and spray velocity for waste cooking oil. They reported a reduction in spray tip penetration with higher ambient pressure. The spray shapes are different for diesel and biodiesel due to cavitation inside the fuel injector nozzle. Recently, Moon et al. [4] used the X-ray phase-contrast imaging technique, which enabled optical excess in the dense region of the fuel spray near the nozzle tip. However, microscopic investigation of spray with the use of experimental methods is still complicated and expensive.

On the other hand, several numerical methods with different spray models have been proposed by the researcher. The most common breakup models are Taylor Analogy Breakup (TAB), Enhanced Taylor Analogy Breakup (ETAB), Wave and Kelvin–Helmholtz Rayleigh–Taylor (KH-RT). However, spray characteristics have a close connection with the internal flow of the fuel nozzle. The linkage between internal flow and spray formation is essential; internal flow is examined by the Eulerian frame of reference, and spray formation is examined by the Lagrangian frame of reference. The development of computational fluid dynamics (CFD) facilitates to conduct spray simulation with the Eulerian–Lagrangian scheme in various commercial software, like KIVA, AVL-Fire, ANSYS Fluent, Converge, Open-FOAM, etc. The atomization of fuel involves the liquid breaking up into droplets through a primary and secondary breakup. Assessment of primary and secondary breakups is important in non-evaporating spray modelling. At the interface of two fluids, the breakup process is controlled by Kelvin–Helmholtz (KH) and Rayleigh–Taylor (RT) instability. High shear forces at the interface cause KH instability, while acceleration in the plane's normal direction causes RT instability (Fig. 1).

Reitz et al. [5, 6] introduced a blob injection model, in which liquid drops with a diameter equal to the nozzle diameter injected and droplet break up followed by KH-RT instability. Sazhin et al. [7] modified the Wave model, taking into account the transient process during the spray injection and implemented it in KIVA II. However, the Wave model, which independently accounts for the breakdown of the liquid core into droplets and their ensuing aerodynamic breakup, was further improved by Turner

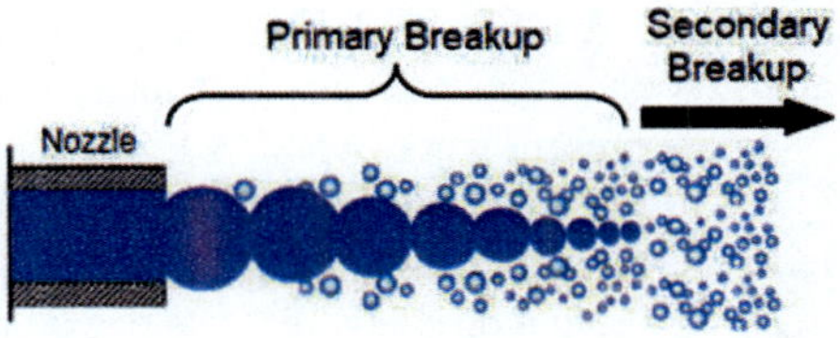

Fig. 1 Primary and secondary spray breakup

et al. [8]. Som et al. [9] develop a new primary breakup model KH-ACT, incorporating the effect of cavitation, turbulence and aerodynamic breakup and implemented it in CONVERGE. They also performed comprehensive validation of a new model for both evaporating and non-evaporating spray. Som et al. [10] performed a detailed 3D simulation for internal nozzle flow with the use of FLUENT and spray simulation in CONVERGE for soy-biodiesel. The two simulations were coupled through the KH-ACT model. Biodiesel exhibits poor atomization characteristics, the STP and the Sauter mean diameter (SMD) are increasing, and, however, cone angle and dispersion are reduced as compared to that of the diesel. Som et al. [11] also investigate the effect of orifice geometry on the inner nozzle flow and spray formation with the use of biodiesel. They reported that the conicity reduces the cavitation and turbulence, which result in an increase in SDM and STP. Battistoni et al. [12, 13] used AVL-Fire to perform a spray simulation with two different nozzles with cylindrical and conical holes. Results of STP, SDM and cone angle have been compared for diesel and SME-biodiesel. Spray characteristics of SME-biodiesel are observed to be less sensitive to the orifice shape, compared to the diesel. Yousefifard et al. [14] performed numerical simulation with the use of Open-FOAM for high injection pressure conditions, similar to that of the Wang et al. [2] experiment. The spray characteristics with different biodiesel (produced from Rapeseed oil) blend (B15, B50) have been investigated by Vajda et al. [15] with the help of AVL-Fire. Agarwal et al. [16] evaluate spray characteristics of Jatropha and Karanja biodiesels through experimental and numerical investigation. They found that Karanja biodiesel provides better atomization and evaporation behaviour compared to Jatropha biodiesel. Mohan et al. [17] used KIVA-IV to investigate spray characteristics of dimethyl ether (DME) and diethyl ether (DEE). The low viscosity and density of the ether fuel result in short STP. Thulasi et al. [18] investigate the effect of injection pressure and chamber temperature of spray characteristics of diethyl ether (DEE) with the use of FLUENT. The spray tip penetration, droplet distribution and vapour formation have been calculated. The STP and SDM were found to be smaller compared to that of diesel and the vapour formation if more compared to diesel because of the lower boiling point. Ishak et al. [19] performed a numerical study on the spray characteristics with different biofuel blends with conical and cylindrical nozzle orifices. They found that diesel is more sensitive to the nozzle geometry in terms of STP, SDM and cone angle compared to biofuels. The fuel properties (i.e. density, viscosity and surface tension) have a significant impact on SDM. Geng et al. [20] conducted an experimental and numerical study of macroscopic and microscopic spray characteristics with a biodiesel-ethanol fuel blend. A high-speed camera captured the STP, and a Malvern laser particle size analyzer determined the size distribution of the droplets and SDM.

In the present work, ANSYS FLUENT was employed to study non-evaporating spray. FLUENT provides a discrete phase model (DPM) to simulate spray characteristics, which is based on the Lagrangian drop Eulerian fluid method. It allows for tracking the fuel droplets (discrete phase) in the Lagrangian frame of reference and the continuous gas phase in the Elerian frame of reference. Furthermore, the Wave breakup model, proposed by Reitz et al. [5], has been used to capture droplet breakup phenomena. The Wave model is suitable for high Weber numbers (We > 100) and high-pressure spray. When the fuel droplets travel in the chamber, they experience the turbulent vortex, which causes the motion of the particles to shift from straight lines to non-smooth curves. The interaction between the droplet and discrete vortex is taken into account by the discrete random walk (DRW) model. The discrete random walk model was used to simulate the turbulent diffusion of the spray droplets.

2 Numerical Modelling of Spray

2.1 Computational Domain and Boundary Condition

The computational geometry used for the present study is based on a single hole injector experimentally studied by Wang et al. [2]. For validation purposes, a nozzle with 0.16 mm diameter and 1 mm in length is used to inject the diesel at the rate of 2.225 gm/s at 300 MPa injection pressure. The fuel injection chamber is symmetrical in shape, hence simulation is carried out with an axi-symmetric computational domain which is created in the ANSYS Design modeller. The discretization of the domain is done with the use of ANSYS ICEM, which allows the creation of fine structural mesh in the centre core of the chamber shown in Fig. 2.

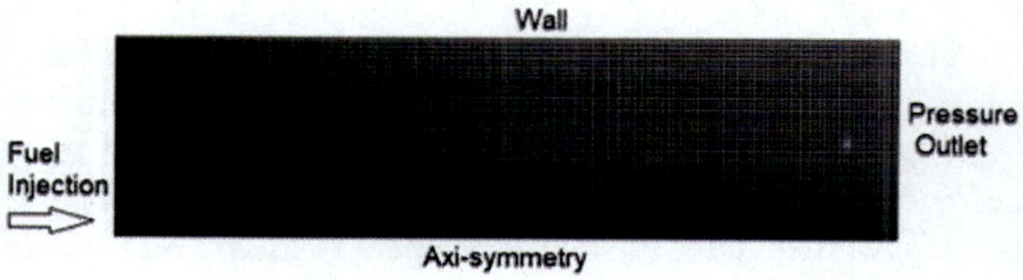

Fig. 2 Boundary condition with detail mesh geometry

2.2 Governing Equations

2.2.1 Turbulent Modelling

The Reynolds Averaged Navier–Stokes equations (RANS)-based model is used to simulate turbulent flow in ANSYS Fluent solver. It takes lesser computational time compared to Large Eddy Simulation (LES). This model considers only time-averaged properties and neglects the instantaneous values:

$$\frac{\partial \rho}{\partial t}+\frac{\partial}{\partial x_i}(\rho \overline{u_i})=0 \tag{1}$$

$$\frac{\partial}{\partial t}(\rho \overline{u_i})+\frac{\partial}{\partial x_j}(\rho \overline{u_i u_j})=-\frac{\partial \overline{p}}{\partial x_i}+\frac{\partial}{\partial x_j}\left[\mu\left(\frac{\partial \overline{u_i}}{\partial x_j}+\frac{\partial \overline{u_j}}{\partial x_i}-\frac{2}{3}\partial_{ij}\frac{\partial \overline{u_l}}{\partial x_l}\right)\right] + \frac{\partial}{\partial x_j}\left(-\rho \overline{u_i\prime u_j\prime}\right) \tag{2}$$

Note the Reynolds stress term $\left(-\rho \overline{u_i\prime u_j\prime}\right)$ which requires additional modelling to close the RANS equations.

The energy equation can be given as

$$\frac{\partial}{\partial t}(\rho E)+\frac{\partial}{\partial x_i}[u_i(\rho E+p)]=\frac{\partial}{\partial x_j}\left(k_{eff}\frac{\partial T}{\partial x_j}+u_i\left(\tau_{ij}\right)_{eff}\right)+S_h \tag{3}$$

where τ_{ij} is stress tensor, andS_h is the source term due to evaporation, momentum and heat transfer.

2.2.2 Discrete Phase Model

The transport equation is solved for continuous flow in the Euler frame of reference, while discrete phase particles are used to simulate spray formation in a Lagrangian frame of reference. The spherical particle introduced into the continuous phase and trajectory are calculated by two-way coupling. The trajectory of the discrete particles is calculated by the force balance on particles given by

$$\frac{\partial u_p}{\partial t}=\frac{3}{4}\frac{\rho_a}{\rho_p}\frac{\left(u_a-u_p\right)\left|\left(u_a-u_p\right)\right|}{d_p}+g_x\frac{\left(\rho_p-\rho_a\right)}{\rho_p} \tag{4}$$

where $\rho_p$andρ_a are the density of particles and air respectively, $u_p$andu_a are the velocity of particles and air, μ is the molecular viscosity of the fluid in MPa and d_p is particle diameter in mm.

2.2.3 Spray Breakup Model

Spray breakup can be resolved by the blob injection model, in which liquid droplet, diameter equal to the nozzle diameter, is experiencing the Kelvin–Helmholtz (KH) and Rayleigh–Taylor (RT) instability. Kelvin–Helmholtz instability is exhibited due to relative velocity between the gas and liquid phases, creating shearing action on the droplet. As a result, the parent droplet breaks into small child droplets. These child droplets experience secondary breakup due to the combined effect of KH and RT. Rayleigh–Taylor instability generates due to the inertia of the denser fluid. The droplet breakup process is similar in both KH and HT. As the Weber number increases, faster aerodynamic effects lead to increases in the formation of the smaller droplets. The Wave model was used to calculate the aerodynamic effects on droplet breakup. The radius of the parent droplet (r) decreases continuously during the break up until it reaches a stable radius (r_{KH}):

$$\frac{dR}{dt} = \frac{r - r_{KH}}{\tau_{KH}} \tag{5}$$

The wavelength of the unstable surface is thought to be precisely proportional to the radius of the stable droplets:

$$r_{KH} = B_o \Lambda_{KH} \tag{6}$$

where Λ_{KH} is the wavelength, $B_o = 0.61$ is model constant and τ_{KH} is the characteristic breakup time of the droplet given as

$$\tau_{KH} = \frac{3.726 B_1 r}{\Lambda_{KH} \Omega_{KH}} \tag{7}$$

B_1 is an adjustable model constant having a range 20 (evaporating) to 60 (non-evaporating), where Λ_{KH} is the wavelength and Ω_{KH} is the frequency.

3 Results and Discussion

3.1 Effect of Mesh Size

The results obtained with the use of computational fluid dynamics always depend on mesh size. So it is essential to refine the mesh, so that solution results are not dependent on mesh size. Moreover, sufficient grid size is essential to simulate discrete phase particles. Four different mesh sizes had been used to perform the grid independence study shown in Fig. 3. With the case, D total number of mesh elements is

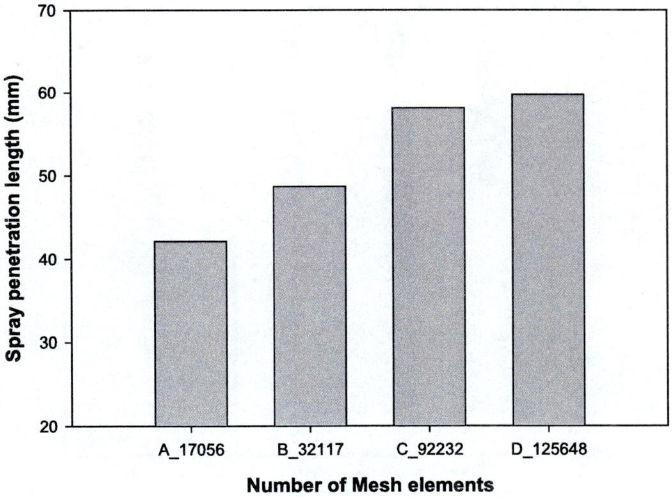

Fig. 3 Effect of mesh elements on spray penetration length

125648 with a minimum element size of 1 μm has been used for further study. The time step of 1×10^{-6} s has been considered to avoid solution instability.

3.2 Effect of the Spray Breakup Model

The atomization of the fuel is a very complex phenomenon, which required a very expensive instrument to diagnostics of its spray characteristics, specifically with high injection pressure. Due to this reason, very little experimental research are available, which will provide the inside detail of spray formation. In the present study, the spray breakup model has been validated with the experimental work carried out by Wang et al. [2]. For validation purposes, a nozzle with a 0.16 mm diameter is used to inject the diesel at the rate of 2.225 g/s at 300 MPa injection pressure. Most of the literature describes the spray tip penetration to study spray characteristics. Spray tip penetration can be defined as the farthest axial distance of the spray boundary from the nozzle tip. The spray tip penetration (STP) of the diesel spray at different injection times has been calculated. The development of spray has been captured from 1 to 10 ms with the time interval of 2 ms shown in Fig. 4.

To compare the capability of different spray models, the SSD, the Wave and the KHRT models have been implemented. STP grows more quickly in the beginning of the injection and less rapidly in the closing stages. The results obtained at the beginning of the fuel injection are quite reasonable with the experimental trend. The KHRT and the Wave model predict the penetration length reasonably well as

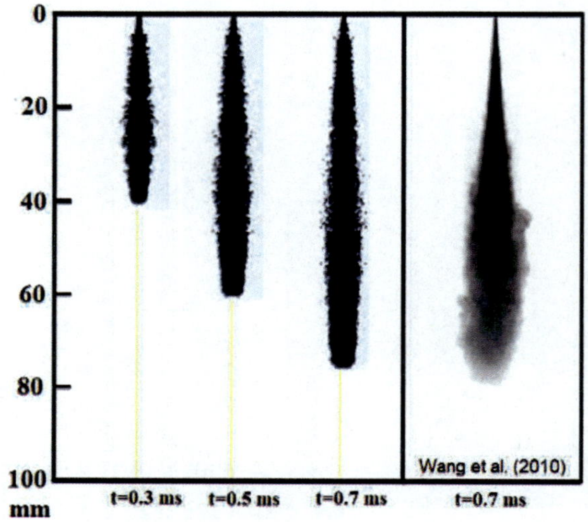

Fig. 4 Spray developments compared with the experimental work of Wang et al. [2] for Diesel

compared to the SSD model shown in Fig. 5. Moreover, the spray shape obtained in the Wave model is closer to the experimental work of Wang et al. [2].

The blob injection model was used to model the primary spray breakup. The liquid droplet, equal to the nozzle diameter, experiences instability and breaks into small droplets. The quality of the spray is described by the droplet size. The diameter of a sphere with the same volume/surface area ratio as the entire spray is referred to as

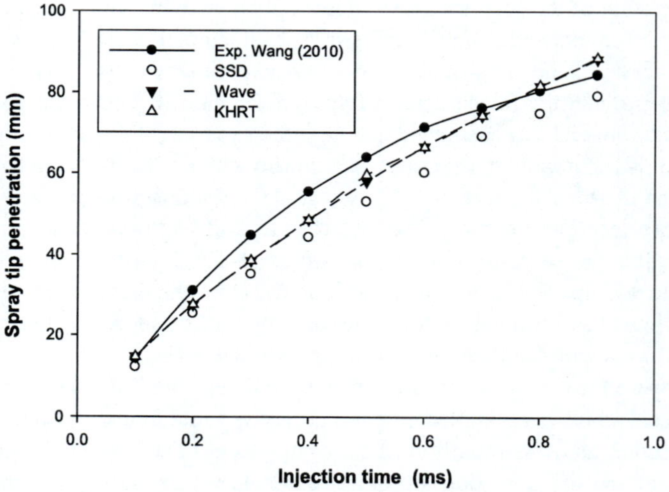

Fig. 5 Comparison of Spray tip penetration of different spray breakup models with experimental data of Wang et al. [2]

Table 1 Particle summary for diesel

SDM	Max. particle diameter	Min. particle diameter
6.73 μm	57.75 μm	0.101 μm

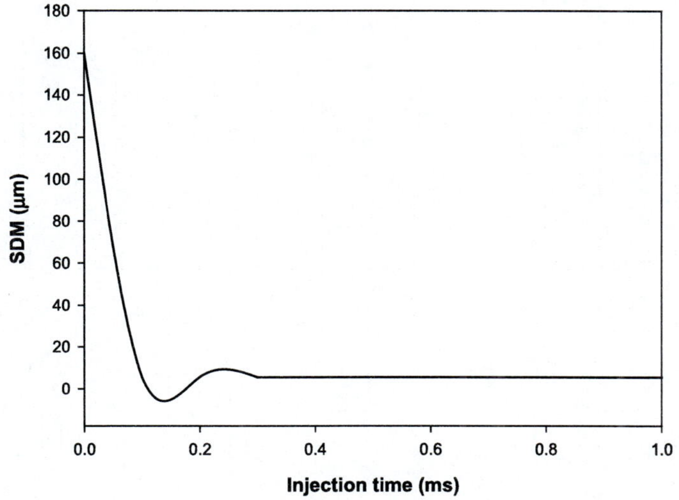

Fig. 6 The Sauter Mean Diameter (SMD) estimation

the SDM. The calculated SDM has been summarized in Table 1 and shown in Fig. 6. At the start of the fuel injection, SMD equals the nozzle diameter then decreases as the atomization of the fuel is proceed.

3.3 Effect of L/D Ratio

The breakup model is governed by the initial diameter of the nozzle. To evaluate the effect of nozzle size, three diameters have been considered (0.16 mm, 0.25 mm and 0.3 mm) with the nozzle length of 1 mm. The STP length has been calculated and compared in Fig. 7 in terms of the L/d ratio. With the decrease in the L/d ratio, the maximum velocity at the nozzle exit will decrease. Moreover, with the decrease in L/d ratio fuel loss the momentum and the STP length would be lower shown in Fig. 7. As the diameter of the nozzle increases (L/d decreases), the initial velocity of the droplet decreases and the breakup of the droplet has been suppressed.

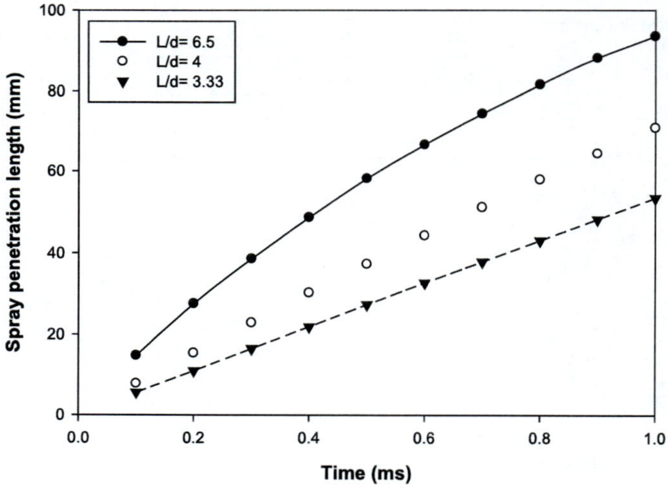

Fig. 7 Spray tip penetration at different L/d ratio

3.4 *Effect of Fuel Property*

The scarcity of petroleum and continuously increasing emissions encourage the use of bioproducts. Biofuels are renewable and that is the major advantage of it. Biodiesel is a greener alternative to diesel, although, due to its fuel properties, its usage is limited. Biodiesel exhibits higher viscosity, which suppresses the instability required for droplet breakup and thus delays the atomization. The development of biodiesel spray is shown in Fig. 8. Although the higher density of the biodiesel leads to better momentum as compared to diesel, the spray tip penetration is slightly higher with biodiesel fuel at a specific time as shown in Fig. 9 (Table 2).

4 Conclusions

Non-evaporating ultra-high injection pressure spray has been simulated by using the discrete phase model. The objective of this work is to evaluate the existing spray breakup model and compare the results with that of Wang et al. [2]. Nozzle with a 0.16 mm diameter is used to inject the diesel at the rate of 2.225 gm/s at 300 MPa injection pressure. The following observation has been derived from this work.

The results obtained at the beginning of the fuel injection are quite reasonable with the experimental trend. The KHRT and the Wave model predict the penetration length reasonably well as compared to the SSD model. The maximum error associated with penetration length calculation is 14% compared to experimental work. The spray shape obtained with the Wave model is more realistic. As the diameter of the nozzle

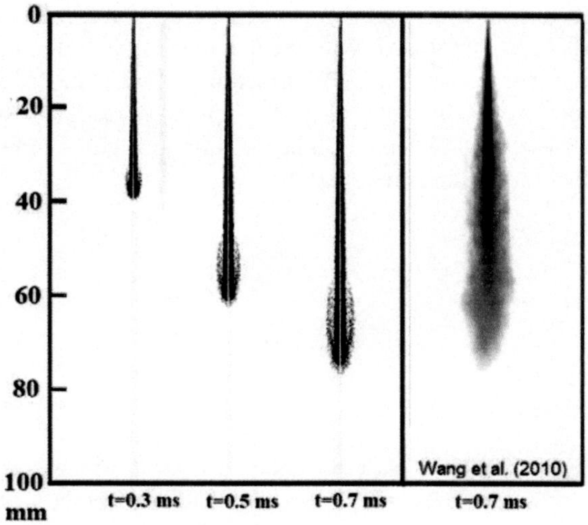

Fig. 8 Spray developments compared with the experimental work of Wang et al. [2] for biodiesel

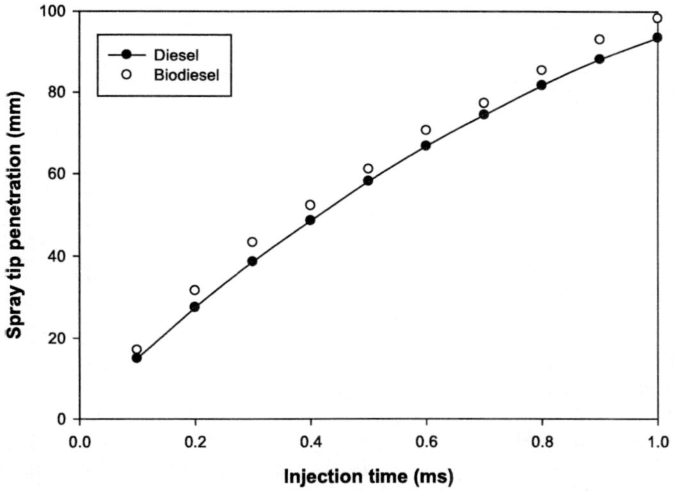

Fig. 9 Comparison of spray tip penetration between diesel and biodiesel

Table 2 Fuel properties

Properties	Diesel	Biodiesel
Liquid phase density (kg/m^3)	830	877
Kinematic viscosity (mm^2/s)	3.37	7.91
Dynamic viscosity (Pa s)	0.0028	0.00694
Surface tension (N/m^2)	0.0255	0.0296

increases (L/d decreases), the initial velocity of the droplet decreases and the breakup of the droplet has been suppressed. With the decrease in the L/d ratio from 6.5 to 4, a 24% reduction has been found in STP. Further decreases in the L/d ratio up to 3.33 reduce STP up to 42%. The higher density of the biodiesel leads to better momentum as compared to diesel, and the spray tip penetration is slightly higher (up to 14%) with biodiesel fuel for the same time instant.

Nomenclature

A	Projected area [m^2]
C_D	Drag coefficient –
d	Diameter [mm]
g	Gravity [m/s]
F	Force [N]
L	Length of nozzle [mm]
r	Droplet radius [mm]
u	Velocity [m/s]
Λ	Wave length [mm]
Ω	Frequency [Hz]
ρ	Density [kg/m^3]
μ	Molecular viscosity [MPa]
τ	Breakup time [s]

References

1. Payri R, Desantes JM, Salvador FJ, Manin J (2009) Influence on diesel injection characteristics and behavior using biodiesel fuels. SAE paper 2009-01-0851
2. Wang X, Huang Z, Kuti OA (2010) Experimental and analytical study on biodiesel and diesel spray characteristics under ultra-high injection pressure. Int J Heat Fluid Flow 31:659–666
3. Mohan B, Yang W, Tay KL, Yu W (2014) Experimental study of spray characteristics of biodiesel derived from waste cooking oil. Energy Convers Manage 88:622–632
4. Moon S, Tsujimura T, Gao Y, Park S, Wang J, Kurimoto N, Nishijima Y, Oguma M (2014) Biodiesel effects on transient needle motion and near-exit flow characteristics of a high-pressure diesel injector. Int J Engine Res 15(4):504–518
5. Reitz RD, Diwakar J (1986) Effect of drop break-up on fuel sprays. SAE technical paper 1986, no 860469
6. Reitz R (1987) Modeling atomization processes in high-pressure vaporizing sprays. At Spray Technol 3:309–337
7. Sazhin SS, Martynov SB, Kristyadi T (2008) Diesel fuel spray penetration, heating, evaporation and ignition: modelling versus experimentation. Int J EngSyst Model 1(1):1–19
8. Turner MR, Sazhin SS, Healey JJ (2012) A breakup model for transient diesel fuel sprays. Fuel 97:288–305

9. Som S, Ramírez AI, Aggarwal SK, Kastengren AL, El-Hannouny EM, Longman DE, Development and validation of a primary breakup model for diesel engine applications. SAE paper no 2009-01-0838
10. Som S, Longman DE, Ramirez AI, Aggarwal SK (2010) A comparison of injector flow and spray characteristics of biodiesel with petrodiesel. Fuel 89:4014–4024
11. Som S, Ramirez AI, Longman DE, Aggarwal SK (2011) Effect of nozzle orifice geometry on spray, combustion, and emission characteristics under diesel engine conditions. Fuel 90:1267–1276
12. Battistoni M, Grimaldi CN (2010) Analysis of transient cavitating flows in diesel injectors using diesel and biodiesel fuels—SAE paper 2010-01-2245. Int J Fuels Lubricants 3(2):879–900
13. Battistoni M, Grimaldi CN (2012) Numerical analysis of injector flow and spray characteristics from diesel injectors using fossil and biodiesel fuels. Appl Energy 2012(97):656–666
14. Yousefard M, Ghadimi P, Mirsalim M (2015) Numerical simulation of biodiesel spray under ultrahighinjection pressure using OpenFOAM. J Braz Soc Mech Sci Eng 37:737–746
15. Vajda B, Lesnik L, Bombek G, Bilus I, Zunic Z, Skerget L, Hocevar M, Sirok B, Kegl B (2015) The numerical simulation of biofuels spray. Fuel 2015(144):71–79
16. Agarwal AK, Som S, Shukla P, Goyal H, Longman D (2015) In-nozzle flow and spray characteristics for mineral diesel, Karanja, and Jatropha biodiesels. Appl Energy 156:138–148
17. Mohan B, Yang W, Yu W (2017) Numerical analysis of spray characteristics of dimethyl ether and diethyl ether fuel. Appl Energy. 185:1403–1410
18. Thulasi V, Karuppa Raj RT (2018) Investigation on the effect of injection pressures on the spray characteristics for diethyl ether and diesel fuel at different chamber temperatures. Front Heat Mass Transf (FHMT) 10:33
19. Ishak MHH, Ismail F, Mat SC, Aziz MS, Abdullah MZ, Abas A (2019) Numerical study on the influence of nozzle spray shape on spray characteristics using diesel and biofuel blends. Biofuels. https://doi.org/10.1080/17597269.2019.1583717
20. Geng L, Xie Y, Wang J, Liu W, Li C, Wang C (2019) Experimental and numerical analysis of the spray characteristics of biodiesel–ethanol fuel blends. Simul: Trans Soc Model Simul Int 1–12. https://doi.org/10.1177/0037549719870244
21. Elkotb MM (1982) Fuel atomization for spray modeling. Prog Energy Combust Sci 8:61–91

Numerical Analysis of Flow Condensation Inside a Horizontal Tube: Current Status and Possibilities

G. R. Mondal, Md N. Hossain, and A. Kundu

Abstract The present paper shows the current status and possibilities of the numerical analysis of flow condensation inside a horizontal tube. The thermohydraulic aspects of flow condensation inside a horizontal tube are complex both from the presence of two-phase and associated heat-transfer phenomena. The issue becomes more complex for the analysis of new eco-friendly refrigerants due to unavailable data in the open literature. On the other hand, different types of tube enhancement make the experimental analysis for different refrigerants more expensive. Hence, the stable numerical solution by a suitable numerical tool can be a smart alternative to the expensive experimental analysis. The present study shows the current status of the available numerical analysis. The paper also presents a numerical approach to the thermohydraulic analysis of flow condensation inside a horizontal tube by the commercial CFD software ANSYS FLUENT (V:14.5). The simulation is carried out under flow condensation of R32 refrigerants which is an eco-friendly refrigerant.

Keywords Flow condensation · Two-phase flow · Void fraction · Condenser tube · Steam quality

Abbreviations

Coeff	Coefficient of relaxation time
d	Liquid droplet diametermm
E	Energy (total)kW
F	Volumetric surface-tension force
Fr	Froude number
g	Gravitational accelerationm/s^2
h	EnthalpykW/kg

G. R. Mondal · M. N. Hossain (✉) · A. Kundu
Department of Mechanical Engineering, Jalpaiguri Government Engineering College, Jalpaiguri, West Bengal, India
e-mail: naimhossain6@gmail.com

S. Das et al. (eds.), *Proceedings of the 1st International Conference on Fluid, Thermal and Energy Systems*, Lecture Notes in Mechanical Engineering,
https://doi.org/10.1007/978-981-99-5990-7_21

h_{lv}	Latent heatkW/kg
k_{eff}	Thermal conductivity
L	Latent heatJ/kg
m_{lv}	Rate of evaporative mass transfer kg/s
m_{vl}	Rate of condensation mass transferkg/s
M	Molecular weightKg/mole
Nu	Nusselt Number
P	Pressurebar
Pr	Prandtl number
q	Heat fluxW/m^2
R	Universal gas constantJ/Kmol
T	TemperatureK
S_h	Volumetric energy source term
t	Times
U	Velocitym/s
β	Accommodation coefficient
ρ	DensityKg/m^3
μ	ViscosityKg/ms
α	Volume fraction
σ	Surface tensionN/m
v	Vapor
l	Liquid
sat	Saturated
$wall$	Tube wall

1 Introduction

Condensation is an important phenomenon for several engineering applications like condenser tubes, boiler condensers, heat exchangers, etc. It is also involved in the heating, ventilation, air-conditioning systems, chemical industries, etc. The condensation process is involved with the mass and the energy transfer between vapor and fluid of the two-phase flow. The flow condensation on the other hand is associated with subcooling. The condensate fluid may become subcooled near the end of the pipe due to the high rate of cooling. The flow regime map and the associated heat transfer may be affected due to the presence of subcooled fluid flow. The difference in temperature between vapor and the solid wall surface is called the driven temperature of condensation. Condensation rate further depends on the fluid properties and the geometry of the tube cross-section, i.e., helical microfin tubes, and grooved tubes. The associated flow regime map of flow condensation is not independent of the two parameters mentioned above. There may present many flow regimes in the flow condensation process, i.e., annular flow, bubble flow, wavy flow, stratified flow, and intermittent flow. For many decades, different researchers focused on the analysis of heat transfer and fluid flow for condensation by experimental analysis.

According to the open literature, there are some established heat-transfer correlations, two-phase frictional factors, and associated flow regime maps for conventional fluids, i.e., water, nitrogen, and conventional refrigerant. However, these data are insufficient for designing the thermal systems (i.e., condenser tubes of refrigeration and HVAC system) run by modern eco-friendly refrigerants. There are two possibilities to acquire proper thermo-physical and thermo-hydraulic data of the modern eco-friendly refrigerants, i.e., experimental analysis or the stable numerical analysis. However, the experimental analysis for different refrigerants with different tube enhancements (i.e., twisted tape tube, microfin tube, helical tube, grooved tube. punched tube, and diamond punched tube) are associated with huge experimental cost. Especially the study of thermo-geometric dependence of the flow condensation is associated with the change of the tube for every case. A proper numerical analysis may be the smart replacement for the costly experimental analysis. A proper numerical model may be utilized to establish proper two-phase heat-transfer correlations, two-phase frictional factors, and flow regime maps for different refrigerants and different tube enhancements.

2 Current Status of Numerical Analysis of Flow Condensation Inside a Horizontal Tube

An extensive literature survey is carried out to find out the past and current scenarios of the flow condensation. Initially, some existing experimental analyses are mentioned. Later the available numerical studies are reflected and further scope of numerical analysis of flow condensation is discussed.

Based on the open literature, it is found that, earlier, researchers have done research work on flow condensation inside the horizontal tubes based on experimental analysis [1–5]. Out of different available data, some important findings are mentioned here.

Rifert et al. [1] carried out the experimental analysis to establish a heat-transfer correlation for flow condensation inside a horizontal tube. The authors found a generalized heat-transfer correlation by experimental data on horizontal flow condensation for different fluids. Rabiee et al. [2] proposed over condensation heat transfer rate by experimental analysis. The authors commented that the local heat-transfer coefficient increased due to the increment of mass flux and vapor quality at the inlet. They also investigated that if the tube length and the mass flux were fixed, the total heat-transfer coefficient increased with the decrease in hydraulic diameter. Yanan et al. [3] proposed a new approach for the heat-transfer analysis for film condensation inside a horizontal tube. The authors verified their model with the existing experimental data (22 different fluids, i.e., water, refrigerants, and organic substances). However, the model was not generic. The model was valid for internal diameters from 2 to 50 mm, the range of pressure from 0.0008 Mpa to 0.91 Mpa, the range of liquid Reynolds number 68–84,827, and for the steam Reynolds number 900–594,373, the vapor quality range from 0.01 to 0.99 and the range of mass flux from 3 to 850 kg/m^2 s. Rahman et al. [4] carried out an experimental analysis on flow condensation and associated heat transfer and the flow pressure drop along the smooth and microfin horizontal tube (2.13 mm of inner diameter). For the horizontal flow condensation, they have proposed that the mass flux and the steam quality decrease with the increasing saturation temperature. Carlos et al. [5] proposed a new generalized model for determining the local flow condensation heat-transfer coefficient inside a plain horizontal tube. The authors considered two types of regimes corresponding to the high mass fluxes and high vapor qualities and for low mass fluxes and low vapor qualities. For different regions, the authors proposed different models which bear a resemblance to the single-phase heat-transfer-coefficient with an equivalent Reynolds number. The authors investigated that the superficial vapor Reynolds number played a significant role in the heat-transfer coefficient. The proposed model was able to predict the heat-transfer coefficient for channels with an equivalent hydraulic diameter of 67 μm and up to 20 mm for different fluids. However, the author observed no significant effect of the channel diameter, shape, and fluid properties on the heat-transfer coefficient.

Apart from the experimental analysis, there is some numerical analysis for flow condensation inside the horizontal pipe for the different working fluids. Juggurnath et al. [6] carried out a three-dimensional numerical simulation for condensing R134a inside the horizontal pipe (smooth tube) with a diameter and length of 8.4 mm and 1.5 m, respectively. The authors used the volume-of-fluid (VOF) approach in their numerical analysis to simulate the void fraction of liquid and vapor phases. The simulations were able to investigate the flow pattern map, i.e., stratified wavy flow. The authors investigated the heat-transfer coefficient, and it was found to be in good agreement with existing data. The authors observed from the simulations that the liquid-phase height at the bottom of the pipe became smaller with the increasing vapor quality, which results in the increment in the heat-transfer coefficient. On the other hand, a thick film and lower heat-transfer coefficient were observed under high saturation temperatures. Juggurnath et al. [7] in another work investigated the local characteristics and the characteristics of condensation of R134a in a horizontal pipe, under mass flux ranges of 100 kg/m^{2}s, 200 kg/m^{2}s, and 400 kg/m^{2}s and undersaturation temperature of 40-degree C. The numerical study was carried out under the mean steam qualities of 0.25, 0.50, and 0.75. The simulation pointed out that there was an increase in condensate film thickness from an angular position. That was on the other hand influenced by the vapor quality of the flow to the bottom of the pipe. The authors also investigated that the heat-transfer coefficient decreases around the circumference of the pipe from bottom to top. The simulation also indicated that in the presence of a specific angular position, the heat-transfer coefficient altered significantly due to the stratified layer of condensate. Also, the velocity profiles of the flow condensation were asymmetrical. The flow velocity and the heat-transfer coefficient reduce along the length of the condensation path. The zonal results predicted that the heat-transfer coefficient was not significantly affected by the film thickness. Ariwibowo et al. [8] investigated the phenomena of condensation of R134a in a horizontal tube numerically [9]. They used the Computational Fluid Dynamics (CFD) with the pseudo-two-phase flow model to investigate the local heat-transfer coefficient of the flow condensation. The authors also investigated the liquid film thickness, the eddy-viscosity, and shear stress variation by the Von Karman universal velocity profiles. The simulation findings showed that the increase in the mass flux and the void fraction increases the two-phase heat-transfer coefficient and the pressure drop. Dong et al. [10] carried out a numerical study to investigate the frictional pressure drop for the flow condensation.

The above-detailed literature survey indicates that, though many experimental analyses for flow boiling exist, detailed numerical simulations are rare. However, there are some numerical simulations for the flow condensation, which indicates the acceptability of the numerical simulation for further investigations. The existing literature survey indicates that the numerical simulations mainly focused on regular refrigerants and plain tubes. Hence, there is a literature gap in the simulations of new eco-friendly refrigerants. The effect of the enhanced tube on the flow condensation is also not explored as in the open literature. Hence, the numerical investigation of different tube profiles, i.e., twisted tube, microfin tube, and grooved tubes for flow condensation of eco-friendly refrigerants, i.e., R410A, R32, and R134zf is still in need of efficient design aspects of HVAC systems and refrigeration plants.

Hence in the present study, a numerical simulation approach of flow condensation along a horizontal plain tube is presented with the VOF model and ANSYS FLUENT (V 14.5) CFD software. The condensation simulation is handled by the evaporation–condensation model of the FLUENT solver. The simulation is mainly carried out to investigate the suitability of numerical analysis with R32 refrigerants which is an eco-friendly new-genre refrigerant. It is also to be noted that the objective of the present study is to investigate the appropriateness of the two-dimensional (2-D) simulation of the flow condensation from the barriers of numerical complexities. This is because the 3-D simulation of the two-phase flow along a horizontal tube required a high level of computing facilities which is beyond the reach of regular researchers. The flow condensation study is carried out under transient mode.

3 Modeling and Simulation of Flow Condensation and the Associated Governing Equations

ANSYS FLUENT (V 14.5) software is used for the condensation simulation in the present paper. The condensation process is simulated by the VOF approach and the existing evaporation–condensation model.

The 2-D simulation domain used in the present simulation is shown in Fig. 1. A tube of 100 mm in length and 10 mm in diameter is used for the present simulation. The tube wall is being cooled under 300 kW/m^2 heat flux. The exit boundary is maintained under atmospheric pressure outlet conditions. It is considered that saturated vapor of R32 refrigerant is entering the tube through the inlet under 0.1 m/s velocity. The simulation is carried out for 3 minute. The boundary of the tube wall is under

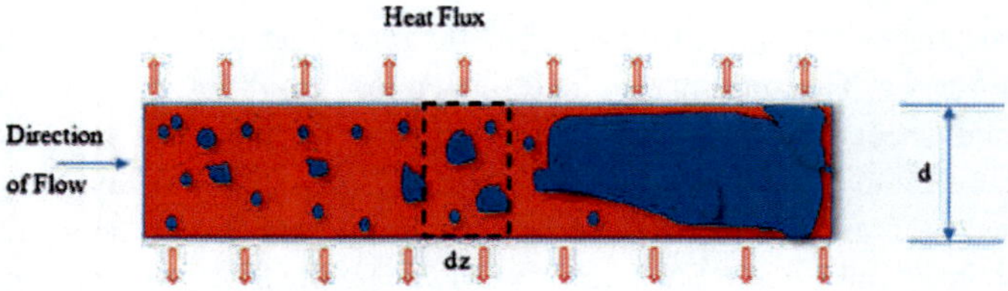

Fig. 1 Schematic diagram of horizontal pipe flow

a no-slip boundary condition. The flow inside the pipe is fully developed flow and initially, the pipe is filled with the saturated vapor of R32.

In the volume of fluid approach, the single set of momentum and energy equations (averaged) of a multiphase flow are solved; these are presented below [11]. The volume fraction for different phases is handled separately.

Momentum equation:

$$\begin{aligned} &\frac{d}{dt}(\rho U) + \nabla(\rho U U) = \\ &-\nabla P + \nabla\left[\mu\left(\nabla U + \nabla U^T\right)\right] + \rho g + F \end{aligned} \tag{1}$$

where U in Eq. (1) is the shared velocity by the different phases [11, 12]. The volume-averaged density and viscosity are presented below

$$\rho = \sum_{i=1}^{j} \alpha_i \rho_i \tag{2}$$

$$\mu = \sum_{i=1}^{j} \alpha_i \mu_i \tag{3}$$

Energy equations [11]:

$$\begin{aligned} &\frac{d}{dt}(\rho E) + \nabla(V(\rho E + P)) \\ &= \nabla\left(k_{eff} \nabla T\right) + S_h \end{aligned} \tag{4}$$

where E is the mass averaged energy. It is presented below

$$E = \sum_{i=1}^{j} \alpha_i \rho_i E_i \Bigg/ \sum_{i=1}^{j} \alpha_i \rho_i \tag{5}$$

The volume-averaged thermal conductivity k_{eff} is presented below

$$K_{eff} = \sum_{i=1}^{j} \alpha_i k_{eff\,i} \tag{6}$$

In the present work, the condensation process is evaluated by the evaporation–condensation model. The mass transfer between the vapor to liquid interphases is estimated by the below volume fraction transport equation expression (Lee model) [11]:

$$\frac{d}{dt}(\alpha\rho_v) + \nabla(\alpha\rho_v\overline{v_v}) = m_{lv} - m_{vl} \tag{7}$$

For the condensation cases,

$$m_{vl} = coeff \times \alpha_v\rho_v\frac{(T_v - T_{sat\,l})}{T_{sat}} \tag{8}$$

For the evaporation case,

$$m_{lv} = coeff \times \alpha_l\rho_l\frac{(T_l - T_{sat})}{T_{sat}} \tag{9}$$

where the $coeff$ term is presented below [11]

$$coeff = \frac{6}{d}\beta\sqrt{\frac{M}{2\pi RT_{sat}}}L\frac{\rho_l}{(\rho_l - \rho_v)} \tag{10}$$

In the present model, it is taken that near the node where the bulk fluid temperature reduces from the saturation temperature, the condensation starts.

It is to be mentioned that the volume fraction of the vapor phase is estimated by Eq. (7). The volume fraction of the liquid phase is estimated by the expression mentioned below

$$\sum_{i=1}^{j} \alpha_i = 1 \tag{11}$$

The turbulence kinetic energy (k) and the dissipation (ε) rate are estimated from the standard $k - \varepsilon$ model [11]. The equations are presented below

$$\frac{\partial}{\partial t}(\rho k) + \frac{\partial}{\partial x_i}(\rho k u_i) = \frac{\partial}{\partial x_j}\left[\left(\mu + \frac{\mu_t}{\sigma_k}\right)\frac{\partial k}{\partial x_j}\right] + G_k + G_b - \rho\varepsilon - Y_M + S_k \tag{12}$$

$$\frac{\partial}{\partial t}(\rho\varepsilon) + \frac{\partial}{\partial x_i}(\rho\varepsilon u_i) = \frac{\partial}{\partial x_j}\left[\left(\mu + \frac{\mu_t}{\sigma_\varepsilon}\right)\frac{\partial \varepsilon}{\partial x_j}\right] + C_{1\varepsilon}\frac{\varepsilon}{k}(G_k + C_{3\varepsilon}G_b) - C_{2\varepsilon}\rho\frac{\varepsilon^2}{k} + S_\varepsilon \tag{13}$$

where G_k and G_b are the generations of turbulent kinetic energy due to mean velocity gradients and buoyancy. The term Y_M is the fluctuating-dilatation effect. σ_k and σ_ε are the Prandtl numbers under turbulent conditions. Their values are taken as 1 and 1.3. Other constant terms, i.e., $C_{1\varepsilon}$, $C_{2\varepsilon}$, and $C_{3\varepsilon}$ are taken at 1.44, 1.92, and 0.99, respectively. μ_t denotes the turbulent viscosity as shown below with C_μ constant:

$$\mu_t = \rho C_\mu \frac{k^2}{\varepsilon} \tag{14}$$

Equations (1), (4), and (7) are calculated by the control volume method with ANSYS FLUENT solver. The pressure and the velocity coupling are done by the "SIMPLE" approach. The criterion of convergence for continuity, velocity, and energy are set to 10^{-5}, 10^{-5}, and 10^{-6}, respectively.

4 Results and Discussion

A grid-independent study and a time-step independent study have been carried out to fix the grid and time steps for the present computational domain. Finally, 90,000 grid and 10^{-4}-time-step numbers have been chosen for the study.

It is to be mentioned here that the present numerical methodology is already compared with the existing numerical findings in our previous works [12, 13]. The adopted properties of R32 used in the preset numerical simulation are taken from NIST-webbook data [14].

Variation of liquid volume fraction, mixture velocity, and total temperature for the condensation process of the R32 vapor is presented in Figs. 2, 3, 4, 5, and 6, respectively. The change in the void fraction of condensate liquid under the different periods of flow (1 s to 2 min) or time of condensation is presented in these figures.

Figure 2a shows the void fraction variation of the liquid phase after 1 s of flow. The figure indicates the beginning of the condensation. The liquid droplets are visible at the later part of the pipe because near the entry point, more saturated vapor is coming continuously. Visible liquid droplets are dispersed in the vapor phase. The phase can be termed a misty flow of mist flow. Figure 2b, c shows the contour of mixture total temperature and mixture velocity, respectively. The velocity profile indicates the fully developed flow. It is to be mentioned that the velocity profile is a little wavy due to the presence of liquid droplet abstraction (Fig. 2c). The temperature contour shows the cooling of the vapor near the wall of the pipe. Also, the local temperature drop near the droplets is visible in the temperature contour (Fig. 2b).

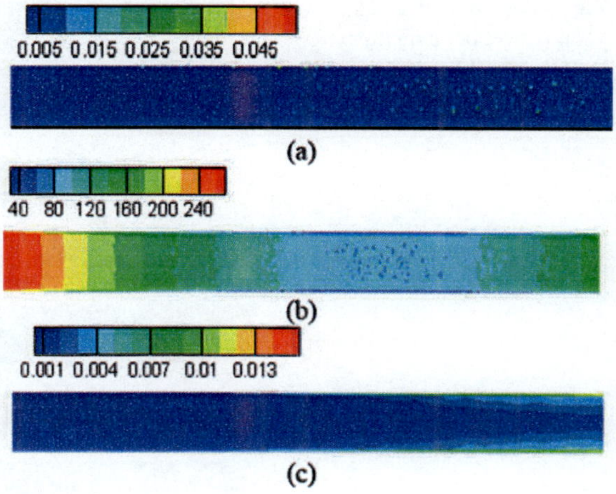

Fig. 2 Variation of **a** liquid volume fraction, **b** mixture temperature (K), and **c** mixture velocity (m/s) after 1 s of flow

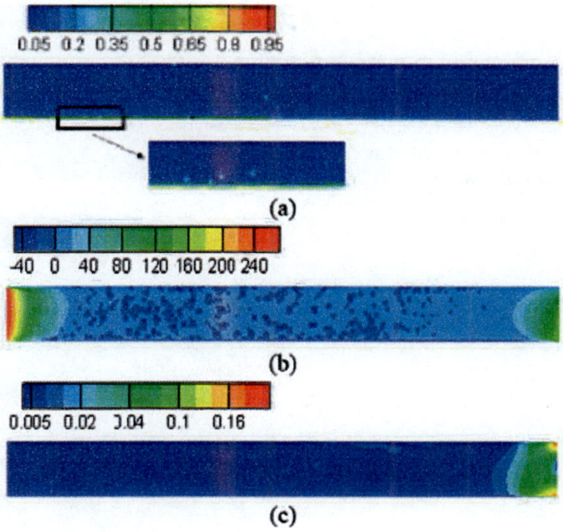

Fig. 3 Variation of **a** liquid volume fraction, **b** mixture temperature (K), and **c** mixture velocity (m/s) after 10 s of flow

Figure 3a shows the void fraction variation of the liquid phase after 10 s of flow. Compared to the previous case, the condensate liquid droplets are falling to the bottom of the pipe (shown in enlarged view). The inlet velocity is spreading the liquid droplets along with the flow. Figure 3b, c shows the variation of mixture total temperature and mixture velocity respectively after 10 s of flow. The velocity profile indicates the presence of turbulence inside the flow and the influence of liquid droplets' obstruction on it. The temperature contour indicates a higher rate of reduction in total temperature near the bottom of the pipe. The accumulation effect of the liquid droplets near the bottom can be attributed to this phenomenon.

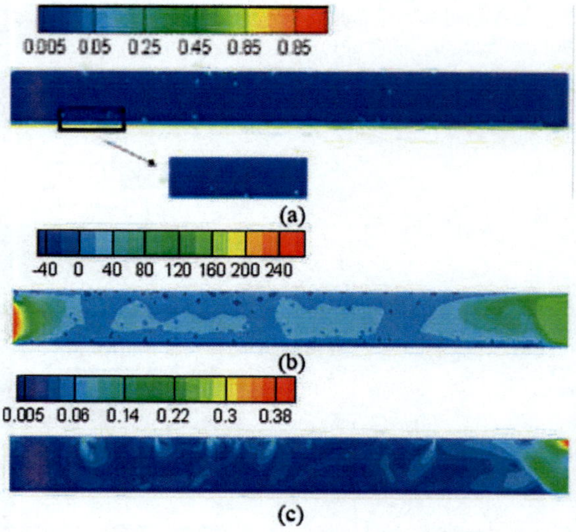

Fig. 4 Variation of **a** liquid volume fraction, **b** mixture temperature (K), and **c** mixture velocity (m/s) after 30 s of flow

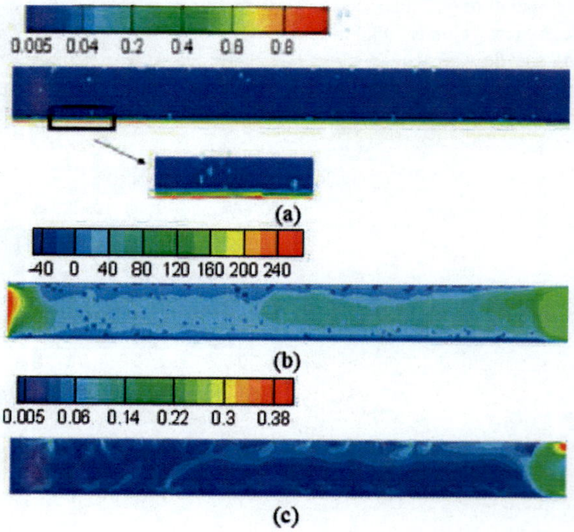

Fig. 5 Variation of **a** liquid volume fraction, **b** mixture temperature (K), and **c** mixture velocity (m/s) after 60 s of flow

Figure 4a shows the void fraction contour of the liquid phase after 30 s of flow. A higher rate of liquid droplet accumulation near the pipe bottom is visible in the figure (shown in the enlarged view). Also, more liquid droplets are forming due to condensation along the pipe.

Figure 4b, c shows the variation of mixture total temperature and mixture velocity respectively after 30 s of flow. A higher rate of turbulence is visible along the pipe. The temperature reduces rapidly near the bottom of the pipe. It influences the condensation near the bottom layer of the liquid.

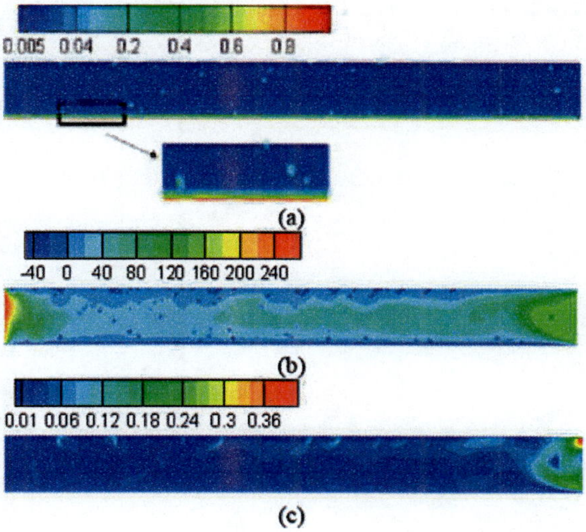

Fig. 6 Variation of **a** liquid volume fraction, **b** mixture temperature (K), and **c** mixture velocity (m/s) after 120 s of flow

The liquid accumulation rate increases compared to the previous case, after 60 s of flow as shown in Fig. 5a. The characteristics of mixture velocity and mixture temperature remain unaltered as shown in Fig. 5b and Fig. 5c, respectively.

However, the rate of liquid accumulation near the pipe bottom does not show a significant increment after 2 min of flow compared to the previous case. This is presented in Fig. 6a. This indicates the steady nature of condensation with a stratified wavy flow pattern. The corresponding temperature and velocity contour are shown in Fig. 6b, c, respectively.

To cross-check the steady pattern of the flow condensation, the average void fraction of liquid along the pipe for different times of flow is identified and presented in Fig. 7. The figure indicates that the average void fraction of liquid accumulation due to condensation increases rapidly from the beginning (1-s case) to the 30 s low. This is identified as a rapid condensation zone in Fig. 7. The increment rate is slow from 30 s to 1 min of flow. This is marked as an intermediate condensation zone. After 1 min to the rest of the flow, the liquid void fraction curve shows diminishing characteristics. Hence, the section is marked as a steady rate of condensation zone in the figure.

The study indicates the steady state of the flow condensation and the stratified wavy form of the flow regime along the pipe.

5 Conclusions

The present paper shows the current status and possibilities of numerical analysis of flow condensation inside a horizontal tube. The detailed literature survey indicates that, though there are many experimental analyses for flow condensation, detailed

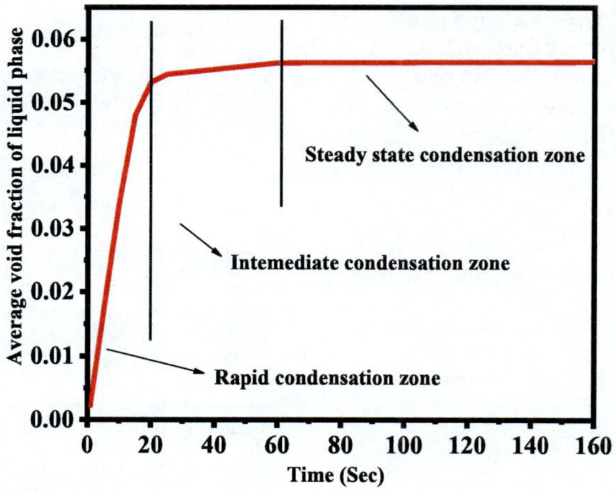

Fig. 7 Average volume fraction variation of liquid phase along the pipe

numerical simulation is rare. The study reveals that a proper numerical analysis of flow condensation may be a smart alternative to an expensive experimental analysis. Flow regime map and heat-transfer coefficients of flow condensation for new eco-friendly refrigerants and enhanced tubes may be simulated by numerical analysis.

The paper also presents a numerical approach to the thermohydraulic analysis of flow condensation inside a horizontal tube by commercial CFD software ANSYS FLUENT (V:14.5). The simulation is carried out under flow condensation of R32 refrigerants which is an eco-friendly refrigerant. Variation of liquid volume fraction and the condensation of the vapor are presented in the present work. The change in the void fraction of condensate liquid under the different periods of flow or time of condensation is presented in these figures. The result shows the visible mist flow, initiation of wavy flow, and the full stratified wavy flow pattern. A further detailed study would be helpful for the detailed flow regime mapping of R32 refrigerants.

References

1. Rifert V, Gorin V, Sereda V, Treputnev V (2019) Improving methods to calculate heat transfer during the condensation inside tubes. J Eng Phys Thermophys 92. https://doi.org/10.1007/s10891-019-01988-6
2. Rabiee R, Désilets M, Proulx P, Ariana M, Julien M, Determination of condensation heat transfer inside a horizontal smooth tube. Department of Chemical Engineering, Université de Sherbrook, Québec, Canada
3. Yanan CM, Abel HGJ, Luis L-O, Ken MC, Oscar MC-F, Osvaldo FG-M (2019) New model for heat transfer calculation during film condensation inside pipes. Int J Heat Mass Transfer 128:344–353
4. Rahman MM, Kudo Y, Bashar MK, Kariya K, Miyara A (2017) Experimental study on condensation heat transfer and pressure drop inside 2.13 mm id smooth and microfin horizontal tube. In: Proceedings of the Asian conference on thermal sciences 2017, 1st ACTS March 26–30, Jeju Island, Korea

5. Carlos AD, Maria F (2018) Simple and general correlation for heat transfer during flow condensation inside plain pipes. Int J Heat Mass Transf 122:290–305
6. Juggurnath D et al (2022) Numerical investigations into the local behavior of condensing R134a flows in a horizontal pipe. J Heat Transfer, ASME 144: 021602-1
7. Juggurnath D, Dauhoo MZ, Elahee MK, Khoodaruth A, Dirker J, Meyer JP, Markides CN (2019) Numerical simulations of condensing R134a flows in horizontal pipes. In: 4th thermal and fluids engineering conference (TFEC) April 14–17, (2019), Las Vegas, NV, USA TFEC-2019
8. Ariwibowo TH, Solihah FH, Harjanto B (2016) A numerical study of condensation heat transfer for R134a in annular flow regime. In: International conference on engineering, science and nanotechnology (ICESNANO 2016) AIP conference proceedings 1788, 030020-1–030020-6. Published by AIP Publishing. 978–0-7354-1452-5/$30.00. https://doi.org/10.1063/1.4968273
9. Olivier SP, Meyer JP, De-Paepe M, Dek K (2016) Measured void fraction and heat transfer coefficients during condensation. In: 11th international conference on heat transfer, fluid mechanics and thermodynamics
10. Qiu G-D, Cai W-H, Li S-L, Wu Z-Y, Jiang Y-Q, Yang Y (2014) Numerical simulation on forced convective condensation of steam upward flow in a vertical pipe. Hindawi Publ Corp Adv Mech Eng 2014, Article ID 589250:11. https://doi.org/10.1155/2014/589250.
11. ANSYS Inc, ANSYS FLUENT User's guide 14.5, 2012.
12. Shahbaz MD, Hossain MdN, Mondal N (2019) Numerical validation of flow boiling analysis in Ansys fluent with VOF approach, TAME_120, 2019. https://doi.org/10.21843/reas%2F2019%2F16-24%2F196162.
13. Hossain MdN, Chakravarty A, Ghosh K (2021) Influence of transient heat input on pool boiling behaviour. In: Proceedings of the 26th national and 4th international ISHMT-ASTFE heat and mass transfer conference December 17-20, 2021, IIT Madras, Chennai-600036, Tamil Nadu, India. https://doi.org/10.1615/IHMTC-2021.3070
14. https://webbook.nist.gov/chemistry/

Effect of Metallic Shell on Energy Storage Characteristics of Macro-Encapsulated PCM System

Sudipta Das, Vidula Athawale, Jegatheesan M, Prasenjit Rath, and Anirban Bhattacharya

Abstract Macro-encapsulated phase change material (PCM) is widely used in energy storage applications for its high energy storing ability without high temperature rise. The effect of shell thickness on energy storage performance of multiple encapsulated PCM systems subjected to flow of hot heat transfer fluid (HTF) is numerically studied by developing a 3D model. The model considers different phases present in the domain such as solid PCM, liquid PCM, metallic shell and HTF. The geometry of the capsules is resolved and the fluid flow between the capsules and also natural convection within the capsules are also captured. Coupled fluid flow and enthalpy model are used to solve the flow of HTF and phase change process of PCM. Melting and energy storage of three different multi-capsule systems with paraffin encapsulated in aluminum shells subjected to heat transfer by flow of high temperature water are studied. The 3 cases have large shell thickness (1 mm), intermediate shell thickness (0.5 mm) and negligible shell thickness (0 mm). The results show that total energy stored in the 0 mm shell system is 5.87 and 14.66% higher than that of 0.5 and 1 mm shell system, respectively. But the melting in 0 mm shell system takes 60.22 and 87.09% more time than that of 0.5 and 1 mm shell system.

Keywords Energy storage · Encapsulated PCM · Shell thickness · Pore-scale model · Packed bed system

Nomenclature

Avg	Volume average
C_1	Large constant
C_2	Small constant
C_P	Specific heat: [J/kgK]
f_{liq}	Liquid fraction

S. Das · V. Athawale · J. M · P. Rath · A. Bhattacharya (✉)
School of Mechanical Sciences, IIT Bhubaneswar, Odisha 752050, India
e-mail: anirban@iitbbs.ac.in

S. Das et al. (eds.), *Proceedings of the 1st International Conference on Fluid, Thermal and Energy Systems*, Lecture Notes in Mechanical Engineering,
https://doi.org/10.1007/978-981-99-5990-7_22

H	Specific enthalpy: [J/kg]
k	Thermal conductivity: [W/mK]
L	Latent energy: [J/kg]
m	Metal
P	Pressure: [N/m^2]
T	Temperature: [K]
T_i	Inlet temperature: [K]
T_m	Melting temperature: [K]
U	Velocity: [m/s]
α	Encapsulation fraction
β	Thermal expansion coefficient: [1/K]
γ	Metal fraction
ρ	Density: [kg/m^3]
μ	Dynamic viscosity: [Ns/m^2]
ω_{in}	Inlet velocity: [m/s]

1 Introduction

There is an increasing interest in research and development on combined latent and sensible energy storage systems using phase change material (PCM). The very narrow temperature range of phase transition of PCM enables charging and discharging of energy at nearly constant temperature. However, the major problem associated with the usage of PCM is conductive heat transfer resistance of PCM and its container. Several studies were carried out to improve thermal conductivity of the overall system using foam, encapsulations, fins and nanoparticles [1, 2] made up of high thermal conductivity materials.

2 Literature Review and Objective

It is found from the literature that providing encapsulations of proper material can serve as thermal conductivity enhancer and protector of the PCM from surrounding media. Liu et al. [3] reviewed different forms of macro-encapsulations and shell material selections for thermal management applications in buildings. Patel et al. [4] numerically investigated the effect of shape of a macro capsule on melting and solidification performance. They found that triangular capsules showed the best thermal performance among cubical, spherical, triangular and plus-shape macro capsules. Ettouney et al. [5] experimentally investigated the effect of capsule diameter on melting and solidification characteristics of paraffin wax in a spherical copper shell. The results showed that natural convection inside liquid PCM increased with capsule diameter. Experimental studies were conducted by Tan et al. [6] on spherical encapsulated PCM for both unconstrained and constrained motions of solid PCM. In unconstrained melting, circulations were generated due to sinking of solid PCM and thus melting of the bottom half was much faster. In constrained melting, melting started concentrically inwards and then the top half melted faster than the bottom half because of natural convection.

In an actual packed bed latent heat storage system, multiple spherical capsules containing PCM are used as primary energy storage media. Nallusamy et al. [7] experimentally investigated flow of water over a packed bed of macro-encapsulated paraffin for charging, batchwise discharging and continuous discharging of the energy storage tank. They found that the combined latent and sensible energy storage system is more efficient than a sensible energy-based storage system. The packed bed of macro-encapsulated PCM was experimentally studied for high-temperature energy storage application in [8]. Ternary carbonate-based hybrid molten salt of melting point 395.1 °C was used in 34 mm diameter and 2 mm thick capsules of stainless steel. They observed that charging and discharging efficiencies increased with heat transfer fluid (HTF) inlet temperature and mass flow rate. They also concluded that the overall performance of the system was better than shell and tube thermal storage systems.

Various numerical models are used to study thermal performance of charging and discharging processes of a latent energy-based packed bed thermal energy storage (TES) system. Single-phase model [9] solves a single temperature field for the entire domain considering instantaneous temperature of the capsule and HTF. The two-phase model treats the capsule and HTF separately. Schumann's model, continuous solid phase model and concentric dispersion model are three different types of two-phase models. Schumann's model [10] neglects thermal diffusion in both HTF and PCM. In the continuous solid phase model [11], the bed is considered as a porous continuous structure and heat diffusion in HTF is considered. Concentric dispersion model [12] calculates thermal gradient inside the capsule. Karthikeyan et al. [13] used three different two-phase models to study packed bed systems filled with spherical encapsulated PCM. They found that the concentric dispersion model showed good agreement with experimental data as it considered the conductive resistance of PCM in each capsule. Although two-phase models consume relatively less computational resources, these models are unable to resolve the actual geometry of the system. Athawale et al. [14, 15] studied a different arrangement of multi-capsule systems using a pore scale model which resolved the geometry of the encapsulated PCM system.

Although there are several studies available on core to coating ratio of micro-encapsulated PCM, only a few studies can be found on the effect of capsule shell thickness on heat transfer and energy storage performance. Wei et al. [16] numerically studied spherical, cylindrical, plate and tubular macro-encapsulated PCM with two different capsule shell thicknesses. The spherical capsule with thick shell had the best performance of heat release among four different shapes, but total energy storage decreased due to less amount of PCM content.

The objective of the present study is to investigate the effect of shell thickness on melting and energy storage characteristics of the encapsulated PCM-based TES system containing multiple capsules. A 3D model is developed and is used to study three different multi-capsule systems having the same capsule outer diameter but different shell thickness while the system is subjected to flow of hot HTF. The model resolves the capsule geometry and captures the phase change of PCM inside the capsules as well as the energy stored in the system.

3 Mathematical Model

A cubical domain of $30 \times 30 \times 30$ mm^3 size containing 18 capsules is considered. The capsules are arranged in a $3 \times 3 \times 2$ manner along X, Y and Z directions. The domain and boundary conditions are shown in Fig. 1. Each capsule has two parts; one is the PCM, the other is the outer shell. Initially, there is no flow and the entire domain is maintained at a constant temperature of 303 K. HTF flows at a constant velocity (ω_{in}) of 1 mm/s, and a constant temperature (T_i) of 373 K is applied at the inlet. The HTF exits the domain via the top face. The side faces of the domain are imposed with adiabatic wall condition. Materials of PCM, shell and HTF are paraffin, aluminum and water, respectively. Table 1 states the properties of different materials. The hot HTF transfers heat to the capsules as it flows over them. The shell transfers heat to the PCM inside it. The PCM gets melted and stores energy in the form of both latent and sensible heat. Some amount of energy is also stored by the metallic shell.

The geometry is created following the technique given in Athawale et al. [15]. But, in the present study, the presence of the shell is also considered. Hence instead of one parameter, here two geometrical parameters α and γ are created by a mapping

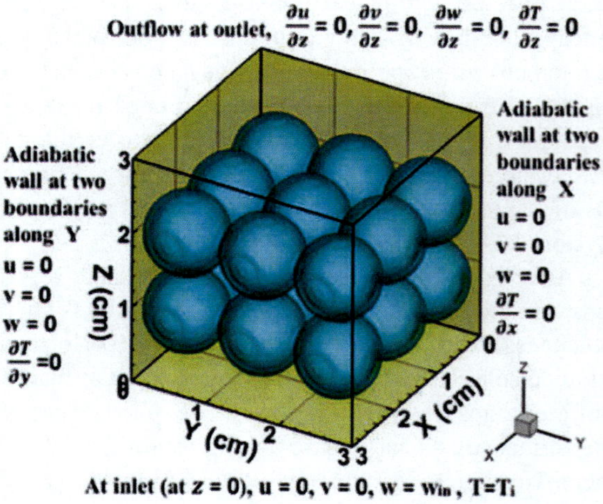

Fig. 1 Domain and boundary conditions

Table 1 Material properties

Parameter	Paraffin	Aluminium	Water
Density (ρ) (kg/m^3)	805	2830	996
Specific heat (C_p) (J/kgK)	2100	910	4178
Thermal conductivity (k) (W/mK)	0.22	205	0.615
Dynamic viscosity (μ) (Ns/m^2)	3.65×10^{-3}	–	0.798×10^{-3}
Melting point (T_m) (K)	333	–	–
Latent heat (L) (J/kg)	169×10^3	–	–

method [17] to include the metal phase present in the domain. The values of α and γ vary from 0 to 1. The value of α for the grids located inside the capsules is 1, and the value of α is 0 for the grids outside capsules. If any grid is located at the capsule and HTF interface, then α value of that grid is between 0 and 1. Similarly, γ is 1 for the grids on shell, 0 for grids outside the shell. At the shell-PCM interface and shell-HTF interface, the value of γ is between 0 and 1. Hence, in PCM region, (α-γ) is 1. Three different geometric configurations are considered by changing the shell thickness but keeping capsule outer diameter and all other parameters same. A thick shell of 1 mm, intermediate thick shell of 0.5 mm and negligible shell of 0 mm are considered while keeping the outer diameter of the capsules equal to 10 mm. The three geometries are shown in Fig. 2.

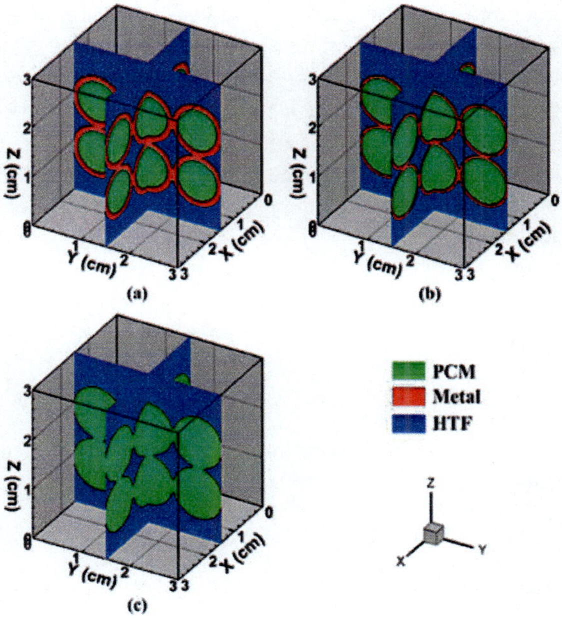

Fig. 2 Geometry with **a** 1 mm, **b** 0.5 mm and **c** 0 mm shell thickness

In the mathematical model, the HTF is considered as a Newtonian fluid and the flow is assumed to be incompressible and laminar. All thermo-physical properties of any distinct phase are considered to remain constant. The model is developed by coupling a phase change model with a fluid flow model. All the properties are volume averaged in terms of α and γ. The volume averaged properties are expressed as follows:

$$\rho_{avg} = \rho_{PCM}(\alpha - \gamma) + \rho_m(\gamma) + \rho_{HTF}(1 - \alpha) \quad (1)$$

$$C_{p_{avg}} = C_{p_{PCM}}(\alpha - \gamma) + C_{p_m}(\gamma) + C_{p_{HTF}}(1 - \alpha) \quad (2)$$

$$k_{avg} = k_{PCM}(\alpha - \gamma) + k_m(\gamma) + k_{HTF}(1 - \alpha) \quad (3)$$

$$\mu_{avg} = \mu_{PCM}(\alpha - \gamma) + \mu_m(\gamma) + \mu_{HTF}(1 - \alpha) \quad (4)$$

In the above-mentioned equations, the density, viscosity, specific heat and thermal conductivity are denoted by ρ, μ, C_p and *k*, respectively. Subscript '*avg*', '*PCM*, '*m*' and '*HTF*' stand for average, PCM, metal and HTF, respectively.

The melting process is simulated using an enthalpy method, which was used by Athawale et al. [15]. Single energy equation expressed in terms of volume average properties is given in Eq. 5. It is discretized using the finite volume method and is solved explicitly to obtain the enthalpy at each control volume. The overall liquid fraction (f_{liq}) of the domain, which is the ratio of amount of liquid PCM present in the domain at any time to the total amount of PCM present, is calculated using Eq. 6. The temperature is calculated depending upon the value of f_{liq}. If the f_{liq} is greater than 0 but less than 1, the temperature at that control volume is equal to T_m. Value of f_{liq} less than or equal to 0 is considered as 0 and greater than equal to 1 is considered as 1. Then the temperature of that control volume is calculated from Eq. 6 using updated *H* and f_{liq} values. The process is repeated until convergence is achieved:

$$\frac{\partial}{\partial t}(\rho_{avg}H) + \nabla \cdot \left(\rho_{avg}\vec{U}H\right) = \nabla \cdot (K_{avg}\nabla T) \quad (5)$$

$$H = C_{p_{avg}}T + (\alpha - \gamma)f_{liq}L \quad (6)$$

The SIMPLER algorithm is used to obtain flow field by solving the continuity equation given in Eq. 7 and the momentum equation given in Eq. 8:

$$\nabla \cdot \vec{U} = 0 \quad (7)$$

$$\rho_{avg}\frac{\partial \vec{U}}{\partial t} + \rho_{avg}\left(\vec{U}\cdot\nabla\right)\vec{U} = -\nabla P$$

$$+\mu_{avg}\nabla^2\vec{U} + (\alpha - \gamma)\rho_{pcm}g\beta(T - T_m) + B\vec{U} \tag{8}$$

The third term on the right-hand side of Eq. 8 is for the buoyancy-driven flow in liquid PCM. The last term of Eq. 8 is used to restrict the flow in the solid PCM and metallic shell. The parameter B in the last term is given as

$$B = -\alpha C_1 \frac{\left(1 - f_{liq}(\alpha - \gamma)\right)}{C_2 + \left(f_{liq}(\alpha - \gamma)\right)^3} \tag{9}$$

In Eq. 9, C1 and C2 are constants having very large and small values, respectively. In the liquid PCM region, both f_{liq} and (α- γ) are 1. So, the value of B is 0. But in the metallic shell and solid PCM region, f_{liq} is 0. Hence, B has a very high value. So, in metallic and solid PCM regions, the velocity is forced to zero due to this term.

4 Results and Discussion

The model was validated for macro-encapsulated PCM with no shell in Athawale et al. [15] with the results given in Khodadadi et al. [18].

For the present study, the liquid fraction contours corresponding to 1 mm shell thickness at 60 s and 80 s are shown in Fig. 3a and b. With time, the melting process proceeds. At 80 s only, the PCM at the central part of the capsule is yet to melt. Similarly, Fig. 3c and d shows the liquid fraction contours for 0.5 mm shell and Fig. 3e and f for 0 mm shell system at 60 and 80 s, respectively. At 60 s, melting of PCM in 1 mm shell is more than that of 0 mm shell, and the melting pattern of the 0.5 mm shell is in between the other two systems.

It can also be seen that in the case of capsules with a metallic shell, the melting starts from the entire periphery and proceeds toward the center almost uniformly. But for the system without any shell, melting starts from the bottom part of the capsule. The PCM which is at the top of the capsule and near the wall is not melted at 60 s. Thus, melting is not uniform. This is because the HTF comes into contact with the bottom part of the capsule first. But in the presence of the shell, the melting starts from the entire periphery. Due to the high thermal conductivity of the shell, heat is transferred from the bottom side to the entire periphery.

Figure 4 shows the corresponding temperature contours for the 3 cases. The capsule temperature increases with time as it takes heat from the HTF. As the capsule temperature increases, the temperature difference between the HTF and capsules decreases which causes less energy transfer from HTF to the capsules. Hence, the drop in HTF temperature decreases with time. The temperature is higher at the top part of the PCM present inside the capsule of 1 mm shell at 80 s because of the presence of natural convection in liquid PCM. The effect of shell thickness is observed from the temperature contours. At 60 s, the 1 mm thick shell system has higher average PCM temperature than 0 mm shell system. The average PCM temperature

Fig. 3 Liquid fraction contours of **a** 1 mm, **c** 0.5 mm and **e** 0 mm shell systems at 60 s and **b** 1 mm, **d** 0.5 mm and **f** 0 mm shell systems at 80 s

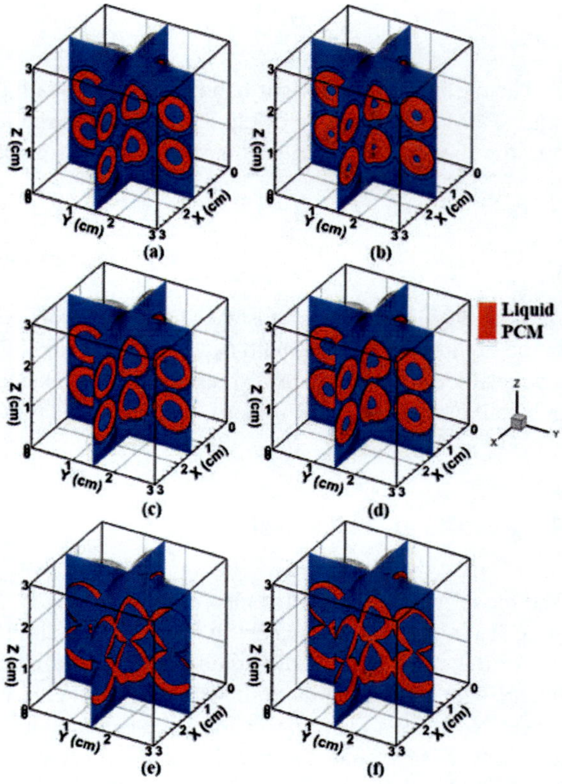

of 0.5 mm shell system is intermediate. Having high thermal conductivity, the shell takes energy from the HTF and transfers it to the PCM at a very high rate.

Figure 5 illustrates the comparison of evolution of overall liquid fraction of the three systems. The start time of melting delays with increase in shell thickness, as the heat needs to cross the metallic shell to melt the PCM. Due to the absence of shell, PCM of 0 mm shell system comes in contact with HTF directly. Therefore, in a 0 mm shell system melting starts at 5.4 s, whereas melting starts at 10.5 and 13.4 s in 0.5 and 1 mm shell systems, respectively. However, thicker metallic shell results in faster melting. The melting completion times of 1, 0.5 and 0 mm shell systems are 93 s, 108.6 s and 174 s, respectively. So, the time required for completing the melting of the 0 mm shell system is 60.22 and 87.09% more than that of the 0.5 and 1 mm shell systems, respectively.

The changes in latent and sensible energy storage in the three systems with time are shown in Fig. 6a and b. After completion of melting, the overall latent energy stored in 1, 0.5 and 0 mm shell systems are 714.41 J, 1017.19 J and 1385.67 J, respectively. This is due to the reduction in PCM volume with increase in shell thickness. The amount of PCM in a thick shell capsule is lower than the amount of PCM in a thin shell capsule. At 180 s, the 1 mm shell thickness system stores highest

Fig. 4 Temperature contours of **a** 1 mm, **c** 0.5 mm and **e** 0 mm shell systems at 60 s and **b** 1 mm, **d** 0.5 mm and **f** 0 mm shell systems at 80 s

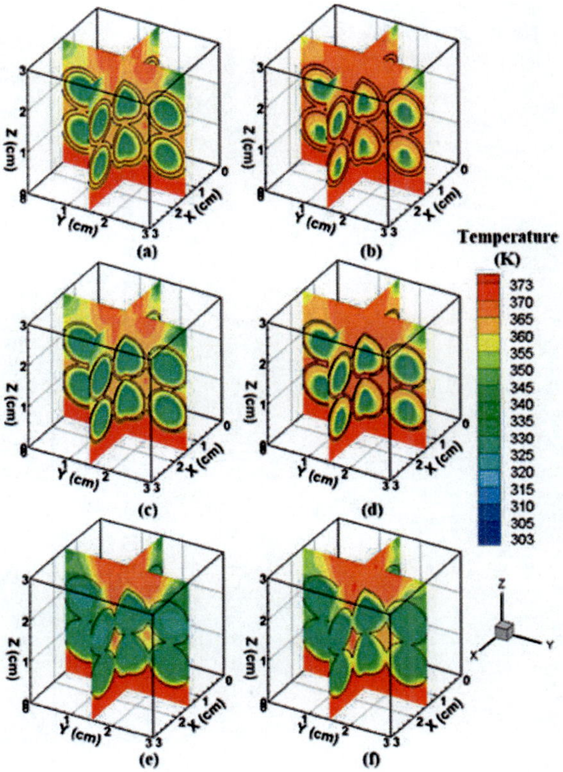

Fig. 5 Variation of overall liquid fraction of the three systems with time

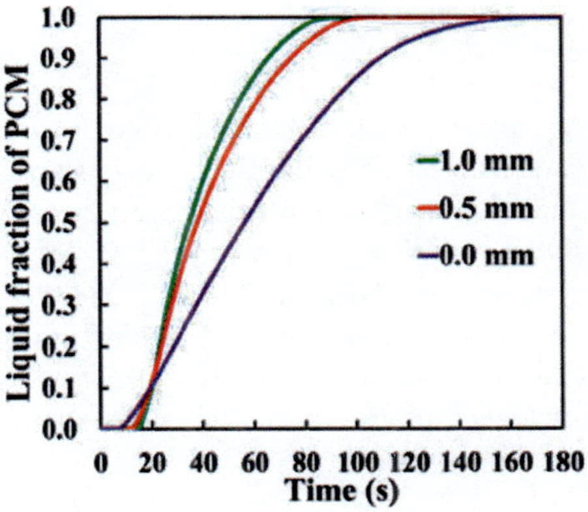

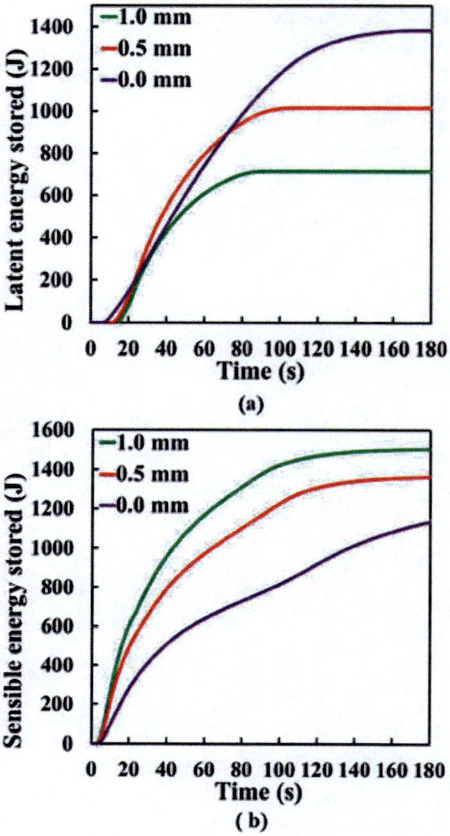

Fig. 6 Comparison of **a** latent and **b** sensible energy stored with time

sensible energy and 0 mm shell system stores highest latent energy. However, in a combined sensible and latent energy storage system, the total energy stored is more important than the individual energy components. The comparison of total energy stored with time is shown in Fig. 7. Total energy stored in 1, 0.5 and 0 mm shell systems after 180 s are 2200.97 J, 2383.69 J and 2523.58 J, respectively. The 0 mm shell system stores 5.87 and 14.66% more energy than that of 0.5 and 1 mm shell systems, respectively. However, the energy storage rates for the thicker shell systems are higher for most of the charging period.

5 Conclusions

The melting and energy storage characteristics during charging of combined sensible and latent energy storage systems containing multiple encapsulated PCM are numerically studied. The metallic shell is also considered, and parametric study is performed

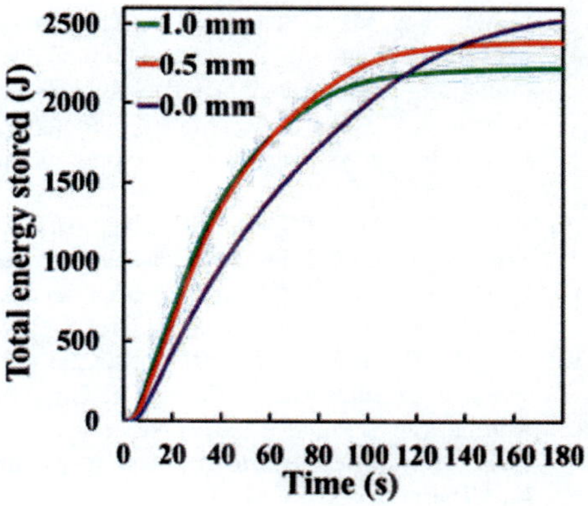

Fig. 7 Comparison of total energy stored with time

by varying the shell thickness to investigate the effect of shell thickness on melting and energy storage of the system. Three different systems having shell thickness 1, 0.5 and 0 mm are studied. The results show that the 0 mm shell system stores 5.87 and 14.66% more energy than that of 0.5 and 1 mm shell systems, respectively. But the 0 mm shell system takes 60.22 and 87.09% more time than the 0.5 and 1 mm shell systems for complete melting. So, use of a thick metallic shell is recommended when melting rate is important. But it reduces PCM content in the system, hence latent and total energy storing ability of the system is reduced.

Acknowledgements This work has been supported by Department of Science and Technology (DST) SERB project CRG/2021/003780.

References

1. Sahoo SK, Das MK, Rath P (2016) Application of TCE-PCM based heat sinks for cooling of electronic components: a review. Renew Sust Energy Rev 59:550–582
2. Salunkhe PB, Shembekar PS (2012) A review on effect of phase change material encapsulation on the thermal performance of a system. Renew Sustain Energy Rev 16(8):5603–5616
3. Liu Z, Yu ZJ, Yang T, Qin D, Li S, Zhang G, Haghighat F, Joybari MM (2018) A review on macro-encapsulated phase change material for building envelope applications. Build Environ 144:281–294
4. Patel JR, Joshi V, Rathod MK (2020) Thermal performance investigations of the melting and solidification in differently shaped macro-capsules saturated with phase change material. J Ener Stor 31:101635
5. Ettouney H, El-Dessouky H, Al-Ali A (2005) Heat transfer during phase change of paraffin wax stored in spherical shells. J Sol Energy Eng 127(3):357–365
6. Tan FL (2008) Constrained and unconstrained melting inside a sphere. Int Commun Heat Mass Transfer 35(4):466–475
7. Nallusamy N, Sampath S, Velraj R (2007) Experimental investigation on a combined sensible and latent heat storage system integrated with constant/varying (solar) heat sources. Renewable Energy 32(7):1206–1227

8. Li MJ, Jin B, Ma Z, Yuan F (2018) Experimental and numerical study on the performance of a new high-temperature packed-bed thermal energy storage system with macroencapsulation of molten salt phase change material. Appl Energy 221:1–15
9. Nagano K, Takeda S, Mochida T, Shimakura K (2004) Thermal characteristics of a direct heat exchange system between granules with phase change material and air. Appl Therm Eng 24(14–15):2131–2144
10. Regin AF, Solanki SC, Saini JS (2009) An analysis of a packed bed latent heat thermal energy storage system using PCM capsules: numerical investigation. Renew Energy 34(7):1765–1773
11. Bhagat K, Saha SK (2016) Numerical analysis of latent heat thermal energy storage using encapsulated phase change material for solar thermal power plant. Renew Energy 95:323–336
12. Yang L, Zhang X, Xu G (2014) Thermal performance of a solar storage packed bed using spherical capsules filled with PCM having different melting points. Ener Build 68:639–646
13. Karthikeyan S, Velraj R (2012) Numerical investigation of packed bed storage unit filled with PCM encapsulated spherical containers–a comparison between various mathematical models. Int J Therm Sci 60:153–160
14. Athawale V, Bhattacharya A, Rath P (2021) Prediction of melting characteristics of encapsulated phase change material energy storage systems. Int J Heat Mass Transf 181:121872
15. Athawale V, Jakhar A, Jegatheesan M, Rath P, Bhattacharya A (2022) A 3D resolved-geometry model for unstructured and structured packed bed encapsulated phase change material system. J Ener Stor 51:104430
16. Wei J, Kawaguchi Y, Hirano S, Takeuchi H (2005) Study on a PCM heat storage system for rapid heat supply. Appl Therm Eng 25(17–18):2903–2920
17. Dinesh BVS, Bhattacharya A (2019) Effect of foam geometry on heat absorption characteristics of PCM-metal foam composite thermal energy storage systems. Int J Heat Mass Transf 134:866–883
18. Khodadadi JM, Zhang Y (2001) Effects of buoyancy-driven convection on melting within spherical containers. Int J Heat Mass Transf 44(8):1605–1608

Heat Exchanger Design for Two-Stage HDH Desalination Unit

Vajeer Baba Shaik, Srinivas Tangellapalli, and Rajeev Kukreja

Abstract Based on the thermal water cycle principle, a two-stage humidification-dehumidification (HDH) desalination cycle has been developed for producing the maximum water output. In the two-stage HDH cycle, a heat exchanger (condenser) plays a major role because it transfers maximum heat, as well as condensation, between the working fluids. This work aims to design a heat exchanger for a two-stage HDH cycle with the heating and humidification process. So in this cycle, the heat exchanger is used as a two-stage air preheater and a dehumidifier to improve the desalinated yield with the optimum area and performance index. In a two-stage plant, the humidifier and air preheater use hot water. For more desalination yield, the first-stage dehumidifier uses chilled water, and an air-cooled evaporator in the VCR is used as a second-stage dehumidifier. So airflow rate, transverse pitch and tube diameter are selected as parameters to optimise the objectives. At 100 kg/h of airflow rate, the second-stage dehumidifier yields 7.62 LPH of freshwater, whereas the first-stage dehumidifier yields only 0.22 LPH of freshwater. The obtained trend demonstrates that high desalinated water production is feasible with optimal heat exchanger area and performance.

Keywords HDH · Desalination · Heat exchanger · Thermal water cycle · Two-stage

Nomenclature

HEx	Heat exchanger
APH-I	First-stage air preheater
WCDH	Water cooled dehumidifier
APH-II	Second-stage air preheater

V. B. Shaik (✉) · S. Tangellapalli · R. Kukreja
Department of Mechanical Engineering, Dr B R Ambedkar National Institute of Technology, Punjab 144011, India
e-mail: vajeerbabashaik@gmail.com

S. Das et al. (eds.), *Proceedings of the 1st International Conference on Fluid, Thermal and Energy Systems*, Lecture Notes in Mechanical Engineering,
https://doi.org/10.1007/978-981-99-5990-7_23

P_{EI}	Performance evaluation index
A	Area: m^2
D	Diameter: m
μ	Dynamic viscosity: Kg/m s
ρ	Density: kg/m^3
u	Flow velocity: m/s
t	Thickness: m
N_u	Nusselt number
Re	Reynolds number
f	Friction factor
m	Mass flow rate: kg/h
N	No. of tubes
S_T	Tubes transverse pitch
N	No. of tubes
S_T	Tubes transverse pitch: m
S_L	Tubes longitudinal pitch: m
F_s	Fin spacing: m
L_2	Fin width: m
L_3	Fin height: m
w	Specific humidity

Subscript

m	Maximum
dw	Desalinated water
f	Fin

1 Introduction

In general, the world's population is increasing, and essentials such as food and drinking water are becoming increasingly scarce. The saline water density is higher as compared to the drinking water density [1]. Reverse osmosis (RO), multi-effect (ME) and vapour compression are the major techniques for converting saline water to desalinated water, but all of these methods require a larger area for installation, are more expensive, require difficulty operating, and mainly consume higher energy sources. However a thermal-based HDH method is an efficient technology for evaporating freshwater from saline water with a lower installation area. A thermal-based HDH desalination cycle works at atmospheric pressure, so there is no need for vacuum pumps [2]. And solar HDH plant has the potential to be installed in small-scale industries also [3].

In the HDH cycle, more researchers used various types of heat exchangers, i.e. shell and tube, helical coil and coil tube heat exchanger for the dehumidification process. M. Abdelgaied et al. [4] explored the hybrid HDH-RO unit with a shell and tube dehumidifier to generate fresh water from the saline water. W. He et al. [5] mathematically investigate the performance of the novel HDD (humidification-dehumidification desalination) system with low-grade waste heat. They recover the low-grade heat from the exhaust gas and utilise it to preheat the atmospheric air before the humidification process at the plate heat exchanger. In an HDD system, a shell and tube cross flow heat exchanger is used as a dehumidifier to maximise the performance yield of the system. And they recommend that the proposed HDD system is far better than old traditional work. W.F. He et al. [6] investigate the performance and entropy rate of the direct contact dehumidifier used in the water-heated HDH desalination cycle. The plate heat exchanger is used as a waste heat recovery application in the plant, and they recommend the waste recovery process in the HDH desalination system as being more efficient. V. Panthalookaran et al. [7] optimise the overall efficiency of the gravity-driven solar HDH distillation plant. A helical coil heat exchanger is used as a dehumidifier and parameters, i.e. coil pitch and diameter are considered as design variables for the simulation process. In the optimisation results, they reveal that the performance yield of the plant significantly improved when the efficiency of the heat exchanger was increased. Albaik et al. [8, 9] numerically investigated the performance of the finned tube heat exchanger used as a finned tube bed in the absorption desalination system. And they reported that a 25% improvement in water production rate was achieved while fin spacing was reduced from 11.5 to 3.6 mm. In this absorption desalination system, a coiled tube heat exchanger is used as a condenser for evaluating the freshwater rate of the system. E. Ahmed et al. [10] focused on optimisation of a finned tube heat exchanger to improve desalination output from a fog desalination system, which is then compared to experimental data. Chiranjeevi and Srinivas [11] simulated and experimented on a cogeneration system, i.e. a VAR-operated HDH unit for generating the water yield and cooling. For an air preheating, condensing and dehumidifying process, a concentric tube HEx was used in the integrated HDH-VAR unit. Y. Zang et al. [12] proposed a novel humidification and dehumidification cycle operating with a heat pump unit. An experimental investigation was carried out into the HDH-HP cycle, and shell and tube HEx were used as dehumidifiers in the two-stage pilot plant. In the experimental observations, the effect of airflow rate is more significant to the performance yield as compared to sea water and air temperature effects. They strongly recommended that air temperatures be insignificant for the productivity of the pilot plant. Srinivas [13] simulated the HP operated HDH system to meet the multiple demands, i.e. water yield, cooling and hot water. In simulation process, VCR's water-cooled condenser and evaporator was used as a humidifier and dehumidifier. And simulation results show that the plant energy performance ratio is increased up to 0.86. Lawal et al. [14] were carried out to analyse the performance of a heat pump operated HDH desalination system with a finned coil tube dehumidifier. And simulation results are reported. Vajeer Baba Shaik and Srinivas Tangellapalli [15] focused on water-cooled dehumidifier design for a two-stage HDH system operated by heat pump unit and simulation results were reported. Y. Zhao et al. [16] are experimentally investigating the performance yield of the four-stage HDH desalination plant using plate heat exchanger and direct contact heat exchanger as a preheater and dehumidifier. M. Li et al. [17] investigate the transient computational fluid dynamics analysis of finned tube heat exchanger performance on an absorption desalination system. Then the authors observed and concluded, the fin tube heat exchangers are the most efficient device for improving the desalination output of the pilot system. However, small-scale water processing industries, including watercraft and islands, require an efficient device with a minimum area for maximising the freshwater yield.

In the two-stage HDH-VCR cycle, a proposed heat exchanger design is highlighted with optimum area and performance. This work aims to evaluate the performance of HEx with the maximum heat transfer rate. Moreover, this work is aimed at analysing the effect of airflow rate, transverse pitch and tube diameter on the desalinated yield of the two-stage HDH plant.

2 Description of Two-Stage HDH-VCR System

Figure 1 shows the two-stage HDH-VCR system operating on an off-grid load for water and cooling generation. Solar energy is used as an energy source for a two-stage HDH-VCR system. Electricity is generated through a solar PV plant, and it drives the fan, pump and VCR. The two-stage HDH plant operates based on the thermal water cycle for water production. At the start, atmospheric air is sucked into the duct through a fan and transferred to the heating and humidification process. But before the heating and humidification processes, atmospheric air is preheated in an air preheater. So a HEx device is used for air preheating process in preheater and coconut coir is packed with a humidifier for the heating and humidification process. A solar water heater is used as a hot water generator and transfers it to the humidifier and air preheater in two stages. The hot water is sprayed into the coir packing through the nozzle sprayer. The temperature and specific humidity of the air are increased in the heating and humidification process and then followed by the dehumidifier. The dehumidifier is also a heat exchanger device, but a first-stage dehumidifier, where cooling and dehumidification processes are used for water production. In the dehumidification process, condensation takes place. After the first stage of the humidification and dehumidification process, the cycle is repeated in the second stage, but the VCR's evaporator is used as the second-stage dehumidifier for water production. R134a is used as a working fluid in the VCR cycle; therefore, water production in the second stage is greater compared to the first stage. For maximum water production, a heat exchanger (condenser type) device is used as an air preheater and a dehumidifier to maintain temperature and specific humidity through sensible heating. As per the norms of WHO (World Health Organization), drinking water contains a minimum of 500 ppm salinity, so minerals are added to desalinated water and then it will be useful for drinking purposes.

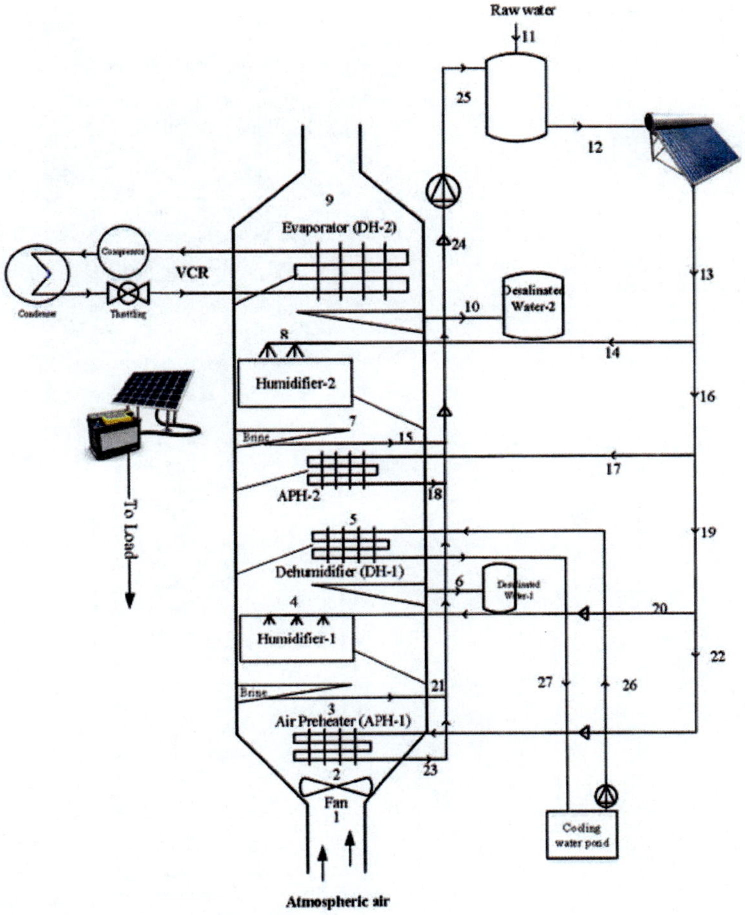

Fig. 1 VCR operated Two-stage HDH cycle

3 Modelling of the Heat Exchanger

The condenser is one of the most effective heat exchanger applications because of its flexibility in design and manufacturing, the wide operating temperature range, and the condensation rate (water from the air) process that also takes place. Figure 2 shows the cross-flow unmixed HEx (condenser type) used for air preheating and dehumidification process in the integrated two-stage HDH-VCR unit. In heat exchanger design, the following assumptions are made, i.e. it operates under steady state conditions, fouling factor is negligible, and properties of air don't change with temperature. The air side non-dimensional factors are calculated [18] [19].

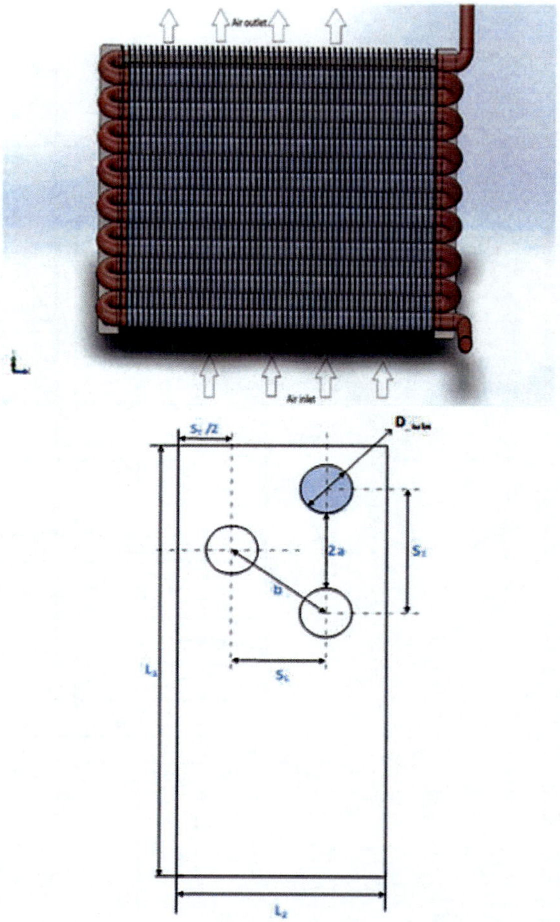

Fig. 2 Heat exchanger in two-stage HDH system

$$Nu = 1.565Re^{0.3414}\left[N\frac{F_s}{D}\right]^{-0.165}\left[\frac{S_T}{S_L}\right]^{0.0558} \tag{1}$$

$$f = 20.713Re^{-0.3489}\left[N\frac{F_s}{D}\right]^{-0.1676}\left[\frac{S_T}{S_L}\right]^{0.6265} \tag{2}$$

$$Re = \left[\frac{\rho D u_m}{\mu}\right] \tag{3}$$

where u_m is the maximum airflow velocity.

$$u_m = u\left[\frac{S_T}{2a}\right] if 2a \leq 2b$$

(or)

$$u_m = u\left[\frac{S_T}{2b}\right] if 2a \geq 2b \tag{4}$$

Then $2a = (S_T - D) - \left[(S_T - D)t_f n_f\right]$,

$$2b = (S_D - D) - \left[(S_T - D)t_f n_f\right]$$

The number of tubes 'N' that are calculated by

$$N = \frac{\mathrm{L_3}}{\mathrm{S_T}}\left[\frac{(\mathrm{L_2/S_L}) + 1}{2}\right] + [(\mathrm{L_3/S_T}) - 1]\left[\frac{(\mathrm{L_2/S_L}) - 1}{2}\right] \tag{5}$$

Using non-dimensional factors, i.e. Nu, friction of the HEx performance is evaluated by[20]

$$\mathrm{PEi} = \left(\frac{\mathrm{Nu}}{\mathrm{f}^{0.333}}\right) \tag{6}$$

During dehumidification process, desalination yield was generated and it was calculated by

$$\mathrm{m_{dw}} = \mathrm{m_{da}}(\mathrm{wi}) - \mathrm{m_{da}}(\mathrm{w_o}) \tag{7}$$

4 Results and Discussion

Figure 3 represents the HEx area and performance index variation at different tube diameters with air volumetric flow rate changes from 100 m^3/h to 250 m^3/h. When the tube diameter is increased by 20%, the performance index increases by a maximum of 16.66% and a minimum of 10.02%. This is due to the air velocity and local air movement at the minimal flow region for a 35 mm transverse tube pitch, which consequently increases the HEx surface area and performance index with a maximum heat capacity rate. And the operating temperatures of the APH, WCDH and evaporator are also different in the system. As a component-wise evaporator, performance is better than a water-cooled dehumidifier (WCDH). The APH effectiveness is 60%, and the temperatures of the inlet water and air are 50 and 28 °C, respectively.

Figure 4 represents the longitudinal pitch effect on heat exchanger area and performance index. At different tube pitches, the HEx area and performance are increased because more air is passed over the fins, which increases the convective air heat transfer coefficient and non-dimensional factors. Due to air velocity and local movement, the maximum flow region for a 10 mm tube diameter, pitch and flow rate changes will consequently increase the heat transfer surface area and performance index with a maximum heat capacity rate. In each heat exchanger design process,

Fig. 3 Variation effect of area and P_{EI} with airflow rate and tube diameter

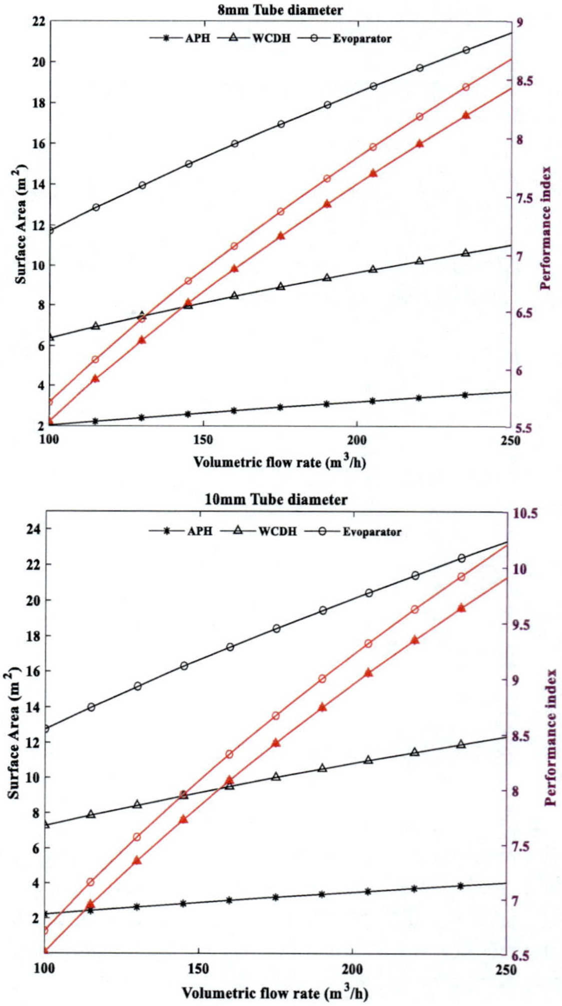

the air properties, i.e. humidity and temperature, are changed because of the heating and cooling process.

Figure 5 represents the role of airflow rate on a specific outcome of a two-stage HDH system. At 100 m^3/h of airflow rate, the first stage's water production is 0.22 L per hour (LPH), and the second stage's is 7.62 LPH. Compared to the first-stage desalinated output, the second stage water production is maximum because refrigerant (R134a) is used as a working fluid and the refrigerant temperature is the evaporator temperature in the HDH system. The increase in airflow rate will increase the amount of water produced in the two-stage HDH system and also cause an increase in longitudinal pitch, causing a decrease in water production but a heat exchanger performance-wise increase because the humid contact surface is decreased.

Fig. 4 Variation of Area and Performance with airflow rate and tubular pitch

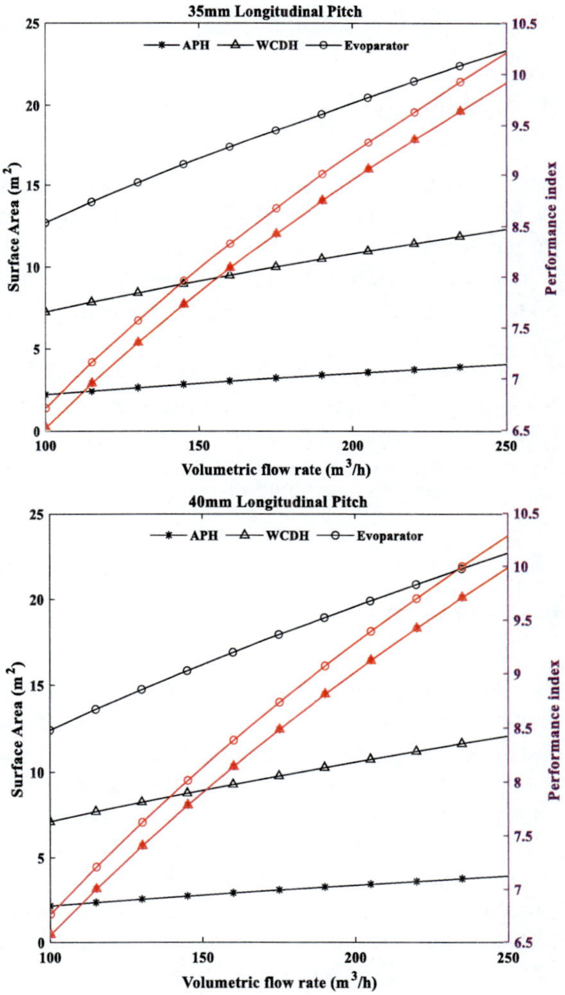

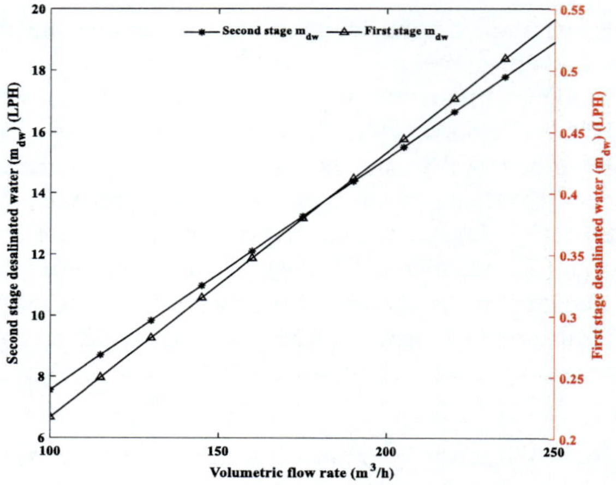

Fig. 5 Variation of 2-stage desalinated outputs with airflow rate

5 Validation

The performance of the two-stage HDH desalinated plant is simulated and compared to the experimental results of Y. Zhao et al. [16]. And they investigated the performance of a four-stage HDH desalination with the experimental setup. In this setup, a direct contact HEx and a plate HEx were used as a dehumidifier and air preheater. So the proposed two-stage HDH plant performance was validated with these experimental values, and it is concluded that the proposed model produces 80% of the desalinated freshwater yield, but only 45.16% of the yield was produced from the four-stage HDH plant. Because in a two-stage HDH desalination plant, the condenser HEx is used as a first-stage preheater, DH (water-cooled DH), second-stage preheater and evaporator (second DH) with the minimum area. So this type of HEx is more efficient in HDH desalination.

6 Conclusions

In heat exchanger design and plant operation, the operational conditions of volumetric flow rate, tube diameter and transverse pitch are thoroughly investigated. A mathematical simulation of the heat transfer area and performance of a heat exchanger is carried out. A two-stage HDH plant performance yield test was also carried out. As seen in the heat transfer surface area and pilot plant performance yield variance as air volumetric flow rate varies from 100 m^3/h to 250 m^3/h, a higher airflow rate increases heat transfer quality and freshwater yield. This heat exchanger's design and operating conditions are considered in off-design conditions. At 10 mm tube bore, heat exchanger area and its performance increased by 42.09% and 16.66%, but just 40% and 10.02% improvement at 8 mm tube diameter. At 100 m^3/h of air mass flow rate, the 7.62 LPH of freshwater yield is produced in the second-stage HDH process, but just 0.22 LPH of freshwater is formed in the first-stage HDH process. Similarly, at 250 m^3/h, the second stage produces 19.5 LPH of water but just 0.53 LPH in the first stage. In vertical ducts, humidifiers will improve the quality (specific humidity) of air. The heat exchangers are placed as an APH, WCDH and evaporator so that the heating and cooling processes take place. The second-stage dehumidifier produces 92.5% of the freshwater yield compared to the first-stage dehumidifier because of the HEx design and its arrangement, working fluid and operational temperatures. This will improve the condensation (air to water).

Acknowledgements The authors acknowledge the project grant of the SERB (Science and Engineering Research Board) under Impacting Research and Technology (IMPRINT-2), PAC Water Resources & River Systems, New Delhi, India (IMP/2019/000444).

References

1. AA, Kazmerski LL (2011) Renewable Energy opportunities in water desalination. *Desalination, Trends Technol*
2. Uche J, Serra L, Herrero LA, Valero A, Turégano JA, Torres C (2003) Software for the anlaysis of water and energy systems. Desalination 156(1–3):367–378
3. Giwa A, Fath H, Hasan SW (2016) Humidification-dehumidification desalination process driven by photovoltaic thermal energy recovery (PV-HDH) for small-scale sustainable water and power production. Desalination 377:163–171
4. Abdelgaied M, Kabeel AE, Kandeal AW, Abosheiasha HF, Shalaby SM (2021) Performance assessment of solar PV-driven hybrid HDH-RO desalination system integrated with energy recovery units and solar collectors : theoretical approach. Energy Convers Manag 239(April):114215
5. He W, Yang H, Wen T, Han D (2019) Thermodynamic and economic investigation of a humidification dehumidification desalination system driven by low grade waste heat. Energy Convers Manag 183(June 2018):848–858
6. He WF, Huang L, Xia JR, Zhu WP, Han D, Wu YK (2017) Parametric analysis of a humidification dehumidification desalination system using a direct-contact dehumidifier. Int J Therm Sci 120:31–40
7. Panthalookaran V, Chettiyadan D, Vadacherry J, Kudakasseril K, Parekkadan V (2016) Design of a helical coil dehumidifier for a novel gravity-driven solar distillation unit. Energy Proc 91:294–302
8. Albaik I, Al-Dadah R, Mahmoud S, Solmaz İ (2020) Non-equilibrium numerical modelling of finned tube heat exchanger for adsorption desalination/cooling system using segregated solution approach. Appl Therm Eng 183(July):2021
9. Albaik I, Al-Dadah R, Mahmoud S, Almesfer MK, Ismail MA (2021) A comparison between the packed and coated finned tube for adsorption system using aluminium fumarate: numerical study. Therm Sci Eng Prog 22(October 2020):100859
10. Abu El-Maaty AE, Awad MM, Sultan GI, Hamed AM (2021) Performance study of fog desalination system coupled with evacuated tube solar collector. Desalination 504(August 2020):114960
11. Chiranjeevi C, Srinivas T (2015) Experimental and simulation studies on two stage humidification-dehumidification desalination and cooling plant. Desalination 376:9–16

12. Zhang Y, Zhu C, Zhang H, Zheng W, You S, Zhen Y (2018) Experimental study of a humidification-dehumidification desalination system with heat pump unit. Desalination 442(May):108–117
13. Tangellapalli S (2021) Humidification-dehumidification and heat pump integration for water purifier and air conditioning. Energy Convers Manag 244:114472
14. Lawal D, Antar M, Khalifa A, Zubair S, Al-Sulaiman F (2018) Humidification-dehumidification desalination system operated by a heat pump. Energy Convers Manag 161(January):128–140
15. Shaik VB, Tangellapalli S (2022) Design and simulation of water - cooled dehumidifier for HDH desalination plant, (April 2022):1–20
16. Zhao Y, Zheng H, Liang S, Zhang N, Long Ma X (2019) Experimental research on four-stage cross flow humidification dehumidification (HDH) solar desalination system with direct contact dehumidifiers. Desalination 467(January):147–157
17. Li M, Zhao Y, Long R, Liu Z, Liu W (2021) Computational fluid dynamic study on adsorption-based desalination and cooling systems with stepwise porosity distribution. Desalination 508(February):115048
18. Wang C, Chi K, Chang C (2000) Heat transfer and friction characteristics of plain ® n-and-tube heat exchangers, part II : Correlation 43:0–7
19. Wang C, Chi K (2000) Heat transfer and friction characteristics of plain ® n-and- tube heat exchangers , part I : new experimental data, 43
20. Xie G, Wang Q, Sunden B (2009) Parametric study and multiple correlations on air-side heat transfer and friction characteristics of fin-and-tube heat exchangers with large number of large-diameter tube rows. Appl Therm Eng 29(1):1–16

Effect of Binary Capsule Size Distribution on Unstructured Packed Bed Encapsulated PCM System

Pranaya Keshari Nahak, Vidula Athawale, Jegatheesan M, Prasenjit Rath, and Anirban Bhattacharya

Abstract In the current circumstances, consistent innovation and improvement in energy storage systems is the need of the hour. In this work, the effect of binary size distribution of PCM capsules on the melting characteristics of encapsulated PCM systems is studied by developing a numerical model which resolves the geometry of the individual capsules in the packed bed. The encapsulated spheres containing PCM are distributed in a cubical domain by implementing an optimized dropping and rolling algorithm. This specific novelty is needed to address the issue of real time packing condition of the spheres inside a packed bed structure in which the spherical PCMs might not be organized in a structured way always. The unstructured packing is done by having spheres of two different sizes mixed together. The ratio of the size of spheres of two kinds is varied in order to study the effect of different kinds of binary distribution on the evolution of melting and the heat transfer characteristics.

Keywords Energy storage · Encapsulated PCM · Pore-scale model · Binary size distribution

Nomenclature

A	Momentum source term coefficient: –
C_p	Specific heat: [J/kgK]
f_l	Liquid fraction: –
H	Enthalpy: [J/kg]
K	Thermal conductivity: [W/mK]
L_f	Latent heat of fusion: [J/kg]
P	Pressure: [N/m^2]
S_B	Boussinesq source term: –
T	Temperature: [K]

P. K. Nahak · V. Athawale · J. M · P. Rath · A. Bhattacharya (✉)
School of Mechanical Sciences, IIT Bhubaneswar, Odisha 752050, India
e-mail: anirban@iitbbs.ac.in

S. Das et al. (eds.), *Proceedings of the 1st International Conference on Fluid, Thermal and Energy Systems*, Lecture Notes in Mechanical Engineering,
https://doi.org/10.1007/978-981-99-5990-7_24

$\vec{U}$ Velocity vector: [m/s]

1 Introduction

From the past several decades, the world has witnessed a huge surge in the consumption of energy. The major percentage of this energy is sought from consumable sources like fossil fuels. Due to the non-replenishing nature of these consumable sources, researchers around the world have started looking for different renewable energy sources which can cater to the energy demand of the future. As most of the renewable sources of energy are intermittent while supplying the required energy, a storage mechanism seems to be a prerequisite. Different kinds of energy storage mechanisms have been developed in the past such as electrochemical energy storage, sensible thermal energy storage systems and latent heat thermal energy storage systems (LTSS). One of the main components of the LTSS system is phase change materials (PCM). This has various advantages when compared to sensible thermal energy storage systems, such as higher energy storage density and operation in very narrow temperature range. In this paper, the research mainly focuses on a type of LTSS in which a novel packed bed structure composed of spherical encapsulated PCMs is arranged following an optimized dropping and rolling (ODR) algorithm [1]. Unstructured arrangements are generated by having two different capsule sizes mixed together in the same packed bed. Thus, the study emphasizes on observing the change in heat transfer and melting phenomena with respect to the change in binary distributions of sizes of PCM containing spheres.

2 Literature Review and Objective

The primary usages of PCM can be found in the domains of solar energy storage, thermal protection component in electrical appliances, recovery of waste heat, etc. [2, 3]. Most of the time, the phase change materials suffer from low thermal conductivity. A lot of researchers proposed the idea of the increasing the heat transfer area by means of the encapsulation process to enhance the heat flux interaction between the HTF and the PCM [4]. Earlier, experiments were conducted on single capsule and a packed bed consisting of multiple capsules by Yagi and Akiyama [5]. The elaborate study of some of the key parameters in the packed bed system is important and owing to this, researchers have studied and developed many numerical models such as the single phase model and two-phase models. In the single phase model [6], the temperature calculated at any instant is treated the same in both the liquid as well as the solid phase and this model is mostly preferred for the PCM having high thermal conductivity and high capacitance with respect to the HTF. On the contrary, the temperature measurement in the two-phase model is distinct and there are again three divisions of this type, i.e. Schuman's model, the continuous solid phase model and the concentric dispersion model. Schumann's model [6] is mainly convection driven model because the thermal conduction is completely neglected and only the exchange of heat fluxes is considered by the convection mechanism. One of its applications in the packed bed system can be found in reference [7]. On the other hand, the continuous solid phase model [6, 8] considers the heat transfer by conduction mode only. The model treats the bed containing all the particles as a continuous solid phase media, not as individual particles. The concentric dispersion model [6, 9] treats each particle individually and hence, this model is able to capture the thermal gradient inside each particle. This model also neglects any kind of heat transfer between the particles while the only kind of heat transfer that persists is between the fluid and bed. More detailed explanations regarding all these models can be found in reference [10].

From literature review, it can be seen that very few studies are present in which the capsule geometry in a packed bed system is resolved. In our previous work, a resolved geometry system was studied in 2D in [11] and subsequently in 3D in [12]. However, although the study presented in [12] considered 3D packing of capsules in an unstructured manner unlike previous studies in literature, the capsule size was taken to be equal and the model for resolving the geometry was accordingly constructed.

In the present study, the model has been improved to include the effect of size variation during 3D packing of spherical capsules in a packed bed system with resolved geometry. The main aim of the current work is to capture the melting and heat transfer characteristics of a packed bed system by changing the radius ratios of the PCM capsules with a binary size distribution considering that the capsules are arranged in unstructured manner in a cubical domain.

3 Mathematical Formulation

In this model, a domain having cubical shape is considered in which the spherical PCMs are distributed following the ODR algorithm. The ODR algorithm [1] starts with initial random positioning of spheres in a (x,y) plane having a range of radius values following binary distribution. Then, the spheres are dropped from a height one after another. The incoming sphere directly contacts the bottom wall if there is no other sphere in its downward path of travel. If it finds any sphere, then it rolls over it. From there, two conditions may arise. If there is no other sphere except the first contacted sphere, then it completely rolls over the first sphere and travels to next bottom layer. But, if it finds another sphere, then it rolls over the two contacted spheres and two conditions may arise again. If it contacts a third sphere, then stability condition is checked to know whether the incoming sphere can be properly positioned above the three spheres. Otherwise, it rolls over all the three spheres to move to next bottom layer. In actual, a complex set of iterative procedures and mathematical formulations are involved in the ODR algorithm, the details of which can be found from reference [1]. The initial domain temperature is set at the normal room temperature and the heat transfer fluid enters through the bottom of the domain with a uniform velocity and temperature. The heat transfer fluid flows over the PCM capsules and starts the melting process gradually. Wall conditions are applied for the momentum equations for the walls. Adiabatic condition is applied for the energy equation for the side walls and the outflow condition is applied at the outlet. The dynamic viscosity value is taken as very high at the interface to model the effect of capsule shell and thus keep the flow of HTF and that in the liquid PCM separate. Some assumptions were made for this model which are described as follows:

1. No thermal resistance is provided by the capsule shell as the thickness is considered to be negligible.
2. For both liquid PCM and solid PCM, the thermo-physical properties are considered as constant.
3. The flow is considered to be laminar and incompressible.
4. At a fixed melting temperature, the phase change occurs.

The mathematical formulation of the mass, momentum and the energy equations are done using the finite volume approach. The packed bed geometry structure having the spheres with unequal size and unstructured arrangement is crucial because the pouring of spheres inside a container will not follow a proper structured arrangement. Hence, to emulate that scenario, the ODR algorithm [1] is used. Additionally, to test the effect of varying size of the spheres, multiple spheres are generated following a binary size distribution. After creation of all the spheres, the region inside the sphere is assumed to be filled with the phase change material and provisions are made for flow of HTF for the region outside of the spheres. To distinguish between the HTF and the PCM, a parameter known as PCM volume fraction 'φ' is defined. The PCM volume fraction is calculated and assigned following a mapping technique, the details of which can be found in reference [13]. The averaged thermo-physical properties are given as follows:

$$\rho_{avg} = \rho_{pcm}\varphi + \rho_{htf}(1-\varphi)$$

$$\mu_{avg} = \mu_{pcm}\varphi + \mu_{htf}(1-\varphi)$$

$$k_{avg} = k_{pcm}\varphi + k_{htf}(1-\varphi)$$

$$C_{p_{avg}} = C_{p_{pcm}}\varphi + C_{p_{htf}}(1-\varphi) \tag{1}$$

Here, the subscript 'avg', 'htf' and 'pcm' mean average properties, properties of heat transfer fluid and phase change material, respectively. The basic equations which govern the flow field are the continuity equation and the momentum equation. The SIMPLER algorithm is used to solve this set of equations in a coupled manner. The formulation is done using the methodology described in [11, 12]. The continuity equation is given by

$$\nabla \bullet \overrightarrow{U} = 0 \tag{2}$$

And the momentum equation is presented as

$$\rho_{avg}\frac{\partial \overrightarrow{U}}{\partial t} + \rho_{avg}\left(\overrightarrow{U} \cdot \nabla\right)\overrightarrow{U} = -\nabla P + \mu_{avg}\nabla^2\overrightarrow{U} + S_B - A\overrightarrow{U} \tag{3}$$

The energy conservation equation is given as

$$\frac{\partial}{\partial t}\left(\rho_{avg}H\right) + \nabla \cdot \left(\rho_{avg}\overrightarrow{U}H\right) = \nabla \cdot \left(k_{avg}\nabla T\right) \tag{4}$$

The energy conservation equation is solved in an explicit way and the 'H' consists of both the latent energy and sensible energy which is given as follows:

$$H = \left[C_{p_{pcm}}\varphi + C_{p_{htf}}(1-\varphi)\right]T + \varphi f_l L_f \tag{5}$$

A cubical domain having each side of 3 cm is considered for numerical analysis. The appropriate boundary condition along with the schematic of the arrangement of the PCM capsules in the cubical domain with binary distribution can be visualized in Fig. 1. The study with the unstructured arrangement for equal size spheres has been earlier conducted by Vidula et al. [12]. However the unstructured arrangement with varying size spheres has been carried out for the first time in this study.

4 Results and Discussion

The arrangement of the encapsulated PCM inside a cubical domain is very crucial, because it changes the melting pattern, the conduction and convection phenomena locally. In the current study, the schematic of the arrangement for a particular binary distribution of spheres can be seen in Fig. 1 in which the HTF enters the domain from the bottom with a velocity of 0.1 cm/s and temperature of 373 K and it flows out from the top face with outflow condition. Paraffin is used as PCM and water as

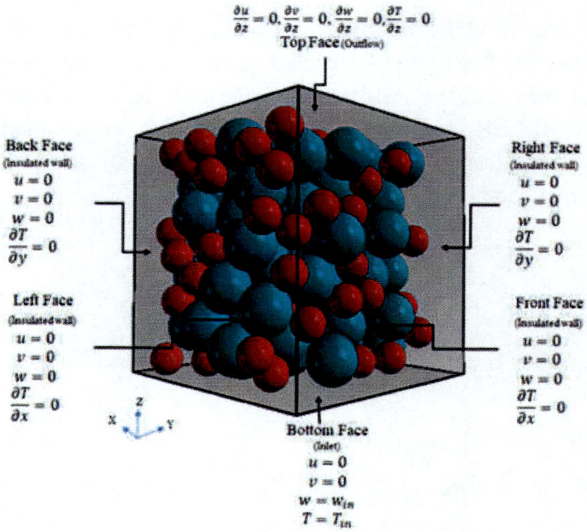

Fig. 1 Schematic of domain with unstructured PCM capsule arrangement

HTF in this study whose thermo-physical properties are given in Table 1. Then this packed bed structure is imported into the earlier validated numerical CFD setup to do the rest of the analysis. The validation was earlier carried out against the findings of Khodadadi and Zhang [14] in which flow and melting pattern of a single spherical capsule was investigated by Vidula et al. [12] which found a very fair agreement.

For the present study, for all the cases, the cubical space is kept constant while there is no constraint on the amount of PCM volume stored. The cubical domain is fully packed with the capsules until no additional capsules can be introduced in the domain. Additionally, the number of small spheres and the number of large spheres are equal for all cases. The small sphere size is kept constant at $r_1 = 0.2$ cm for all cases, but, the size of large sphere is allowed to vary according to the ratio of 1:1, 1.5:1 and 2:1 for the three different cases.

In Fig. 2, the variation of melt fraction of the PCM at two discrete times is studied and its variation with the change in the radius of large spheres is also depicted. The

Table 1 Thermophysical properties [11]

Parameter	PCM	HTF
Thermal conductivity (W/mK)	0.22	0.615
Specific heat (J/kgK)	2100	4178
Density (kg/m^3)	805	996
Viscosity (Ns/m^2)	3.6×10^{-3}	0.79×10^{-3}
Latent heat (J/kg)	169×10^3	-
Melting temperature (K)	333	-
Coefficient of thermal expansion (1/K)	3.1×10^{-4}	-

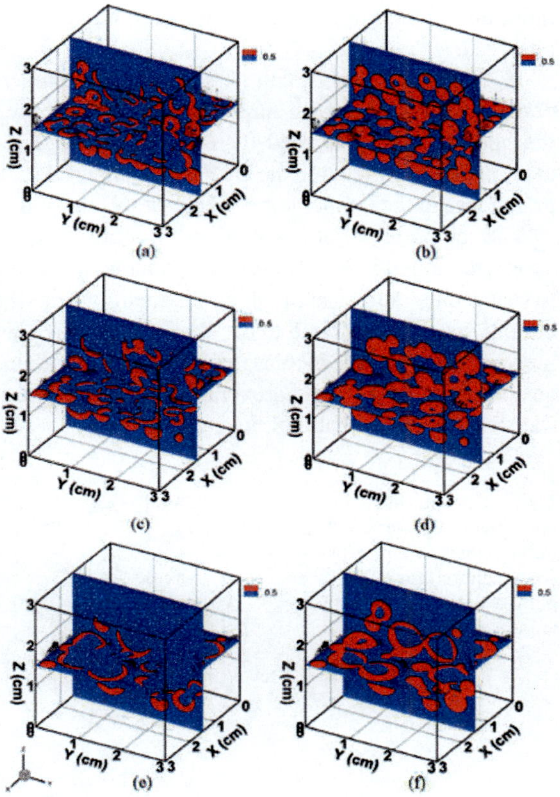

Fig. 2 Contour of liquid fraction at times t = 30 s (left column) and t = 60 s (right column) for radius ratio = 1, 1.5 and 2 (top to bottom)

content of PCM melted mainly depends upon the amount of surface area of the encapsulated PCM exposed to the HTF.

From Fig. 2, it can be clearly discerned that the PCM near the surface of the spherical capsules melt initially and then the melting gradually moves to the inside of the spheres. Hence, the bigger capsules suffer from slow penetration of the heat flux from the HTF and subsequently it leads to lesser fraction of melting. On an aggregate, this gives rise to lesser liquid fraction of the whole system at any particular time for the larger radius ratio case.

The velocity field also regulates the effective flow and convection of heat in the PCM capsules. As the flow field becomes stronger, the heat transfer rate also becomes higher. This phenomenon can also be observed in Fig. 2. When the radius ratio is small, the flow passage between the capsules also narrows down and subsequently, the HTF accelerates in these regions and improves the melting more.

The heat transfer characteristics can also be noticed from the contours of temperature in Fig. 3. At the time of 30 s, the cold region inside the PCM is very less in the case of lowest radius ratio. While a lot of patches of cold region can be observed in the case of the highest radius ratio. Similar is the case at 60 s also. In the lowest radius ratio, due to more effective heat transfer, higher temperature inside the capsules is observed.

In the HTF region, it can be seen that the temperature in the upper half of the domain starts to rise with time. Initially, most of the sensible energy coming from the bottom of HTF is used for melting the PCM. Later, as the time evolves, the amount of energy leaving the HTF and being used for the PCM decreases. This leads to increase in the temperature of HTF at the top of the domain with increasing time.

The temporal variations of liquid fraction of all capsules with time for all the cases are plotted in Fig. 4. It can be seen that the overall melting time is least for the lowest radius ratio case in which the melting time is found to be 101.2 s while for other two cases it is found to be 105.9 s and 144.7 s with increase in radius ratio. As mentioned earlier, the increase in melting time with increasing radius ratio is mainly attributed to the decreasing overall heat transfer area and decreasing effective heat transfer rate to the centre of the capsules.

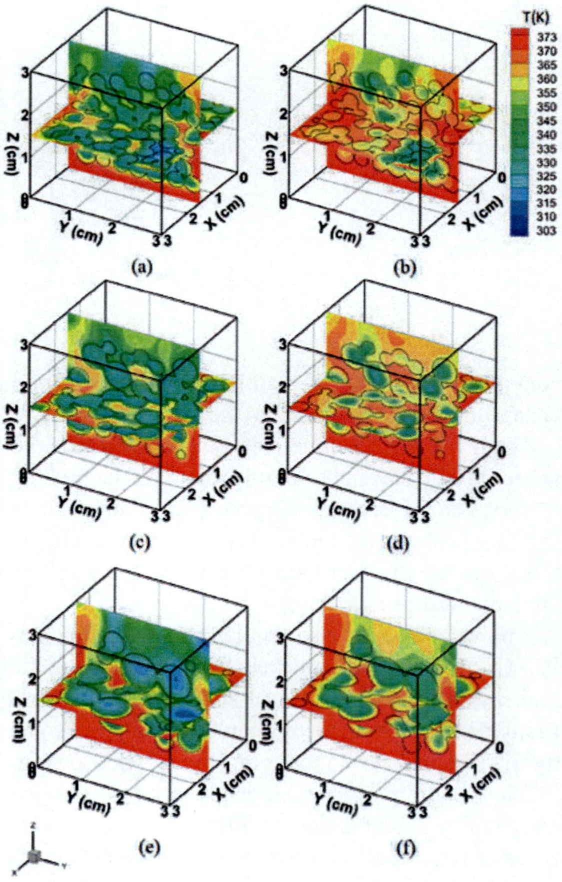

Fig. 3 Contour of temperature at times t = 30 s (left column) and t = 60 s (right column) for radius ratio = 1, 1.5 and 2 (top to bottom)

Fig. 4 Line plots of liquid fraction v/s time

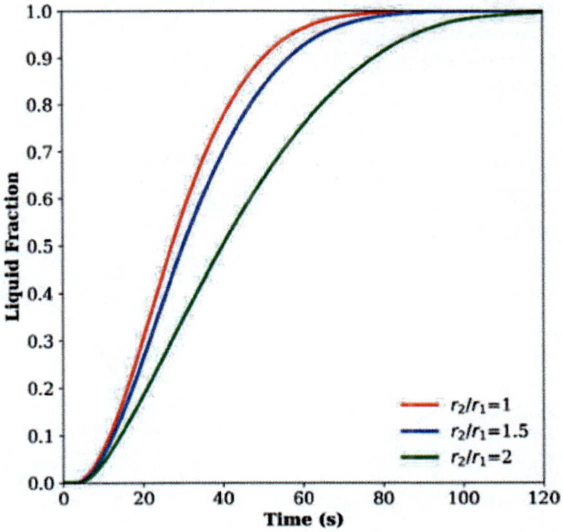

One more thing to note here is that the melting rate, i.e. the slope of the line plot decreases with the increase of radius ratio and the gap between the line plots also widens for the consecutive cases. This is because, the increase in the radius follows arithmetic progression, but the volume manifests as the cube of the radius value for each case. Hence, as the radius of a single capsule increases, the amount of remote PCM (i.e. far from the HTF region) becomes large which leads to increase in the melting time. Figure 5a and b shows the temporal evolution of sensible and latent energy content of the entire packed bed system considering all the capsules for all the three cases. It is seen that initially, the slope is more for each case and gradually the slope decreases. This change is mainly attributed to the way in which the system consumes the thermal energy from the HTF.

In this study, the cubical domain volume is held constant for all cases while the entire domain is fully packed with PCM capsules. Thus the PCM volume can vary for different cases depending on the effectiveness of packing. Hence, there is a difference in the total amount of latent energy stored in each of the cases. The latent energy content is highest for the smallest radius ratio case with a value of 1.56 kJ. For other two cases, it is found to be 1.54 kJ and 1.42 kJ, respectively. It can be observed that when the radius ratio increases to 1.5, the latent energy stored decreases by 1.53%, while when the radius ratio increases to 2, the latent energy stored decreases by 9.15%.

The variations of the total energy with time for the 3 radius ratio cases are plotted in Fig. 6. As the total energy is the sum of sensible and latent energy, the amount of total energy storage also follows a similar trend.

Fig. 5 Line plots of sensible energy v/s time (top) and latent energy v/s time (bottom)

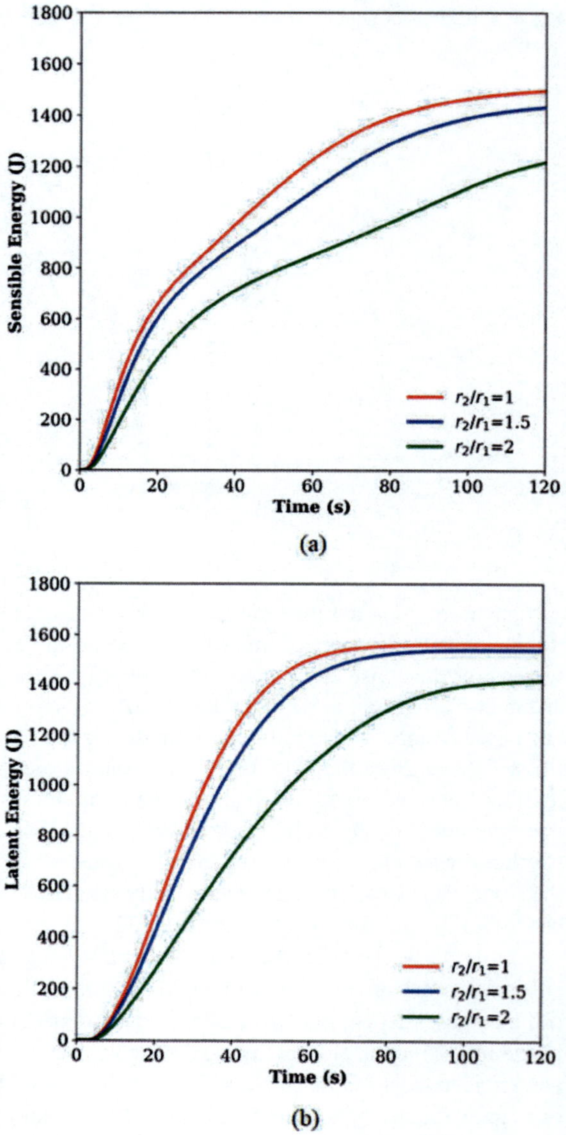

5 Conclusions

The main highlight of this research work is to construct an unstructured geometrical packing of spherical PCM capsules following a binary size distribution. As the cubical space is fixed, the amount of PCM volume stored is allowed to vary and it is observed that with the highest radius ratio, the amount of total latent energy stored is the lowest

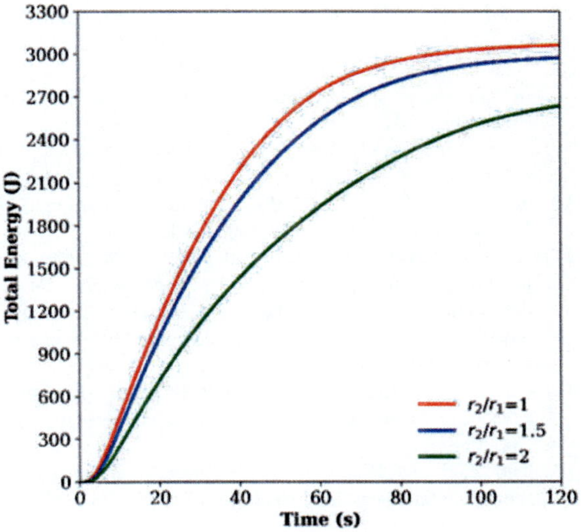

Fig. 6 Line plots of total energy v/s time

and this is mainly attributed to the more void spaces present in the highest radius ratio cases. It is also observed that with the increase in the radius ratio of the spheres, the melting time increases and this is due to the considerable rise of volume of remote PCM away from the proximity of the HTF. Quantitatively, considerable change in melting time can also be seen. When the radius ratio is raised to 1.5, the melting time increases by 4.64%, but by doubling the radius ratio, the melting time is found to increase by 42.98%.

Acknowledgements This work has been supported by DST (Department of Science and Technology) SERB project CRG/2021/003780.

References

1. Hitti K, Bernacki M (2013) Optimized dropping and rolling (ODR) method for packing of poly-disperse spheres. Appl Math Model 37(8):5715–5722
2. Sharif MA, Al-Abidi AA, Mat S, Sopian K, Ruslan MH, Sulaiman MY, Rosli MAM (2015) Review of the application of phase change material for heating and domestic hot water systems. Renew Sustain Energy Rev 42:557–568
3. Regin AF, Solanki SC, Saini JS (2008) Heat transfer characteristics of thermal energy storage system using PCM capsules: a review. Renew Sustain Energy Rev 12(9):2438–2458
4. Alam TE, Dhau JS, Goswami DY, Stefanakos E (2015) Macroencapsulation and characterization of phase change materials for latent heat thermal energy storage systems. Appl Energy 154:92–101
5. Yagi J, Akiyama T (1995) Storage of thermal energy for effective use of waste heat from industries. J Mater Process Technol 48(1–4):793–804
6. Ismail KA, Stuginsky R Jr (1999) A parametric study on possible fixed bed models for pcm and sensible heat storage. Appl Therm Eng 19(7):757–788

7. Sanderson TM, Cunningham GT (1995) Performance and efficient design of packed bed thermal storage systems. Part 1, Appl Energy 50(2):119–132
8. Raul A, Jain M, Gaikwad S, Saha SK (2018) Modelling and experimental study of latent heat thermal energy storage with encapsulated PCMs for solar thermal applications. Appl Therm Eng 143:415–428
9. Wu M, Xu C, He YL (2014) Dynamic thermal performance analysis of a molten-salt packed-bed thermal energy storage system using PCM capsules. Appl Energy 121:184–195
10. de Gracia A, Cabeza LF (2017) Numerical simulation of a PCM packed bed system: a review. Renew Sustain Energy Rev 69:1055–1063
11. Athawale V, Bhattacharya A, Rath P (2021) Prediction of melting characteristics of encapsulated phase change material energy storage systems. Int J Heat Mass Transf 181:121872
12. Athawale V, Jakhar A, Jegatheesan M, Rath P, Bhattacharya A (2022) A 3D resolved-geometry model for unstructured and structured packed bed encapsulated phase change material system. J Energy Stor 51:104430
13. Dinesh BVS, Bhattacharya A (2019) Effect of foam geometry on heat absorption characteristics of PCM-metal foam composite thermal energy storage systems. Int J Heat Mass Transf 134:866–883
14. Khodadadi JM, Zhang Y (2001) Effects of buoyancy-driven convection on melting within spherical containers. Int J Heat Mass Transf 44(8):1605–1618

Effect of Capsule Shape on Melting and Energy Storage Rates for Encapsulated PCM-Based Systems

Tanmayee Kopparthi, Vidula Athawale, Prasenjit Rath, and Anirban Bhattacharya

Abstract Latent heat thermal energy storage using phase change materials (PCM) has become a topic of interest as it has the advantages of high energy storage density. Encapsulated PCM energy storage systems are being widely studied as they have large heat transfer area available for charging by the heat transfer fluid (HTF). The shape of the capsules can considerably affect the melting rate and thus the energy charging characteristics. In this work, a numerical analysis of encapsulated PCM system with a single capsule but with different shapes has been carried out by developing a 3D finite volume-based model. The developed model is used to predict the melting and energy storage rates of spherical, cuboidal, cylindrical and triangular prism-shaped capsules keeping the PCM volume constant. It is observed that the time required for melting is reduced by 3, 10.39, 22.64, 11.68% for cubical, horizontal cylinder, horizontal prism and vertical prism capsules as compared to the spherical capsule having the same volume. Thus, the horizontal prism capsule has the best melting and energy storage rates among all the considered shapes.

Keywords Encapsulated phase change material · Heat transfer fluid · Latent energy storage · Melting rate

Nomenclature

C_p	Specific heat (J/kgK)
f_l	Volume fraction of liquid PCM
H	Volume-averaged enthalpy (J/kg)
K	Thermal conductivity (W/mK)
P	Pressure (N/m^2)
R	Resistance source coefficient
S_B	Boussinesq source term

T. Kopparthi · V. Athawale · P. Rath · A. Bhattacharya (✉)
School of Mechanical Sciences, IIT Bhubaneswar, Odisha 752050, India
e-mail: anirban@iitbbs.ac.in

S. Das et al. (eds.), *Proceedings of the 1st International Conference on Fluid, Thermal and Energy Systems*, Lecture Notes in Mechanical Engineering,
https://doi.org/10.1007/978-981-99-5990-7_25

t	Time
T	Temperature
T_{in}	Inlet temperature
$\vec{U}$	Velocity (m/s)
u, v, w	x, y, z Velocity component (m/s)
w_{in}	Inlet velocity
x,y,z	X, Y, Z coordinate (m)
ρ	Density (kg/m^3)
μ	Viscosity (Ns/m^2)

1 Introduction

The depletion of conventional energy sources like coal and petroleum and rapid increase in energy demands have led to the use of renewable energy sources. Although solar energy is a viable option, it is intermittent in nature which calls for the need of developing efficient thermal energy storage (TES) systems. Phase change materials (PCM) are used to store this large amount of thermal energy in the form of latent heat. They are advantageous as they have high energy density and they operate at constant temperature. Organic PCMs are generally preferred due to their operating temperature ranges and stability at room temperature. But a major drawback is that they have very low thermal conductivity which gives rise to low melting and heat transfer rates. To address this issue, many thermal enhancement techniques have been developed like adding metal fins [1], adding metal foams [2], nanoparticle addition [3] and encapsulation [4]. Encapsulation is the process of incorporating PCM in capsules made of materials having high thermal conductivity. This increases the melting rate by increasing the effective heat transfer area between the HTF and PCM. Also, encapsulation ensures enhanced life of the system by reducing the reactivity with surroundings and also curtails the issues of corrosion and leakage.

2 Literature Review and Objective

Encapsulated PCM TES systems consisting of spherical capsules have been thoroughly studied by researchers. Tan et al. [5] studied PCM melting in a sphere in two different modes. In one mode, called constrained melting, the solid is fixed and thus cannot move in the molten PCM. In the other mode, called unconstrained melting, the solid PCM can sink or float depending on its density difference with the molten PCM. Khodadadi and Zhang [6] analysed PCM melting in a spherical capsule using finite volume method, taking into account the effect of natural convection. Veerappan et al. [7] analytically studied the effects of capsule size, PCM material, HTF temperature and PCM temperature on rate of solidification. Considerable research has been done to increase heat transfer rates for encapsulated PCM systems. Narasimhan et al. [8] examined the melting pattern of encapsulated PCM system with PCM distributed with high conductivity particles. Fan et al. [9] studied the effect of fin height on melting rate in a spherical capsule with circumferential fins. Melting was found to be faster as fin height increased. It is noteworthy that finned capsules are difficult to manufacture and interior fins block the PCM flow within the capsule, reducing heat transfer rates.

Encapsulated PCM systems with spherical capsules are not the best option as spheres have low surface area to volume (SA/V) ratio which reduces melting/ solidification rates. Thus, some researchers studied the performance of capsules of other shapes having higher SA/V ratio. Cheng et al. [10] studied melting in red blood cell-shaped capsule. Its charging rate was observed to be 2.12 times that of a spherical capsule. Awasthi et al. [11] investigated the effect of irregularities on the capsule by creating sinusoidal wavy surfaces on the circular cross-section of a cylindrical capsule. Patel et al. [12] studied melting in spherical, cubical, triangular and plus-shaped capsules with constant temperature boundary condition.

From literature review, it is observed that capsules of shapes other than sphere have not been explored to their full potential. Furthermore, a comparative study between different shapes has not been reported yet. Patel et al. [12] did a comparative study as discussed above considering only isothermal boundary condition. However, the nature of HTF flow over the capsules strongly affects the melting characteristics, which was not considered in the aforementioned study. The study presented in this paper aims to examine the melting and energy storage characteristics of PCM capsules of different shapes. We compared the melting and energy storage rates of spherical, cubical, horizontal cylinder, vertical cylinder and triangular prism-shaped capsules, taking into consideration the HTF flow over the capsules and keeping the capsule volume constant.

3 Numerical Model

A 3D cubical domain with a side of 2 cm is taken with a single PCM capsule having a volume of 1 cm^3 placed at the centre. Paraffin wax is chosen as the PCM. Water is considered as the HTF. Table 1 lists their thermo-physical properties. Initially, the domain is at 303 K. HTF flows from the bottom boundary with a velocity of 0.1 cm/s at 373 K and leaves the domain from the top boundary. All other walls are insulated. The thickness of the shell is considered to be negligible. The interface between the PCM capsule and the HTF does not require a separate boundary condition as heat transfer from the HTF to the PCM is assumed to be continuous with negligible thermal resistance. A very high viscosity value (10^{20} Ns/m^2) is artificially assigned to the interface region, which prevents HTF flow in PCM region and keeps the natural convection in PCM separate from the forced convection in the HTF. This value is arbitrarily taken and any large value which suppresses the flow velocity to zero at the interface cells can be used. The domain of interest and the boundary conditions are given in Fig. 1.

An encapsulation fraction parameter φ is defined to denote the quantity of PCM in each control volume. A value of 1 corresponds to PCM region and a value of 0 corresponds to HTF zone. A value ranging from 0 to 1 refers to PCM-HTF interface region. A mapping technique given in [13] is used to calculate the encapsulation factor. The thermo-physical properties are volume averaged in terms of φ, as given in Eq. 1.

$$\rho_{avg} = \rho_{pcm}\varphi + \rho_{htf}(1 - \varphi)$$

$$k_{avg} = k_{pcm}\varphi + k_{htf}(1 - \varphi)$$

$$C_{p_{avg}} = C_{p_{pcm}}\varphi + C_{p_{htf}}(1 - \varphi)$$

Table 1 Properties used for simulation [14]

Property	Paraffin	Water
Density (kg/m^3)	805	996
Viscosity (Ns/m^2)	3.6×10^{-3}	0.79×10^{-3}
Thermal conductivity (W/mK)	0.22	0.615
Specific heat (J/kgK)	2100	4178
Latent enthalpy (J/kg)	169×10^3	-
Melting point (K)	333	-
Thermal expansion coefficient (1/K)	3×10^{-4}	-

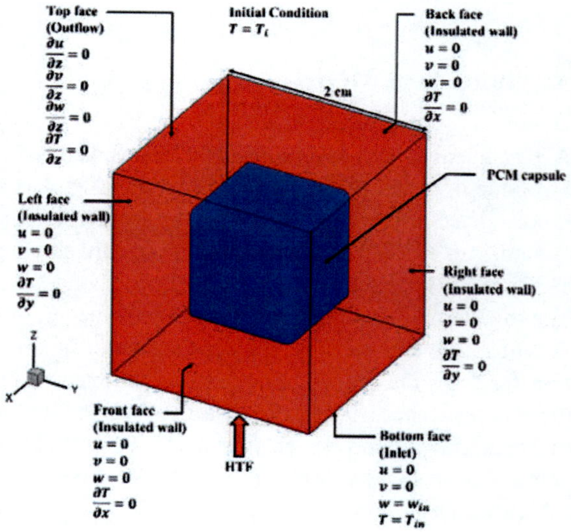

Fig. 1 Schematic of domain with cubical PCM capsule and boundary conditions

$$\mu_{avg} = \mu_{pcm}\varphi + \mu_{htf}(1-\varphi) \tag{1}$$

Here, the thermo-physical properties of PCM and HTF are denoted by their respective subscripts.

Based on the nature of the problem, some assumptions are made:

- Phase change of PCM takes place at isothermal conditions.
- HTF and PCM have constant thermo-physical properties.
- Capsule shell is of infinitesimally small thickness and thus, thermal resistance is negligible.
- Natural convection within the liquid PCM region is enabled by considering Boussinesq approximation.

The governing equations comprises the continuity, momentum and energy equations as given in Eqs. 2, 3 and 4.

$$\nabla \bullet \vec{U} = 0 \tag{2}$$

$$\rho_{avg}\left[\frac{\partial \vec{U}}{\partial t} + \left(\vec{U}\cdot\nabla\right)\vec{U}\right] = \mu_{avg}\nabla^2\vec{U} - \nabla P - R\vec{U} + S_B \tag{3}$$

$$\frac{\partial}{\partial t}(\rho_{avg}H) + \nabla\cdot\left(\rho_{avg}\vec{U}H\right) = \nabla\cdot(k_{avg}\nabla T) \tag{4}$$

Here, P is the pressure and $\vec{U}$ is the velocity, T represents temperature and H denotes the volume-averaged enthalpy given by $H = \left[C_{p_{pcm}}\varphi + C_{p_{htf}}(1-\varphi)\right]T + \varphi f_l L_f$ where f_l is PCM liquid fraction and L_f is latent heat of melting. Explicit finite volume method (FVM) is used to discretise the energy equation. The following steps are used for finding the evolution of melting in each control volume.

- By solving the energy equation, the updated value of H can be found, which is then used to calculate the updated liquid fraction value as $f_{l_{new}} = \left[H - (C_{p_{pcm}}\varphi + C_{p_{htf}}(1-\varphi))T_m\right]/\varphi L_f$ where T_m is melting temperature of PCM.
- If $f_{l_{new}} < 0$, melting has not begun and $f_{l_{new}}$ value in that cell is set to 0. Temperature, which is less than T_m is thus calculated as $T = H/(C_{p_{pcm}}\varphi + C_{p_{htf}}(1-\varphi))$
- If $f_{l_{new}} > 1$, melting has completed and $f_{l_{new}}$ value in that cell is set to 1. Temperature which is greater than T_m is thus calculated as $T = H/(C_{p_{pcm}}\varphi + C_{p_{htf}}(1-\varphi) + \varphi L_f)$
- If $f_{l_{new}}$ is between 0 and 1, the control volume is in the melting interface and T is given the value of T_m

Natural convection in molten PCM region is taken into account as a source term with Boussinesq approximation defined as $S_B = \rho_{pcm}\vec{g}\beta\varphi(T - T_m)$ where β is thermal expansion coefficient and $\vec{g}$ is gravitational acceleration. The term $R\vec{U}$ is used to drive the velocity to 0 in the solid PCM region and is defined as a function of f_l as given in [15].

4 Results and discussion

The present FORTRAN code is an extension of an existing in-house code. The existing program has been validated by Vidula et al. [16] with the results of Khodadadi and Zhang [6]. For validation, they have considered a spherical capsule with constant temperature boundary condition. They also considered the effect of natural convection within the molten PCM. The results almost match with the results obtained in [6] with an error of 3.92% in dimensionless melting time. Here, we studied the melting pattern and energy storage rate of encapsulated TES systems consisting of spherical, cubical, horizontal and vertical cylinder, and horizontal and vertical triangular prism-shaped capsules. The PCM volume was kept constant for all the cases. The geometries of the aforementioned systems are shown in Fig. 2.

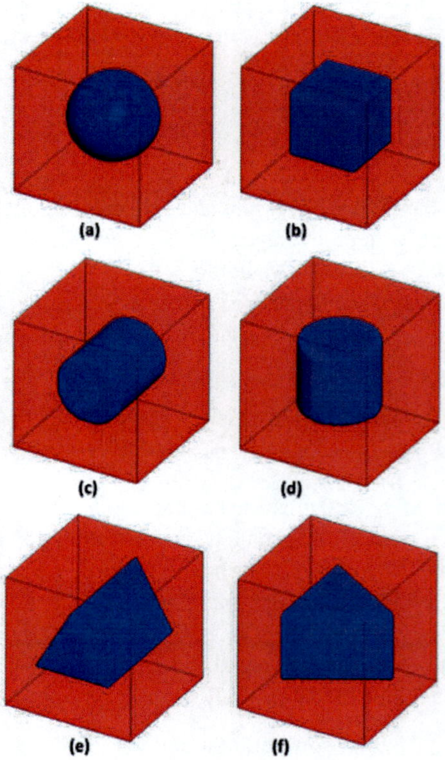

Fig. 2 Shapes considered: **a** sphere **b** cube **c** horizontal cylinder **d** vertical cylinder **e** horizontal prism **f** vertical prism

When HTF flows over the PCM capsule, heat transfer takes place from HTF to PCM due to the temperature difference between them. Initially, conduction is the predominant mode of heat transfer, resulting in increase in the solid PCM temperature. When the PCM temperature reaches its melting point, it starts melting.

Buoyancy-driven natural convection currents are generated in the molten PCM significantly affecting the heat transfer characteristics. Figure 3 illustrates the liquid fraction contours for different shaped capsules at 120 s. It is observed that melting is the fastest in horizontal triangular prism-shaped capsule and the slowest in the vertical cylinder capsule. This is because horizontal prism has very high SA/V ratio and also larger base area available for transfer of heat from HTF to PCM. The temperature contours for the same time are shown in Fig. 4. Again, it is found that the temperature in the horizontal prism is highest due to the earlier completion of melting resulting in further sensible heating.

Figure 5 shows the liquid fraction versus time plots for all the six capsules. Initially, there is large temperature gradient between HTF and PCM, resulting in heat transfer by conduction. As a result, the temperature of PCM increases, which reduces the temperature difference between PCM and HTF. Thus, the effect of conduction mode of heat transfer decreases with time. When PCM starts melting, the volume of molten PCM increases and convection currents are established due to thermal buoyancy.

Fig. 3 Liquid fraction contours **a** sphere **b** cube **c** horizontal cylinder **d** vertical cylinder **e** horizontal prism **f** vertical prism at 120 s

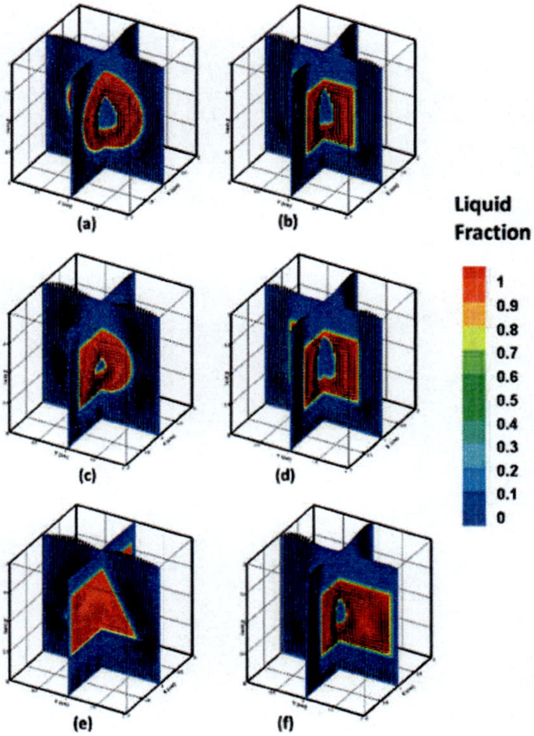

Fig. 4 Temperature contours **a** sphere **b** cube **c** horizontal cylinder **d** vertical cylinder **e** horizontal prism **f** vertical prism at 120 s

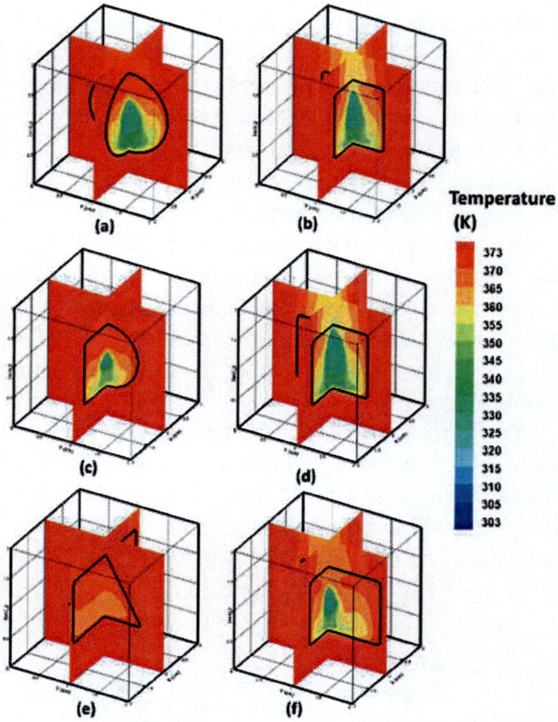

Fig. 5 Liquid fraction versus time plot for capsules of various shapes

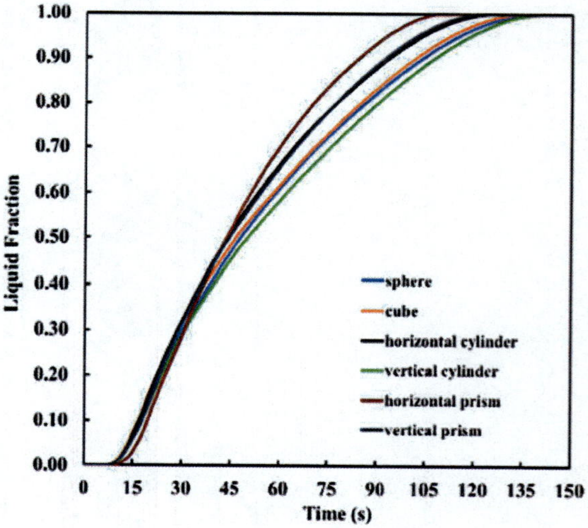

The total PCM melting time is 139.6, 135.5, 125.1, 142.1, 108, 123.3 s for spherical, cubical, horizontal cylinder, vertical cylinder, horizontal prism and vertical prism capsules, respectively. This means that the total melting time is 3, 10.39, 22.64, 11.68% lower in cubical, horizontal cylinder, horizontal triangular prism and vertical triangular prism-shaped capsules than a spherical capsule having the same volume. In the study performed by Patel et al. [12], the total melting time of horizontal triangular prism capsule is 26.78% lower than the spherical capsule. The vertical cylinder performs the worst with a melting time higher than the spherical capsule.

The rate of change of sensible energy, latent energy and total energy for all the six capsules are shown in Figs. 6a, b and 7, respectively. During the charging process, PCM stores energy as sensible and latent heat. It can be observed that the horizontal prism stores the highest amount of sensible and total energy as it melts faster. Thus, horizontal prism has the best melting and energy storage rates.

Figure 8 shows the change of latent energy fraction with time for all the six capsules. Latent energy fraction refers to the ratio of latent energy to the total energy stored in the system and can be considered as the effectiveness of the system in storing energy as latent energy. It is observed that the horizontal prism case shows the highest peak value of 0.7 which is slightly higher than all the other cases. This is because melting rate is higher in this case, which results in increase in the latent energy during the intermediate period of melting. But after melting has completely occurred, sensible energy increases, which reduces the latent energy fraction for this case. For the other cases, the variation of latent energy fraction is almost similar with a slightly flatter curve.

5 Conclusions

The melting and energy storage characteristics of an encapsulated PCM TES system subjected to HTF flow with sphere, cube, horizontal and vertical cylinder, and prism-shaped capsules with the same PCM volume are simulated using a numerical model. It is observed that initially the melting is conduction dominated and natural convection becomes more significant with increasing liquid fraction of PCM. Also, the capsule geometry significantly affects the melting rate. It is found that melting time is decreased by 3%, 10.39%, 22.64%, 11.68% for cubical, horizontal cylinder, horizontal prism and vertical prism capsules as compared to the traditional spherical capsule of same volume while the melting time increases for the vertical cylinder case. Thus, horizontal prism capsule has the highest melting and energy storage rates among all the considered shapes. This is due to the higher surface area for a fixed volume and higher base area available for transfer of heat from HTF to PCM. The developed model can be extended in future to study packed bed systems with multiple PCM capsules of different shapes.

Fig. 6 Variation of **a** sensible energy and **b** latent energy with time

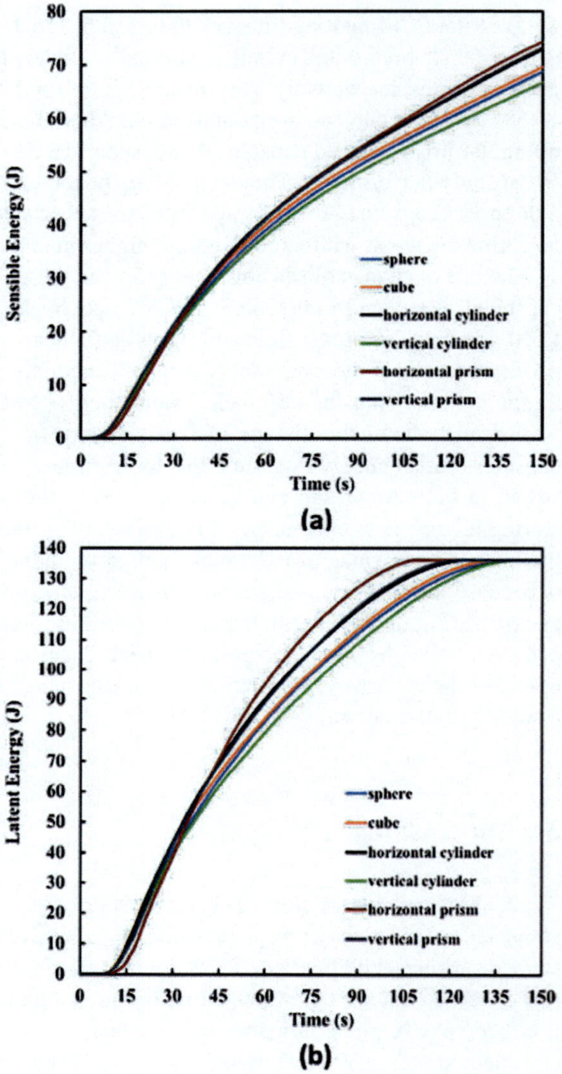

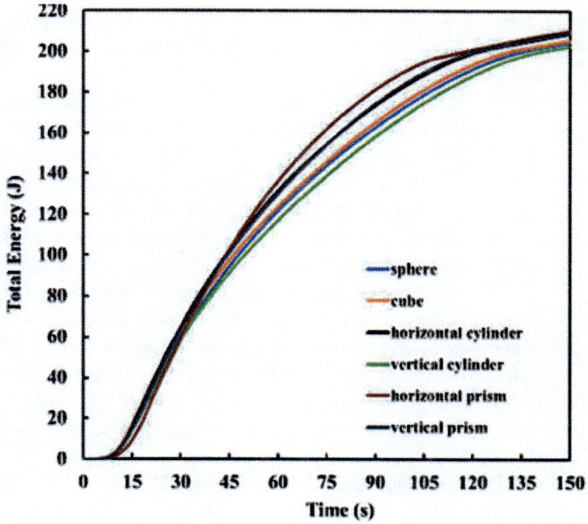

Fig. 7 Total energy versus time plot for capsules of various shapes

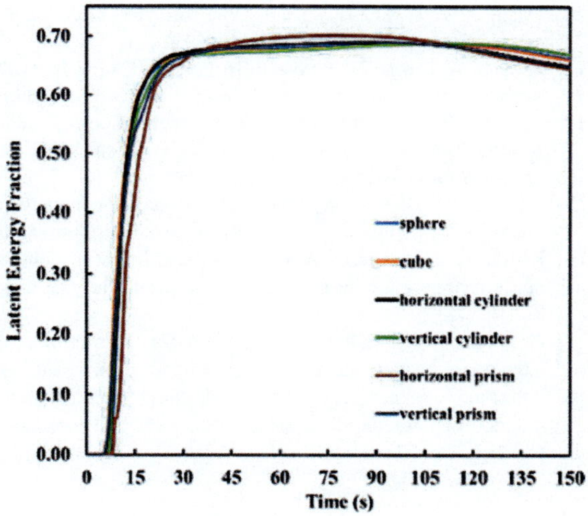

Fig. 8 Latent energy fraction versus time plot for capsules of various shapes

Acknowledgements This work has been supported by DST (Department of Science and Technology) SERB project CRG/2021/003780.

References

1. Zhang Y, Faghri A (1996) Heat transfer enhancement in latent heat thermal energy storage system by using the internally finned tube. Int J Heat Mass Transf 39(15):3165–3173

2. Dinesh BVS, Bhattacharya A (2019) Effect of foam geometry on heat absorption characteristics of PCM-metal foam composite thermal energy storage systems. Int J Heat Mass Transf 134:866–883
3. Ho CJ, Gao JY (2009) Preparation and thermophysical properties of nanoparticle-in-paraffin emulsion as phase change material. Int Commun Heat Mass Transfer 36(5):467–470
4. Bellan S, Gonzalez-Aguilar J, Romero M, Rahman MM, Goswami DY, Stefanakos EK, Couling D (2014) Numerical analysis of charging and discharging performance of a thermal energy storage system with encapsulated phase change material. Appl Therm Eng 71(1):481–500
5. Tan FL (2008) Constrained and unconstrained melting inside a sphere. Int Commun Heat Mass Transfer 35(4):466–475
6. Khodadadi JM, Zhang Y (2001) Effects of buoyancy-driven convection on melting within spherical containers. Int J Heat Mass Transf 44(8):1605–1608
7. Veerappan M, Kalaiselvam S, Iniyan S, Goic R (2009) Phase change characteristic study of spherical PCMs in solar energy storage. Sol Energy 83(8):1245–1252
8. Narasimhan NL, Bharath R, Ramji SA, Tarun M, Arumugam AS (2014) Numerical studies on the performance enhancement of an encapsulated thermal storage unit. Int J Therm Sci 84:184–195
9. Fan LW, Zhu ZQ, Xiao SL, Liu MJ, Lu H, Zeng Y, Yu ZT, Cen KF (2016) An experimental and numerical investigation of constrained melting heat transfer of a phase change material in a circumferentially finned spherical capsule for thermal energy storage. Appl Therm Eng 100:1063–1075
10. Cheng X, Zhai X (2017) Thermal performance analysis of a novel PCM capsule in red blood cell shape. Appl Therm Eng 120:130–137
11. Awasthi A, Kumar B, Nguyen HH, Lee SS, Chung JD (2020) Effect of sinusoidal cylindrical surface of PCM on melting performance. J Mech Sci Technol 34(8):3395–3402
12. Patel JR, Joshi V, Rathod MK (2020) Thermal performance investigations of the melting and solidification in differently shaped macro-capsules saturated with phase change material. J Energy Stor 31:101635
13. Dinesh BVS, Bhattacharya A (2020) Comparison of energy absorption characteristics of PCM-metal foam systems with different pore size distributions. J Energy Stor 28:101190
14. Parida A, Bhattacharya A, Rath P (2020) Effect of convection on melting characteristics of phase change material-metal foam composite thermal energy storage system. J Energy Stor 32:101804
15. Athawale V, Bhattacharya A, Rath P (2021) Prediction of melting characteristics of encapsulated phase change material energy storage systems. Int J Heat Mass Transf 181:121872
16. Athawale V, Jakhar A, Jegatheesan M, Rath P, Bhattacharya A (2022) A 3D resolved-geometry model for unstructured and structured packed bed encapsulated phase change material system. J Energy Stor 51:104430

Performance Optimization of Hybrid Silica Gel-Based Desiccant Dehumidifier: Experimental Investigation

Zafar Alam and Taliv Hussain

Abstract Experimental exploration of the functioning of the hybrid dehumidifier desiccant wheel system is signified in this work with actual consideration of the performance deviation as a function of the flow rates of air at the process and recovery division of the desiccant wheel for hot and damp environments. The desiccant material is regenerated by an electric heater fitted in the regeneration section. Airflow rates have been varied (i.e., 2.8, 3.5, 4.5, and 5.2 ms^{-1}) at the process subdivision and regeneration section of the DW. At various process inlet temperatures (i.e., 30, 31, 32, 33, and 35 °C) and at different airflow rates, performance parameters like latent COP, sensible COP, Total COP, dehumidification effectiveness, Regeneration effectiveness, and Thermal effectiveness have been analyzed and found that optimal values of these performance parameters in this work for the same process and regeneration air inlet velocity are found at 3.5 ms^{-1} inlet speed of air at process and Regeneration section and 31°C process air inlet temperature.

Keywords Dehumidifier · Temperature · Humidity · Effectiveness · Regeneration

Terminology Used

c_p	Specific heat air (kJ kg^{-1} K^{-1})
h	Specific enthalpy (kJ kg^{-1})
V	Volumetric flow rate (m3 h^{-1})
COP	Coefficient of performance
DW	Desiccant wheel
RMSE	Root mean square error
T	Temperature (°C)

Z. Alam (✉) · T. Hussain
Department of Mechanical Engineering, AMU Aligarh, Aligarh, UP 202001, India
e-mail: zafaralamalig003@gmail.com

S. Das et al. (eds.), *Proceedings of the 1st International Conference on Fluid, Thermal and Energy Systems*, Lecture Notes in Mechanical Engineering,
https://doi.org/10.1007/978-981-99-5990-7_26

Greek signs

Δh_{vs}	Water latent heat of vap. (kJ kg^{-1})
ρ	Density of air (kg m^{-1})
ω	Air humidity fraction (kg kg^{-1})
η	Effectiveness (dimensionless)

Subscripts

deh	Dehumidification
reg	Recovery/regeneration
P	Process
tol	Total
sen	Sensible

1 Introduction

Refrigeration and cooling (RAC) play very significant part in our everyday working for cooling and warming prerequisites. Refrigeration frameworks are broadly utilized for controlling temperature in working/living climate to give comfort to humankind. The utilization of refrigeration innovation for cooling isn't new; notwithstanding, cooling innovation is needed to advance because of numerous natural guidelines. One of the significant parts of cooling is to give the comfort climate to the humankind in every one of the conditions. Comfort conditions improve human living as well as improve the exhibition. Comfort is essentially settled by temperature and mugginess. For comfort feeling, the general dampness should be inside indicated range, for example, relative mugginess 55 to 60%, and dry bulb temperature 22 to 26 °C [1, 2]. Common forced air systems are used to maintain comfortable temperatures and humidity levels. Air dehumidification is performed by lowering the temperature in the cooling loop below the dew point, which concentrates water droplets. One of the least expensive ways to do it is to use a desiccant dehumidifier.

2 Literature Review and Objective

Angrisani et al. [3] broke down the exhibition by shifting five working conditions: rotational speed, volume wind stream rates, recovery temperature and temperature of open air and dampness proportion. A few presentation boundaries, in view of electric, warm and essential energy, are explained. Jani et al. [4] have given various sorts of mathematical simulations of turning desiccant type dehumidifier, mechanical improvement, implementation measurement of desiccant cooling structure. Hussain et al. [5] recommendations for the ideal airflow rate and matching cooling duration enhanced the precooling of spherical food products. They recommended that cooling be enhanced equal to max speed, or 2 m/s, for the spherical product [5]. Cellulose and steel wire mesh condensers were used by Hussain et al. [6] to improve VCRS system's coefficient of performance. They claimed that evaporative cooling caused the COP of cellulose condensers to be higher than that of steel wire mesh condensers [6]. Alam and Hussain [7, 8] improved dehumidifier (desiccant) operation in accordance with ambient process air temperature, process air flow rates, renewal rate, and other critical factors based on performance [9]. The majority of the researchers experimented with using natural gas, electric heaters, solar energy, and other heat sources to recover desiccant material at the solid desiccant dehumidifier's recovery side. Few of them focused on and compared the various performance criteria for solid desiccant dehumidifiers to other types of desiccant dehumidifiers, such as liquid or composite desiccant dehumidifiers. However, none of them concentrated on optimizing these parameters for the Indian climate (Aligarh) based on inlet temperatures, recovery, and process air flow rates and for both the recovery and process sides. In light of ambient temperatures and airflow rates, the present study intends to improve and optimize the solid desiccant dehumidifier's functioning.

3 Experimental Unit

Process air is processed via blower and then passed over the DW structure. After passing through the DW, process air temperature rises and the process air humidity decreases, and the process air passes over the evaporator section of the hybrid VCRS arrangement as presented in Figs. 1, 2 and 3 display pretreatment section and the line illustration of investigational setup.

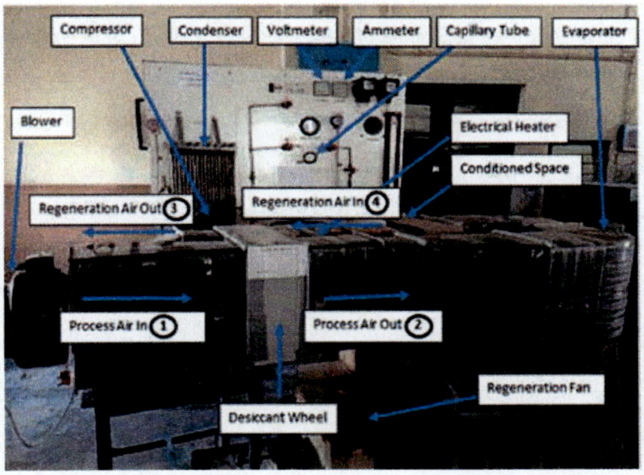

Fig. 1 Experimental setup

Fig. 2 Pre-treatment section

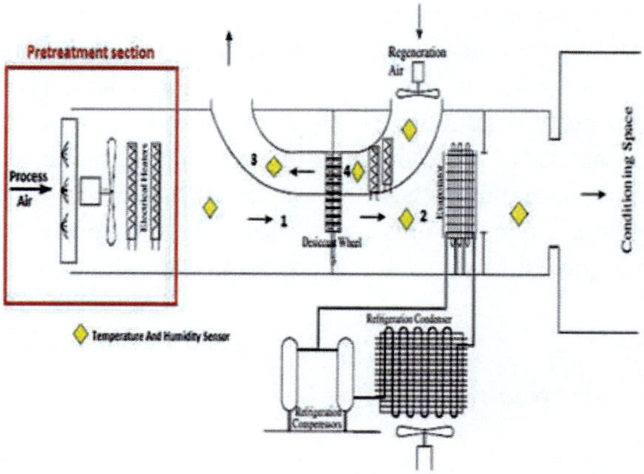

Fig. 3 Line illustration of investigational setup

Table 1 Shows the measuring instruments

Equipment	Voltage	Sampling rate	Range	Accuracy
Tweex data logger	Single phase 220 V standard supply	5 s to 255 min	−10 to + 85 °C 0 to 100 %RH	± 0.5°C ± 2.5%
Digital anemometer	3.0 V DC	–	0 to 30 m/s	± 5%

Relative humidity of air and Dry bulb temperature is noted using Data Logger. Air velocity is assessed via anemometer as shown in Table 1.

3.1 Properties of Components

See Tables 2, 3 and 4.

Table 2 Shows the specification of desiccant wheel and silica gel properties

Aspect	Magnitude
DW ratio of Volume (Φ)	0.48
Thermal conductivity (silica gel), k_d (W/mK)	0.175
Density (silica gel), ρ_d(kg/m^3)	1129
Elevation of passing route, 2a (m)	0.0018
Wheel's dia, D (m)	0.37
Pitch of flow route, 2b (m)	0.0032
Radius of Pore, *r* (m)	11×10^{-10}
Wheel's Length, Lw (m)	0.1
Matrix materials Specific heat,c_m (J/kgK)	1030
Silica gel's specific heat, c_d (J/kgK)	921
Matrix material density of, ρ_m (kg/m^3)	625
Permeability, ε	0.4
Wall Size of channel, δ (m)	0.00034
Complete area of one track (Ratio), (Ar)	0.844

Table 3 Indicates the description of blower

Voltage (V)	Frequency (Hz)	Power (W)	Speed (RPM)	Flow rate of Air (m^3 h^{-1})	Noise (dB)	Operating Temperature (°C)
115/120	60	220	2600	1050	68	−30 to + 60

Table 4 Indicates the design of duct

Material	Thickness of sheet	Padding Material	Padding thickness
Iron sheet (Galvanized)	18/1.214 mm Gauge	Foam (PU)	2.5 cm

4 Methodology

The setup for hot and humid conditions yielded investigational data on dry bulb temperature and relative humidity. Data is being collected by varying process and renewal inlet air velocities. All the experimental data is measured at inlet and outlet of regeneration and process region of the DW. Range of process air inlet velocities taken into consideration is due to the limitation of experimental setup. The temperature range that was taken into consideration for this study was 30 – 35 °C because it only examined hot and humid conditions. Air flow rates were changed at the process section and on the desiccant wheel's regeneration section (i.e., 5.2, 4.5, 3.5, and 2.8 ms^{-1}).

4.1 Performance Parameters

4.1.1 The Dehumidification Effectiveness ($\eta_{\mathbf{deh}}$)

Symbolizes the ratio between the actual and the ideal dehumidification ability of the DW [10].

$$\eta_{deh} = \frac{\omega_1 - \omega_2}{\omega_1} \tag{1}$$

4.1.2 Regeneration Effectiveness ($\eta_{\mathbf{reg}}$)

Indicates the desiccant wheel's latent load in terms of the regeneration thermal power necessary for the desiccant wheel's adsorption operation [10].

$$\eta_{reg} = \frac{\omega_3 - \omega_4}{\omega_4} \tag{2}$$

4.1.3 Thermal Effectiveness (ηth)

Represents the proportion of heat transmission rate to the greatest heat transfer rate feasible [10].

$$\eta_{th} = \frac{T_2 - T_1}{T_4 - T_1} \tag{3}$$

4.1.4 Latent COP ($\mathbf{COP_{Lat}}$)

Signifies the ability to diminish air moisture dependent on the quantity of power used to replenish it [11].

$$COP_{Lat} = \frac{\dot{V}_P \times \Delta h_{vs} \times (\omega_1 - \omega_2)}{\dot{V}_{reg} \times (h_4 - h_1)} \tag{4}$$

4.1.5 Sensible COP (COP_{Sen})

Fluctuations in temperature of air and sensible heat going via a DW are measured with this element [11].

$$COP_{sen} = \frac{\dot{V}_P \times C_p \times (T_2 - T_1)}{\dot{V}_{reg} \times (h_4 - h_1)} \tag{5}$$

4.1.6 Total COP (COP_{tol})

Is the proportion of the process air total heat change to the power delivered for the recovery procedure [11].

$$COP_{tol} = \frac{\dot{V}_P \times (h_1 - h_2)}{\dot{V}_{reg} \times (h_4 - h_1)} \tag{6}$$

It is observed from the Fig. 4 that the dehumidification effectiveness of process section is straightly linked to the process inlet temperature. With rise in Inlet Temperature of Process Air dehumidification effectiveness surges. When the ambient moisture increases it is observed that effectiveness also increases and the effectiveness drops when the process outlet moisture rises. Dehumidification effectiveness reaches its peak at a process and air intake velocity of 3.5 ms^{-1} on the recovery side and an air inlet temperature of 31°C on the process side.

Figure 5 reveals the deviation of thermal effectiveness with different process air inlet temperature. It decreases with decrease in deviation in temperature at the outlet and inlet of DW process side.

Figure 6 demonstrated regeneration efficacy rises as air flow rate and air inlet temperature rise on the desiccant wheel's process side.

Figure 7 makes it abundantly evident that COP_{latent} increases first before decreasing. The humidity proportion at the process side's air intake rises as COP_{latent} does. In this instance, the process and recovery intake air velocities are 3.5 ms^{-1} and the inlet temperature of air is 31°C at the process side of the desiccant wheel, respectively.

Sensible COP decreases with reduction of temperature difference at the process outlet and inlet of the DW as shown in Fig. 8.

As shown in Fig. 9, as the enthalpy change at the process inlet and the outlet increase, total COP also increase and decrease with decrease in enthalpy change at the process exit and inlet division of DW. At 3.5 m/s process and regeneration

Table 5 Shows measured values for 2.8 m/s process inlet velocity

T_1	ϕ_1	ω_1	T_2	ϕ_2	ω_2	T_4	ϕ_4	ω_4	T_3	ϕ_3	ω_3
30	59	0.0157	33.1	42.5	0.0135	55.2	14	0.0139	28	62	0.0149
31	57	0.0164	33.1	42	0.0133	56.3	12	0.0126	27.7	60	0.0141
32	54	0.0163	33.5	42.4	0.0137	58.2	11	0.0125	28.1	60	0.0145
33	50	0.016	34	43	0.0143	59.5	10	0.0122	27.7	60	0.0141
34	46	0.0155	35	40.5	0.0143	59.9	9	0.0111	27.8	61	0.0145
35	42	0.0148	36	37.5	0.014	60	8	0.0099	27.7	61	0.0144

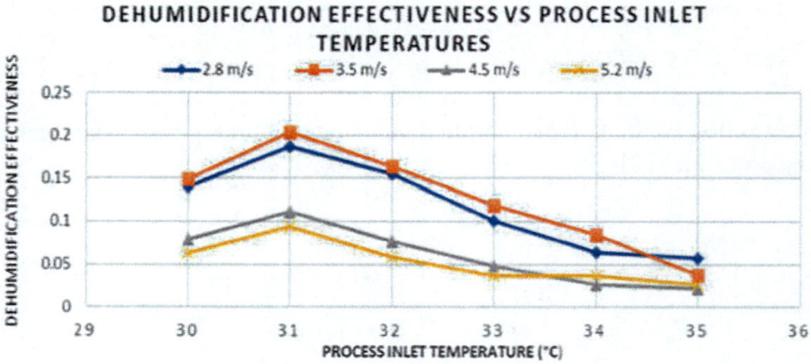

Fig. 4 Shows the deviation of Dehumidification effectiveness

air intake velocity and 31°C process air inlet temperature, the maximum value of COP_{total} is attained (Table 5).

5 Results and Discussion

See Figs. 4, 5, 6, 7, 8 and 9.

6 Error Analysis

The root mean square error (RMSE) approach detailed by Kline and McClintock [12] is used in this study's error investigation, as shown in the table below (Table 6).

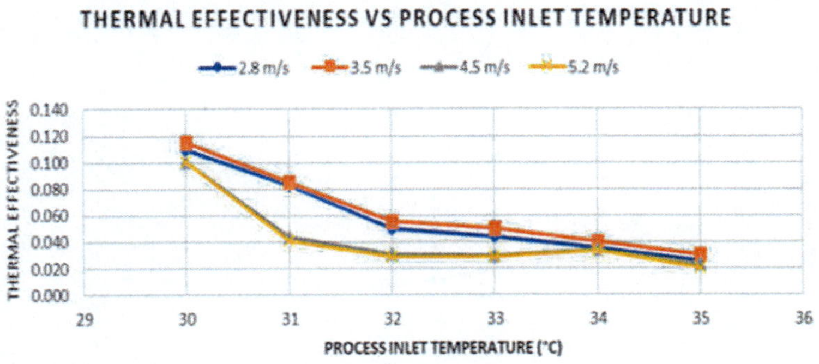

Fig. 5 Illustrates the deviation of Thermal effectiveness

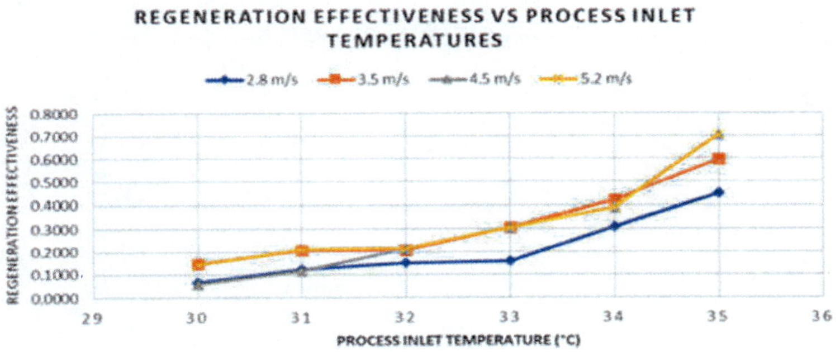

Fig. 6 Illustrates the change of Regeneration effectiveness

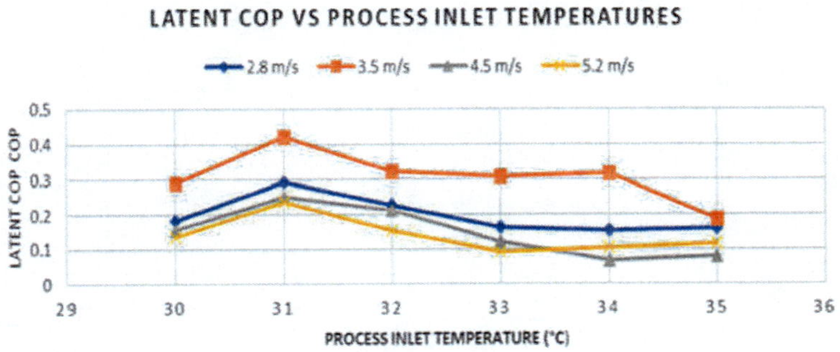

Fig. 7 Indicates the deviation of Latent COP

7 Conclusions

- Dehumidification effectiveness is exactly proportional to the process inlet temperature. With rise in process inlet moisture, dehumidification effectiveness surges and vice versa.

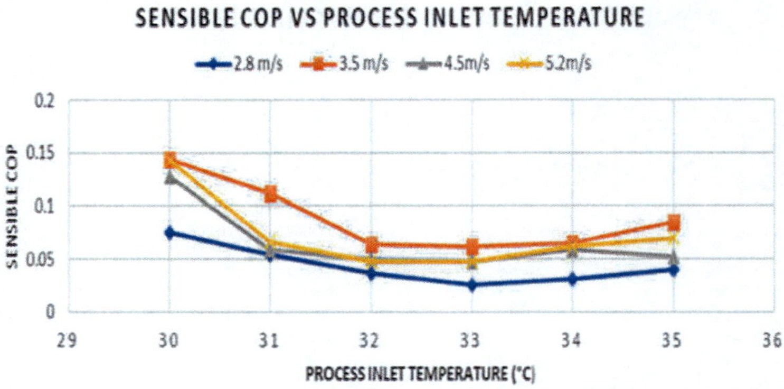

Fig. 8 Signify the deviation of Sensible COP

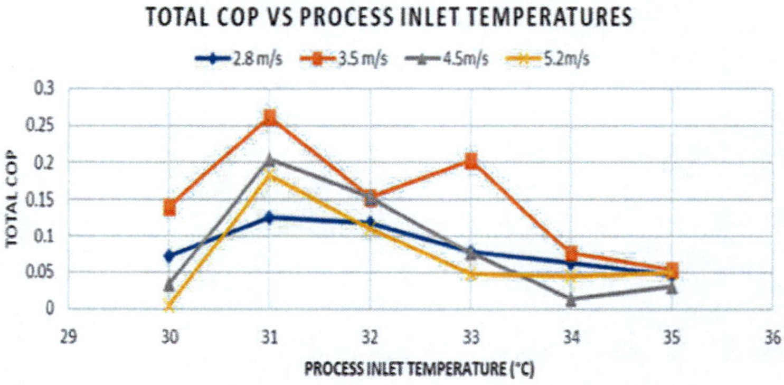

Fig. 9 Shows the deviation of Total COP

Table 6 Shows the uncertainty of considered parameters

Factors	Uncertainty (%)
Dehumidification Effectiveness (η_{deh})	7.70
Regeneration Effectiveness (η_{reg})	5.27
Thermal effectiveness (ηth)	2.37
Latent COP (COP_{latent})	2.02
Sensible COP ($COP_{sensible}$)	2.99
Total COP (COP_{total})	3.11

- Enthalpy change at the desiccant wheel's process input and outflow is directly related to total COP. The temperature difference between the desiccant wheel's process input and output, on the other hand, directly affects the latent coefficient of performance (COP).
- Regeneration effectiveness is inversely related to the growth in ambient dampness because of the growth in vapor pressure of renewal air, which causes this one to decrease through increased ambient dampness. Recovery effectiveness is related to recovery temperatures and the recovery velocities at air inlet.
- The maximum values of these factors of performance were obtained at 3.5 m/s recovery and process air inlet velocity and 31 °C.

Acknowledgements Government of India's TEQIP-III program provided funding for the current experimental investigation.

References

1. ASHRAE (1997) Handbook—Fundamentals, American Society of Heating, Refrigeration and Air-conditioning Engineers, Atlanta, GA, USA
2. ASHRAE (1995) Handbook—HVAC Applications, American Society of Heating, Refrigeration and Air- conditioning Engineers, Atlanta, GA, USA
3. Angrisani G, Roselli C, Sasso M (2015) Experimental assessment of the energy performance of a hybrid desiccant cooling system and comparison with other air-conditioning technologies. Appl Energy 138:533–545
4. Jani DB, Mishra M, Sahoo PK (2016) Solid desiccant air conditioning–a state of the art review. Renew Sustain Energy Rev 60:1451–1469
5. Hussain T, Kamal MA, Alam Z, Hafiz A, Ahmad A (2021) Experimental and numerical investigation of spherical food product during forced convection cooling. Measur Food 3:100006
6. Hussain T, Singh AK, Mittal A, Verma A, Alam Z (2020) Performance evaluation of vapor compression refrigeration system by varying air flow rates in air-cooled and evaporatively cooled condensers. Intern J Energy Clean Environ 21(1)
7. Alam Z, Hussain T (2022) Experimental exploration on performance advancement of solid desiccant dehumidifier. IOP Conf Series: Mater Sci Eng 1224(1):012011. https://doi.org/10.1088/1757-899x/1224/1/012011
8. Alam Z, Hussain T (2022) Experimental analysis and performance evaluation of solid desiccant dehumidifier using silica gel. Int J Anal Exp Finite Elem Anal 9(1)
9. Alam Z, Hussain T (2022, October) Experimental investigation on heat transfer performance of rotatory desiccant dehumidifier. In: IOP conference series: materials science and engineering, vol 1259, no 1, p 012003. IOP Publishing. https://doi.org/10.1088/1757-899X/1259/1/012003
10. Mandegari MA, Pahlavanzadeh H (2009) Introduction of a new definition for effectiveness of desiccant wheels. Energy 34(6):797–803
11. Angrisani G, Capozzoli A, Minichiello F, Roselli C, Sasso M (2011) Desiccant wheel regenerated by thermal energy from a microcogenerator: experimental assessment of the performances. Appl Energy 88(4):1354–1365

12. Goodarzia G, Thirukonda N, Heidari S, Akbarzadeh A, Date A (2017) Performance evaluation of solid desiccant wheel regenerated by waste heat or renewable energy. Energy Procedia 110:434–439
13. Kline SJ (1953) Describing uncertainty in single sample experiments. Mech Eng 75:3–8

Numerical Study of PCM-Based Energy Storage System Using Finite Difference Method

C. Abhijith and Ranjith Maniyeri

Abstract Energy storage systems incorporating phase change material (PCM) are becoming the answer to intermittent energy availability in the area of solar cooking vessels and solar room heating systems. These thermal energy storage systems are efficient, reliable and can reduce running costs and investments. The present work investigates the melting of n-octadecane using enthalpy formulation method by using finite difference method-based discretization. Accordingly, a numerical model is developed in MATLAB which is validated first by comparing with previous works. Further using the developed model studies on wall materials, thickness, different PCMs and temperatures are carried out. It is found that metal containers with higher thermal conductivities provided significant boost in energy storage. The increased thickness of walls adds significant change only if containers are of lower thermal conductivity such as steel. Different PCMs are analysed, and their operating temperature and energy storage capabilities are studied. In addition, enthalpy equation in cylindrical coordinates is solved since practical PCM storages are commonly cylindrical in shape. At the end, the numerical results are compared with an experimental case with paraffin wax as PCM.

Keywords Energy storage · Enthalpy formulation · Finite difference method · Melt fraction · Phase change materials

1 Introduction

Thermal energy storage systems are increasingly becoming attractive since many renewable energy supplies must deal with inconsistent demand and supply. Latent heat storage systems store energy while melting and solidification of phase change materials (PCM) as latent heat. PCM-based systems have the advantage of high

C. Abhijith · R. Maniyeri (✉)
Department of Mechanical Engineering, National Institute of Technology Karnataka (NITK), Surathkal, Karnataka Manglore-575025, India
e-mail: mranji1@nitk.edu.in

S. Das et al. (eds.), *Proceedings of the 1st International Conference on Fluid, Thermal and Energy Systems*, Lecture Notes in Mechanical Engineering,
https://doi.org/10.1007/978-981-99-5990-7_27

storage capacity and relatively constant operating temperature. Such systems can be employed to bridge the mismatch between supply and demand of renewable systems such as solar cooking vessels and room heating systems. Two-dimensional (2D) numerical heat transfer studies of PCM materials have been done by Chen et al. [1] and Costa et al. [2] using enthalpy formulation method. Research works were reported on the studies of Melting characteristics with respect to temperature and storage vessel materials. It is understood that melting characteristics such as melting and solidification time, latent energy stored, wall temperature and container material should be studied in detail. Most of the studies considered rectangular coordinate system for the simulation but in reality PCM containers are of cylindrical shape and hence solving enthalpy equations in cylindrical coordinates is of practical significance. Inspired by this, we propose finite difference-based simulation model to study PCM-based energy storage system under different wall temperatures, metal containers and wall thicknesses. We also aim to see how our numerical model relate with that of experimental works on solar box cooker embedded with a PCM developed by Anilkumar et al. [3].

2 Literature Review and Objective

Several researchers have studied the melting and energy storage capabilities of PCM materials. The challenging part of a phase change problem exists due to the moving boundary between phases. Costa et al. [2] used enthalpy formulation method to simulate PCM melting in 1D and 2D domains. Enthalpy formulation solves one single enthalpy equation for both liquid and solid domains and hence the interface does not need any extra conditions. Costa et al. [2] used finite difference method and an additional melt fraction update equation to simulate the melting of n-octadecane PCM material.

Many researchers have studied to incorporate the PCM into useful applications. Buddhi and Sahoo [4] developed a latent heat storage-based box solar cooker to demonstrate feasibility of using a phase change material (stearic acid) in Indian climatic conditions. They demonstrated the variation of temperature with time. Sharma et al. [5] studied the operation of a cylindrical PCM storage system that uses erythritol as PCM. They studied temperature and melt fraction variation during operation. Halawa et al. [6] used enthalpy formulation to investigate the freezing characteristics of thin slabs of PCM with variation in wall temperatures. Chen et al. [1] studied melt fraction variation of PCM storage system with time and wall conduction using enthalpy formulation in 2D domain.

Zivkovic and Fujii [7] compared rectangular and cylindrical PCM storage melting and validated the enthalpy formulation method by experiment. Vyshak and Jilani [8] expanded upon this work to compare rectangular, cylindrical and cylindrical shell PCM storages and concluded that cylindrical shell configuration provided more favourable energy storage rates.

Recently, Anilkumar et al. [3] conducted experiments on a cooking pot of solar box cooker integrated with a cylindrical PCM storage system and analysed its ability to retain favourable temperatures. Paraffin wax was used for analysis and optimum amount of PCM was calculated and different PCMs were analysed and best PCM for the application was recommended to be erythritol.

Deriving inspiration from previous studies and to contribute further the present work aim to develop a computational model using finite difference method to predict PCM melt fraction in thermal energy storage systems and to study the effect of parameters such as PCM materials, container material, container thickness, boundary condition on melting and energy storage. In addition, melting and solidification in a cylindrical storage to compare with an experimental case is also studied.

3 Materials and Methods

3.1 Mathematical Model

Enthalpy formulation is employed as the mathematical model to solve the given problem where the variable is enthalpy. Assumptions involved in enthalpy formulation method are.

1. PCM thermal properties are not dependent on temperature. Thermal properties of liquid and solid phases can be different.
2. Initial condition of PCM is solid.
3. PCM is assumed to be isotropic and homogenous.
4. PCM is subjected to only conduction heat transfer.

For any phase change problem, the total volumetric enthalpy can be represented by

$$\frac{\partial H}{\partial t} = \Delta(k_k(\Delta T)) \tag{1}$$

k_k denotes the thermal conductivity of phase k. Total volumetric enthalpy represented in terms of sensible and latent heat is

$$H = h(T) + \rho\lambda f(T) \tag{2}$$

ρ denotes the density, f(T) is the melt fraction and λ is the latent energy of the PCM. Sensible heat h can be written as

$$h = \int_{T_m}^{T} \rho_k c_k dT \tag{3}$$

The melt fraction f at different temperature is

$$f = \begin{cases} 0 \ if \ \ T < T_m \ (solid) \\ 0 \ to \ 1 \ \ if \ \ T = T_m \ (mushy) \\ 1 \ \ if \ \ T > T_m \ (Liquid \end{cases}$$

Enthalpy at each phase can be written as

$$H = \begin{cases} \int_{T_m}^{T} \rho_s c_s dT T < T_m \ (solid) \\ \rho_l f \lambda T_m \ (mushy) \\ \int_{T_m}^{T} \rho_l c_l dT T > T_m \ (liquid) \end{cases}$$

Value of temperature can be obtained from enthalpy as

$$T = \begin{cases} T_m + \frac{H}{\rho_s c_s} H < 0 \ (solid) \\ T_m 0 < H < \rho_l \lambda \ (mushy) \\ T_m + \frac{H - \rho_l \lambda}{\rho_l c_l} H > \rho_l \lambda \ (liquid) \end{cases}$$

From (2) and (3), two-dimensional heat transfer equation while melting of a PCM is given as

$$\frac{\partial h}{\partial t} = \frac{\partial}{\partial x}\left(\alpha \frac{\partial h}{\partial x}\right) + \frac{\partial}{\partial y}\left(\alpha \frac{\partial h}{\partial y}\right) - \rho_l \lambda \frac{\partial f}{\partial t} \quad (4)$$

3.2 Numerical Solution

The fully implicit finite difference approach can be used to solve Eq. (4). For uniform grids the discretisation leads to

$$h_p = h_p^0 + \alpha R\left[h_E - 4h_p + h_S + h_N + h_S\right] + \rho_l \lambda [f_p^0 - f_p^k] \quad (5)$$

$$a_E h_E + a_W h_W + a_P h_P + a_N h_N + a_S h_S = Q \quad (6)$$

where

$$a_E = a_W = a_N = a_S = -\alpha R,$$

$$a_P = 1 - a_E - a_W - a_N - a_S$$

$$Q = h_p^0 + \rho_l \lambda [f_p^0 - f_p^k],\ R = \frac{dt}{dx^2}$$

Here superscript 0 denotes the preceding time step and k means the kth iteration for updating the melt fraction Voller [9] derived the formula as

$$f_p^k = f_p^0 + \frac{-a_E h_E - a_W h_W - a_N h_N - a_S h_S + h_p^0}{\rho_l \lambda} \tag{7}$$

The algorithm developed by Voller [9] is.

1. Compute the coefficients.
2. To constrain sensible enthalpy h_p while melting or solidification, a_p value is set to 10^{15} during phase change.
3. Solve for enthalpy and update liquid fraction at each iteration. Along with under and over correction for the melt fraction

$$f = \begin{cases} 0\ if\ f < 0 \\ 1\ if\ f > 1 \end{cases}$$

4. Repeat calculation until convergence. Large a_p value is retained until phase change is completed or convergence is reached.

Convergence is reached at a time step where total enthalpy is less than the given tolerance

$$\frac{ABS(H^k - H^{k+1})}{\rho_k c_k} < TOL$$

Tolerance for the present work is taken as 10^{-4}.

Interface Properties

Because properties of the two phases differ, the domain will have non-uniform thermal conductivity, Specific heat and density As a result, the equivalent properties of the mushy control volume must be calculated. As an example thermal conductivity of mushy control volume is

$$k_{mushy} = k_s(1 - f) + k_f f \tag{8}$$

Boundary and Initial Conditions

Initial temperature of PCM is fixed at 5^0C lower than the melting point of the PCM. Temperature at the left boundary is kept as 10^0C (or 15^0C) and other boundaries are given adiabatic condition. All the equations listed above have been solved by developing a code in MATLAB.

4 Results and Discussion

4.1 Grid Independence Study and Validation

Grid independence study is carried out by choosing 32 × 32 grid and 64 × 64 grid and melt fraction vs time graph is plotted as shown in Fig. 1. There is very little difference between both grids hence 32 × 32 grid is computationally much efficient. But while plotting melt front progression 62 × 62 grid provides much clearer picture hence 62 × 62 grid is selected as optimum grid for further simulations.

Then, by comparing our results with that of Costa et al. [2] the developed code is validated as depicted in Fig. 2. The PCM used is n-octadecane in an aluminium container. From the figure, it can be inferred that there is good agreement with our results with that of Costa et al. [2] results which proves the reliability of our code. Further, the progression of melt front with time is also recorded at 2,10,30,60,150 and 240 min to illustrate the melting of PCM which also shows excellent agreement with that of Costa et al. [2] studies as represented in Figs. 3 and 4 (Table 1).

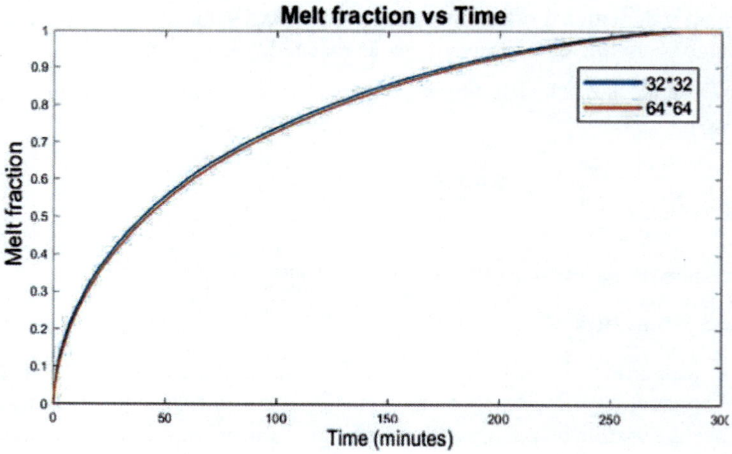

Fig. 1 Melt fraction versus time for two different grids

Table 1 Properties of n-octadecane

N-octadecane	
Melting temperature	300.7 K
Latent heat of fusion	243.5 kJ/kg
Density (liquid)	777.0 kg/m^3
Thermal conductivity	
Liquid	0.148 W/mK
Solid	0.358 W/mK
Thermal diffusivity	
Liquid	$8.64*10^{-8}$ m^2/s
Solid	$2.14*10^{-7}$ m^2/s

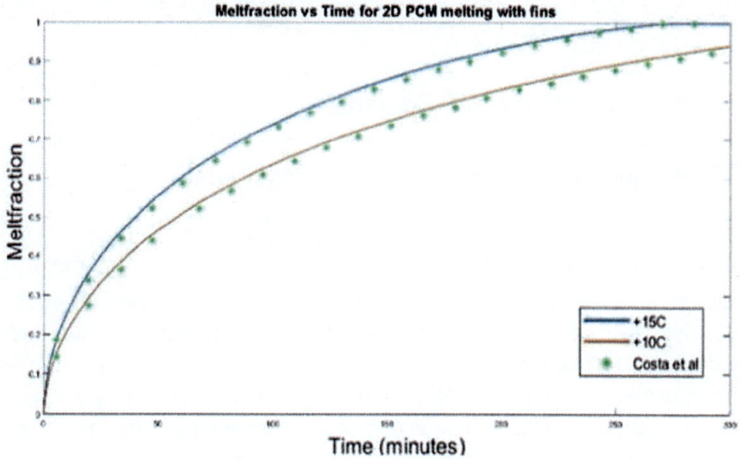

Fig. 2 Melt fraction vs time for current code versus Costa et al. [2] results

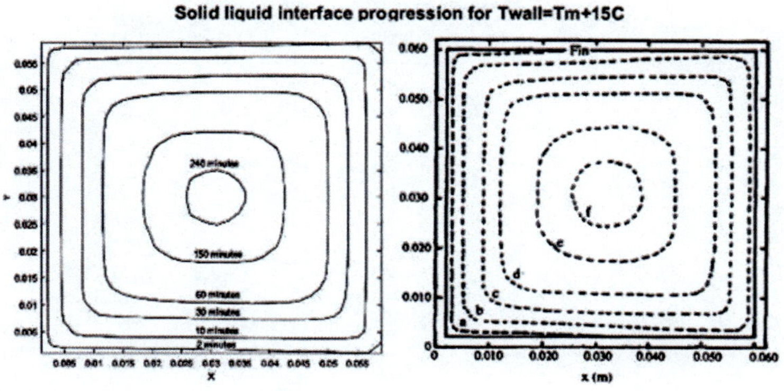

Fig. 3 Solid liquid interface progression with wall temperature 15 °C for **a** Current code (left) **b** Costa et al. [2] (right)

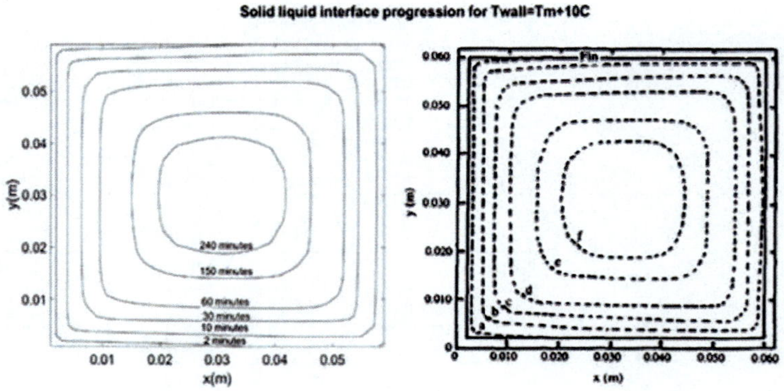

Fig. 4 Solid liquid interface progression with wall temperature 10 °C for **a** Current code (left) **b** Costa et al. [2] (right)

4.2 *Factors Affecting Heat Transfer*

4.2.1 Heat Transfer with Different Container Materials

Numerical simulations are performed for various container materials like copper, tin, stainless steel and glass to see the change in melting characteristics. The thermal conductivity values of these metals result in different melting characteristics.

While investigating the melt front propagation, we can see significant changes in low-conductivity materials such as steel and glass since they take significantly more time for melting. With metals such as copper and tin which have appreciable conductivity change in melting time is minimal. There is very little improvement in using copper in place of aluminium. Figure 5 depicts these observations (Fig. 6).

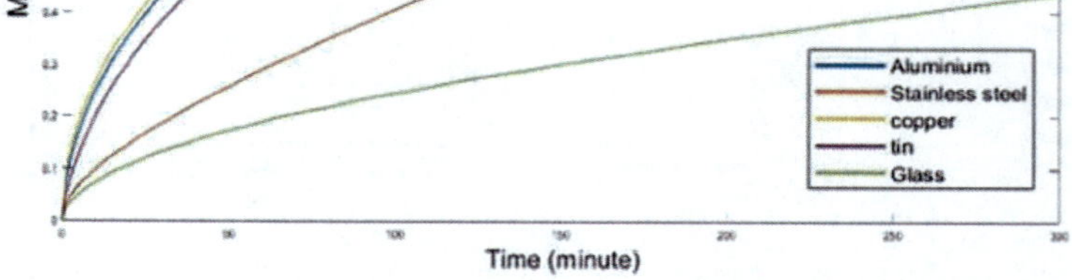

Fig. 5 Melt fraction versus time for different container materials

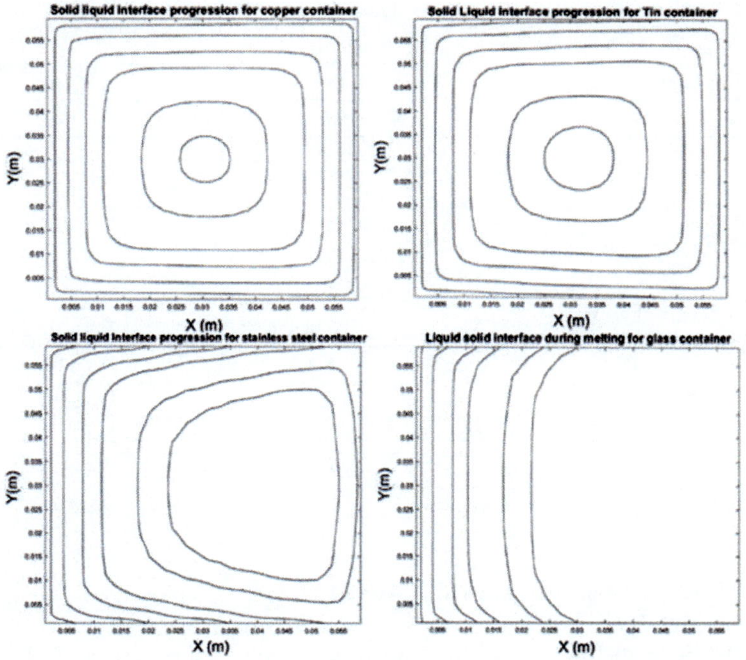

Fig. 6 Solid liquid interface progression for different fin materials **a** Copper **b** Tin **c** Stainless steel **d** Glass

The melting in copper and tin are very similar to aluminium container but for lower thermal conductivity metals such as steel more melting is done near the heated wall and the right end melts slower than the other side. In the glass container, there is very little heat conduction hence melt front is essentially flat progressing from left to right. It is apparent that glass does not provide any heat transfer benefit and only acts as a container for PCM.

4.2.2 Heat Transfer with Different Thickness of Metal Container

The previously mentioned simulations are done for 2 mm metal containers. To see the effect of change in thickness of the metal containers the simulations are repeated for 1mm, 2 mm, 3 mm and 4 mm thicknesses. Total melting time computed in each case shed light on the effectiveness of melting.

From the total melting time values shown in Table 2, with higher thickness heat transfer is improved and time for melting is decreased. The effect is more pronounced in metals with lower thermal conductivity such as steel where a decrease of 38.2% in time is observed when thickness is increased from 1 to 4 mm while in copper this change is merely 2.46%.

Table 2 Total melting time for different metal thickness

	Total melting time (minutes)			
Thickness (mm)	Aluminium	Copper	Tin	Steel
1mm	291.6	279.9	328.2	706.3
2mm	282	275.3	302.8	537.9
3mm	279	273.7	294.9	472.7
4mm	277.4	273.05	289.4	436.4

Table 3 Total melting time for different PCMs

PCM	Latent heat (kJ/Kg)	Time (minutes)
n-Octadecane	243.5	282.033
Stearic Acid	160	100.3667
Acetamide	263	210.4
Acetanilide	222	186.35
Erythritol	339.8	368.3

4.2.3 Melting Characteristics of Different PCMs

Different PCM materials operate at different melting temperatures as shown in Table 3. PCMs such as n-octadecane, stearic acid, acetamide, acetanilide and erythritol are subjected to melting at excess temperature of 15 °C from their respective melting temperature. Melt fraction as well as latent energy stored per unit length with respect to time is plotted for each PCM which is illustrated in Figs. 7 and 8.

Even though erythritol seems to store more energy during its melting, fixing PCM for a particular usage is heavily dependent upon the operating temperature available.

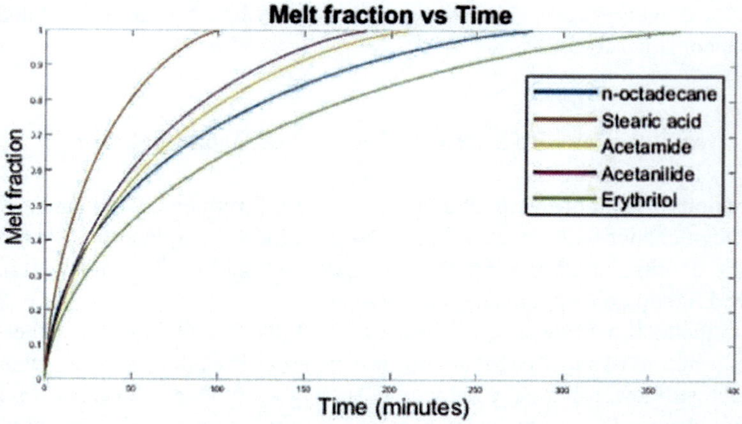

Fig. 7 Melt fraction versus time for different PCMs

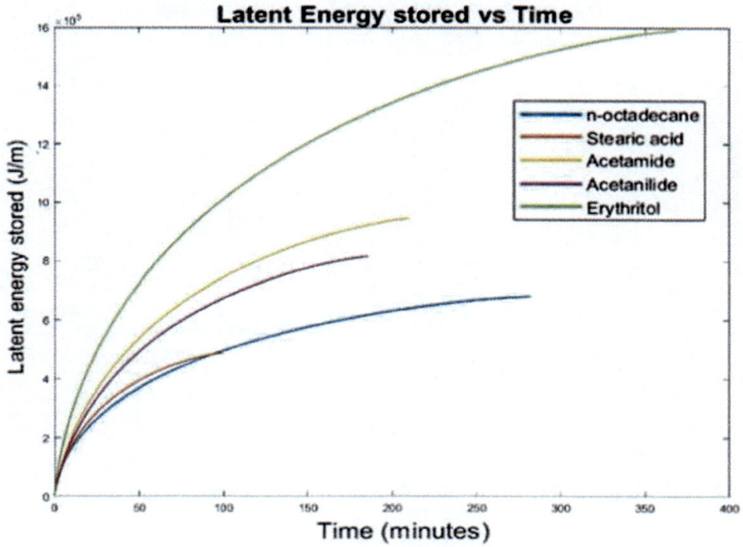

Fig. 8 Latent energy stored during melting for different PCMs

So PCMs with high energy storage may not be suitable since they also require high operating temperature. It is important to select PCM based on both energy storage and operating temperature requirements.

4.3 Cylindrical PCM Geometry

In the next stage, we extend the developed model for the case of cylindrical geometry. Accordingly, the governing equation describing the problem in cylindrical coordinates is given by

$$\frac{\partial h}{\partial t} = \frac{1}{r}\frac{\partial}{\partial r}\left(\alpha \frac{\partial h}{\partial r}\right) - \rho_l \lambda \frac{\partial f}{\partial t} \tag{9}$$

Again, the code in cylindrical coordinate system is first validated by comparing with that of Zivkovic and Fujii [7] where the melting of a cylinder calcium chloride hexahydrate ($CaCl_2.2H_2O$) with radius 20 mm kept in air temperature of 60 °C is studied. Outer surface of the cylinder is given a convective boundary condition. Temperature at the centre of the cylinder is plotted to monitor the process which is illustrated in Fig. 9. Both results are in good agreement.

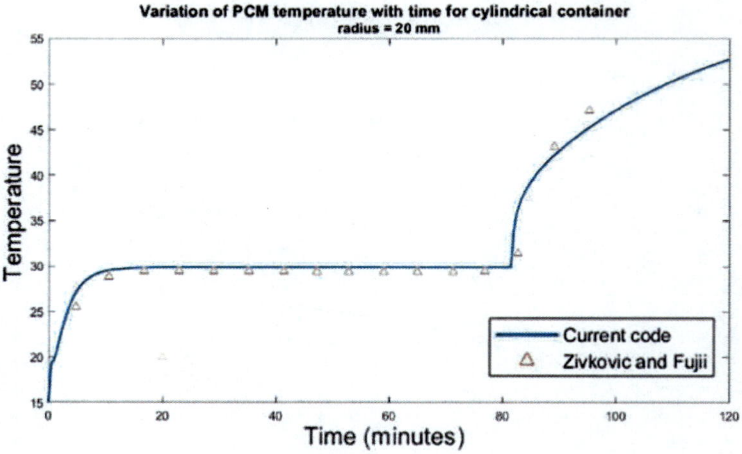

Fig. 9 Centreline temperature for current code versus Zivkovic and Fujii [7] results

Cylindrical Thermal Energy Storage Unit Experiment

In the final stage, we attempt to implement our developed computational model for the study reported by Anilkumar et al. [3] in which an experiment consisting of cylindrical PCM unit filled with paraffin wax is performed for a solar box cooker. Cylinder has diameter of 16 cm and height of 18 cm. To ascertain the PCM's ability to retain specific temperature a 6-h experiment is done by filling the cavity with hot water and measuring the lateral surface temperature of the unit. The unit is partially heated and PCM is partially melted at the start of the experiment and 78 °C water is filled and temperature is measured throughout the span of experiment. The present code is used to analyse this case by employing convective boundary condition at both water and air side of the cylinder.

Figure 10 shows the comparison of lateral surface temperature variation with time for present work and that of Anilkumar et al. [3]. From the plot, it can be seen that for both experimental and numerical works the temperature is very close to the melting temperature of the PCM and is successful in maintaining that temperature. More variation in experimental result is due to the isothermal assumption taken while in reality paraffin wax has melting range of 55 °C–60 °C. Hence more drop in temperature is expected. Both results show that PCM units are successful in maintaining desired temperature over the duration of operation. Maximum error between Experimental and Numerical study was found at the 1-h mark where the deviation is found to be 4 °C.

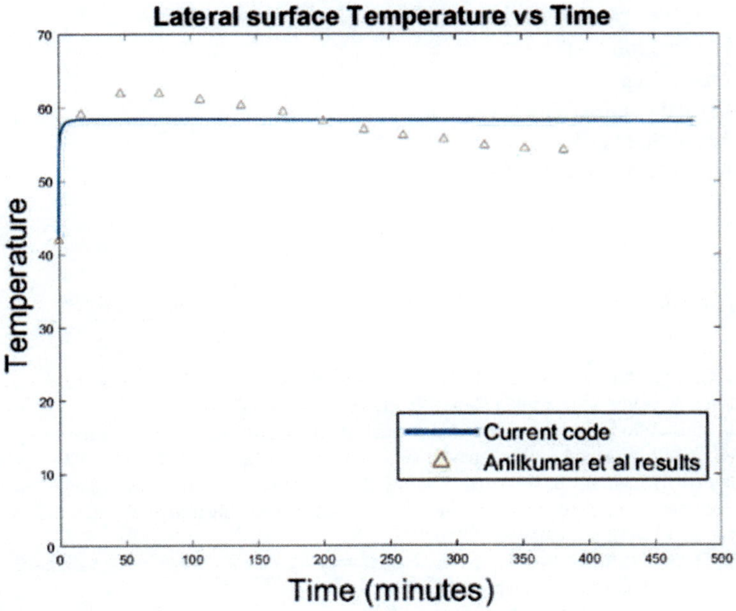

Fig. 10 Lateral surface temperature (°C) for current code versus Anilkumar et al. [3] experiment

5 Conclusion

In the present work, numerical simulation of melting of PCM is done through enthalpy formulation and fully implicit finite difference method. Initially, the developed code is validated. Then, the changes in melting characteristics for different container materials, wall thickness, wall temperature and using different PCMs are observed. The increasing of wall thickness imparted little improvement in melting except for material with low thermal conductivity. Similarly, changing container material does not have much influence on melting unless it is a material with low conductivity. PCMs operating at different melting temperatures gives different amount of latent energy stored and different total melting time. These characteristics of the system can be manipulated for optimal conditions according to a specific use. A model for enthalpy formulation in cylindrical coordinate is also developed and validated. The ability of paraffin wax PCM to maintain desired temperature is analysed using developed code and compared with available experimental results.

Nomenclature

h Sensible enthalpy (J/Kg)
T Temperature (°C/K)

λ Latent energy (J/Kg)
f Melt fraction (–)
ρ Density (Kg/m^2)
k Thermal conductivity (W/mK)
C Specific heat (J/KgK)
α Thermal diffusivity (m^2/s)

References

1. Chen CR, Atul Sharma, Tyagi SK, Buddhi D (2008) Numerical heat transfer studies of PCMs used in a box-type solar cooker Renew Energy 33 1121–1129
2. Costa M, Buddhi D, Oliva A (1996) Numerical simulation of a latent heat thermal energy storage system with enhanced heat conduction. Energy Convers Manag 39(3/4):319–330
3. Anilkumar BC, Maniyeri R, Anish S (2021) Optimum selection of phase change material for solar box cooker integrated with thermal energy storage unit using multi-criteria decision-making technique. J Energy Storage 40:102807
4. Buddhi D (1997) And L. K Sahoo, Solar cooker with latent heat storage: design and experimental testing, Energy Convers Manag 38:493–498
5. Sharma SD (2005) Takeshi Iwata, Hiroaki Kitano, Kazu Nobu Sagara, Thermal performance of a solar cooker based on an evacuated tube solar collector with a PCM storage unit. Sol Energy 78:416–426
6. Halawa E, Bruno F, Saman W (2005) Numerical analysis of a PCM thermal storage system with varying wall temperature. Energy Convers Manag 46:2592–2604
7. Zivkovic and Fujii (2001) An analysis of isothermal phase change of phase change material within rectangular and cylindrical containers. Sol Energy 70:51–61
8. Vyshak NR, Jilani G (2005) Numerical analysis of a PCM thermal storage system with varying wall temperature. Energy Convers Manag 46:2592–2604
9. Voller VR (1990) Fast implicit finite-difference method for the analysis of phase change problems. Numer Heat Transf, Part B: Fundam: Int J Comput Methodol 17(2):155–169

CFBC Technology—An Essential Alternative for Present Indian Coal Power Sector

S. Naga Kishore, T. Venkateswara Rao, and M. L. S. Deva Kumar

Abstract Pulverized coal combustion (PCC) is a dominant technology in Indian power sector and has occupied 51.69% in the grand total installed capacity. Though circulating fluidized bed combustion (CFBC) technology is economically sound, technically feasible and environmentally friendly than PCC technology, CFBC technology utilization in India is still in the nascent stage. In the present paper, two present difficulties (retirement of huge thermal capacity and coal import from other countries) of PCC boilers were explained and need of better technology which is required immediately to come out from these difficulties of PCC boilers was discussed. The objective of this paper is to encourage CFBC technology in India for (i) constructing new boilers and for renovating and revamping old PC fired boilers to extend their life up to 25 years and (ii) improving local low grade coal utilization in present power generation of Indian coal power sector. This paper also provides comparison between CFBC and PCC boilers on three main features such as type of fuel, emissions and cost which are required for manufactures to establish CFBC technology for Indian coal power sector.

Keywords CFBC · PCC · Boiler · Coal · Emissions · Cost

S. N. Kishore (✉)
Department of Mechanical Engineering, Anurag University, Hyderabad 500088, India
e-mail: nagakishore.surisetty@gmail.com

T. V. Rao
Department of Mechanical Engineering, DBS Institute of Technology, Kavali 524201, India

M. L. S. D. Kumar
Department of Mechanical Engineering, JNT University, Anantapur 515002, India

S. Das et al. (eds.), *Proceedings of the 1st International Conference on Fluid, Thermal and Energy Systems*, Lecture Notes in Mechanical Engineering,
https://doi.org/10.1007/978-981-99-5990-7_28

1 Introduction

Coal-based sub-critical PC fired power plants have been used as key technologies for power generation in India. Even though the power generation by renewable sources is being increased, the coal utilization for power generation has been increased in India every year to reach demand. From the statistics of Central Electricity Authority (CEA), the average growth in coal consumption for last five years in Indian power plants is observed as 3.273% [1]. The share of coal consumption for electricity generation will remain unchanged for the foreseeable future due to rising cost of oil imports and difficulties encountered in importing natural gas from neighbouring countries [2].

Despite India possessing immense potential of power generation from hydel, natural gas, oil and coal resources, the progress in energy production in the India has not taken place on a scale commensurate with the possibilities. As a result, there exists a gap between availability and demand. The installed capacity of power has reached 392107.24 MW at the end of November 2021. From the report of CEA, India, it was identified that power of 5,875 MW was in deficit at the end of November 2021 [1].

Electricity shortage of coal fired plants in India is a big problem which is based on rise of imported coal prices and retirement of huge capacity [3, 4]. Many coal fired plants in India were installed in early sixties. Number of power plants having capacity of 22,716 MW are ready to stop working due to retirement condition during the period of 2017–2022 [1, 4]. Therefore, there is an immediate need for installing more new power plants in India to meet demand. At the same time, better alternative for old PCC boilers is essential to revamp and renovate with less cost, to improve local coal utilization, to reduce cost at similar environmental performance compared to traditional PCC technology.

2 Difficulties of PCC Boilers

At present, two main difficulties of PCC boilers in Indian coal power sector are described below.

2.1 Coal Import

Although the whole world is looking towards oil, natural gas and renewable sources for power generation, still many countries have used coal as primary source for generating power. Some of the countries in that list are South Africa (71.02%), China (56.55%), India (51.69%), Kazakhstan (52.09%) and Vietnam (51.3%) [1, 5].

Coal reserves value in India is reached to 3,44,021 MT at the end of March 2020 [6]. Even though huge availability of coal, still India is one of the top five coal importers. The coal used in power plants is a combination of Indian and imported coal with blend ratio as 70:30 [7]. There are two main reasons to import coal for power generation. One is that Indian subbituminous coals have high ash content and another reason is that power plants in India are designed to operate with high-quality coals.

The ash content in Indian subbituminous coal varies from 30 to 50% and high ash coals are not possible to be used directly in all cases. One of the major problems involved in any coal fired plant is solid waste disposal. When high ash coal is burned in the furnace, the discharging of solid waste and other problems of ash disposal further increased. Huge land in hundreds of acres should be provided for high ash coals in case of ash pond [8, 9]. Another problem in the operation with the usage of high ash coals is fouling and slagging due to high temperatures in the furnace. To mitigate this problem, usually boiler design has to be altered including furnace dimensions. Strong fly ash is also caused to exacerbate some more problems such as the corrosion problems on walls, reduction in heat transfer rate, resistance in gas flow through boiler [10].

The coal quantity in million tonnes (MT) which is imported for last nine years from neighbouring countries is represented in Fig. 1. It is noticed that the value of coal imported from neighbouring countries was raised by 5.64% than previous year and reached 248.53 MT for the year 2019–20. The amount of money spent for this imported coal was 1527320.55 million rupees. Mainly, Indonesia and South Africa coals are predominantly used to be mixed with Indian coal in power plants [6].

Figure 2 shows that the quantity of coal in MT which is imported from neighbouring countries in the year 2019–20 [6]. To reduce the cost of coal import from other countries, an effective technology which supports to burn low grade fuels must be required.

2.2 *Retirement of Large Capacity*

Another difficulty to be discussed is that most of thermal PC fired power plants in India have reached an average age of 30 years and ready for modernization or revamping due to shortage of power and capital. Power units having capacity of 16,789 MW are going to retirement due to insufficient space for flue gas desulphurization (FGD) system and not following the latest guidelines of ministry of environment and forest (MOEF). Twenty seven thermal units from various states were gone for shut down during the period 2019–2020 and as a result, 2,463 MW power was in deficit throughout India [4].

Figure 3 shows the capacity number of thermal plants which are expected for retirement at the end of the year 2022 due to age limit [4]. In India, many PC fired power plants of 210/200 MW capacity have been identified for renovation and needed to be modernized prior to 2022.

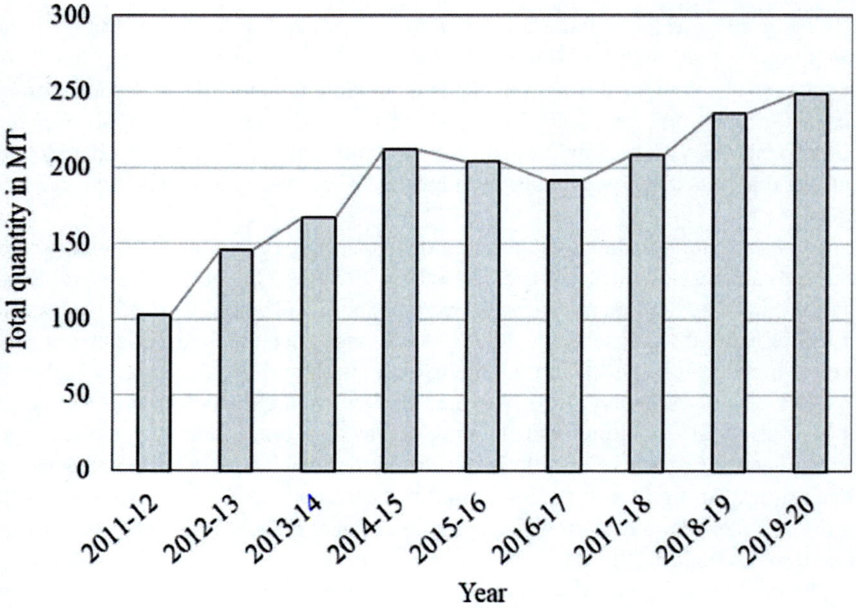

Fig. 1 Yearly import of coal of India [MT] [6]

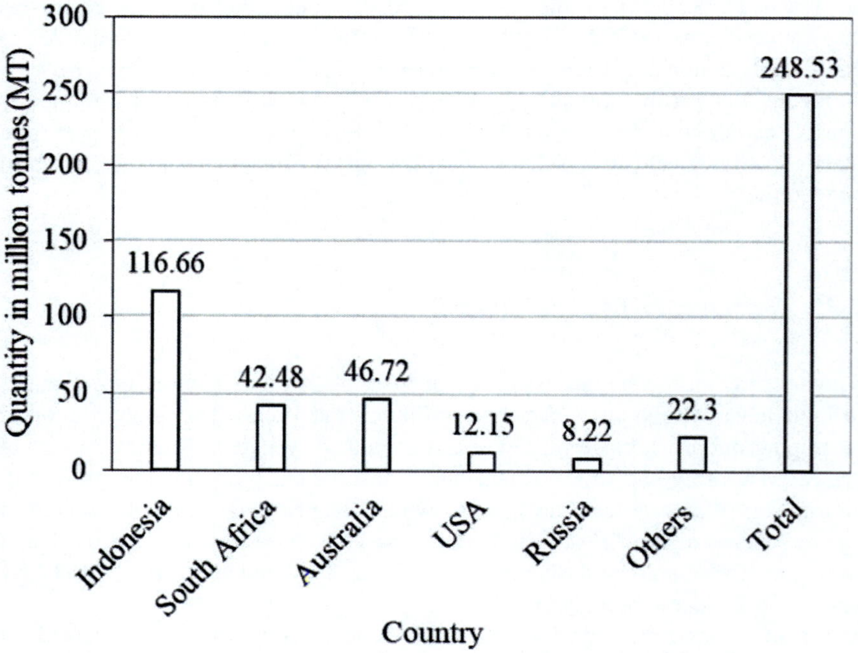

Fig. 2 Countrywise import of coal of India for the year 2019–20 [6]

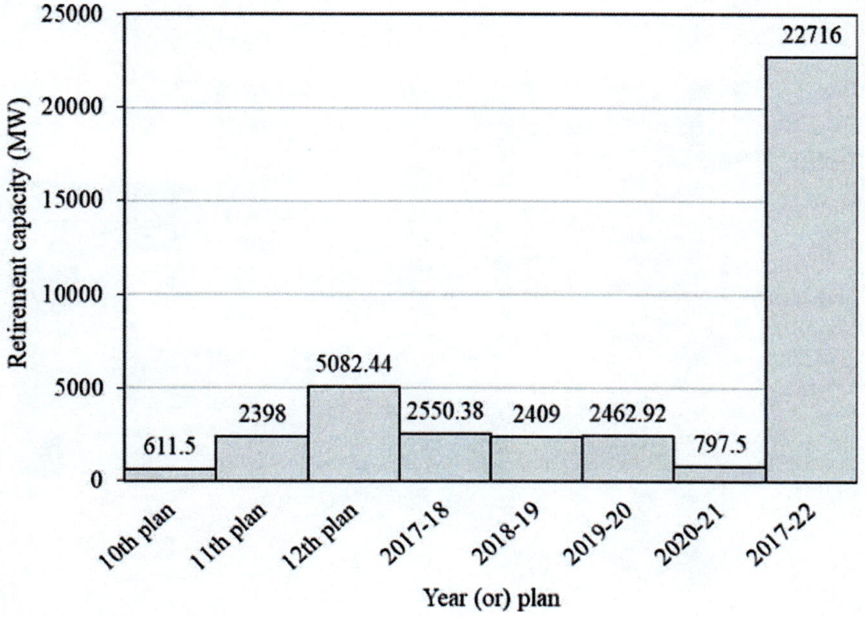

Fig. 3 Capacity (MW) of thermal plants for retirement [4]

The PCC boiler refurbishment with the same technology would not give good results today in India. Definitely, it is required to refurbishment of old PCC boilers with better technology which must be environmentally friendly and this technology should also improve thermal and environmental performance.

3 Need of CFBC Technology for Indian Coal Power Sector

The better alternative for two problems mentioned above is CFBC technology. The reasons to show this CFBC technology as immediate alternative in current situation for Indian coal power sector are mentioned below.

(i) .The coal ministry of India has planned to reduce import coal by one-third in coming five years by increasing the local coal utilization. Hence, to reduce importing of coal from other countries and to utilize local low grade coals (high ash Indian subbituminous coal and Indian lignite) in effective manner, CFBC technology is the best alternative for power generation. Fuel flexibility is the biggest advantage which keeps CFBC boiler in the market as popular. CFBC boiler burns different types of fuels (low and high quality fuels) and releases less emissions without adding any selective catalytic reduction (SCR) and FGD system which are essential for PC fired boiler for same environmental

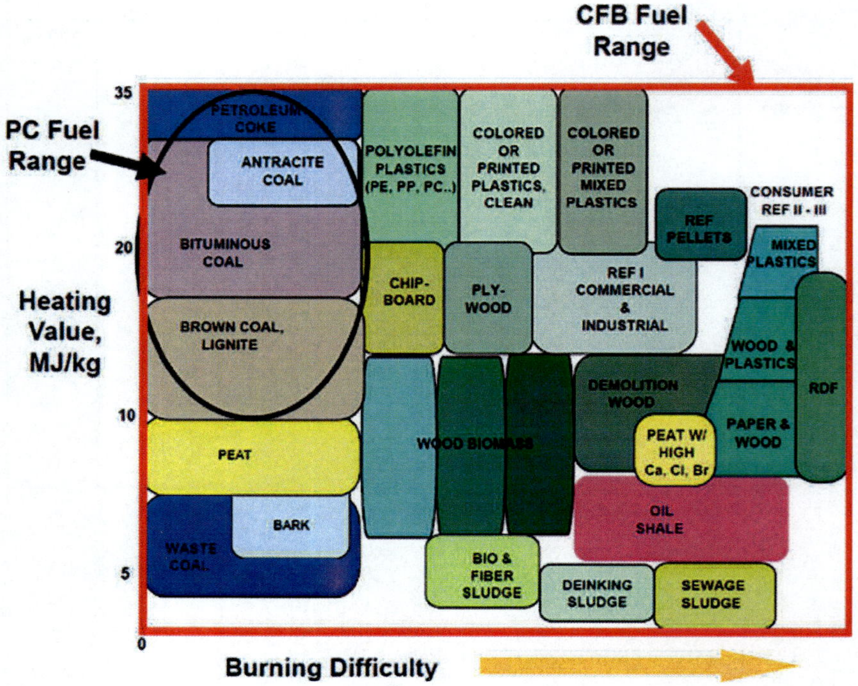

Fig. 4 Comparison of list of fuels burnt in PCC and CFBC boilers [12]

performance [11]. The following Fig. 4 represents the range of fuels which can be burned by PCC and CFB technology based on calorific value and burning difficulty.

Indian lignite is an effective alternative to be used in power plants for reducing coal import from other countries and avoiding problems of high ash Indian coals. Even though large amounts of lignite resources are available in India, lignite usage for power generation is very less. The statistics from the Ministry of power, India, [13] said that the percentage of lignite-based power generation was 1.8% in the year 2020 from total installed capacity. To increase the usage of this lignite for Indian power plants, CFBC technology is the best choice to prefer. The Indian lignite coals have high moisture content (40–55%), volatile matter (45–55%), sulphur content (0.3–6%) and low heating value [14]. The lignite is mainly concentrated in the states of Tamil Nadu, Gujarat and Rajasthan. The total quantity in million tonnes available in various states of India for the year 2019–20 is 46,019 MT [6] in which 79.29% is from Tamil Nadu lignite (Neyveli), Rajasthan lignite has 13.8% and Gujarat lignite share is 5.91% as shown in Fig. 5.

The composition of three lignite samples which are popularly available and used in commercial operations in India are shown in Table 1.

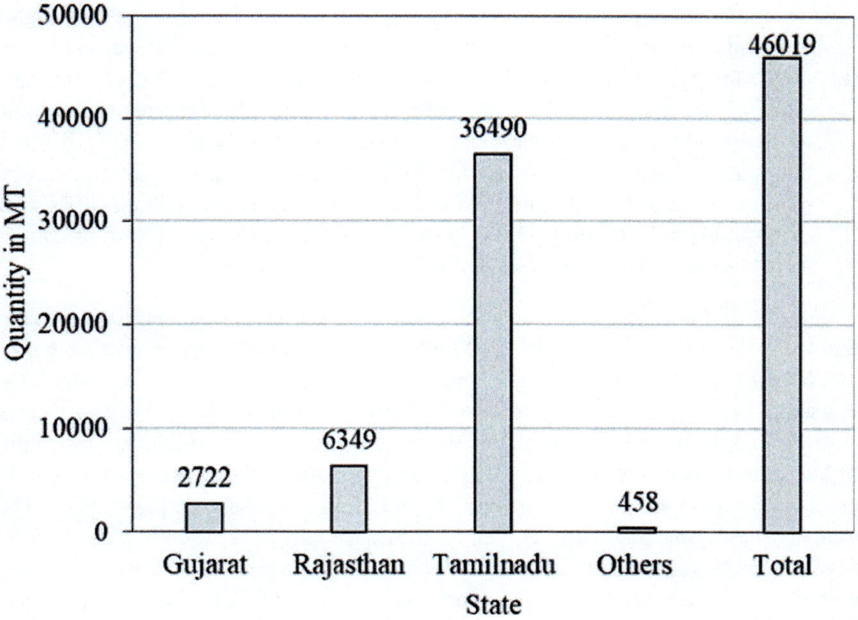

Fig. 5 Statewise inventory of geological reserve of lignite as on 1 April 2020 [6]

Table 1 Ultimate analysis of three popular Indian lignite samples [15, 16]

Ultimate analysis (% by weight)	Tamil Nadu (Neyveli) lignite	Gujarat (Surat) lignite	Rajasthan (Giral) lignite
Carbon (%)	27.5	43.3	30.09
Hydrogen (%)	2.2	3.14	2.24
Sulphur (%)	0.7	1.2	6.0
Oxygen (%)	10.4	8.55	6.27
Nitrogen (%)	0.2	0.81	0.4
Moisture (%)	50.5	24.0	40.0
Ash (%)	8.5	19.02	15.0
HHV (kcal/kg)	2650	4240	3000

(ii) .There are two ways for rebuilding the retirement of huge thermal capacity. One is refurbishment of old plants and second one is constructing new plants by demolishing old plants. Building of new power plants in specified areas is not at all supportable due to environmental issues. Not only environmental problems but also huge amount of capital cost and time are required for developing infrastructure of new power plants. Hence, manufacturers see modernization of old plants instead of new plants. Sen et al. [17, 18] and Kavidass et al. [19] proved that CFBC technology is the best option to choose compared to other alternative technologies in case of revamping of PC fired boilers. In case of revamping of old boilers with CFB technology, most of the equipment which was existed in PCC unit can be maintained same without changes. Small changes and new equipment are required for CFB boiler such as auxiliary equipment, limestone crushers, secondary coal crushers, distributed control system (DCS), dry ash handling system and bag house filters. With small changes of equipment, the cost of revamping of PCC unit reduces by 52.3% to be changed into CFBC unit without the need of replacing a new a PCC or CFBC unit and life of old boiler can be extended to another 25 years [20].

Even in the construction of new plants, CFBC boilers are economically sound and technically feasible than PC fired boilers. The operation and maintenance cost of CFB unit is 5–10% lower than PC fired boiler due to less moving equipment. SO_2 emissions in PC fired boiler are removed by FGD units which are expensive with more than 10% of the overall cost of plant [21]. However, there is a significant absorption of SO_2 emissions in the CFB furnace that takes place due to introduction of limestone along with coal. As a result, the lower SO_2 emissions are possible without FGD. The temperature during combustion in the CFB furnace ranges between 800 and 900 °C at which the NO_X emissions are not formed much and no need of SCR system. Many authors proved for similar environmental performance of NO_2 and SO_2 emissions that the FGD plant and SCR system are not required for CFBC unit. As a result, the capital cost is 8–10% lower than PC fired boiler [22]. The cost comparison between PCC and CFBC boilers in case of construction of plants is shown in Table 2.

4 CFBC Technology Development

CFBC technology has a long history of development. The first CFBC development started in 1955 in Germany for the use as roaster for sulphite ores. Subsequently, CFBC technology has been used for many applications including for power generation and can be considered as commercial proven technology. The first boiler with CFBC technology was built in 1982, Germany, by Lurgi for the application of electricity generation. This plant generated 84 MW in which 9 MW was for generating electricity, 31 MW for processing steam and 44 MW for melting molten salt by burning low grade coal washery residues with the addition of limestone. The CFBC technology was completed 40 years on 31 May 2016 and has entered into 5th decade in the area of power plant applications. In these 40 years, it has owned many milestones [27]. The development or growth of CFBC technology in 40 years is explained in given Fig. 6.

Table 2 Comparison of cost between commercial PCC and CFBC units

Reference	Unit	Plant size (MW)	Parameter	PCC boiler	CFBC boiler
Subramaniam and Patil [23]	Cr/MW	150	Cost of emission control	0.63–0.8	0.05 (Max)
			Equipment (FGD + SCR)		
Belaid [21]	$	125	Annual O&M cost	3300.00	3000.00
John [24]	$/kW	N/A	Capital cost for similar	803	678
			Environmental performance		
Wheeldon and Thimsen [25]	$/kW	150	Total plant cost	4390	3990
	$/MWh	150	Total cost of electricity	109.6	101.1
	$/kW	500	Total plant cost	2680	2660
	$/MWh	500	Total cost of electricity	74.5	72.8
Reddy and Sethi [26]	N/A	N/A	Capital cost/kW	1.03–1.19	1.0

N/A—Not available

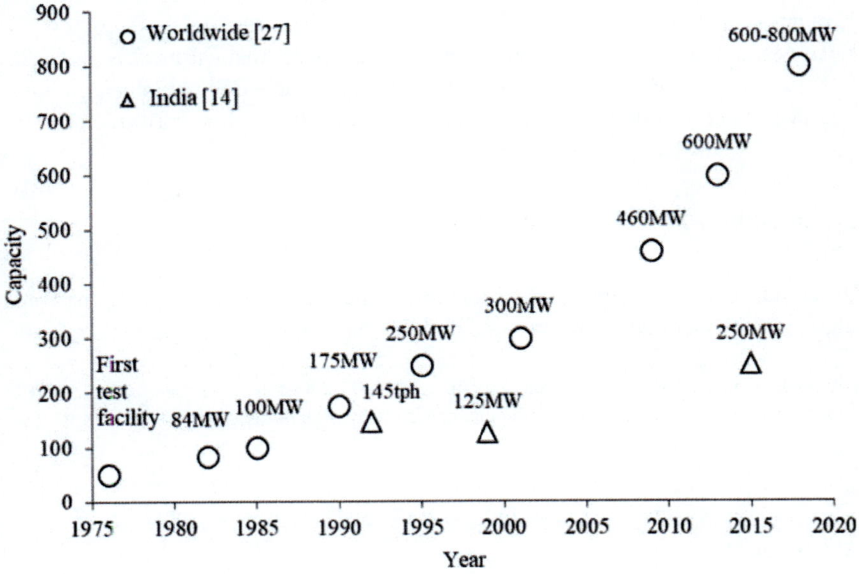

Fig. 6 Development of CFBC technology (Capacitywise) worldwide and India

4.1 Development of CFBC Boiler in India

Already some CFBC systems have started working with other low grade fuels like lignite and pet coke. In India, 2 × 145 tph CFB boilers-based cogeneration plant was set up by Indal at Hirakud in 1992. Later, the first major plant of 125 MW CFBC boiler with two units was installed by Gujarat Industries Power Company Ltd (GIPCL) in 1999. The major developments of CFBC technology in India are shown in Fig. 6. Owing to successful results of this technology, some more plants installed in Gujarat, Rajasthan and Tamil Nadu [14]. Still, India has necessary operating experience for CFBC power plants.

5 Conclusions

- Two major issues of PC fired boilers in present Indian thermal power plants were discussed with proved literature.
- It can be concluded that the CFBC is promising technology for current Indian coal power sector. Renovation, modernization and life extension of old PCC boilers in India are always cost-effective options compared to CFBC technology.
- To achieve the policy introduced by Government of India for improving local coal utilization and reducing huge value of import coal cost, CFBC technology is the best option to burn low grade fuels such as Indian lignite effectively for power plants.
- To encourage CFBC technology in India, more mathematical models on CFBC boilers should be required to analyse the design of new CFBC boilers, optimize processes in operation and predict thermal and environmental performance.

6 Declaration of Interest

The authors declare that they have no known competing financial interests or personal relationships that could have appeared to influence the work reported in this paper.

Acknowledgements The authors are grateful to the Shri. Kali Basu, Retired DGM, BHEL, Hyderabad, for his valuable support and guidance in submission of the paper.

References

1. Central Electricity Authority (2021) Executive summary on power sector, ministry of power. Government of India
2. https://www.weforum.org/agenda/2017/10/fossil-fuels-will-dominate-energy-in-2040/

3. Coal shortage: will your electricity bill go up? Times of India, October 13, 2021. https://timesofindia.indiatimes.com/business/india-business/coal-shortage-will-your-electricity-bill-go-up/articleshow/86984085.cms. Last Accessed Mar 2022
4. Central Electricity Authority, CEA annual report 2019–2020, ministry of power. Government of India
5. BP (2021) Statistical review of world energy 2021, 70th edn.
6. Coal Controller's Organisation (2021) Provisional coal statistics 2019–20, ministry of coal. Government of India
7. https://powerline.net.in/2016/08/26/optimum-ratios/. Last Accessed Mar 2022
8. Chandra A, Chandra H (2004) Impact of Indian and imported coal on Indian thermal power plants. J Sci Ind Res 63:156–162
9. Review of Land Requirement for Thermal Power Stations (2010) Central Electricity Authority, New Delhi. https://cea.nic.in/wp-content/uploads/2020/04/land_review_report.pdf. Last Accessed Mar 2022
10. Bhatt MS (2006) Effect of ash in coal on the performance of coal fired thermal power plants. Part II: capacity and secondary energy effects. Energy Sources Part Recover Util Environ Eff 28(1):25–41
11. Basu P (2015) Circulating fluidized bed boilers-design, operation and maintenance. Springer
12. Kostamo M (2012) Foster wheeler CFB technology–enabling fuel flexibility for power and CHP generation. In: The European Power Generation Strategy Summit, Prague, Czech Republic, 3rd–5th December 2012. http://www.europeanpowergeneration.eu/files/2012Presentations/MarkkuKostamo_FosterWheelerEnergia.pdf. Last Accessed Mar 2022
13. Ministry of Power (2020) Annual report 2019–20. Government of India
14. Lakshminarasimhan M, Ravi Kumar B, Lawrence A, Muthukrishnan M (2013) High sulfur lignite fired large CFB boilers-design and operating experience. In: Knowlton TM (ed) Proceedings of 10th international conference on circulating fluidized beds and fluidization technology, PSRI. ECI symposium series
15. Balasubramanian S, Sudhindra K, Bhat S (2022) An insight into advanced technology in circulating fluidised bed combustion steam generators. TATA Consulting Engineers Ltd., Bangalore. https://www.academia.edu/34881373/An_Insight_into_Advanced_Technology_in_Circulating_Fluidised_Bed_Combustion_Steam_Generators?auto=download. Last Accessed Mar 2022
16. https://www.scribd.com/presentation/19666291/CFB-Presn-NTPC11-07-06. Last Accessed Mar 2022
17. Sen AN, Miller L, Basu P, Dutta A (2003) Revamping of 4 × 58 MWTH pulverized coal fired boilers with circulating fluidized bed firing. In: Proceeding of FBC 2003, 17th international fluidized bed combustion conference, May 18–21, Jacksonville, Florida, USA
18. Dutta A, Sen A, Basu P (2005) Revamping of 2 × 32 MWe pulverized coal fired boilers with circulating fluidized bed firing. In: Proceedings of FBC 2005, 18th international fluidized bed combustion conference, May 22–25, Toronto, Ontario, Canada
19. Kavidass S, Walker DJ, Nortan Jr GS (1999) IR-CFB repowering: a cost effective option for older PC fired boilers. In: Power-gen international 99, 30th November–2nd December, New Orleans, Louisiana, USA
20. Dutta A, Dutta S, Basu P (2006) Techno-economic evaluation of revamping of coal-fired power plants in India. Int J Glob Energy Issues 26(1/2):6–42
21. Belaid M, Falcon R, Vainikka P (2014) Pulverized coal versus circulating fluidized-bed boilers perspectives and challenges for South Africa. S Afr J Chem Eng 19(3):72–81
22. Lockwood T (2013) Techno-economic analysis of PC versus CFB combustion technology. IEA Clean Coal Centre
23. Subramaniam CR, Patil AR (2017) CFB technology, an overview on emission moderation to meet new regulation. In: 16th National workshop on technological innovation & efficient operations of industry boilers, 2nd–3rd February 2017
24. John T (2013) Circulating fluidized bed technology for Indian and other coals. In: Proceedings of conference on clean coal and carbon capture and storage technologies, STEP-TREC, Trichy, Dec 2013. http://www.carboncap-cleantech.com/. Last Accessed Mar 2022

25. Wheeldon JM, Thimsen D (2013) Economic evaluation of circulating fluidized-bed combustion power generation plants. In: Fluidized bed technologies for near-zero emission combustion and gasification. Woodhead Publishing Limited
26. Reddy DN, Sethi VK (2002) Renovation and refurbishment of old PC power plants using CFBC boilers–technology status in India. In: 8th European roundtable on cleaner production, Cork, Ireland Oct 9–11, 2002
27. Leuschke F, Heirmann G, Beckner T (2017) 40 years of circulating fluidized bed (CFB) power plant technology

Design and Development of Linear Compressor for Miniature Vapor Compression Refrigeration System

S. Satheesh Kumar, G. Kumaraguruparan, S. Raja Rajeshwari, and M. M. Devarajan

Abstract In this paper, solenoid motor-based miniature linear compressor is developed for miniaturized vapour compression (MVC)-based cooling system. This linear compressor is developed for the cooling capacity range of 100–150 W. This research work is divided into three major parts. First, theoretical analysis is done for the various forces relating on the compressor movement. Then, 3D simulation of the compressor component is performed and discussed the relationship of the displacement of mass and velocity with respect to time using 20 sim software. In latter part of this work, the fabrication of compressor components is done and it is controlled by the microcontroller. The 20 sim simulation results shows that the piston force is directly proportional to piston velocity and displacement. The enhancement of piston velocity and displacement of 300% for the piston force is increased from 2.5 N to 10 N. The prototype analyse shows that the pressure ratio of 3 is obtained with the help of solenoid valve actuation. This pressure ratio is useful to produce the cooling capacity of 150 W at the evaporator and condenser temperature of 10 °C and 30 °C respectively. It is concluded that the system with highly efficient linear compressor is developed with lesser weight and required capacity.

Keywords Linear compressor · Miniaturization · Teflon · Theoretical analysis · Piston model · Vibration

S. S. Kumar (✉)
Department of Mechanical Engineering, M. Kumarasamy College of Engineering, Karur-639113, India
e-mail: sateeskanna@gmail.com

G. Kumaraguruparan · S. R. Rajeshwari · M. M. Devarajan
Department of Mechatronics Engineering, Thiagarajar College of Engineering, Madurai-625015, India
e-mail: gkgmech@tce.edu

S. Das et al. (eds.), *Proceedings of the 1st International Conference on Fluid, Thermal and Energy Systems*, Lecture Notes in Mechanical Engineering,
https://doi.org/10.1007/978-981-99-5990-7_29

1 Introduction

The development of the compressor is focused due to their demand, high efficiency and energy consumption rate. The some of the applications of highly efficient compressor are household, electronics cooling system, aircraft cooling system, automobile and HVAC systems. The Refrigeration system comprises the compressor, condenser, expansion valve and evaporator. The compressor is the only one moving component and consumes the input energy in larger amount. Thus, by improving the compressor efficiency or making even a small modification in the compressor design and control, it will be delivered the high overall performance of the system.

Nowadays, linear compressor is focusing due to miniaturization and higher performance of vapour compression-based cooling system. Linear compressors have various advantages over rotary compressors which include lesser friction, lesser number of moving points, lesser noise and higher efficiency. And for our application the lesser weight, lesser friction and lesser number of parts helps when the system needs to be easily portable. A linear compressor can be driven by various actuators or linear motors which based on three types. They are moving iron type motor, moving coil type motor and moving magnet type motor.

2 Literature Review and Objective

Bijanzad et al., [1] done the analytical model for linear compressor. They made compressor system which is an oil free and used for household refrigerators. The linear compressor is actuated by solenoid system. The dynamic models of the system were validated with the simulation results as well as experimental setup results. They achieved the higher efficiency when excited at natural frequency. The result showed that natural frequency and work done of the system are 19 Hz and 1.3 J respectively.

Bradshaw et al., [2] discussed about the model of linear compressor with small size for the electronics cooling application. Linear compressor is interesting for refrigeration application in the electronics cooling field. They validated their compressor model results with the experimental data. The overall displacement and average stroke of the compressor are 3 cm and 0.6 cm respectively. They built the prototype compressor and utilized it with the omission of the springs and linear motor. Their model result showed a good agreement with experimental result and the predicted values of the volumetric efficiency and the overall isentropic efficiency are 24% and 31% respectively.

Liang et al., [3] performed and fabricated the linear compressor based on moving magnet. They made the compressor design with clearance seal and flexure bearing. It is useful for a refrigerating compressor system. They used a compact type heat exchanger and they made the system needed for CPU cooling. Liang [4] discussed yet another paper about linear compressors design with clearance seal and flexure bearing. They made the compressor design for space applications. They presented and constructed the moving magnet motor in a novel design. They discussed the use of flexure spring, linear motor, valves and their incorporation. The companion paper covers the experimental evaluation of the linear compressor. They used nitrogen for experiments, with a constant pressure ratio and showed that the linear motor system under resonance achieved high overall efficiency. They found that the motor efficiency of 74% is achieved when pressure ratio is 3.0. They also found that the overall adiabatic and the isothermal efficiencies are 54% and 46% respectively. They concluded that these efficiency values are within a reasonable range for a small compressor.

Bailey et al., [5] done a research work about a novel designed oil-free type linear compressor based on a moving magnet type motor. They used clearance seal and flexure spring in the linear compressor and they incorporated into vapour compression-based refrigeration system for electronics cooling system. They fabricated the compressor that has the stroke and piston diameter of 14 mm and 9 mm respectively and it is operated using novel designed moving magnet type motor. They done the work for every operating condition using stroke, discharge and suction pressure to predict the resonant frequency. They done the experiment for several strokes under individually pressure ratios. They done the experiment with fixed outlet condenser temperature (50°C) and evaporator temperature (Range: 6 °C–27 °C). They result showed that the COP is 3.0 for the pressure ratio of 2.5.

Also, the LG Electronics [6] developed the oil free piston type linear compressor for household refrigerator applications which is an energy efficient. It has no crank-driven mechanism. Its piston motion is brought by using linear motor (forward motion) and helical coil spring (backward motion). LG established the compressor delivering the outstanding energy efficiency due to the following reasons.

- Due to the simple mechanism, its mechanical loss is considerable lesser than conventional crank-driven type compressors.
- Motor efficiency has been achieved more than 90% while using the moving magnet type linear oscillating motor.
- With the simple electronics, the system cooling capacity can be obtained for better system efficiency. They concluded that LG linear compressor is 20% to 30% higher efficient than current conventional crank-driven type compressors.

LG Electronics used the free piston mechanism for better compressor efficiency by lowering the frictional loss and better flow path of the piston. Additionally, by adjusting the simple voltage of the electronic circuits, the stroke of the free piston can be changed and therefore, the cooling capacity of the system can be obtained.

Jomde [7] developed a new oil free moving coil linear compressor for refrigeration system. R134a is used as working fluid. Rulon material is used for piston surface to reduce the friction between the cylinder surface and piston body. They found that COP of cooling system with linear compressor is 18.6% more than the reciprocating compressor.

Liang [8] reviewed the linear compressor for vapour compression refrigeration system with applications of electronics cooling, aerospace cooling and household refrigerator applications. They found that cost is a major factor for linear compressor fabrication. They suggested that linear motor has to be designed at cost effective.

Alzoubi [9] experimentally investigated the energy efficient VCR system using linear compressor. The developed the VCR system with linear compressor achieved high COP of 4.5.

From the above literature study, linear compressor is mainly focused in the area of miniature refrigeration system and it also produced the required cooling capacity compared to conventional reciprocating compressor. The miniature linear compressor is a promising area to develop the energy efficient VCR system. So, highly efficient motor design is focused on this paper. The objective of this work is to develop an oil free and free piston solenoid controlled linear compressor and to achieve high system performance.

3 Linear Compressor Design

The linear compressor consists piston, connecting rod, cylinder, spring and linear motor. The piston is freely moved from forward motion using the force exerted by the linear motor and backward motion is by using springs. The schematic diagram of oil free and free piston solenoid controlled linear compressor is illustrated in Fig. 1.

The external forces involving linear compressor is taken for this study and the expressions are

$$\sum f_{ext} = m_{mov}\, a = \; - k_{sp}\Delta x - \; c_{fr}.\, v - \; f_g(t) + \; f_{mag} \tag{1}$$

$$m_{mov}\, a + \; k_{sp}\Delta x + \; c_{fr}.\, v + \; f_g(t) = \; f_{mag} \tag{2}$$

where m_{mov} is moving mass, k_{sp} is mechanical spring stiffness, c_{fr} is viscous damping coefficient of friction, Δx is displacement of the piston, a and v are the acceleration and velocity respectively.

The instantaneous gas force is calculated by

$$f_g(t) = \left(P_{cy}(t) - P_{suc}\right) * A_{cr} \tag{3}$$

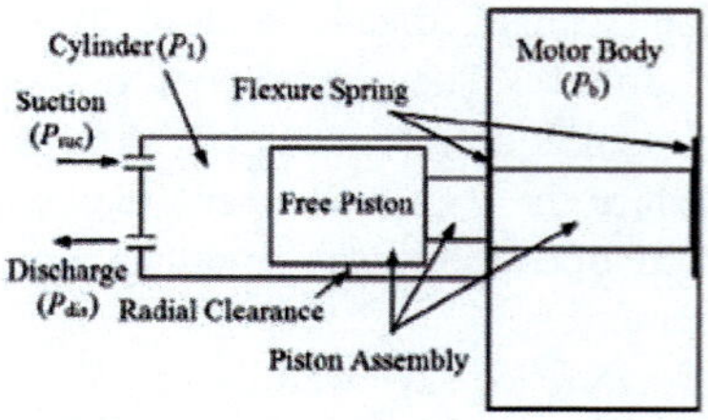

Fig. 1 Schematic diagram of linear compressor [10]

where P_{suc} is the suction pressure, $P_{cy}(t)$ is instantaneous cylinder pressure and A_{cr} is the Piston cross-sectional area.

3.1 *Simulation of Linear Compressor Using 20 sim*

20 sim is a software tool which provided us to analyse the modelling and simulation process for various mechatronic system. 20 sim software enables the model to choose graphically. Using these models, we can simulate and analyse the behaviour of the dynamic system and create the control systems. It can also be able to generate C code and simulate it using the hardware.

20 sim software programs enable us to create the model easily, quickly and intuitively. Also, it can help to create the models using equations, physical components, block diagrams and bond graphs. Various tool boxes help in building models, simulate and analyse its performance. Iconic diagrams or components are the building blocks of physical systems as shown in Fig. 2.

20 sim software is used to analyse the motion of piston. This motion depends on the force acting on the piston or moving system. The moving system comprises the armature and the piston connected to it. The motion is caused due to various forces including the electromagnetic force induced by the solenoid, gas force acting opposite to the moving piston due to the compression (and this gas force can in turn be modelled as a spring damper system), the force caused by the spring, force caused due to friction (i.e. damping) and the force caused due to the inertia (because of the moving mass). The assumptions of this model are given below [11].

- Assume constant spring stiffness
- Mass of the piston is 0.1 kg
- At fixed world velocity is zero
- Force by the solenoid is one directional
- Spring retraction force is considered in one directional.
- Assume constant gravitational force.

In the simulator we can also add a 3D plot, say 3D animation and then add the physical components with the dimensions as per the model either using the inbuilt

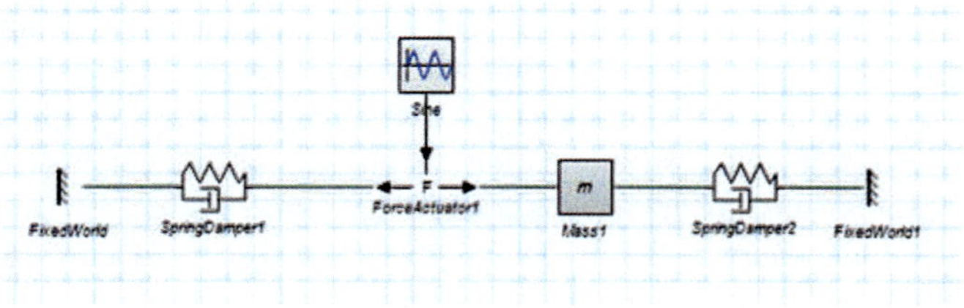

Fig. 2 Iconic diagram modelled in 20 sim

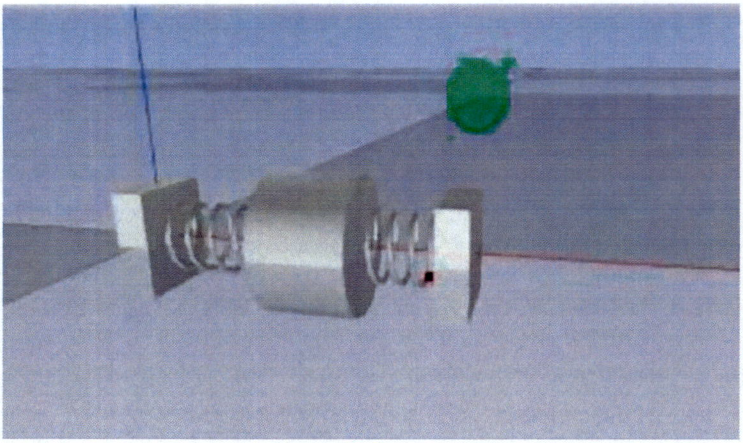

Fig. 3 3D simulation window

structures or import a 3D CAD file. The three-dimensional model of an oil free and free piston solenoid controlled linear compressor is presented in Fig. 3.

3.2 Fabrication of Linear Compressor

The cylinder is made with a stainless-steel rod which is initially turned, drilled and then bored for the required dimensions. The piston is made with Teflon material by turning and drilling process in the lathe. The piston is connected with the AC solenoid valve to move the forward motion. The schematic diagram of the piston rod is shown in Fig. 4. The cylinder head is fabricated with aluminium material and a reed valve acts as an inlet and exit port of the refrigerant. The schematic diagram of cylinder case/head is shown in Fig. 5. Input and output ports of the copper tubes are connected with the pressure gauge to measure the inlet and discharge pressures after intake and compression processes respectively. The picture of reed valve and inlet and outlet copper tube location is illustrated in Figs. 6 and 7.

Steel rod has been used for machining of the cylinder. The valve assembly is attached to one end of the cylinder and a piston is let through the other end. The piston is made using Teflon as a material and the reason behind using Teflon for this application is that it has very low friction when compared to other metals and also Teflon coating over the piston gets easier. This has high lubricating properties and there is no need of the external lubrication oil. Hence, oil-free compression can be achieved and due to this advantage, this compressor can be worked at any orientation (0°–360°). AC power supply can be directly given to the solenoid but we control the compression through Arduino. Hence, we convert the AC supply into the required voltage that an Arduino can accept. This is done using a relay module which can

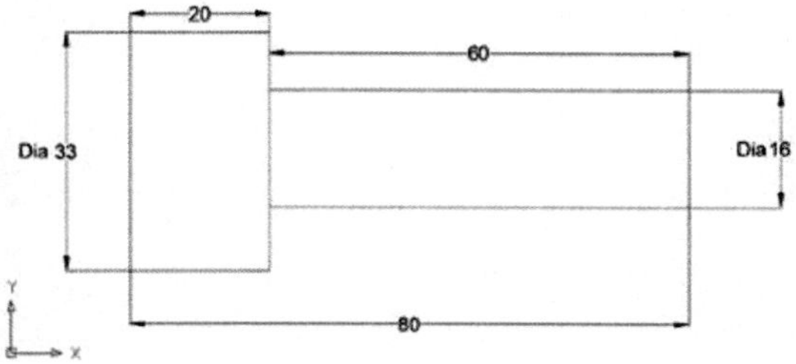

Fig. 4 Schematic diagram of piston rod

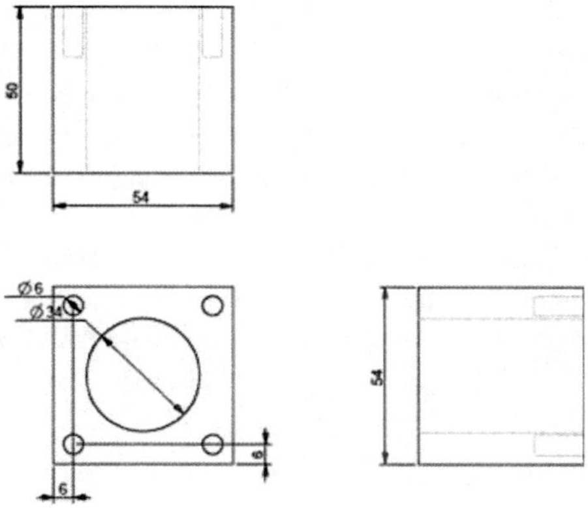

Fig. 5 Schematic diagram of cylinder head

Fig. 6 Picture of reed valve

Fig. 7 Picture of inlet and outlet copper tube location

Fig. 8 Prototype of developed oil free linear compressor

convert the 220 V into the required 5 V. The control code is written with Arduino and it is explained in the next section. Now whenever the solenoid is given a supply and is excited, the piston moves forward and backwards is obtained by the springs and thus compression is done inside the cylinder and the compressed refrigerant is passed on to the next stage.

The prototype of the linear compressor has been developed with the basic requirements that have been stated as above. This includes AC Power supply, AC Solenoid, Arduino, Relay module, Teflon piston, Cylinder, Valve assembly and Pressure gauge. The prototype of an oil free and free piston solenoid controlled linear compressor is shown in Fig. 8.

4 Results and Discussion

The motion of piston is analysed for the various piston forces of 2.5 N, 5 N and 10 N. This analysis is conducted for the time of 200 s. The mass of the piston is taken for the simulation as 100 g and the damping coefficient is 500 Ns/m. The relationship between time with displacement and velocity of piston for several piston forces is illustrated in Fig. 9. The simulation results show that the piston is moved with velocity of 0.005 m/s and piston displacement of 5 mm is obtained for the piston force of 2.5 N. The piston velocities of 0.01 m/s and 0.02 m/s is attained for the piston force of 5 N and 10 N respectively. Also, for these forces, the achieved displacement are 10 mm and 20 mm respectively. The piston velocity and displacement are directly proportional to the piston force which is also confirmed by Fig. 9. When the piston force is increased from 2.5 N to 10 N, the piston velocity and displacement has been enhanced at the percentage of 300% respectively. The 20sim simulation is also performed for various spring damping coefficients of 10 Ns/m, 50 Ns/m and 100 Ns/m respectively. Its results showed that the damping coefficient is decreased with increase of the displacement and piston velocity. For the damping coefficient of 10 Ns/m, the achieved piston velocity and displacement are 2 m/s and 2 m respectively. This displacement value is not in our range. The linear compressor is modelled for the stroke of 10–15 mm. Hence, it is concluded that the spring has been selected for

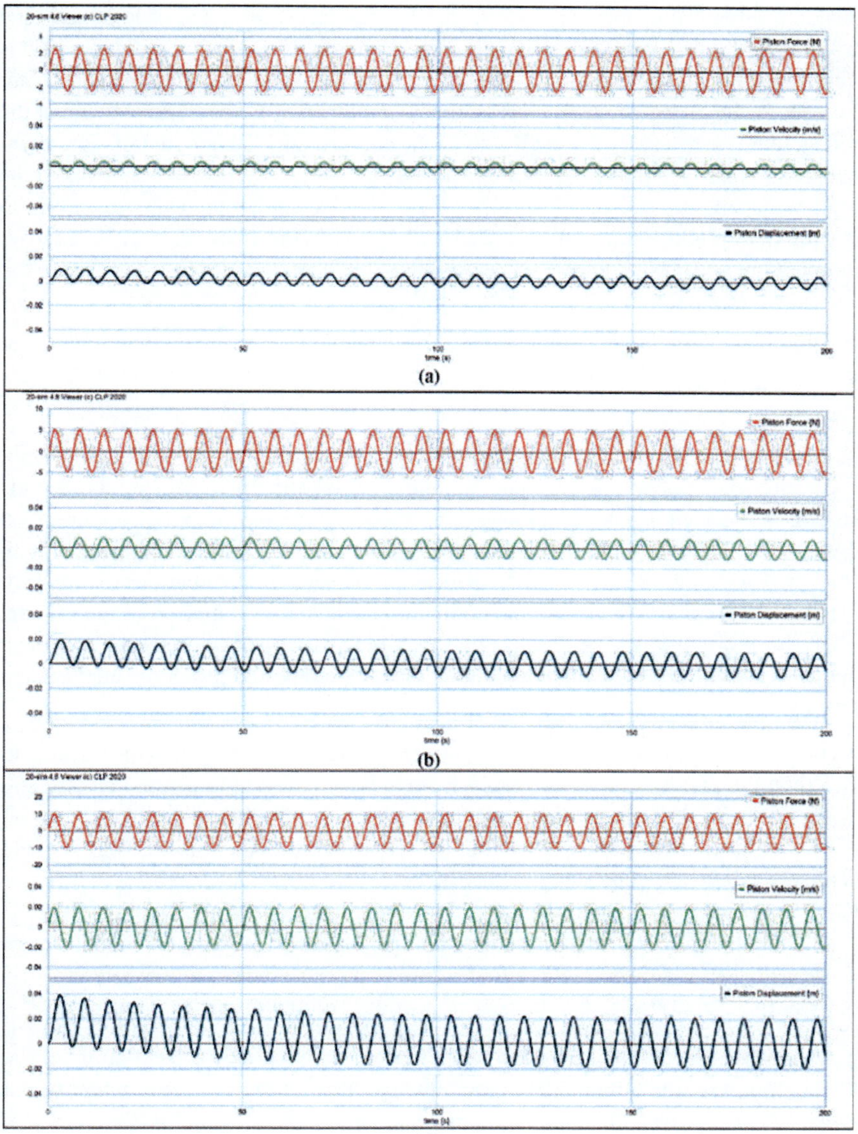

(c)

Fig. 9 Relationship of the piston force, velocity and displacement of the piston with respect to time for various piston forces **a**) Piston force = 2.5 N, **b**) Piston force = 5 N and **c**) Piston force = 10 N

the damping coefficient of 500 Ns/m and piston force of 5 N is enough to move the piston at the stroke of 10 mm.

Using the simulation results, the prototype model of linear compressor is analysed for the cooling capacity of 100–150 W. The solenoid is actuated by the programmable

Arduino setup. This is actuated at the various displacement range of 5–20 mm and translational speed of 0.005–0.015 m/s. This result shows that the pressure ratio of 3 is obtained for the maximum range of displacement and velocity of 20 mm and 0.02 m/s respectively. Here, the suction pressure is maintained at 1.5 bar. The leakage of the refrigerant is arrested by the enclosure of the compressor. The amount of refrigerant charged to this system is 80 g. From this analysis, linear compressor has worked the maximum pressure ratio of 3 with the solenoid valve actuation. This pressure ratio is enough to produce the cooling capacity of 150 W at the evaporator and condenser temperature of 10°C and 30°C respectively.

5 Conclusions

In this work, an oil-free and free piston type solenoid controlled linear compressor is developed. The compression rate is controlled with the help of a solenoid. This system was developed for a miniaturized vapour compression-based refrigeration system. The various forces that act on the system and the relationship between them are found and studied. 20 sim has been used for the software simulations to represent the entire system and to develop an animation to indicate the motion of the system. Then the prototype was developed by machining processes and coating has been done to reduce the friction. The results of the simulation and prototype analysis are given below.

- The piston force is directly proportional to the piston velocity and displacement. When the piston force is increased from 2.5 N to 10 N, the piston velocity and displacement has been enhanced at the percentage of 300% respectively.
- When the spring damping coefficient is decreased, the piston velocity and displacement are enhanced drastically.
- The piston force of 5 N is enough to move the piston stroke of 10 mm.
- With help of solenoid actuator, the prototype linear compressor obtained the pressure ratio of 3. This pressure ratio is useful to produce the cooling capacity of 150 W at the evaporator and condenser temperature of 10 °C and 30 °C respectively.

In future work, we will be focused to reduce the vibration of linear compressor, spring design optimization. Also, fabrication of moving magnet types linear motor using permanent magnet is to be focused. The other refrigerants of R1234yf, R445, R456, R515 and mixture of R134a + R1234yf and R134a + R1234ze will be used for better performance of the miniature vapour compression-based cooling system.

Acknowledgements Author thanks to the Department of Science and Technology for financial support under the scheme of the Instrumentation Development Programme (F. No: IDP/IND/11/2015).

Nomenclature

Acr	Cross-sectional area of piston [m^2]
k_{sp}	Mechanical spring stiffness [N/m]
a	Acceleration [m/s^2]
v	Velocity of the piston [m/s]
m_{mov}	Moving mass [kg]

References

1. Bijanzad A, Hassan A, Lazoglu I (2016) Analysis of solenoid based linear compressor for household refrigerator. Int J Refrig 74:116–128
2. Bradshaw CR, Groll EA, Garimella SV (2013) Sensitivity analysis of a comprehensive model for a miniature-scale linear compressor for electronics cooling. Int J Of Refrig, 36(7), pp 1998–2006
3. Kun Liang, Mike Dadd, Paul Bailey, Clearance seal compressors with linear motor drives. Part 2: Experimental evaluation of an oil-free compressor, J Power and Energy, 2013, pp 1–9
4. Liang K, Stone R, Davies G, Dadd M, Bailey P (2014) Modelling and measurement of a moving magnet linear compressor performance. Energy 66:487–495
5. Liang K, Stone R, Dadd M, Bailey P (2014) A novel linear electromagnetic-drive oil-free refrigeration compressor using R134a. Int J Refrig 40:450–459
6. Linear and Reciprocating compressor, compressor technology for refrigeration, 2019, pp 1–22
7. Jomde A, Anderson A, Bhojwani V, Shreyans Kedia, Nitish Jangale, Kshitij Kolas, Khedkar P (2018) Modeling and measurement of a moving coil oil-free linear compressor performance for refrigeration application using R134a, Int J Refrig, 88, pp 182–194
8. Liang K (2017) A review of linear compressors for refrigeration. Int J Refrig 84:253–273
9. Mahmoud A (2015) Alzoubi, TieJun Zhang, Characterization of energy efficient vapor compression cycle prototype with a linear compressor. Energy Procedia 75:3253–3258
10. Liang K, Stone R, Dadd M, Bailey P (2016) Piston position sensing and control in a linear compressor using a search coil. Int J Refrig 66:32–40
11. Kleijn C, Groothuis MA (2017) Getting started with 20-sim 4.6, Enschede, Controllab Products B.V

Effect of Internal Flow Channels on Encapsulated PCM with Constant Volume

Shivam Kumar Pandey, Kartik Tewari, Vidula Athawale, and Anirban Bhattacharya

Abstract A numerical study is presented to analyze the effect of internal flow channels on latent heat thermal energy storage (LHTES) systems with encapsulated phase change materials (PCMs) while maintaining constant PCM volume. A 3D cubical domain is modeled with a PCM capsule located at its center subjected to hot heat transfer fluid (HTF). Internal channels are made into the PCM capsules to improve the melting rate. Maintaining a constant volume makes the total latent energy stored constant although the rate of energy storage varies with change of internal radius. The model also captures the natural convection within the capsules. Coupled fluid flow and enthalpy model is used to solve the flow of HTF and phase change process of PCM. Capsules with base radii of 4 mm and 8 mm are studied under four different configurations with the ratio of the channel to the base radius (r/R_0) varying as 0.250, 0.333, 0.416, and 0.500. The results show that the melting duration reduces by 42.58% for the capsule with a 4 mm base radius when r/R_0 is 0.500. Similarly, the melting duration reduces by 51.15% with an 8 mm base radius when r/R_0 is 0.500. For all r/R_0 ratios, it is found that the effect of internal channels on energy storage rates and melting time are considerably more for the larger capsule size.

Keywords Energy storage · Encapsulated PCM · Internal channels · Heat transfer enhancement

1 Introduction

Phase Change Material (PCM), a latent energy storage technique, allows energy storage while showing a negligible change in the temperature of the material used. PCMs are stable chemically and safe for the environment. To increase the thermal conductivity of PCMs and improve the latent energy storage ability, various techniques are available in the literature. A few of these include encapsulation of

S. K. Pandey · K. Tewari · V. Athawale · A. Bhattacharya (✉)
IIT Bhubaneswar, School of Mechanical Sciences, Bhubaneswar, Odisha-752050, India
e-mail: anirban@iitbbs.ac.in

S. Das et al. (eds.), *Proceedings of the 1st International Conference on Fluid, Thermal and Energy Systems*, Lecture Notes in Mechanical Engineering,
https://doi.org/10.1007/978-981-99-5990-7_30

PCM, including metal fins, and using carbon or silver nanoparticles. Besides solar energy storage, they are also widely implemented in textiles, buildings, and thermal management systems for batteries.

2 Literature Review and Objective

The literature shows that significant work has been conducted on encapsulated PCM-based energy storage systems. However, very few works have highlighted the effect of enhancing latent energy storage rate by making minor modifications to the PCM capsule geometry.

A computational investigation by Khodadadi and Zhang involved analyzing the melting of PCM within an encapsulation under the influence of combined conduction and buoyancy-driven convection [1]. Regin et al. [2] conducted studies to determine the role of natural convection and phase change temperature on the heat transfer of a cylindrically encapsulated PCM with varying capsule radii. Three stages of melting were analyzed both numerically and experimentally.

Kalaiselvam et al. [3] conducted an analytical study to determine the location of the transient interface for solidification and melting. The time for attaining steady-state and solidification characteristics were both explained using a non-dimensional heat generation parameter. Agyenim et al. [4] reviewed materials and container geometries proposed earlier to store PCMs. Container geometries, resembling day-to-day engineering systems, such as long thin heat pipes, cylindrical or rectangular containers, were used to simulate the melting of PCMs. The lowest heat losses were minimal from the shell and tube systems. Hlimi et al. [5] conducted a numerical investigation on encapsulated gallium placed inside a horizontal cylinder. The key observation was that in the later stages of the process, unsymmetrical melting occurred along the horizontal axis of the capsule. Liu et al. [6], presented a review of the macro-encapsulated PCMs for buildings. The paper compared the merits and limitations of micro and macro encapsulation methods. It also emphasized the importance of geometry and material selection on PCM's overall phase change characteristics.

Energy storage characteristics of PCMs installed within a water tank were analyzed numerically and experimentally by Bony and Citherlet [7]. A variety of geometries and material combinations for the PCMs were considered. A study on spherically encapsulated PCM was conducted by Sattari et al. [8] with varying operating conditions and PCM container diameters. The results indicated that the change in surface temperature has more influence on melting duration, melting rates, and overall heat transfer compared to other process parameters such as the container's diameter and the initial PCM temperature.

Our present work analyzes the energy storage characteristics of encapsulated PCMs with internal channels through the development of a 3D finite volume-based model. For each case, the total volume of the PCM with an internal channel is maintained constant and equal to the volume of the corresponding capsule with no channel. Thus, by choosing fixed set of values for the channel radius, the value of PCM's outer diameter is determined. It was found from the literature that the central region of a spherically encapsulated PCM takes a long time to melt completely. In order to improve the rate of energy storage, the inclusion of such channels into the PCMs is proposed.

3 Mathematical Model

All the simulations are performed in a three-dimensional cubical domain of size 20 × 20 × 20 mm^3, which along with boundary conditions, is represented in Fig. 1. Simulations are conducted on a paraffin wax-based single capsule. The heat transfer fluid, i.e., water, flows from the bottom to the top across the domain, with a constant inlet temperature of 373 K. The domain is initially maintained at 303 K. As soon as the heat transfer fluid enters the domain, the temperature and thus sensible energy of PCM rises. Once the melting temperature is reached, the capsules begin to melt, and any further energy is stored in the PCM as latent heat. To ensure the complete melting of capsules, all the simulations were run for a sufficiently long duration. The material properties used for the simulations are specified in Table 1.

The following are the assumptions made in the present analysis:

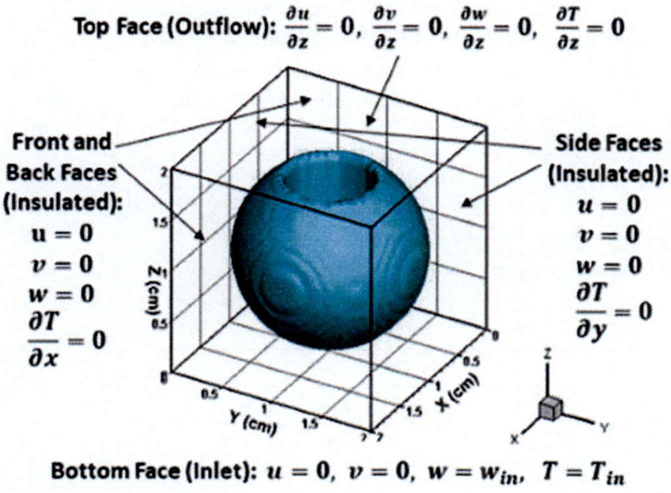

Fig. 1 Domain and boundary conditions

Table 1 Material properties

Parameter	Paraffin	Water
Density (ρ) (kg/m^3)	805	996
Dynamic viscosity (μ) (Ns/m^2)	3.65×10^{-3}	0.798×10^{-3}
Thermal conductivity (k) (W/mK)	0.22	0.615
Specific heat (C_p) (J/kgK)	2100	4178
Coefficient of thermal expansion (1/K)	3.08×10^{-4}	–
Melting point (T_m) (K)	333	–
Latent heat (L) (J/kg)	169×10^3	–

- Properties of paraffin and water are constant.
- Natural convection is implemented using Boussinesq approximation.
- The melting temperature is constant.
- The shell for PCM capsules is considered infinitely thin. Negligible thermal resistance is considered for the shell.
- The interface between the PCM and the HTF has been modeled by using a high artificial viscosity value of 10^{16} Ns/m^2 for the interface cells, thus suppressing the interfacial velocity to zero. This value of viscosity is arbitrarily chosen and any other high value which suppresses the velocity to zero at the interface can also be used.
- The channels pass precisely through the geometric center of individual capsules.

Internal channels are modeled within the PCM geometry by modifying the existing in-house codes used in previous research work by our group [9]. The encapsulation fraction parameter denoted using φ is defined to account for the fraction of PCM material present inside each control volume. Mathematically speaking, $0 \leq \varphi \leq 1$. Table 2 indicates the modified values of outer radius of PCM capsule for increasing values of internal channel radius; all determined while maintaining constant total volume, equal to corresponding base capsule volumes without any internal channels.

Table 2 External radius of capsules for constant PCM volume

Base capsule radius(R_0) in mm	Modified outer radius of capsules (R_0) in mm			
	$r/R_0 = 0.250$	$r/R_0 = 0.333$	$r/R_0 = 0.416$	$r/R_0 = 0.500$
4.000	4.083	4.146	4.227	4.324
8.000	8.165	8.292	8.454	8.648

For the present study, all the properties, such as density (ρ_{avg}), , thermal conductivity (k_{avg}) and viscosity (μ_{avg}), are volume-averaged. Each of them can be expressed in terms of the encapsulation fraction (φ) as

$$\rho_{avg} = \rho_{PCM}\varphi + \rho_{HTF}(1 - \varphi) \tag{1}$$

$$k_{avg} = k_{PCM}\varphi + k_{HTF}(1 - \varphi) \tag{2}$$

$$\mu_{avg} = \mu_{PCM}\varphi + \mu_{HTF}(1 - \varphi) \tag{3}$$

The subscripts "avg," "PCM," and "HTF" in the Eqs. (1, 2, and 3) stand for average, PCM, and HTF, respectively.

Considering the heat transfer model, a volume-averaged enthalpy-based energy conservation formulation, as indicated by Eq. (4) below, is used to describe the melting of the PCM

$$\frac{\partial}{\partial t}(\rho_{avg}H) + \nabla \cdot \left(\rho_{avg}\vec{v}H\right) = \nabla \cdot \left(K_{avg}\nabla T\right) \tag{4}$$

where

$$H = \left[C_{P_{PCM}}\varphi + C_{P_{HTF}}(1 - \varphi)\right]T + \varphi f L_f \tag{5}$$

In the above equation, f signifies the liquid fraction of PCM, which is the ratio of the amount of liquid PCM in a control volume to the total PCM present in the control volume, and L_f is the latent heat of melting. The energy Eq. (4) is discretized explicitly using the finite volume method, in which the values for velocity ($\vec{v}$) and temperature (T) are used from the previous time step for each new iteration. The discretization and numerical implementation schemes are similar to our previous work [10]. The discretized energy equation can be solved to obtain the new value of enthalpy (H_{new}). Upon substituting the value of H_{new} into Eq. (5) and choosing the value of temperature at the HTF-PCM interface as the melting temperature of PCM (T_m), the updated value of liquid fraction (f_{new}) is attained.

In continuation, f_{new} is utilized to obtain the temperature of the cell (T_{cell}). The value of T_{cell} varies with the value of the updated liquid fraction (f_{new}), for either of which, the following interpretations can be made:

- If the value of the liquid fraction obtained is less than zero, i.e., $f_{new} \leq 0$ then $T_{cell} = \frac{H}{[C_{P_{PCM}}\varphi + C_{P_{HTF}}(1-\varphi)]}$. This indicates that the melting of PCM has not yet begun. The value of the liquid fraction assigned to the solver is 0.
- If liquid fraction obtained from (4) lies between 0 and 1, i.e., $0 < f_{new} < 1$ then the value of $T_{cell} = T_m$. This indicates that cells are at an interface of solid and liquid phases, and the value of the liquid fraction assigned to the solver is f_{new}.
- Finally, if the liquid fraction obtained from $f_{new} \geq 1$, $T_{cell} = \frac{H-\varphi L_f}{[C_{P_{PCM}}\varphi + C_{P_{HTF}}(1-\varphi)]}$. This further indicates that the cells have completely melted, and the value of the liquid fraction assigned to the solver is 1.

Similarly, considering the fluid flow model, the fluid flow is considered in the heat transfer fluid and the liquid PCM. As shown in Eqs. (6) and (7), the continuity and momentum conservation equations are modified to incorporate the effect of different phases. The equations are solved by using the SIMPLER algorithm.

$$\nabla \cdot \vec{V} = 0 \tag{6}$$

$$\rho_{avg}\frac{\partial \vec{V}}{\partial t} + \rho_{avg}\left(\vec{V}\cdot\nabla\right)\vec{V} = -\nabla P + \mu_{avg}\nabla^2\vec{V} + \vec{S_B} + \tau\vec{V} \tag{7}$$

In Eq. (7), P refers to the pressure at the center of each control volume. The source term due to Boussinesq approximation is given by: $\vec{S_B} = \rho_{PCM}\vec{g}\,\alpha\varphi(T_{cell} - T_m)$. $\vec{g}$ and α denote acceleration due to gravity, and the thermal expansion coefficient, respectively. $\vec{S_B}$ is non-zero only if $T_{cell} > T_m$. The term $\tau\vec{V}$ forces the velocity to become zero in the solid PCM region [10].

4 Results and Discussion

The entire formulation is implemented by modifying the in-house code used previously for the study presented in Athawale et al. [11]. The validation of the code was carried out by Athawale et al. [11] with the existing work of Khodadadi and Zhang [1].

Liquid fraction contours for PCM with 4 mm and 8 mm radii at 50 s are shown in Figs. 2 and 3, respectively. It is observed that, upon increasing the channel size, while maintaining constant volume, a quicker melting is observed due to higher effective heat transfer surface area.

Corresponding temperature contours at 50 s are shown in Figs. 4 and 5, respectively. It can be seen that it is a case of non-uniform heat transfer since the melting starts from the bottom, where the HTF comes into contact with the capsules first. The HTF then proceeds through the periphery and the internal channels and melts the PCM completely. For the lower radius channels, the temperature in the PCM is lower indicating less heat transfer to the internal parts of the capsule. This is more prominent for the 8 mm cases. Faster melting at the bottom and thermal stratification

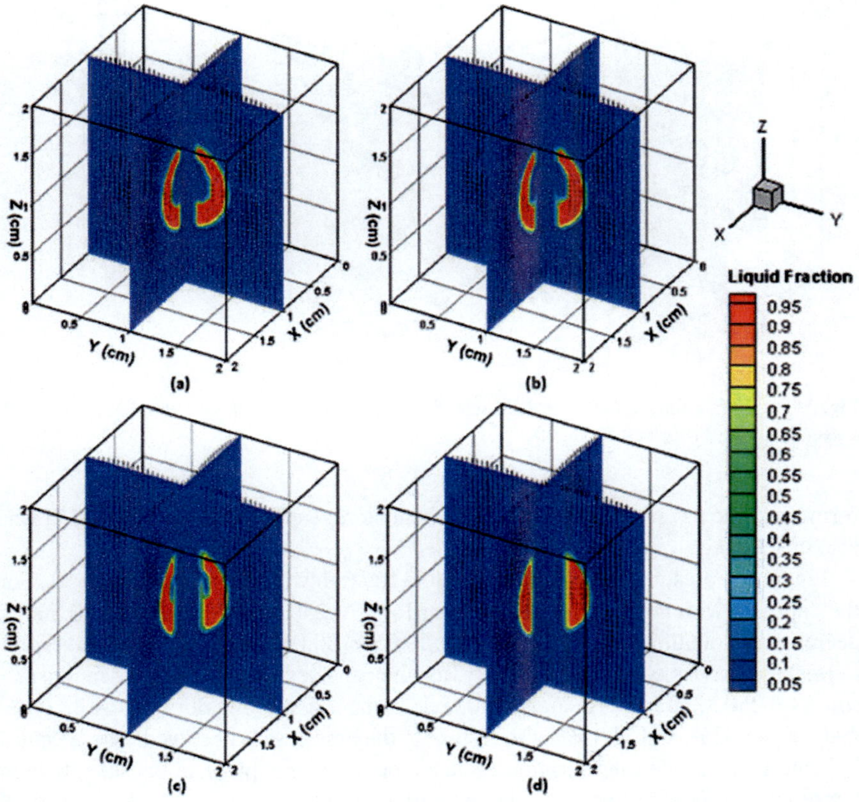

Fig. 2 Liquid fraction contours for PCM with 4 mm base radius **a** $r/R_0 = 0.250$, **b** $r/R_0 = 0.333$, **c** $r/R_0 = 0.416$, **d** $r/R_0 = 0.500$ at 50 s

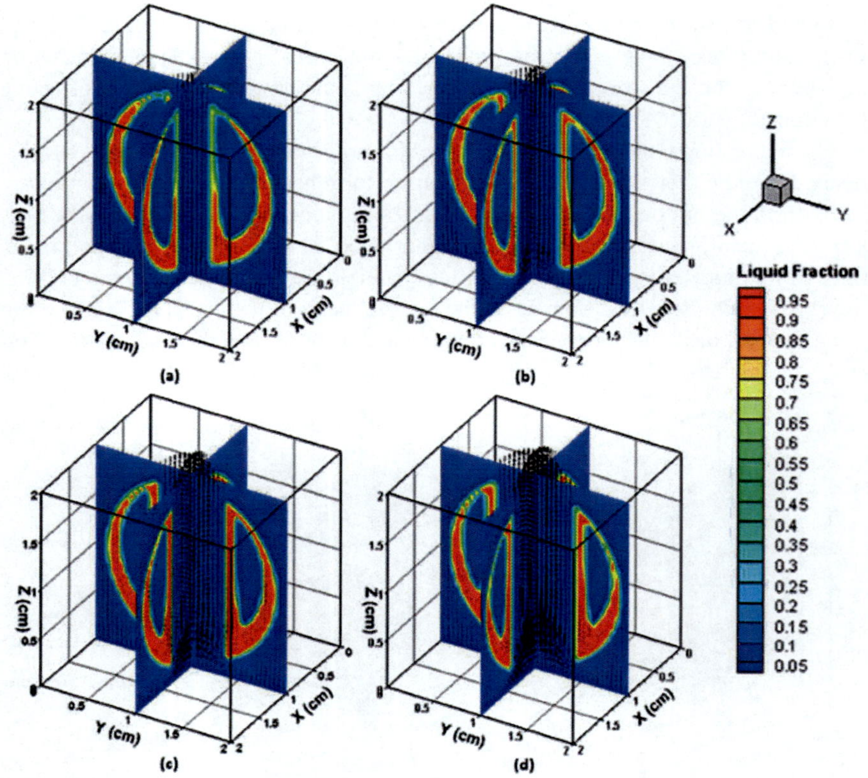

Fig. 3 Liquid fraction contours for PCM with 8 mm base radius **a** $r/R_0 = 0.250$, **b** $r/R_0 = 0.333$, **c** $r/R_0 = 0.416$, **d** $r/R_0 = 0.500$ at 50 s

forming at the top resulting in increased heat storage efficiency similar to [12] was also observed.

Figure 6 represents the variation of liquid fraction with time for all the cases. For the capsules with 4 mm base radius and r/R_0 as 0.250, 0.333, 0.416, and 0.500, a decrement in melting duration by 13.02%, 19.75%, 30.05%, and 42.58% with respect to the base case is observed. Similarly, for the capsules with 8 mm base radius and r/R_0 as 0.250, 0.333, 0.416, and 0.500, a decrement in melting duration by 33.64%, 42.08%, 47.93%, and 51.15% with respect to the base case is observed. These values indicate that the melting duration reduces much more rapidly in capsules having larger internal flow channels and larger external radii.

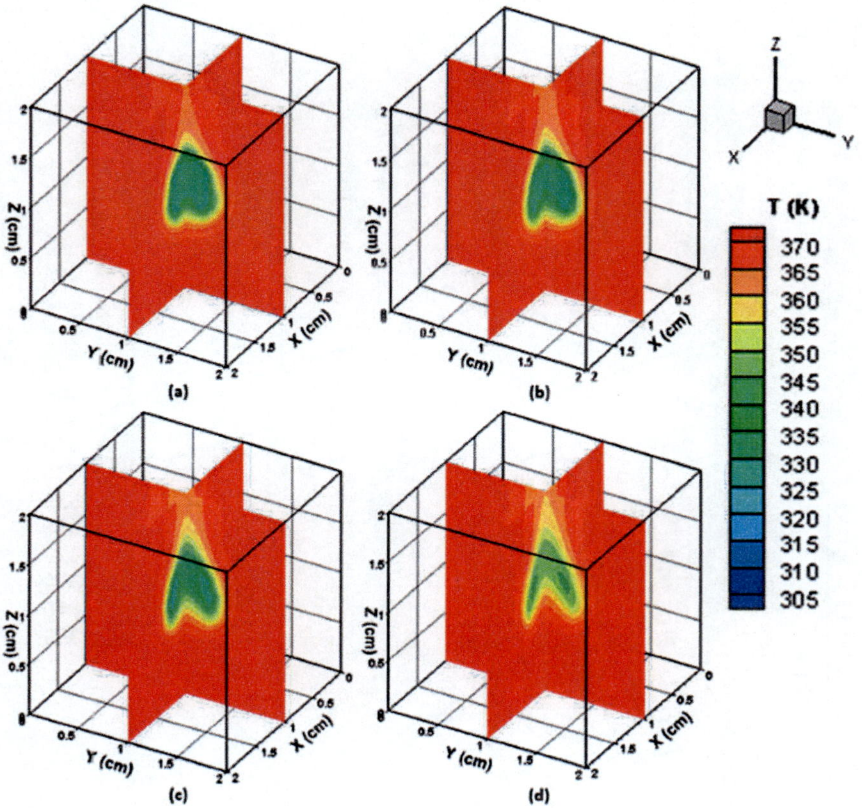

Fig. 4 Temperature contours with 4 mm base radius **a** $r/R_0 = 0.250$, **b** $r/R_0 = 0.333$, **c** $r/R_0 = 0.416$, **d** $r/R_0 = 0.500$ at 50 s

Figures 7 and 8 indicate the variation of latent, sensible, and total energies for the capsules with 4 mm and 8 mm base radii, respectively. The latent energy for all the 4 mm cases settles at 36.47 J. The sensible energy increases with an increase in channel radius, because of the PCM's higher effective heat transfer area, arising due to an increasing nature of internal channel size. The latent energy for all the 8 mm cases settles at 291.76 J. The sensible energy trend for the 8 mm cases can be understood using similar arguments proposed for Fig. 7b.

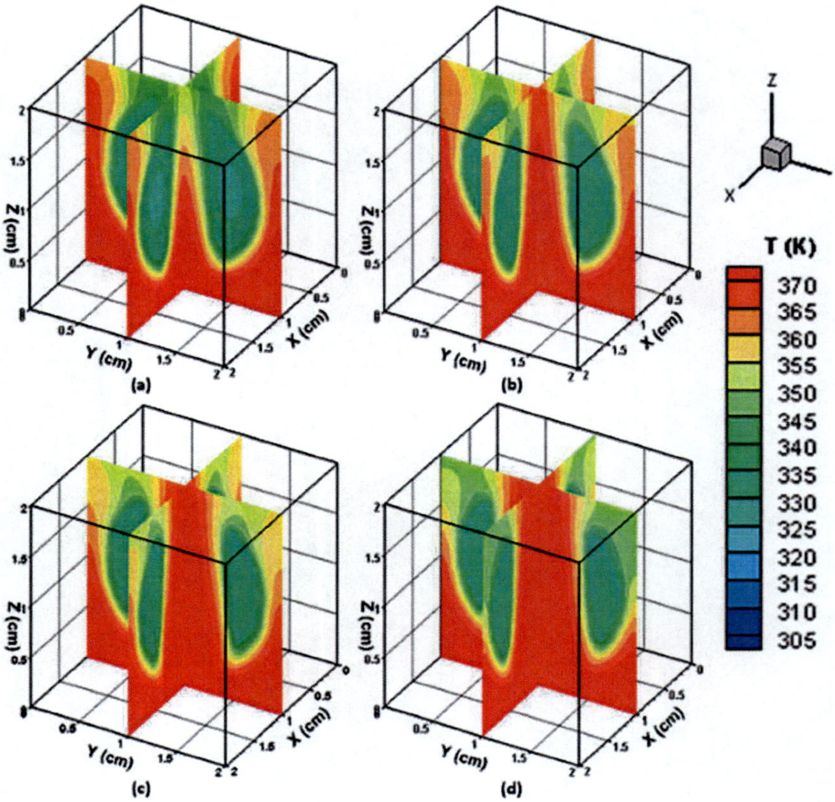

Fig. 5 Temperature contours with 8 mm base radius **a** $r/R_0 = 0.250$, **b** $r/R_0 = 0.333$, **c** $r/R_0 = 0.416$, **d** $r/R_0 = 0.500$ at 50 s

Figure 9 represents the latent energy fraction (latent energy / total energy) plotted against time for all the cases combined. From the plot, the latent energy fraction of cases corresponding to 4 mm capsules begins to rise after a lag compared to the 8 mm capsules. Since the capsules are located at the center of the cubical domain, the HTF comes into contact with the capsules for the 8 mm cases earlier than the 4 mm cases. The 4 mm cases peak earlier and the peak values are higher than the 8 mm cases, owing to their lesser volume, resulting in quicker complete melting. Finally, the latent energy fraction for the 8 mm PCM cases is higher indicating lesser sensible energy storage due to longer melting period. Additionally, the base cases have lower sensible energy content due to less effective heat transfer when compared to PCMs having internal channels. Thus their latent energy fraction is higher upon complete melting although the trends for the peak values are reversed.

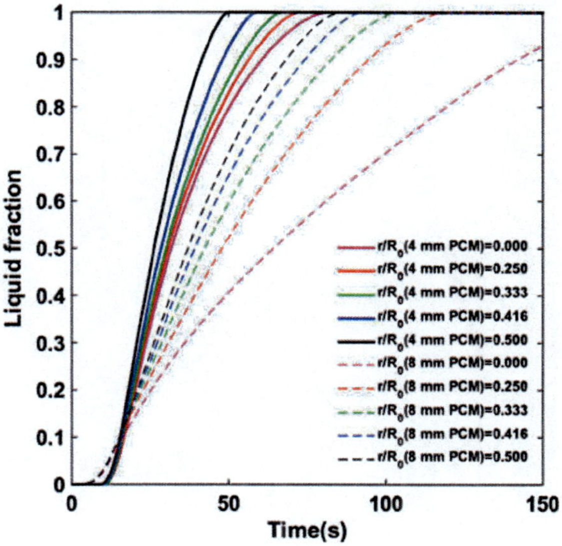

Fig. 6 Variation of liquid fraction with time

5 Conclusions

The energy storage characteristics, in terms of liquid fraction, latent, sensible, and total energies for spherical encapsulated PCMs with internal cylindrical channels and constant total volume, have been analyzed in this paper. The melting duration reduces by 13.02%, 19.75%, 30.05%, and 42.58% with respect to the capsule having a base radius of 4 mm. Similarly, melting duration reduces by 33.64%, 42.08%, 47.93%, and 51.15% with respect to the capsule having a base radius of 8 mm. These values show that the melting duration reduces faster with increase in channel radius ratio for larger capsules. Thus, having internal channels for macro-encapsulated PCM systems with larger capsules is more effective. The work can be extended to a packed bed system containing multiple capsules arranged in a structured manner for further analysis.

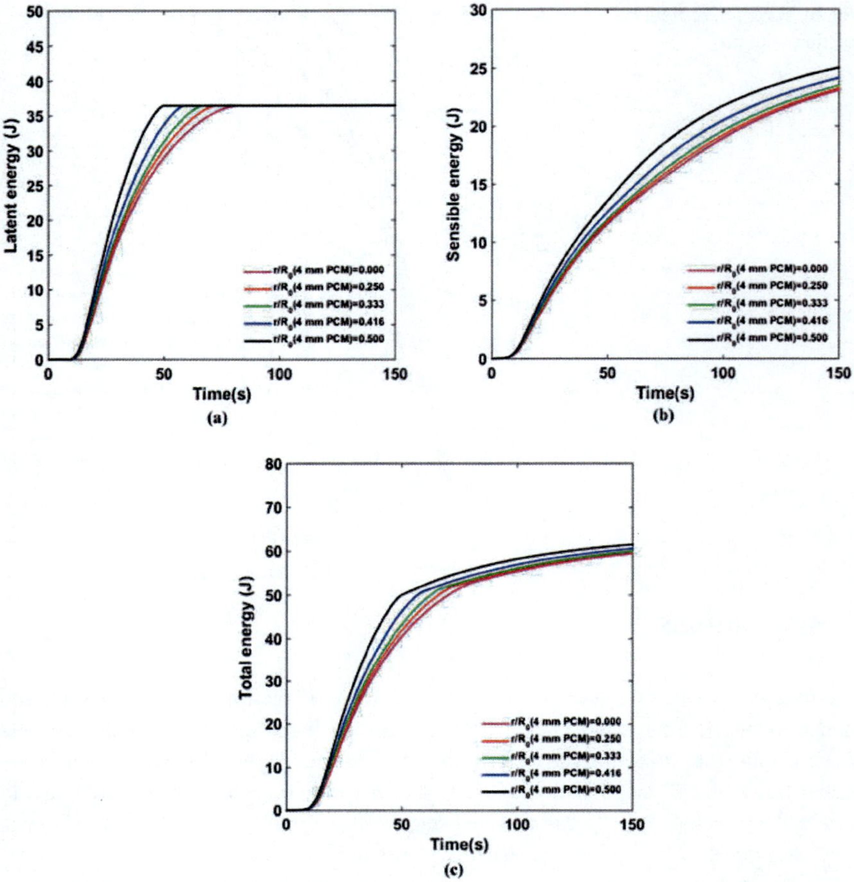

Fig. 7 Variation of **a** latent energy, **b** sensible energy, **c** total energy of PCM with time for 4 mm base radius

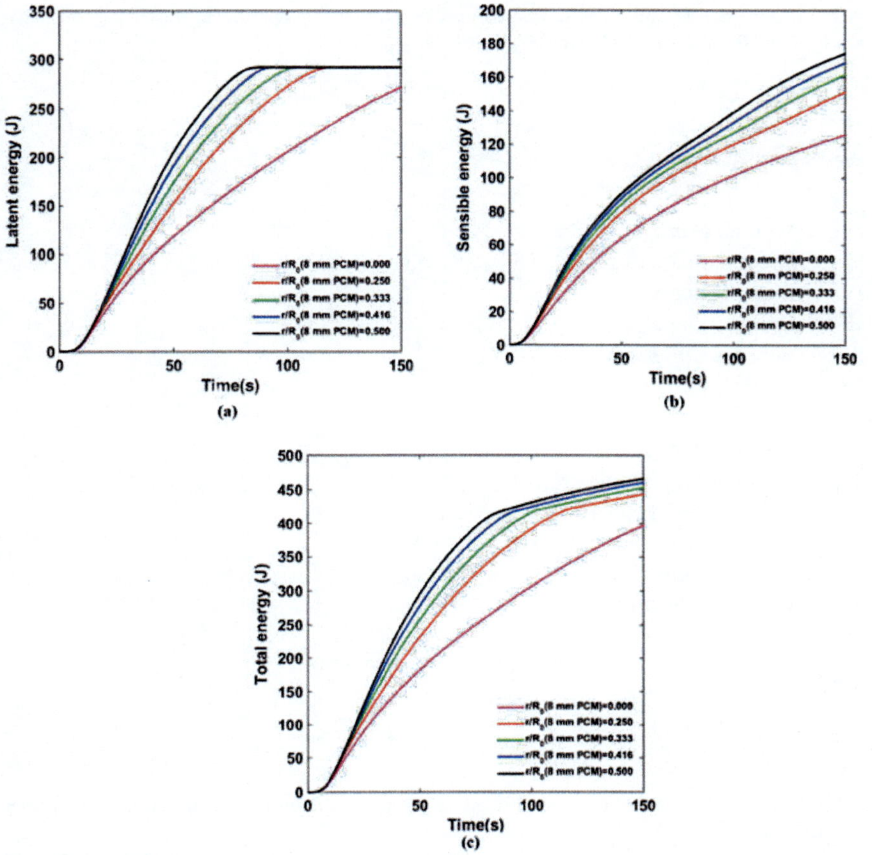

Fig. 8 Variation of **a** latent energy, **b** sensible energy, **c** total energy of PCM with time for 8 mm base radius

Fig. 9 Variation of latent energy fraction with time

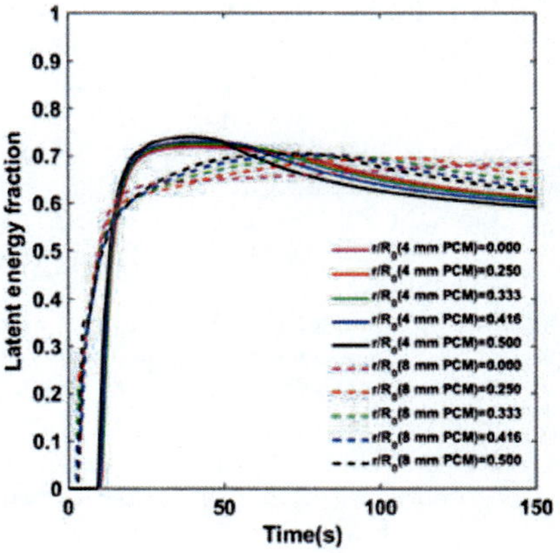

Acknowledgements This work has been supported by DST (Department of Science and Technology) SERB project CRG/2021/003780.

Nomenclature

avg	Volume average
C_p	Specific heat [J/kgK]
f	Liquid fraction
H	Specific enthalpy [J/kg]
K	Thermal conductivity [W/mK]
L	Latent energy [J/kg]
P	Pressure [N/m^2]
T	Temperature [K]
T_m	Melting temperature [K]
v	Velocity [m/s]
α	Coefficient of thermal expansion [1/K]
ρ	Density [kg/m^3]
μ	Viscosity [Ns/m^2]

References

1. Khodadadi JM, Zhang Y (2001) Effects of buoyancy-driven convection on melting within spherical containers. Int J Heat Mass Transf 44(8):1605–1608
2. Regin AF, Solanki SC, Saini JS (2006) Latent heat thermal energy storage using cylindrical capsule: Numerical and experimental investigations. Renew Energy 31(13):2025–2041
3. Kalaiselvam S, Veerappan M, Aaron AA, Iniyan S (2008) Experimental and analytical investigation of solidification and melting characteristics of PCMs inside cylindrical encapsulation. Int J Therm Sci 47(7):858–874
4. Agyenim F, Hewitt N, Eames P, Smyth M (2010) A review of materials, heat transfer and phase change problem formulation for latent heat thermal energy storage systems (LHTESS). Renew Sustain Energy Rev 14(2):615–628
5. Hlimi M, Hamdaoui S, Mahdaoui M, Kousksou T, Msaad AA, Jamil A, El Bouardi A (2016) Melting inside a horizontal cylindrical capsule, Case Studies in Thermal. Eng 8:359–369
6. Liu Z, Yu ZJ, Yang T, Qin D, Li S, Zhang G, Haghighat F, Joybari MM (2018) A review on macro-encapsulated phase change material for building envelope applications. Build Environ 144:281–294
7. Bony J, Citherlet S (2007) Numerical model and experimental validation of heat storage with phase change materials. Energy Build 39(10):1065–1072
8. Sattari H, Mohebbi A, Afsahi MM, Yancheshme AA (2017) CFD simulation of melting process of phase change materials (PCMs) in a spherical capsule. Int J Refrig 73:209–218
9. Dinesh BVS, Bhattacharya A (2019) Effect of foam geometry on heat absorption characteristics of PCM-metal foam composite thermal energy storage systems. Int J Heat Mass Transf 134:866–883
10. Athawale V, Bhattacharya A, Rath P (2021) Prediction of melting characteristics of encapsulated phase change material energy storage systems. Int J Heat Mass Transf 181:121872

11. Athawale V, Jakhar A, Jegatheesan M, Rath P, Bhattacharya A (2022) A 3D resolved-geometry model for unstructured and structured packed bed encapsulated phase change material system. J Energy Storage 51:104430
12. Soni V, Kumar A, Jain VK (2018) Modeling of PCM melting: Analysis of discrepancy between numerical and experimental results and energy storage performance. Energy 150:190–204

Simulation of a Single Zone Building in Meghalaya Using Solar Air Heater in TRNSYS Software

Noel M. Rajive, Ashish B. Khelkar, and Rajat Subhra Das

Abstract The global demand for renewable energy is increasing daily. A substantial portion of this renewable energy will be provided by solar energy. In this scenario, solar air heaters are a suitable replacement for electric heaters. Solar air heaters (SAHs) find wide application in space heating, crop drying, and process heating, among many others. The simulation of SAH for various locations can be carried out by the Transient System Simulation software, commonly known as TRNSYS. The current study focuses on the simulation of a SAH system for a building in the location of Cherrapunjee in the state of Meghalaya in India. A wet and cold location is explicitly chosen to determine the application of the solar air heater in cold climates for space heating applications. A single zone building located at the NIT Meghalaya main campus in Cherrapunjee is considered the building whose outlet temperature is determined. A significant increase in the output temperature is obtained using SAH. The temperature of 40.52 °C was obtained at 5821 h at a mass flow rate of 1000 kg/hr. The humidity ratio obtained was 0.028 at 5920 h.

Keywords Solar energy · TRNSYS · Solar air heater · Building simulation · Meghalaya

1 Introduction

The World Resource Institute 2017 released a report which stated that India accounts for 6.65% of total global carbon emissions and is ranked fourth in the world. While some of this can be attributed to the large population size of the country, the reality of the threat of climate change still looms large. When the energy consumption of large economies is considered, India will hold the position of top contributor by the year 2040. Coal will be meeting most of this increase in demand, followed by renewable energy. In 2020, renewables overtook gas and oil to become the second

N. M. Rajive (✉) · A. B. Khelkar · R. S. Das
Department of Mechanical Engineering, NIT Meghalaya, Shillong-793003, India
e-mail: noelrajive94@gmail.com

S. Das et al. (eds.), *Proceedings of the 1st International Conference on Fluid, Thermal and Energy Systems*, Lecture Notes in Mechanical Engineering,
https://doi.org/10.1007/978-981-99-5990-7_31

most important source of domestic power production in India. Renewables and their demand in India will significantly grow to 256 million tons in 2040 from 17 million tons in 2016, with the annual increase being 12% (Kumar et al., 2020).

With this scenario in mind, the push for renewables becomes more critical. Renewables can be introduced for commercial and industrial use only if they have comparable to or more efficient than traditional power sources. The earth captures approximately 1.8×10^{11} MW of energy from the sun annually [1]. This value is so much larger than the current level of consumption on the planet of all commercial energy sources. Thus, in theory, solar energy could supply the entire world's present and future energy needs for a very long time. Therefore it becomes critical that the available solar equipment improve efficiency to meet the market requirements.

The solar air heater is equipment that finds wide use in cold locations. Research has been carried out to see if it is a viable alternative to the electric heater. The bulk of the work that has been carried out assumes a steady state of heat flux falling on the absorber plates. But, in practical cases, this is not true. The solar energy falling on a system varies throughout the day. It increases as the day gets older, peaks during a particular time, and again comes down. But most computational works do not consider this. Constant heat flux is assumed for analysis. This is where the TRNSYS simulation comes in handy. The Transient System Software, otherwise known as TRNSYS, is to be used in the computational study of the SAH. TRNSYS program software has various libraries in it. Parts used in a thermal system such as solar collectors, pipes, pumps, and fans are found in this library. Each part is designated as a type. This type consists of mathematical modeling or a group of governing equations in the TRNSYS simulation program. These types are further divided into groups. Each group has a different number of types represented by specific applications such as HVAC, solar, hydronic, and building load. TRNSYS simulation allows the user to create studio projects as per requirement. The part needs to be selected from the available library in the TRNSYS environment.

Works have been carried out in TRNSYS by researchers, and these works are studied and presented in the literature review section.

2 Literature Review and Objective

The solar air heater is one such device that needs significant performance enhancement. Although many different proposals have been put forward, they are yet to take off in the commercial market. The suitability of SAH for every weather condition is an area of study. Much work has been done on the solar water heater and its applications [2–4]. Works have also been carried out in solar air heaters, and their application is space heating [5]. The introduction of turbulators and vortex generators increases the heat transfer from the absorber plate to the air stream. This has also increased the pumping power requirement of the system. Still, it has been demonstrated that the gains in heat transfer outweigh the increase in pumping power consumption. Another novel idea presented by researchers is to modify the fluid path that flows over the absorber plate. The SAH has been modified in different ways. Absorber plates roughened with ribs, vortex generators, and turbulators have increased the heat transfer between the plates and the air over the heated surface [6–8]. This includes the modifications in duct shape or the airflow passage.

Shandilya et al. [9] have done a TRNSYS simulation of a building located in New Delhi. The energy demand for a New Delhi climatic zone building was simulated using TRNSYS 18. Two retrofitting scenarios were considered to save energy: thermal envelope and high-efficiency windows. They concluded that these methods could save a maximum amount of Rs.68,000, including using 2-pane windows, glazed windows, and walls with extruded polystyrene. Egrican et al. [10] have simulated an office building located in Istanbul, Turkey. The results showed that a double PV façade is significant when designing energy-efficient buildings. The southeast façade of the building is covered with PV panels integrated into the building.

The objective of this work is the TRNSYS simulation of a single-room building located on the National Institute of Technology campus in Cherrapunjee, Meghalaya. The solar air heater operates in the weather location of Cherrapunjee, Meghalaya. The weather location is available in the TRNSYS 18 weather library. The building is heated using a solar air heater. The output temperature of the building is plotted, and the results are analyzed.

3 Trnsys System Description

3.1 TRNSYS System

The connection between the different components is made in the TRNSYS software simulation studio, shown in Fig. 1. The location of the single zone building is in Cherrapunjee, (25.2702° N, 91.7323° E), Meghalaya. The weather data file is imported into the Type 15 Weather component. The Meteonorm file format of weather data is imported from the library to the weather data component.

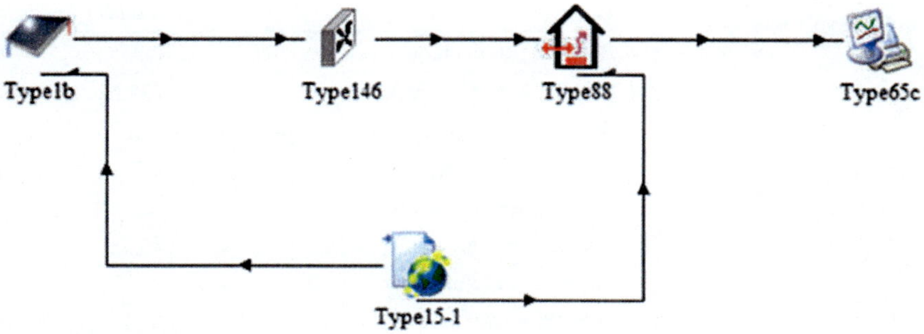

Fig. 1 TRNSYS system configuration

The weather component is connected to the flat plate solar collector (FPC) (Type 1b). The dry bulb temperature of the environment is given as the input to the inlet and the ambient temperature of the solar collector. All other corresponding input components from the weather data are connected to the flat plate solar collector. The input components connected include the ambient temperature, total radiation on horizontal, sky diffuse radiation on horizontal, total radiation on a tilted surface, angle of incidence for tilted surface, and slope of an inclined surface.

The solar collector's output is passed on to a Type 146 (Constant speed fan). The outlet temperature of the FPC is given as the inlet air temperature input of the constant speed fan. The outlet flow rate of the collector is also the inlet flow rate of the constant speed fan. This fan is then connected to the load.

The load considered here is a Type 88 lumped capacitance building with internal gains. All the building parameters that are considered are inputted into the model.

The collector area is taken as $5m^2$ for the analysis. The heat capacity of air, which is the working fluid, was set to 1.007 kJ/kgK considering an operating temperature of 300 K. The default values of various parameters of FPC are taken for the study.

3.2 Building Description

A small building on the main campus of the National Institute of Technology, Cherrapunjee, Meghalaya, is chosen for this study. The dimensions of the building are such that it has a length of 2.3 m, 2.3 m in width, and 2 m high. The total area and volume of the room are calculated for the cuboidal building. The whole building/room is considered to be a single zone in the simulation study. The roof and walls are made of plywood. The floor, meanwhile, is made of concrete. The thickness is uniformly 30 mm. The heat loss coefficients of the corresponding materials are considered while calculating the overall heat loss coefficient and the thermal capacitance of the building. The parameters used for the building in TRNSYS type are given in the table below (Table 1).

Table 1 Parameters of building in TRNSYS

Parameters	Values
Building loss coefficient (kJ/hr/m^2/K)	0.95
Building Capacitance (kJ/K)	838.626
Specific heat of building air (kJ/kgK)	1.007
Density of building air (kg/m^3)	1.2
Building surface area (m^2)	5.29
Building volume (m^3)	10.58
Initial Temperature (°C)	15
Initial humidity ratio	0.01
Latent heat of vaporization (kJ/kg)	2260

4 Results and Discussion

The results of the simulation are shown in the following figures. The weather-related parameters throughout the year for the location are initially plotted. The location of Cherrapunjee (25.2702° N, 91.7323° E) located in Meghalaya state in India is analyzed. The wind velocity of the location is shown in the Fig. 2.

It can be seen that the location experiences winds having velocities that peak at 7 m/s. The wind plays a significant role in the calculation of convection coefficient

$$\mathrm{h_w} = 5.7 + 3.8\mathrm{W}\ \left(\mathrm{W/m^2K}\right) \tag{1}$$

In turn, the convection coefficient is important in calculating the overall heat transfer coefficient.

$$U_{L,j} = \frac{3.6}{\frac{N_G}{\frac{C}{T_{P,j}}\left[\frac{(T_{av,j}-T_a)}{N_a+f}\right]^{0.33}+\frac{1}{h_w}}} + \frac{3.6\sigma(T_{av,j}^2 + T_a^2)}{\frac{1}{\varepsilon_p+0.05N_G(1-\varepsilon_p)} + \frac{2N_G+f-1}{\varepsilon_g} - N_G} + U_{be} \tag{2}$$

This coefficient, in turn, is used in calculating the outlet temperature of the solar air heater, which is calculated by the following equation.

$$T_o = \frac{AF_{R,j}(I_T(\lambda\alpha) - U_{L,j}(T_{i,j} - T_a))}{N_s\,\dot{m}_c C_{pc}} + T_i \tag{3}$$

The humidity ratio of the exact location is also shown. The ambient temperature and the wet-bulb temperature of the location are also plotted annually. It can be seen that the location has a very low temperature annually. The humidity ratio is relatively high. It can be clearly seen that the weather is wet and cold compared to other locations in India (Figs. 3, 4).

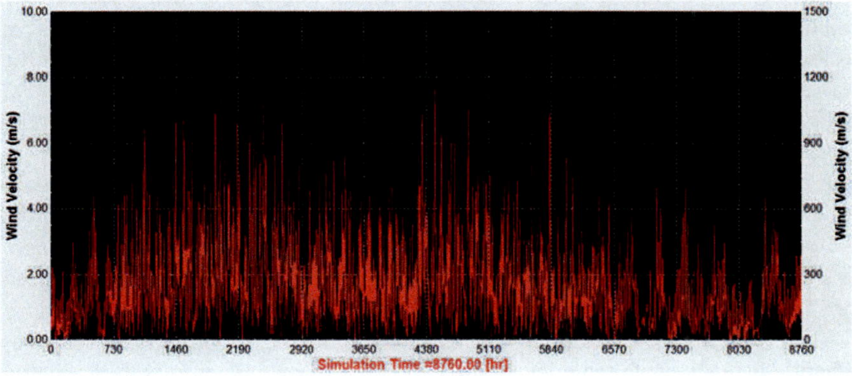

Fig. 2 Wind velocity of cherrapunjee

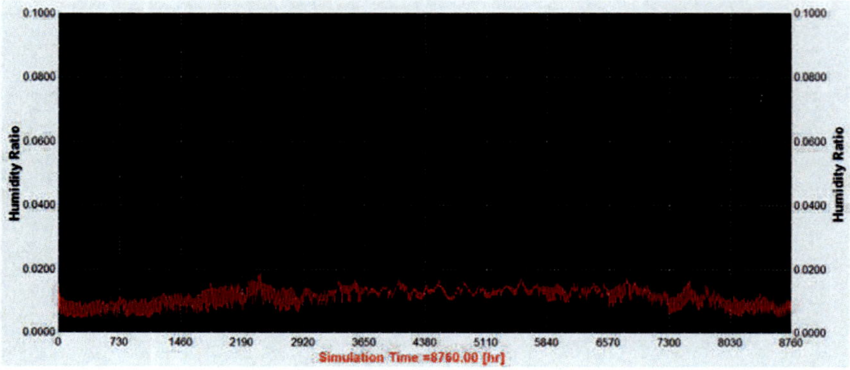

Fig. 3 Humidity ratio of Cherrapunjee

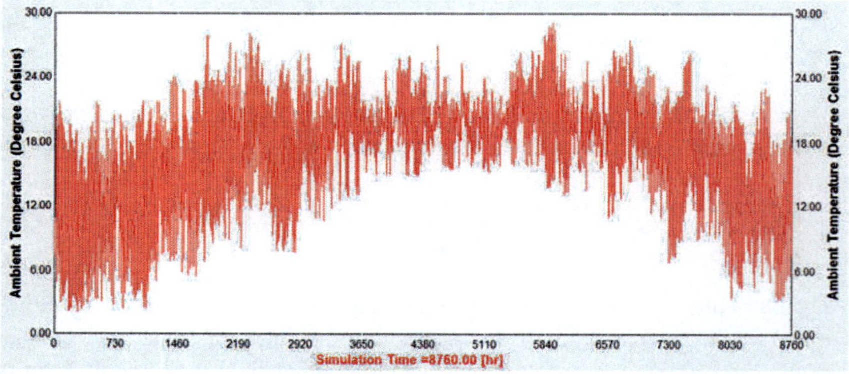

Fig. 4 Ambient temperature of Cherrapunjee

The analysis of the pattern of ambient temperature variation shows that the summer months, as expectedly, show higher values than other months, and the lowest values are recorded during the winter months. Winter and summer months occur as per the seasonal changes of any location in the northern hemisphere. December, January, and February record the lowest temperatures, and months like April and September record high temperatures.

It can be seen that the average temperature, even during the peak of summer, does not go above 25 °C. The temperatures dip to an average of below 10 °C during the winter months. The high elevation of the location contributes to this phenomenon. Cherrapunjee stands at around 1430 m above sea level. This clearly shows the need for space heating for this location.

The incident solar radiation of the location is plotted using TRNSYS. The solar radiation data shows that the location receives a healthy amount of solar radiation despite its cold and wet climate. The highest values are recorded above 1000 W/m^2.

The minimum values are recorded as expectedly during the rainy season when there is a lot of cloud cover. Therefore the months of June and July receive the lowest amount of solar radiation (Fig. 5).

The outlet temperature of the single zone building that is being heated is plotted. The figure shows the temperature versus time graph for various mass flow rates for one year (Figs. 6, 7, 8, 9).

When the mass flow rate is 1000 kg/hr, the average temperature of the heated building is above 20^0C throughout most of the year. The maximum value reaches almost 40.52^0C at the beginning of September, i.e., 5821 h. As expected from the solar radiation data, the summer months give higher output temperatures. The winter months of January and December have comparatively low output values (Fig. 10).

The humidity ratio variation throughout the year for the building is shown in Fig. 10. The maximum value obtained is 0.028 at 5920 h.

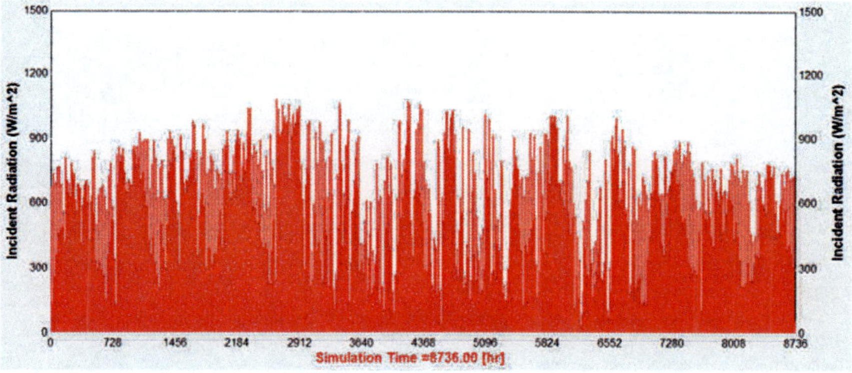

Fig. 5 Incident radiation of Cherrapunjee

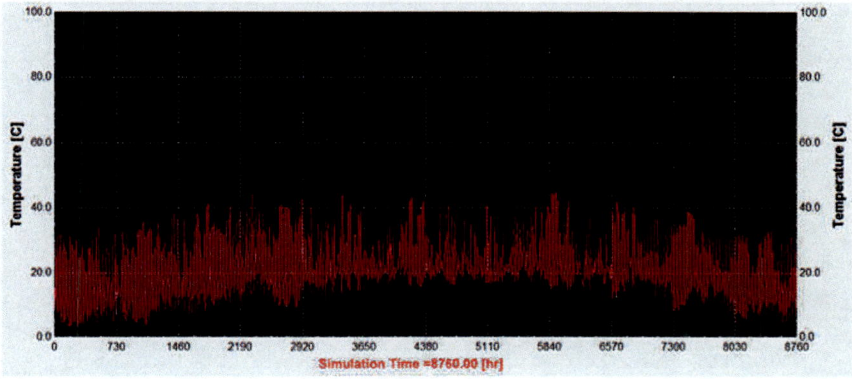

Fig. 6 Outlet temperature of building for 1000 kg/hr

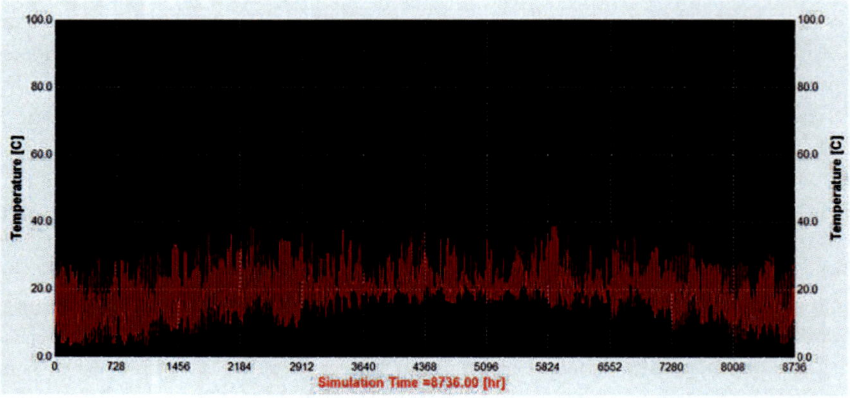

Fig. 7 Outlet temperature of building for 1500 kg/hr

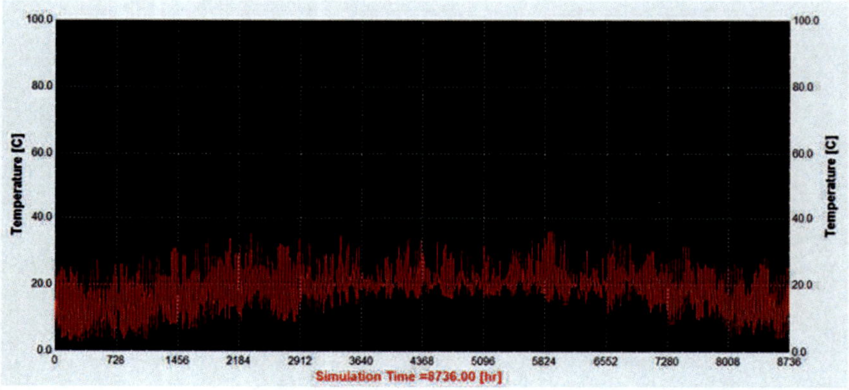

Fig. 8 Outlet temperature of building for 2000 kg/hr

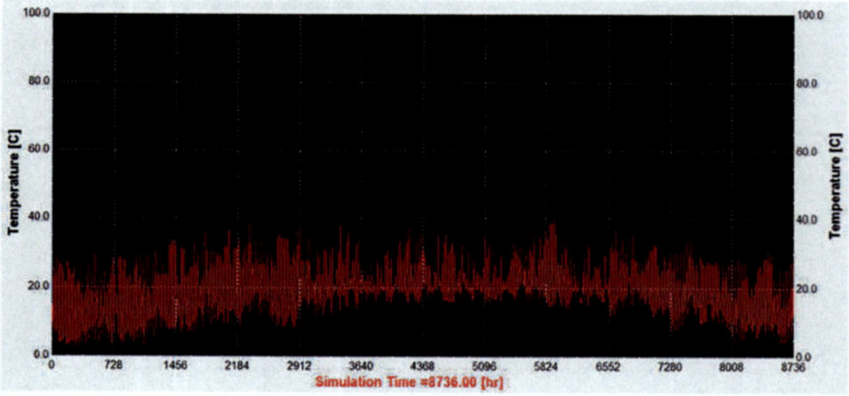

Fig. 9 Outlet temperature of building for 2500 kg/hr

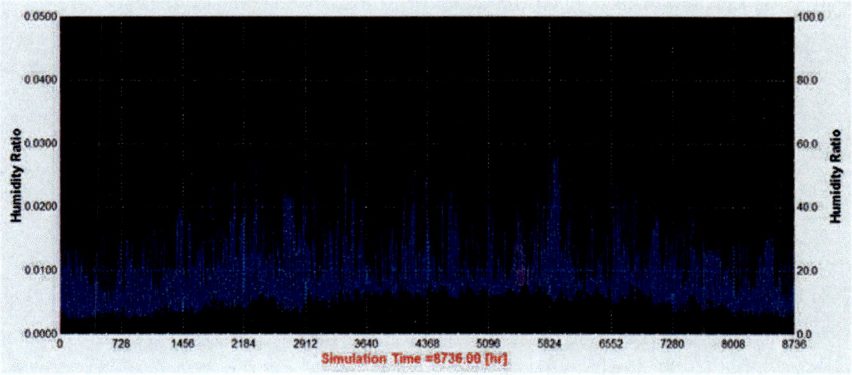

Fig. 10 Humidity ratio of building for 1000 kg/hr

5 Conclusions

The results clearly show significant heating of the building using the solar air heater. It can be seen that the temperature of the single zone building has increased significantly from the ambient temperature. Taking the case of the month of September for a mass flow rate of 1000 kg/hr, there is an increase of 15^{0}C from ambient values. This is significantly more than what is required for a summer month. This can be reduced or adjusted by reducing the airflow rate in the solar air heater. Also, the air heater will be employed during the cold winter months, when such a temperature rise will be useful. The month of December experiences sub-twenty temperatures, and the solar air heater will prove to be very useful during this time. For the study, the values of 1000 kg/hr, 1500 kg/hr, 2000 kg/hr, and 2500 kg/hr were chosen. The values can be varied as per requirement. As can be observed, as the mass flow rate is increased, the outlet temperature of the building is reduced. The humidity ratio of the room shows a rise in value from ambient values, with a maximum of 0.028 recorded during September.

The solar radiation intensity received by the location is sufficient to develop an efficient solar air heater capable of heating a room. Although the location remains cold and wet throughout the year, this study proves that it receives enough solar radiation to run an efficient space heating system. Human comfort lies around 25 °C, and it can be seen that temperatures above 40 °C are obtained for a mass flow rate of 1000 kg/hr during the summer months. The extra heat received during summer can be stored to be used during non-sunshine hours. The variation in the outlet temperature can be explained by the interpretation of solar radiation. Since there is no solar radiation during the night time, the outlet temperature comes down. This problem can be resolved by employing Phase-change Materials (PCMs) in conjunction with a solar air heater. The usage of a thermal storage wall as a substitute for Phase-change Materials (PCMs) can be considered. This helps further increase the average outlet temperature and can overcome the intermittent nature of the solar energy problem.

Nomenclature

A	Total collector array aperture or gross area in m^2
β	Collector slope in degrees
C_p	Heat capacity at constant pressure
C_{pf}	Specific heat of collector fluid in kJ/kgK
D_h	Hydraulic diameter
F'	Fin efficiency of collector
f	Friction factor
h	Heat Transfer coefficient in W/m^2K
h_w	Heat transfer coefficient due to wind in W/ m^2K
I	Global (total) horizontal radiation in kJ/h m^2
I_d	Diffuse horizontal radiation in kJ/h m^2
I_T	Global radiation incident on the solar collector in kJ/h m^2
I_T	Global radiation incident on the solar collector in kJ/h m^2
k_f	Thermal conductivity of fluid in W/mK
m	Mass flow rate in kg/s
m_c	Mass flow rate through the collector in kg/h
N_s	Number of identical collectors in series
N_G	Number of collector covers(or glazings)
Nu	Nusselt number
P	Pitch length in m
Re	Reynolds Number
T_i	Initial temperature of fluid in Kelvin/^{0}C
T_o	Final temperature of fluid in Kelvin/^{0}C
T_a	Ambient air temperature in ^{0}C
T_{av}	Average collector fluid temperature in ^{0}C
T_i	Inlet temperature of fluid to collector in ^{0}C
$\tau\alpha$	Transmittance-absorptance product of the collector
U_L	Overall loss coefficient from the collector defined per unit top surface area in kJ/hr m^2K
W	Wind velocity in m/s

References

1. Sukhatme S, Nayak J (2008) Solar energy. 4th ed. Chennai: McGraw Hill education private limited; 2008, pp 31–32
2. Kalogirou SA (2001) Use of TRNSYS for modeling and simulation of a hybrid PV- thermal solar system for Cyprus, Renew Energy, 23, pp 247–260
3. Selvadhurai M, Ponmurugan V, Vijayakumar R, Dillibabu V (2019) Creating a TRNSYS model of spiral flow flat plate solar water heater (SFSWH) and comparing unique results with parallel flow flat plate collector. Int J Innov Technol Explor Eng (IJITEE), ISSN: 2278–3075, 8:12S

4. Mohammed MN, Alghoul MA, Abulqasem K, Mustafa A, Glaisa K, Ooshaksaraei P, Yahya M, Zaharim A, Sopian K (2011) TRNSYS simulation of solar water heating system in Iraq, Recent Researches in Geography, Geology. Energy, Environ Biomed
5. Zhao DL, Li Y, Dai YJ, Wang RZ (2011) Optimal study of a solar air heating system with pebble bed energy storage. Energy Convers Manag 52 2392–2400
6. Bezbaruah PJ, Das RS, Sarkar BK (2019) Thermo-hydraulic performance augmentation of solar air-duct using modified forms of conical vortex generators. Heat Mass Transf 55–1387–1403
7. Bezbaruah PJ, Das RS, Sarkar BK (2019) CFD analysis of solar collector with truncated hemispherical vortex generators. In: 14th International Conference on Heat Transfer, Fluid Mechanics and Thermodynamics
8. Bezbaruah PJ, Das RS, Sarkar BK (2020) Solar air heater with finned absorber plate and helical Flow path: A CFD Analysis, Appl Sol Energy 2020, 56(1), pp 35–4
9. Shandilya A, Streicher W (2017) Performance and cost analysis of retrofit strategies applied to a sample single family house located in New Delhi assisted by TRNSYS Energy Simulations Tool- A Case Study. Int J Eng Res Technol (IJERT), ISSN:2278–0181, 6(11)
10. Egrican N, Akguc A (2011) Modeling of the energy efficient office building using TRNSYS Simulation Program. Int Cent Appl Thermodyn (ICAT)
11. Jani DB, Bhabhor K, Dadi M, Doshi S, Jotaniya PV, Ravat H, Bhatt K (2020) A review on use of TRNSYS as simulation tool in performance prediction of desiccant cooling cycle. J Therm Anal Calorim 140:2011–2031

Numerical Analysis of Phase Change Material ($SiCH_3$) Used for Energy Storage

Manish Kumar Prasad and Shambhunath Barman

Abstract The high energy storage density of Phase Change materials is one of the primary reason for their widespread application in the energy storage due to its constant phase change temperature. In this paper analysis of PCM has been shown using ANSYS Fluent software. Here, the PCM has been used for the analysis is Methyl-Silylidine ($SiCH_3$). Here, in this paper the analysis has done in cylindrical model as in future it can be used for energy storage unit model. The outer pipe is declared with 511 K temperature due to which the PCM inside the annular region will melt and when it solidifies the heat is being released which is carried out by the water flowing inside the inner pipe. For first 100 s the value of liquid fraction was taken which was 0.690 and similarly up to 600 s the data was taken and the decrement in the liquid fraction was observed due to the increment in temperature. Also the outlet temperature for the PCM material was taken starting from 100 to 600 s and the decrement in the temperature was observed which indicates that when the phase change material melts and solidifies, it releases heat. A good number of research has been done and still going on the Phase Change Material which is being widely used for energy storage. Due to the energy crisis in the day to day life, it has been a concern and also motivated researchers to do research on different Phase Change Material (PCM) which are being widely used for the energy storage systems.

Keywords Phase Change Material (PCM) · Thermal Energy Storage (TES) · Temperature dependent properties · Methyl-Silylidine (SiC) · Liquid Fraction

M. K. Prasad · S. Barman (✉)
Department of Mechanical Engineering, NIT Sikkim, Namchi, South Sikkim-737139, India
e-mail: shambhunath.barman@nitsikkim.ac.in

S. Das et al. (eds.), *Proceedings of the 1st International Conference on Fluid, Thermal and Energy Systems*, Lecture Notes in Mechanical Engineering,
https://doi.org/10.1007/978-981-99-5990-7_32

1 Introduction

Since eighteenth century, Phase Change Materials (PCMs) have been used for the various applications in the thermal energy storage and lots of commercial products have been developed till the date. Basically depending on the field of applications PCMs are being selected based on their temperature and the phase-transition enthalpy. Since many years a lot of research works for the energy storage and also its methodologies are going on worldwide due to the energy crisis and the environment pollutions [1]. It is seen that how the demand of energy is increasing and moreover fossil fuels have adverse effects on the environment so it is necessary to move on to renewable sources of energy for proper fulfillment of the required energy. As the solar energy is available in plentifully, but it is discontinuous, unpredictable, and available in the day time, so efficient Thermal Energy Storage (TES) systems are required for storing the excess energy available during the day time and which can be used during night and in required time [4]. So, Phase Change Materials (PCM) can be used for Thermal Energy System (TES) systems. The thermal energy storage systems (TES) give lots of alternatives for the efficient energy use and for conservation [2]. Thermal energy storage (TES) is termed as the temporary storing of thermal energy in the form of hot or cold substances for using it later in the required field for utilization [2]. Phase Change Materials (PCMs) are the kinds of materials which are having a very unique ability to catch and save energy in the periods of plentiful heat, and that releases it during required necessary periods [3].

2 Literature Review and Objective

Phase Change Material (PCM) type thermal energy storage plays an excellent role in the field of energy storage by the fact that of its high energy storage and constant phase change temperature [1]. PCMs are latent thermal storage materials and use chemical bonds for storing and releasing the heat [6]. The transfer of thermal energy happens when the phase change took place in the materials, i.e., from solid to liquid or vice-versa. An alternative system for efficient energy use and its conservation happens by using the Thermal energy storage (TES) system. TES main types are sensible and latent. TES complete process includes a of minimum three steps: charging, storing, and discharging [2]. But in practical systems, few steps occur at the same time (for example, charging and storing) and every step can occur more than once in each storage cycle. PCMs can generally use in building applications in two ways: direct impregnation into the building materials and another by encapsulation [5]. A model based on 2-D rectangular cavity is used for investigating transport phenomena during the solidification of PCM {CCHH $(C_aCl_2, 6H_2O)$} [1]. Vikash et al. [8] used the ANSYS (Fluent) 15.0 software for modeling and simulation for melting of PCM. Nitin Kukreja et al. [9] performed an intensive numerical study inside the shell and tube heat exchanger for finding the melting performance of PCM. They also show that the heat transfer fluid inlet temperature shows a very significant role for completing the charging and discharging process of the phase change material, as the inlet temperature increases and same the charging time reduces. Anica Trp et al. [10] developed a transient heat transfer phenomenon during the charging and discharging of the shell and tube latent thermal energy storage system.

The PCMs that are employed in the developing of thermal storage systems should have acceptable thermophysical, chemical, and kinetics properties, some examples of which are listed below [11]:

(i) Have suitable phase-transition temperature.
(ii) Be characterized by a high latent heat of transition.
(iii) Having an effective transfer of heat.

Whenever PCM is used for a particular application, the operating temperature of heating and cooling must be matched with the transition temperature of PCM. Latent heat must be as high as possible, and that is also in a volumetric basis for minimizing the physical size of the heat store. Wei-Wei Wang et al. [12] studied about the effect of the temperature difference between the inlet of heat transfer fluid and the melting point of PCM. Effect of inlet heat transfer fluid temperature, the mass flow rate on the characteristics of charging and discharging processes are insight.

3 Materials and Methods

In this paper, the Material which has been consider for analysis is methyl-silylidine ($SiCH_3$), which is having temperature dependent properties like the melting temperatures and density variations at various temperature zones. For this analysis, ANSYS software have been used. For solving five steps were involved they are Geometry, Meshing, Setup, Solution, and finding the results. Each process has been explained in this paper.

An outer cylindrical pipe having a depth of 1000 mm, and inside that another cylindrical pipe of depth 1000 mm but small in dimension as compared to the outer cylinder pipe has been constructed using ANSYS software for the analysis. So in this model two fluid zones are present which have been separated.

In the annular region, Phase Change Material (PCM) will be filled and inside the small pipe the water will be flowing. The model consists of four parts, i.e. outer pipe, inner fluid (i.e., inside pipe where water will flow), inner pipe, and PCM (Fig. 1).

Geometry (Fig. 2):

Meshing (Fig. 3):

Meshing is that process in which the disintegrating of geometry into simple elements from a large integrated model is done to performing the calculations (Table 1).

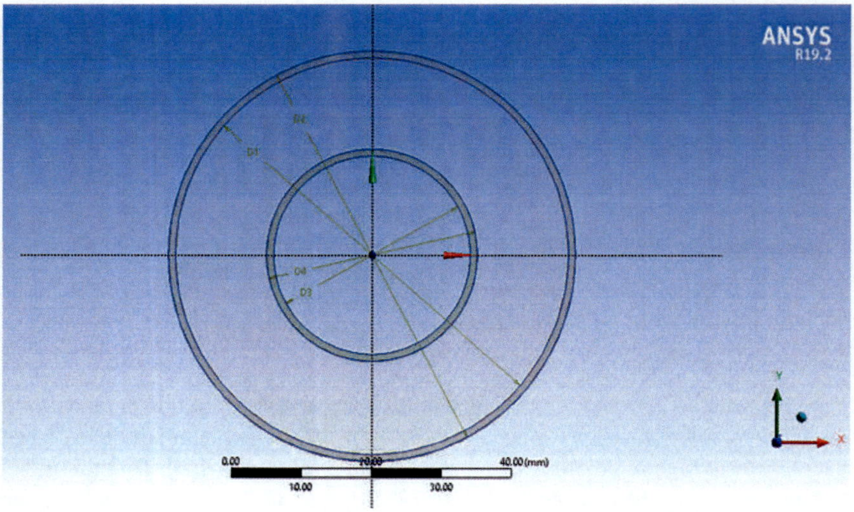

Fig. 1 Schematic diagram of the model (front view)

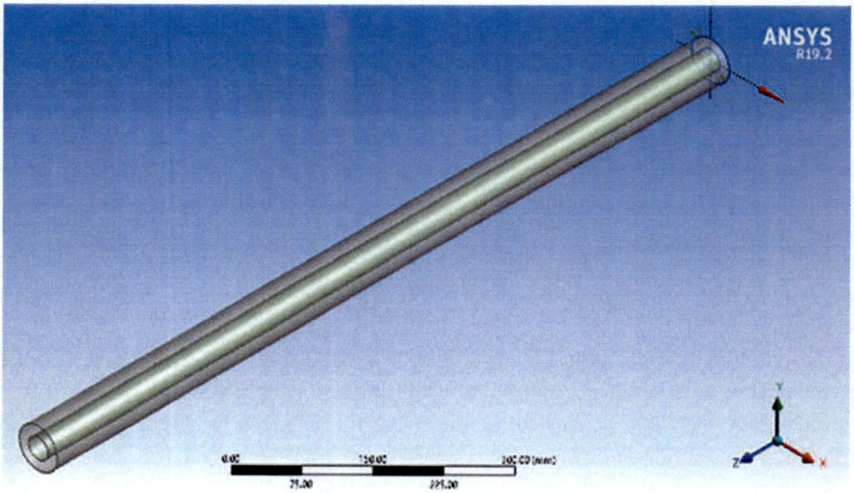

Fig. 2 Geometry

Setup.

The analysis was carried out in unsteady mode. The energy equation has been solved.

Energy Equation

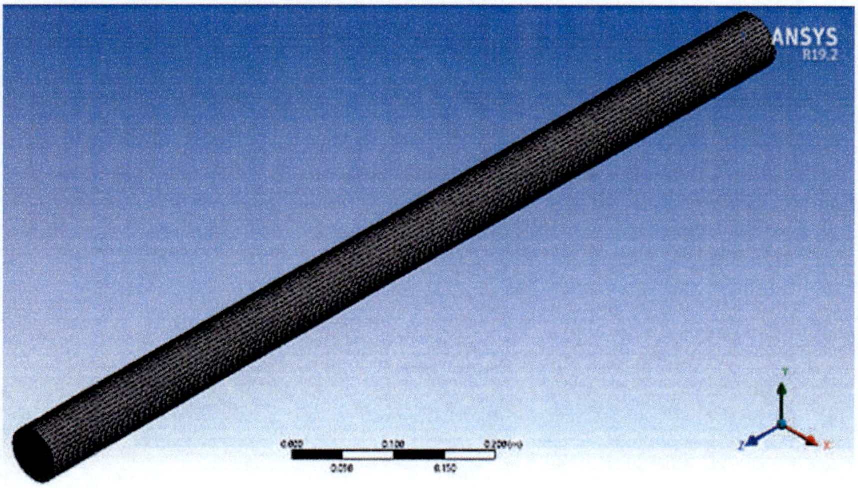

Fig. 3 Meshing view of geometry

Table 1 Mesh Statistics:

	Number
Nodes	153140
Elements	145524

$$\frac{1}{r}\frac{\partial}{\partial r}\left(kr\frac{\partial T}{\partial r}\right)+\frac{1}{r^2}\frac{\partial}{\partial \phi}\left(k\frac{\partial T}{\partial \phi}\right)+\frac{\partial}{\partial z}\left(k\frac{\partial T}{\partial z}\right)+\dot{q}=\rho c_p\frac{\partial T}{\partial r} \quad (1)$$

Material:

Methyl-silylidine (as PCM), water-liquid, air], Copper (Table 2).

When the temperature increases the phase change material behavior constantly changes with respect to temperature. So, for the better analysis of the PCM effective values have been taken [13] like Density, Specific heat, and the Thermal conductivity (Table 3).

Boundary Conditions:

Mass flow rate = 0.05 kg/sec.

Inlet total temperature = 300 K.

Table 2 Thermo-physical properties of Copper

Density	8.94 Mg/m^3
Melting point	1355 K
Specific heat	388 J/Kg-K
Thermal conductivity	370 W/m.K

Table 3 Thermo-physical properties of PCM (Methyl-silylidine) [13]

Point	Temperature (K)	Density (Kg/m^3)	Specific heat (J/Kg.K)
1	500	1945.628	1486.499
2	550	1913.828	1495.089
3	600	1882.028	1503.679
4	650	1850.228	1512.269
5	700	1818.428	1520.859
6	750	1786.628	1529.449
7	800	1754.828	1538.039
8	850	1723.028	1546.629
9	900	1691.228	1555.219

Outer pipe wall temperature = 511 K.

Here the outer pipe temperature is assigned as 511 K, due to which the adjacent PCM in annular space will melt and the water in the inner pipe will carry away the heat from the PCM.

Solution Methods:

For finding solutions Pressure velocity coupling has been used. SIMPLE scheme has been considered. Green Gauss cell based has been taken into consideration for linear interpolation. To compute pressure second order has been considered. For Momentum Second order upwind and for Energy First order upwind has been used.

4 Results and Discussion

Since the experimental results of PCMs are not available, so in this paper a numerical analysis has been shown and the authors are preparing the experimental setups for the validation of the numerical analysis. In this current paper, only numerical analysis has been shown successfully. Decrement in the liquid fraction was recorded due to the increment in the temperature. The outlet temperature of phase change material was found to be decreased (figure number 6), which indicates that when the phase change material melts and after being solidified at constant temperature it releases sufficient heat which can be used in various applications.

Here, in this analysis after doing with all modeling, meshing, and setups, at first for 100 s the liquid fraction and temperature for PCM was obtained. So, similarly up to 600 s the data has been taken to analyze the liquid fraction and temperature with respect to time.

In Fig. 4, it can see how the density decreases with increase in the temperature. It's showing the relation that the density is inversely proportional to temperature (Table 5).

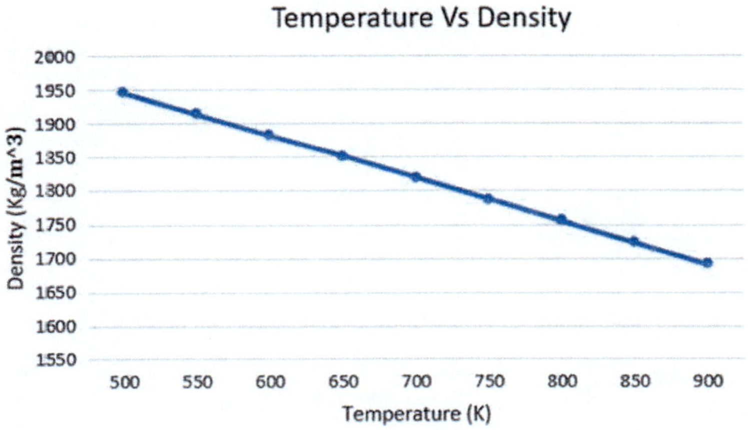

Fig. 4 Temperature Versus Density plot of PCM

Table 4 Thermo-physical properties of PCM (Methyl-silylidine) [13]

Point	Temperature (K)	Thermal conductivity (W/m.K)
1	500	0.486300
2	550	0.496126
3	600	0.505896
4	650	0.515660
5	700	0.525436
6	750	0.535206

The above Table 4 and graph of Fig. 5 give the idea that how the liquid fraction decreases with respect to time. As with respect to time the temperature has been increased, so it has been observed that decrease in the liquid fraction (Table 6).

From Table 5 and the graph of Fig. 6, it can be observed that the outlet temperature is more as compared to the inlet temperature and the reason for that is due to the heat transfer from the melted PCM to the flowing water inside the small pipe.

Table 5 Liquid fraction with respect to time

Time (s)	pcm	Net
100	0.69020735	0.69020735
200	0.5049174	0.5049174
300	0.3421868	0.3421868
400	0.24495045	0.24495045
500	0.15468354	0.15468354
600	0.085066138	0.085066138

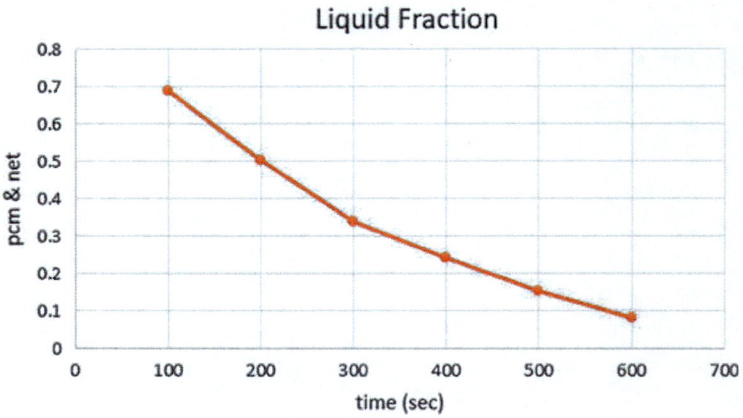

Fig. 5 Liquid fraction of PCM

Table 6 Temperature w.r.t. time

Time (s)	Inlet (K)	Outlet (K)	Net (K)
100	300	320.84048	310.42024
200	300	316.07795	308.03898
300	300	313.25032	306.62516
400	300	311.62195	305.81098
500	300	310.52133	305.26067
600	300	310.02314	305.01157

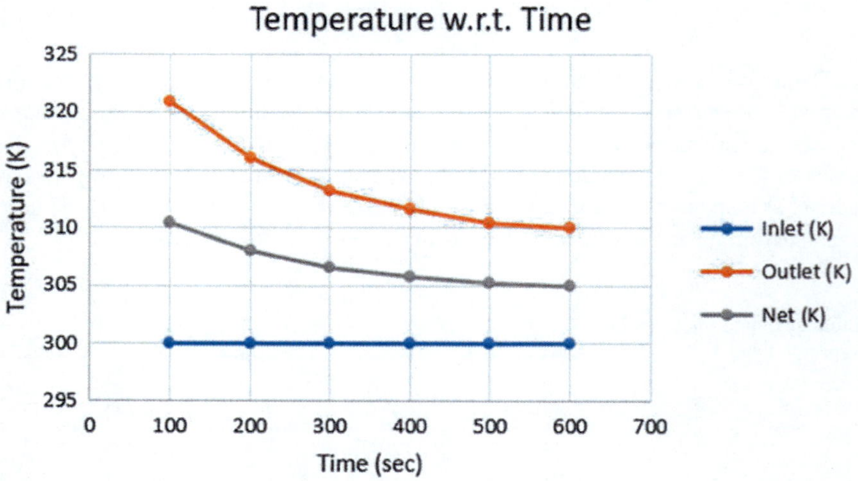

Fig. 6 Temperature w.r.t. time

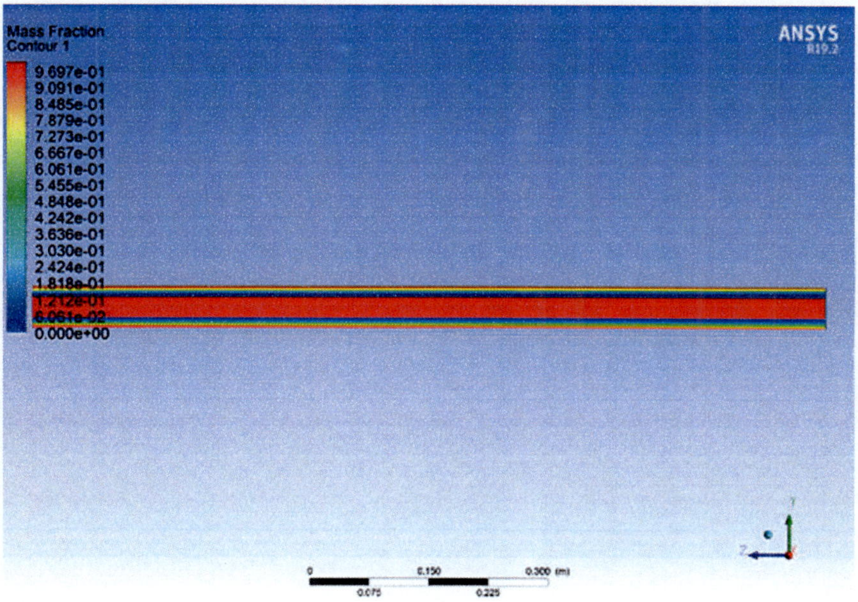

Fig. 7 Contour plot for mass fraction

Figure 7 shows the contour plot for the Mass fraction, the red coloring indicates the amount of mass of water inside the small pipe. Figure 8 shows the contour plot for temperature in which it can be seen that the red coloring in the outer pipe indicated the maximum temperature as already mentioned in this paper the outer pipe has been declared with maximum temperature so that the PCM will melt and heat will be transfer to the flowing fluid. Figure 9 shows the contour plot for velocity in which it can be that the red color indicating the maximum velocity at those areas and that that is due to the excitation of the water molecules after heat observed by the molecules from the PCM. Figure 10 shows the contour plot for the velocity streamline.

5 Conclusions

The present paper shows the analysis of PCM (methyl-silylidine). It has been shown that when the phase change material melts and solidifies in the annular region, it releases heat and the heat is carried out by the water which is flowing inside the pipe. Decrement in the liquid fraction was observed due to the increment in temperature. The decrement in the outlet temperature of phase change material was observed, due to the heat released from the phase change material after it is being solidified. This numerical analysis gives a brief idea about the phase change material, regarding its behavior and various applications where it can be used. For better analysis effective

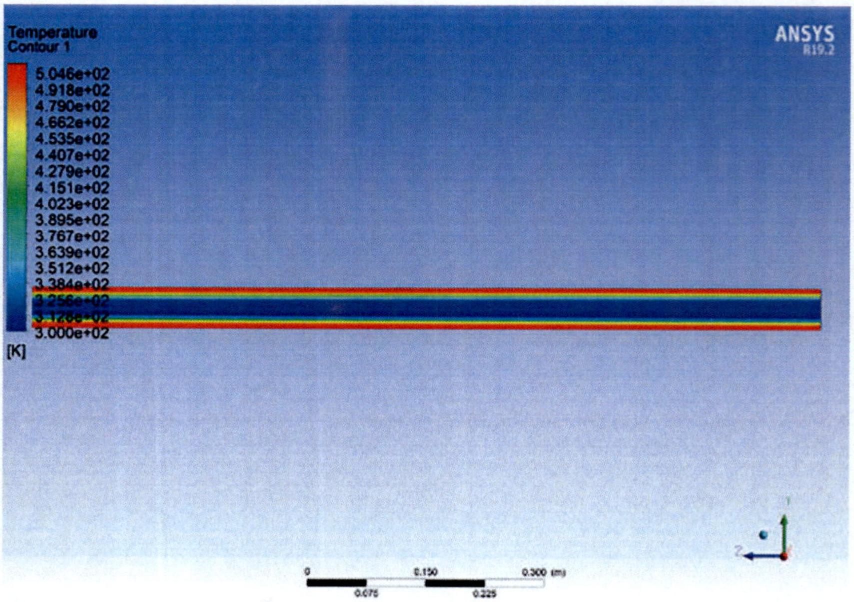

Fig. 8 Contour plot for temperature distribution

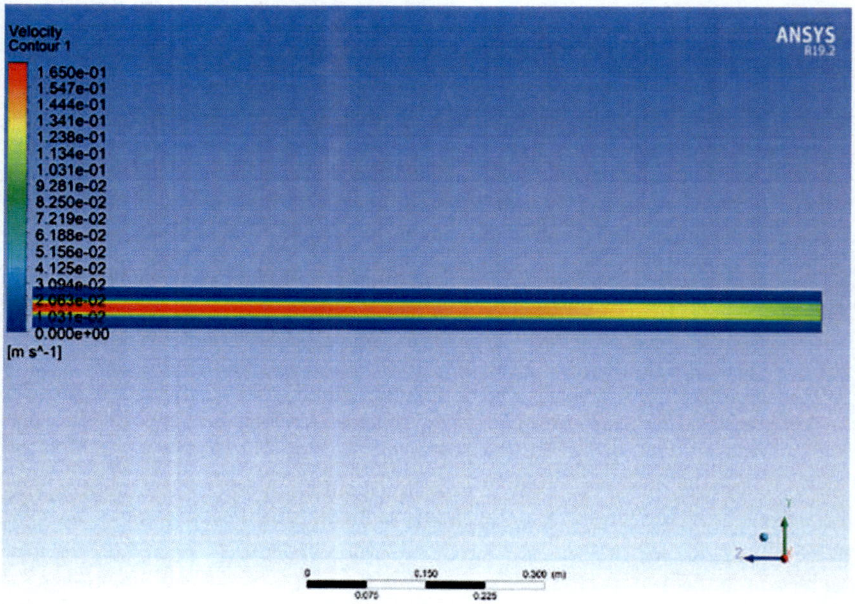

Fig. 9 Contour plot for velocity distribution

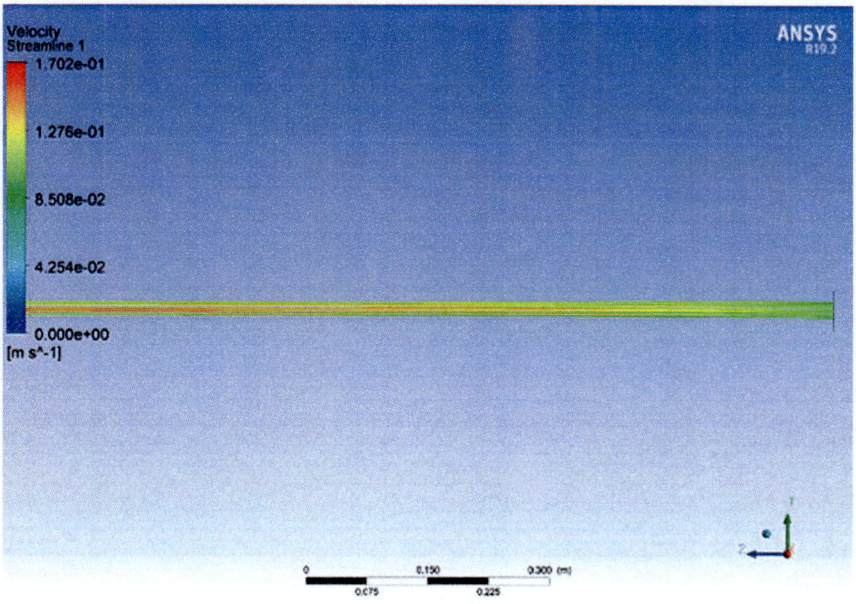

Fig. 10 Contour plot for velocity streamline

properties values had been taken [13] like density, specific heat, and thermal conductivity, etc. Phase Change Material can be used for thermal energy systems which are also known to be thermal energy storage devices. Large amount of thermal energy can be stored by using PCMs which help in fulfillment of energy crisis in ongoing time. Huge applications of phase change materials in buildings, construction, and energy storage devices will drive the overall market and also air pollution can be decreased which protects our environment. More numerical analysis can be performed for some other emerging phase change materials. The authors will also be working for an experimental setup based on energy storage unit.

Acknowledgements Authors deeply acknowledge NIT SIKKIM for utilizing the computational laboratory of Department of Mechanical Engineering for working on license approved ANSYS software.

Nomenclature

k	Material Conductivity ($W.m^{-1}.k^{-1}$)
q_v	Rate at which energy generated per unit volume of the medium ($W.m^{-3}$)
c_p	Specific heat capacity ($J.kg^{-1}.k^{-1}$)
K	Temperature (K)

References

1. Samanta H, Roy PC, Barman N (2015) Modeling of solidification of CCHH (CaCl2, 6H2O) in a Shell-and-Tube PCM based heat storage unit. Procedia Eng, 127, pp 816–823
2. Socaciu LG (2012) Thermal energy storage with phase change material. Leonardo Electron J Pract Technol
3. Kylili A, Theodoridou M, Ioannou I, Fokaides PA (2016) Numerical heat transfer analysis of phase change material (PCM)—enhanced plasters. COMSOL Inc., Munich, Germany, pp 1–7
4. Vinod Kumar Naidu P, Ajay M Nair (2018) Analysis of paraffin wax a phase change material. Int J Curr Eng Technol
5. Gordana Toplicic-Curcic, Aleksandar Kekovic, Dusan Grdic, Snezana Doric- Velkkovic, Zoran Grdic (2018). Phase change material (PCMs)—innovative materials for improvement of energy efficiency of buildings. Contemp Achiev Civ Eng
6. Nitin .D Patil, Karale SR (2012) Design and analysis of phase change material based thermal energy storage for active building cooling: a Review. Int J Eng Sci Technol (IJEST)
7. John A Noel, Samer Kahwaji, Louis Desgrosseilliers, Dominic Groulx, Mary Anne White (2016) Phase Change Materials
8. Vikas, Ankit Yadav, Soni SK (2017) Simulation of melting process of a Phase Change Material (PCM) using ANSYS (Fluent). IRJET
9. Kukreja N, Gupta SK, Rawat M (2020) Performance analysis of phase change material using energy storage device. In: 10th International conference of materials processing and characterization
10. Trp A, Lenic K, Frankovic B (2006) Analysis of the influence of operating conditions and geometric parameters on heat transfer in water-paraffin shell-and-tube latent thermal energy storage unit 16:1830–1839
11. Sharma A, Tyagi VV, Chen CR, Buddhi D (2009) Review on thermal energy storage with phase change materials and applications 13(2):318–345
12. Wei-Wei W, Kun Z, Liang-Bi W, He, Ya-Ling (2013) Numerical study of the heat charging and discharging characteristics of a shell-and-tube phase change heat storage unit. Appl Therm Eng. 58(1–2), pp 542–553.
13. Naresh G, Sagar KV, Reddy LS, Reddy EN, Reddy NS, Vohra M (2021) Battery thermal management using fins in PCM (Phase Change Material), 2021 JETIM, 8:5
14. Ahmet kurklu (1997) Energy storage applications in greenhouses by means of phase change materials (PCMs); a review., Renewable Energy, 1, 1998. pp 89–103
15. Ziye Ling, Jiajie Chen, Xiaoming Fang, Zhengguo Zhang, Tao Xu, Xuenong Gao, Shuangfeng Wang (2014). Experimental and numerical investigation of the application of phase change materials in a simulative power batteries thermal management system. Appl Energy 121, pp 104–113
16. Akhil Johnson, Hussain Dr. J (2016) Numerical analysis of thermal storage systems using phase change material. Int J Mech Ind Technol
17. Li Y, Liu S (2014) Experimental study on thermal performance of a solar chimney combined with PCM. Appl Energy 114:172–178
18. Halawa E, Saman W (2011) Thermal performance analysis of a phase change thermal storage unit for space heating. Renewable Energy 36:259–264

Estimation of the Settling Velocity and Drag Coefficient of Sphere Settling in a Vertical and Inclined Channel at Different Temperatures

Mohammad Hussain and Basudeb Munshi

Abstract The settling of particles in a fluid is one of the most common phenomena in nature and industrial applications like pharmaceutical, chemical, oil and gas production. The particle terminal velocity (V), Coefficient of drag (C_D) and the effect of the wall plays a significant role in the multiphase flows. The terminal velocity (V) and Coefficient of drag (C_D) are measured experimentally in the same Newtonian fluid at two different temperatures for different spheres and the Diameter ratio (0.06–0.32) at different angles (*90°–30°*). As the temperature rises, the viscosity decreases, increasing the terminal velocity (V) of the spherical particle and influencing the drag coefficient (C_D) of the particles. The wall effect on the particle is also studied by considering the channel of different diameters. The empirical relation for the wall factor in terms of diameter ratio is developed. The predicted and the experimental results show good agreement with each other.

Keywords Settling velocity · Wall factor · Drag coefficient · Diameter ratio · Reynolds number

1 Introduction

The concept of particle settling is very interesting in many industrial applications, such as pneumatic conveying, mill ducts in power plants, fluidized beds, food processing, solid particle transportation in the fluid channel, and mineral and chemical processing. The drag force on the particles is created by the relative motion of the settling particle and fluid, and the drag force is used to calculate the drag coefficient by dividing the drag force by the projected area. Both friction and inertia forces are considered to evaluate the drag coefficient in incompressible fluids. For decades, researchers have been fascinated by the motion of a spherical particle in

---

M. Hussain · B. Munshi (✉)
Laboratory of Transport Phenomenon, Department of Chemical Engineering, NIT Rourkela, Rourkela 769008, Odisha, India
e-mail: basudebm@gmail.com

S. Das et al. (eds.), *Proceedings of the 1st International Conference on Fluid, Thermal and Energy Systems*, Lecture Notes in Mechanical Engineering,
https://doi.org/10.1007/978-981-99-5990-7_33

a fluid channel filled with Power-law fluid and Newtonian. Clift et al. [1], Khan and Richardson [2], Uhlherr and Chhabra [3], and Chhabra [4] published qualitative works that discussed the Coefficient of drag experienced by the spherical particle while passing through different fluids. The studies were conducted in order to calculate the drag coefficient (C_D) for settling in an inclined channel filled with Newtonian fluid. The previous literature focused primarily on the spherical particle. There is a wealth of literature on spherical particles flowing in Newtonian and non-Newtonian fluids. Jan and Chen [5], Jalal and Ganji [6], Chhabra and Ferreira [7], Chhabra et al. [8], and Unnikrishnan and Chhabra [9] also developed a set of C_D-*Re* correlations by varying the diameter ratio (*d/D*) and angle of the channel. The change in temperature of the working fluid influences the behavior of particle, and it becomes critical to study the settling behavior of particle at different temperatures. However, no work is communicated to investigate the behavior of spherical particles settling in Newtonian fluid at various temperatures. This research aims to calculate the Coefficient of drag of spherical particles moving in Newtonian fluid in the inclined and vertical channels at different temperatures. The effect of the wall on the spheres at two different temperatures is also discussed in this work. The diameter ratio is described as the ratio of the particle's diameter to the flow channel's inner diameter. Some of the goals of this work are to fill gaps in the literature.

- To calculate the Drag Coefficient (C_D) of spherical particles at two different temperatures.
- To determine how the channel angle and diameter ratio (*d/D*) affect the measured particle settling terminal velocity (V).
- To calculate the wall factor on the spheres related to the flow channel.

2 Physics Around Sphere Particle

A diagrammatic representation of the particle is shown in Fig. 1. The net bodyweight of the particle (F_B sinθ), solid–solid resistance (F_R), and drag force (F_D) are the various forces acting on a spherical particle as it settles in an inclined channel filled with test fluid; at equilibrium, the force balance equation is

$$F_D + F_R = F_B \sin\theta \tag{1}$$

The drag force expressed as [10]

$$F = \frac{1}{2} C_D \rho_F V^2 A \tag{2}$$

The net drag force acting on the sphere is represented by the non-dimensional term drag coefficient. In an inclined channel, the drag coefficient (C_D) is expressed as

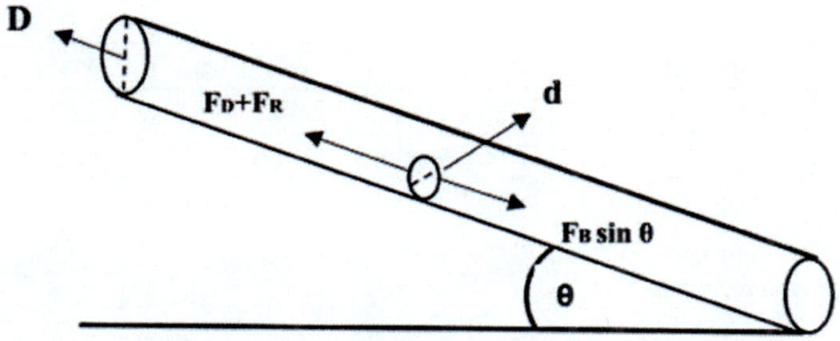

Fig. 1 The experimental setup is represented in diagram form

$$C_D = \frac{4gd}{3V^2}\left(\frac{\rho_S - \rho_F}{\rho_F}\right)\sin\theta \tag{3}$$

Equation (3) is simplified for the vertical channel because the angle is 90° (sin *90* = *1*).

$$C_D = \frac{4gd}{3V^2}\left(\frac{\rho_S - \rho_F}{\rho_F}\right) \tag{4}$$

The blockage ratios (*d/D*) and channel angle are significant constraints to consider when studying the drag coefficient (C_D) in the inclined channel; thus, the Coefficient of drag (C_D) can be expressed as.

$$C_D = f(Re, d/D, \theta) \tag{5}$$

3 Experimental

3.1 Materials

Steel spheres ranging in diameter from 3 to 8 mm and densities varying from 7675.2 to 8056.3 kg/m^3 are used. A 1500 mm long perplex tube with 24.8, 50, and 68.3 mm diameters is used as the flow region for all spheres. At two different temperatures, pure glycerin solutions were used as working fluids. A Rheometer (Rheolab QC, Anton Paar, Germany) and a pycnometer are used to measure fluid viscosities and densities, respectively. Table 1 lists all of the required physical properties of working fluids. The experimental setup with the testing region is shown in Fig. 2.

Table 1 Properties of working fluids

S. No.	Temperature (°C)	Fluids	Density (kg/m^3)	Viscosity (Pa s)
1	23	100% glycerin	1258.9	1.0782
2	30	100% glycerin	1254.6	0.59792

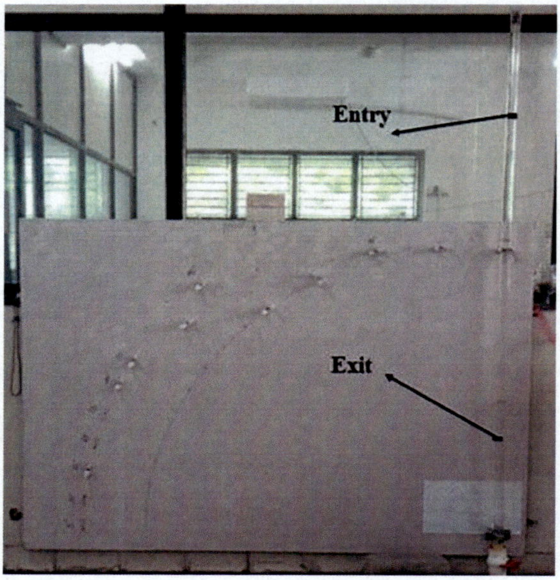

Fig. 2 The experimental setup with the testing region

3.2 Experimental Procedure and Setup

The experimental setup consists of transparent tubes with diameters of 24.8, 50.0, and 68.3 mm, and a length of 1500 mm. The tube is supported by a wooden frame, with one end clamped to the frame and the other end set free to be adjusted to various angles. The test fluid is loaded into the channel 24 h before starting the experiment to ensure thermal stability and bubble-free [5, 7]. A forceps is used to hold the particle, and a stopwatch measures the particle's settling time. The testing region in the tube is chosen so that the particle's velocity changes at a constant rate. Each settling time measurement process is repeated four times, and the average value is used to calculate the settling velocity (V) of the particle. Finally, the Reynolds number and Coefficient of drag (C_D) are calculated using the settling velocity.

4 Results and Discussions

4.1 Impact of Diameter Ratio (d/D) on Terminal Velocity (V)

The terminal velocity (V) is an important parameter to calculate the Reynolds number and Coefficient of drag, but the terminal velocity depends on the particle shape, size, and diameter ratio [9, 11]. Figure 3 shows the terminal velocity (V) variation of particles with diameter ratio at two different temperatures. The terminal velocity is observed to increase as the particle diameter increases because the particle's weight increases. Figure 3 depicts the variation of particle terminal velocity with temperature. The rise in temperature of the working fluid decreases the fluid's viscosity. Since the viscosity is decreasing, the particle's viscous forces are also decreasing, which causes the particle's terminal velocity to increase.

The impact of the channel angles (90°–30°) on the settling velocity (V) is shown in Fig. 4. Figure 4a, b depicts the increasing trend of the settling velocity (V) with a d/D ratio for all channel angles by keeping the diameter of the flow channel at 68.3 mm. It is discovered that the settling terminal velocity (V) of the particles is maximum when the flow channel is vertical and decreases as the angle of the incline decreases. It occurs due to reducing body force's effect on the settling velocity (V) by varying the flow channel angles, and variation of particle terminal velocity (V) with temperature is also studied. The increase in terminal velocity is observed by increasing the temperature for all channel angles. The same trends are observed by varying the diameter of the flow channels in Fig. 4c–f.

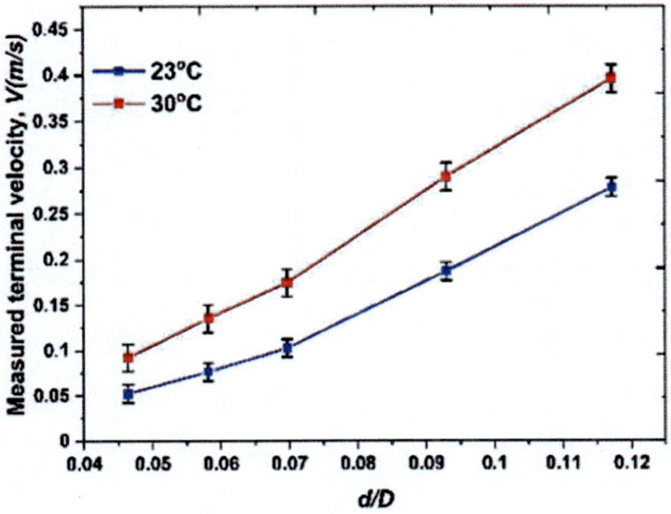

Fig. 3 Terminal velocity (V) variations with diameter ratio (d/D) in 100% glycerin and 90° channel angle

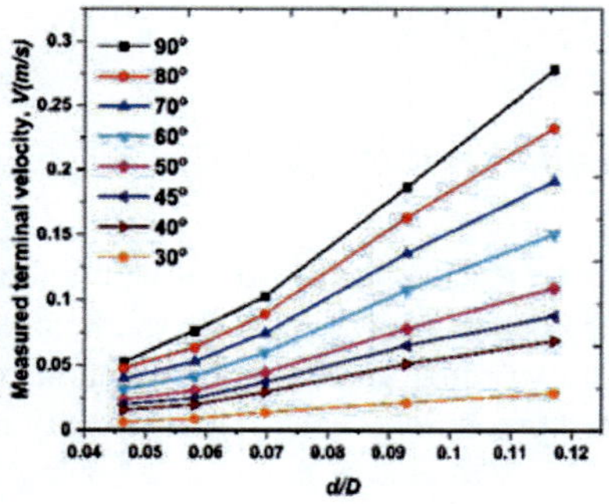

(a) D=68.3mm at 23°C

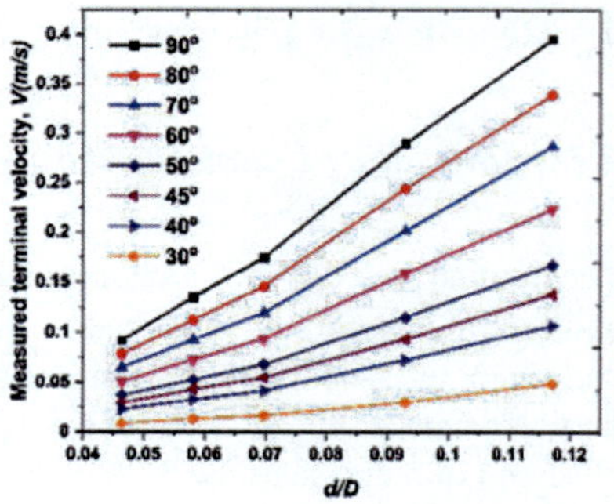

(b) D=68.3mm at 30°C

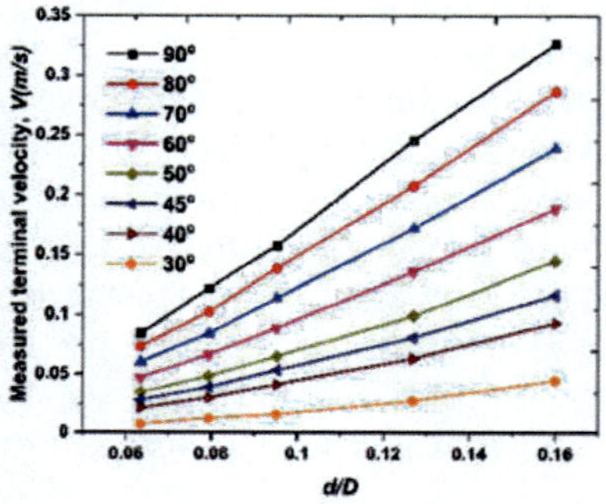

(c) D=50.0mm at 23°C

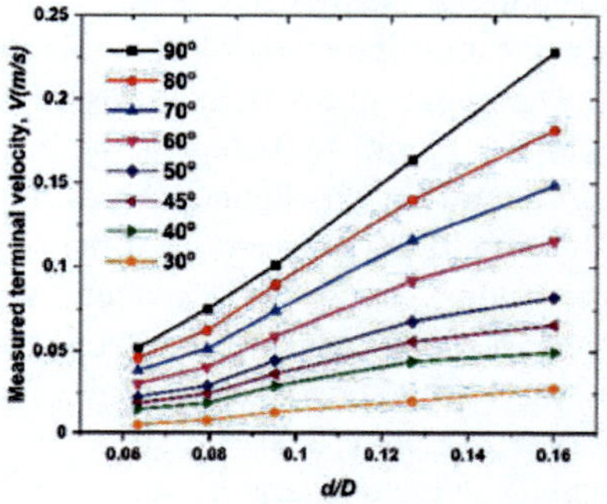

(d) D=50.0mm at 30°C

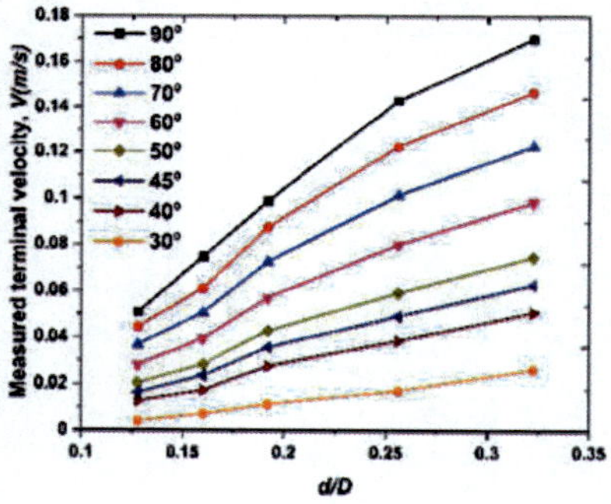

(e) D=24.8mm at 23°C

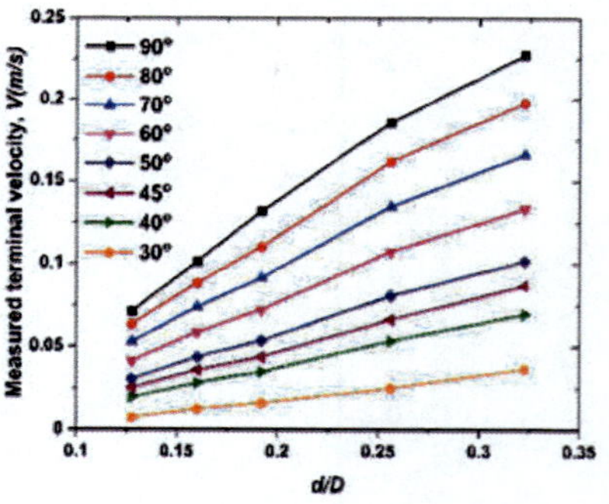

(f) D=24.8mm at 30°C

Fig. 4 Terminal velocity (*V*) variations of the particle with diameter ratio (*d/D*) at different temperatures

4.2 Wall Effect

Figure 5 depicts the effect of changing the flow channel diameter on the terminal velocity of each particle. The increase in the terminal velocity is observed by increasing the diameter of the flow channel. The extrapolation method is used to calculate the terminal velocity in an unbounded region by considering (*d/D* = zero). When *d/D* = 0, the gap between the wall of the flow channel and particle is very large; therefore, the wall does not impact the particle's terminal velocity, and it is considered the terminal velocity of a particle in the unbounded region. Figure 5 also shows the variation of terminal velocity (*V*) with the different temperatures of the working fluid. The calculated terminal velocity in the unbounded region is used to calculate the wall factor, *f*, the terminal velocity (V) in the confined flow region is divided with the velocity in unbounded flow region to calculate the wall factor.

The influence of the diameter ratio (*d/D*) on the wall factor (*f*) is depicted in Fig. 6. The wall coefficient decreases as the diameter ratio increases because of the space between the particle and the wall boundary narrows. The wall factor is also affected by the temperature of the working flow. The wall factor increases with an increase in the temperature of the working fluid because the temperature affects the fluid's viscosity.

The experimental results are compared with the Haberman and Sayre [12] data. It shows that the present spherical particle experiences less wall effect than literature data because of the properties and temperature of the working fluids. The increasing trend of wall factor with decreasing diameter ratio is the same as the literature data. When the diameter ratio increases, the gap between the wall of the flow channel and particle decreases, so the particle experience additional resistance from the wall. The equations are developed to predict the experimental wall factor. Equations (6) and (7) predict the experimental data with an average error of 4.25%, a maximum error of 8.58% for working fluid at 23 °C temperature, an average error of 2.59%, and a maximum error of 8.15% for working fluid at 30 °C temperature.

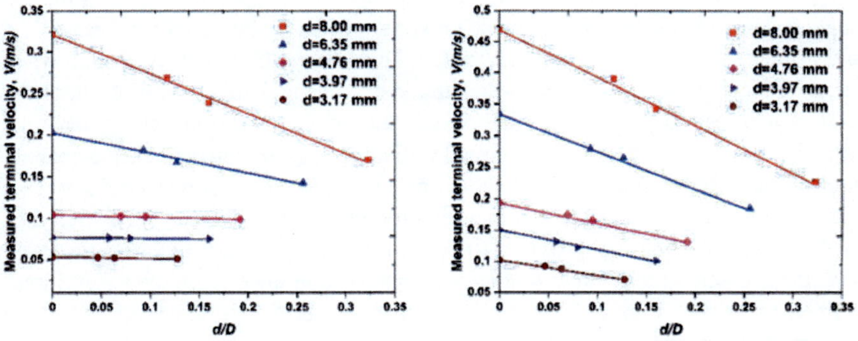

(a) Temperature=23°C (b) Temperature=30°C

Fig. 5 Impact of diameter ratio (*d/D*) on the terminal velocity (*V*)

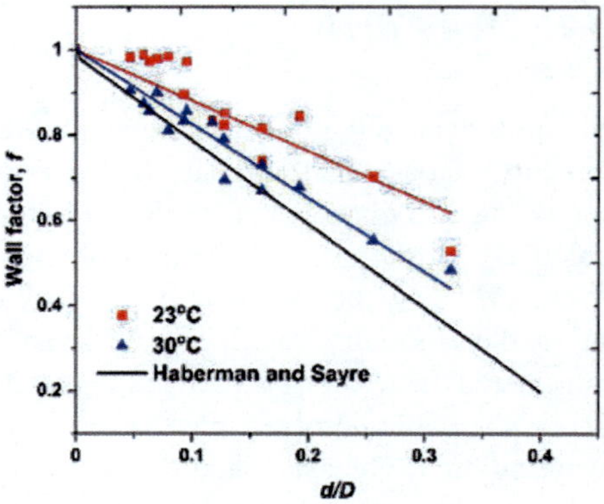

Fig. 6 Wall factor versus diameter ratio

$$f = 1 - 1.1832\left(\frac{d}{D}\right) \quad (6)$$

$$f = 1 - 1.7351\left(\frac{d}{D}\right) \quad (7)$$

4.3 *Relation Between Reynolds Number (Re) and Drag Coefficient (C_D)*

For different channel angles and diameter ratios (*d/D*), the Coefficient of drag (C_D) and Reynolds number (*Re*) is calculated by means of the properties of the spheres, fluid, and the settling velocity (*V*) of the spherical particles.

Figure 7 depicts the relationship between the Reynolds number (*Re*) and the drag coefficient (C_D) As previously observed by researchers, the expected decreasing trend is observed. The Reynolds number and drag coefficient are inversely propositional, so the Reynolds number is found maximum when the flow channels are vertical and decreases with a decrease in the angles. The drag coefficient is maximum when the channel is inclined and decreases with a rising channel angle. As a result, as the angle decreases, the drag coefficient increases. At 90°, the maximum value of terminal velocity is obtained. The settling terminal velocity decreases with decreasing channel angle for 80°–30° inclinations [10]. From Eqs. (3) and (4), it can be observed that (C_D) is inversely proportional to the terminal velocity square (V^2). The drag force increases by decreasing the angle of channel (θ) because the drag coefficient varies inversely with the projected area. Small-diameter particles experience more drag than large-diameter particles. Figure 7a, b explain how the experimental results of (C_D) and *Re* vary with temperature. At 30 °C, the fluid's viscosity decreases, causing particle velocity and the Reynolds number to increase. The Reynolds number is inversely related to the Coefficient of drag (C_D), and as the Reynolds number (*Re*) increases,

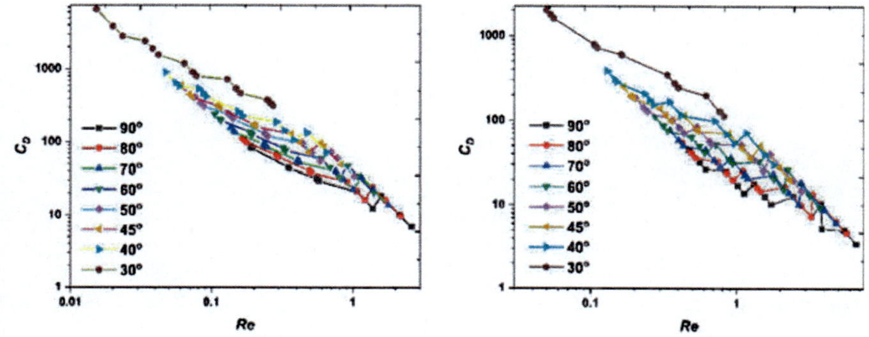

(a) Temperature=23°C

(b) Temperature=30°C

Fig. 7 Relation between Reynolds number (*Re*) and Drag coefficient (C_D)

the particle's Coefficient of drag decreases. When the two results are compared, the trend is the same at all temperatures.

5 Conclusions

The Coefficient of drag of spherical particles settling in both the inclined and vertical planes is obtained experimentally in this work. When the temperature of the working fluids changes, the effect of channel, diameter ratio, wall effect, angle of inclination, and Reynolds number (*Re*) on the coefficient of drag (C_D) of the particle in Newtonian fluids are discussed. The decrease in temperature of the working fluid increases the viscosity, so the particle's Coefficient of drag increases, and the Reynolds number decreases. The angle of inclination and *d/D* ratios also affect the Coefficient of drag because the terminal velocity decreases as the angles change from 90° to 30°. The different diameter channels were used to demonstrate the impact of the wall on particle terminal velocity (*V*). The effect of wall boundaries on the spherical particle velocities is represented in terms of the wall factor (f). The wall factor increases when the gap between the particle and the wall increases, and the particle experiences a high wall effect when the gap is reduced. The empirical relation is developed to calculate the wall factor at different temperatures. The predicted and the experimental results show good agreement with each other.

Nomenclature

d Diameter of particle (mm)
ρ_S Density of solid (kg/m^3)

ρ_F	Density of fluid (kg/m^3)
V	Settling velocity (m/s)
D	Flow channel diameter (mm)
Re	Reynolds number (–)
C_D	Drag coefficient (–)
f	Wall factor (–)

References

1. Clift RR, Grace JR, Weber ME (1978) Bubbles drops, and particles
2. Khan AR, Richardson JF (1987) The resistance to motion of a solid sphere in a fluid. Chem Eng Commun 62(1–6):135–150
3. Uhlherr PHT, Chhabra RP (1995) Wall effect for the fall of spheres in cylindrical tubes at high Reynolds number. Can J Chem Eng 73(6):918–923
4. Chhabra RP (2006) Bubbles, drops, and particles in non-Newtonian fluids. CRC Press
5. Jan C-D, Chen J-C (1997) Movements of a sphere rolling down an inclined plane. J Hydraul Res 35(5):689–706
6. Jalaal M, Ganji DD (2010) An analytical study on motion of a sphere rolling down an inclined plane submerged in a Newtonian fluid. Powder Technol 198(1):82–92
7. Chhabra RP, Ferreira JM (1999) An analytical study of the motion of a sphere rolling down a smooth inclined plane in an incompressible Newtonian fluid. Powder Technol 104(2):130–138
8. Chhabra RP, Kumar M, Prasad R (2000) Drag on spheres in rolling motion in inclined smooth tubes filled with incompressible liquids. Powder Technol 113(1–2):114–118
9. Unnikrishnan A, Chhabra RP (1991) An experimental study of the motion of cylinders in Newtonian fluids: wall effects and drag coefficient. Can J Chem Eng 69(3):729–735
10. Samantaray SK et al (2018) Newtonian flow past a hollow frustum in the vertical and inclined plane: an experimental observation for terminal velocity and drag coefficient. Powder Technol 326:114–122
11. Sharma MK, Chhabra RP (1991) An experimental study of free fall of cones in Newtonian and non-Newtonian media: drag coefficient and wall effects. Chem Eng Process 30(2):61–67
12. Haberman WL, Sayre RM (1958) Motion of rigid and fluid spheres in stationary and moving liquids inside cylindrical tubes. David Taylor Model Basin Washington DC

Experimental and Numerical Investigation on Hydrodynamics of Swirling Fluidised Bed Reactor

Dekketi G. C. Vikram Reddy, Kiran Malieakkal Eyyanan, V. Sri Pratibha Reddy, and Panneerselvam Ranganathan

Abstract The performance of chemical reactors depends on the transport behaviours of the bed, i.e. heat and mass transfer. As reported in the literature fluidised bed reactors have very efficient transport properties, but problems like channelling and entrainment can be seen due to the axial flow of air. To further improve the reactor performance, process intensification strategy can be used. One of intensified approaches is the swirling motion in the fluidised bed. In this reactor, due to its non-axial movement of air, it avoids the problems which usually occur in the conventional fluidised bed. In this work, we aim to study the hydrodynamics of swirling fluidised bed reactor experimentally and numerically. First, the experimental work related to the swirling fluidised bed is conducted and the results are used to validate the present numerical model. Then, the numerical modelling of single and multiphase phase flows in a swirling fluidised bed is studied. The developed model is validated with the literature experimental data. Finally, the effect of model parameters and operating conditions on the hydrodynamics of the swirling fluidised bed reactor is investigated.

Keywords Fluidised bed reactor · Swirling fluidised bed reactor · Entrainment · Validation · Experimental setup

1 Introduction

Fluidisation is a method to cause the solid granular particles to behave like a fluid to maximize the interaction between the particles. In a conventional fluidisation, when velocity of the gas is greater than the minimum fluidisation velocity, slugging occurs and when it is greater than the terminal settling velocity (the velocity enough to carry away the solid particles out of the reactor), the problems like elutriation occurs [1].

D. G. C. V. Reddy · K. M. Eyyanan · V. Sri Pratibha Reddy · P. Ranganathan (✉)
Department of Chemical Engineering, National Institute of Technology Calicut, Kozhikode 673601, Kerala, India
e-mail: pranganathan@nitc.ac.in

S. Das et al. (eds.), *Proceedings of the 1st International Conference on Fluid, Thermal and Energy Systems*, Lecture Notes in Mechanical Engineering,
https://doi.org/10.1007/978-981-99-5990-7_34

These limitations of the conventional fluidised bed reactor have captivated the necessity for an intensified reactor. One of the intensified reactor designs is a swirling fluidised bed (SFB), which incorporates a tangential inlet entry and a spiral distributor plate. In SFB, the fluidising gas enters the reactor through the tangential inlet, passes through the blades of the distributor plate which are fixed at a desired angle These form a trapezoidal gap that leads to an annular region where three velocity components are generated—axial, radial and tangential, the latter component which is due to the tangential entry of the gas.

In literature, few studies are reported on the swirling fluidised bed, focusing on its hydrodynamics using both experiments and computational fluid dynamics (CFD) simulation. Miin et al. [2] determined the effects on the velocity profile with varying blade angle, particle shape and size, and bed weight using PIV technique in a swirling fluidised bed reactor. Sirisomboon and Arromdee [3] conducted a computational fluid dynamics study of the gas–solid distribution of geldart B particles in SFB and found that the RNG k-ε model was the best fit among standard k-ε, RNG k-ε and RSM k-ε models. Othman et al. [4] have compared different turbulence models for air velocity in the plenum chamber using CFD techniques and validated the results with PIV data to find the best turbulence model. Venkiteswaran [5] performed fundamental (cold bed model) study to get to know the physics behind the phenomena in the system and the author mentioned that blade inclination in the distributor plate has a major influence on the bed hydrodynamics as it affects the bed pressure drop and distributor pressure drop as well. Mohideen et al. [6] proposed the swirling fluidisation technique to dry the oil palm frond (OPF) and concluded that drying of OPF using swirling fluidised bed is effective for both continuous and batch drying and the ability to fluidise the highly irregular shapes of OPF at relatively low energy consumption using the swirling fluidisation technique unlike other methods of fluidised bed drying.

Though few modelling studies are reported in the literature, a detailed multiphase flow modelling of swirling fluidised bed has not reported. Thus, the present study focuses on the hydrodynamics of SFB using experimentally and numerically.

2 Methods

2.1 Experimental

The setup of the SFB reactor consisted of a distributor, a cone, a central hub and a plenum chamber, all of which was enclosed in an acrylic column which is shown in Fig. 1. The distributor consisted of two pairs of inner and outer rings to which the blades were attached. The cone, hub and rings were fabricated using Delta WASP 4070 industrial 3D printer available at TBI NITC and Raise 3D printer at DIC NITC. The material used for 3D printing is polylactic acid (PLA). 50 blades were fabricated from acrylic sheets using a laser cutting machine, which was then mounted onto the

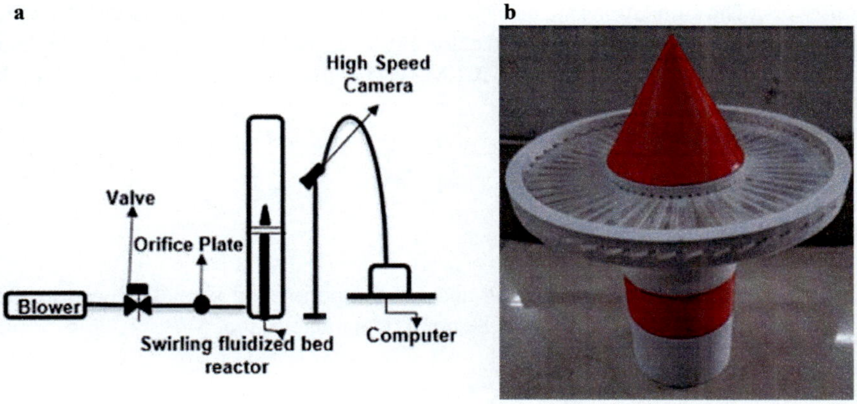

Fig. 1 **a** Schematic diagram of experimental setup. **b** Fabricated distributor plate

pair of rings. The blades and rings were designed to have a 10° declination with a blade overlap angle of 9°.

Plenum height is considered to be 50 mm and an extension of 50 mm was kept to avoid excessive backflow as well as to resemble an actual system of reactor on top of the plenum [4]. The total domain height is now at 100 mm as shown in Fig. 1. Air was chosen as the fluid single phase medium. The experiments are performed to measure bed expansion height for different operating conditions including particle size and density, air velocity and blade angle and inclination.

2.2 *Modelling*

CFD simulations are performed to predict hydrodynamics of a swirling fluidised bed. Ansys Single-phase flow is simulated inside the plenum chamber to observe the velocity profiles and contours to visually verify the swirling flow. RNG k-ε with swirl dominated flow which belongs to the RANS (Reynolds Averaged Navier Stokes) turbulence models is considered for this study. Air inlet velocity is taken as the boundary conditions at the inlet, 12.3 m/s for plenum without hub and 12.8 m/s for plenum with hub [4]. A zero Pa is used as the pressure outlet boundary condition, and with the no-slip condition at the walls. The $Y+$ value used at the boundary walls is about 42.27.

For multiphase flow modelling, Eulerian-Eulerian method is used to model gas–solid flows in SFB. We assume that the bed particles follow the kinetic theory of granular flow in which particles resemble to be as gas. The gas phase is modelled with the RNG k-ε turbulence model assuming no-slip condition at the walls. Inlet BCs are considered to be the inlet velocity condition and the outlet BC with zero Pa pressure outlet. Wen Yu's model is used to account for interphase drag force in this work. The simulations are carried out for different bed densities to visualize the flow and bed behaviour inside the reactor. Fluent is used to analyse the flow pattern in SFB. The meshing is done using ICEM-CFD, with unstructured tetrahedral mesh with over 100,000 nodes.

3 Results and Discussion

In this work, we will be discussing about the experimental studies as well as the validation and parametric study of the numerical modelling of SFB.

3.1 Experimental Results

One of the aspects that is investigated in this work is the effect of operation conditions such as the particle size, shape and density on the bed expansion height and these results are shown in Table 1. Initially, 200 g of each particle was added into the reactor and a flow rate of 200 m^3/h was provided. It was noted that the bed expansion was larger for smaller and less denser particles as expected. The larger denser particles had a more sluggish movement when compared to the less dense ones making it more difficult to fluidise and mix.

Table 2 shows the result of the effect of cone size on fluidisation. Two different sizes of cone with diameters of 100 and 140 mm were used for this purpose. Due to the trapezoidal shape of the blades, the area for air opening in the region surrounding the cone was lesser when compared to the other parts of the distributor in the case smaller size of cone. This provided a huge resistance to the air flow with particle loading resulting in a dead zone in the region surrounding the cone for low flow rates of air. It is noted that the air resistance due to small opening can be reduced by using a larger cone size above the distributor plate which compromises the area of the distributor but increases the uniformity in fluidisation and mixing.

Table 1 Influence of particle size and density on bed height

Type	Size	Density	H_{initial}	H_{final}
Beads (sphere)	9.6	610	12	17
Beads (sphere)	3.7	200	16.5	20.51
Beads (elliptical)	11/9	370	13	20.5
Brown chick pea	8	1350	11	15
Soy seeds	5	715	8.7	12
Soy bean	12	650	1	3

Table 2 Effect of cone size on fluidisation

Size of cone	H_{initial} (mm)	$H_{\text{final(min)}}$ (mm)	$H_{\text{final(max)}}$ (mm)
Big cone (140 mm)	30	38	47.9
Small cone (100 mm)	19	26	37

3.2 Single-Phase Flow Model Validation

A single-phase flow of air in SFB has been simulated with commercially available computational fluid dynamics software (Ansys 19.2). The flow behaviour inside the plenum chamber of a swirling fluidized bed reactor is studied and validated with the literature experimental data of Othman et al. [4]. The authors measured air velocity in a swirling fluidised bed using PIV. A single flow inside the plenum chamber of a swirling fluidised bed with and without the cylindrical hub (central body) is simulated. Figure 2 shows three planes as considered in the plenum chamber. These planes are considered at heights of 66 mm, 250 mm and 486 mm respectively from the bottom of the reactor.

Figure 3 depicts the comparison of air velocity between the CFD model and literature data at height of 250 mm. It is shown in figure that CFD model prediction matches with the literature experimental data. Also it is shown that the rise in velocities near the walls is due to the action of centrifugal forces as the flow is dominated by swirling action. RNG k-ε with swirl dominated flow was used as the turbulence model in this work. Though the Reynolds stress model (RSM) and RNG k-ε models

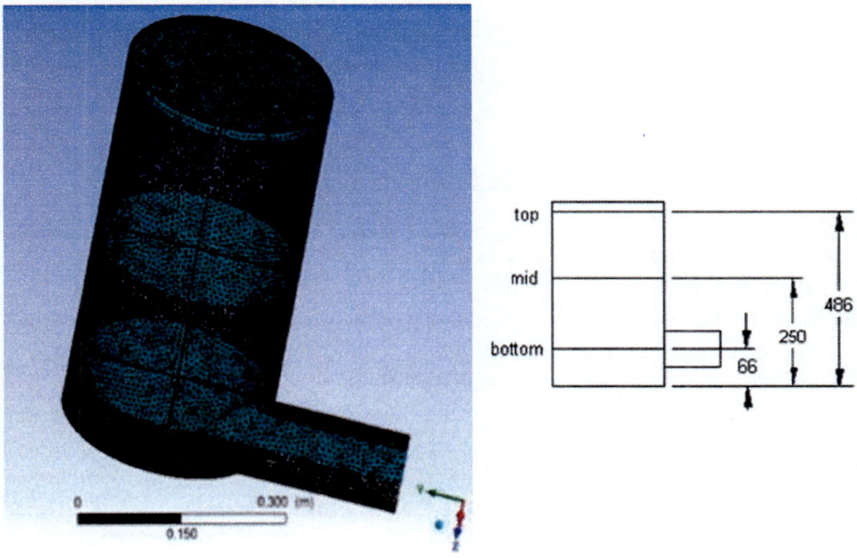

Fig. 2 **a** Mesh of plenum chamber **b** various planes considered in the plenum

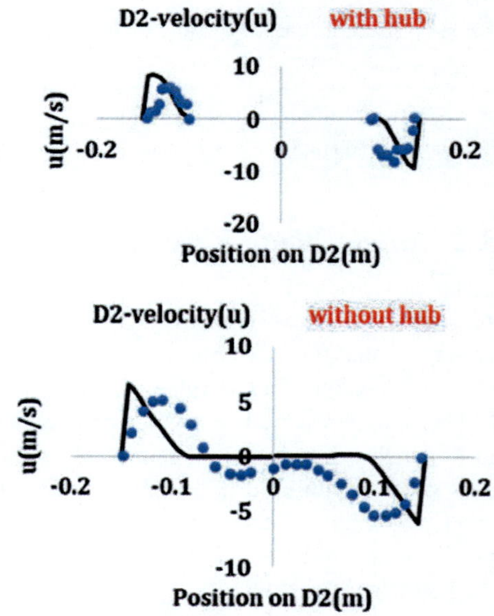

Fig. 3 Comparison of air velocity of the plenum between experiments and CFD simulation at the mid plane of D2

are mostly used to predict the flow, RSM was ignored in this study due to its high computational power.

Figure 4 shows the velocity vector profiles in plenum of swirling fluidised bed at the bottom plane.

It is found that the axial motion is less dominant when the hub is incorporated into plenum chamber as the hub maintains the swirling flow, whereas the domination of axial flow near to the centre of the column is found in case of without hub. Also, the high velocity at the bottom plane of plenum is observed as expected.

3.3 *Multiphase Flow Model Validation*

The result of multiphase modelling is discussed in this section. For multiphase flow modelling, the whole swirling fluidised bed reactor including plenum, distributor and bed is considered. The conditions used in the model are based on the literature experimental work of [5]. The validation of the present model is performed by comparing the bed pressure drop with experimental data. Figure 5 shows the comparison plot of pressure drop between the experimental and model prediction. It is found that the model prediction matches experimentally derived values with a slight deviation. These deviations can be accountable to the capabilities of present model including the interphase drag force and turbulence model. This needs to be improved which can be considered for future work.

Figure 6 shows the bed expansion of 4 mm particles with 857 kg/m^3 plastic beads with bed voidage of 0.44 at the gas velocity of 12 m/s and the initial bed height of

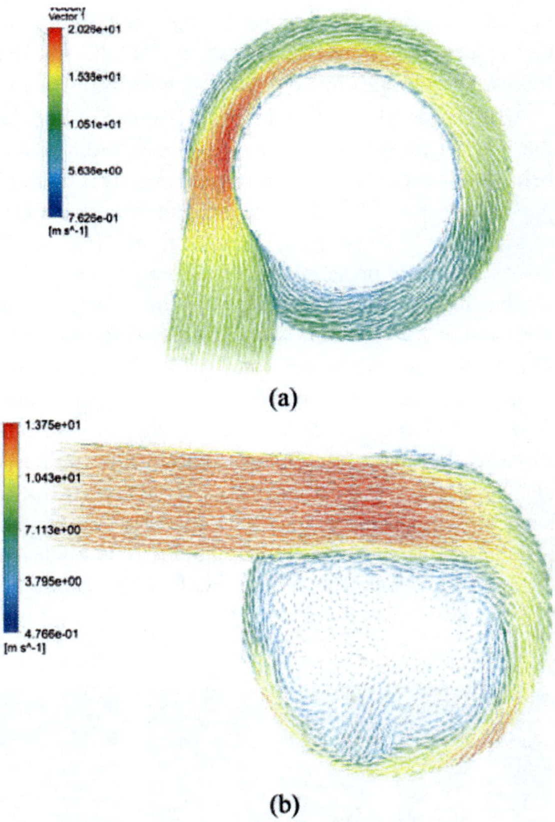

Fig. 4 The profiles of velocity vectors at the bottom plane of the plenum chamber **a** with hub **b** without hub

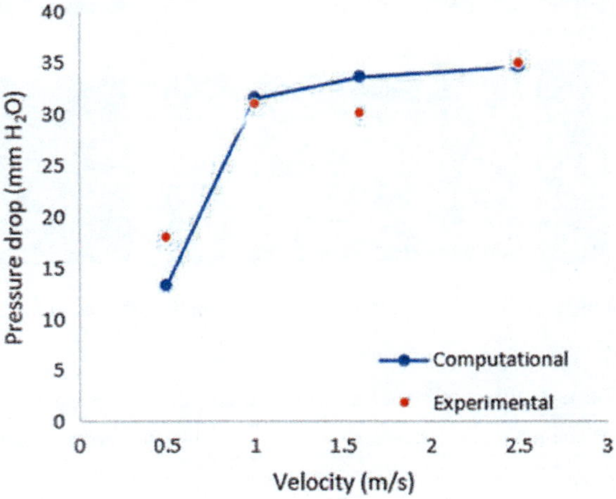

Fig. 5 Comparison plot of bed pressure drop at different superficial velocities between CFD and experiment

80 mm. It is shown that as time progresses, the mixing of solid particles in the bed takes place though the bed expansion is small. Also, it is found that due to centrifugal forces, the particles move towards the walls of the reactor.

Distributor plate pressure drop is one of the important parameters to study the hydrodynamics of the swirling distributor plate, which influences fluidisation behaviour and energy consumption. Agarwal et al. [7] mentioned that a minimum pressure drop of 350 mm water of distributor plate is required for proper fluidisation in a shallow bed system of a conventional fluidised bed. Also, the lower minimum distributor plate pressure drop may lead to improper distribution. Sreenivasan and Raghavan [8] mentioned in the case of a swirling fluidised bed fluidisation can be achieved at low pressure drop compared to that in the conventional fluidised bed. This is the main advantage for swirling fluidised bed.

Influence of various parameters on the pressure drop across the distributor plate is studied. Table 3 indicates the change in distributor plate pressure drop at the superficial velocity of 12.5 m/s with different blade inclination angles of 10°, 15° and 20°. From this table, we can observe that 20° inclination blade distributor blades have the least pressure drop. This can be explained due to the fact that the flow resistances offered by the blade angles. The blades with 20° inclination offer least resistance as compared to other blade inclinations.

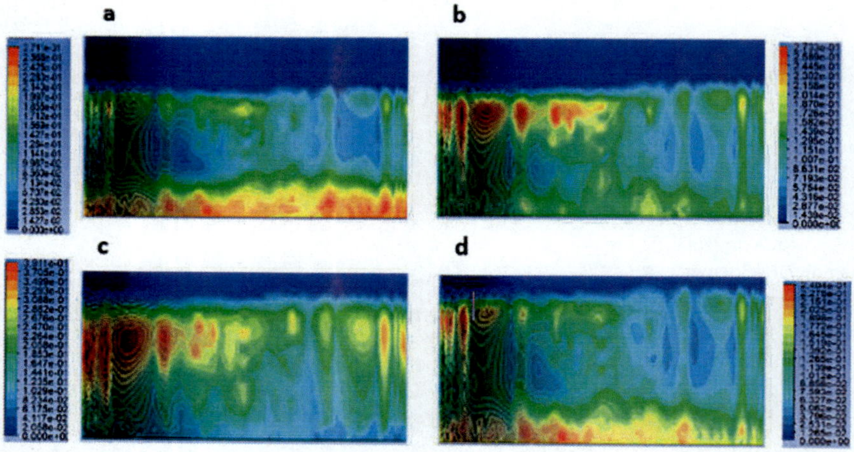

Fig. 6 Solid volume fraction contours of bed at time **a** 1 s **b** 1.5 s **c** 2 s **d** 2.3 s

Table 3 Effect of blade inclinations on the distributor plate pressure drop

Blade inclination (degrees)	Distributor pressure drop by CFD (Pa)
10	850
15	310
20	135

Table 4 indicates the change in distributor plate pressure drop at the superficial velocity of 12.5 m/s with different blade overlap angles of 9°, 12° and 15°. A large overlap angle defines larger overlap width and creates a longer path of flow in between the blades. The overlap angle is chosen instead of the overlap length as the length varies with the distance from the inner to outer, whereas the overlap angle remains constant.

Table 5 shows that variation of bed pressure drops for the variation in blade inclination angle at the superficial velocity of 12.5 m/s. It is observed that bed pressure drop increases with the decrease in the blade inclination angle. The reason may be due to the fact that the increase in the blade inclination angle increases the percentage of opening in the distributor. As the percentage of area opening increases, it leads to the decrease in the resistances at the distributor and thus, decreases distributor pressure drop.

Table 6 shows the change in bed pressure drop for change in blade overlap angle at the superficial velocity of 12.5 m/s. It is observed that blade overlap angle has a significant influence on the bed pressure drop. The overlap angle governs the velocity pattern at the distributor outlet. A small overlap angle means a smaller width or a smaller flow development length which leads to underdeveloped flow with low velocity at the distributor outlet. The gas entry angle into the bed is lower as we increase the blade overlap angle, with higher horizontal component of the gas velocity, due to which horizontal momentum increases and consequently, vigorous swirl has been observed in the bed.

Table 4 Effect of blade overlaps on the distributor plate pressure drop

Blade overlap (degrees)	Distributor pressure drop by CFD (Pa)
9	850
12	853
15	855

Table 5 Influence of blade inclinations on bed pressure drop

Blade inclination (degree)	Bed pressure drop (Pa)
10	340
15	260
20	240

Table 6 Effect of blade overlap angle on the bed pressure drop

Blade overlap (degrees)	Bed pressure drop (Pa)
9	320
12	410
15	430

Figures 7 and 8 show bed expansion behaviours of particle in SFB at different times. It is found that more mixing takes place outside of the cone. Also, it is found that the particle concertation is more close to the cone region due to improper mixing.

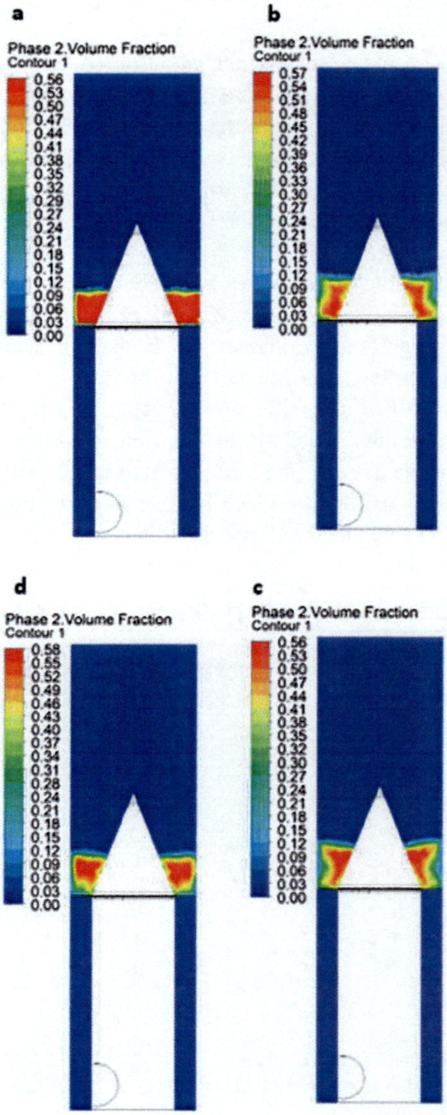

Fig. 7 Bed expansion behaviour in SFB (density: 971 kg/m^3 and 6 mm particles) at times **a** $t = 0$ s **b** $t = 0.5$ s **c** $t = 1$ s **d** $t = 2$ s

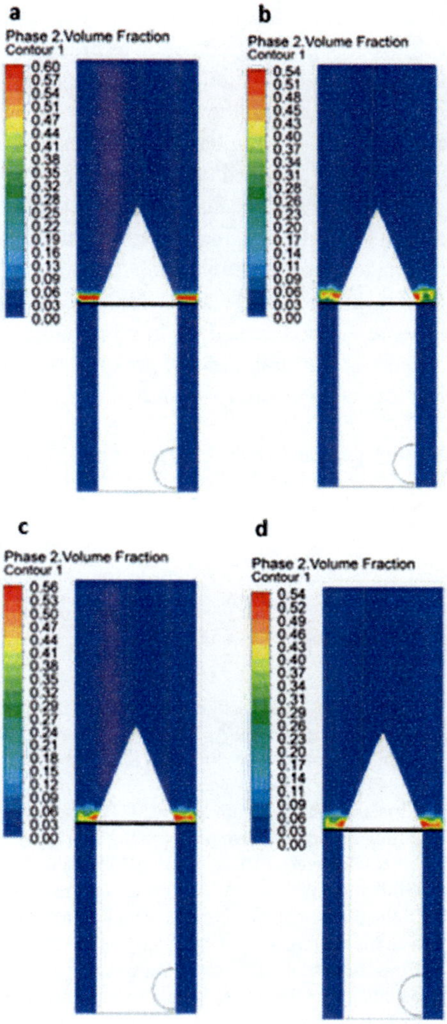

Fig. 8 Bed expansion behaviour in SFB (density: 2306 kg/m^3 size: 2 mm particles) at time **a** $t = 0$ s **b** $t = 0.5$ s **c** $t = 1$ s **d** $t = 2$ s

4 Conclusions

In this work, a lab-scale swirling fluidised reactor was developed to study the bed expansion behaviour through both experimental and numerical modelling. Furthermore, a study was conducted on the effect of different parameters of distributor plate such as blade overlap angle, blade inclination angle and some other related parameters studied on the hydrodynamic behaviour of the swirling fluidized bed reactor. It was observed through modelling that blade inclination plays a crucial role in the bed hydrodynamics as it affects the distributor as well as the bed pressure drop.

Regarding CFD modelling, the model validation of the single-phase flow has been achieved. A good agreement was predicted between the numerical and experimental flow patterns and velocities. The slight deviation observed in the flow and velocity is acceptable due to the high complexity of the flow in the plenum chamber. Further, the validated single-phase model was extended to study multiphase flow modelling in a swirling fluidised bed to obtain hydrodynamic parameters.

In this work, the fluidisation of Geldart D particles was performed since it is difficult to fluidise using conventional fluidised bed reactors. As compared to various parameters, the velocity of the fluidising medium is the most dominant factor which is evident from the results. No significant effect of various parameters has been observed in bed expansion in swirling fluidization regime.

This is the first step towards the design and scale-up of swirling fluidised bed reactors and can apply to high temperature and high pressure works.

Acknowledgements The authors would like to thank Design Innovation Centre (DIC), Technology Business Incubator (TBI) and Department of Mechanical Engineering at NIT Calicut for their support and guidance throughout the work.

References

1. Bamido A (2018) Design of a fluidized bed reactor for biomass pyrolysis, p 98
2. Miin CS, Sulaiman SA, Raghavan VR et al (2015) Hydrodynamics of multi-sized particles in stable regime of a swirling bed. Korean J Chem Eng 32:2361–2367. https://doi.org/10.1007/s11814-015-0151-6
3. Sirisomboon K, Arromdee P (2021) A computational fluid dynamics study of gas–solid distribution of Geldart Group B particles in a swirling fluidized bed. Powder Technol 393:734–750. https://doi.org/10.1016/j.powtec.2021.08.020
4. Othman S, Wahab AA, Raghavan VR (2010) Validation by PIV of the numerical study of flow in the plenum chamber of a swirling fluidized bed. CFD Lett 2:85–96
5. Venkiteswaran VK (2014) Study of hydrdynamic properties of swirling fludiised bed. PhD. thesis, Universiti Teknologi Petronas
6. Mohideen MF, Faiz M, Salleh H et al (2011) Drying of oil palm frond via swirling fluidization technique. In: Proceedings of world congress on engineering, WCE 2011, vol 3, pp 2375–2380
7. Agarwal JC, Davis WL, King DT (1962) Fluidized bed coal dryer. Chem Eng Prog 58(11):85
8. Sreenivasan B, Raghavan VR (2002) Hydrodynamics of a swirling fluidised bed. Chem Eng Process 41:99–106. https://doi.org/10.1016/S0255-2701(00)00155-0

Asymptotic Analysis to Viscoelastic Fluid Flow Over a Stretching Sheet

Amit Kumar Pandey and Abhijit Das

Abstract In this paper, a global asymptotic solution of the second-order, incompressible fluid flow due to stretching of a plane elastic surface is obtained using a simple direct method called Homotopy Renormalization Method (HTR). Comparisons of the asymptotic solutions, with those available in the literature and also with those obtained using the Homotopy Analysis Method (HAM) show that our explicit solutions are valid over the entire domain.

Keywords HTR · Viscoelastic fluid · Stretching · HAM

1 Introduction

In a recent work, Liu [7] proposed a renormalization method (TR Method) depending on the Taylor series expansion for the asymptotic analysis. Although the standard renormalization group method (RG) [5] can be derived from the TR method, there exist some weakness (see Sect. 5.1 of [7]) in both RG and TR methods. In order to overcome these weaknesses Liu [7] further proposed the HTR method by combining the topological deformation with the TR method. The HTR method provides liberty in selecting the first approximate solution in perturbation expansion, which has the ability to improve the global approximation solution. Even if equations do not include a small parameter HTR can still be used to obtain the global asymptotic solution. The effectiveness and validity of the method is evident from the number of works devoted to it, see [6, 10] and the references therein.

A. K. Pandey (✉) · A. Das
Department of Mathematics, NIT, Trichy, Tamil Nadu 620015, India
e-mail: amitpandey754703@gmail.com

S. Das et al. (eds.), *Proceedings of the 1st International Conference on Fluid, Thermal and Energy Systems*, Lecture Notes in Mechanical Engineering,
https://doi.org/10.1007/978-981-99-5990-7_35

2 Literature Review and Objective

Our objective is to apply the HTR method to the equations governing the flow of a second-order viscoelastic fluid caused by the stretching of a plane surface. The fluid flow of a stretching surface has a wide range of applications in science and engineering; hence it has gotten a lot of attention, see [4, 6, 9] and the references therein.

The constitutive equation of an incompressible second-order fluid provided by Coleman and Noll [3] is

$$T = -pI + \mu A_1 + \alpha_1 A_2 + \alpha_2 A_1^2, \tag{1}$$

where T and p indicate stress tensor and pressure respectively, and μ, α_1, α_2 are material constants such that $\alpha_1 < 0$, and

$$A_1 = (\nabla v) + (\nabla v)^T,$$
$$A_2 = \tfrac{d}{dt} A_1 + A_1 \cdot (\nabla v) + (\nabla v)^T \cdot A_1.$$

Consider a fluid flow, confined to $y > 0$, that obeys Eq. (1) past a flat sheet coinciding with the plane $y = 0$. Keeping the origin fixed, the wall is stretched by the application of two equal and opposite forces along the x-axis. For this fluid, Beard and Walters [1] obtained the steady incompressible 2-D boundary layer equations as

$$\frac{\partial u}{\partial x} + \frac{\partial v}{\partial y} = 0, \tag{2}$$

$$u\frac{\partial u}{\partial x} + v\frac{\partial u}{\partial y} = \nu\frac{\partial^2 u}{\partial y^2} - k\begin{bmatrix} \frac{\partial}{\partial x}\left(u\frac{\partial^2 u}{\partial y^2}\right) \\ +\frac{\partial u}{\partial y}\frac{\partial^2 v}{\partial y^2} + v\frac{\partial^3 u}{\partial y^3} \end{bmatrix}, \tag{3}$$

where

$$\nu = \frac{\mu}{\rho},\ k = -\frac{\alpha_1}{p}. \tag{4}$$

And the corresponding boundary conditions are

$$u = cx,\ v = 0 \text{ at } y = 0;\ c > 0\,(c \text{ constant}) \tag{5}$$

$$u \to 0 \text{ as } y \to \infty. \tag{6}$$

It must be noted that, while deriving the above equations, both ν and k were found to be of $O(\delta^2)$, where δ is the boundary layer thickness.

Since the free stream velocity is zero, the flow is only due to the stretching of the wall. Using the following similarity variables [4]

$$u = cxF'(\eta), \ \ v = -\sqrt{\nu c}F(\eta), \tag{7}$$

$$\text{where} \quad \eta = \sqrt{\frac{c}{\nu}}y, \tag{8}$$

in (3), we have

$$F'^2 - FF'' = F''' - k_1\Big(2F'F''' - F''^2 - FF^{iv}\Big), \tag{9}$$

where prime indicates the differentiation with respect to η and

$$k_1 = \frac{kc}{\nu}. \tag{10}$$

Equations (5) and (6) become

$$F'(0) = 1, \ \ F(0) = 0, \ \ F'(\infty) = 0. \tag{11}$$

3 Asymptotic Solution of Viscoelastic Fluid Flow Over a Stretching Sheet

Consider a general non-linear equation and a general linear operator as

$$\mathcal{N}\big(t, y, y', \cdots\big) = 0, \tag{12}$$

$$\mathcal{L}(y) = 0. \tag{13}$$

We construct the following homotopy equation,

$$(1-\epsilon)L(y) + \epsilon N(t, y, y', \cdots) = 0, \tag{14}$$

where the parameter ϵ is such that $0 \le \epsilon \le 1$. As ϵ changing from 0 to 1, the homotopy equation changes from linear to the aim equation. Expand the solution of the homotopy equation as a power series of ϵ,

$$y(t, \epsilon) = \sum_{k=0}^{\infty} y_k(t)\epsilon^k, \tag{15}$$

where $y_k(t)$'s are unknown functions. Equations (14) and (15) yield

$$\mathcal{L}(y_{k+1}) = \mathcal{L}(y_k) - \mathcal{N}(t, y_k, y_{k'}, \cdots) \tag{16}$$

where $k = 0, 1, \cdots$. Solving (15) and (16), expanding every y_k as a power series at some point t_0, then rearranging the solution and at the end taking $\epsilon = 1$ gives the solution of original equation as

$$y(t, t_0) = \mathcal{Y}_0(t_0, A, B, \cdots) + \mathcal{Y}_1(t_0, A, B, \cdots)$$

$$(t - t_0) + \cdots + O(1) \tag{17}$$

where A, B, etc., are parameters. From (17), the solution is

$$y(t) = \mathcal{Y}_0(t, A, B, \cdots). \tag{18}$$

Consider that these parameters are functions of t_0. Then, determine them by the renormalization equation

$$\frac{\partial \mathcal{Y}_0}{\partial t_0} = \mathcal{Y}_1. \tag{19}$$

Through a suitable choice of the closed equations system, we solve these integral constants such as A and B and substitute them into the expression of $\mathcal{Y}_0$ to obtain the asymptotic global solution.

The above proposed method is called the homotopy renormalization method based on the Taylor expansion, for simplicity, the HTR method.

In the frame of HTR method, we take homotopy equation as

$$F'' + F' = \epsilon \begin{bmatrix} F'' + F' - F'^2 + FF'' + F''' \\ -k_1\left(2F'F''' - F''^2 - FF^{iv}\right) \end{bmatrix}. \tag{20}$$

Now expand F as

$$F(\eta) = \sum_{m=0}^{\infty} F_m(\eta)\epsilon^m. \tag{21}$$

On substituting (21) into (20), we obtain

$$F_0'' + F_0' = 0 \tag{22}$$

$$\begin{aligned} F_1'' + F_1' = F_0'' + F_0' - F_0'^2 + F_0 F_0'' + F_0''' \\ - k_1\left(2F_0' F_0''' - F_0''^2 - F_0 F_0^{iv}\right). \end{aligned} \tag{23}$$

On solving (22) we obtain,

$$F_0 = A + Be^{-\eta}. \tag{24}$$

Using (24) in (23) we get F_1 as

$$F_1 = B[1 - (1 + k_1)A]\big[(\eta - \eta_0)e^{-\eta} + e^{-\eta}\big]. \tag{25}$$

Thus, the first-order solution is

$$\begin{aligned} F &= F_0 + \epsilon F_1 \\ &= A + Be^{-\eta} + \epsilon\{[1 - (1 + k_1)A]\, B \\ &(\eta - \eta_0)e^{-\eta} + [1 - (1 + k_1)A]Be^{-\eta}\big\}. \end{aligned} \tag{26}$$

Assuming A and B as a function of η_0, and using the renormalization equation

$$\begin{aligned} \frac{\partial}{\partial \eta_0}(F(\eta_0)) = &- Be^{-\eta_0} + \epsilon[\{1 - (1 + k_1)A\} \\ &Be^{-\eta_0} - \{1 - (1 + k_1)A\}Be^{-\eta_0}\big], \end{aligned} \tag{27}$$

$$\begin{aligned} A' &= 0 \\ B' &= \tfrac{1-(1+k_1)A}{2-(1+k_1)A} B, \end{aligned} \tag{28}$$

we get the solution of closed equations system as

$$\begin{aligned} A &= A_0 \\ B &= B_0 \exp\Big(\tfrac{1-(1+k_1)A_0}{2-(1+k_1)A_0}\eta_0\Big), \end{aligned} \tag{29}$$

where A_0 and B_0 are the constants of integration that can be obtained using boundary conditions. On substituting (29) into (26) at $\eta_0 = \eta$ and $\epsilon = 1$, we get

$$F = A_0 + [2 - (1 + k_1)A_0]B_0$$

$$exp\left(\frac{-\eta}{2 - (1 + k_1)A_0}\right),$$

which is the global asymptotic solution. On using the boundary conditions, we will get

$$F = \frac{1}{r}[1 - exp(-r\eta)], \tag{30}$$

where $r = \frac{2+k_1}{2}$.

Table 1 Values of τ

k_1	HTR	HAM	Numerical solution [8]
0.005	−0.997487	−0.997497	−0.9975
0.01	−0.99495	−0.994987	−0.9949
0.03	−0.98455	−0.984886	−0.9846
0.05	−0.97375	−0.9746679	−0.9738

4 Results and Discussion

The elastic parameter k_1 defined by (10) for a second-order fluid is generally small and hence, r is positive. Thus, a remarkably simple and exact analytical solution of (9) that satisfies (11) is obtained from (30). Therefore, from (7) we get the velocity components as

$$u = cxe^{-\eta r}$$
$$v = -\frac{\sqrt{\nu c}}{r}(1 - e^{-\eta r}). \tag{31}$$

From (30), the wall shear stress τ is obtained as

$$\tau = (1 - k_1)F''(0) = -\frac{(1 - k_1)(2 + k_1)}{2} \tag{32}$$

That is, τ vanishes when $k_1 = 1$. But this is impossible as the validity of the model given in (1) depends on the tacit assumption that $k_1 \ll 1$\$ and further when $k_1 = 1$, (14) and (12) show that $r \to \infty$, and hence $F'(\eta) = 0$, which is inconsistent with (11).

Thus, we infer that skin friction can never vanish, which contradicts with the result provided in [2].

In Table 1, a comparison of the values of τ for different values of k_1 is presented with $10th$ order HAM solutions and also with the results of [8]. It can be observed that the values of τ decrease with increasing values of the elastic parameter k_1, indicating that the power expenditure in stretching the sheet decreases for higher values of k_1.

Table 2 shows the variation of $F'(\eta)$ with η for different values of k_1. We see that for increasing values of k_1, u decreases. These findings are inline with the results given in [8].

Table 2 Variation of $F'(\eta)$

k_1	η								
	0.28			2.47			8.38		
	HTR	HAM	Numerical solution [8]	HTR	HAM	Numerical solution [8]	HTR	HAM	Numerical solution [8]
0.005	0.7553	0.7552	0.7535	0.0841	0.0840	0.0842	0.00022	0.00022	0.0002
0.01	0.7547	0.7529	0.7529	0.0835	0.0835	0.0837	0.00022	0.00021	0.0002
0.05	0.7505	0.7503	0.7487	0.0795	0.0793	0.0795	0.00018	0.00018	0.0001

5 Conclusion

This paper deals with the asymptotic analysis of the viscoelastic fluid over a stretching sheet using the HTR method. By considering the proper homotopy equation, we obtain the asymptotic solution which agrees with the boundary conditions. Moreover, our results indicate that the solution obtained using the HTR method is very accurate and agrees well in comparison with the available numerical solution and the HAM solution. From these it is also evident that the solutions we obtained behave quite well on the local scale; that is, for relatively small η. Hence, they are not just "asymptotic" solutions, but can be treated as global solutions. The following can be drawn from the obtained results:

1. The solution obtained in this paper is easy to apply due to its simple form. We can simply calculate the physical parameters like skin friction coefficient.
2. Comparison with the HAM and numerical solution shows that the obtained solution is not only valid asymptotically as their local properties are great. That is, they are nearly accurate solutions.
3. Since the changes in the boundary conditions as well as parameters do not influence the computational process of the HTR method significantly, it is of practical use.

Acknowledgements I express my sincere gratitude to my institute, National Institute of Technology Trichy, for providing me with the necessary facilities as and when required.

References

1. Beard DW, Walters K (1964) Elasticoviscous boundary-layer flows I. Two-dimensional flow near a stagnation point. Math Proc Cambridge Philos Soc 60(3)
2. Bujurke NM, Biradar SN, Hiremath PS (1987) Second-order fluid flow past a stretching sheet with heat transfer. Zeit Für angew Math Und Phy ZAMP 38(4)
3. Coleman BD, Noll W (1974) An approximation theorem for functionals, with applications in continuum mechanics. Found Mech Therm Dyn

4. Dandapat BS, Gupta AS (1989) Flow and heat transfer in a viscoelastic fluid over a stretching sheet. Intl J Non-linear Mech 24(3)
5. Gell-Mann M, Low FE (1954) Quantum electrodynamics at small distances. Phy Rev 95(5)
6. Kai Y, et al (2019) Exact and asymptotic solutions to magnetohydrodynamic flow over a nonlinear stretching sheet with a power-law velocity by the homotopy renormalization method. Phy Fluids 31(6)
7. Liu C (2017) The renormalization method based on the Taylor expansion and applications for asymptotic analysis. Nonlinear Dyn 88(2)
8. Rajagopal KR, Na TY, Gupta AS (1984) Flow of a viscoelastic fluid over a stretching sheet. Rheol Acta 23(2)
9. Wahid NS, et al (2020) Magnetohydrodynamic (MHD) slip Darcy flow of viscoelastic fluid over a stretching sheet and heat transfer with thermal radiation and viscous dissipation. CFD Lett 12(1)
10. Wang C, Gao W (2017) Asymptotic analysis of reduced navier-stokes equations by homotopy renormalization method. Reports on M.P 80.1

Impact of Relative Size and Rotation Rate of an Upstream Cylinder on the Hydrodynamics of a Stationary Cylinder in Tandem

G. Krithin, D. Sathish Kumar, J. Ramarajan, and S. Jayavel

Abstract This work aims to numerically investigate the hydrodynamics of the flow past two circular cylinders in tandem using the finite volume method. The upstream cylinder rotates at a non-dimensional rotation rate, $0 \leq \alpha \leq 1.6$, while the main cylinder is stationary. The ratio of the diameter of the upstream cylinder, d, to the diameter of the main cylinder, D, varies as 0.25, 0.5 and 0.75. The distance of separation between the cylinders is fixed at $L = 0.8D$ to study the flow past the cylinders in the co-shedding regime. The cylinders are exposed to an incoming fluid at $Re =$ 500 with reference to the main cylinder. The behaviours of the root-mean-square lift (C_l^{RMS}) and drag coefficients (C_d^{RMS}) of the main and upstream cylinders are investigated. The drag coefficient of the upstream cylinder interestingly decreases when $\alpha = \omega D/2U_\infty$ increases, while that of the main cylinder increases when α increases. The lift coefficients of both the cylinders increase with the increase in α. The Strouhal number variations are analysed, and the vortex-shedding frequency of the upstream cylinder is found to be lesser than that of the main cylinder.

Keywords Circular cylinder · Hydrodynamic forces · Rotation rate · Tandem arrangement · Varying diameter ratio

1 Introduction

A circular cylinder is a highly symmetric structure. Yet it proves to be sufficient to make an incoming symmetric flow asymmetric. The wake structure and vorticity downstream of the cylinder has inspired a great volume of historical research since the time of Ludwig Prandtl and Theodore von Karman.

Berger and Wille [1] and Williamson [2] had studied the flow structures and the periodic phenomena in the wake of a cylinder exposed to an incoming flow. Improvements to the fundamental study were made through methods such as altering the

G. Krithin (✉) · D. Sathish Kumar · J. Ramarajan · S. Jayavel
Department of Mechanical Engineering, IIITDM, Kancheepuram 600127, Tamil Nadu, India
e-mail: krithing@gmail.com

S. Das et al. (eds.), *Proceedings of the 1st International Conference on Fluid, Thermal and Energy Systems*, Lecture Notes in Mechanical Engineering,
https://doi.org/10.1007/978-981-99-5990-7_36

geometry of the cylinder, changing the flow regime of the incoming fluid and rotating the cylinder. Flow past a rotating cylinder has drawn the attention of researchers since Glauert [3] and evidently earlier. Mittal and Kumar [4] reported that a rotating cylinder exhibits two vortex-shedding regimes and an intermediate no-shedding regime with respect to its rotational velocity.

The study of the flow past more than one cylinder also came as an advancement which attracted researchers further, owing to the complexity of the wake downstream of the cylinders. Two cylinders could be arranged in tandem, side-by-side or staggered configuration as explained by Sumner [5]. Cylinders in tandem involve interesting phenomena because the upstream cylinder behaves like a turbulence generator for the second, while the main cylinder acts as a wake stabilizer for the first. Since the configuration of two cylinders in tandem and the rotation of a cylinder independently cause three vortex-shedding regimes, it is fascinating to explore the combination of these regimes.

Vortex shedding was observed by Mittal and Kumar [4] for $\alpha < 1.91$ and $4.34 < \alpha < 4.70$. These two vortex-shedding regimes are separated by a no-shedding regime. Mittal et al. [6] performed numerical studies on the change in the behaviour of the flow over a pair of identical cylinders in tandem by varying the longitudinal spacing between their centres as $2.5D$ and $5.5D$ for $Re = 100$ and $Re = 1000$. Meneghini et al. [7] performed simulations for the flow over two identical cylinders in orthogonal arrangements for spacing between $1.5D$ and $4D$ between the centres of the cylinders for $Re = 200$ using FEM. They could broadly observe the classification of the co-shedding, interference and two-wake flow regimes. Recently, the flow over two cylinders of non-identical diameters in tandem has witnessed substantial exploration. Song et al. [8] provided extensive numerical results of the hydrodynamic force coefficients for both the cylinders and also introduced new parameters called total force coefficients based on the sum of diameters and the spacing between the cylinders. The computations were performed for $Re = 800$ for the larger cylinder while varying the spacing and d/D. A distance of $1.5D$ between the backward stagnation point of the upstream cylinder and the forward stagnation point of the main cylinder is said to be the critical point. Zhao et al. [10] investigated the flow past cylinders of different diameters in tandem to compare the total RMS forces with that of an isolated cylinder. Extensive experimental studies have been performed by Alam and Zhou [12] to study the flow structure interactions for two cylinders in tandem. Wang et al. [13] experimentally investigated the consequence of varying the diameter of the upstream cylinder relative to the main cylinder. Significant contributions include further division of the co-shedding regime into the lock-in and no-lock-in regions depending on *St*.

From the literature survey, it is seen that the consequences of varying relative size of the upstream cylinder and its position relative to the main cylinder have been studied for stationary cylinders. However, flow past rotating cylinders in tandem hasn't seen as much investigation. The present work aims to computationally investigate the flow over a rotating cylinder upstream of a stationary cylinder in tandem. The d/D ratios of 0.25, 0.5 and 0.75 are explored while the non-dimensional rotation rate of the smaller cylinder is incremented from 0 to 1.6 in steps of 0.2. *Re* is fixed at

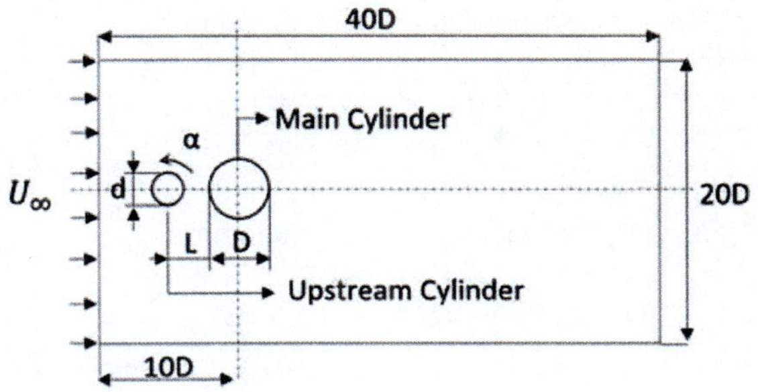

Fig. 1 Computational domain

500 with respect to the main cylinder. This work aims to expose both the cylinders to a uniform flow of low *Re*. Hence, *Re* with respect to the upstream cylinder is 125, 250 or 375. The flow past a rotating cylinder has been rigourously investigated for such cases, and this helps compare the present system of cylinders in tandem with an isolated cylinder. The spacing between cylinders is defined as the distance between the centre of the upstream cylinder and the forward stagnation point of the main cylinder in accordance with Wang et al. [13].

2 Numerical Methodology

2.1 Problem Statement

The schematic for the present study is shown in Fig. 1. The system comprises an infinitely long cylinder of diameter d rotating counter-clockwise at a constant angular velocity ω and an infinitely long stationary cylinder of diameter D placed in tandem. The distance between the centre of the upstream cylinder and the forward stagnation point of the main cylinder is constant L

The cylinders are exposed to the incoming fluid moving with constant freestream velocity U_∞ with linear velocity profile in the positive x direction.

2.2 Governing Equation

ANSYS Fluent solver is employed to solve the non-dimensional Navier–Stokes equation.

Continuity Equation

$$\frac{\partial U}{\partial X} + \frac{\partial V}{\partial Y} = 0$$

Navier–Stokes equation in x direction

$$\frac{\partial U}{\partial \tau} + \frac{\partial (UU)}{\partial X} + \frac{\partial (VU)}{\partial Y} = -\frac{\partial p}{\partial X} + \frac{1}{Re}\left(\frac{\partial^2 U}{\partial X^2} + \frac{\partial^2 U}{\partial Y^2}\right)$$

Navier–Stokes equation in y direction

$$\frac{\partial V}{\partial \tau} + \frac{\partial (UV)}{\partial X} + \frac{\partial (VV)}{\partial Y} = -\frac{\partial p}{\partial Y} + \frac{1}{Re}\left(\frac{\partial^2 V}{\partial X^2} + \frac{\partial^2 V}{\partial Y^2}\right)$$

2.3 Computational Domain and Mesh

The two-dimensional computations are performed on a mesh developed in the $40D \times 20D$ rectangular domain, as shown in Fig. 2. The dimensions have been chosen after extensive literature review indicating to allow adequate distance downstream of the larger cylinder. The origin coincides with the main cylinder's centre, at a distance of $10D$ from the left, bottom and top boundaries and $30D$ from the right boundary. The smaller cylinder is placed longitudinally in front of the main cylinder based on the spacing L. Due to the definition of L, the centre of the upstream cylinder will be at a distance of $L + D/2$ from the origin for all cases. The difference is accommodated with the reduction in the region between the cylinders, as d increases. The incoming fluid, fixed to be air, enters the domain with constant inlet velocity normal to the left boundary, corresponding to $Re = 500$ and exits through the right boundary.

The mesh used is an O-type grid in the neighbourhood of the cylinders and a Cartesian grid farther away as shown in Fig. 2. This technique has provided more accurate results compared to a completely Cartesian mesh owing to better orthogonal

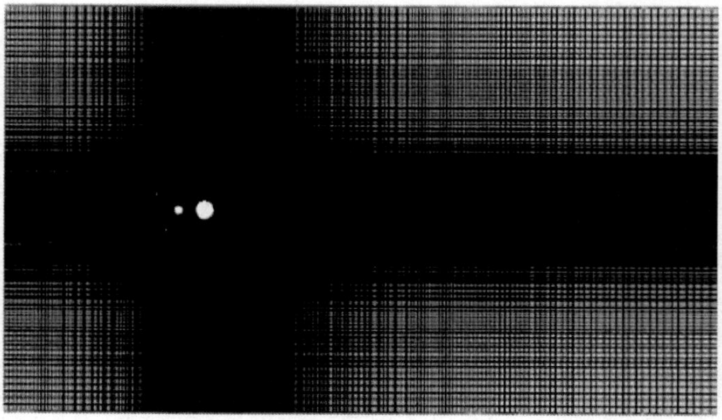

Fig. 2 Structured mesh for the computational domain

quality and aspect ratio of the cells. The defining characteristic of the mesh is the number of points on the cylinder. The entire mesh is then generated to ensure smooth transition of the size of elements increasing from the cylinders to the boundaries. The mesh consists of 1,76,610 elements and 1,75,714 nodes for *d/D* = 0.5. The number of grid points varies slightly for different *d/D* owing to the need for uniform fine meshing between the cylinders and change in geometry. However, the number of nodes on one quarter of the surface of the main cylinder is maintained at 90.

2.4 Solution Methodology

Second order upwind scheme is used for discretization of the momentum terms, while the second order scheme is used for the discretization of the pressure term. Second order implicit scheme is employed for temporal discretization. The governing equations are solved using the semi-implicit method for pressure linked equations (SIMPLE). The computations are performed for a non-dimensional time step $\tau =$ 0.01. The convergence criterion for velocity and continuity is set at 10^{-5}.

The boundary conditions are as follows:

- Left boundary: inlet, $U = 1$, $V = 0$.
- Right boundary: outlet, $\frac{\partial U}{\partial X} = 0$, $\frac{\partial V}{\partial Y} = 0$.
- Top and bottom boundaries: symmetry,

$$< U, \Delta V > - < V, \Delta U > = 0$$

- Cylinders: wall

$$U = -\alpha \cos \phi, \quad V = -\alpha \sin \phi$$

The upstream cylinder and main cylinder are rotating and stationary wall boundaries, respectively.

The hydrodynamic force coefficients are calculated from the lift force, F_L, and drag force, F_D, on the two cylinders as follows:

$$C_d = \frac{F_d}{0.5\rho U_\infty^2 DH} \quad C_l = \frac{F_l}{0.5\rho U_\infty^2 DH}$$

H is the characteristic length of the structure and is taken to be 1 for this two-dimensional study.

$C_l^{RMS} and C_d^{RMS}$ are computed using MATLAB once the time series of the lift and drag forces has attained dynamic stability. The vortex-shedding frequency, *f*, is characterized by the Strouhal number calculated as

Table 1 Grid independence test details

Grid	No. of elements	C_{l1}^{RMS}	Deviation (%)
Very coarse	54,040	0.663	
Coarse	100,246	0.703	5.93
Standard	180,634	0.726	3.39
Fine	399,119	0.725	0.13

$$St = \frac{fD}{U_\infty}.$$

2.5 Grid Independence Test

Four grids with increasing number of elements have been used for the grid resolution study. The grids have been chosen based on the characteristic count of grid points on the surface of the cylinder.

The case for the incoming flow at *Re* = 500, longitudinal spacing L^* between the cylinders of 0.8D, *d/D* = 0.25 and $\alpha = 0.2$ is tested. C_{l1}^{RMS} is compared across the meshes and the observed comparison is presented in Table 1.

It is seen that the standard grid offers significantly more accurate results against the coarse grid. However, the fine grid offers only negligible deviation from the standard grid for an increase of more than 2 lakh elements. Hence, the standard grid is chosen for the problem statement.

3 Results and Discussion

3.1 Validation

To validate the numerical model and methodology, the desired coefficients of a single rotating cylinder are compared with Kang et al. [9]. The computations for the validation were performed on a $40D \times 20D$ rectangular domain similar to that as shown in Fig. 1. The mesh consisted of 66,240 elements and 65,700 nodes. The methodology and the convergence criterion were the same. The non-dimensional time step was fixed at 0.025. *St* and the means and amplitudes of C_l and C_d were compared for *Re* = 160 steady flow. The results are presented in Table 2.

The present results are found to agree with the results reported by Meneghini et al. [7] for flow past a single cylinder.

The methodology is also validated through simulation of flow past two cylinders of dissimilar diameters in tandem for d/D = 0.25 and the distance between the

Table 2 Flow past cylinder at $Re = 160$

α	C_d^{mean} from present results	C_d^{mean} from Kang et al. [9]	C_l^{mean} from present results	C_l^{mean} from Kang et al. [9]
0	1.35	1.32	0	0
1	1.13	1.10	−2.47	−2.43
1.5	0.81	0.78	−3.93	−3.88

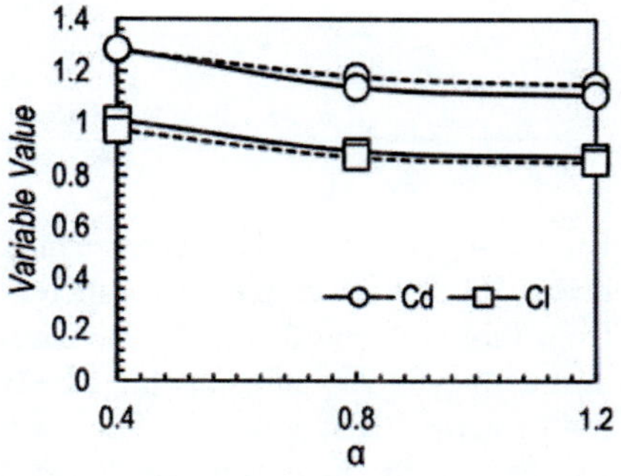

Fig. 3 Validation for flow past two cylinders in tandem. Dashed line: present work. Solid Line: Song's et al. [8] results

backward stagnation point of the upstream cylinder and the forward stagnation point of the main cylinder varying as 0.4D, 0.8D and 1.2D.

Figure 3 shows the comparison of C_{l1}^{RMS} and C_{d1}^{RMS}. The force coefficients of the upstream cylinder were observed to agree with Song's et al. [8] results.

3.2 Flow Past an Isolated Rotating Cylinder

The flow past an isolated rotating cylinder at $Re = 500$ and $0 < \alpha < 1.6$ is studied to compare the behaviour of the hydrodynamics of the tandem arrangement with that of the isolated cylinder. The flow past a stationary cylinder for $Re = 125$, 250, 375 and 500 is also studied since the Re with respect to the upstream cylinder varies in these values. The results are presented in Table 3.

C_l^{RMS} and C_d^{RMS} and St increase for increase in Re when the cylinder is stationary. C_d^{RMS} and St decrease for increase in α for a fixed Re. C_l^{RMS} increases with increase in α for a fixed Re.

3.3 Force Coefficients of the Main Cylinder

C_{d1}^{RMS} And C_{l1}^{RMS} for varying α and d/D are presented in Fig. 4a and b, respectively. C_{d1}^{RMS} increases with increase in α for a constant d/D. Kang et al. [9], Mittal and

Table 3 Flow past a single cylinder

Re	α	C_d^{RMS}	C_l^{RMS}	*St*
125	0	1.369	0.301	0.182
250		1.398	0.598	0.205
375		1.441	0.758	0.221
500	0	1.472	0.854	0.229
	0.4	1.436	1.329	0.227
	0.8	1.332	2.213	0.223
	1.2	1.197	3.259	0.216
	1.6	0.956	4.290	0.201

Kumar [4] had reported an increase in C_d^{RMS} for increase in α up to $\alpha = 1.5$ at low *Re* below 200. However, the rate of increase varied for different *Re*. The present result from Table 1 shows that C_d^{RMS} decreases for increase in α. Thus, the presence of the upstream cylinder causes the increase in drag on the main cylinder. C_{d1}^{RMS} is smaller for a greater d/D at a given α. Song et al. [8] and Wang et al. [13] had also reported greater drag coefficients for smaller *d/D*.

C_{l1}^{RMS} increases with increase in α. $\alpha = 0.6$ is seen to be a critical point after which the rate of increase of C_{l1}^{RMS} is greater. This change is more pronounced for greater *d/D* up to $\alpha = 1$. Furthermore, C_{l1}^{RMS} is smaller for greater *d/D*. The reported dependence of the force coefficients of the main cylinder on *d/D* could be explained as the main cylinder is more submerged in the upstream cylinder's wake for greater *d/D*.

3.4 Force Coefficients of the Upstream Cylinder

C_{d2}^{RMS} And C_{l2}^{RMS} for varying α and *d/D* are presented in Fig. 4c and d, respectively. C_{d2}^{RMS} decreases with increase in α for a constant *d/D* up to $\alpha = 1$. For $\alpha > 1$, C_{d2}^{RMS} for $d/D = 0.75$ increases while it continues decreasing at a different rate for $d/D = 0.25$ and $d/D = 0.5$. C_{d2}^{RMS} is lesser C_d^{RMS} of an isolated cylinder subjected to an incoming flow of the same identical *Re*. Meneghini et al. [7] and Song et al. [8] had reported the presence of a downstream cylinder caused reduction in drag on the upstream cylinder. C_{d2}^{RMS} is greater for larger *d/D* as has been reported for two stationary cylinders in tandem. Greater *d/D* implies that *Re* of the incoming flow at the upstream cylinder is greater which in turn implies greater drag on the cylinder, in accordance with Kang et al. [9]. The result of increasing *Re*, shown in Table 3, supports this inference.

C_{l2}^{RMS} steadily increases for increase in α across all *d/D*. C_{l2}^{RMS} is greater for greater *d/D* at a given α. This behaviour is in accordance with the results of lift force on an isolated cylinder. However, C_{l2}^{RMS} is lesser than C_l^{RMS} of an isolated cylinder subjected to an incoming flow of the same *Re*.

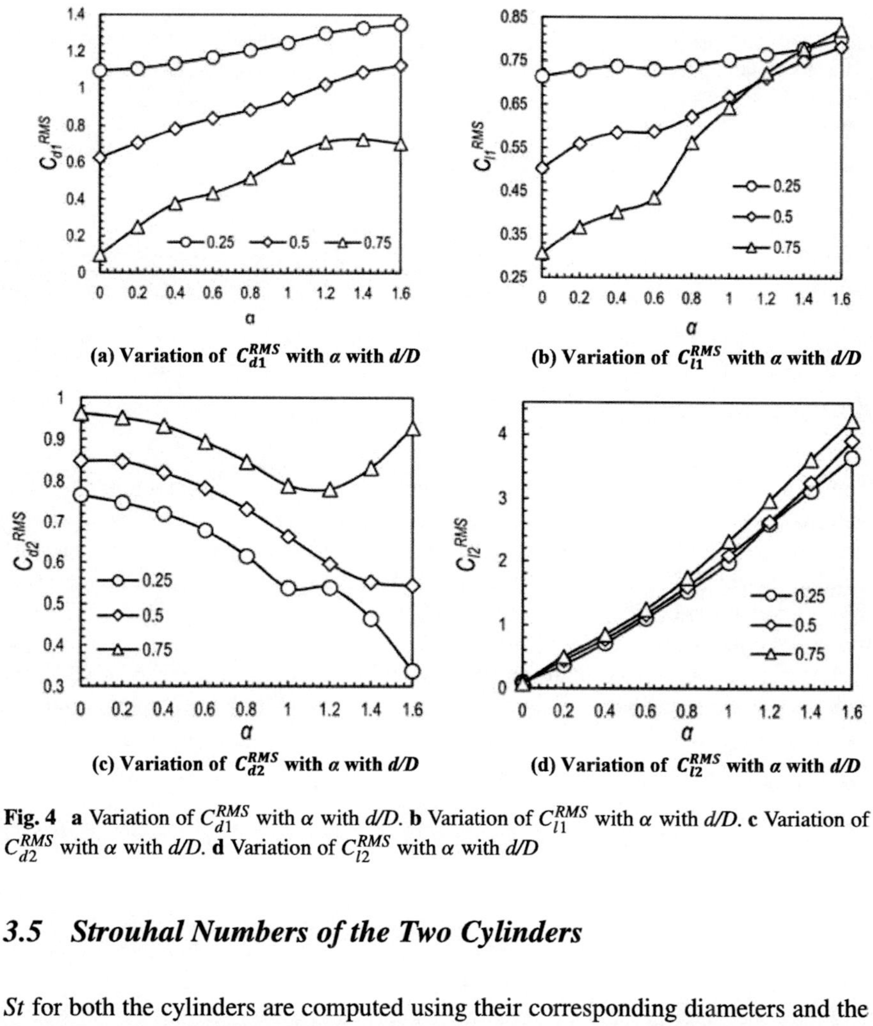

Fig. 4 **a** Variation of C_{d1}^{RMS} with α with *d/D*. **b** Variation of C_{l1}^{RMS} with α with *d/D*. **c** Variation of C_{d2}^{RMS} with α with *d/D*. **d** Variation of C_{l2}^{RMS} with α with *d/D*

3.5 Strouhal Numbers of the Two Cylinders

St for both the cylinders are computed using their corresponding diameters and the frequency of variation of the lift force after the system has attained dynamic stability. The variation of *St* of cylinders with α and *d/D* is shown in Fig. 5. *St* of both the cylinders is lesser for greater *d/D*. St_2 is lesser than *St* of an isolated cylinder subjected to an incoming flow of identical *Re*. This is expected since the spacing between the cylinders is fixed within the co-shedding regime where there is only one wake for both the cylinders. If the upstream cylinder sheds vortices due to the magnitude of *Re* with respect to *d*, the main cylinder would definitely obstruct the wake thereby causing a change in the frequency of vortex shedding.

The upstream cylinder's presence causes a slight reduction in St_1. The periodicity of the lift and drag forces also doesn't exhibit a perfect curve as in the case of an isolated cylinder. The power spectral graphs would show more than one peak as against the behaviour exhibited by an isolated cylinder.

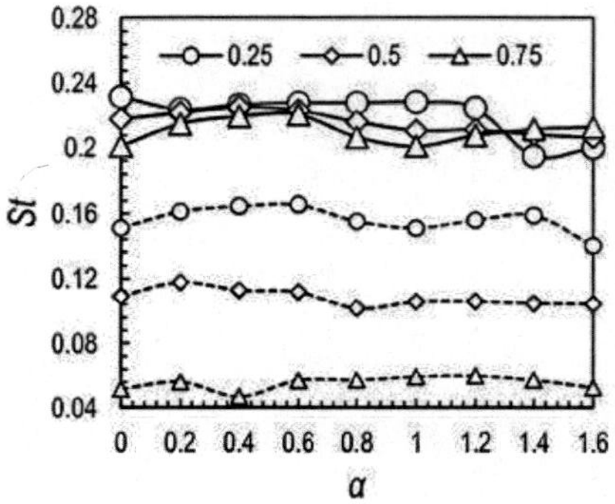

Fig. 5 Variation of St with α and d/D for both the cylinders. Solid line: main cylinder. Dashed line: upstream cylinder

4 Conclusions

The hydrodynamic force coefficients of the flow past a rotating cylinder with a stationary cylinder in tandem have been studied. The reduction in the force coefficients of the upstream cylinder validates the tendency of the main cylinder to act as a wake stabilizer. The presence of the upstream cylinder is felt by the main cylinder through reduction in C_{l1}^{RMS} and C_{d1}^{RMS}. $\alpha = 1.2$ is seen as a critical point for the drag forces for both cylinders.

The present work has explored the first vortex-shedding regime of a rotating cylinder combined with the co-shedding regime of two cylinders in tandem. Future studies could involve larger Re of incoming flow and varying L^* to study the other regimes for flow over cylinders in tandem.

Nomenclature

C_d	Drag coefficient	–
C_l	Lift coefficient	–
D	Diameter of main cylinder	(m)
d	Diameter of upstream cylinder	(m)
L	Spacing between the cylinders	(m)
L^*	Non-dimensional spacing between the cylinders (L/D)	–
Re	Reynolds number	–
St	Strouhal number	–
U_∞	Freestream velocity of incoming fluid	(m/s)
U	Stream-wise velocity	(m/s)
U	Non-dimensional stream-wise velocity	–
V	Cross-stream velocity	(m/s)
V	Non-dimensional cross-stream velocity	–
α	Non-dimensional rotation rate of upstream cylinder	–
μ	Dynamic viscosity of fluid	(kg/m · s)

Φ	Angle of velocity vector at forward stagnation point	(rad)
ρ	Density of fluid	(kg/m^3)
τ	Non-dimensional time	–
ω	Angular velocity	(rad/s)
1	Subscript for main cylinder	–
2	Subscript for upstream cylinder	–

References

1. Berger E, Wille R (1972) Periodic flow phenomena. Ann Rev Fluid Mech 4(1):313–340
2. Williamson CH (1996) Vortex dynamics in the cylinder wake. Ann Rev Fluid Mech 28(1):477–539
3. Glauert MB (1957) The flow past a rapidly rotating circular cylinder. In: Proceedings of the royal society of London. Series A. Mathematical and physical sciences, vol 242, no 1228, pp 108–115
4. Mittal S, Kumar B (2003) Flow past a rotating cylinder. J Fluid Mech 476:303–334
5. Sumner D (2010) Two circular cylinders in crossflow: a review. J Fluids Struct 26(6):849–899
6. Mittal S, Kumar V, Raghuvanshi A (1997) Unsteady incompressible flows past two cylinders in tandem and staggered arrangements. Int J Numer Meth Fluids 25(11):1315–1344
7. Meneghini JR, Saltara F, Siqueira CL, Ferrari JA Jr (2001) Numerical simulation of flow interference between two circular cylinders in tandem and side-by-side arrangements. J Fluids Struct 15(2):327–350
8. Song Z, Duan M, Gu J (2017) Hydrodynamics and vortex shedding characteristics of two tandem cylinders of different diameters in steady flow. Ships Offshore Struct 12(1):86–99
9. Kang S, Choi H, Lee S (1999) Laminar flow past a rotating circular cylinder. Phys Fluids 11(11):3312–3321
10. Zhao M, Cheng L, Teng B, Liang D (2005) Numerical simulation of viscous flow past two circular cylinders of different diameters. Appl Ocean Res 27(1):39–55
11. Alam MM, Moriya M, Takai K, Sakamoto H (2003) Fluctuating fluid forces acting on two circular cylinders in a tandem arrangement at a subcritical Reynolds number. J Wind Eng Ind Aerodyn 91(1–2):139–154
12. Alam MM, Zhou Y (2008) Strouhal numbers, forces and flow structures around two tandem cylinders of different diameters. J Fluids Struct 24(4):50526
13. Wang L, Alam MM, Zhou Y (2018) Two tandem cylinders of different diameters in cross-flow: effect of an upstream cylinder on wake dynamics. J Fluid Mech 836:5–42

A Numerical Investigation on Turbulent Convective Flow Characteristics Over Periodic Grooves of Different Curvatures

Auronil Mukherjee and Arnab Chakroborty

Abstract The effect of curvatures on extended surfaces found many applications in heat transfer devices since a decade. So, understanding the underlying flow physics and heat transfer is the prime importance of the present work. In this work, numerical simulations of turbulent forced convection are conducted by introducing curvatures at sharp corners of the periodic groove at the lower wall. A constant heat flux is supplied in the lower wall while the upper wall is insulated. Computations were carried out using k- $\in$ (RNG) model in RANS formulation implemented in ANSYS Fluent 19.2®. The simulations are performed over varying Reynolds numbers (Re) of 6000–36,000 where ratio of groove width to channel height, groove pitch ratio, and depth ratio are kept constant as 1, 2, and 0.5, respectively. Assessment of coefficient of heat transfer, frictional losses, and magnitude of heat enhancement are systematically carried out over varying radii of curvatures on grooves. The insertion of curvatures improves 10–12% in overall heat transfer, while overall 5% heat transfer enhancement is observed with increment in curvature radius, Reynolds number, and friction factor. The present study proposes several optimal parameters for the enhancement of heat transfer in practical thermal devices.

Keywords Grooved Channel · Curvature · Turbulence · CFD · Heat transfer enhancement

1 Introduction

The performance of major engineering applications relies on improved convective heat transfer in order to reduce heat exchanger size, weight, and cost. Extended surfaces such as ribs, groove, and rib with helical geometry have been used for a long

A. Mukherjee (✉)
Department of Applied Mechanics, IIT Madras, Chennai 600036, India
e-mail: am20s013@smail.iitm.ac.in

A. Chakroborty
Department of Mechanical Engineering, IIT Madras, Chennai 600036, India

S. Das et al. (eds.), *Proceedings of the 1st International Conference on Fluid, Thermal and Energy Systems*, Lecture Notes in Mechanical Engineering,
https://doi.org/10.1007/978-981-99-5990-7_37

time to influence the fluid motion leading to formation of transverse and longitudinal vortices that improve performances of heat exchangers. Channels where ribs are fitted orthogonally to flow direction induce complex flow scenarios to investigate the influence of inserts on heated turbulent convective flows. The phenomenon of heat transfer and associated frictional losses in flow domains with ribs in perpendicular direction have been the subject of numerous experimental research since a long time.

Donne and Meyer [1] have reported an extensive investigation and measurement of heat transfer and coefficients of pressure drop for different rib designs. Several studies are reported on heat transfer and fluid flow with very less geometric parameter modifications. Martin and Bates [2] reported the velocity field and turbulence structure in a rectangular asymmetric ribs channel. Hong and Hsieh [3] studied the effect of alignments of ribs on forced convection in a rectangular or square channel at various alignments. Baughn and Yan [4] studied thermal transmission at mutually opposite walls of a square channel introducing orthogonally oriented square ribs. The coefficient of heat transfer was measured in the upper region of the ribs. Liou and Hwang [5] used holographic interferometry technique in order to assess heat transfer coefficients locally along with the capturing frictional loss in an asymmetrical channel with square ribs at different pitch-to-height ratios. Lorenz et al. [6] investigated the distribution patterns of coefficient of heat transfer and frictional losses on the walls within an asymmetric grooved channel with ribs for turbulent flow and heat transfer. Yang and Hwang [7] investigated thermal characteristics in rectangular ducts where slit and ribs were installed on one side of the wall. Mukherjee and Saha [8] studied the effect of fillets in contracted and expanded rectangular grooves for a laminar incompressible flow. They demonstrated an improvement in heat transfer using fillets of an arbitrary radius. However, the study was restricted to low Reynolds number flow, and the effect of the fillet radius on flow physics was not reported. Smith and Pongjet [9] studied the effect of different turbulent models and groove width to channel ratio in a rectangular grooved channel. They claimed a 158% increment in heat transfer rate in a grooved channel in comparison to a smooth channel. However, detailed investigations of effect of curvature of varying radii in a one-sided expanded grooved channel in a high-speed turbulent convective flow is not present in the existing literature.

The present study numerically investigates the role of curvature radius at the sharp corners in a grooved channel and its effect on heat transfer for a unidirectional convective flow of air using finite volume-based commercial package ANSYS Fluent 19.2®. The associated pressure loss due to the introduction of the curvatures of different radii are also studied along with respective visualization the heat transfer enhancement factor over a range of *Re* varying from 6000 to 36,000.

2 Methodology

2.1 Description of the 2D Numerical Domain

The geometry of the present work consists of a horizontal grooved channel as illustrated in Fig. 1 (Fig. 1a, b: Grooves without curvature and with curvature respectively). The grooves are located in the lower region of the domain. The geometric dimensions are kept the same as mentioned in Eiamsa-ard and Promvonge [9], with expanded region height, $H = 40$ mm. Curvatures of various radii ranging from 3 mm to 9.5 mm are incorporated in the sharp corners of the initial rectangular geometry. The top wall is insulated while constant heat flux is applied in the lower wall as stated by Lorenz et al. [6]. Air is used as the working fluid, and it travels unidirectionally along the specified direction. The flow is assumed to be steady and incompressible [9].

The basic dimensions of the groove are shown in Fig. 2a and in Fig. 2b. The mean inlet velocity ranges from 1.19 to 3.58 m/s, and pressure gradient is absent at the exit, with no-slip boundary conditions at the wall [9]. Chaube et al. [10] reported that the results obtained using a 2D flow model are in reasonable agreement with a 3D model. Hence, 2D geometry is used in the present work to save more computer memory, processing time, and the computational costs.

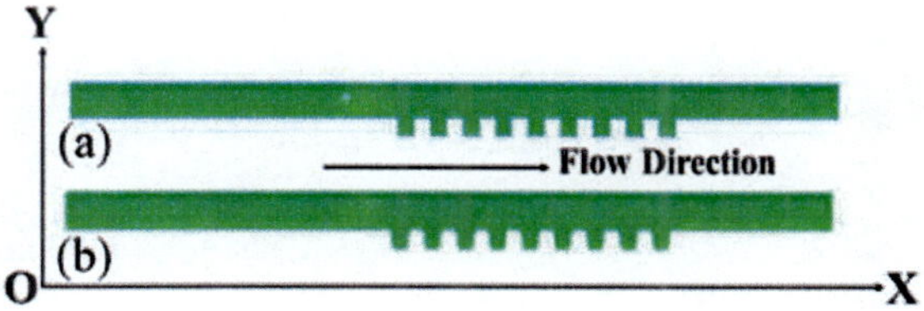

Fig. 1 Schematic of the computational domain: **a** without curvature at corner of fillets, and **b** with curvature at corner of fillets

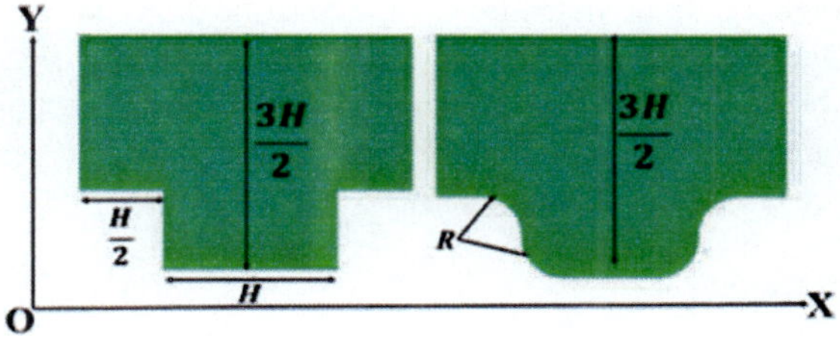

Fig. 2 Geometric details of a single groove: **a** without curvature at corner of fillets, and **b** with curvature at corner of fillets

2.2 Governing Equations of Fluid Flow

2D steady equations of continuity, incompressible Navier–Stokes equations, and the energy equation regulate dynamics of the fluid flow and thermal transmission through the flow domain. The equations in the tensor systems are as follows:

Continuity Equation:

$$\nabla \cdot \vec{U} = 0 \tag{1}$$

Momentum Equation:

$$\rho \frac{D\vec{U}}{Dt} = \rho \vec{g} - \nabla P + \mu \nabla^2 \vec{U} \tag{2}$$

Energy Equation:

$$\rho c_p \frac{DT}{Dt} = \rho \dot{q} + k \nabla^2 T + \varphi \tag{3}$$

The governing equations for the RNG k-ϵ model are

$$\frac{\partial}{\partial t}(\rho k) + \frac{\partial}{\partial x_i}(\rho k u_i) = \frac{\partial}{\partial x_j}\left(\alpha_k \mu_{eff} \frac{\partial k}{\partial x_j}\right) + G_k + G_b + \rho \varepsilon - Y_M + S_k \tag{4}$$

$$\begin{aligned} \frac{\partial}{\partial t}(\rho \varepsilon) + \frac{\partial}{\partial x_i}(\rho \varepsilon u_i) = & \frac{\partial}{\partial x_j}\left(\alpha_\varepsilon \mu_{eff} \frac{\partial \epsilon}{\partial x_j}\right) + C_{1\varepsilon} \frac{\varepsilon}{k}(G_k + C_{3\varepsilon} G_b) \\ & - C_{2\varepsilon} \rho \frac{\epsilon^2}{k} - R_\varepsilon + S_\varepsilon \end{aligned} \tag{5}$$

The generation of turbulence kinetic energy due to mean velocity gradients is denoted by G_k in the Eqs. 4 and 5, while the generation of turbulence kinetic energy due to buoyancy is given by G_b. The contribution of variable dilatation in compressible turbulence to the overall dissipation rate is represented by Y_M. The inverse effective Prandtl numbers for k and ϵ are α_k and α_ε, respectively. S_k and S_ε represent the user-defined source terms.

2.3 Meshing and Grid Independence Study

Figure 3 depicts the variation of coefficient of heat transfer at the bottom wall over five different mesh sizes. The laminar sub-layer is resolved by rectangular grids using face meshing techniques, with grid adoption for y + ≈2 at a near wall zone [9]. To establish the optimal mesh size for numerical simulations, a grid sensitivity

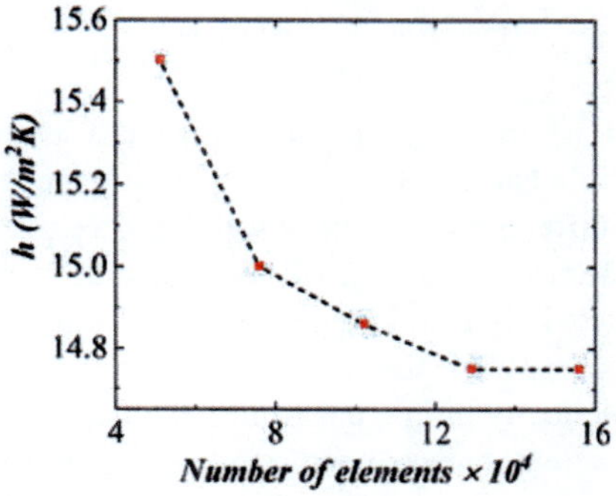

Fig. 3 Plot of mesh independent study at different grid sizes

analysis is further performed. The number of cells used to attain grid independence is varied between 50,900 and 155,992 rectangular types of meshes, and the heat transfer coefficient is calculated for each case at a Reynolds number of 12,000. The optimum grid is obtained, keeping residual less than 10e-03.

2.4 CFD Model Validation

The results obtained through numerical simulations were validated for a steady incompressible flow at a *Re* of 12,000 using experimental results of Lorenz et al. and Eiamsa-ard et al. [6, 9], respectively. The variation of the local Nusselt number (Nu) is investigated along the wall of a single groove as shown in Fig. 4. The results of the present simulation are observed to satisfy a reasonable match with literature.

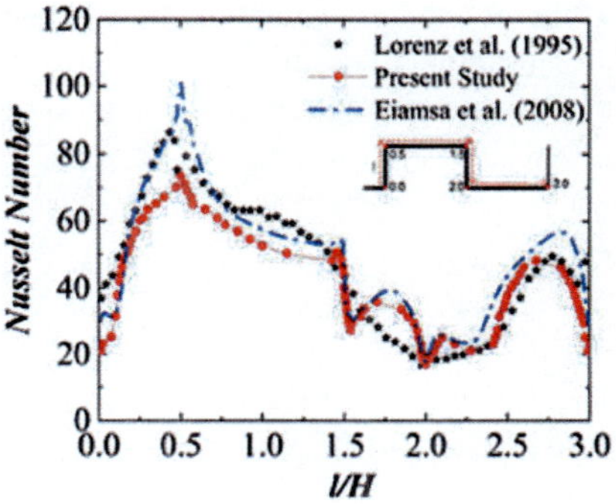

Fig. 4 The local Nusselt number *(Nu)* distribution over different l/H

2.5 Numerical Methods

The finite volume method solver was used to discretize both the steady incompressible Navier–Stokes equations and the RANS turbulence model. QUICK and central differencing schemes were used respectively to discretize convective and diffusion terms. Discrete nonlinear equations were implicitly implemented using pressure–velocity coupling algorithm. Pressure field was evaluated using the SIMPLE algorithm while the second order upwind scheme is used to model the energy. A constant velocity profile has been established at the inlet. The channel wall has an impermeable boundary condition, while the lower wall of the computational domain has a supply of fixed heat flux. At the inlet, the turbulence intensity was fixed at 10% [9].

3 Results and Discussions

3.1 Fluid Flow Regimes and Heat Transfer

Figure 5 depicts the variation of mean Nusselt number with different curvature radii over varying Reynolds numbers of 6000–36,000. The heat transfer coefficient increases with an increase in curvature radius at any particular Reynolds number. This clearly demonstrates that the intensification of the formation of the longitudinal vortices is directly proportional to the curvature radius of the fillet that further causes the heat transfer enhancement with increase in Reynolds Number.

Figure 6 depicts a histogram of the amount of heat transfer enhancement has been observed due to the insertion of these curvatures on the simple rectangle fillet. The maximum magnitude of Nusselt number is obtained for a curvature radius of 9.5 mm. The comparison of the heat transfer between a grooved channel, without and with an optimum curvature, is obtained using this configuration. The average percentage of heat transfer enhancement due to the introduction of this optimum curvature is found to be around 9%.

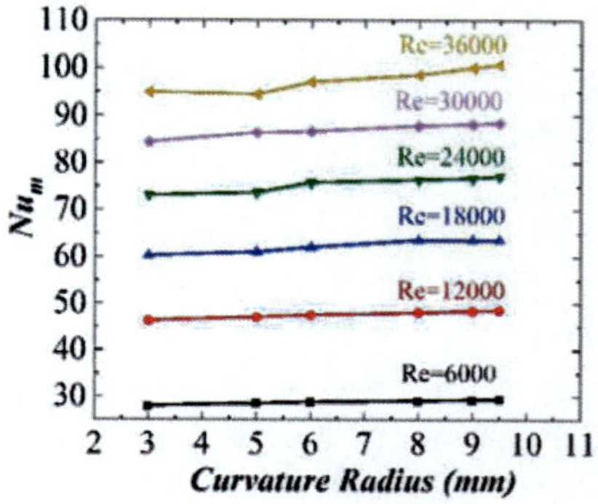

Fig. 5 Plot of Nu_m with curvature radius at Re in range 6000–36,000

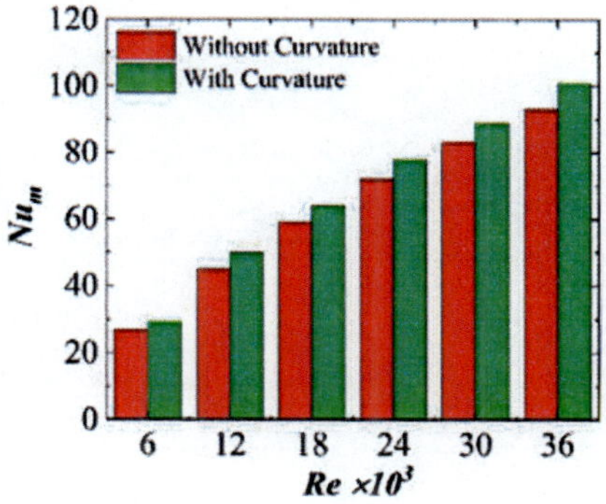

Fig. 6 Comparative histogram of heat transfer enhancement due to the insertion of a curvature on the fillet

The flow structure of a time-averaged flow inside the grooved channel can be observed by looking into the vorticity and stream function plots for both without and with curvature configurations, as shown in Fig. 7a, b. From Fig. 7, it could be interpreted that the curvature groove region is having the largest re-circulation zone with respect to the normal rectangular groove as addition of a curvature at the sharp corners results in increased formation of longitudinal vortices in the near wall region proximal to the groove. Longitudinal vortices are more effective in enhancing heat transfer than transverse vortices [13]. Previous researchers found a similar observation of high association between the vorticity flow and the Nusselt number at the vortex production zones [11, 12]. However, the curvature effects further improve the effective mixing area of the flow which is indicated by the shifting of the high-velocity streamlines towards the lower wall region of the groove. Figure 7c illustrates the temperature variation inside the channel. It is observed that addition of curvatures increases the mean temperature distribution proximal to the wall region which enhances the heat transfer rate.

3.2 *Frictional Losses*

Figure 8 illustrates a trend of approximate values of the friction factor with curvature radius of the fillet using the k-∈ turbulence model for unit groove width-to-height ratio.

The fluid on entering the flow domain suffers by a finite magnitude of pressure drop caused due to a loss in energy in corresponding flow fields. However, it could be observed that the losses incurred due to friction reduce with increase in *Re,* as the sub-layer zone of the viscous medium is suppressed. Also, the magnitude of frictional losses rises moderately with increment in curvature radius.

A comparative plot of the friction factors for both the configurations at different *Re* is depicted in Fig. 9. The frictional losses have moderately increased due to the introduction of the curvatures on the fillet. This increment in friction factor is due to curved surfaces; the contact region is asymmetrical which results in a shear due

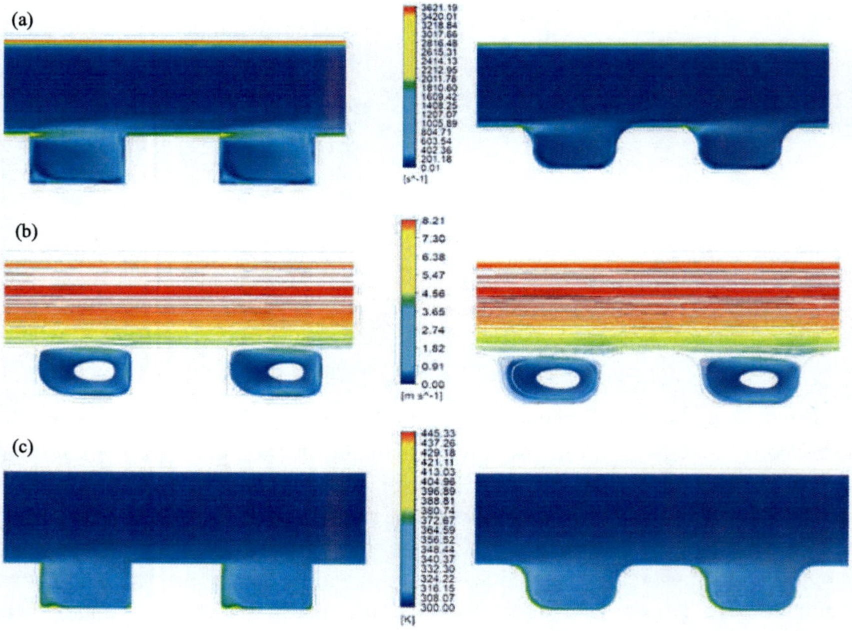

Fig. 7 The contours and streamlines for without (Left column) and with (Right) curvature at $Re =$ 36,000 for 6th and 7th groove: **a** Vorticity, **b** Velocity streamline, and **c** Temperature

Fig. 8 Trend of frictional losses with curvature radius with *Re*

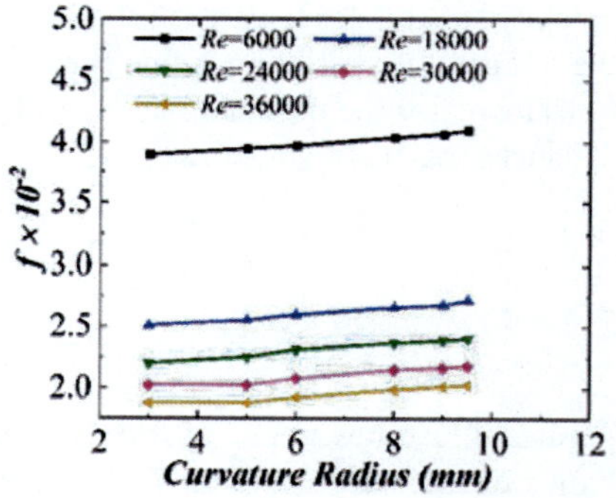

to friction. It contributes an additional force along the motion of the fluid particles which increases the frictional pressure drop. As velocity increases, the magnitude of the shear force acting on the curved surface decreases; as a result, the friction factor values show a decreasing trend in higher Reynolds number regime.

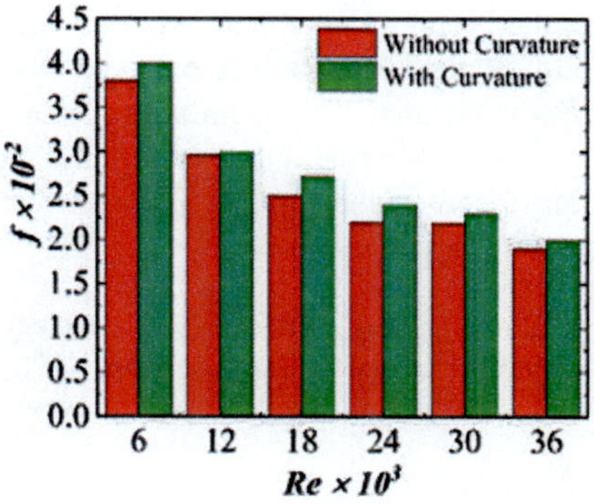

Fig. 9 Comparison of friction factor due to the insertion of curvature on grooves

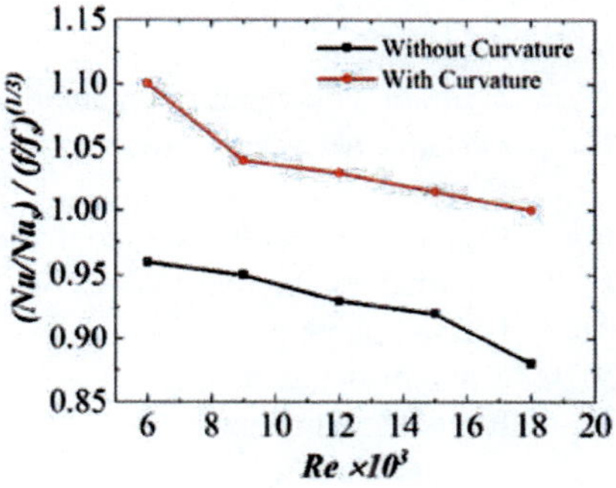

Fig. 10 Comparison of trend of heat transmission improvement with ***Re*** for both configurations

3.3 *Heat Transfer Enhancement*

Figure 10 presents a comparative analysis of the trend in magnitude of heat transmission improvement with *Re* for both the configurations without and with a curvature of 9.5 mm.

Here, Nu_s and f_s is the corresponding Nusselt number and friction factor for a smooth channel in agreement with reported literature [9]. The heat transfer enhancement factor is observed to improve for a grooved channel with curvature, and it further reduces with increment in *Re*.

4 Conclusions

Turbulent convective thermal characteristics and flow dynamics behaviours across a periodic groove channel are numerically studied in this work. The primary goal of this research is to investigate the role of curvatures of different radii in heat transfer enhancement and the corresponding frictional loss in turbulent high-speed flow. The numerical results are validated by comparing it to prior experimental results reported in the literature conducted under a similar set of conditions. It is observed that the introduction of curvature enhances the heat transfer with minimal rise in friction factor. Further, the magnitude of heat transfer enhancement shows an increasing trend with increment in the curvature radius. Overall, the average percentage of heat transfer enhancement due to the introduction of these curvatures is found to be around 10%.

Acknowledgements The authors are thankful to the PG Senapathy Computing Resources at IIT Madras. The authors sincerely acknowledge IIT Madras for providing a quality education based on which they are able to pursue the present research independently.

Nomenclature

D_h Hydraulic diameter of channel = $2H$ mm
H Distance between grooves mm
f Friction factor
h Convective heat transfer coefficient W m^{-2} K^{-1}
k Thermal conductivity W m^{-1} K^{-1}
Nu Nusselt number
p Static pressure Pa
Re Reynolds number
T Température K
U Mean velocity ms^{-1}
u′i Fluctuation velocity components ms^{-1}
μ Dynamic viscosity m^{-1} kg s^{-1}
μ_{eff} Effective viscosity m^{-1} kg s^{-1}
ω Turbulent specific dissipation rate s^{-1}
ε Turbulent dissipation rate m^2 s^{-3}
ρ Density kg/m^3

References

1. Dalle Donne M, Meyer L (1977) Turbulent convective heat transfer from rough surfaces with two-dimensional rectangular ribs. Int J Heat Mass Transf 20(6):583–620
2. Martin SR, Bates CJ (1992) Small-probe-volume laser Doppler anemometry measurements of turbulent flow near the wall of a rib-roughened channel. Flow Meas Instrum 3(2):81–88
3. Hong YJ, Hsieh SS (1993) Heat transfer and friction factor measurements in ducts with staggered and in-line ribs
4. Baughn JW (1992) Local heat transfer measurements in square ducts with transverse ribs. Enhanc Heat Transf ASME 202:1–7
5. Liou TM, Hwang JJ (1992) Developing heat transfer and friction in a ribbed rectangular duct with flow separation at inlet

6. Lorenz S, Mukomilow D, Leiner W (1995) Distribution of the heat transfer coefficient in a channel with periodic transverse grooves. Exp Thermal Fluid Sci 11(3):234–242
7. Yang YT, Hwang CW (2004) Numerical calculations of heat transfer and friction characteristics in rectangular ducts with slit and solid ribs mounted on one wall. Numer Heat Transf, Part A: Appl 45(4):363–375
8. Mukherjee A, Saha S (2021) Effect of fillet on fluid flow and heat transfer characteristics in a grooved channel: a numerical study. In: Proceedings of the 26th national and 4th international ISHMT-ASTFE heat and mass transfer conference December 17–20, 2021, IIT Madras, Chennai-600036, Tamil Nadu, India. Begel House Inc.
9. Eiamsa-ard S, Promvonge P (2008) Numerical study on heat transfer of turbulent channel flow over periodic grooves. Int Commun Heat Mass Transfer 35(7):844–852
10. Chaube A, Sahoo PK, Solanki SC (2006) Analysis of heat transfer augmentation and flow characteristics due to rib roughness over absorber plate of a solar air heater. Renew Energy 31(3):317–331
11. Lemenand T, Habchi C, Della Valle D, Peerhossaini H (2018) Vorticity and convective heat transfer downstream of a vortex generator. Int J Therm Sci 125:342–349
12. Habchi C, Lemenand T, Della Valle D, Peerhossaini H (2010) On the correlation between vorticity strength and convective heat transfer. Int Heat Transfer Conf 49378:377–382
13. Fiebig M (1998) Vortices, generators and heat transfer. Chem Eng Res Des 76(2):108–123
14. Matsson JE (2021) An introduction to ANSYS fluent 2021. SDC Publications
15. Fox RW, McDonald AT, Mitchell JW (2020) Fox and McDonald's introduction to fluid mechanics. Wiley
16. ANSYS Inc (2015) ANSYS fluent theory guide 12.0
17. White FM (2011) Fluid mechanics. McGraw-Hills, New York, USA

Numerical Analysis of Supersonic Flow Past a Double Cavity Scramjet Combustor

Vaisakh S. Nair and K. Muraleedharan Nair

Abstract The fundamental physics involved in cavity flows are of great academic interest. The cavity flows are characterized by an unstable shear layer, vortex formation, its mutual interactions, acoustic generation and its propagation. Dual cavities have lots of applications in flame holders of combustors due to their remarkable potential to stabilize the process of combustion. The present study gives an understanding of the fundamental flow features when flow past a dual flame holder cavity is of aspect ratio 3.0 for a Mach number of 3.0. The results were then compared with the flow features around a single cavity of similar inflow and geometric conditions to study the influence of change in the shape of the geometry on flow dynamics. It is observed that combustor geometry has a considerable influence on flow dynamics and the acoustic behavior of fluid flow in the combustor. All the simulations were done using ANSYS FLUENT 19.2 using k-epsilon model. Most of the research work till date is limited to single cavity flow analysis and flow control methods. However, there are limited research works found on flow dynamics around a scramjet combustor. The present simulation resolves hydrodynamic and acoustic flow features in and around during supersonic flow past a dual cavity.

Keywords Supersonic flow · Aspect ratio · k-epsilon turbulence model · Mach number · Dual cavity

1 Introduction

Testing and evolution of air-breathing hypersonic flight technology has been an intriguing factor since the 1980s because of its tremendous applications. NASA's Hyper-X program and triumphant aviation of X-43A are some of the promising

V. S. Nair (✉) · K. M. Nair
Department of Mechanical Engineering, Mar Baselios College of Engineering and Technology, Thiruvananthapuram, Kerala 695015, India
e-mail: vaisakh.s@mbcet.ac.in; vaisakh.nair28@gmail.com

S. Das et al. (eds.), *Proceedings of the 1st International Conference on Fluid, Thermal and Energy Systems*, Lecture Notes in Mechanical Engineering,
https://doi.org/10.1007/978-981-99-5990-7_38

flights in this regime. With extensive studies in the hypersonic domain, the operations of supersonic combustion chambers are subjected to tremendous variations in thermodynamic conditions. The pressure change related to the combustion process is often highly related to the chamber geometry. Effective combustion cannot be initiated until uniform mixing has been achieved.

The air enters the diffuser and undergoes compression process which highly depends on the vehicle speed. The fuel mixes with the compressed air, ignites and burns in the chamber with the aid of a flame holder that stabilizes the flame. Though the operations are simple, actual application is often limited by extreme technical challenges. Further, maintaining combustion in a supersonic flow regime adds more complexity due to the fact that the events like fuel injection, mixing, ignition and burning should happen within milliseconds. Hence, research on supersonic combustion in a scramjet engine is a very demanding topic nowadays. The cavity in the flame holder and thus the formed shear layer and the recirculation flow help an effective and sustained combustion. This turbulent mixing, shock interaction and associated heat liberation make it a complex phenomenon.

Supersonic propulsion systems have increasing demands in military industry and in the high-speed non-combatant transportation sector. To develop such systems, it is necessary to understand the fundamentals of mixing and combustion occurring in the engine. In the present work, we tried to understand the fundamental behavior of fluid flow around a combustor at supersonic inflow condition. The shear layer, compressive wave formations and their mutual interactions, Mach and temperature distribution were resolved.

2 Literature Review and Objective

The flow dynamics and oscillation mechanism of cavity flow were examined [1] employing optical interferometer and Schlieren visualization. It was found that the fundamental frequency of oscillation heavily depends on inflow Mach number. The increasing aspect ratio of the cavity leads changes in the mode of oscillation initially but after certain length, the shear layer loses its ability to oscillate in a particular mode and develops fluctuations on a large scale which increases the drag on cavity [2]. The investigation using implicit large eddy simulation for supersonic cavity flows past an open cavity for aspect ratio of 2.0 observed that a thick inflow boundary layer changes the oscillation mechanism for an open cavity [3]. The shear-layer/wake mode in supersonic flow past cavities [4] proposed that the compressive waves move along with shed vortex downstream and get deflected from the trailing edge and cause the generation of acoustic wave within the cavity. The analysis on the influence of sidewall and cavity width on flow characteristic [5] illustrates the confinement of flow passage due to the interaction of wall vortex. The occurrence of self-sustained oscillation is strong, and becomes coherent for open cavities rather than closed cavities [6].

The Mach wave formed as a result of the supersonic progress of the turbulent eddies interacts with the trailing edge. The generated reflected wave propagates upstream and excites the shear layer causing feedback mechanism. The numerical studies [7–11] concentrated on well-defined features of cavity flow such as the production and interaction of compressive wave and shear layer. The influence of fuel injection temperature and pressure on the combustor performance, and the performance of a dual cavity combustor over various flight operating and fuel injection conditions were investigated [12, 13].

3 Methodology

The computational domain depicted in Figs. 1 and 2 shows the applied boundary conditions. The finite-volume method is utilized to resolve the governing equations. To reduce the computing time and cost, two-dimensional compressible flow past a cavity flame holder with L/D ratio 3.0 is solved in ANSYS FLUENT 19.2 with realizable k-epsilon turbulence model, most suitable for the simulation of combustion regime, and helps to understand the flow physics. The solver employed SIMPLE scheme with second order accuracy.

Velocity is provided at the inlet boundary and at the fuel injection port. Non-reflective condition following the characteristics is given as outflow. Symmetric modeling is used with no-slip condition at other wall regions of the domain. Pressure-based solver is implemented for the analysis (Table 1).

3.1 Grid Independent Test

Three grid levels were used for the grid independent test, namely Grid 1(500 × 600), Grid 2(580 × 780) and Grid 3(600 × 780). Grid distribution is made in such a way that more number of grid elements is pumped at the wall vicinity and stretched away from the wall; the computational grid is indicated in Fig. 3a.

Figure 3b shows that increase in the number of grid elements has no significant effect on the solution from grid 2 onwards. The simulation attempted with the higher grid elements yielded no variation in pressure along streamwise locations. Hence, grid 2 has been considered for the subsequent calculations.

Table 1 Flow conditions

Variables	Air	Fuel (hydrogen)
Mach number	3.0	1.0
Total temperature	1500 K	300 K
Pressure	1.0 MPa	0.7 MPa

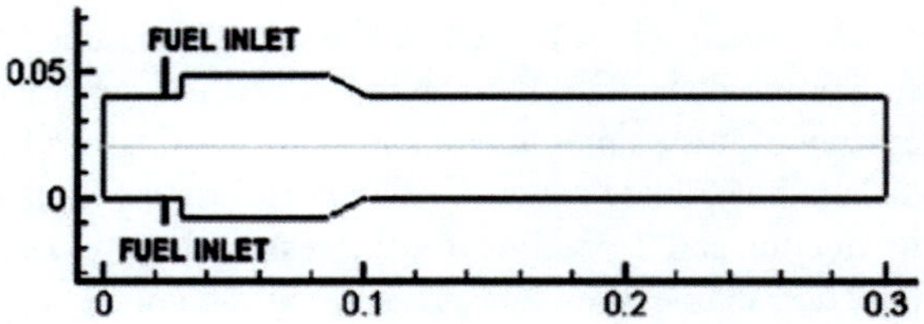

Fig. 1 Computational domain

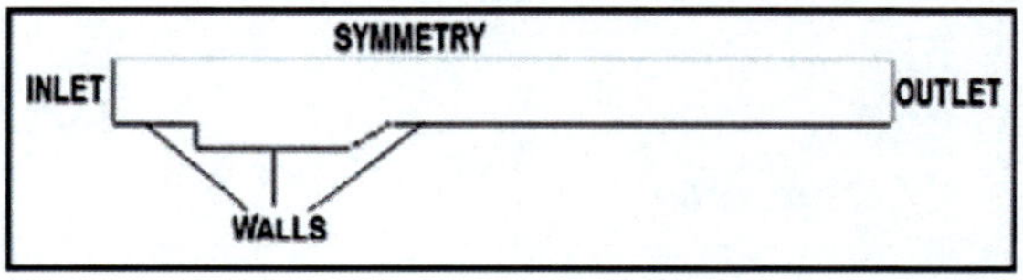

Fig. 2 Imposed boundary conditions

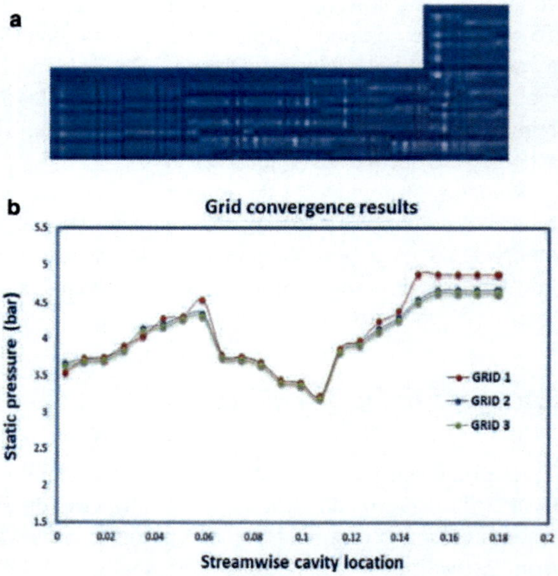

Figure 3 **a** Computational grid. **b** Grid Independent test

3.2 *Validation*

The present solver is validated using experimental data of [11].

Figure 4 depicts the non-dimensionalized streamwise mean velocity (U/U∞) along the normalized cavity length (Y/L). Profiles of U/U∞ depict small deviation from the experimental data along the shear layer in the second half of the structure. The dissipation in validation graph occurs because of the overpredicted growth rate in that cavity region. However, the trend of the profile shows good agreement with that of experimental data.

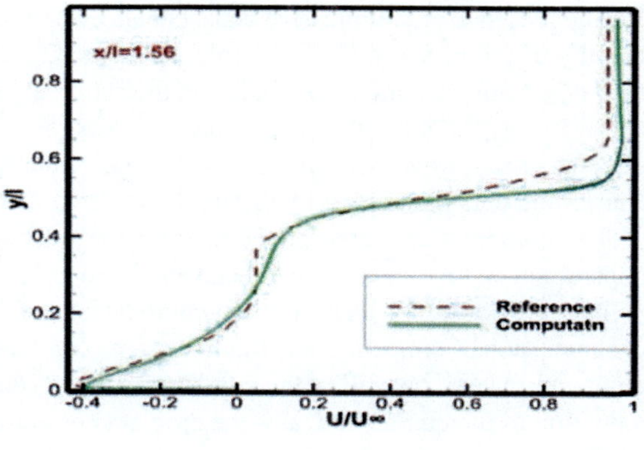

Fig. 4 Validation

4 Results and Discussion

The flow features around a dual cavity scramjet combustor were investigated using k-epsilon model. The numerical scheme accurately simulated the combustion flow field that helped to predict the formation and status of the shock wave. It successfully resolved the influence of inflow and geometric parameters on the hydrodynamics properties of the flow around the cavity. All the results were taken once the flow becomes dynamically stable, and then it is allowed to complete 10 to 12 vortex shedding cycles. As the high-speed flow approaches the forefront of the cavity structure, on account of the immediate geometry change, a velocity gradient exists among the main flow and that near the fore edge of the cavity structure. This in turn forms the shear layer and associated vortex.

The interaction between the incoming shear layer and the fuel injection along with the recirculation interactions leads the formation of compressive waves within the combustor. The formed compressive waves travel downstream, and these waves mutually interact with each other and with the shear layer enhancing the fuel air mixing within the combustor. Iso-contours of density gradient in Fig. 5 demonstrated the presence of compressive waves and its movement.

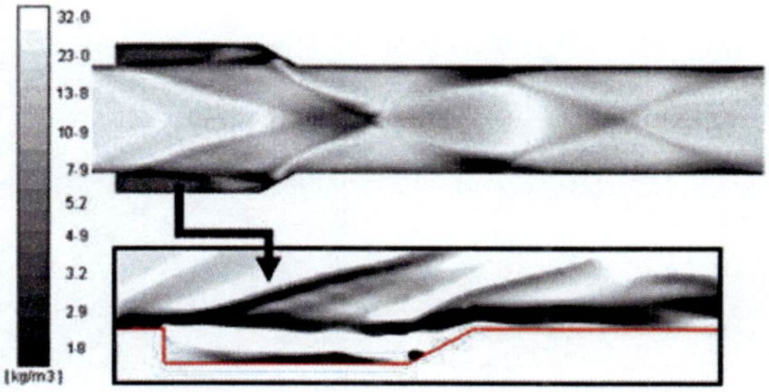

Fig. 5 Iso-contours of the density gradient

The presence of compressive wave enhances pressure along the flame holder and thereby the compression of air. At the downstream of the combustor, the reflection of waves from the combustor walls and their mutual interaction are evident (Fig. 5), and it makes the flow dynamics more complex. The formation and the movement of compressive waves within the combustor eventually enhance the oxidation of the charge and thus the mixing of air and the injected fuel.

The compressive wave reflection from the wall and their interactions are also apparent in static pressure contour given in Fig. 6.

It is also observed form the iso-contours of the temperature that a significant concentration of temperature is found near the combustor wall. The increasing static temperature near the wall of the combustor (Fig. 7) due to the significant amount of oxidation of the charge indicates the efficient combustion process in that region.

The presence of cavity in upper and lower walls of the combustor enhances the effectiveness of the fuel air mixing. This may be due to the fact that supersonic flow stays for more time within the cavity due to the existing vortices at the cavity mouth and along the shear layer indicating its interaction with the recirculation eddies as shown in the iso -contours of vorticity (Fig. 8).

From the Mach number contours depicted in Fig. 9, it is further observed that subsonic flow within the cavity bottom wall region helps to enhance effectiveness of flame holder during the combustion process.

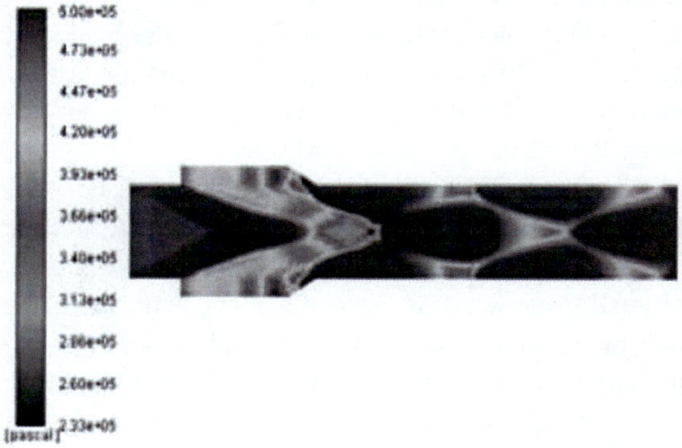

Fig. 6 Iso-contours of the pressure

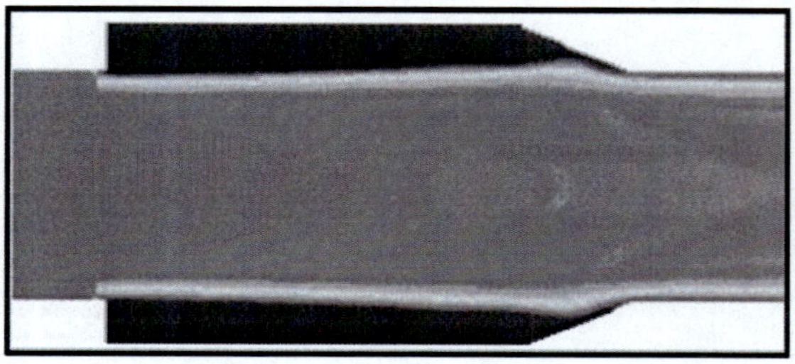

Fig. 7 Iso-contours of temperature

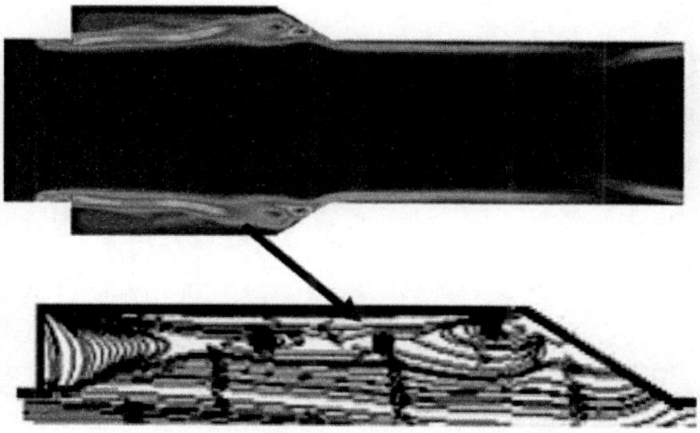

Fig. 8 Iso-contours of vorticity

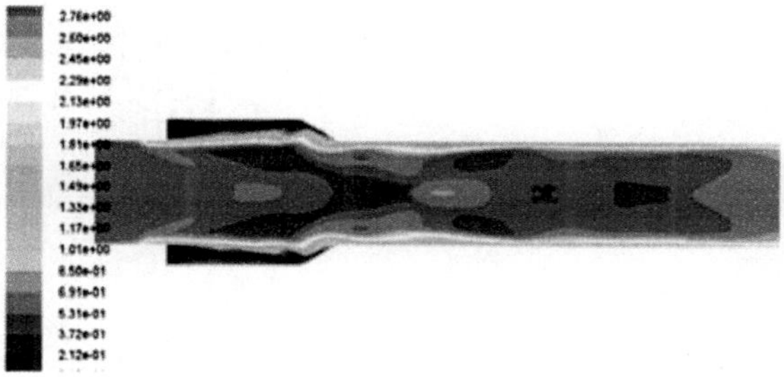

Fig. 9 Iso-contours of Mach number depict the propagation of waves in the streamwise direction

However, when supersonic flow approaches a single cavity of similar geometric and inflow conditions as shown in Fig. 10, it is observed that the velocity gradient prevails between the free stream and flow inside the cavity and shear layer is formed. The shear layer near the fore wall of the cavity gains energy from the recirculation eddies and is separated from the cavity's leading edge. The separated flow convects toward the trailing edge and produces huge acoustic disturbances once they impinge on the rear edge of the cavity which may cause damage to the structures.

The partially clipped vortices in turn move along with recirculation eddies upstream and interact with the shear layer near the front wall causing self-sustaining flow oscillations.

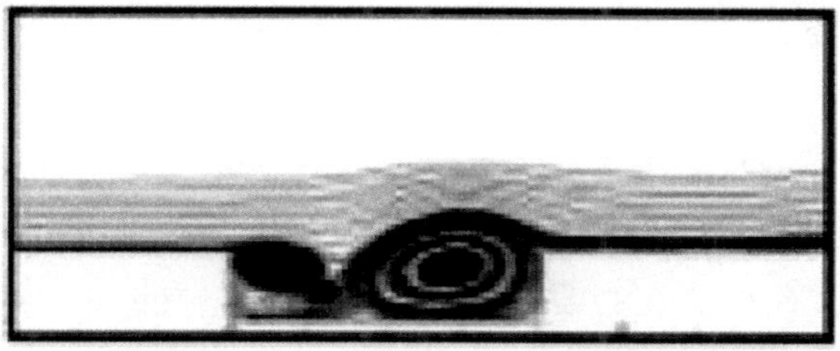

Fig. 10 Iso-contours of the vorticity for single cavity of L/D = 3

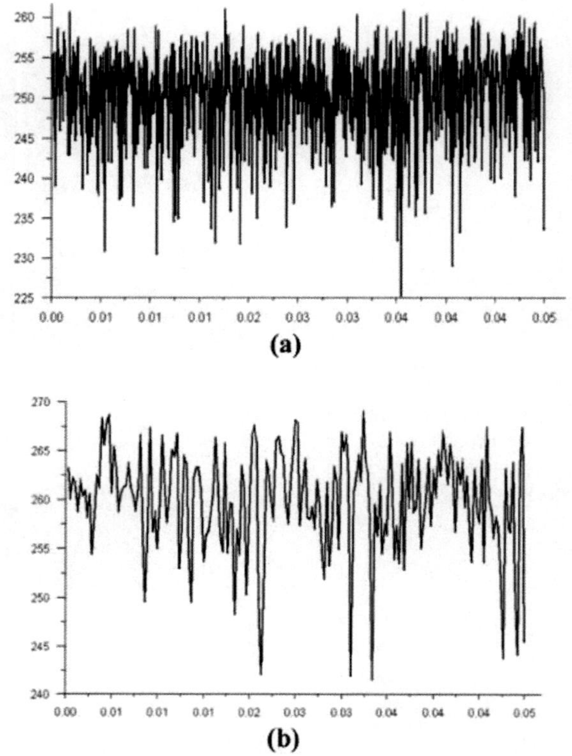

Fig. 11 Pressure fluctuations along cavity location—**a** Inside cavity and **b** Downstream of the cavity

4.1 Acoustic Analysis

The pressure fluctuations across the flame holder cavity (Fig. 11) and its influence on flow inside and downstream of the combustor are analyzed.

It is observed from the OASPL (dB) verses streamwise locations of the combustor (Fig. 12b) that the strength of acoustic disturbances increases as the flow moves downstream, inside the cavity. It may be due to the series of reflections of compressive waves from the combustor rear walls and their mutual interaction. However, the oscillations' strength reduces further downstream of the combustor.

5 Conclusions

Cavities play a significant role in various applications like weapon bays, aircraft fuel ports, combustion chambers and so on. In the current work, compressible flows past flame holder dual cavity structure for an inflow Mach number 3.0 have been analyzed. CFD simulations are performed and the features of flow oscillation were found to be

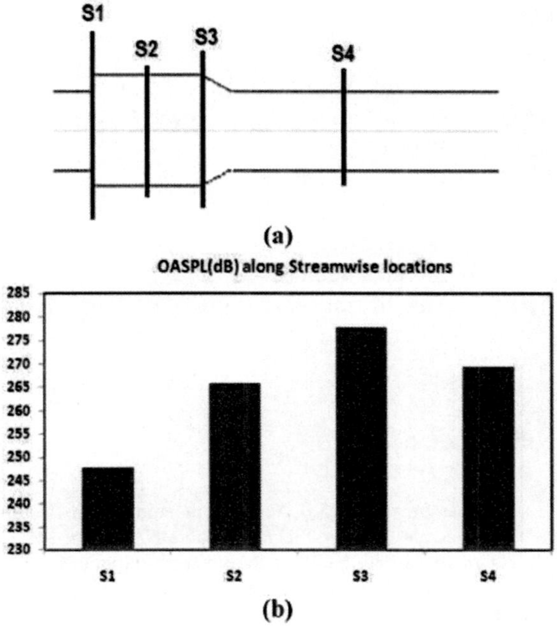

Fig. 12 **a** Station points, and **b** Overall sound pressure level across the flame holder locations

different from that of flow past a single cavity. The k-epsilon model helps to simulate the mean flow features for turbulent conditions, with good prediction of flow field vortices within the cavity structures with respect to the formation of shear layer and compression waves within the dual cavity, recirculation eddies and their interaction with the formed shear layer. This in turn helped to understand the fluctuations within the structure. The presence of dual cavity configuration of the combustor enhances the effectiveness of the fuel air mixing. This may be due to the fact that supersonic flow stays for more time within the cavity due to the presence of vortices and their interaction with the recirculation eddies and compressive waves. It is further observed that the acoustic disturbances vary as the flow moves downstream due to the series of reflections of compressive waves from the combustor rear walls and their mutual interactions. The subsonic flow within the cavity bottom wall region helps to enhance the effectiveness of flame holder during the combustion process.

Acknowledgements The authors would like to express their deepest gratitude to the authors whose previous research provided the motivation to pursue this research and also Mar Baselios College of Engineering and Technology for the facilities provided during the course of the work. We would also like to extend our gratitude to the Centre for Engineering Research and Development (CERD) and APJ Abdul Kalam Technological University for the support received throughout the work.

Nomenclature

ρ	Density of air [kg/m^3]
OASPL	Overall sound pressure level [dB]
Pa	Pascal –
D	Depth of cavity [m]
L	Length of cavity [m]
AR	Aspect ratio –
St	Strouhal number –
$U\infty$	Free stream velocity [m/s]

References

1. Krishnamurty K (1955) Acoustic radiation from two-dimensional rectangular cut outs in aerodynamic surfaces. Technical report, NACA TN-3487
2. Gharib M, Roshko A (1987) The effect of oscillations of cavity drag. J Fluid Mech 177:501–530
3. Sun Y, Zhang Y, Taira K, Cattafesta L, George B, Ukeiley L (2016) Width and sidewall effects on high speed cavity flows. In: 54th AIAA aerospace sciences meeting, AIAA paper 2016-1343
4. Sridhar V, Gai SL, Kleine H (2016) Oscillatory characteristics of shallow open cavities in supersonic flow. AIAA J 54(11):3495–3508
5. Pandian S, Desikan SLN, Niranjan S (2018) Experimental investigation of starting characteristics and wave propagation from a shallow open cavity and its acoustic emission at supersonic speed. Phys Fluids 30:016104
6. Baysal O, Stallings R (1988) Computational and experimental investigation of cavity flow fields. AIAA J 26(1):6–7
7. Rizzetta DP (1988) Numerical simulation of supersonic flow over a three-dimensional cavity. AIAA J 26:799–807.9. Zhang (1995)
8. Tam C-J, Orkwis PD, Disimile PJ (1996) Algebraic turbulence modelsimulations of supersonic open-cavity flow physics. AIAA J 34(11):2255–2260
9. Colonius T, Basu AJ, Rowley CW (1999) Numerical investigation of the flow past a cavity. In: AIAA paper, pp 99–1912; 10–12
10. Rowley CW, Colonius T, Basu AJ (2002) On self-sustained oscillations in two- dimensional compressible flow over rectangular cavities. J Fluid Mech 455:315–346
11. Larchevêque L, Sagaut P, Le T-H, Comte P (2004) Large-eddy simulation of a compressible flow in a three-dimensional open cavity at high Reynolds number. J Fluid Mech 516:265–301
12. Forestier N, Jacquin L, Geffroy P (2003) The mixing layer over a deep cavity at high-subsonic speed. J Fluid Mech 475:101–145
13. Huang W, Bin Luo S, Liu J, Wang ZG (2010) Effect of cavity flame holder configuration on combustion flow field performance of integrated hypersonic vehicle. Sci China Technol Sci 53:2725–2733
14. Collatz MJ, Gruber MR, Olmstead DT, Branam RD, Lin K, Tam C (2009) Dual cavity scramjet operability performance and study 1–15
15. Muraleedharan Nair K, Nair VS, Vishnu Prasad S, Kartha G (2021) Numerical analysis of supersonic flow past tandem deep cavities. In: Proceedings of the 48th national conference on fluid mechanics and fluid power (FMFP). BITS Pilani, Pilani Campus, India
16. Muraleedharan Nair K, Vishnu Prasad S, Nair VS, Kartha G (2019) Analysis of supersonic flow past deep cavity with different aspect ratios. In: AIP conference proceedings 2134, 020011

Pressure Drop in Concurrent Gas–Liquid Upflow Through Packed Beds

A. V. Raghavendra Rao, Bhaskar Bethi, G. Bhanu Radhika, and R. Kishore Kumar

Abstract Separate correlations were proposed to predict a two-phase frictional pressure drop for the three identified flow regimes, namely bubble flow, pulse flow, and spray flow for the concurrent gas–liquid upflow through the packed bed from the large experimental data. In this work, the gas–liquid systems studied are air–water, air-56% glycerol, and air-monoethanolamin; the column packing investigated are Raschig rings, Intalox saddles, and ceramic spheres of two sizes. Liquid flow rates were varied from 3.53 to 43.92 kg/(m^2s) and air flow rates from 0.265 to 1.768 kg/(m^2s). It is noted that pressure drop increases with an increase in gas and liquid flow rates, which is consistent with literature findings. It is observed that pressure drop is higher for the system with high liquid viscosity, and it is also observed that pressure drop is influenced by the combined effect of packing particle size, shape, and porosity of the beds. These correlations predicted the pressure drop of the current experimental data (1150 data points) and literature data (770 data points) with an RMS deviation of 12.8%.

Keywords Two-phase flow · Pressure drop · Concurrent upflow · Packed bed · Flow regimes

1 Introduction

The determination of a two-phase pressure drop (ΔP) for gas–liquid concurrent upflow through a packed bed is a very important design parameter. The ΔP depends on the flow regimes in the bed [1], phase flow rates [2–7], fluid physical properties [8], and packing characteristics [9]. Bubble flow (BF), pulse flow (PF), and spray flow (SF) are the flow regimes that occur when gas and liquid flow concurrently

A. V. R. Rao (✉) · B. Bethi · G. B. Radhika
Department of Chemical Engineering, B V Raju Institute of Technology, Narsapur 502313, India
e-mail: raghavendrarao.av@bvrit.ac.in

R. K. Kumar
INM Technologies Private Limited, Bangalore 560074, India

S. Das et al. (eds.), *Proceedings of the 1st International Conference on Fluid, Thermal and Energy Systems*, Lecture Notes in Mechanical Engineering,
https://doi.org/10.1007/978-981-99-5990-7_39

upwards through a packed bed. These flow regimes manifest sequentially at constant liquid flow rate with increasing gas flow rate.

2 Literature Review and Objective

Some authors investigated pressure drop in all three flow regimes [2–7] but limited their studies to one G-L system (mostly A-W) and a single type of packing (either CS or cylinders, either of a single size or of different sizes). Some authors studied pressure drop in BF or BF extended to PF [9–13]. Multiple G-L systems were used to study pressure drop [14–17], but only a single type of packing was used in their studies. Multiple G-L systems and multiple bed packing were used in one study [8], but their flow regimes were limited to bubble flow to dispersed BF and PF regimes. Only one author [1] investigated pressure drop using multiple G-L systems and multiple bed packing in all three flow regimes. Pulse flow and spray flow were also termed as slug flow [5] and surge flow [6]. Some authors [9, 10] termed their flow regimes as single- and two-phase pore flow. A model was proposed [18, 19] for predicting the ΔP. The proposed model [19] is a modified version of the reported model [20] in which the constants in Ergun's equation, interfacial drag and relative permeability, and passability were modified. Pressure drop was experimentally studied [21] in an inclined packed bed, and it was found that with an inclination towards the horizontal position, the ΔP is reduced.

It was found in one study [22] for pressure drop measurements for A-W upflow through a bed of cylindrical particles and found that the Reeds model [23] predicted friction well at a high air rate, and the Schulenberg and Muller model [24] predicted well at low gas rates. The increase in the flow rate of the gas and liquid reduces the bubble travelling time and increase the pressure drop [25]. Pressure drop and flow regimes were measured and analysed [26] under microgravity, and found that pressured drop depends on wetting characteristics under a viscous-capillary regime ($1 < Re_l < 10$, $0.1 < Re_g < 1$). Effect of G-L distributor on gas and liquid holdup, back mixing, and pressure drop was investigated [27], and found that the pressure drop and back mixing are lower for distributors with multi-step type.

The investigators presented their ΔP findings either as plots or as correlations. Most of the existing correlations [2, 7, 9, 10, 15, 16], which include all three flow regimes, are based on one G-L system and a single type of packing. The correlations proposed based on multiple gas–liquid systems and packing were limited in their study to pulse FR only [8]. Such correlations do not adequately represent the effect of the characteristics of the bed packing, physical properties of the fluids, and all flow regimes. There is only one study [1] covering all three flow regimes, multiple G-L systems, and different bed packings. Hence, there is a necessity to add more data with multiple G-L systems and multiple packing particles and develop robust correlations for predicting the ΔP in all three flow regimes.

The objectives of the present investigation are (1) to experimentally study the effect of gas and liquid flow rates, physical properties of fluids, type and size of packing on the two-phase pressure drop and (2) to propose separate correlations to predict ΔP_f for the three identified flow regimes, namely BF, PF, and SF.

3 Experimental

Figure 1 depicts the computer-controlled experimental setup schematic diagram, the details of which are mentioned in earlier work [28, 29].

For different gas and liquid flow rates flowing concurrently upwards through a packed bed of 10 cm id and 100 cm height, a computer records the pressure at different locations along the bed's height, which was transmitted by the Rosemount Pressure Transmitters 3051 of Emerson make with a performance accuracy of 0.04% and the calibration ranges from 0.25 to 276 mbar. The accuracy of the experimental results was checked by measuring the ΔP of single-phase flow and comparing it with Ergun's equation. The match was very good.

Pressure drop data was recorded for the three G-L systems (A-W, A-G, and air-MEA) and four types of packing (ceramic CS (3.87, 6.9 mm), RR (6 mm), and IS

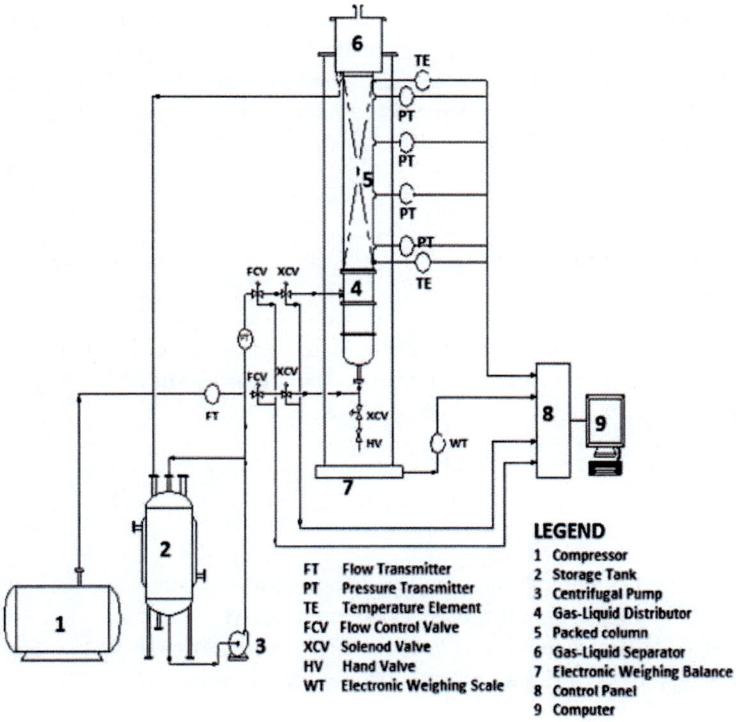

Fig. 1 Experimental setup schematic diagram

Table 1 Bed characteristics

Bed characteristics			
d_c (m)	d_p (m)	ε	ϕ
0.1	0.069 (CS)	0.354	1
	0.00387 (CS)	0.352	1
	0.006 (RR)	0.44	0.58
	0.0072 (IS)	0.536	0.514

Table 2 Physical properties of the fluids

Fluid	ρ (kg/m^3)	μ (kg/(m.s))	σ (N/m)
Water	1000	0.001	0.072
56% Glycerol	1165	0.007	0.069
MEA	1020	0.015	0.049
Air	1.165	0.000018	

(7.92 mm)). For each G-L system, data was recorded for four-column packings. For each column packing and for each G-L system, liquid flow rates were increased stepwise 10 times, and for each liquid flow rate, the gas rate was increased stepwise 10 times. The liquid flow rates ranged from 3.53 to 43.92 kg/(m^{2}s) and air flow rates ranged from 0.265 to 1.768 kg/(m^{2}s). A total of 1150 ΔP data points were obtained in the present study. Apart from the present experimental data, 770 literature data is obtained. Table 1 depicts the bed characteristics in the present study, and Table 2 gives the physical properties of the fluids used.

4 Results and Discussion

The experimentally measured pressure drop is the total ΔP. It is a combination of ΔP due to friction and ΔP due to liquid holdup. The ΔP_f is estimated by the following formula:

$$\frac{\Delta P_f}{h} = \frac{\Delta P_t}{h} - \rho_B g \quad (1)$$

where, $\rho_B = \beta_t \rho_l + (1 - \beta_t)\rho_g$

The total liquid holdup is measured by the weighing method. In the present work, the measured liquid holdup is used in calculating the ΔP_f.

The liquid holdup in the above equation can also be estimated from the correlation [29], which predicts the liquid hold up with an RMS deviation of 5.25%.

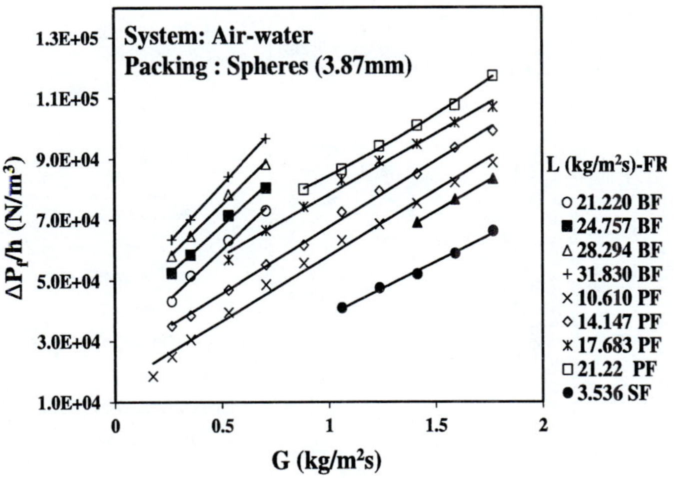

Fig. 2 ΔP_f variation with G and L

4.1 Effect of Gas and Liquid Flow Rates

Figure 2 depicts the variation of ΔP_f with gas and liquid flow rates in all observed flow regimes for spherical particles (3.87 mm) for the A-W system. It is observed that with an increase in flow rates of phases, the ΔP_f increases. The same trend is observed for other G-L systems in the present study (A-G, Air-MEA), in all observed flow regimes, and for all packings (CS, RR, IS). All the data in the literature [1–7] also show that with an increase in flow rates of phases, the ΔP_f increases.

4.2 Effect of Physical Properties of the Fluids

Figure 3 depicts the variation of ΔP_f with the physical properties of the G-L systems for IS in all the three flow regimes. It is observed that the system with high liquid viscosity exhibits a higher ΔP_f than the system with low liquid viscosity [1, 8]. A similar trend is observed in all packings.

Since the ΔP_f for flow through the porous medium is a combination of viscous forces, inertial forces, and interfacial drag, it can be stated that as the viscosity of the liquid increases, the ΔP_f increases. In this work, the influence of liquid surface tension on a ΔP_f could not be established independently. Due to the creation of a large interfacial area in pulse flow and spray flow regimes, its role may not be trivial.

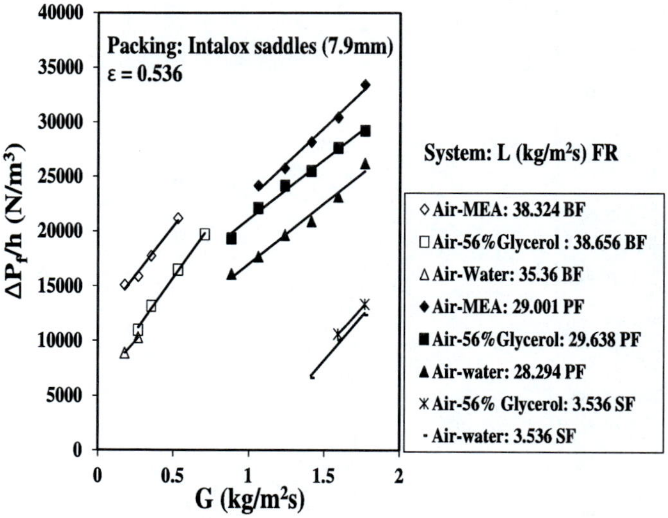

Fig. 3 ΔP_f variation with gas–liquid system

4.3 *Effect of Type of Packing*

Figure 4 depicts the effect of packing characteristics on the ΔP_f in all flow regimes for the A-G system. It is noted that the ΔP_f in the bed of 3.87 mm CS was higher than that of 6.9 mm CS, though the porosity of both packings was almost the same. The same trend is observed for all G-L systems. It indicates that for the particles of the same shape, the ΔP_f increases with decreasing particle size [9]. The present data for CS and IS shows that ΔP_f increases with a decrease in the bed porosity [3].

But the present data of RR does not follow this trend in all the flow regimes and for all the G-L systems. The bed of 6.9 mm RR has a porosity of 0.44 and the beds of 6 and 3.87 mm CS have equal porosity of 0.35. It can be seen from Fig. 4 that RR bed exhibits a higher pressure drop than the 6.9 mm CS bed but less pressure drop than the 3.87 mm CS bed. In one case, RR exhibits the trend of increasing ΔP_f with increasing bed porosity [8], and, in the other case, it exhibits the trend of increasing ΔP_f with decreasing bed porosity [1, 3]. This anomaly of the RR bed with respect to the beds of different sizes of CS may be due to the generation of more interfacial area in the bed of small CS. In a bed of smaller particles, the probability of ascending bubbles hitting the packing and breaking into smaller bubbles is greater than in a bed of larger particles. More interfacial area is generated when larger bubbles break into smaller ones. Additional energy consumed to generate more interfacial area in a bed of smaller particles may be the cause of the higher pressure drop. It may be explained by noting that though the particles used by them are of different shapes, they are almost of the same size. So, it can only be said that the combined effect of particle size, shape, and bed porosity influences the ΔP_f.

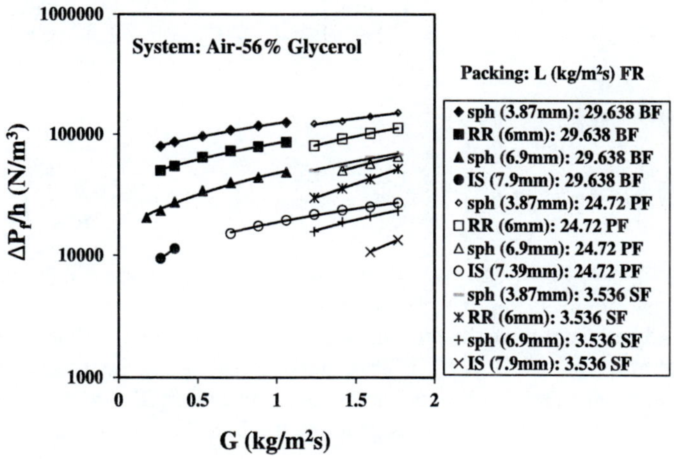

Fig. 4 ΔP_f variation with type of packing

4.4 Correlation of the Experimental Data

In the present work, it is observed that ΔP_f depends on phase flow rates, G-L system physical properties, and packing characteristics [1, 8].

It is attempted to correlate ΔP_f to these variables in terms of dimensionless parameters.

The two-phase flow friction factor f is defined as follows:

$$f = \frac{\Delta P_f}{h} \frac{d_e}{2\rho_g {u_g}^2} \tag{2}$$

The friction factor is correlated with gas and liquid Reynolds numbers, the sphericity of the packing particles, and the porosity of the bed. The Morton number [30] is included to develop the present correlation in spray FR also, which is a gas continuous phase.

Regression analysis of the ΔP_f data obtained using three G-L systems and four different packing gives the following correlations for each identified flow regimes. Flow regimes were identified visually while measuring the pressure drop [28].

Bubble flow (BF)

$$f = 6.82\mathrm{x}10^5 {Re_l}^{0.85} {Re_g}^{-1.596} \left(\frac{d_e}{d_c}\right)^{0.025} \left(\frac{\varepsilon}{1-\varepsilon}\right)^{-2.147} \phi^{-1.51} \left(\frac{g {\mu_l}^4}{\rho_l {\sigma_l}^3}\right)^{0.226} \tag{3}$$

Pulse flow (PF)

$$f = 1.3x10^5 Re_l^{0.49} Re_g^{-1..33} \left(\frac{d_e}{d_c}\right)^{0.041} \left(\frac{\varepsilon}{1-\varepsilon}\right)^{-1.73} \phi^{-1..27} \left(\frac{g\mu_l^4}{\rho_l \sigma_l^3}\right)^{0.135} \quad (4)$$

Spray flow (SF)

$$f = 3.7x10^4 Re_l^{0.166} Re_g^{-1..16} \left(\frac{d_e}{d_c}\right)^{0.192} \left(\frac{\varepsilon}{1-\varepsilon}\right)^{-1..35} \phi^{-1..12} \left(\frac{g\mu_l^4}{\rho_l \sigma_l^3}\right)^{0.044} \quad (5)$$

The appropriate FR for the given conditions can be determined from the equations [28] of transition lines from one FR to other FR.

Figures 5 and 6 compare the present experimental data in BF, PF, and SF regimes and the literature data [1, 3, 7, 16, 17] respectively with predicted values. The match between the experimental values and predicted values is good.

The root mean square deviation is given by the following equation:

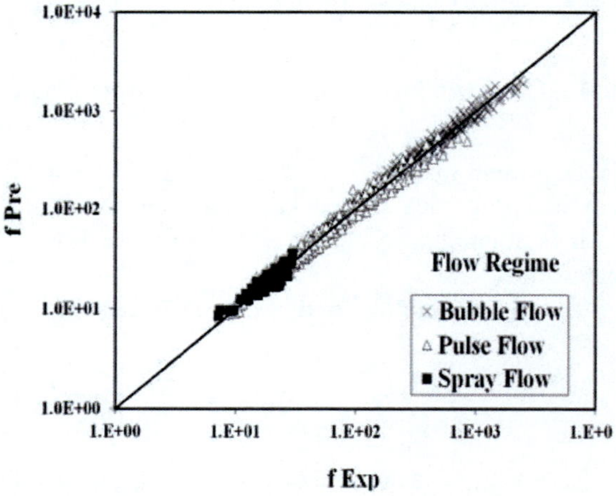

Fig. 5 Comparison of present data with the data predicted

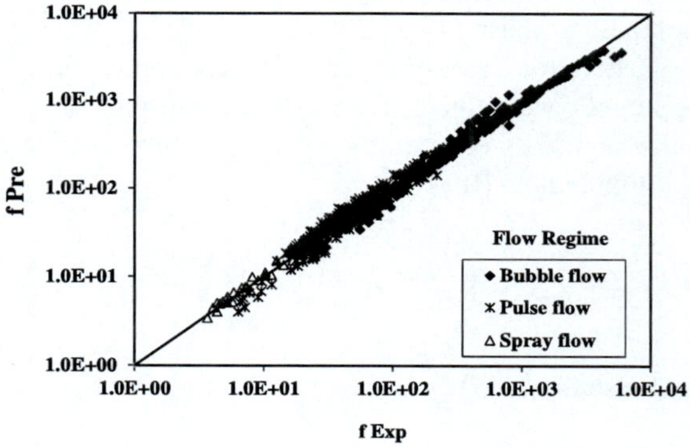

Fig. 6 Comparison of literature data with the data predicted

$$RMS = \sqrt{\frac{1}{n}\sum_{1}^{n}\left(\frac{x_{expn} - x_{pren}}{x_{expn}}\right)^2}$$

5 Conclusions

Two-phase pressure drop was measured experimentally for concurrent upflow of gas and liquid through a packed bed using three G-L systems (A-W, A-G, and air-MEA) and four types of packing (ceramic CS (3.87, 6.9 mm), RR (6 mm), and IS (7.92 mm)). The liquid flow rates ranged from 3.53 to 43.92 kg/(m^2 s) and air flow rates ranged from 0.265 to 1.768 kg/(m^2 s). The fluid flow rates cover the BF, PF, and SF regimes. It is observed that with an increase in gas and liquid flow rates, ΔP_f increases. It is observed that with an increase in liquid viscosity, the ΔP_f increases. It is also observed that ΔP_f was influenced by the combined effect of packing size, shape, and porosity of the bed. From the present experimental data (1150 points) and literature data (770) points, three separate correlations were proposed to predict the ΔP_f in three observed flow regimes in terms of the liquid physical properties, system variables, and operating variables. The correlations proposed predicted the literature and present data with an RMS deviation of 12.8%.

Nomenclature

d_c	Column diameter [m]
d_e	Equivalent particle diameter (2/3)(ϕ d_p) (ε/1-ε) [m]
d_p	Particle diameter [m]
f	Two-phase friction factor –
g	Acceleration due to gravity [m/s^2]
G	Gas flow rate [kg/(m^2 s)]
h	Distance between the pressure tapping [m]
L	Liquid flow rate [kg/(m^2 s)]
MO	Morton number ($g\mu_l^4/\rho_l\sigma_l^3$) –
n	Number of observations
ΔP	Two-phase pressure drop [N/m^2]
Re	Reynolds number –
Rms	Root mean square deviation –
u	Velocity [m/s]

Greek Symbols

ε	Bed porosity –
ϕ	Sphericity –
μ	Viscosity [kg/(m.s)]
ρ	Density [kg/m^3]
ρ_B	Bulk density [kg/m^3]
σ	Surface tension [N/m]

Abbreviations

A-W	Air-water
A-G	Air-56% Glycerol
BF	Bubble flow
CS	Ceramics spheres
Exp	Experimental
FR	Flow regime
G-L	Gas-liquid
Id	Internal diameter
IS	Intalox saddles
MEA	Monoethanolamine
PF	Pulse flow
Pre	Predicted
RR	Raschig Rings
SF	*Spray flow*
Sph	Ceramics spheres

Subscripts

B	Bulk
f	Frictional
g	Gas
l	Liquid
t	Total

References

1. Khan A (1998) Flow pattern of the phases, phase holdup and pressure drop on concurrent gas-liquid upflow through packed beds. PhD thesis, Department of chemical Engineering, IIT Madras, India
2. Turpin JL, Huntington RL (1967) Prediction of pressure drop for two-phase, two component concurrent flow in packed beds. AIChE J 13(6):1196–1202
3. Specchia V, Sicardi S, Gianetto A (1974) Absorption in packed towers with concurrent upward flow. AIChE J 20(4):646–653
4. Sato Y, Hirose T, Ida T (1974) Kagaku Kogaku upward cocurrent gas liquid flow in packed beds, vol 38, pp 534–543
5. Smith D, Reznik A, Pontello A (1979) Pitsburgh energy research centre quarterly reports. In: Shah YT (ed) Personnel communication cited in gas-liquid-soild reactor design, p 231. McGraw-Hill Book, New York
6. Mazzarino I, Sicardi S, Baldi G (1987) Effectiveness in an upflow multiphase reactor. Chem Eng J 36:151–160
7. Goto G, Gaspillo PD (1992) Multiple hydrodynamic states in gas-liquid upflow and downflow through beds of small packings. Ind Eng Chem Res 31:629–632
8. Murugesan T, Sivakumar V (2002) Pressure drop and flow regimes in cocurrent gas-liquid upflow through packed beds. Chem Eng J 88:233–243
9. Saada MY (1975) Fluid mechanics of co-current two phase flow in packed-beds: pressure drop and liquid hold-up studies. Period Polytech Chem Eng 19:317–337
10. Ford LH (1960) Multiphase flow through porous media, with special reference to the turbulent regime. PhD thesis, University of London, UK
11. Sicardi S, Baldi G, Specchia V, Mazzarino I (1984) Hydrodynamics of fluid bed reactors with cocurrent upward flows. Quad Ing Chim Ital 20:66–69
12. Wu Y, Khadilkar MR, Al-Dahhan MH, Dudukovi MP (1996) Comparison of upflow and down-flow two-phase flow packed-bed reactors with the without fines: experimental observations. Ind Eng Chem Res 35:397–405
13. Molga EJ, Westerterp KR (1997) Experimental study of a cocurrent upflow packed bed bubble column reactor: pressure drop, holdup and interfacial area. Chem Eng Process 36:489–495
14. Yang XL, Wild G, Euzen JP (1992) A comparison of hydrodynamics of packed-bed reactors with cocurrent upflow and downflow of gas and liquid. Chem Eng Sci 47:1323–1325
15. Srinivas KV, Chabra RP (1994) Pressure drop in two phase cocurrent upward flow in packed beds: air/non-Newtonian liquid systems. Can J Chem Eng 72:1085–1091
16. Larachi F, Wild G, Laurent A, Midoux N (1994) Influence of gas density on the hydrodynamics of cocurrent gas-liquid upflow fixed bed reactors. Ind Eng Chem Res 33:519–525
17. Iliuta I, Thyrion FC, Muntean O (1996) Hydrodynamic characteristics of two-phase flow through fixed beds: air/Newtonian and non-Newtonian liquids. Chem Eng Sci 51:4987–4995
18. Sindhu H, Sai PST (2003) A model for pressure drop and liquid saturation in gas-liquid co current upflow through packed beds. Chem Eng J 93:135–142
19. Park JH, Park HS, Lee M, Moriyama K (2018) Modelling of pressure drop in two-phase flow of mono-sized spherical. Int J Heat Mass Transfer 127:986–995
20. Schmidt W (2003) Influence of multidimensionality and interfacial friction on the coolability of fragmented corium. University of Stuttgart, p 13
21. Bouteldja H, Hamidipour M, Larachi F (2013) Hydrodynamics of an inclined gas-liquid concurrent upflow packed bed. Chem Eng Sci 102:397–404
22. Li L, Zhang S, Wang K, Wang H (2020) Pressure drops and dryout heat fluxes of packed beds with cylindrical particles. Heat Transf Eng 41:1014–1025
23. Reed AW (1982) PhD thesis, Massachusetts Institute of Technology, USA
24. Schulenberg T, Muller U (1987) An improved model for two phase flow through beds of coarse particles. Int J Multiph Flow 13:87–97

25. Xu L, Yu B, Wang C, Jiang H, Liu Y, Chen R (2022) Particle-resolved CFD simulations of local bubble behaviors in a mini-packed bed with gas–liquid concurrent flow. Chem Eng Sci 41:85–108
26. Motil BJ, Ramé E, Salgi P, Taghavi M, Balakotaiah V (2021) Gas-liquid flows through porous media in microgravity: the international space station packed bed reactor experiment. AIChE J 67
27. Yu K, Wang W, Zhang T, Yong Y, Yang C (2019) Effects of internals on phase holdup and backmixing in a slightlyexpanded-bed reactor with gas–liquid concurrent upflow. Chin J Chem Eng 27:227–2283
28. Raghavendra Rao AV, Kishore Kumar R, Sankarshana T, Khan A (2011) Identification of flow regimes in concurrent gas liquid upflow through packed beds. Chem Eng Technol 34:1909–1917
29. Kumar RK, Rao AR, Sankarshana T, Khan A (2012) Liquid holdup in concurrent gas liquid upflow through packed column with random and corrugated structured packing. In: Proceedings of the world congress on engineering and computer science, vol II
30. Ettiyappan P, Babu J, Arunagiri A, Murugesan T (2006) Prediction of two-phase pressure drop and liquid holdup in co-current gas-liquid downflow of air-newtonian systems through packed beds. J Chem Technol Biotechnol 81:70–81

Numerical Analysis of Fluid Flow and Enhancement of Heat Transfer Over a Square Cylinder with Semicircle Attached at Front and Back Ends

Kunwar Pal Singh Dhania, Prabhat Kumar, Ajoy Kumar Das, and Subrata Kumar Ghosh

Abstract A numerical study of enhanced fluid flow and heat transmission around a square cylinder with semicircles affixed at the front and rear ends was conducted. Both the Reynolds (Re = 100) and Prandtl (Pr = 0.7) numbers are low. By examining both circumstances of a cylindrical square with a semicircle located at the front end and a semicircle linked to the rear end of a square cylinder, a variety of characteristics have been calculated, including the drag coefficient (C_D), lift coefficient (C_L), and Nusselt number (N_u). The square with a semicircle attached at the front end reduced fluid forces and boosted the heat transfer rate when compared to a square with a semicircle added at the back end.

Keywords Square cylinder · Semicircle · Heat transfer enhancement · Steady flow · Fluid flow

1 Introduction

Bluff bodies due to their significant engineering importance in heat exchangers, solar heating systems, natural circulation boilers, nuclear reactors, dry cooling towers, electronic cooling, vortex flow metres and flow dividers, probes and sensors, and a broad variety of other applications are among the products available. For a long time, significant research has been conducted on flow and heat transmission across narrow cylindrical bluff bodies. Under normal conditions, a bluff body is one in which the flow separates from a considerable area of the surface, resulting in a wide wake zone downstream. The interplay of three shear layers, namely a boundary layer, a separating free shear layer, and a wake, all of which have unique or related processes

K. P. S. Dhania (✉) · P. Kumar · A. K. Das · S. K. Ghosh
Department of Mechanical Engineering, NIT Agartala, Tripura 799046, India
e-mail: kpsinghdhania@gmail.com

S. Das et al. (eds.), *Proceedings of the 1st International Conference on Fluid, Thermal and Energy Systems*, Lecture Notes in Mechanical Engineering,
https://doi.org/10.1007/978-981-99-5990-7_40

of creating instabilities as the Reynolds number grows, has presented a substantial fluid mechanics challenge. Such flows are used in a variety of technical applications. In the great majority of inquiries, standard square, round, and triangular cylinders are used.

2 Literature Review and Objective

A square cylinder's corner roundness is quantitatively examined and the heat transmission quantity is greatest for the nano fluid volume fraction at r = 0.51, although fluid forces are least for all volume fractions at r = 0.50 [1]. The inverse and direct proportionality of C_D and C_{Lrms} to the Φ, with maximum and minimum values of r = 0.71 (square) and 0.50 (circle), respectively. Dey and Das [2] according to the research, heat transmission improves gradually and can be enhanced by 14 per cent by employing a new cylinder shape with a corner radius of r = 0.51 and a non-dimensional corner radius of r = 0.50, 0.71 with low Reynolds number, Re = 100, utilising Prandtl number, Pr = 0.01–1,000. Dey and Das [3] an expanded triangular solid (thorn) placed on it, as compared to a square cylinder friction, drag, and lift were decreased by 16 and 46 per cent for Re = 100 and 22 and 60 per cent for Re = 180, respectively (without thorn). When the thorn is positioned at the front stagnation point rather than the back stagnation point, the drag and lift force fluctuation is minimised, and there is a pretty large variety in drag and lift. Investigating an unstable flow via a triangle extended solid coupled to a square cylinder. Bhattachary and Mahapatra [4] when the buoyancy effect is taken into account, the wake's centreline symmetry is lost, and the cylinder experiences a downward lift. When the cylinder is exposed to a horizontal cross-flow, vortex shedding is not suppressed. When the Reynolds number and Grashof number grow, heat transmission from the cylinder increases. Dey and Das [5] the numerical study of an unstable two-dimensional laminar forced convection heat transfer around a square cylinder with rounded corner edges for Re = 80–180 and non-dimensional corner radius, r = 0.50–0.71 at Pr = 0.71 (Air). Gene expression programming (GEP) forecasts heat transmission qualities, and particle swarm optimisation uses the Nuavg explicit equation obtained by GEP to optimise corner radii for maximum heat transfer rate. A novel cylinder design may increase the heat transfer rate of a circular cylinder by 12% with a corner radius of 0.51. Dhiman et al. [6] Reynolds and Prandtl numbers in cross-flow were used to evaluate the heat transport features of an isolated square cylinder under conditions ranging from 1 Re 45–0.7 Pr 4,000. (The highest value of Peclet number being 4,000). As the Reynolds and/or Prandtl numbers grow, the local Nusselt number at each corner of the square cylinder increases. The front cylinder surface has the greatest average Nusselt number, followed by the centre and back. Akhilesh et al. [7] in the unstable two-dimensional periodic flow regime, the influence of the Reynolds and Prandtl numbers on the rate of heat transfer from a square cylinder is numerically examined; $60 \leq$ Re ≤ 160 and $0.7 \leq$ Pr ≤ 50 are the values (the maximum value of Peclet number being 4000). The local Nusselt number on each face of the barrier grows as the Reynolds and Prandtl numbers climb. On the front surface of the cylinder, the surface average Nusselt number has the maximum value. Sen et al. [8] at low Reynolds numbers, the flow through a square cylinder was investigated. In the constantly separated flow regime, the wake is closed and symmetric around the x-axis, and it stays connected to the cylinder's back surface. The separation bubble lengthens linearly as Re increases. Res = 1.15 is determined for the unbounded flow using the linearity of bubble length with Re. Res of 1.16 is predicted from the surface vorticity distribution. Sharma and Eswaran [9] researchers looked at heat and fluid flow across a square cylinder in a two-dimensional linear flow and discovered that the cylinder average Nusselt number climbs monotonically as the Reynolds number increases.

In this study, a semicircle is attached to a square cylinder at both the front and back ends, significantly improving heat transfer and reducing fluid forces.

3 Numerical Analysis

Previously, square cylinders, square cylinders with triangular thorns attached at the front end, and square cylinders with round corners had been studied; even so, in this study, a semicircle is attached to a square cylinder at both the front and back ends, significantly improving heat transfer and reducing fluid forces.

3.1 Geometrical Structure

An arithmetical model is created to investigate the fluid flow. In Fig. 1a and b, a square cylinder with semicircles (r = 0.5D) attached at the front and back ends is positioned inside a rectangle domain. A dimensionless side of the cylinder equals "D." $L_U = 10D$ and $L_D = 40D$ are the lengths of the cylinder's left and right sides, respectively. H represents the vertical space between the top and bottom walls and is set at 20D to simulate a 0.05 obstruction (D/H) (Fig. 2).

3.2 Governing Equations

The suggested numerical solution's equations are based on incompressible laminar flow in a two-dimensional steady flow regime.

Continuity equation:

$$\frac{\partial u}{\partial x} + \frac{\partial v}{\partial y} = 0 \tag{1}$$

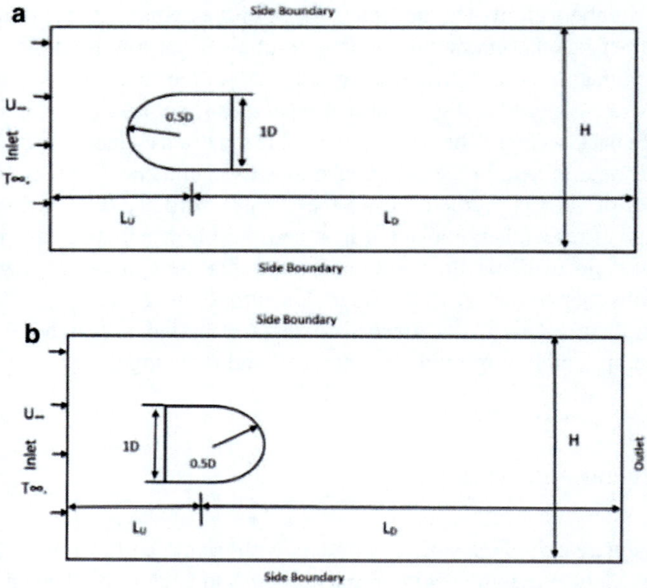

Fig. 1 **a** The computational field is shown graphically: a square cylinder with a semicircle (r = 0.5D) attached at the front end. **b** The computational field is shown graphically; a square cylinder with a semicircle (r = 0.5D) attached at the back end

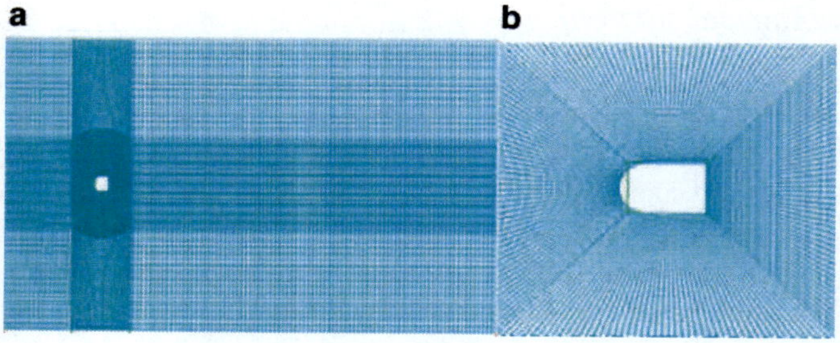

Fig. 2 **a** Computational grid sharing square cylinder with a semicircle (r = 0.5D) attached at the front end; **b** a closer look at the model's grid sharing

X-Momentum equation:

$$\frac{\partial u}{\partial t}+\frac{\partial uu}{\partial x}+\frac{\partial vu}{\partial y}=-\frac{\partial p}{\partial x}+\frac{1}{Re}\left(\frac{\partial^2 u}{\partial x^2}+\frac{\partial^2 u}{\partial y^2}\right) \tag{2}$$

Y-Momentum equation:

$$\frac{\partial v}{\partial t}+\frac{\partial uv}{\partial x}+\frac{\partial vv}{\partial y}=-\frac{\partial p}{\partial y}+\frac{1}{Re}\left(\frac{\partial^2 v}{\partial x^2}+\frac{\partial^2 v}{\partial y^2}\right) \tag{3}$$

Energy Equation:

$$\frac{\partial \theta}{\partial t}+\frac{\partial (u\theta)}{\partial x}+\frac{\partial (v\theta)}{\partial y}=\frac{1}{RePr}\left(\frac{\partial^2 \theta}{\partial x^2}+\frac{\partial^2 \theta}{\partial y^2}\right) \tag{4}$$

The following are some of the non-dimensional variables:

$$u=\frac{u}{U_\infty},\ v=\frac{v}{V_\infty},x=\frac{x}{D},\ y=\frac{y}{D},\ t=\frac{U_\infty t}{D},\ p=\frac{p}{\rho U_\infty^2},\ \theta=\frac{T-T_\infty}{T_{W}-T_\infty} \tag{5}$$

3.3 Grid Independency Test

The numerical simulation used the grid pattern of the current model. When square cylinder with a semicircle (r = 0.5D) attached at the front end is depicted in Fig. 3a. The cylinder part is exposed in more detail in Fig. 3b. Grid structure of the square cylinder with a semicircle (r = 0.5D) attached is shown at the rear end in Fig. 4a and the model itself in Fig. 4b.

The present numerical analysis allocates and displays the structured non-uniform grid over the full computational field in Figs. 3 and 4. Nine sub-blocks make up the entire computational domain. The four sub-blocks connected to the core block are given a better mesh to precisely portray the flow behaviour in both directions. As you go closer to the wall, the finer meshes get coarser. The largest grid size is 0.5D, which is constructed in an irregular pattern from the smallest grid size of 0.07D. To test how mesh influences the incidence of warmth transmission, grids with numbers

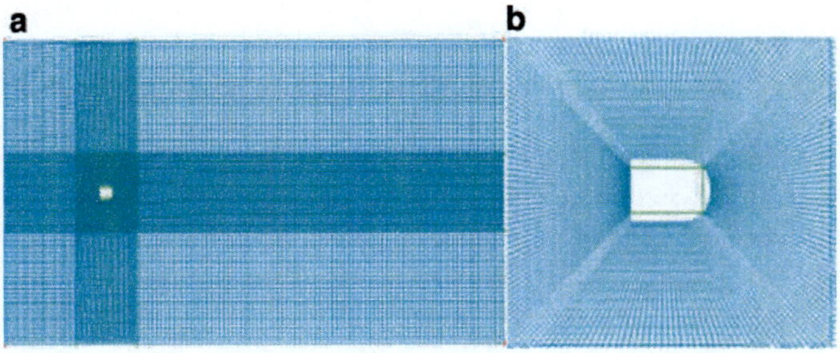

Fig. 3 **a** Computational grid sharing square cylinder with a semicircle (r = 0.5D) attached at the back end; **b** a closer look at the model's grid sharing

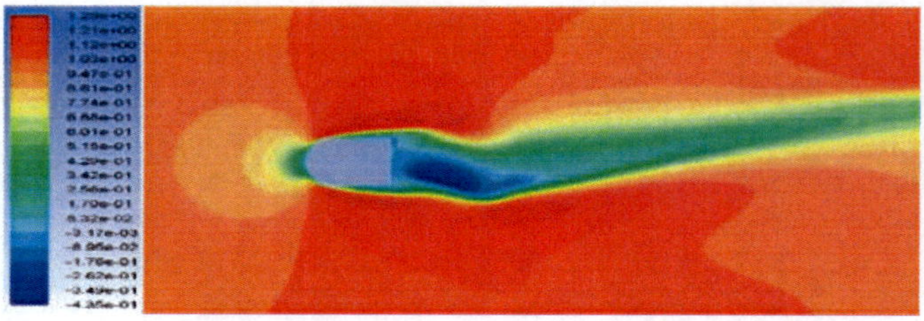

Fig. 4 X-Velocity distribution profile

Table 1 Test for mesh independence

Grid Independence test						
No. of grids	Cdavg	% Change	Clrms	% Change	Nusselt-No	% Change
15,000	1.3284	–	0.0079	–	4.4805	–
25,000	1.3451	1.2436	0.0665	88.1950	4.4098	1.5789
40,000	1.3428	0.1702	0.0100	84.9378	4.3705	0.8906

of 15,000, 25,000, and 40,000 were employed. Between the 25,000 and 40,000 grids, there are minor differences in the results. The findings are summarised in Table 1. It is clear that there are no substantial differences in the outcomes between Grid2 and Grid3. As a result, a 25,000-grid-size grid (Grid2) is found to be sufficient having peak values of Nu number and after which it starts decreasing Dey and Das [10]. The mesh independence and mesh diversity criteria time limit for computation.

3.4 Domain Independency Test

Upstream and downstream domain lengths are utilised to double-check the numerical simulation results. As a result, it's vital to investigate the impact of domain lengths, and a variety of trial and error samples have been calculated to accomplish so as done by Dey and Das [2]. The results are listed in Table 2. With the goal of having optimum domain independence, a rigorous domain independence analysis was completed at downstream distance (Lu) and upstream distance (Lu). Several hit-or-miss combinations of L_U and L_D were examined by taking an initial value of $L_U = 10D$ and $L_D = 40D$, and the outcomes are listed in Table 2. The values have changed very little of C_{Davg}, C_{Lrms}, and Nusselt number for the upstream length ($L_U = 10D$) having peak value of calculated parameters and minimum percentage variation so upper side have been chosen. And for downstream length, there is negligible change in C_{davg}, C_{Lrms}, and Nusselt number so for downstream length ($L_D = 40D$) is selected having peak value of calculated parameters and minimum percentage variation so upper side have been chosen. After evaluating the value of L_U and L_D is selected as 10D and 40D, respectively.

Table 2 Test for domain independence

Domain independence test						
Cdavg		% Change	Clrms	% change	Nu number	% Change
Lu (at Ld = 40D)						
5D	1.3586	–	0.0090	–	4.3697	–
10D	1.3383	1.4988	0.0072	19.7948	4.3926	0.5226
15D	1.3245	1.0331	0.0092	21.3345	4.3849	0.1763
Ld (at Lu = 10D)						
20D	1.3518	–	0.0083	–	4.4859	–
30D	1.3285	1.7226	0.0084	1.4546	4.3822	2.3130
40D	1.3383	0.7277	0.0072	13.7460	4.3926	0.2387

3.5 Boundary Conditions

i. For the "velocity inlet" at the inlet border, a boundary condition is assigned (left face) with free stream velocity, $U\infty$ and temperature $T\infty$, and Neumann boundary condition for pressure is used as $(\partial p/\partial x)$.
ii. At the cylinder's surface, the standard no-slip boundary condition is applied, i.e. $u = 0$, $v = 0$ conditions for pressure ($\nabla p.n = 0$, where n is the unit normal vector).
iii. The normal slip boundary condition is applied to the top and bottom surfaces of the computational domain, i.e. $u = U$, $v = 0$.
iv. The outlet is assigned to the computational domain's extreme right surface. For the "pressure outlet" at the exit, a boundary condition is used.
v. Pressure boundary condition of Dirichlet type, $p = 0$. Additionally, there is a completely formed flow scenario.

4 Validation of Current Numerical Data Using Previous Research

When a semicircle was connected to the front end with a square cylinder, the drag and lift coefficients were lowered, according to the data. And, as indicated in Table 3, when the semicircle is connected to the rear end, the drag coefficient is lowered while the lift coefficient is raised by 2–3 times.

By comparing the data to previously published studies, the Nusselt Number was confirmed and found that when the semicircle ($r = 0.5D$) was attached at the front end, it saw an increase of Nusselt number. The semicircle ($r = 0.5D$) attached at the rear end has seen a minute decrease in the Nusselt number. The Nusselt numbers when analysed numerically at $Re = 100$ and $Pr = 0.7$ are given in Table 4.

Table 3 Validation of drag and lift coefficients with previous study

Source	Square with semicircle attached at front end				Square with semicircle attached at rear end			
	Cdavg	% Change	Clrms	Clrms	Cdavg	% Change	Clrms	Clrms
Present study	0.7249	–	0.1253	–	1.2951	–	0.7183	–
Dey and Das [3]	1.5299	52.6158	0.1698	35.4736	1.5299	18.1288	0.1698	323.0048
Sahu et al. [4]	1.4878	51.2750	0.188	49.9943	1.4878	14.8781	0.188	282.0544
Sharma and Eswaran [5]	1.4936	51.4642	0.1922	53.3453	1.4936	15.3259	0.1922	273.7056

Table 4 Validation of average Nusselt number with previous study

Source	Square with semicircle attached at front end		Square with semicircle attached at back end	
	Avg. Nusselt no	% Change	Avg. Nusselt no	% Change
Present study	5.4283	–	4.3000	–
Dey and Das [3]	3.8443	29.1805	3.8443	11.8551
Sahu et al. [4]	4.0252	34.8581	4.0252	6.8281
Sharma and Eswaran [5]	4.0439	34.2345	4.0439	6.3341

5 Different Profiles When Semicircle (R = 0.50D) is Attached at Front End of a Square Cylinder

(See Figs. 5 and 6).

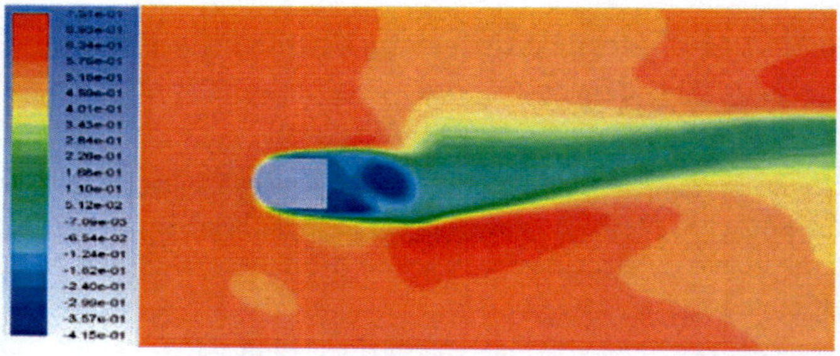

Fig. 5 Pressure distribution profile

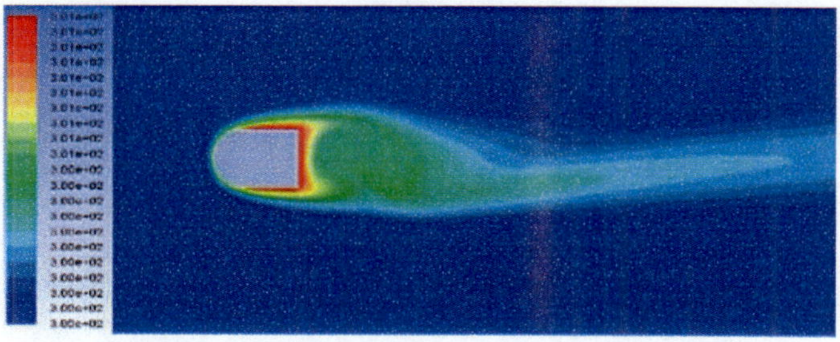

Fig. 6 Temperature distribution

6 Different Profiles When Semicircle (R = 0.50D) is Attached at Back End to a Square Cylinder

(See Figs. 7, 8 and 9).

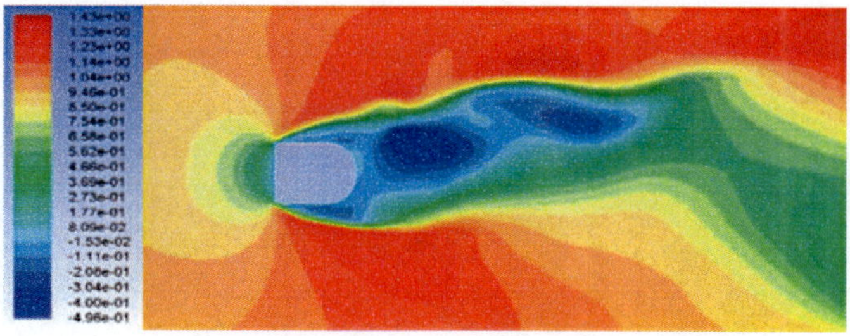

Fig. 7 X-Velocity distribution profile

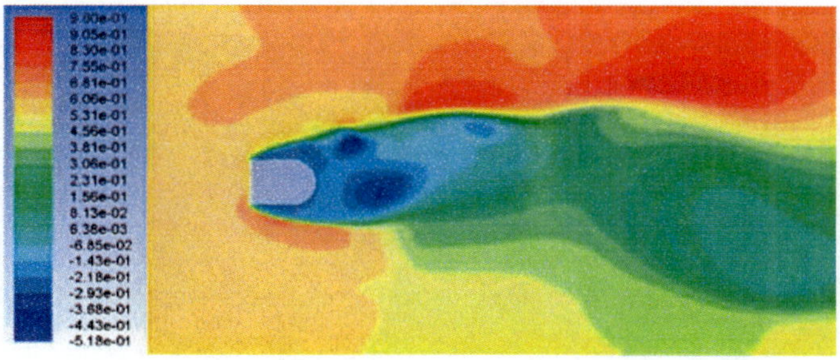

Fig. 8 Pressure distribution profile

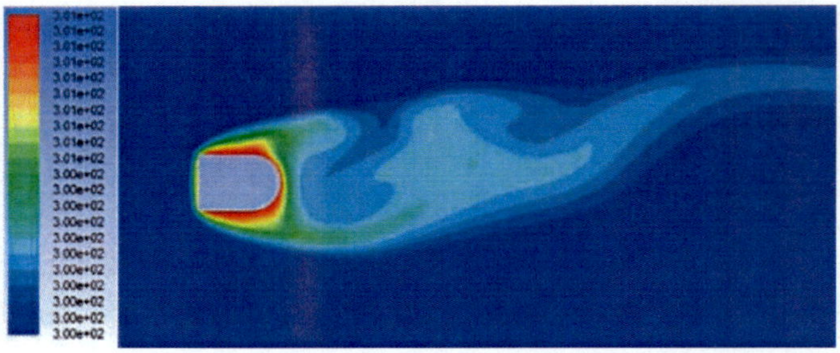

Fig. 9 Temperature distribution profile

7 Different Graphs When Semicircle (R = 0.50D) is Attached at Front End of a Square Cylinder

(See Figs. 10, 11, 12 and 13).

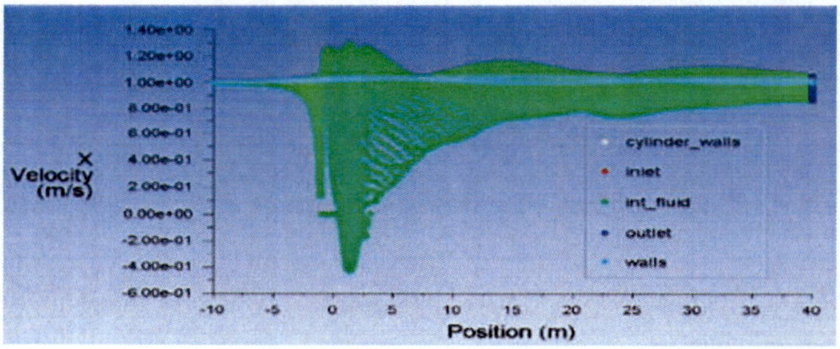

Fig. 10 X-Velocity distribution variation

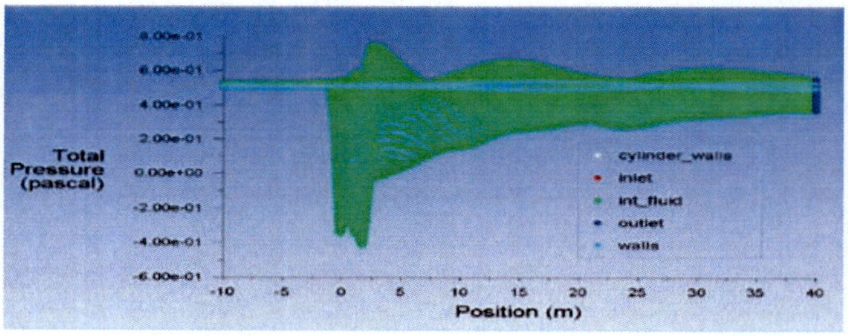

Fig. 11 Pressure distribution variation

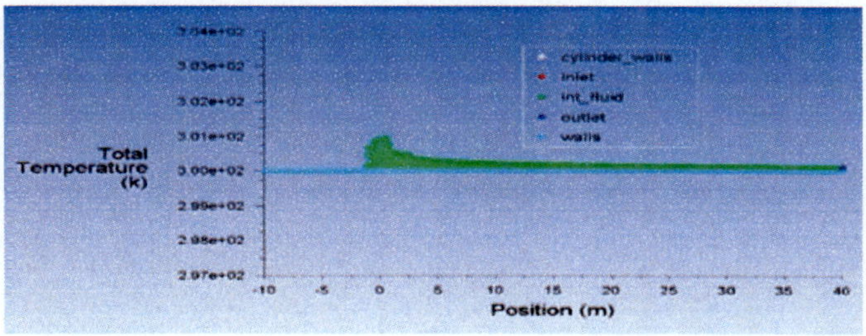

Fig. 12 Temperature distribution variation

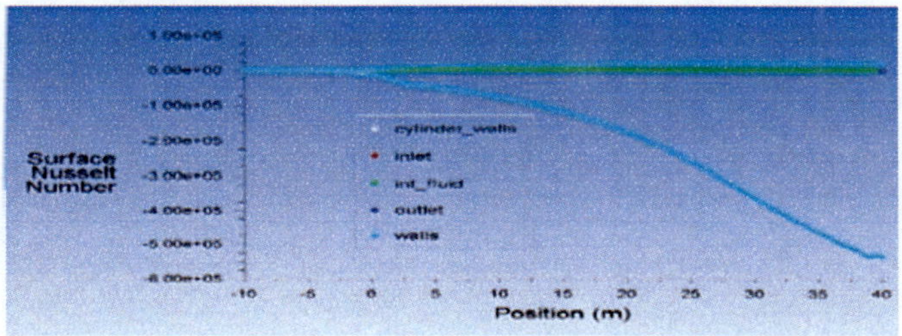

Fig. 13 Nusselt number distribution variation

8 Different Graphs When Semicircle (R = 0.50D) is Attached at Back End of a Square Cylinder

(See Figs. 14, 15, 16 and 17).

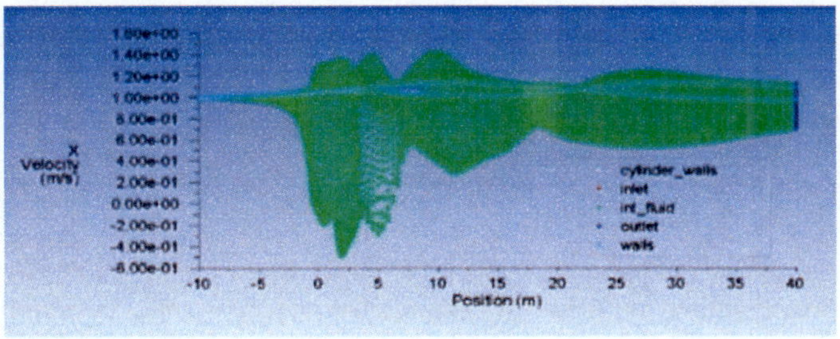

Fig. 14 X-Velocity distribution variation

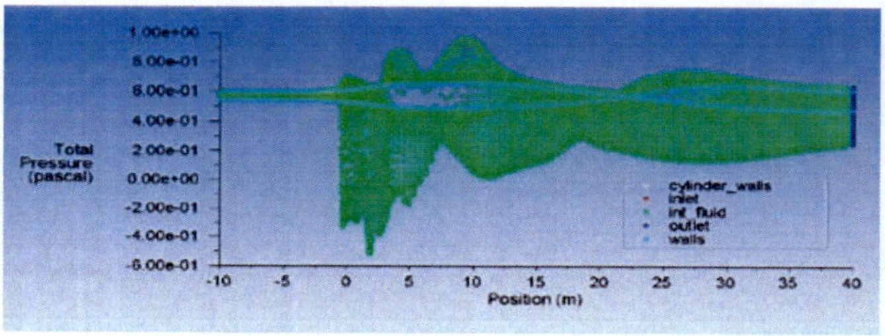

Fig. 15 Pressure distribution variation

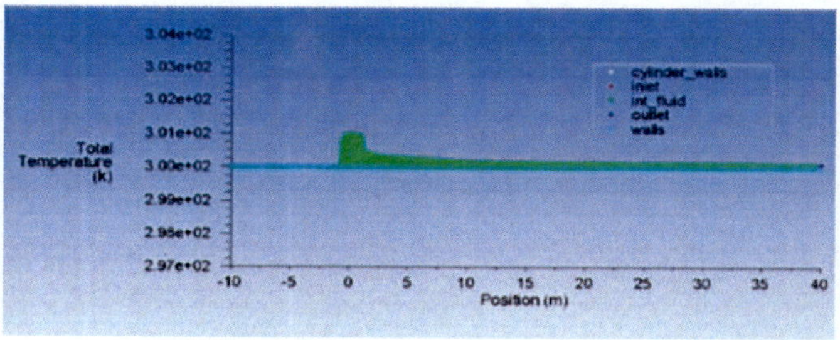

Fig. 16 Temperature distribution variation

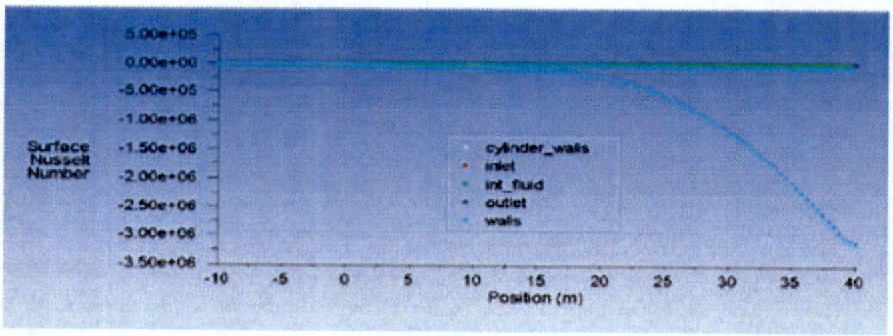

Fig. 17 Nusselt number distribution variation

9 Results and Discussion

In steady flow regimes, the Reynolds number Re = 100; the low Reynolds number has been taken because the present study is based on the steady flow, and 2-D analysis at higher Reynolds number of the flow can be turned into 3-D. Prandtl number (Pr = 0.7) has been utilised to minimise fluid forces throughout the flow. Semicircle of radius 0.50D attached at the front end of a square cylinder and semicircle of radius 0.50D attached at the back end were the two methods employed by using Ansys Fluent software because you'll have more time to experiment and enhance your product's performance using Ansys Fluent. Put your simulation findings in the hands of software that has been well tested across a broad variety of applications. With Ansys Fluent, you can create complex physics models and investigate a variety of fluid phenomena, things in a flexible and user-friendly environment [11]. The drag and lift coefficients have been illustrated.

9.1 When Semicircle is Attached at the Front End, the Results Are as Follows

a. In the present study, drag coefficient, i.e. drag force is reduced to a large extent when compared with the result reported by Dey and Das [3], i.e. 52.61%, when compared with Sahu et al. [7], i.e. 51.27%, and when compared with Sharma and Eswaran [9], i.e. 51.46%.
b. The lift coefficient of the present study is reduced when compared with the result reported by Dey and Das [3], i.e. 35.47%, by Sahu et al. [7], i.e. 49.99% and by Sharma and Eswaran [9], i.e. 53.34%.
c. On comparing both cases to one another, the square cylinder with semicircle r = 0.50D has seen a high reduction in the forces as compared to the square with semicircle R = 0.50D attached to the rear end.

The reasons for decrease in drag and lift coefficients is cause of when the semicircle attached at the front end of a square cylinder the combined geometry becomes the similar shape of aerodynamic cause frontal area has been reduced and flow becomes more smooth and streamlined. Therefore, the coefficient of drag (C_D) and coefficient of lift (C_L) have been reduced. Figure 4 shows the x-velocity variation when the semicircle of radius r = 0.5d is attached at front end, and Fig. 10 shows the variation of x-velocity along the domain length. In both the cases as shown in Figs. 4 and 10, near the cylinder the velocity has increased. As shown in Figs. 5 and 11, velocity gradient and the pressure near the cylinder surface decreased. Therefore, the boundary layer over the surface is more smooth which reduced the drag and lift forces drastically.

9.2 When Semicircle of Radius 0.5D is Attached to the Rear End, the Results Are as Follows

1. The drag coefficient, i.e. drag force is reduced when compared with the result reported by P. Dey and Das [3], i.e. 18.12%, by Sahu et al. [7], i.e. 14.87% and by Sharma and Eswaran [9], i.e. 15.32%.
2. But the lift coefficient, i.e. lift force is increased by 323.0%, 282.05%, 273.70% as compared with the result reported by Dey and Das [3], Sahu et al. [7], and Sharma and Eswaran [9], respectively.

The drag and lift coefficients are reduced because for the semicircle attached at the back end of a square cylinder, the combined geometry becomes a similar shape of aerodynamic in flow in the opposite direction. As shown in Figs. 7 and 14 the x-velocity distributions are shown near the cylinder at the front side. The velocity gradient is more and as the flow passes the cylinder, it decreases so the boundary layer separation takes place and the flow deflects out of the cylinder which increases the coefficient of drag (C_D). On the other hand, due to deflection of flow out of the cylinder the lift forces increase so the coefficient of lift (C_L) increases to a large extent.

When the square with semicircle r = 0.50D is attached at the front end, the Nusselt number increases compared to the result reported by Dey and Das [3], Sahu et al. [7], and Sharma and Eswaran [9] by 29.18%, 34.85%, and 34.23%, respectively.

As we have seen, the two cases: the first is when a square cylinder with semicircle r = 0.50D is attached at the front end and the second is when a square cylinder with semicircle r = 0.50D is attached at the back end. The increase in Nusselt number is when the semicircle r = 0.50D attached at front end and at the back end. Of both cases, the first case has shown the better result, as shown in Figs. 6 and 13. It is due to the semicircle attached at the front end that the flow becomes more streamlined and boundary layer over the cylinder forms so that the total heat generated is taken away by the fluid. As shown in Figs. 9 and 17, Nusselt number variability is throughout the domain length; Fig. 17 depicts this. When the semicircle r = 0.5D is attached at the back end, the flow has deflected out of the cylinder, the boundary layer is separated, and the flow becomes less streamlined near the cylinder surface as it was in the first case so the Nusselt number reduces. The Nusselt number variation as well as domain length are shown in Fig. 17. It has been noted that in both cases, Nusselt number increased as compared to that previously studied as shown in Table 4.

10 Conclusions

At low Reynolds numbers (Re = 100) and Prandtl numbers (Pr = 0.7), a numerical study of fluid flow and heat transfer enhancement around a square cylinder with semicircle r = 0.5D was performed. Several features including drag coefficient (C_{Davg}), lift coefficient (C_L), and Nusselt number have been discovered by analysing both the front and back ends of a square cylinder with semicircle r = 0.50D. When a semicircle was affixed to the front end of a square cylinder, fluid forces were minimised and heat transfer rate was increased:

- Drag coefficient, i.e. drag force is reduced to a large extent when compared with the result reported by Dey and Das [3], i.e. 52.61%.
- The lift coefficient of the present study is reduced when compared with the result reported by Dey and Das [3], i.e. 35.47%.

- The square cylinder with semicircle r = 0.50D has seen a high reduction in the forces as compared to the square with semicircle R = 0.50D attached to the rear end.
- With semicircle R = 0.50D attached to the rear end, the drag coefficient, i.e. drag force is reduced when compared with the result reported by Dey and Das [3], i.e. 18.12%.
- With semicircle R = 0.50D attached to the rear end, the lift coefficient, i.e. lift force is increased by 323.0% as compared with the result reported by Dey and Das [3].
- When the square with semicircle r = 0.50D is attached at the front end, the Nusselt number is increased compared to the result reported by Dey and Das [3] by 29.18%.
- With semicircle R = 0.50D attached to the rear end, the Nusselt number is increased compared to the result reported by Dey and Das [3] by 11.85%.

Nomenclature

C_P	Specific heat (J/kg k)
D	Area of cross-section in the flow direction (m)
h	Coefficient of local convective heat transfer (W/m^2 k)
H	Height of the domain (m)
k	Thermal conductivity (W/m K)
L_d	Distance downstream from the cylinder centre (m)
L_u	Upstream distance from the cylinder centre (m)
N_u	On the cylinder surface, there is a local Nusselt number (dimensionless)
N_{uavg}	Averaged Nusselt number over the cylinder surface (dimensionless)
P_r	Prandtl number (=(μcp)/k)
Re	Reynolds number (=ρ_(fU_∞ D)/μ_f)
μ_f	Fluid Viscosity in a Dynamic State
ρ_f	The fluid density
t	Time (dimensionless)
T	Temperature (K)
U_∞	Free stream velocity (m/s)
x, y	Cartesian coordinate
P_∞	The pressure of a free-flowing stream u, v = Components of velocity in the x and y directions (dimensionless).

Greek Symbols

α	Thermal diffusivity (m^2/s)

μ	Dynamic viscosity (kg/m s)
ρ	Density (kg/m^3)
θ	Dimensionless temperature
υ	Kinematic viscosity (m^2/s).

References

1. Dey P, Das AK (2017) Analysis of fluid flow and heat transfer characteristics over a square cylinder: effect of corner radius and nanofluid volume fraction. Arab J Sci Eng 42:1687–1698
2. Dey P, Das AK (2017) A numerical study on effect of corner radius and Reynolds number on fluid flow over a square cylinder, pp 1155–1165
3. Dey P, Das AK (2015) Steady flow over triangular extended solid attached to square cylinder–a method to reduce drag
4. Bhattacharyya S, Mahapatra S (2005) Vortex shedding around a heated square cylinder under the influence of buoyancy. Heat Mass Transf 41:824–833
5. Dey P, Das AK (2016) A utilization of GEP (gene expression programming) metamodel and PSO (particle swarm optimization) tool to predict and optimize the forced convection around a cylinder. Energy 95:447–458. ISSN 0360-5442
6. Dhiman AK, Chhabra RP, Sharma A, Eswaran V (2006) Effects of Reynolds and Prandtl numbers on heat transfer across a square cylinder in the steady flow regime
7. Sahu AK, Chhabra R, Eswaran V (2009) Effects of Reynolds and Prandtl numbers on heat transfer from a square cylinder in the unsteady flow regime. Int J Heat Mass Transf 52(3):839–850
8. Sen S, Mitta S (2010) Flow past a square cylinder at low Reynolds numbers
9. Sharma A, Eswaran V, Sharma A, Eswaran V (2004) Heat and fluid flow across a square cylinder in the two dimensional laminar flow regime. Numer Heat Transf Part A Appl 45(3):247–269
10. Dey P, Das AK (2016) Prediction and optimization of unsteady forced convection around a rounded cornered square cylinder in the range of Re
11. Li (2015) Comparison between three different CFD software and numerical simulation of an ambulance hall (master of science thesis) EGI-2015-017MSC

A Flow-Visualization Study of an Elevated Jet in Crossflow

Jyoti Gupta and Arun K. Saha

Abstract This paper presents the physics of a jet issuing from a circular stack orthogonal to crossflow. A circular pipe of thickness 2 mm and outer diameter of 12 mm, having an aspect ratio $AR = 9$, is modeled as a stack. The experiments are conducted for a fixed crossflow Reynolds number of $\mathrm{Re}_D = 632$ with a velocity ratio of $R = 0.5$. The quantitative measurements have been taken using Laser Doppler Velocimetry (LDV). Two types of flow-visualization techniques, namely (i) bulk flow visualization and (ii) Laser-Induced Fluorescence (LIF), have been used. The bulk flow visualization gives the volumetric visualization in the projected XY plane, where the clockwise rotating windward jet shear layer morphs into the shape of hairpin vortices in the downstream direction via a stretching mechanism. The LIF visualization in the XZ plane is done by traversing the laser sheet in various span-wise locations. The evolution of flow inside the stack gives rise to the formation of a concentric ring that transforms into counter-rotating vortex pair near the stack free-end. The flow field captured in this plane (XZ) reveals the switching of vortex shedding from an alternate to an arch-type vortex.

Keywords Elevated jet · Stack · Stack-wake · Flow visualization · Vortex shedding

1 Introduction

The study of a jet in crossflow has been widely studied in the past due to large industrial applications. According to the source of emission, the jet can be categorized into two types: jet exiting from a wall opening called a wall-mounted jet and jet exiting in crossflow from a pipe/stack of a certain height called an elevated jet. Wall-mounted jet in crossflow, also popularly known as jet in crossflow (JICF), has been studied

J. Gupta (✉) · A. K. Saha
Department of Mechanical Engineering, Indian Institute of Technology Kanpur, Kanpur 208016, India
e-mail: jyotika@iitk.ac.in

S. Das et al. (eds.), *Proceedings of the 1st International Conference on Fluid, Thermal and Energy Systems*, Lecture Notes in Mechanical Engineering,
https://doi.org/10.1007/978-981-99-5990-7_41

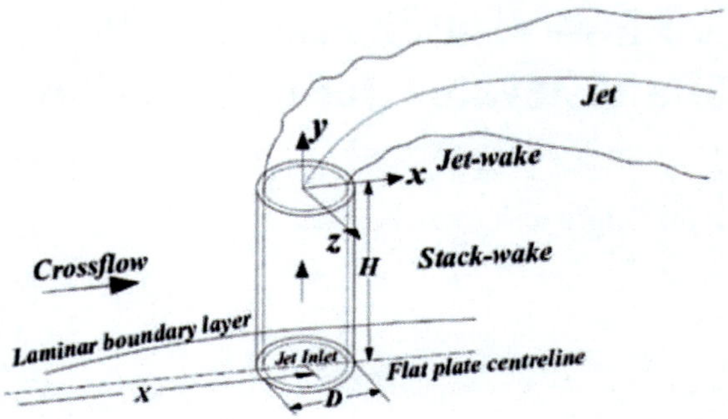

Fig. 1 Definition sketch and co-ordinate system

extensively because of numerous applications encountered in film cooling of turbine blades and compressor, thrust control of a propulsive system, S/VTOL aircraft [1], power plant combustors for optimum mixing ratio [2], and fuel injection in high-speed air-breathing engines [3]. Recently, researchers have been interested in the study of vortex dynamics of an elevated jet in crossflow (EJICF) because of various applications in the environmental field, such as exhaust of polluted air from the stack into the atmosphere, stack flares [2], and wastewater disposal into deep oceans [3]. Apart from the industrial and environmental application, a biological application of JICF has been studied by Delavan et al. [4]. They observed the variation in the excurrent jet velocity of bivalve clams as a function of different environmental factors. The wall-mounted jet is subjected to crossflow, jet, jet-wake, and wall-boundary layer interaction. On the other hand, the stack-issued jet in crossflow involves the interaction among crossflow, jet, jet-wake, and stack-wake behind the stack, as shown in Fig. 1.

The interactions between jet and crossflow have been studied in the past, mainly focused on a wall-mounted jet in crossflow. The complex interaction of the jet with the crossflow leads to the formation of different vortical structures, namely (i) jet shear layer vortices, (ii) horseshoe vortices, (iii) wake vortices, and (iv) counter-rotating vortex pair (CRVP). In an elevated jet, there is an additional wake present in the flow field known as stack-wake. The vorticity contained in these vortical structures has vorticity generated from the wall surface. Therefore, in the present study of EJICF, the only source of vorticity is within the stack or outside the stack surface. The vorticity created near the wall (on which the stack is mounted) has a negligible effect on the main flow as the jet exit is far from the wall. On the other hand, in the case of JICF, the near-wall vorticity affects the jet-wake vortices [5, 6]. Andreopoulos and Rodi [5] observed two pairs of counter-rotating vortex pairs (CRVP), a stronger one in the deflected jet and a weaker one in the wake of the jet. The key observation is that the vortical structures in the jet's wake are not shed by the jet itself but are formed from vorticity, which originates in the wall boundary layer of the crossflow [6]. There are some other authors who advocate that jet can shed some of its vorticity [7, 8]. The vortices formed in jet-core, jet-wake, and stack-wake regions are shed in a downstream direction in either a symmetric or anti-symmetric manner.

The jet shear layer includes leading-edge vortices and lee-side vortices formed at the upstream and downstream of the jet exit, respectively. The formation of shear layer vortices results from a Kelvin–Helmholtz (K–H) instability at the jet shear layer and crossflow interface [6–9]. The cylindrical vortex sheet exiting from the stack is the source vorticity of these two rows of vortices [10]. Fric and Roshko [6] emphasized that vorticity must be formed at the interface of a solid and fluid rather than at the interface of two homogenous fluids of the same density. The present study of EJICF is partly dedicated to confirmation that the only source of jet vortices is from the boundary layer of the stack's inner walls. The shedding of jet fluid in stack-wake, jet-wake, and jet region is also reported in the second part of the study.

2 Experimental Methods

2.1 *Water Tunnel*

The experimental work is carried out in a low-speed closed-circuit water tunnel. All the visualization is undertaken in a water tunnel with a uniform test section of 0.4 m high, 0.54 m wide, and 1.5 m long. A series of mesh screens and honeycomb are installed upstream of the contraction channel to reduce the turbulence in the test section, which is within 3% at the maximum speed of 0.5 m/s. On the test section floor, an acrylic flat plate of 1.3 m in length and 0.36 m in width is kept, with a distance of 70 mm between the upper surfaces of the plate and the floor, which act as a false floor for the elevated jet.

2.2 *Visualization Apparatus*

The flow visualization has been carried out in two phases. In the first phase, the bulk/volume visualization is done in an oblique plane to capture the three-dimensional structures. The visualization apparatus consists of an LED light panel of 24 W (to illuminate the background), a video series light of 150 W (to brighten dye streaks), and a camera of 500 frames per second (shown in Fig. 2). The color ink is used for bulk flow visualization to capture the deflected jet. To make the dye neutrally buoyant, it is first combined with alcohol and then diluted with water. Also, the dye is prepared freshly at room temperature, and the tunnel is refilled with fresh water at the same room temperature to minimize the temperature variation.

In the second phase, the visualization is done by illuminating different planes, parallel to the bottom plate using the laser-induced fluorescence (LIF) technique. A typical LIF visualization system consists of four main components: (i) a solid-state laser source (5 W), (ii) optical lenses to convert the laser beam into a sheet, (iii)

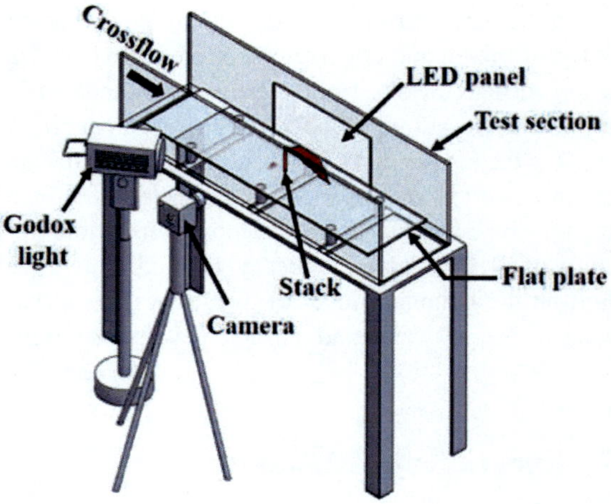

Fig. 2 Setup for bulk flow visualization

sodium fluorescence dye, and (iv) a camera. A beam compressor is used to compress the diameter of the laser beam to get a thin laser sheet of 0.5 mm.

2.3 Velocity Measurement

Laser Doppler Velocimetry has been used for measuring the flow velocity at selected locations. In the present study, two-dimensional preliminary flow measurements are performed in order to characterize the flow. The focal length of the two laser beams is 500 mm with wavelengths of 532 and 485 nm. The laser beam has a diameter of 1.8 mm and a nominal power of 2.5 W. The silver-coated hollow glass spheres with a mean diameter of 12 μm have been used as the seeding particles. Approximately 3000 samples are taken at the jet exit, whereas a low data rate is found during the measurement of crossflow velocity due to the lower number of seeding particles reaching the measurement volume. LDV measurements at the jet exit ($y = 0.042D$) are carried out at a predetermined flow rate with no crossflow to ascertain definite velocity profiles. Figure 3 shows the centerline jet velocity profile to be nearly fully developed parabolic.

2.4 Test Model

The test model or stack is constructed from a perspex tube of a wall thickness of 2.0 mm with an outer diameter of 12.0 mm and aspect ratio (AR = H/D) of 9.0, where H is the height and D is the outer diameter of the stack. The free end of the stack is sufficiently below ($16D$) the free surface to reduce its interaction with the

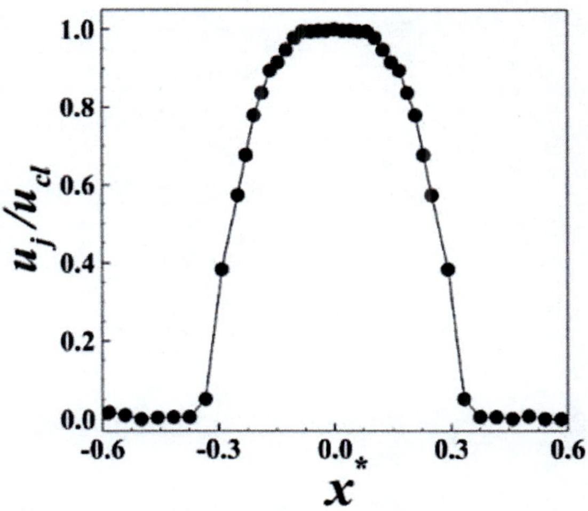

Fig. 3 LDV measurement of mean centerline jet exit velocity profile

jet. The flow setup from axisymmetric pipe jet (stack) includes (i) false floor used to mount the stack (described earlier), (ii) settling chamber (to provide eddy free supply to jet), (iii) flow meter (to control the jet flow rate), (iv) pump, and (v) header tank (a separate water reservoir for jet flow). A few small holes of 0.9 mm diameter are made on the circumference of the stack at a distance of 1.4*D* below the false floor, for allowing the dye to be released slowly under gravity onto the inner surface of the stack to mark the jet shear layer, as shown in Fig. 4. For smooth operation of pump, a separate constant voltage transformer (CVT) has been used.

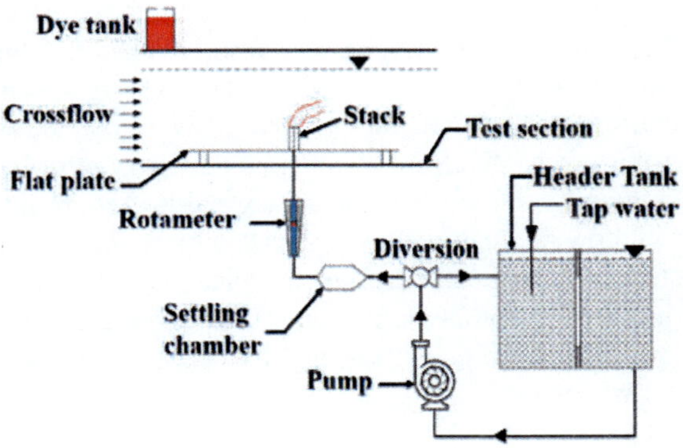

Fig. 4 Setup for jet flow arrangement

Table 1 List of parameters

Parameters	Value
Free stream velocity (m/s)	0.047
Boundary layer thickness (mm)	15
Shape factor	2.1
Reynolds number (Re_X)	3.2×10^4
Flow type	Laminar
Stack diameter (mm)	12
Reynolds number (Re_D)	632
Velocity ratio (R)	0.5

2.5 Flow and Geometrical Parameters

In the present study, the stack is partially immersed in a laminar boundary layer, as shown in Fig. 1. The characteristics of the boundary layer are measured, using LDV, and is shown in Table 1. The same fluid (water) was used for both the jet and the crossflow in this study. Therefore, the jet to crossflow momentum flux is simplified as velocity ratio R, which is defined as $R = \overline{u}_j/u_\infty$, where $\overline{u}_j$ is the mean jet velocity at the exit of the stack and u_∞ is the freestream velocity. The Reynolds number, based on freestream velocity and outer diameter of the stack (D), is $Re_D = u_\infty D/\nu$, where ν is the kinematic viscosity of water.

3 Results and Discussion

The study of a jet in crossflow results in a three-dimensional complex flow field with several vortices. These vortices change significantly when the jet and crossflow momentum flux alter and are captured by the bulk flow-visualization techniques. The dye is discharged from the internal periphery of the stack in this investigation via a quiescent reservoir constructed around it. As a result, the dye should be supposed to follow the fluid path and precisely mark the vorticity within the stack. Since the stack is transparent, therefore, the relationship between the shear layer's vorticity and the vorticity within the stack may be provided. As the crossflow fluid is not marked upstream of the pipe, therefore, the vorticity generated upstream of the stack tip is not witnessed.

3.1 Bulk Flow Visualization

The jet shear layer of the emerging jet has been captured at a fixed Reynolds number $Re_D = 632$ and velocity ratio $R = 0.5$. The bulk flow visualization has been carried

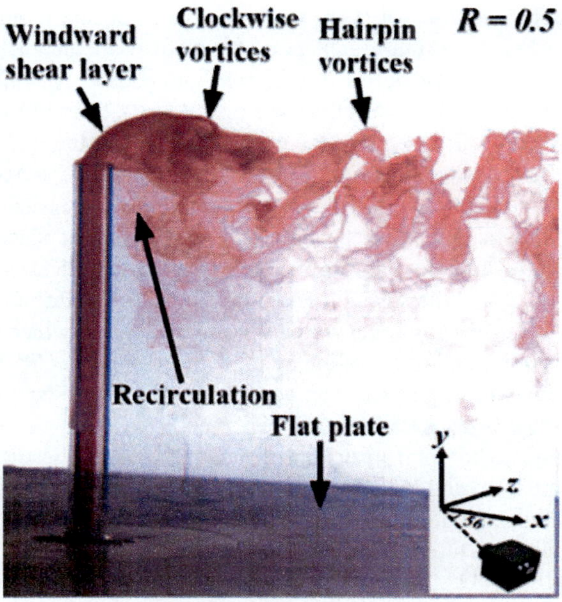

Fig. 5 Streak image of clockwise downwash vortices at $R = 0.5$ and $Re_D = 632$, captured in a plane at an angle of 56° from the XY plane

out at an angle of 56° from the vertical XY plane, shown in Fig. 5. In the present study, the crossflow velocity is greater than the jet velocity as $R < 1.0$; therefore, the jet bends and spreads quickly. The perspective view of the flow described that the mass of fluid ejecting from the jet is entrained into stack-wake and further advanced in the downstream direction of the stack. Initially, at the jet exit, no vortices appear on the windward side of the jet shear layer. At some downstream distance from the jet exit, the windward shear layer rotates in the clockwise direction. On the other hand, the leeside of the shear layer engulfed into the low-pressure region of the stack-wake. While moving downstream, the lateral parts of these shear layer vortices got stretched into pairs of the thin filament, connected at the upper end, referred to as the head of the shear layer. Therefore, the clockwise rotating jet shear layer morphs into the shape of hairpin vortices in the downstream direction via a stretching mechanism. The jet's oblique view (with respect to the streamwise direction) clearly brings out both the side legs of hairpin vortices. The bottoms of these hairpin structures are initially inclined toward the stack. On moving further downstream, they slowly rotate toward a vertical orientation. Behind the stack, a relatively large longitudinal recirculating bubble forms.

3.2 *LIF Visualization*

LIF flow visualization has been carried out to study the structures at different XZ planes in a spanwise direction. In the bottom parallel plane (XZ) at the spanwise location $y^* = -1.0$, the vortex lines are concentric rings within the stack, as shown in Fig. 6a. As the laser plane is traversed toward the stack-free end, i.e., at $y^* = -0.66$, the concentric vortex ring morphs into counter-rotating vortex pair (Fig. 6b), which is more prominent at the stack tip ($y^* = 0$), clear from Fig. 6c. This counter-rotating vortex pair resembles the core of the jet. As the laser plane is further traversed in the spanwise direction ($y^* = 0.33$), the jet-core shifts from the center of the stack (Fig. 6d), as well as the counter-rotating vortex pair becomes weaker in this plane which signifies the bending and realignment of vorticity from existing XZ plane. In the three folding processes of a cylindrical sheet, it is elaborated that the folding process at the sides of the jet column gives contribution to the formation of CRVP [10]. The cylindrical vortex sheet at the jet exit, which contains circular vortex lines, is prevented from rolling up into vortex rings by CRVP. The present study is also in line with the above statement. To clearly understand the folding process of the jet shear layer, Kelso et al. [9] performed an experiment at Re = 940 and $R = 2.3$. They injected dye far upstream of the wall-mounted jet in the crossflow that surrounds the front and sides of the jet, named as hovering vortex. The hovering vortices and associated folding process together appear to constitute the hanging vortex (a tube-like structure formed at the lateral sides of the jet) and the eventual evolution of the CRVP [11].

Some authors believe that a jet can shed vorticity in its wake, while others believe it cannot shed vorticity like a solid cylinder. The present study of EJICF investigates the possibility of the existence of jet vortices in both the jet and stack-wake. Since the jet fluid is only illuminated in the laser plane, therefore, the image captured in any of the regions viz, jet-core, jet-wake, and stack-wake will show the shedding of jet vortices. The flow in the wake region is unsteady; therefore, flow patterns differ slightly from image to image. Figures 7, 8, and 9 show the flow patterns in

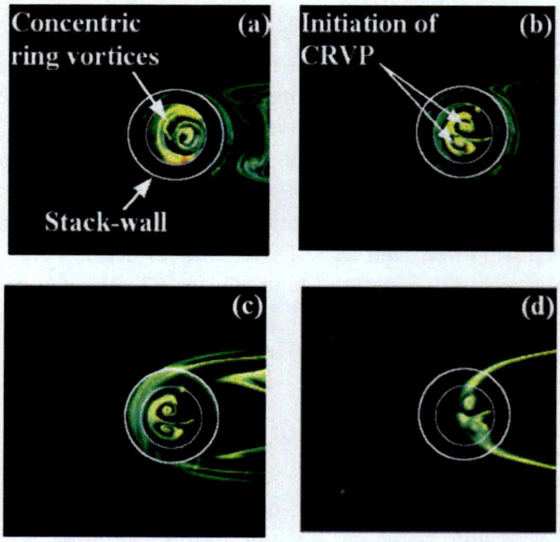

Fig. 6 A sequence of photographs showing the vorticity inside the stack and at the jet exit for $Re_D = 632$, $R = 0.5$ and spanwise locations **a** $y^* = -1.0$, **b** $y^* = -0.66$, **c** $y^* = 0$, and **d** $y^* = 0.33$

the XZ plane covering the jet-wake, jet-core, and stack-wake regions, respectively. From Fig. 7, the dye streaks are predominantly in the jet-wake region, covering the spanwise location $0 < y^* \leq 0.67$. At $y^* = 0$, the shed jet fluid follows the separation event from the lateral stack walls, while a little dye present in the immediate wake of the stack tip represents the recirculation of jet fluid from the stack-wake region. After the recirculation region up to $x^* = 2.8$, the jet fluid is shed in alternate nature, as shown in Fig. 7a. The solid curved lines indicate their directions of rotation. While moving the laser plane in the jet-wake region (Fig. 7b and c), the very near-wake region of the jet is open, i.e., no dye is present and bounded by the streak lines shedding from the jet fluid in its wake. The near-wake region shows the symmetric nature of the shedding, while in the far-wake, the shedding is of alternate nature. These vortices extract their vorticity from the shear layer that surrounds the jet as it issued from the stack. The width of jet-wake (Δz^*) for these two cases is 1.4 and 1.2 respectively at downstream location $x^* = 1.8$, which is decreasing in nature while traversing from stack tip to jet region. From Fig. 8a and b, while traversing the laser plane $0.67 < y^* \leq 1.0$, the very near-wake region of the jet is closed, i.e., dye is present, which resembles the jet-core region. Since the jet fluid is entrained in the stack-wake region, therefore, in the XZ plane at $y^* = 0.83$ and 1.0, a small portion of the jet region persists in the downstream direction. The shedding of jet-core fluid is symmetric in nature.

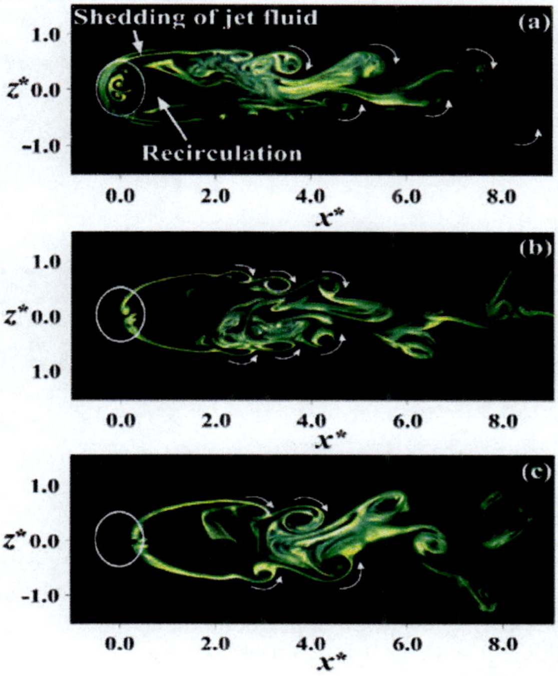

Fig. 7 Cross-sectional view of near flow field of jet-wake region at $Re_D = 632$, $R = 0.5$, and **a** $y^* = 0$, **b** $y^* = 0.42$, and **c** $y^* = 0.67$

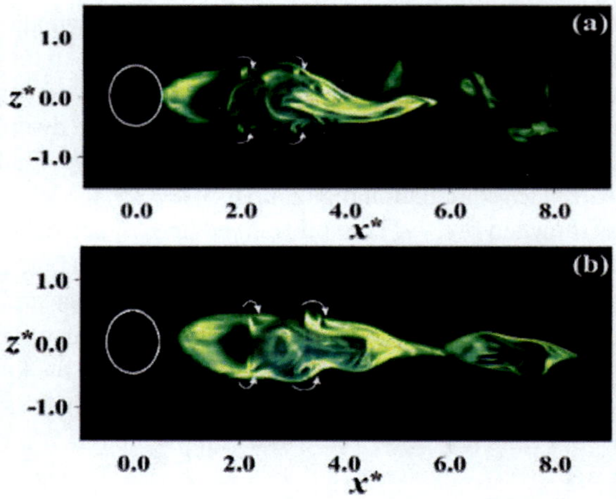

Fig. 8 Cross-sectional view of near flow field of jet-core region at $\mathrm{Re}_D = 632$, $R = 0.5$, and **a** $y^* = 0.83$ and **b** $y^* = 1.0$

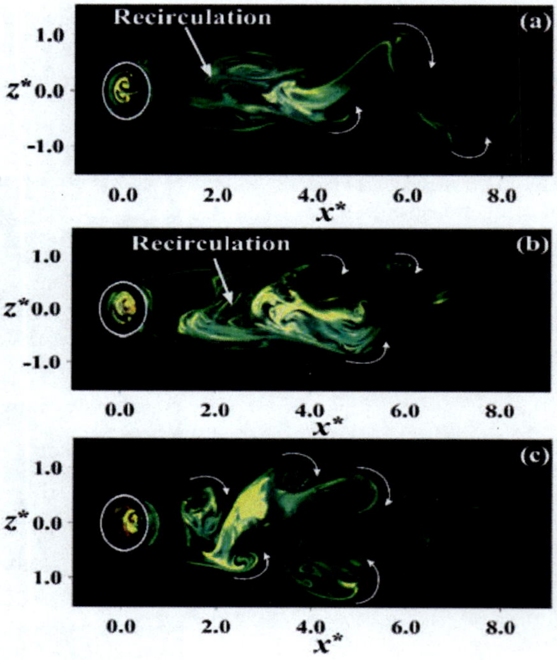

Fig. 9 Cross-sectional view of near flow field of stack-wake region at $\mathrm{Re}_D = 632$, $R = 0.5$, and **a** $y^* = -0.67$, **b** $y^* = -1.3$, and **c** $y^* = -3.0$

The bulk flow-visualization image (Fig. 5) clearly shows that the jet fluid is entrained in the stack-wake region and forms the recirculation region behind the stack. The downwash of the jet fluid into the stack-wake marks the vortices shed in this region. Figure 9 shows another separation event within 9.0 stack diameters of the stack-wake. At spanwise location $y^* = -0.67$, a recirculation bubble (in the immediate stack-wake region) is observed that lasts up to downstream distance $x^* = 3.2$.

Beyond this distance, the fluid is shed in an alternate sense, shown in Fig. 9a, where the three patches of dye streak near the wake centerline are observed as footprints of stack-wake vortices. Traversing the laser plane down in the stack-wake region at $y^* = -1.3$, the downstream distance of the recirculation bubble decreases up to $x^* = 2.7$, and the fluid is shed in an alternate sense like before. This is so because the jet fluid engulfed into stack-wake moves toward the stack outer rear wall, clear from Fig. 5. Traversing down further in a spanwise direction, the size of the recirculation zone becomes negligible, and the jet fluid is shed in an alternate sense, as shown in Fig. 9c. The size of the alternate vortices at this plane is more prominent than in the earlier cases shown in Fig. 9a and b, which signifies that on moving down in the stack-wake region, it becomes stronger in nature.

4 Conclusions

The current research is based entirely on the qualitative examination of flow structures utilizing various visualization techniques. The three-dimensional clockwise rotating jet shear layer structure is captured via bulk flow visualization that morphs into a hairpin structure downstream via stretching mechanism. The study uses LIF visualization to describe the connection between the shear layer's vorticity and the vorticity within the circular stack of finite height. The vortex line at the stack wall changes from concentric rings to counter-rotating vortex pairs near the stack-free end. The vortex shedding from the stack, jet-wake, and the jet-core region has been described. At the specified region in the XZ plane, two types of vortex-street are formed namely, arch-type (in-phase) shedding and alternate (out-of-phase) shedding. The former one appears in the jet-wake and the jet-core region while the latter occurs near the stack's free end and in the stack-wake region.

Acknowledgements The authors gratefully acknowledge the financial support received from FIST program of Department of Science and Technology, Government of India, in developing the facility. Also, the financial assistance received from SERB, Government of India, for the execution of this work is gratefully acknowledged.

Nomenclature

D	Outer diameter [mm]
H	Free stream velocity [m/s]
u_∞	Free stream velocity [m/s]
u_j	Centerline mean jet velocity [m/s]
u_{cl}	Mean peak centerline jet velocity [m/s]
$\overline{u}_j$	Mean jet velocity [m/s]
ν	Kinematic viscosity of water [m^2/s]

x	Streamwise co-ordinate [m]
x^*	Non-dimensional streamwise co-ordinate, x/D –
y^*	Non-dimensional spanwise co-ordinate, y/D –
z^*	Non-dimensional transverse co-ordinate, z/D –
R	Velocity ratio, $\overline{u}_j/u_\infty$ –
AR	Aspect ratio, H/D –
Re_X	Local Reynolds number, $u_\infty x/\nu$ –
Re_D	Reynolds number, $u_\infty D/\nu$ –
Re_j	Jet Reynolds number, $\overline{u}_j D/\nu$ –
Δz^*	Non-dimensional jet-wake width –

References

1. Karagozian AR (2010) Transverse jets and their control. Prog Energy Combust Sci 36:531–553
2. Andreopolous J (1982) Measurements in a jet-pipe flow issuing perpendicularly into a cross stream. J Fluid Eng 104:493–499
3. Koh RCY, Brooks NH (2005) Fluid mechanics of waste water disposal in the ocean. Annu Rev Fluid Mech 187–211
4. Delavan SK, Webster DR (2011) Unsteadiness of Bivalve Clam Jet Flow According to Environmental Conditions. Aquat Biol 13:175–191
5. Andreopolous J, Rodi W (1984) Experimental investigation of jets in a crossflow. J Fluid Mech 138:93–127
6. Fric T, Roshko A (1994) Vortical structure in the wake of a transverse jet. J Fluid Mech 279:1–47
7. Moussa Z, Trischka JW, Eskinazi S (1977) The near field in the mixing of a round jet with a cross-stream. J Fluid Mech 80(1):49–80
8. Smith S, Lozano A, Mungal M, Hanson R (1933) Scalar mixing in the subsonic jet in crossflow. Comput Exp Assess Jets Crossflow 13:28003 07–34
9. Kelso RM, Lim TT, Perry A (1996) An experimental study of round jets in crossflow. J Fluid Mech 306:11–144
10. Lim TT, New TH, Luo S (2001) On the development of large-scale structure of a jet normal to a crossflow. Phys Fluids 13(3):770–775
11. Yuan LL, Street RL, Ferziger JH (1999) Large-eddy simulations of a round jet in crossflow. J Fluid Mech 379:71–104

Numerical Study on the Aerodynamic Characteristics of Supersonic and Hypersonic Missiles

Parv Kumar Goyal and Jayahar Sivasubramanian

Abstract Aerodynamics plays an important factor in the design of a missile. In this paper, a study of the aerodynamic characteristics has been carried out for a 2D missile geometry using CFD simulations. The missile chosen for the analysis was the first modern ballistic missile, V2 Rocket. Parametric study at supersonic and hypersonic flow regimes at 0° and 5° angle of attack was carried out. Two different types of computational grids were generated with one of them resolving the boundary layer. The grid and computational domain were generated using the ANSYS Workbench environment. The numerical simulation was carried out using ANSYS Fluent.

Keywords Hypersonic · Aerodynamics · Missiles · CFD · ANSYS fluent

1 Introduction

Earlier, the only way to study aerodynamics was through experimental setups and theoretical approximations. Now, with the development in science and technology, in particular with the advent of computers, we can study the aerodynamic features for a whole flight vehicle using Computational Fluid Dynamics (CFD). CFD is an important and popular method to study and understand aerodynamic characteristics through numerical simulations. The study of aerodynamics has an important role in the development of the missile technology. The aim of this paper is to study the aerodynamic characteristics of a missile geometry using CFD simulations.

Missiles can be defined as a self-propelled guided air weapon, capable of carrying a payload with explosives such as warheads, and can cause mass destruction [11]. Missiles play a crucial role in the warfare and are very important for the security of a country. The first modern cruise missile, 'V-1 Rocket', was built by the Germans in

P. K. Goyal (✉) · J. Sivasubramanian
Department of Aerospace Engineering, Ramaiah University of Applied Sciences, Bangalore 560058, India
e-mail: parvgoyal@gmail.com

S. Das et al. (eds.), *Proceedings of the 1st International Conference on Fluid, Thermal and Energy Systems*, Lecture Notes in Mechanical Engineering,
https://doi.org/10.1007/978-981-99-5990-7_42

1944 during World War II. Also, the first ballistic missile 'V-2 Rocket' was developed and operated by the Germans in 1944 during the war [8].

The development of V2 missile started a new era of modern warfare, where science and technology was required to design and build such weapons. The V2 had played a significant role during WW2. The first operational V2 Rocket was launched against London in September of 1944 and over the next year, around 3600 V2 weapons were used [12].

The aerodynamic characteristics can affect the design, altitude, usage and the range of a missile. Through numerical simulations, aerodynamic properties such as C_L and C_D can be obtained for the missile geometry. However, there is very limited data available on the research of missiles in the open domain.

Many previous researchers including Meng et al. [1], Ishak et al. [2], Gaonkar et al. [3], Dol [4], Pandie and Prabantara [5], Divakaran et al. [6] and Wee [7] have numerically investigated the aerodynamics of the missile using analytical, numerical and wind tunnel test methods.

In addition to playing a major role in any country's defence strategy, modern-day missiles are also a relevant geometry for academic research, as students can understand and develop strong fundamental aerospace engineering knowledge by studying them. The aim of this study is to predict the aerodynamic characteristics of the first ballistic missile. As the CFD techniques have become a popular tool for understanding the flow features for such geometry, ANSYS Workbench environment will be used for the study. ANSYS Workbench provides a good environment for meshing and creating the computational domain for the simulations.

2 Literature Review and Objective

Gaonkar et al. [3] performed numerical simulation over a missile in supersonic regime, and the study was carried out in order to optimize the C_L/C_D ratio by changing the angle of attack (AoA) and keeping the Mach number (M) same using ANSYS Fluent; the study found that the AoA giving the maximum C_L is 12° with a C_L as 2.15058 and with a maximum C_L/C_D ratio of 2.5. The results were validated with the analysis reported in the literature; the observation was made that the C_L and C_D values are similar with an error percentage of 4% to 5%. It was observed that by increasing AoA from 0° to 12°, there was a rise in the value of C_L which interprets that the turbulence of the flow increases with AoA. At the exit of the missile, the maximum velocity was observed with increasing AoA from 0° to 12°.

Dol [4] conducted a numerical study using ANSYS Fluent for different fins at various AoA of 0°, 20° and 40° at a velocity of 150 m/s. From the results, it can be observed that there is a relation between the drag coefficient and the dimensionless parameters of the fins. Experimental verification of the results have been done using a wind tunnel and re-scaling a single fin.

Meng et al. [1] numerically studied aerodynamics of a missile and concluded that as strong shock waves can be observed in supersonic flow, the air velocity may decrease while the warhead pressure may increase. There is a decrease in C_D as there is an increase in free stream Mach number from 2 to 4.5, and the strength of the shock wave increases correspondingly. It can be observed that the strength of the shock wave has been increased as there is a decrease in the radius of curvature of head, and with increase in the slenderness ratio, the C_D shows a gradual trend of decrease. The stagnation point position may change.

with change in the AoA. Strong compressed air around the warhead is observed. Hence, the missile temperature is affected by shock waves and also from the viscous forces, and a vortex was observed in the rear end of the missile.

Ishak et al. [2] studied the aerodynamic properties of a curved fin rocket. The method used for this study was a semi-empirical method with numerical simulation. The USAF DATCOM tool used a semi-empirical method for reference. For the numerical technique, the ANSYS Fluent environment was chosen. The configuration of the rocket comprises a cylindrical body, conical nose and four curved fins. The study was conducted at various Mach numbers of 0.15, 0.4, 0.6 and 0.8 for subsonic flow. For supersonic flow, the Mach numbers were 1.2, 1.4 and 2.0. The AoA changes from 0° to 25° for every velocity, with 5° increments. The results were compared using the wind tunnel test data, USAF DATCOM data, numerical simulation data and the data from the literature. It can be observed from the results that every method shows a similar rocket aerodynamic characteristic trends.

Pandie and Prabantara [5] performed simulations to understand the aerodynamic features around the missile body when simulated using the ANSYS environment which shows various characteristics of pressure and velocity with change in AoA given. It can be seen that the change in the AoA is directly proportional to the pressure value. The velocity value varies relative to each variation in AoA.

Divakaran et al. [6] performed numerical simulations on a ballistic missile. Pressure far-field boundary conditions were used for the simulation using ANSYS environment to predict the performance and the aerodynamic features of the missile in a 2D plane. The study has been conducted at various Mach numbers ranging from 2–3.5 at various angles of attack changing from 0 to 16°. It can be observed that if the AoA increases then the air density over the missile is not affected, and the coefficient of pressure decreases with an increase in temperature.

It can be observed in the literature that only limited research data are available for geometries related to missiles. Also, there is a need for research, especially in the high-speed regime, to understand the aerodynamics of the missile. A better understanding of the aerodynamics characteristics will help identify the criteria for optimizing the geometry for high-speed applications in future research. Furthermore, for the first modern ballistic missile 'V2 Rocket', aerodynamic studies can lead to better understanding of the flow physics, in particular at high Mach numbers.

The objective of this work is to study the aerodynamic characteristics of a missile geometry using Computational Fluid Dynamics.

3 Methodology

The aerodynamic analysis was performed on the German V2 missile. 2D model of the missile was built using the CATIA V5 software (Fig. 1). The dimensions of the missile geometry were taken from the literature [8]. The domain and meshing were constructed in the ANSYS Workbench environment (DesignModeler). The study was conducted using the Finite Volume Method in ANSYS Fluent. A parametric study was conducted for two flow regimes at supersonic and hypersonic Mach numbers. A C-Domain with computational grid was used for the numerical simulation. The study was conducted at 1.2 km altitude from the sea level at a temperature of 280.2 K and pressure of 25,636.79 Pa with the fluid taken as air (ideal-gas).

Pressure far-field boundary conditions were used for the analysis. The study was conducted at Mach number 4.5 and 7. The CFL number was taken as 0.7, and the number of iterations took to converge varied according to the flow regime. Discretization scheme used was Second Order Upwind as it provides more accurate results. Implicit method with density-based solver was used for the simulation. The turbulent flow field was solved using the k-omega (SST) model. The convergence criteria were set as 1E-5.

The meshing was done using the ANSYS Workbench environment. Two different types of computational grid were generated to understand and study the importance of the mesh quality, element types and the size.

A hybrid structured mesh was generated for the first case (Fig. 2), where a structured mesh was generated around the body and away from the body; it was unstructured with the total number of cells as 53,403.

In the second case (Fig. 3), a fully structured mesh was created to resolve the boundary layer with the total number of cells as 97,826.

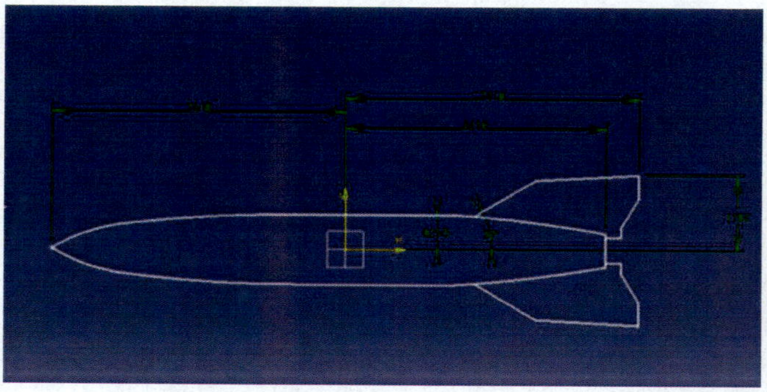

Fig. 1 The 2D sketch of the V2 missile

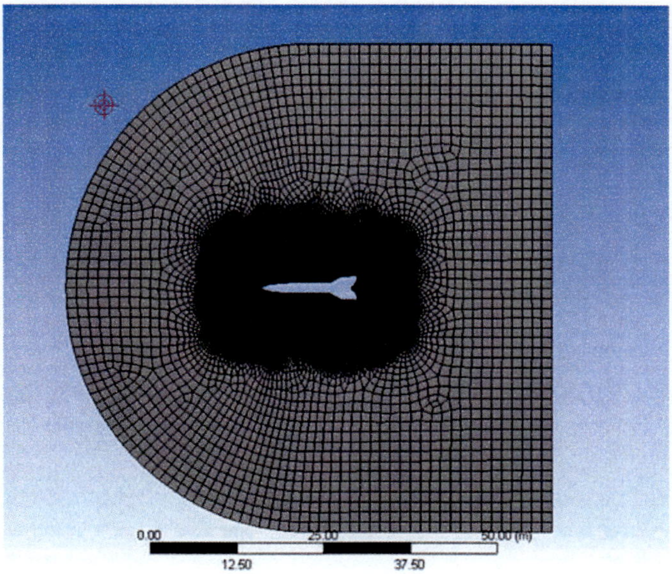

Fig. 2 Hybrid grid

4 Results and Discussion

The aerodynamic analysis was carried out on the 2D geometry of the missile body for two different computational grids. The flow contours of Mach number, pressure, density and temperature were extracted (Figs. 4, 5, 6, 7, 8, 9, 10, 11) (Tables 1, 2).

We can interpret from the results that there is a significant difference in C_L for two different grids. The computational grid is very important for numerical simulations as it can affect the quality of the results. It can be observed that by changing hybrid grid to fully structured grid, there was a change in the value of C_L and thus, improvement in the results. It was observed that at 0° AoA (Figs. 8, 9, 10, 11), the flow shows a symmetrical behaviour on the bottom and top surfaces of the missile. The free stream flow encounters a strong compression around the missile nose leading to the formation of shock waves. Bow shock wave is generated at the nose of the missile and also a series of expansion waves occurs. The flow tends to separate from the body due to the formation of shock waves. The normal shock wave interaction can be observed at the tail section of the missile. The flow contours of Mach number, pressure, density and temperature at the bottom and the top of the missile show identical features because theoretically at zero angle of attack, the air flow around the missile body is symmetrical. At 5° AoA (Figs. 4, 5, 6, 7), the temperature, pressure and density at the bottom of the missile are higher than the upper part of the missile, resulting in higher velocity at the top surface. Therefore, it can be interpreted that the areas with high temperature, pressure and density have a lower Mach number flow.

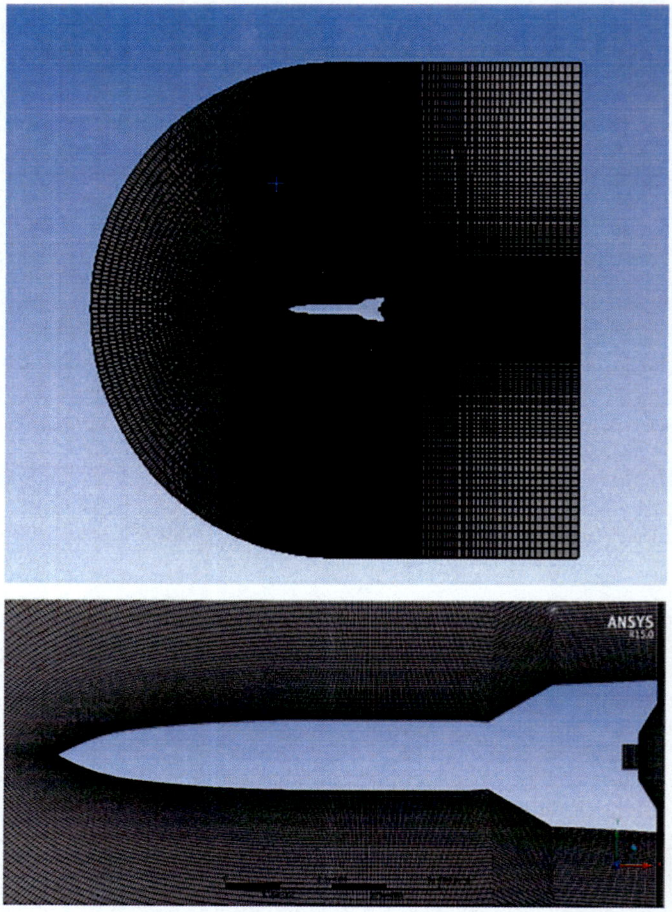

Fig. 3 Fully structured grid

At the upper side, expansion waves and at the lower surface, oblique shock waves can be observed.

The maximum C_L/C_D ratio was obtained for the fully structured grid and the ratio for the hybrid grid was comparatively less at 5° angle of attack. Aerodynamically, a missile will be more efficient at 5° angle of attack compared to the 0° angle of attack. The highest pressure region can be observed at the nose section, and the second highest at the fin section for both hypersonic and supersonic regimes. The maximum density regime can be observed at the nose and the fin section for the supersonic flow but in case of hypersonic flow, the highest density regime can be seen at the nose section only. High temperature can be noted in the fin section of the missile. The lower half experiences a high temperature, pressure and density because of the formation of the expansion waves. At the fin section, the Mach number contour shows

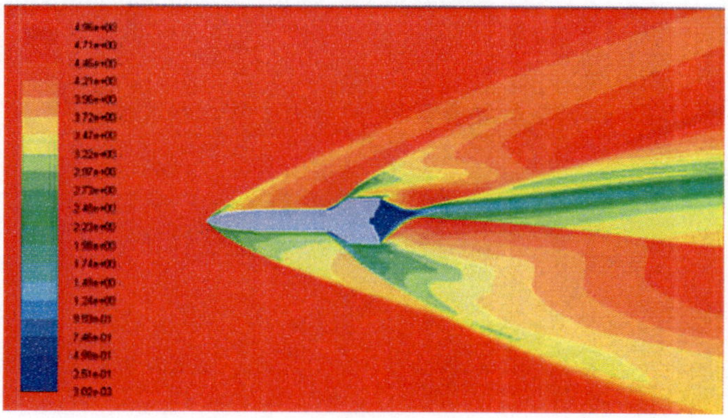

Fig. 4 Contour of mach number at M = 4.5 and AoA = 5°

Fig. 5 Contour of density at M = 4.5 and AoA = 5°

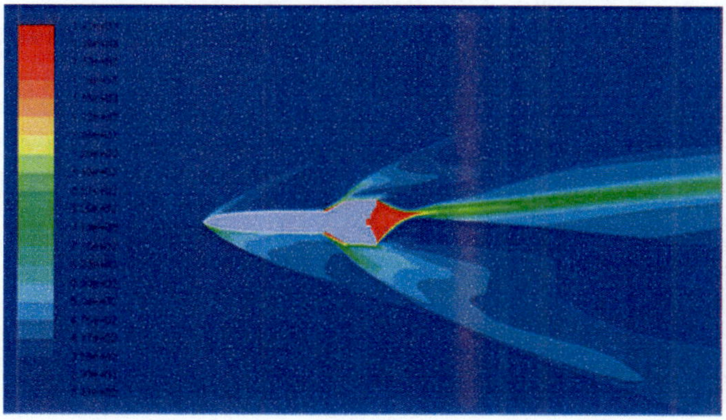

Fig. 6 Contour of temperature at M = 4.5 and 5°

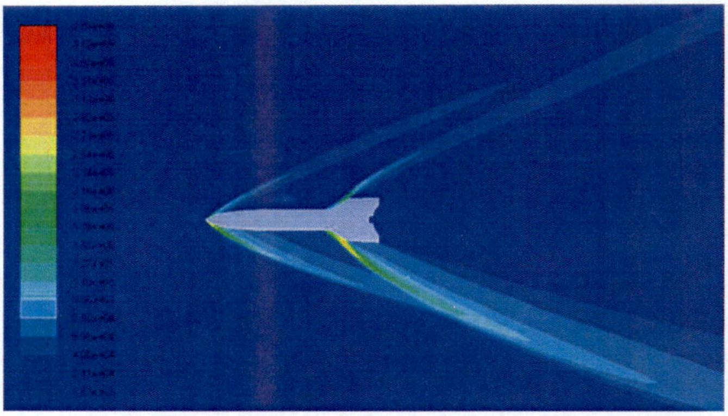

Fig. 7 Contour of pressure at M = 4.5 and AoA = 5°

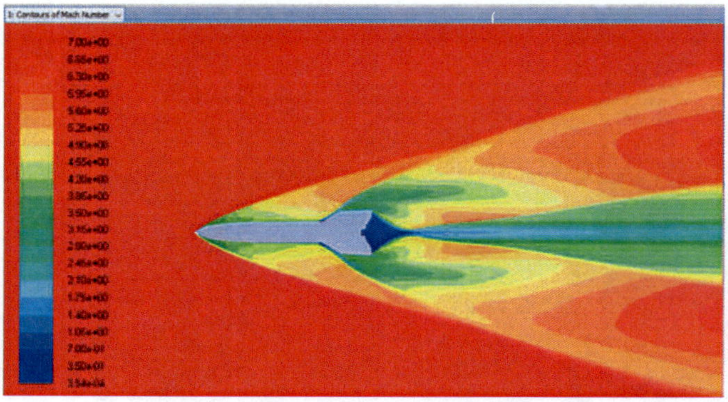

Fig. 8 Contour of mach number at M = 7 and AoA = 0°

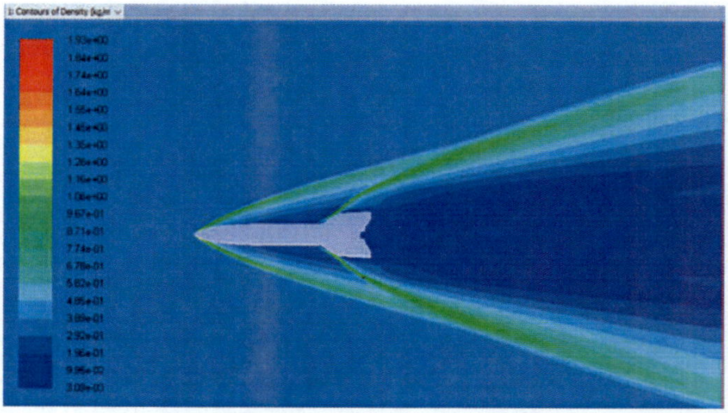

Fig. 9 Contour of density at M = 7 and AoA = 0°

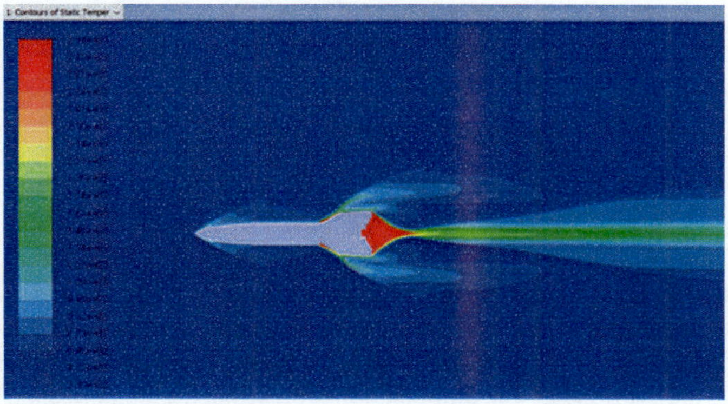

Fig. 10 Contour of temperature at M = 7 and AoA = 0°

Fig. 11 Contour of pressure at M = 7 and AoA = 0°

Table 1 Numerical results (hybrid grid)

Mach number	C_D	C_L	$\alpha^{(o)}$
4.5	1.690999	1.8E-3	0
7	1.257848	8.6E-4	0
4.5	1.885775	2.42345	5
7	1.462041	2.09742	5

Table 2 Numerical results (fully structured grid)

Mach number	C_D	C_L	$\alpha^{(o)}$
4.5	1.536241	1.2E-4	0
7	1.738449	1.4E-4	0
4.5	1.746315	2.35740	5
7	1.562561	2.51234	5

a subsonic regime. Near the body and nose of the missile, supersonic flow can be observed at zero angle of attack.

5 Conclusions

The first modern ballistic missile (V2) was modelled and investigated numerically using ANSYS Fluent. A parametric study was carried out for two different types of computational mesh at two different Mach numbers to predict the aerodynamic characteristics of the 2D missile geometry. The fully structured grid shows better results than the hybrid grid. The flow contours were studied in order to understand the aerodynamics features of the missile geometry. The missile is more aerodynamically efficient at hypersonic regime with angle of attack of 5° compared to 0°. The C_L obtained for the 5° angle of attack is 2.51234 with C_L/C_D ratio as 1.607834. The formation of shock wave resulted in a decrease of flow parameters like density, Mach number, pressure and temperature.

Acknowledgements We would like to thank the management of Ramaiah University of Applied Sciences, India, for providing us with the resources needed for conducting this research. Moreover, we would like to thank the ICFTES committee for providing us with this opportunity to publish this paper.

Nomenclature

C_L	Lift coefficient –
M	Mach Number –
C_D	Drag coefficient –
α	Angle of attack (°)

References

1. Meng Y-S, Yan L, Huang W, Tong X-Y (2020) Numerical investigation of the aerodynamic characteristics of a missile. IOP Conf Ser: Mater Sci Eng 887:012001
2. Ishak IS, Mohd RN, Dahalan MdN (2017) Aerodynamic study of air flow over a curved fin rocket. J Adv Res Fluid Mech Therm Sci 40(1). ISSN: 2289-7879
3. Gaonkar AA, Menon P, Srinivas G (2019) Aerodynamic performance enhancement of supersonic 2D missile using ANSYS. Univers J Mech Eng 7(6A):1–11. https://doi.org/10.13189/ujme.2019.071401
4. Dol SS (2021) Aerodynamics analysis of grid fins inner lattice structure in cruise missile. Wseas Trans Fluid Mech. https://doi.org/10.37394/232013.2021.16.9
5. Pandie AR, Prabantara PL (2020) Analysis of characteristics of guided missile's aerodynamics using CFD (ANSYS R15.0) software. AIP Conf Proc. https://doi.org/10.1063/5.0002284

6. Divakaran R, Barretto N, Srinivas G (2020) Aerodynamic performance enhancement of missile using numerical techniques. In: First international conference on advances in physical sciences and materials. https://doi.org/10.1088/1742-6596/1706/1/012222
7. Wee HC (2011) Aerodynamic analysis of a canard missile configuration using ANSYS-CFX. Master's thesis, Naval Postgraduate School (USA)
8. Aerodynamic configuration design of a missile. http://www.v2rocket.com/
9. https://www.oldrocketplans.com/narts/GERMANV2_102.pdf
10. Nielsen JN, Missile aerodynamic. Mcgraw-Hill, USA
11. Sethunathan P, Sugendran RN, Anbarasan T (2015) Aerodynamic Configuration design of a missile. Int J Eng Res & Technol (IJERT) 4(03). ISSN: 2278-0181
12. Leroy Spearman M, Historical review of missile aerodynamic developments. NASA Langley Research Center

Numerical Studies to Investigate Flow Behaviour of Two-Phase Liquid–liquid Flow in Microchannels

C. Bersha Angelin Christal and Aarathi Pradeep

Abstract Microfluidic liquid–liquid extraction devices have unique advantages including lower fluid requirements and high extraction efficiency. In this work, we study the flow behaviour of two immiscible fluids in a cross-flow microchannel. The effect of different geometry that can influence the flow pattern is evaluated. Towards this, we compared the flow through a simple cross-flow microchannel with the flow through a microchannel modified with ridges. The ridges are arranged alternately as well as oppositely, and their effect is studied. Simulation studies were performed with COMSOL Multiphysics with diethyl ether and water as the two immiscible model fluids. Phase-field method coupled with the multiphase laminar flow was used to solve the velocity profile, pressure field and the dispersed phase fraction in the microchannels. The simulation results revealed the formation of slug flow within the microchannel. The presence of ridges will lead to a higher dispersion of the fluids leading to enhanced mass transfer between the phases.

Keywords Multiphase flow · Cross-flow microchannels · Triangular ridges · Immiscible flow · Phase-field method

C. Bersha Angelin Christal
Department of Chemical and Material Science Engineering, Amrita Vishwa Vidyapeetham, Coimbatore 641112, India

A. Pradeep (✉)
Amrita Biosensor Research Lab, Amrita Vishwa Vidyapeetham, Coimbatore 641112, India
e-mail: p_aarathi@cb.amrita.edu

Department of Sciences, Amrita Vishwa Vidyapeetham, Coimbatore 641112, India

S. Das et al. (eds.), *Proceedings of the 1st International Conference on Fluid, Thermal and Energy Systems*, Lecture Notes in Mechanical Engineering,
https://doi.org/10.1007/978-981-99-5990-7_43

1 Introduction

In recent years, many microfluidic devices have emerged due to their superior heat and mass transfer properties [1]. Microfluidic devices have several advantages like easy flow control, smaller size, smaller sample and reagent requirement, faster analysis and less power consumption [1]. They have explicit applications in several fields including biomedicine, solvent extraction and also in chemical synthesis. Liquid–liquid extraction systems are used to separate components from one phase to the other based on differences in solubilities or chemical potential of the species in either phase [2]. Microfluidic liquid–liquid extraction systems are advantageous for separating compounds that are expensive as they typically require only low sample volumes and have higher efficiency of extraction due to higher diffusion as they have an increased surface-to-volume ratio. They can be used for developing lab-on-a-chip devices for the selective extraction and quantification of different radio-nucleotides or metabolites. For example, cholesterol is an important biomarker that provides information on cardiovascular diseases. Non-enzymatic cholesterol biosensors are promising for the development of point-of-care diagnostic devices for the quantification of cholesterol. However, the oxidation potential of cholesterol for non-enzymatic electrochemical sensing in aqueous solutions typically requires a high potential, which may lead to interference with other molecules. Therefore, cholesterol detection in organic solvents like diethyl ether or acetonitrile will be able to improve the selectivity of the biosensor [3]. In situations like this, it is essential to extract cholesterol from the sample to an organic medium like diethyl ether.

In order to design an efficient microfluidic device for liquid–liquid extraction process, it is essential to understand the microfluidic flow behaviour in two-phase systems. Since the flow is in microscale, the movement of liquid is not influenced by the force of gravity but they depend upon the surface tension, and viscous and inertial forces [4]. The dimensionless numbers like Capillary number (Ca), Reynolds number (Re), Weber number (We), Bond number (Bo) and Ohnesorge number (Oh) are used to understand the relationship between various forces acting on the liquid. The flow pattern is influenced by different parameters like the microchannel material, hydrodynamics of microchannel, inlet configurations and fluid properties [5]. In a typical two-phase liquid flow, different flow patterns like parallel flow and segmented flow can be observed. This depends on the geometry of the microchannels as well as the ratio of flow velocities of the aqueous and the dispersed phase [4]. In a parallel flow, the transfer of solute occurs at the interface, and it will be easy to separate the phases after extraction, whereas, in a segmented flow, the mass transfer is enhanced due to an increase in the interfacial area. Therefore, the geometry and the flow velocity must be carefully optimized in a two-phase microfluidic device to generate the desired flow pattern and optimum performance.

2 Literature Review and Objective

The droplet-based extraction of phenol from oil was studied by Das et al. (2015), in an acrylic-based microchannel reactor. The target molecules migrated from highly viscous phase (silicone oil) to less viscous liquid (phenol). It was seen from their work that the extraction efficiency increased when the formation of droplets takes longer time because the phenolate ions occur at slower rate. Hence, a low flow rate ratio was favoured [6].

Marsousi et al. (2019) studied the extraction of calcium using ionic liquids in spiral microfluidics. They found in their study that in conventional method, only 10% of total mass is transferred during the droplet formation but in the case of Y-shaped microchannel have 89% and Spiral-shaped microchannel it is 94%. The increase in crown ether concentration in ionic liquid increases efficiency. They also observed that spiral microchannels have high efficiency due to centrifugal force creating the formation of elongated droplets which increases the contact area and mass transfer. This force increases as flow rate increases, so does the extraction efficiency [7].

Han and Chen (2018) have done a systematic study on the 3-dimentional simulation of the microfluidic channel for the generation of micro-droplets and also studied the influence of the droplet diameter with different parameters such as the flow rate ratio, continuous phase viscosity, interfacial tension and the contact angle. It is observed that when the interfacial tension and flow rate ratio increase, the effective droplet diameter also increases and it decreases when contact angle and continuous phase viscosity increase [8].

Raad et al. (2021) have studied the mechanism of droplet formation and splitting in an immiscible liquid/liquid two-phase flow in the branched T-junction microchannel. Due to the effect of pressure gradients at the outlet branches, there is a formation of asymmetric split at the T-junction. They infer that, as the length ratio of the outlet branches increases, the diameter of the droplets after split up increases. They also observe that when the length of the outlet branches is equal, the droplet diameter after splitting up is also of the same size [9].

Takada et al. (2014) did a study on the slug droplet formation in the T-junction microchannel with square cross section for an immiscible liquid–liquid two-phase extraction with hydrophilic solid walls. For a silicone oil-pure water system, the ratio of the volumetric flow rate of the dispersed phase to continuous phase is fixed to be 1 with low Ca, Re and We numbers. They infer that the droplets of both phases decrease steadily at constant length difference as their flow rates increased [10].

Bi et al. (2017) have done a simulation as well as experimental validation of a 3-dimentional microfluidic channel using the phase-field method. This paper points out the relationship between mobility and the formation of droplets as well as its shape and size. They define the characteristic mobility as the correct relaxation time of the interface. They have performed studies by maintaining the characteristic mobility ($M = \chi \varepsilon^2$) constant and varying interfacial thickness (ε) to capture the physical process of droplet formation. They also mention that the value of interfacial thickness is selected based on the mesh size [11].

Liu (2022) studied the extraction of molybdenum-99 by using the parallel multiphase flow microfluidic liquid–liquid solvent extraction inside a double Y-shaped microfluidic channel. They found that the slug length is linearly proportional to the Pe^{-1} number and the Cn number when the Ca number of the two phases is smaller than 1×10^{-3}, and it is described by the empirical correlation: $L_{slug} = 0.72\ Cn\ Pe^{-1} + 3.86$. To obtain a parallel flow, at first they started with the influence of the mobility parameter (M) and capillary width (ε) and infer that increasing the value of mobility parameter (M) and capillary width (ε) pushes the critical Re value high. Complete phase separation is hard to achieve in a parallel flow pattern, and there will be a leakage from one phase to the other, hence preferential wall wettability is done to avoid leakage at the outlet [12].

The main objective of this work is to study the flow patterns of two immiscible fluids in microfluidic devices by multiphase simulation studies. By studying the flow pattern, microfluidic devices for the extraction of biologically relevant component from body fluids using selective solvents can be designed.

3 Materials and Methods

3.1 Design

In order to study the effect of geometry on the flow pattern of two immiscible liquids, different cross-flow microchannels were designed. Figure 1a shows the geometry of a simple cross-flow microchannel. The inlets have a length of 2000 μm and are 100 μm in width and are joined to a common microchannel having a length (*l*) of 12,000 μm and a width (*w*) of 100 μm. The length and width are maintained constant for all other geometrical shapes.

Figure 1b shows the microchannel modified with triangular-shaped ridges arranged alternately with the distance (*d*) between two ridges as 500 μm each. The ridges project inwards into the microchannel to a length of 50 μm and height of 50 μm. Figure 1c shows the opposite arrangement of triangular ridges in the microchannel, where the ridges are arranged on either side of the microchannel, with a distance of 500 μm between the ridges.

3.2 Numerical Simulation

The simulation studies were carried out using the CFD tool, COMSOL Multiphysics (version 5.5), to study the flow behaviour of two-phase liquid–liquid flow in microchannels. The multiphase problem was solved using the phase field interface coupled with Multiphase Laminar flow and was used to track the flow behaviour

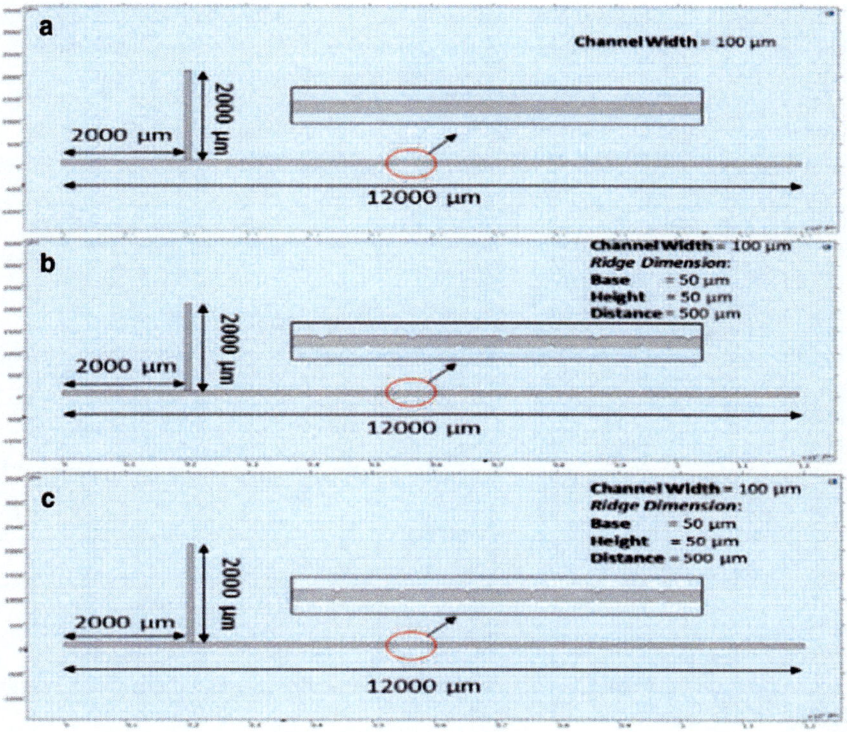

Fig. 1 Different geometries of cross-flow microchannels; **a** straight microchannel; **b** microchannels modified with alternate triangular ridges; and **c** microchannels modified with opposite triangular ridges

in the microchannel. The model fluids were chosen to be diethyl ether and water at 20 °C. The fluid properties are tabulated in Table 1.

The inlet boundary condition for both the fluids was given at a velocity of 1.39 × 10^{-2} m s^{-1} with inlet pressure of 1 atm at 20 °C. At the outlet, a pressure of 0 Pa was defined. The dispersed phase was taken as diethyl ether with the initial volume

Table 1 Model fluids used in study

Parameters	Fluid 1	Fluid 2
Dynamic viscosity (μ)	1×10^{-3} Pa s	2.24×10^{-4} Pa s
Density (ρ)	1000 kg m^{-3}	713.4 kg m^{-3}
Initial volume fraction (Φ)	1	−1
Initial velocity (u)	1.39×10^{-3} m s^{-1}	1.39×10^{-3} m s^{-1}
Inlet pressure (P)	101,325 Pa	101,325 Pa
Outlet pressure (P_o)	0 Pa	0 Pa

fraction, $\Phi = -1$, and water was taken as the continuous phase with the initial volume fraction, $\Phi = 1$. The Navier–Stokes equation for momentum balance and Continuity equation are used for solving incompressible liquids. Equation of Continuity is given by Bai et al. [13]

$$\nabla \cdot \vec{u} = 0 \tag{1}$$

Navier-Stroke's equation for incompressible liquids is given by Bai et al. [13]

$$\rho\left[\frac{\partial \overrightarrow{u}}{\partial t} + (u.\nabla)u\right] = -\nabla \rho + \nabla \cdot \mu\left(\nabla u + \nabla u^T\right) + G\nabla\varnothing + f_{ext} \tag{2}$$

For phase-field method, the Cahn–Hilliard equation is given by Bai et al. [13]

$$\frac{\partial \varnothing}{\partial t} + u \cdot .\nabla\varnothing = M\nabla^2 G \tag{3}$$

$$G = -K\nabla^2\varnothing + f\prime(\varnothing) \tag{4}$$

$$f\prime(\varnothing) = u\varnothing^3 - r\varnothing \tag{5}$$

The relation between the surface tension coefficient (σ) and the mixing energy density (K) is given by Bai et al. [13]

$$\sigma = \frac{2\sqrt{2}}{3}\frac{K}{\varepsilon} \tag{6}$$

When the interfacial thickness approaches zero, the Cahn–Hilliard equation changes to Bai et al. [13]

$$\frac{\partial \varnothing}{\partial t} + u \cdot \nabla\varnothing = \frac{MK}{\varepsilon^2}\nabla^2\Psi \tag{7}$$

$$\Psi = -\varepsilon^2\nabla^2\varnothing + \varnothing(\varnothing^2 - 1) \tag{8}$$

In terms of surface tension coefficient and interfacial thickness, the chemical potential is expressed as Bai et al. [13]

$$G = \frac{3\varepsilon\sigma}{2\sqrt{2}}\left(-\nabla^2\varnothing + \frac{\varnothing(\varnothing^2 - 1)}{\varepsilon^2}\right) \tag{9}$$

As mentioned earlier, the surface tension is one of the important forces responsible for the flow of microfluids in a microscale. The contact angle of water-diethyl ether interface is calculated using the Young–Laplace equation.

The Young–Laplace equation for air-water interface, air-diethyl ether interface and water-diethyl ether interface is given by [13]

$$\cos\theta_{WD} = \frac{\sigma_{AD}\cos\theta_{AD} - \sigma_{AW}\cos\theta_{AW}}{\sigma_{WD}} \tag{10}$$

where the surface tension coefficient of air–water interface (σ_{AW}) is 0.072 N m^{-1} [14] and contact angle (θ_{AW}) is 77° [14], the surface tension coefficient of air-diethyl ether interface (σ_{AD}) is 0.017 N m^{-1} [11] and contact angle (θ_{AW}) is 0°, and the surface tension coefficient of water-diethyl ether interface (σ_{WD}) is 7.36×10^{-3} N m^{-1} [15]. By substituting the known parameters in Eq. (6), we get contact angle of water-diethyl interface (θ_{WD}), which is 83.73°.

Time-dependent studies were performed to analyse the flow of the two immiscible fluids in the microchannel. Mesh convergence studies were performed and physics-controlled 'normal' meshing was done in the entire domain.

4 Results and Discussion

The 2-dimensional simulation results obtained for the different geometries analysed are presented here. It was seen from the studies that for an inlet velocity (u) of 1.39×10^{-2} m s^{-1}, the average droplet size is more than the hydraulic diameter of the microchannels. Hence, the flow pattern observed in the microchannels is predominantly slug flow. The velocity ratio of both inlets is taken as 1, hence both inlets have the same velocity (1.39×10^{-2}), with the Capillary number of continuous phase (Ca_c) and Weber number of dispersed phase (We_d) found to be 1.83×10^{-4} and 2.68×10^{-4}. Since Weber number (We) and Capillary number (Ca) greatly influence the flow regime of droplet formation, for this We and Ca numbers the flow regime observed is squeezing regime. Here, slug length is defined as the length of the aqueous phase (Water) dispersed in the organic phase (Diethyl Ether). The following section presents the simulation results obtained for different geometries analysed. Here, the blue colour represents the volume fraction of water, whereas red colour represents that of diethyl ether and the intermediate colours at the interface between two droplets indicate the diffusive layer of that droplet.

4.1 Straight Microchannel with No Modifications

The volume fraction and velocity profile of the organic phase and the aqueous phase obtained in the simple straight microchannel are shown in Fig. 2. Figure 2a represents the volume fraction of two immiscible fluids in a simple microchannel, and Fig. 2b shows velocity profile of the microchannel. At the junction, the regime formed is squeezing flow and the droplet size of 240 μm for water and 220 μm for diethyl

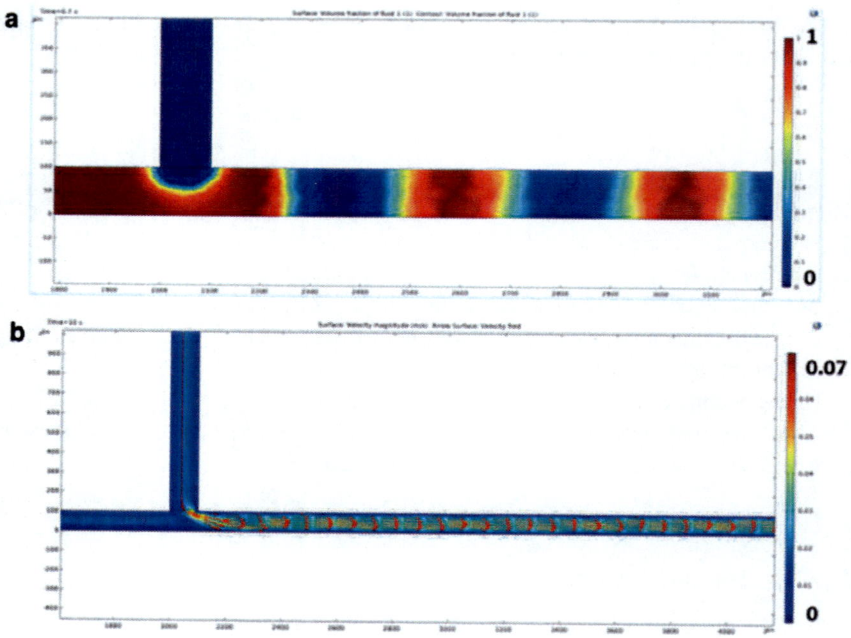

Fig. 2 **a** Volume fraction of the two immiscible fluids observed at different time intervals in a straight microchannel and **b** velocity profile in the microchannel

ether. The fluid flow follows a parabolic velocity profile in the straight cross-flow microchannel with maximum velocity at the centre of the microchannel (Fig. 2b). It is to be noted that the velocity inside the droplets is higher leading to chaotic advections within the droplets that can promote mixing within the droplets. It is also observed that velocity is higher at the junction due to the droplet formation and hence it increases mass transfer rate.

4.2 Alternate Triangular Ridges Modified Microchannel

The effect of placing ridges in an alternate manner within the microchannel in two-phase liquid flow was analysed, and the results are presented in Fig. 3a. Similar behaviour as in the case of straight microchannel was observed. However, the slug lengths were different from that of the earlier geometry. Due to the presence of ridges, the droplet size was of 230 μm for water and 230 μm for diethyl ether. Figure 3b shows the velocity profile inside the microchannels modified with alternate ridges. It can be seen that at the junction of ridges, the velocity is higher leading to better advections within the channel. This can promote the mass transport between streams as well as enhance mixing within the droplets.

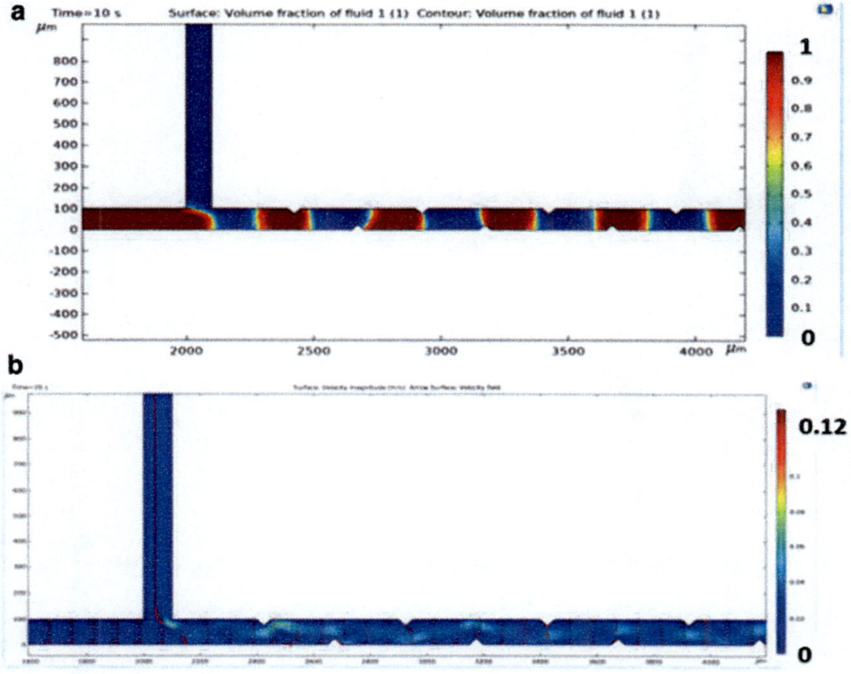

Fig. 3 **a** Volume fraction of the two immiscible fluids observed at time intervals in a microchannel modified with triangular ridges alternately; **b** Velocity profile of the corresponding microchannel

4.3 Opposite Triangular Ridge Modified Microchannel

Figure 4 shows the volume fraction of the two phases of fluid obtained in microchannels having triangular-shaped ridges arranged on either side of the microchannel. Figure 4a shows the volume fraction of the organic phase and the aqueous phase at $t = 0.4$ s, and Fig. 4b shows the velocity profile in the microchannel. The droplet size was of 220 μm for water and 210 μm for diethyl ether. The presence of the ridges leads to irregular sizes of droplets throughout the flow. Due to the presence of ridges in the microchannel, the fluid undergoes expansion and contraction leading to enhanced generation of chaotic advections and mixing between the fluid streams. However, the pressure drop incurred in this microchannel will be higher than the straight channel or the microchannel modified with alternate ridges.

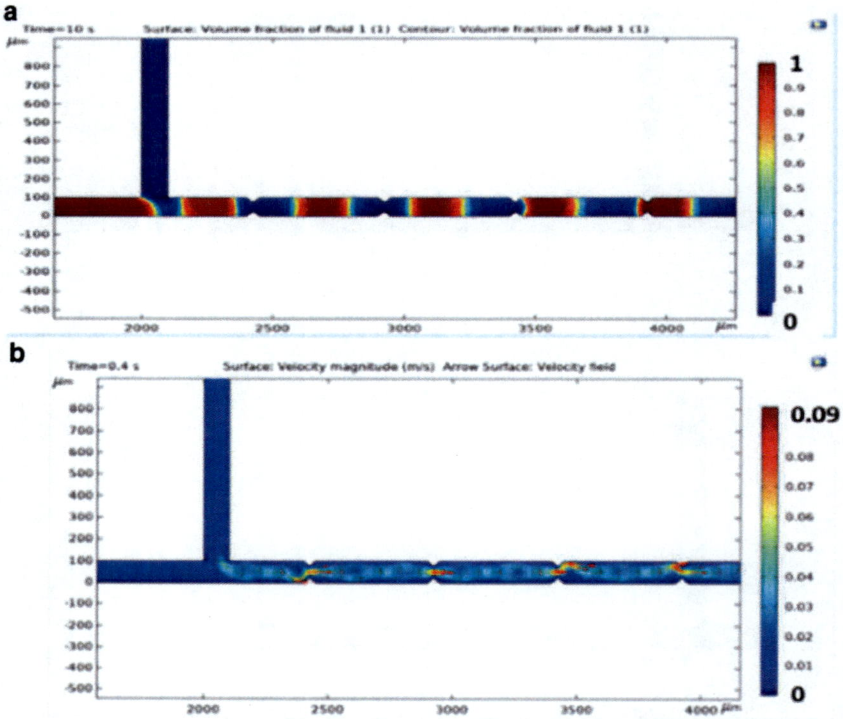

Fig. 4 **a** Volume fraction of the two immiscible fluids observed at time intervals in a microchannel modified with triangular ridges in opposite arrangement; **b** Velocity profile of the corresponding microchannel

4.4 Comparison of Flow Behaviour and Mass Transfer in the Microchannels

The velocity profile along a cut-line in the three different microchannel geometries is shown in Fig. 5. As seen from the figure, at the junction of the microchannel where the aqueous and the organic phases meet, the velocity is high due to mixing of two immiscible fluids. In the microchannels modified with ridges, chaotic advections are introduced in the microchannel. In the microchannel modified with alternate ridge channel, the velocity fluctuates from 0.02 to 0.047 m/s. In the microchannels modified with opposite ridges, constant high and low peaks in velocity are seen which are due to the arrangements of ridges opposite each other. The sudden decrease in the channel width from 100 to 50 μm leads to localized increase in velocity from 0.025 to 0.072 m/s, contributing to chaotic advections.

The mass transfer characteristics in each of the microfluidic channel were also analysed (Fig. 6). It can be seen from the simulation results that faster dispersion of the solute happens in the microfluidic channels with ridges, which can be attributed

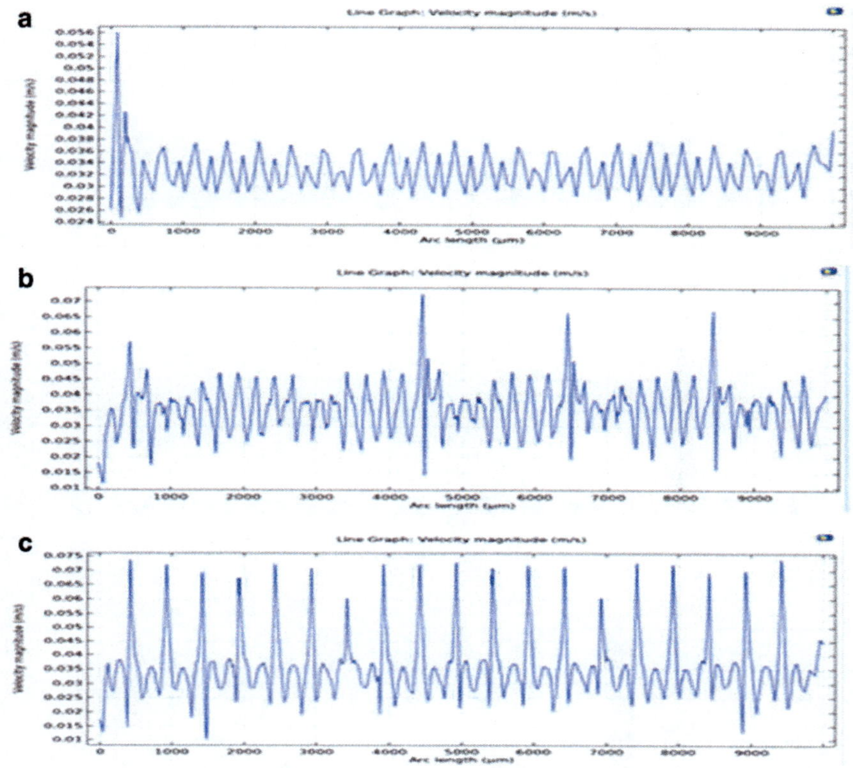

Fig. 5 Cut-line analysis of velocity in the different microchannels. **a** Straight microchannel, **b** channels modified with alternate ridges and **c** channel modified with opposite arrangement of ridges

to the chaotic advections in these microchannels. The vorticity in the droplets in ridge-modified channels leads to faster dispersion of the solute from the aqueous phase to the organic phase and within the organic phase slugs.

5 Conclusions

In this study, the flow patterns of two immiscible fluids on three different microfluidic geometries were analysed using simulation studies. In order to have maximum extraction efficiency on the microfluidic device, the effective interfacial area between the continuous and the dispersed phase must be higher. It was seen from the analysis that by modifying the microchannel with ridges in the microchannel, there was no significant change in the frequency of droplet generation. However, the chaotic advections in the ridge-based microchannels can enhance the transport of the solute from the

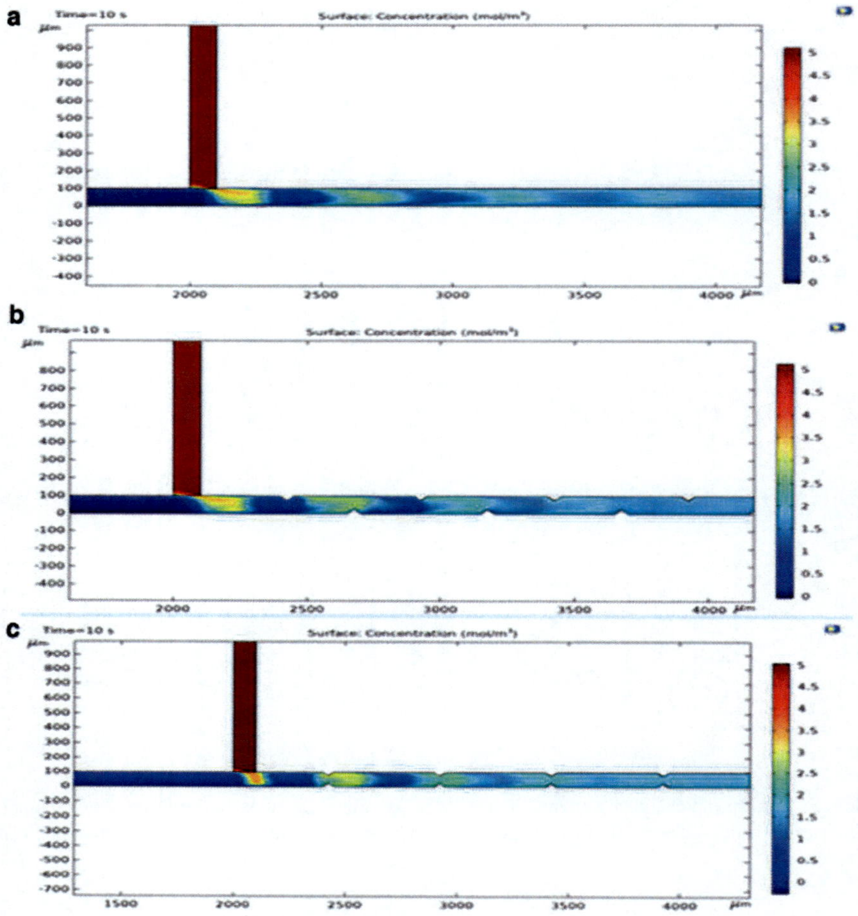

Fig. 6 Concentration profile in the microchannels. **a** Straight microchannel, **b** channels modified with alternate ridges and **c** channel modified with opposite arrangement of ridges

dispersed phase to the continuous phase as well as mixing of the solute within the continuous phase. Further studies will be done to study the effect of different inlet velocities, and the extraction efficiencies will be evaluated for the different geometries. Such microfluidic devices for liquid–liquid extraction can be potentially used for developing point-of-care testing devices that will be able to analyse the sample without any complex sample preparation steps.

Nomenclature

μ	Dynamic viscosity	(Pa · s)
Φ	Initial volume fraction	–

u	Initial velocity	(m/s)
P	Inlet pressure	(Pa)
P_0	Outlet pressure	(Pa)
ρ	Density of liquids	(kg/m^3)
σ	Surface tension coefficient	(N/m)
θ	Contact angel	(^{o})
t	Time	(s)
l	Length	(μm)
w	Width	(μm)
d	Distance between two ridges	(μm)

References

1. Su Y, Chen G, Kenig EY (2015) An experimental study on the numbering-up of microchannels for liquid mixing. Lab Chip 15(1):179–187
2. Handlos AE, Baron T (1957) Mass and heat transfer from drops in liquid-liquid extraction, 3(1):127–136
3. Punnakkal N, Raveendran J, Vasu SP, Nair BG, Satheesh Babu TG. Highly sensitive and wide range non-enzymatic electrochemical detection of cholesterol using pencil lead electrodes. J Electrochem Soc 168:047515
4. Al-Azzawi M, Mjalli FS, Husain A, Al-Dahhan M (2021) A review on the hydrodynamics of the liquid—liquid two-phase flow in the microchannels. Ind Eng Chem Res 60:5049–5075
5. Pradeep A, Nair BG, Suneesh PV, Satheesh Babu TG (2021) Enhancement in mixing efficiency by ridges in straight and meander microchannels. Chem Eng Process: Process Intensification 159:108217
6. Das D, Duraiswamy S, Yi Z, Chan V, Yang C (2015) Continuous droplet-based liquid-liquid extraction of Phenol from oil. Sep Sci Technol 50:1023–1029
7. Marsousia S, Karimi-Sabetb J, Moosaviana MA, Aminic Y (2019) Liquid-liquid extraction of calcium using ionic liquids in spiral microfluidics. Chem Eng J 356:492–505
8. Han W, Chen X. Numerical simulation of the droplet formation in a t-junction microchannel by a level-set method. Austr J Chem 71(12)
9. Raad M, Rezazadeh S, Jalili H, Fallah DA (2021) A numerical study of droplet splitting in branched T-shaped microchannel using the two-phase level-set method. Adv Mech Eng 13(11):1–13
10. Takada N, Matsumoto J, Matsumoto S (2014) Numerical simulation of two-phase fluid motion in microchannel based on phase-field model. In: 11th world congress on computational mechanics, pp 3895–3903
11. Bi S, Zhao G, Wu J (2010) Surface tension of diethyl ether, diisopropyl ether, and dibutyl ether. J Chem Eng Data 55(4):1523–1526
12. Liu Z (2022) Purifying radionuclides with microfluidic technology for medical purpose: simulating multiphase flows inside a microfluidic channel with the phase field method. https://doi.org/10.4233/uuid:e1bebcdd-185a-4515-b352-76d68f65ace8
13. Bi F, He X, Yang X, Zhou R, Wang C (2017) Three dimensional phase-field investigation of droplet formation in microfluidic floe focusing devices with experimental validation. Int J Multiph Flow 93:130–141

14. Zhao Y, Chen G, Yuan Q (2007) Liquid-liquid two-phase mass transfer in the t-junction microchannels. AIChE J 53(12)
15. Shahid MZ, Usman MR, Akram MS, Khawaja SY, Afzal W (2017) Initial interfacial tension for various organic–water systems and study of the effect of solute concentration and temperature. J Chem Eng Data 62(4):1198–1203

Optimization of Phase Shift Mechanism in Pulse Tube Cryocooler

Sarthak Srivastava, Gagan Agarwal, Numan Ahmad, Naimesh Patel, Shaunak R. Joshi, and B. A. Modi

Abstract Pulse Tube Refrigeration is considered as an appropriate refrigeration technology for the space, military, and other industries, leading to extensive studies on increasing and optimizing its thermodynamic efficiency. By controlling the phase difference between pressure and mass flow rate, the acoustic power of a pulse tube can be reduced. In this study, electrical analogy for thermodynamic properties is used, which iteratively computes the compliance, inertance, and resistance of the inertance tube by discretization of the tube in a large number of nodes. The numerical model is further validated by the existing published experimental results. In order to get required phase shift from inertance tube, the geometry is optimized and further effect of a varied cross-sectional tube is studied that leads to lower acoustic power.

Keywords Inertance tube · Pulse tube · Cryocoolers · Optimization · Electrical analogy

1 Introduction

Cryocooler is a refrigeration system that can approach a cryogenic temperature. Usually small and compact coolers are extensively used in space technology, rockets, and many more. Figure 1 shows an inertance tube pulse tube refrigerator, which consists of a compressor, after cooler (AC), regenerator, cold heat exchanger (CHX), pulse tube, hot heat exchanger (HHX), inertance tube, and reservoir.

When compared to the Stirling cryocooler, the pulse tube has the advantage of having no moveable parts in the cold finger, resulting in high dependability, extended life, and less vibration at the cold tip. An appropriate phase shift at the regenerator

S. Srivastava (✉) · B. A. Modi
Institute of Technology, Nirma University, Ahmedabad 380065, India
e-mail: sarthak.srivastava1201@gmail.com

G. Agarwal (✉) · N. Ahmad · N. Patel · S. R. Joshi
Space Application Centre, ISRO, Ahmedabad 380015, India
e-mail: agl.gagan@gmail.com

S. Das et al. (eds.), *Proceedings of the 1st International Conference on Fluid, Thermal and Energy Systems*, Lecture Notes in Mechanical Engineering,
https://doi.org/10.1007/978-981-99-5990-7_44

Fig. 1 Schematic of inertance pulse tube refrigerator

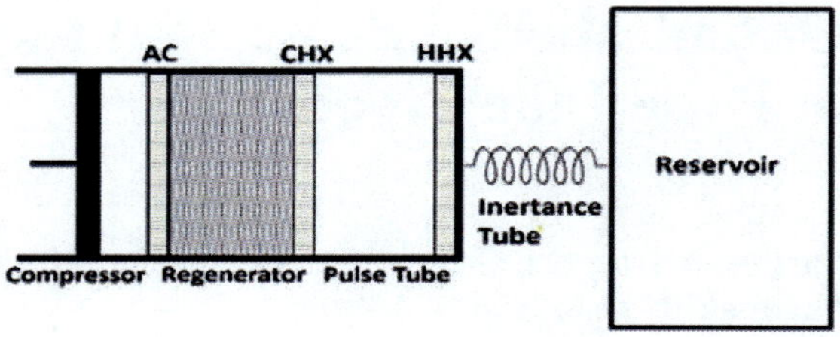

produces the maximum refrigeration effect per unit of working fluid mass in an Inertance Pulse Tube Refrigerator (IPTR). The phase shift needed in the inertance tube is increased due to losses in the pulse tube. Therefore, as a result of optimizing the phase shift mechanism (PSM), the cryocooler's thermodynamic efficiency improves.

The inertance tube provides resistance to the fluid which along with inertia and compliance of the reservoir causes pressure to either lag or lead the mass flow by a desired phase angle.

2 Literature Review and Objective

Various investigations on the design and development of phase shift mechanisms for pulse tube cryocoolers have been conducted previously. Mikulin [1] for orifice pulse tube and Zhu et al. [2] for double inlet pulse tube initiated pulse tube in cryocooler which have allowed the cryocooler to be more efficient and added a new dimension to manipulate in order to get better performance than a Stirling cryocooler. In the orifice pulse cryocooler, an orifice is provided with a calculated diameter in order to cause a phase shift necessary for maximizing the performance; however, orifice pulse tube has a limitation that it only allows phase shift when mass flow rate is leading pressure. Thus, Kanao et al. [3] later worked by introducing inertance tube that doesn't have such limitation like in Orifice Pulse Tube Refrigerator (OPTR); by optimizing and precise calculation of the dimension of inertance tube, the required phase shift can be achieved.

De Boer [4] provides a detailed mathematical model of an IPTR that directly solves the sinusoidal driving pressure's linearized governing equations. Luo et al. [5] employed a transmission line analogy to investigate phase shift in an inertance tube for laminar and turbulent flows, and the model was further confirmed with experimental data. Radebaugh et al. [6] optimized a transmission line model in order to obtain the necessary complex impedance for the best geometry.

Previous studies have primarily focused on low frequency pulse tube cryocoolers. The phase shift needed at the inertance tube is larger with high frequency PTC. This study involves an elaborate optimization technique as well as a parametric study in order to attain a higher phase shift. This study also includes design plots to examine the efficiency of different parameters on phase and acoustic power of an ITPR, as well as the influence of varying cross sections of inertance tube, resulting in lower acoustic power.

3 Model Development

The model is designed such that it converts thermodynamic properties into electrical components where the pressure is analogous to voltage and mass flow rate is analogous to current. The inertia of the fluid inside the tube is analogous to the inertance of the circuit. As the flow propagates across the tube, some fraction of the total mass flow will remain static on every point of the tube which is analogous to the capacitance of the circuit. The friction from the inner wall of the tube, contraction, and expansion lead to resistance which is taken as the resistance component of the circuit [6].

This model converts inertance tube as a circuit of 'n' number of nodes, each node consisting of Inertance (I_i), Compliance (C_i), and Resistance (R_i). At the end, inertance tube connects to the reservoir which acts as a capacitor and has a certain compliance (C_{res}) along with the resistance (R_{res}) due to expansion of the fluid from the tube end resulting in a complex impedance of reservoir which connects with the circuit of inertance tube as shown in Fig. 2. The inertance and resistance are connected in series, while capacitance is connected in parallel at each node. The compliance behavior would intend to store a certain amount of mass flow, resulting in lowering the mass flow rate for the next node input. This model has discrete node implementation giving the advantage of incorporating the loss due to contraction/ expansion in the tube ends; this loss causes a significant decrease in the mean pressure across the time period.

Since resistance in the circuit is a function of mass flow rate, make it vary spatially across the nodes. In the case of a uniform cross-sectional tube, capacitance and inertance of each node are constant throughout the tube, while in case of a varying cross-sectional inertance tube, the impedance associated with capacitance and inertance changes for every node. The distributed model using an implicit method can account for this effect by using an iterative method that updates the resistance values dependent on the local mass flow rate. The model is designed for a turbulent flow in

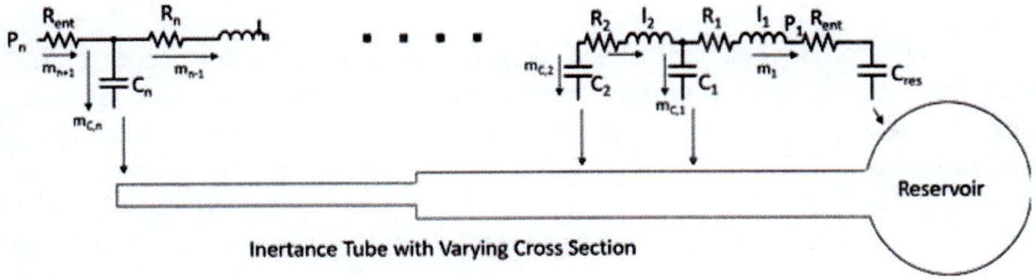

Fig. 2 Electrical analogy used in inertance tube modeling

rough pipe, and accordingly Fanning friction of each node is calculated [10]:

$$f_i = -3.6\log\log\left[\frac{6.9}{Re_i} + \left(\frac{\in /D_i}{3.7}\right)^{10/9}\right] \quad (1)$$

Based on the Fanning friction value, the resistance (R_i) of each node is estimated [7]:

$$R_i = \frac{64 f_i \dot{m}_i (L/n)}{\pi \rho D_i^5} \quad (2)$$

where $\dot{m}_i$ is mass flow rate flowing across ith node, ρ is the mean density of the gas inside that node, L is the total length of the uniform cross-sectional tube, and D is the diameter of the tube. The change in mass flow rate leads to change in Reynold number, leading to different friction factors on every node. With the varied diameter of the tube, the resistance will vary as well, leading to different values of resistance for every node. The inertance and compliance of ith node are shown in Eqs. (3) and (4) [7]:

$$C_i = \frac{\pi D^2 (L/n)}{4\gamma R_g T} \quad (3)$$

$$I_i = \frac{4L}{n\pi {D_i}^2} \quad (4)$$

Here, γ is the heat capacity ratio, T is the average temperature of the gas in the node, and R_g is the gas constant for helium gas. The compliance and inertance of a system are determined by its length and diameter. The compliance and inertia of a system will alter as its diameter changes. The model proceeds to calculate the impedance. The resistance or impedance due to the sudden contraction and expansion of fluid at the inlet of the reservoir can be significant in many cases, and thus it could be determined using orifice equation with contraction coefficient as given in Eq. (5) [10]:

$$\Delta p = k\frac{v^2}{2} \quad (5)$$

The total impedances are calculated using the already defined impedances of each node, starting from the inlet of the reservoir and stopping at the end of the inertance tube. The impedances corresponding to inertance, compliance, and resistance of each node are estimated [7]:

$$Z_{I,i} = j\omega I_i \quad (6)$$

$$Z_{C,i} = \frac{1}{j\omega C_i} \tag{7}$$

$$Z_{R,i} = R_i \tag{8}$$

where j is imaginary number, and ω is angular frequency. The circuit is designed as the combination of several impedances in a cascade manner. The innermost impedance is the reservoir, and the outermost impedance is determined by the total impedance of the system of inertance tube and reservoir. The nodes have components connected in series and parallel, shown by subscripts s and p. The first node considers the reservoir's impedance as shown in Eqs. (9) and (10):

$$Z_{res} = Z_{\text{Rent}} + Z_{C,res} \tag{9}$$

$$Z_{s,1} = Z_{R,1} + Z_{res} + Z_{I,1} \tag{10}$$

$$Z_{s,i} = Z_{R,i} + Z_{l,i} + Z_{p,i} i = 2, \ldots, n \tag{11}$$

$$Z_{p,i} = \left[\frac{1}{Z_{c,i}} + \frac{1}{Z_{s,i}}\right]^{-1} i = 2, \ldots, n \tag{12}$$

$$\dot{m}_{n+1} = \frac{P_a}{Z_{\text{circuit}}} \tag{13}$$

$$\theta = -\angle Z_{\text{circuit}} \tag{14}$$

Equations (9)–(12) show the computational pattern in the model [7], beginning from the reservoir and approaching the end of the inertance tube. The entrance effect at the pulse tube end may be accounted for in this model by combining the resistance value of the entrance effect with the tube's final impedance. The mass flow rate at the intake may be computed using the electrical analogy as the ratio of mean pressure at the pulse tube end to the absolute value of complex total impedance of the inertance tube (Eq. 13). Similarly, as demonstrated in Eq. (14) [7], the phase shift between mass flow rate and pressure may be attributed to the angle of the complex total impedance. The phase and mass flow rate related to each node can be computed using the same equations for ith terms. The Acoustic Power is calculated by Eq. (15) [8]:

$$W = R_g T \frac{m_{\text{inlet}} P_a}{2P} \cos\theta_m \tag{15}$$

3.1 Model Validation

The model is validated by the existing experimental results of uniform cross-section inertance tube by Luo et al. [5]. Figure 3 shows the comparison of the predicted phase difference between mass flow rate and pressure as a function of length for the 0.5 inch OD (11.07 mm ID) tube operating at a compressor frequency of 15.3 Hz. Figure 4 shows the plot between phase difference and length for the 0.375 inch OD (7.09 mm ID) tube operating at a compressor frequency of 15.3 Hz. It is noted that the phase at the inlet of the inertance tube agrees very well with the inferred experimental data.

The experiment work by Luo et al. [5] has uncertainty under 5–10%. This may be due to fluctuation in the input parameters like temperature and mass flow rate during experimentation. Though the numerical estimated curves are very similar to experimental number for almost all tube lengths, however, for some length values, there is a minor deviation, which is under the uncertainty limit.

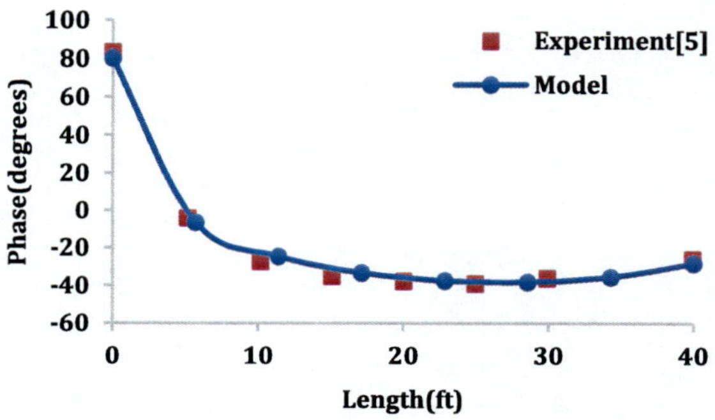

Fig. 3 Phase as a function of length for the 0.5 inch OD (11.07 mm ID) tube at flow frequency of 15.3 Hz

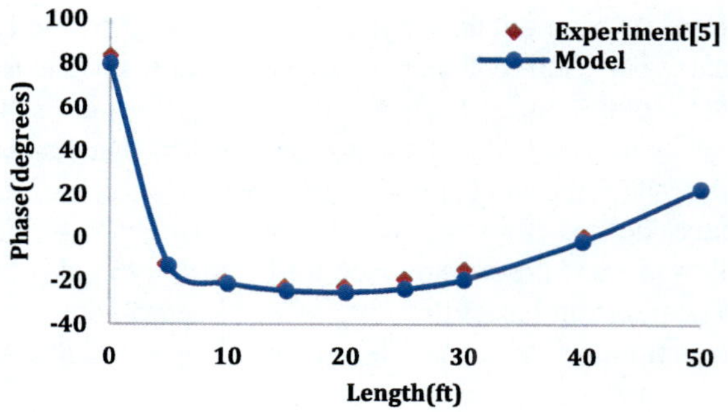

Fig. 4 Phase as a function of length for the 0.375 inch OD (7.09 mm ID) tube at flow frequency of 15.3 Hz

4 Results and Discussion

The validation of the model in the previous section makes it capable to be used as a tool for further analysis and optimization of PSM. The frequency was taken as 100 Hz, mean pressure was 30 bar, mean temperature was 300 K, and Pressure Ratio was 1.2. Dimensionless parameters were introduced for length as ratio of actual length to wavelength: $\overline{L} = L/\lambda$, whereλ is ratio of speed of sound to operating frequency. Dimensionless diameter is the ratio of actual diameter to viscous penetration depth: $\overline{D} = D/\delta$, where δ is square root of ratio of kinematic viscosity to operating frequency. Dimensionless volume: $\overline{V} = V/\lambda A$, where λ is wavelength and A is cross-sectional area of input port of reservoir. Firstly, the effectiveness of various parameters is analyzed in order to understand the effect of length, diameter, and volume of reservoir on phase shift and the total acoustic power. Figure 5 shows the phase shift behavior with the variation in total length; the phase damps out as the length increases. The graph has been plotted from the data generated from the model while keeping the volume of reservoir constant. Figure 6 depicts the curve of acoustic power as a function of length for different diameter and constant volume of reservoir. It can be understood from Figs. 5 and 6 that Phase and Acoustic Power for lower lengths behave like a wave and at higher length they become saturated. In addition, the acoustic power with respect to length is more for higher tube diameters. Figures 7 and 10 demonstrate the behavior of phase and acoustic power with change in volume of reservoir; it is noted that effectiveness of volume of reservoir is only in some specific region of the plot after which the curves saturate and the reservoir becomes obsolete for higher length of the tube, hereby such volume should be taken of reservoir that leads to maximum effectiveness of it and the system can be miniaturized (Fig. 8).

Fig. 5 Phase versus total length of the inertance tube

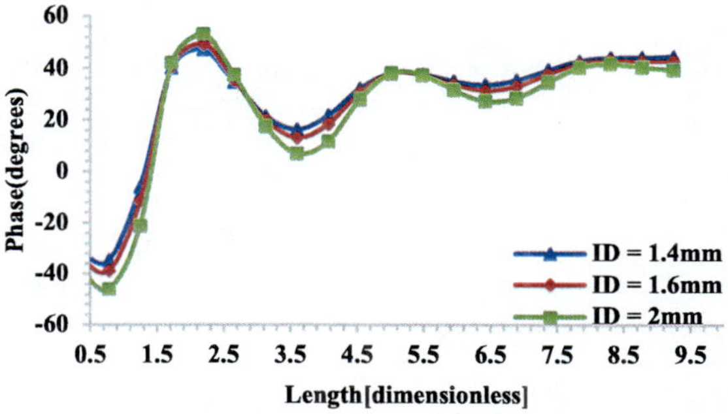

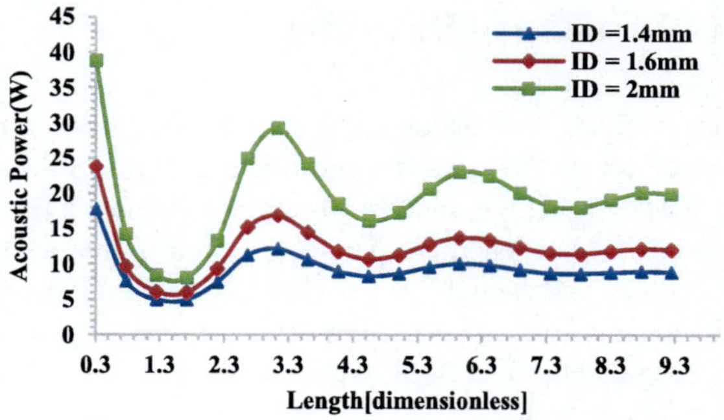

Fig. 6 Acoustic power versus total length for specific I.D. tubes

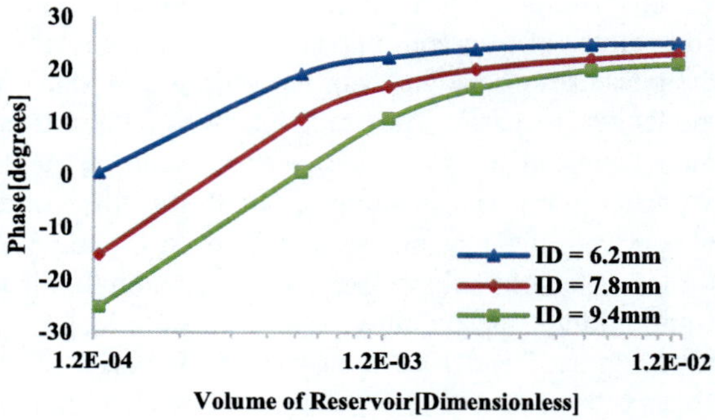

Fig. 7 Phase at the pulse tube end as a function of volume of reservoir for different I.D.

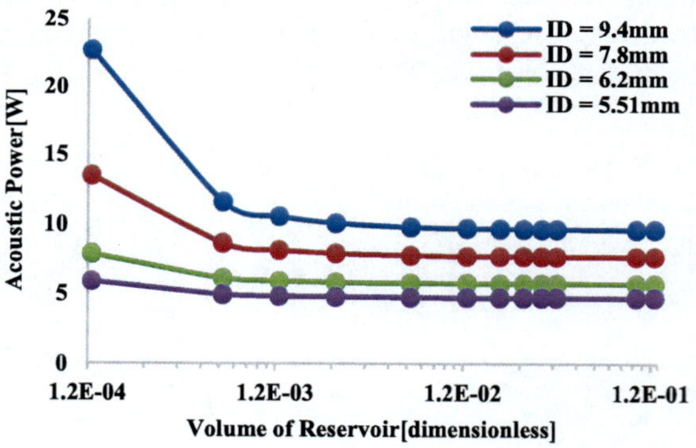

Fig. 8 Acoustic power versus volume of reservoir for four different I.D. of the inertance tube

4.1 Optimization

The final requirement from the model is to yield a geometry of the inertance tube in order to get −45° phase shift and with minimum power (Fig. 9). Table 1 shows the different ID values and the corresponding lengths and volumes of reservoir along with acoustic power. Figure 10 is a three-dimensional graph that shows the relation of acoustic power with phase shift and length from an inertance tube of different

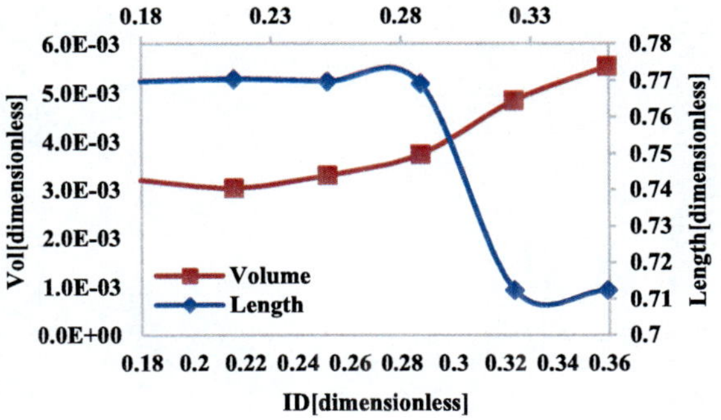

Fig. 9 Optimized volume of reservoir and inertance tube length as a function of ID of the tube to obtain −45° phase shift

diameters. Figure 11 shows the process of optimization, where tubes of two different cross section are used, and their length is distributed in n percentage of the total length. The optimization study with various combination of available tube IDs, length, and volume is carried out using Simplex optimization technique in EES software [9].

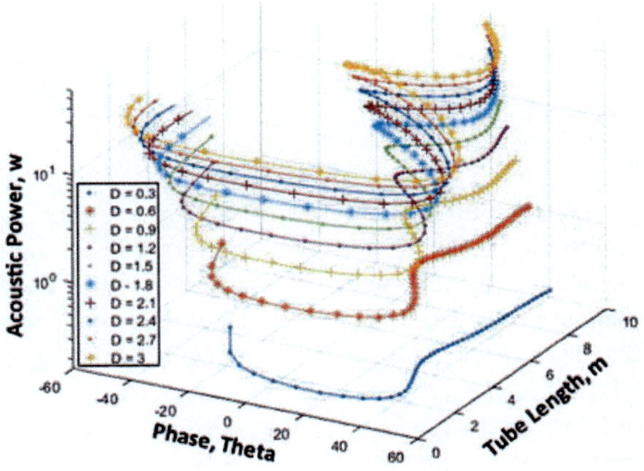

Fig. 10 Acoustic power, phase, and tube length for 100 Hz, mean pressure 40 bar

Table 1 Optimized dimensions for the desired phase shift from the inertance tube

ID tube [mm]	Vol. of res [cc]	Length [mm]	Acoustic power [W]	Phase at tube inlet
5	253.3	1249	114.4	−45°
6	240.2	1250	185.7	−45°
7	261.2	1249	277.1	−45°
8	296	1248	389.9	−45°
9	383.4	1156	549.7	−45°
10	439.1	1156	711.4	−45°

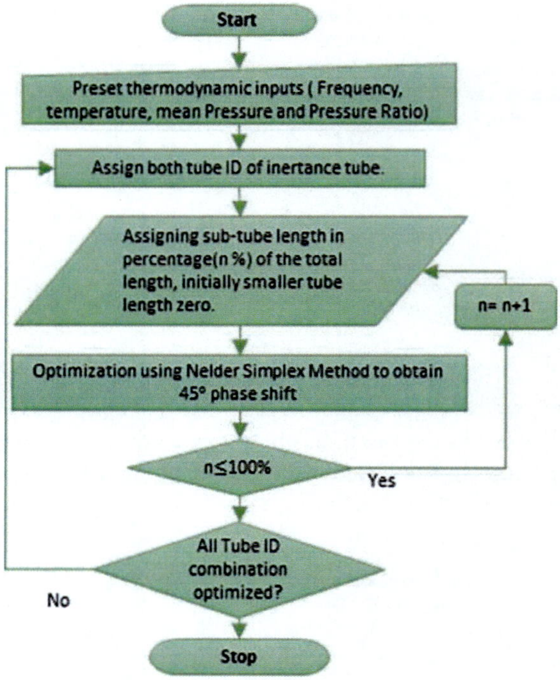

Fig. 11 Flowchart of optimization strategy

Initially uniform cross section of tube was taken, and best combination of inner diameter (ID), length, and volume of reservoir are found to achieve −45°, and corresponding acoustic power is estimated as shown in Table 1. It is noted that the requirement of acoustic power is very high with the minimum being 114.4 W for the inner diameter of tube being 5 mm, volume of reservoir 2.5E-4 m^3, and length of tube 1.25 m. Figure 12 shows the trend of acoustic power with respect to ID of tube in order to get 45° phase shift. To reach much lower acoustic power, variation in the cross section of inertance tube was explored, and its effect was analyzed. It was found that a slightly larger cross section in the tube near the reservoir end yields a much less acoustic power.

The tube was converted into two different diameter tubes with varying length and volume of reservoir in order to find the optimum combination of geometry. The length per unit node was kept constant to avoid error in computation, and optimization was done by varying both the length and volume of the reservoir, keeping the bounds of the geometry in tube area where the reservoir and inertance tube are effective and sensitive to the phase shifting mechanism. After extensive optimization, the dimensions were computed by the model, giving the total length of 1.761 m with 1.233 m long tube of 3.34 mm ID tube and 0.5284 m long tube of 1.65 mm ID tube, the volume of reservoir being 3e-4 m^3 with a minimum acoustic power of 11.26 W. Figure 13 demonstrates the effectiveness of the smaller diameter tube that can be

Fig. 12 Optimized acoustic power as a function of ID of the tube to obtain −45° phase shift

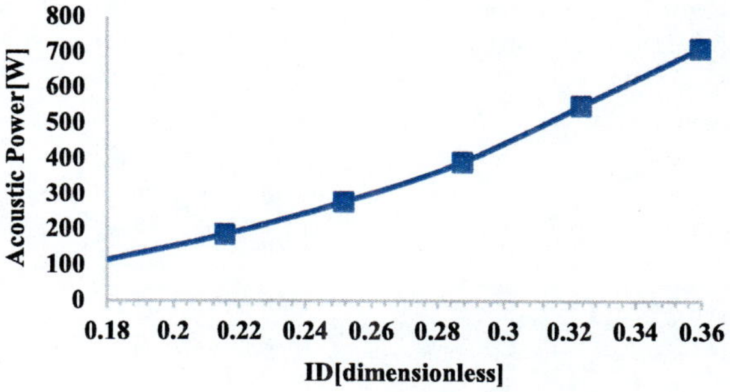

adjusted to get optimum phase shift after the realization of the component. Subsequently, a study is carried out to estimate the effect of non-geometrical parameters (P_o and frequency). As shown in Figs. 14 and 15 with increase in P_o and frequency, acoustic power and phase decrease.

Fig. 13 Phase and acoustic power versus smaller ID tube length

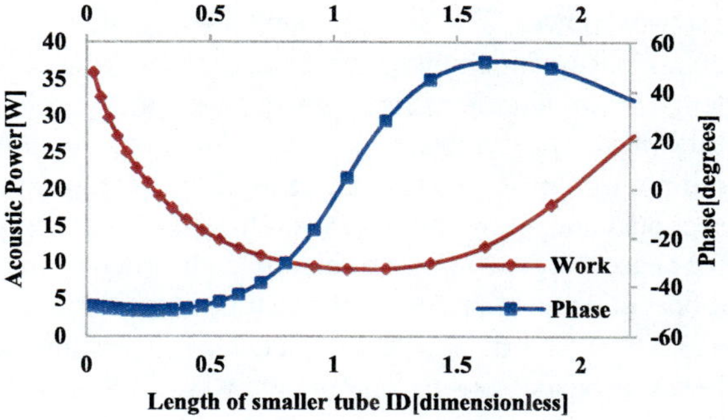

Fig. 14 Mean pressure effect over phase shift and acoustic power

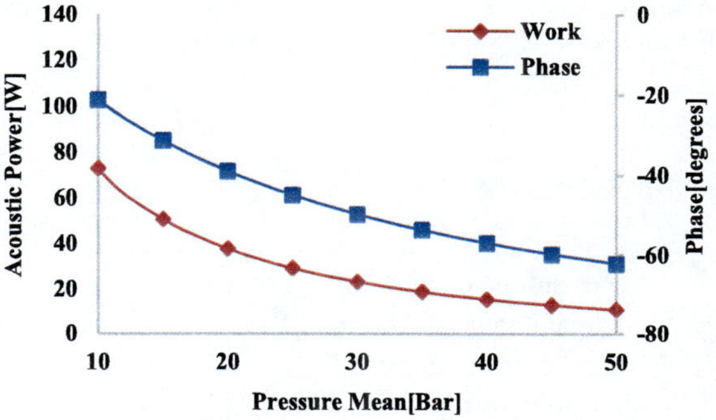

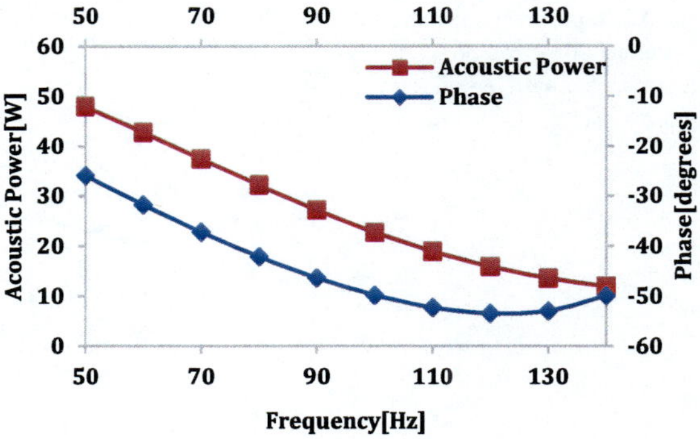

Fig. 15 Phase and acoustic power versus frequency

5 Conclusions

The numerical model presented in this paper gives a detailed analysis about different geometric parameters and its effectiveness toward phase shifting mechanism and acoustic power. The model shows the significance of inertance, resistance, and compliance of inertance tube and reservoir in the PSM. The cross-sectional area, length, and volume of reservoir can be optimized in a way that we get the necessary phase shift; however, such combination of geometrical dimensions that gives minimum acoustic power can be used. Further the advantage of using varied cross-sectional area of inertance tube is studied. The behavior and effect of inertance tube for phase shift in PTC can be clearly understood by the plots shown in this paper. The model concluded that an inertance tube combination of 1.23 m length of 3.34 mm ID and 0.53 m of 1.65 mm ID connected to a reservoir of volume 0.0003 m^3 will lead to lowest acoustic power requirement of 11.26 W providing us with an optimum phase shift of −45°.

Acknowledgements The authors are grateful to Director and Technology Development Committee-SAC, ISRO, for their encouragement and support.

Nomenclature

R_i	Resistance of ith node [ohm]
C_i	Capacitance of ith node [F]
I_i	Inertance of ith node [H]
T	Average temperature [K]
ρ	Density of helium [kg/m^3]
L	Length of the tube [m]
ω	Operating frequency [rad/s]
γ	Ratio of specific heat
s	Series connection

p	Parallel connection
P	Mean pressure [Bar]
V_{res}	Reservoir volume [m^3]
W	Acoustic power [W]
$\dot{m}$	Mass flow rate [kg/s]
Θ_m	Inlet phase of mass flow [deg]
R_{ent}	Resistance due to entrance [Ω]
P_a	Pressure amplitude [Bar]

References

1. Mikulin EI, Tarasov AA, Shrebyonock MP (1984) Low-temperature expansion pulse tube. Adv Cryog Eng 29:629–637
2. Zhu S, Wu P, Chen Z (1990) Double inlet pulse tube refrigerators: an important improvement. Cryogenics 30:514–520
3. Kanao K, Watanabe N, Kanazawa Y (1994) A miniature pulse tube refrigerator for temperature below 100 K. Cryogenics 34(supplement):167
4. De Boer PCT (2002) Performance of the inertance pulse tube. Cryogenics 42:209–221
5. Luo E, Radebaugh R, Lewis M (2004) Inertance tube models and their experimental verification. In: AIP conference proceedings, vol 710, no 1. American Institute of Physics
6. Radebaugh R et al (2006) Inertance tube optimization for pulse tube refrigerators. In: AIP conference proceedings, vol 823, no 1. American Institute of Physics
7. Swift GW (200) Thermoacoustics: a unifying perspective for some engines and refrigerators. Condensed Matter and Thermal Physics Group, Los Alamos National Laboratory
8. Xiao JH (1995) Thermoacoustic heat transportation and energy transformation Part 1: formulation of the problem. Cryogenics 35(1):15–19
9. Klein SA, Alvarado FL (2002) EES—engineering equation solver, F-Chart Software
10. Ali S (2001) Pressure drop correlations for flow through regular helical coil tubes. Fluid Dyn Res 28(4):295

Empirical and Machine Learning Approaches for Turbulent Thermal Convection in Rectangular Enclosures Tilted at Acute Angles

N. Sen, A. S. Pisharody, and U. Madanan

Abstract The present study proposes both an empirical and a machine learning approach to predict Nusselt numbers for turbulent thermal convection in rectangular enclosures inclined at acute angles for a given range of Rayleigh numbers (1.85×10^6–1.04×10^{11}) and aspect ratios (1, 3, 6, 10). Nusselt numbers predicted using the empirical correlation and the machine learning model are found to yield a mean absolute percentage error of 5 and 7%, respectively, when compared with the corresponding experimental values. Nusselt numbers predicted using both the approaches are also compared with the corresponding reported values from the literature and found to agree reasonably well.

Keywords Inclined enclosure thermal convection · Nusselt number · Rayleigh number · Aspect ratio · Machine learning

1 Introduction

Turbulent (or, high-Rayleigh-number) thermal convection in enclosures with differentially heated walls has been an area of significant research interest owing to its natural occurrence in various fields such as meteorology, oceanic convection, geophysics, and astrophysics. It also finds numerous engineering applications, such

N. Sen and A. S. Pisharody are the joint first authors.

N. Sen
Department of Mechanical Engineering, Jadavpur University, Kolkata 700032, West Bengal, India

A. S. Pisharody
Department of Computer Science and Engineering, IIT Kanpur, Kanpur 208016, Uttar Pradesh, India

U. Madanan (✉)
Experimental Thermofluids Laboratory, Department of Mechanical Engineering, IIT Kanpur, Kanpur 208016, Uttar Pradesh, India
e-mail: umadanan@iitk.ac.in

S. Das et al. (eds.), *Proceedings of the 1st International Conference on Fluid, Thermal and Energy Systems*, Lecture Notes in Mechanical Engineering,
https://doi.org/10.1007/978-981-99-5990-7_45

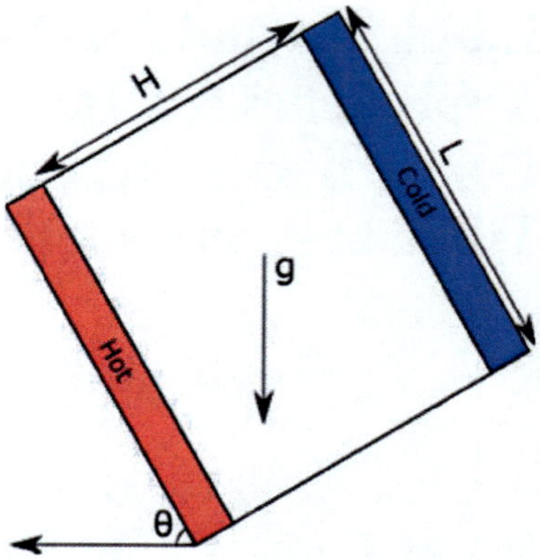

Fig. 1 Schematic of an inclined rectangular enclosure

as solar energy flat-plate collectors, nuclear reactors, crystal growth in metallurgy, and electronics cooling. Interestingly, most of these applications can be modeled as thermal convection in an enclosure that is tilted at an acute angle with respect to the horizontal (refer to Fig. 1 for details). Thus, characterizing the heat transfer rate for such inclined enclosures becomes paramount. A basic scaling analysis suggests that, in addition to the angle of inclination, θ, the non-dimensional heat transfer coefficient, i.e., Nusselt number (Nu), is a function of dimensionless parameters such as Rayleigh number (Ra), Prandtl number (Pr), and aspect ratio (AR). However, in most of these applications, the working fluid is fixed and, consequently, Prandtl number becomes invariant. Therefore, developing a correlation or a machine learning model that can predict Nusselt number as a function of remaining significant dimensionless parameters ($Nu = f(Ra, AR, \theta)$), which is valid for the entire range of acute angles, can be very useful.

2 Literature Review and Objective

Thermal convection in enclosures inclined at various acute angles can be broadly classified into Rayleigh-Bénard convection (i.e., when temperature gradient is anti-parallel to gravity ; $\theta = 0°$) and convection within enclosures inclined at a non-zero acute angle (i.e., $0° < \theta < 90°$). Rayleigh-Bénard convection is of significant theoretical relevance due to its pivotal role in helping us understand turbulence and is equally important from a practical perspective due to its prevalence in many engineering situations and natural phenomena. For this reason, review of the literature is broken down into separate subsections to address both these categories.

Table 1 Select scaling relations for Rayleigh-Bénard convection ($\theta = 0°$)

Authors	Ra range	Nu–Ra scaling
Niemela and Sreenivasan [1]	1.1×10^6–2.9×10^{12}	$0.088Ra^{0.32}$
Chong and Xia [2]	3×10^4–1×10^{11}	$1 + 0.106Ra^{0.308}$
Waleffe et al. [3]	1×10^7–1×10^9	$0.115Ra^{0.31}$

2.1 Turbulent Rayleigh-Bénard Convection

A vast majority of research for turbulent Rayleigh-Bénard convection suggest that Nusselt number is independent of aspect ratio. This immediately yields a relation of the form $Nu = CRa^n$ since Nusselt number is then dependent only on Rayleigh number for $\theta = 0°$ [1–3]. Some of these significant scaling relations for turbulent Rayleigh-Bénard convection are presented in Table 1.

2.2 Turbulent Thermal Convection in Enclosures Inclined at Non-zero Acute Angles

Inaba [4] experimentally studied thermal convection in an inclined rectangular cavity of varied aspect ratios filled with air. He outlined a criterion for characterizing a fully turbulent regime for thermal convection in an inclined enclosure, i.e., $Ra\cos\theta > 4 \times 10^5$. Based on his experiments, he proposed a relation of the form $Nu = 0.0785(Ra\cos\theta)^{0.330}$ for acute angles of inclination. Goldstein et al. [5] employed an electrochemical mass transfer technique to study thermal convection in rectangular enclosures and proposed correlations for Nusselt number following two different forms: (i) $Nu = f(Ra, AR)$ for a given angle of inclination and (ii) $Nu = C(Ra\cos\theta)^n$ for a given range of acute angles (as can be inferred from Table 2). Dropkin and Somerscales [6] conducted experiments for a wide range of Prandtl numbers and reported that a relation of the form $Nu = CRa^{1/3}Pr^{0.074}$ exists if one needs to account for the effect of Prandtl number. They further stipulated that the value of C reduces with increase in angle of inclination (refer to Table 2). More recently, Madanan and Goldstein [7, 8] performed an experimental study of high-Rayleigh-number thermal convection ($1.85 \times 10^6 \leq Ra \leq 1.04 \times 10^{11}$) using compressed gases in rectangular enclosures of varied aspect ratios ($AR = 1, 3, 6, 10$) at different angles of inclination ($\theta = 0°, 30°, 60°, 90°, 120°, 150°$). They observed that Nusselt number decreases monotonically with angle of inclination due to a decrease in the buoyant force in the flow direction. Also, for a given Rayleigh number and inclination angle, Nusselt number was found to follow a decreasing trend with an increase in aspect ratio. This effect was found to be negligible at $\theta = 0°$ and gradually amplified with increase in angle of inclination. They proposed a scaling relationship of the form $Nu = 0.067(Ra\cos\theta)^{0.331}$ to predict the heat transfer rate for $\theta \leq 60°$.

Table 2 Select scaling relations for inclined enclosure convection ($0° \leq \theta \leq 60°$)

Authors	*Ra*	*AR*	*Nu–Ra* scaling
Goldstein et al. [5]	2×10^8–1.6×10^{12}	1.8–36	$0.066(Ra\cos\theta)^{1/3}$
Dropkin and Somerscales [6]	5×10^4–7.17×10^8	4.41–16.56	$CRa^{1/3}Pr^{0.074}$ $C = 0.069(0°)$, $0.065(30°)$, $0.057(60°)$

2.3 Machine Learning in Turbulent Convection

Recent years have seen rising popularity in the use of machine learning modeling in the field of thermal sciences [9]. Ability to identify trends and patterns, ease of handling large amount of data, and applicability to a wide variety of problems are a few reasons for its gaining reputation in this field. Many contemporary studies in the field of thermal convection have featured machine learning approaches to challenges such as modeling turbulent heat flux for thermal convection in low Prandtl number fluids [10], predicting temperature in mantle convection using known relevant parameters [11], analyzing heat transport by turbulent superstructures in a Rayleigh-Bénard convection layer [12], and providing improved predictions for Reynolds and Nusselt numbers in turbulent Rayleigh-Bénard convection [13]. However, to the best of the authors' knowledge, most studies in the literature had an abundance of data (obtained either experimentally or computationally) and therefore used neural networks for building these machine learning models. Unfortunately, for problems that may have only a limited amount of data available, this is not a feasible option. Therefore, the present study attempts to explore an alternate machine learning approach that may offer reasonable results with small-sized data.

Based on a careful review of the literature pertaining to high-Rayleigh-number thermal convection in acute angled rectangular enclosures, it is apparent that most of the proposed correlations for Nusselt number are either for a particular angle of inclination or do not include all the relevant parameters of interest, i.e., Ra, AR, and θ for a given fluid or Prandtl number. The present study therefore aims to develop a unified approach for determining Nusselt number given a set of Rayleigh number, aspect ratio, and angle of inclination for turbulent thermal convection in enclosures inclined at acute angles. This is achieved using two techniques:

I. An empirical correlation based on experimental data from the literature [7, 8] taking into account the heat transfer physics communicated in the past literature.
II. A machine learning model based on the eXtreme Gradient Boosting (XGBoost) algorithm trained using the acute angle experimental data from the literature [7, 8].

3 Methodology

3.1 Development of Empirical Correlation

The present study proposes a generic correlation that considers the various heat transfer physics and trends reported by Madanan and Goldstein [7, 8]. This correlation is established using the experimental data presented in Table 3 and takes into account the combined effect of three independent variables, i.e., Rayleigh number, aspect ratio, and angle of inclination, on Nusselt number. Two physics-based adjustments are carried out on the *Nu*–*Ra* relation for Rayleigh-Bénard convection ($\theta = 0°$) to achieve this correlation for all the acute angles and a given range of aspect ratios.

Firstly, acceleration due to gravity, g, is replaced with $g\cos\theta$ in the expression for Rayleigh number to consider the effect of angle of inclination on the net buoyant force. Thus, the correlation takes a modified form: $Nu = C(Ra\cos\theta)^n$, where C and n are arbitrary constants. Following this, the effect of aspect ratio, which has a gradually increasing influence on Nusselt number with an increase in angle of inclination (with almost no discernible effect for Rayleigh-Bénard convection), is also

Table 3 Dataset used for the present study [7, 8]

Ra	*AR*	*Nu*		
		0°	30°	60°
1.85×10^6	10	8.87	7.02	6.66
3.25×10^6	10	10.39	8.32	7.63
6.14×10^6	10	12.80	10.01	9.25
8.77×10^6	10	14.33	12.01	10.99
1.05×10^7	10	14.83	12.64	11.87
1.28×10^7	10	15.93	13.77	12.76
3.98×10^7	6	21.78	18.64	16.31
6.33×10^7	6	24.22	22.72	20.99
9.76×10^7	6	29.71	27.67	25.23
2.89×10^8	3	41.33	36.23	34.27
7.90×10^8	3	57.32	54.67	49.26
1.00×10^9	3	62.14	58.93	54.19
1.82×10^9	3	78.11	76.22	69.99
3.96×10^9	1	96.78	88.77	83.19
9.23×10^9	1	133.22	124.55	113.15
1.62×10^{10}	1	161.55	152.77	141.38
4.92×10^{10}	1	219.19	207.81	195.67
7.60×10^{10}	1	256.12	239.23	228.11
1.04×10^{11}	1	291.12	283.11	269.23

incorporated into this correlation. This increasing influence of aspect ratio on Nusselt number is negative in nature, i.e., for a given acute angle of inclination, Nusselt number decreases with increase in aspect ratio. These observations suggest that the functional relationship between Nu and AR should be of the form:$Nu = \left(AR^{-m}\right)^{r}$, where m is an arbitrary positive constant and r is a function such that $r \approx 0$ for $\theta = 0°$ and $r \approx r_{max}$ as the angle of inclination approaches 90°. A simple logic hints that r must be a function of $\sin p\theta$ and can be expressed as $r = A\sin p\theta$ with A and p being positive constants. It is also known that for turbulent thermal convection, Nusselt number decreases monotonically with an increase in angle of inclination [7, 8]. However, since the effect of angle of inclination has already been included while establishing the independent relation of Nusselt number with both Ra and AR, additional individual effect of angle of inclination is ignored for this correlation. Finally, the authors use a hit and trial method that minimizes the root mean square error between the predictions and the corresponding experimental values of Nusselt number [7, 8] to obtain the aforementioned constants in the correlation.

The final form of the proposed correlation becomes

$$Nu = 0.069(Ra\cos\theta)^{0.33}\left(\sin\theta\left(1 + \frac{1}{AR}\right)\right)^{0.3\sin\theta} \tag{1}$$

This correlation is valid for all values of acute angles, given $1.85 \times 10^6 \leq Ra \leq 1.04 \times 10^{11}$ and $1 \leq AR \leq 10$.

3.2 Machine Learning (ML) Model

The ML model developed in the present paper is constructed using three input features: Rayleigh number, aspect ratio, and angle of inclination. The complete dataset consists of 57 experimental data points (as shown in Table 3), which is randomly split using a 70–30 ratio for model building, i.e., 70% (or, 38 data points) for determining the model parameters and 30% (or, 19 data points) for testing robustness of the model to unseen data. The range of input features used is Rayleigh number (1.85×10^6–1.04×10^{11}), aspect ratio $(1, 3, 6, 10)$, angle of inclination $(0°, 30°, 60°)$.

Due to the size limitations of the dataset used, a major concern when choosing a machine learning algorithm is the model overfitting to the training data. Keeping this in mind, simple non-linear algorithms such as Support Vector Machine and K Nearest Neighbor are attempted initially. However, these are found to be unsuitable for the present dataset. Ensemble learning algorithms are less prone to overfitting and among the algorithms available presently, eXtreme Gradient Boosting algorithm (XGBoost) [14] is known to deliver superior performance for structured data. Although more popularly used for larger datasets, its robustness against overfitting and high predictive ability are the main motivation behind choosing this algorithm for the present problem statement. Through optimized system and algorithmic performance, XGBoost additively builds an ensemble of weak learners (in this case, decision trees), such that each consecutive learner learns from previous learner's errors and combines their individual prediction results to build an improved strong learner. This is achieved by optimizing the objective function with each new tree addition. Here, the objective function is of the form:

$$obj(f) = L(f) + \Omega(f) \tag{2}$$

where L is the loss function, Ω is the regularization term, and f consists of parameters to be determined for optimization. The loss function aids the predictive capability of the model, and the regularization term ensures that the model does not overfit on the training data (or, it fails to generalize to unseen data).

The *scikit-learn API* of the XGBoost algorithm, *XGBRegressor*, is used for training the present model. GridSearchCV (from the *scikit-learn* library) and Bayesian Optimization [15] methodologies are utilized to tune the hyperparameters of the model. Some important hyperparameters to be tuned include maximum depth of each decision tree (*max_depth*), regularization terms (*reg_alpha* and *reg_lambda*), and number of boosted decision trees (*n_estimators*). The loss function used for the model is *squared error* (or, the root mean squared error). The final model is chosen based on how they performed against the test dataset on three metrics: R-squared (R^2), Root Mean Squared Error (RMSE), and Mean Absolute Percentage Error (MAPE).

4 Results and Discussion

Nusselt numbers estimated using the proposed empirical correlation are compared with those reported by Madanan and Goldstein [7, 8]. This is done by plotting them against Rayleigh number on a decadic log–log scale as shown in Fig. 2. As can be inferred from the figure, they are in very good agreement for all the three tested angles of inclination. To assess the performance of the correlation, mean absolute percentage error, established using the experimental values as true values, is determined to be within 5% for all the three angles of inclination. This clearly underscores the correlation's ability to predict Nusselt numbers with reasonable accuracy for the investigated range of aspect ratios and angles of inclination.

To determine how well the proposed correlation follows the reported trends of Nusselt number with aspect ratio for various tested angles of inclination [7, 8], normalized Nusselt numbers, defined as $\frac{Nu}{Nu_{AR=10}}$, are plotted against aspect ratio for all the tested angles at a fixed Rayleigh number of 4.4×10^7 (refer to Fig. 3). As depicted in the figure, normalized Nusselt numbers are found to follow a similar trend with aspect ratio as reported in the literature [7, 8]. However, there is a slight offset in the values predicted.

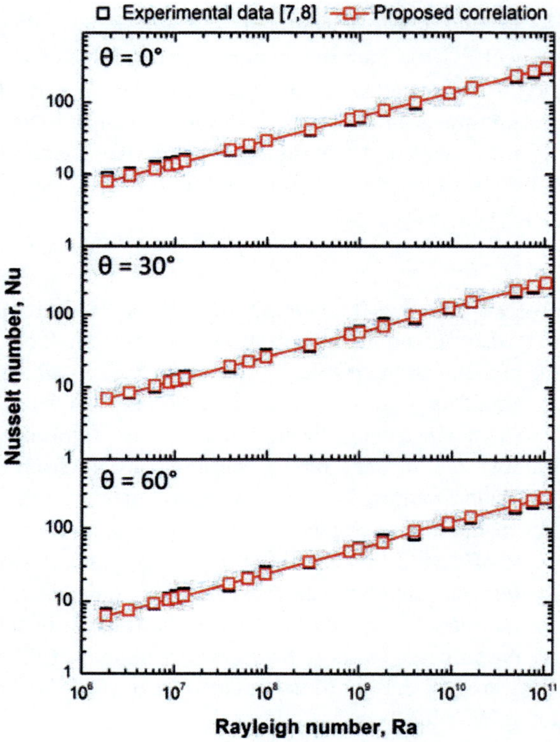

Fig. 2 Experimental data versus correlation

The machine learning model is initially used for predicting Nusselt numbers for the test dataset reserved at the beginning. These predictions are found to be in reasonable agreement with experimental Nusselt numbers (MAPE ~ 7%, RMSE ~ 3.76). The model predictions are plotted against experimental Nusselt numbers (as shown in Fig. 4) and present a reasonably accurate *R*-squared value of 0.997 for the test dataset. The model is also cross-validated for robustness using a tenfold cross-validation system, wherein the model is fitted a total of ten times on a training set consisting of 90% random samples, with the remaining 10% used as test set. This cross-validation yielded a reasonable mean absolute percentage error of 6.79%, with a standard deviation of 4.65%. Although a more in-depth analysis is required to gain deeper insights, a preliminary examination of feature importance given to each input feature by the model shows that Rayleigh number is given the highest importance, followed by angle of inclination, and finally aspect ratio (as can be seen from Fig. 5) when building decision trees during the model training phase.

Finally, correlations proposed in the literature for 0° [1–3], 30° [5, 6], and 60° [5, 6] are used to estimate Nusselt numbers corresponding to the reserved dataset. These estimated Nusselt number values along with the predictions from the correlation and the ML model are then plotted against Rayleigh number on a decadic log–log scale

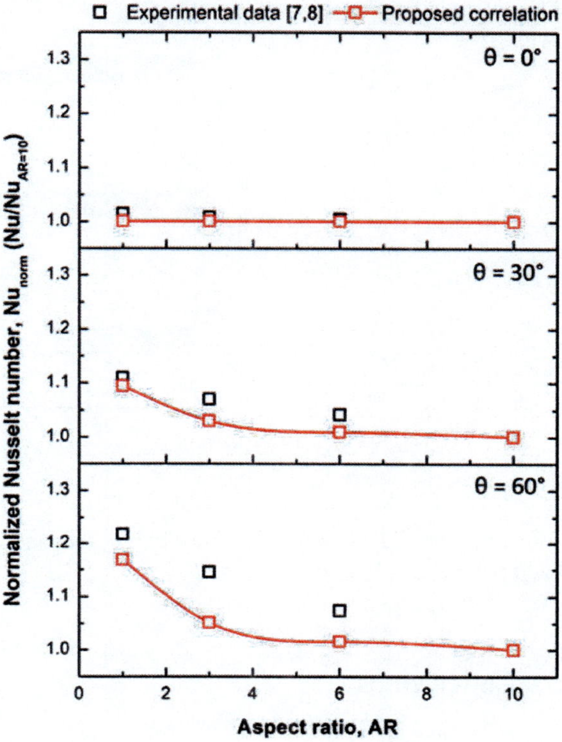

Fig. 3 Influence of aspect ratio on Nusselt number for varied angles of inclination at $Ra = 4.4 \times 10^7$

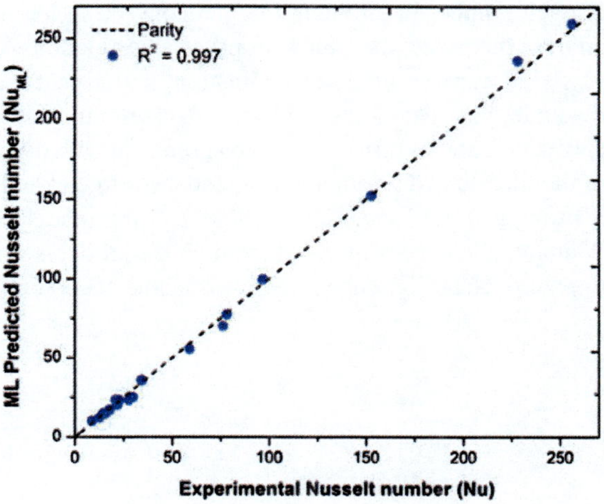

Fig. 4 Performance of ML model

Fig. 5 Feature importance in ML model

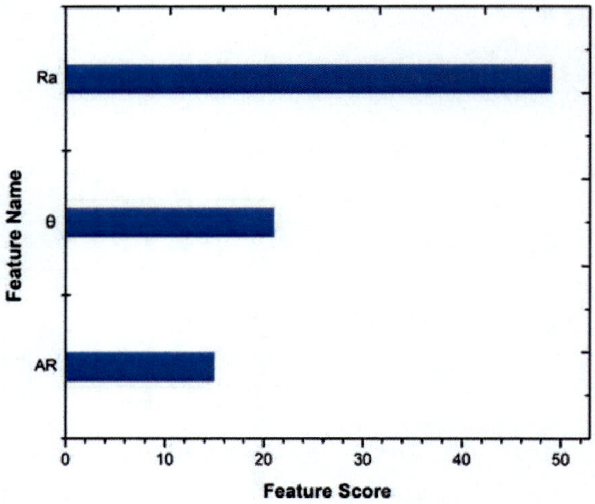

for comparison. As can be seen from Fig. 6, they are in reasonable agreement with each other.

5 Conclusions

The present study attempts to address the unavailability of a unified correlation for Nusselt number in turbulent thermal convection for acute angles of inclination. To address this issue, this study proposes a correlation for Nusselt number in terms of Rayleigh number, angle of inclination, and aspect ratio, taking into consideration the underlying heat transfer physics. Nusselt numbers estimated using the empirical correlation are found to be in good agreement with the corresponding values reported in the literature. The authors also endeavor to tackle this problem through machine learning and put forward a machine learning model using the XGBoost algorithm. Although the dataset used for building the model is small-sized, the model is found to predict Nusselt numbers with reasonable accuracy.

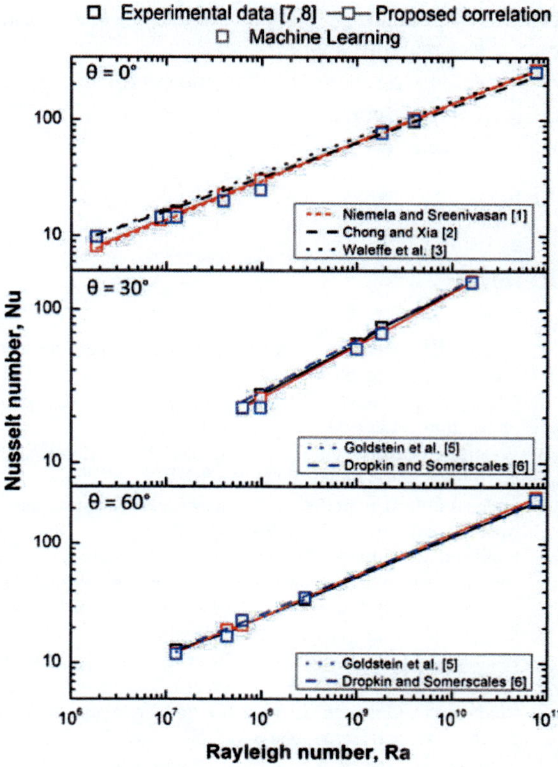

Fig. 6 Comparison with the literature

Acknowledgements The author U. M. would like to take this opportunity to acknowledge Heat Transfer Laboratory at the University of Minnesota Twin Cities where he acquired the experimental data used in the present study.

Nomenclature

AR	Aspect ratio $= L/H$
C	Arbitrary constant
Gr	Grashof number ($Gr = g\beta\Delta T\ H^3/\nu^2$)
g	Acceleration due to gravity [m/s^2]
H	Spacing between the hot and cold plates [m]
h	Average heat transfer coefficient [W/m^2 − K]
k_f	Thermal conductivity of the fluid [W/m − K]
L	Extent of the square hot (or cold) plate [m]
Nu	Average Nusselt number $= hH/k_f$
Pr	Prandtl number $= \nu/\alpha$
Ra	Rayleigh number ($Ra = g\beta\Delta T\ H^3/\nu\alpha$)

ΔT	Temperature difference between hot and cold plates [K]
α	Thermal diffusivity [m^2/s]
β	Volumetric thermal expansion coefficient [K^{-1}]
θ	Angle of inclination [°]
ν	Kinematic viscosity [m^2/s]

References

1. Niemela JJ, Sreenivasan KR (2006) Turbulent convection at high Rayleigh numbers and aspect ratio 4. J Fluid Mech 557:411–422
2. Chong KL, Xia KQ (2016) Exploring the severely confined regime in Rayleigh-Bénard convection. J Fluid Mech 805:R4–R1
3. Waleffe F, Boonkasame A, Smith LM (2015) Heat transport by coherent Rayleigh-Bénard convection. Phys Fluids 27(5):051702
4. Hideo I (1984) Experimental study of natural convection in an inclined air layer. Int J Heat Mass Transf 27(8):1127–1139
5. Goldstein RJ, Chiang HD, Sayer E (1987) Natural convection mass transfer in an inclined enclosure at high Rayleigh number. In: Proceedings of 2nd international symposium on transport phenomena in turbulent flows, October 1987, Tokyo, Japan, pp 229–255
6. Dropkin D, Somerscales E (1965) Heat transfer by natural convection in liquids confined by two parallel plates which are inclined at various angles with respect to the horizontal. J Heat Transf 87(1):77–82
7. Madanan U, Goldstein RJ (2020) High-Rayleigh-number thermal convection of compressed gases in inclined rectangular enclosures. Phys Fluids 32(1):017103
8. Madanan U, Goldstein RJ (2019) Experimental investigation on very-high-Rayleigh-number thermal convection in tilted rectangular enclosures. Int J Heat Mass Transf 139:121–129
9. Pandey S, Schumacher J, Sreenivasan KR (2020) A perspective on machine learning in turbulent flows. J Turbul 21(9–10):567–584
10. Fiore M, Koloszar L, Mendez MA, Duponcheel M, Bartosiewicz Y (2022) Physics-constrained machine learning for thermal turbulence modelling at low Prandtl numbers. arXiv preprint arXiv:2201.06301
11. Agarwal S, Tosi N, Kessel P, Breuer D, Montavon G (2021) Deep learning for surrogate modeling of two-dimensional mantle convection. Phys Rev Fluids 6(11):113801
12. Fonda E, Pandey A, Schumacher J, Sreenivasan KR (2019) Deep learning in turbulent convection networks. Proc Natl Acad Sci 116(18):8667–8672
13. Bhattacharya S, Verma MK, Bhattacharya A (2022) Predictions of Reynolds and Nusselt numbers in turbulent convection using machine-learning models. arXiv preprint arXiv:2201.03200
14. Chen T, Guestrin C (2016) XGBoost: a scalable tree boosting system. In: Proceedings of the 22nd ACM SIGKDD international conference on knowledge discovery and data mining, August 2016, San Francisco, California, USA, pp 785–794
15. Nogueira F (2014) Bayesian optimization: open source constrained global optimization tool for Python. https://github.com/fmfn/BayesianOptimization

Investigation on Performance Characteristics of Convergent Fin Heat Sink Under Forced Convection Using CFD

Anurag Arjunan and P. Kumaresan

Abstract Majority of the existing studies and literature, in the field of Electronic cooling, primarily focus on increasing the surface area of the heat sinks to achieve better heat transfer rates. This study focuses on how to enhance the average surface heat transfer coefficient by converging the flow of fluid. Inspiration has been taken from the convergent nozzle. Fins on the heat sink are placed at various angles relative to each other. This is done in order to mimic the convergent behaviour under forced convection. Simulations are done for a number of test cases by varying the following parameters—Air velocity, Fin angle, and heat input. Analysis of the CFD results shows that the average heat transfer coefficient (HTC) for convergent angled fins is higher compared to straight fins by ~15%. Similarly, average surface heat flux is also higher when compared to straight fin heat sinks. Also, as a consequence of the fin orientation, almost 11% mass reduction is seen in the angled fin heat sinks. The flow profile in the wake of the convergent angled fins proved to be suitable for restricted ventilation spaces.

Keywords Forced-convection · Convergent fins · Boundary layer · Average HTC · Weight reduction

1 Introduction

As we know there has been significant progress in the field of Electronics over the past few decades and despite the fact that the components are being miniaturized, the amount of heat dissipated by them is still off the charts. It is for this purpose, that the design modifications and development of heat sinks are becoming highly substantial with each passing day.

Existing research and studies focus on increasing the exposed area of the heat sink to maximize heat dissipation in forced convection scenarios. This led to the

A. Arjunan (✉) · P. Kumaresan
PDIC, Bharat Electronics Ltd., Bangalore, Karnataka 560013, India
e-mail: anuragiitdh2020@gmail.com

S. Das et al. (eds.), *Proceedings of the 1st International Conference on Fluid, Thermal and Energy Systems*, Lecture Notes in Mechanical Engineering,
https://doi.org/10.1007/978-981-99-5990-7_46

development of complex geometries like wavy fins, branched fins, flared fins [1, 2], perforated pin fins [8], etc. Experiments on these heat sinks have shown better performance than traditional heat sinks. But manufacturing of these heat sink requires complex processes like EDM, Wire cutting, and metal 3D printing. As a result, these heat sinks become less preferred option.

2 Literature Review and Objective

Various studies have been carried out to improve the performance of heat sinks. Many of them focus on increasing exposed surface area, while others focus on varying convective airflow (natural or convective). As per existing research, heat sink orientation and fin angles are the parameters to enhance the flow.

ATS has conducted experiments to show that maxiFLOW™ flared-fin performs better than other standard configurations. It had significant improvement in thermal resistance and HTC [1]. Similarly, one of the investigations also proved, experimentally, that flared heat sinks have less thermal resistance and is further enhanced with an increase in the number of fins [2]. S.K. Farooq studied the effect of fin inclination with the base surface on base temperature distribution and HTC in the case of natural convection [3]. Challenges in thermal management of photovoltaic cells and improvement studies strengthen the role of convergent-divergent construction in microchannel heat sink design which enhances the heat transfer rate compared to straight microchannel heat sink design [4]. The thickness of the fin influences the performance of the heat sink, lesser the thickness of the fins, the better the heat dissipation by fluid flowing over the surface and lesser the weight of the heat sink [5]. In addition, the optimization of plate-fin heat sinks with variable fin thickness along the direction normal to fluid flow reduces the thermal resistance compared to uniform fin thickness [6]. This literature has proved that Heat sink orientation, flared fins, flow channel construction, and fin thickness can significantly improve the thermal performance of the system. However, there is negligible data on the effect of fin angle variation, on HTC, across the flow.

In this study, an attempt has been made to improve the straight fin heat sink by varying the fin angle along the axis parallel to airflow. Inspiration has been taken from the convergent nozzle as the primary objective is to enhance the airflow usage efficiency. Emphasis has been given to mass reduction too. The machining process complexity is the same across, the straight fin heat sink and angled heat sinks, which makes it useful for industrial applications. Since the concept is novel, the scope of numerical investigation is limited.

3 Materials and Methods

3.1 Geometry

As per the literature survey, Flat plate inclinations had a significant effect on boundary layer thickness and heat distribution [11]. Also, a convergent nozzle leads to an increase in velocity (By continuity equation). The effect on heat transfer, due to convergence, was studied in the case of forced convection. To incorporate these improvements, the following two types of geometries were modelled (Refer Fig. 1, 2, and Table 1).

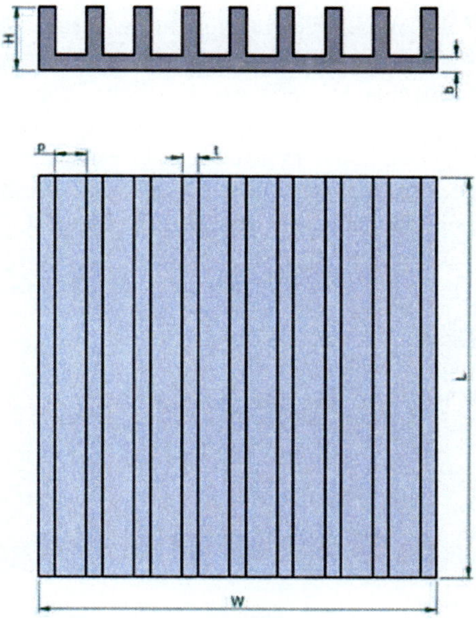

Fig. 1 Straight fin geometry parameters

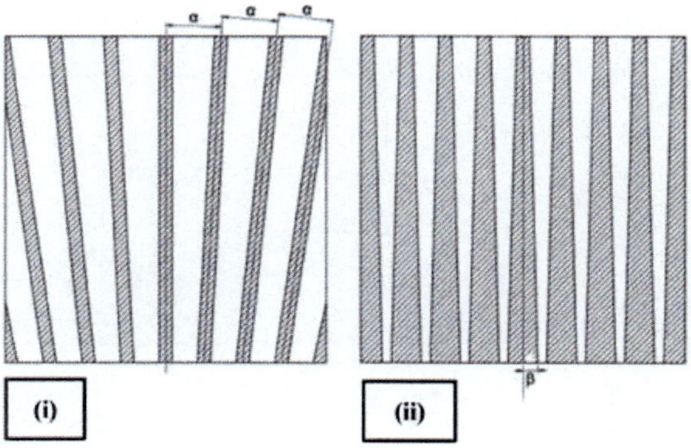

Fig. 2 Two types of convergent heat sinks (top view) (i) angled fins, (ii) longitudinal tapered fins (α—Fin orientation angle, β—Taper angle)

Table 1 Geometry details

W	*L*	*H*	*p*	*t*	*b*
50 mm	50 mm	8 mm	4 mm	2 mm	2 mm

Control parameter	Reference	Angled fins			Tapered fins		
	Straight fins	$\alpha = 1°$	$\alpha = 2°$	$\alpha = 3°$	$\beta = 0.5°$	$\beta = 1.0°$	$\beta = 1.5°$
Surface area (exposed to fluid) $\times 10^{-3}$ m^2	0.852	0.79	0.76	0.75	0.856	0.86	0.864

3.2 Simulation Setup

Solid Works Flow Simulation was used to simulate all the test cases obtained by varying 'α', 'β', air velocity, flow direction, and heat input. The material defined for the heat sinks was AA6061-T6. It is used extensively in Electronic cooling systems due to its comparatively higher thermal conductivity (155 W/m^2-K at 300 K).

The ambient room temperature was kept at 25 °C. For heat input, typical FC PBGA package was used. The contact between the heat sink and IC package was defined as a Chomerics thermal pad (Therm-A-Gap, thermal resistance: 5.8×10^{-4}m^2-K/W) (Fig. 3).

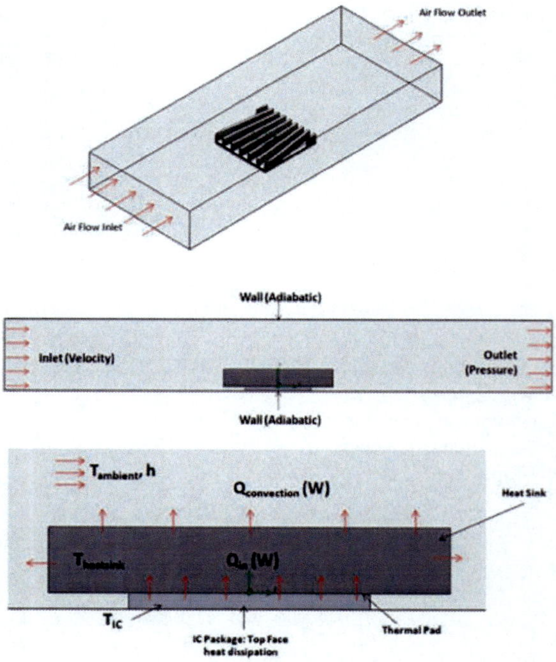

Fig. 3 Computational domain and boundary conditions

3.3 Governing Equations

CFD Analysis for heat sink under forced convection will involve governing equations of both fluid mechanics and heat transfer.

$$\nabla \cdot V = 0 \tag{1}$$

$$\rho\left(\frac{DV}{Dt}\right) = -\nabla p + \mu \nabla^2 V \tag{2}$$

$$\rho C_p\left(\frac{DT}{Dt}\right) = k\nabla^2 T \tag{3}$$

where V, ρ, p, μ, Cp, k, and T represent the velocity vector, fluid density, pressure, dynamic viscosity, specific heat, thermal conductivity, and temperature, respectively.

4 Validation

Pressure drop across the heat sink is used to validate the correctness of the simulations. The pressure drop calculation method is mentioned in an electronic cooling article [9]. This method has been devised by using equations from Culham and Muzyhka [10].

$$\Delta P = \left(K_c + 4 \cdot f_{app} \cdot \frac{L}{D_h} + K_e\right) \cdot \rho \frac{V^2}{2} \tag{4}$$

$$D_h = \frac{4 \cdot A_c}{P} \tag{5}$$

$$K_c = 0.42 \cdot (1 - \sigma^2) \tag{6}$$

$$K_e = (1 - \sigma^2)^2 \tag{7}$$

$$\sigma = 1 - \frac{N_{fin} \cdot t}{w} \tag{8}$$

$$f_{app} = \frac{\left[\left(\frac{3.44}{\sqrt{L^*}}\right)^2 + (f.Re)^2\right]^{1/2}}{Re} \tag{9}$$

$$L^* = \frac{L}{D_h \cdot Re} \tag{10}$$

Fig. 4 Pressure drop versus Inlet air velocity

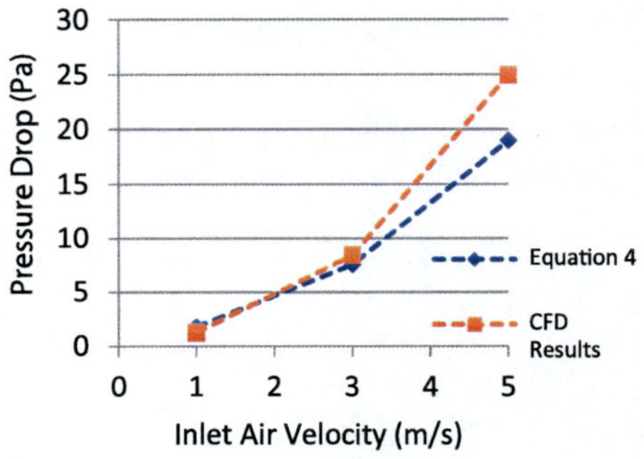

$$Re = \frac{\rho \cdot V \cdot D_h}{\mu} \tag{11}$$

$$f = (24 - 35.527 \cdot \lambda + 46.721 \cdot \lambda^2 - 40.829 \cdot \lambda^3 + 22.954 \cdot \lambda^4 - 6.089 \cdot \lambda^5) \tag{12}$$

In Fig. 4 Pressure drop values (for straight fins) are compared resulting from Eq. (4) and CFD results. Observed deviation can be due to the fact that the equation is modelled for fully developed laminar flow.

5 Results and Discussion

5.1 *Velocity Profile and Boundary Layer*

No slip condition gives rise to a boundary layer just above the surface. Observing Fig. 5, for straight fins, flow is exactly parallel to the surface thus boundary layer is very well developed. A significant amount of momentum is lost due to this. For angled fins, the angle of attack from the inlet air is nonzero. There exists a component of velocity that is perpendicular to the fin surface. Thus, boundary layer thickness is reduced compared to straight fins. As a result, the momentum loss is reduced. This behaviour is in accordance with a numerical investigation by Oguz Turgut [11].

Also, because of the convergent nature of the heat sink, the area of the airflow path keeps on reducing. Therefore, velocity increases as per 'Continuity Equation: $A_i V = A_o\ V_o'$.

Due to these two phenomena, heat sinks with angled fins have higher velocity sustained in between the fins (Refer to Fig. 6). With the increase in orientation angle, the boundary layer thickness further reduces.

For longitudinal tapered fins, fin spacing reduces to an extent where flow gets restricted. It negatively affects the heat dissipation.

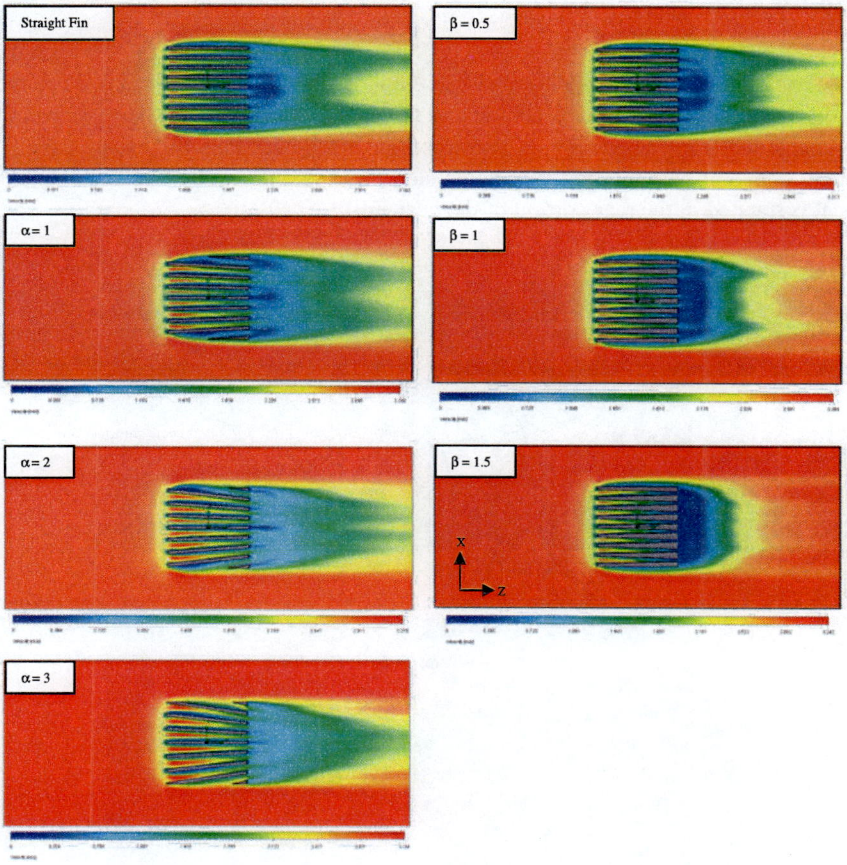

Fig. 5 Velocity contours for various geometries. Top view, cut plot at midpoint of fins (Inlet velocity: 3 m/s, heat input: 10W)

Fig. 6 Velocity along *Z*-axis, between two adjacent fins

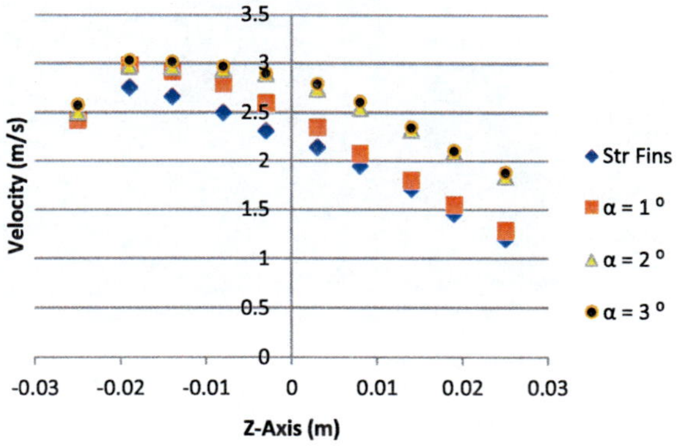

5.2 Temperature Profiles

Figure 7 shows the temperature distribution of the fluid. Angled fins have the most favourable temperature contour as far as the electronic cooling application is concerned. The wake region is converged and elongated. This will ensure the efficient escape of heat through ventilation slots. Tapered fins hinder the airflow, thus heat is trapped in the rear portion of the heat sink.

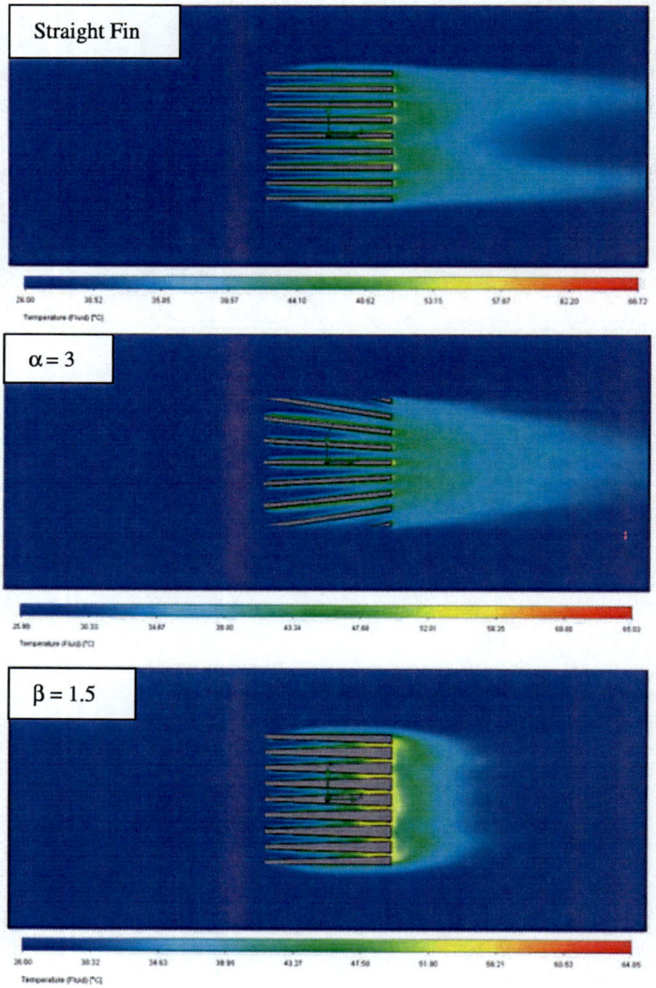

Fig. 7 Fluid temperature contour for 3 major geometries (Inlet velocity: 3 m/s, heat input: 10 W)

5.3 *Average Surface HTC and Heat Flux*

In Fig. 8, it is seen that the average HTC increases with an increase in orientation angle. This behaviour becomes significant when air velocity goes higher. HTC has a direct correlation with speed [7]. Also, for higher heat inputs, average HTC is higher since the heat sink reaches higher temperatures relative to ambient air temperature. This promotes better heat transfer. It is seen that the greater the orientation angle, the better the surface average HTC.

Fig. 8 Variation of Avg. HTC with air velocity for various heat inputs

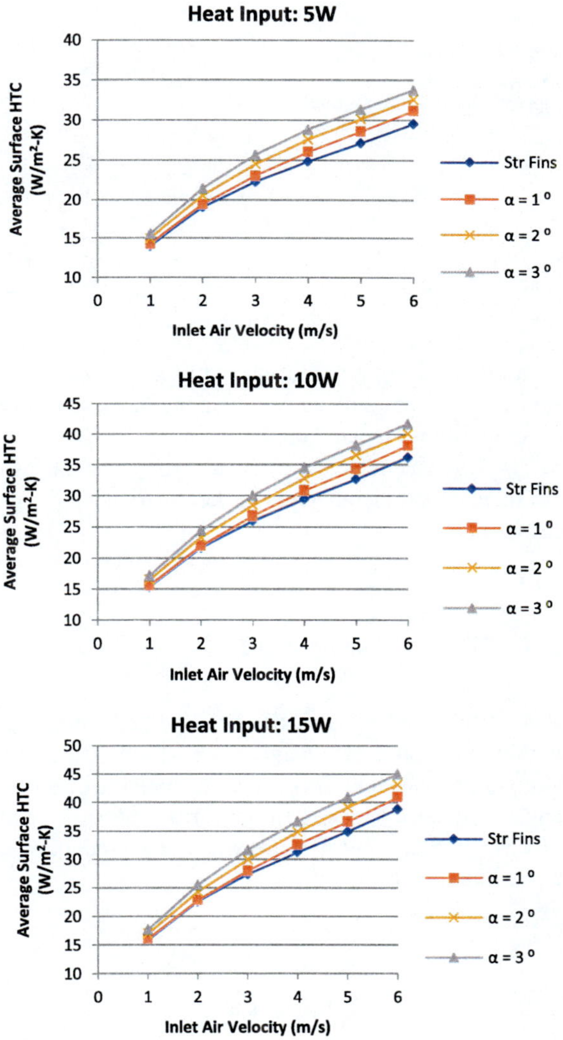

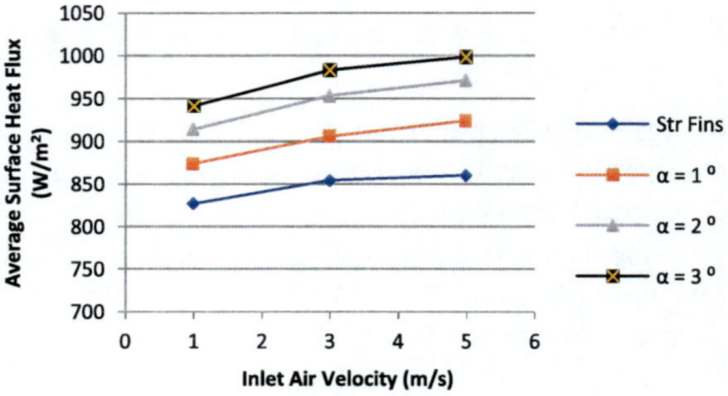

Fig. 9 Variation of Avg. heat flux with air velocity

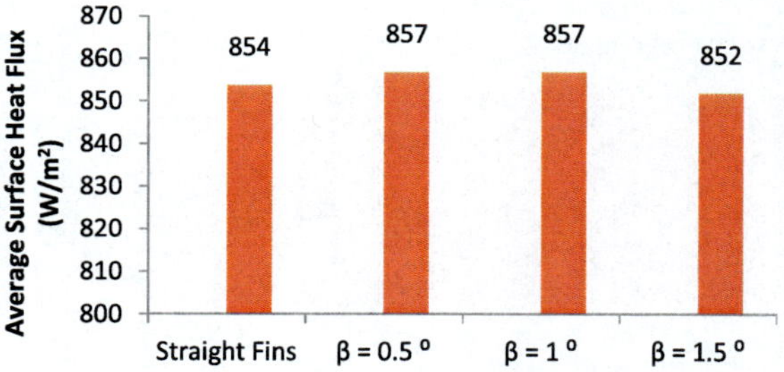

Fig. 10 Comparison of average surface heat flux for tapered fins and straight fins

Similarly, average surface heat flux increases for angled fins (Refer to Fig. 9). The greater the orientation angle, the higher the average surface heat flux.

Longitudinal tapered fins have almost the same heat flux as the straight fins (Refer to Fig. 10). The thickness of the fins keeps increasing which limits its effectiveness. An increase in tapered angle further restricts the air path and thus negatively affects its performance.

5.4 *Weight Comparison*

Weight of each heat sink has been plotted (refer Fig. 11). Heat sinks having fins with orientation angle, weigh less by 11.3% (in case of $\alpha = 3°$) when compared to straight fin heat sink. This is due to chopping of the end fins. Fins with longitudinal angle increase the cross-sectional area, adding more mass (up to 30.2% in case of $\beta = 1.5°$). In Electronic systems where mass reduction is highly favoured, convergent angled fins can be opted.

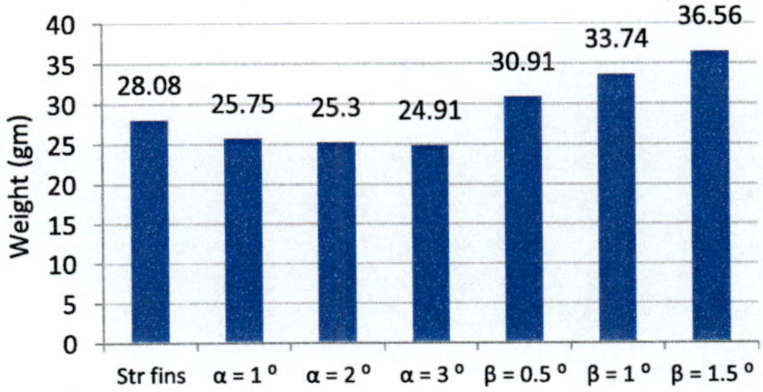

Fig. 11 Weight versus type of heat sink

6 Conclusions

In this study, performance parameters of angled and longitudinal tapered fins were compared with straight fin heat sink. It is concluded that:

i. Angled fins have a better velocity profile. Momentum loss is reduced. As a result avg. HTC is more
ii. With the increase in orientation angle, the average HTC increases (~15% for $\alpha = 3°$), which implies efficient use of airflow.
iii. An increase in average surface heat flux implies better use of exposed surface area.
iv. Maximum heat sink temperature for angled fins is a couple of degrees less than straight fins, that too with ~11% less mass.
v. Surface average HTC as well as heat flux have a direct correlation with Inlet air velocity and heat input.

Acknowledgements Sincere thanks to Mr. Jagadesh Puvvala M.Tech (Manager, PDIC) and Mr. Sreevalsen M.Tech (Sr. Engineer, PDIC) for their guidance and support.

Nomenclature

A	Total exposed area of fins [m^2]
A_c	Cross sectional area [m^2]
A_i	Inlet area between two fins [m^2]
A_o	Exit area between two fins [m^2]
D_h	Hydraulic diameter [m]
H	Heat input [W]
HTC	Heat transfer coefficient [W/m^2-K]
N_{fin}	No. of fins
P	Wetted perimeter [m]
ΔP	Pressure drop [Pa]
Re	Reynolds No.

V	Inlet air velocity [m/s]
V_o	Outlet air velocity [m/s]
α	Orientation angle (°)
β	Longitudinal taper angle (°)
λ	Aspect ratio

References

1. How Air Velocity affects the thermal performance of heat sinks: a comparison of Straight fin, Pin fin and maxiFLOW^{TM} Architectures by ATS. Qpedia, Oct 2008
2. Luo Q, Li P, Cai L, Chen X, Yan H, Zhu HX, Zhai P, Li P, Zhang Q (2019) Experimental investigation on the heat dissipation performance of flared-fin heat sinks for concentration photovoltaic modules. Elsevier
3. Farooq SK (2017) Effect of fin inclinations on natural convection heat transfer by using CFD. 9(3). https://doi.org/10.21817/ijet/2017/v9i3/1709030258
4. Ali AYM, Abo-Zahhad EM, Elqady HI, Rabie M, Elkady MF, Ookawara S, El-Shazly AH, Radwan A (2020) Thermal analysis of high concentrator photovoltaic module using convergent-divergent microchannel heat sink design. Appl Therm Eng. https://doi.org/10.1016/j.applthermaleng.020.116201
5. Mehmedagic I, Krug J (2015) Heat sink design and optimization. Technical report ARMET-TR-15036
6. Kim D-K, Jung J, Kim SJ (2010) Thermal optimization of plate-fin heat sinks with variable fin thickness. Elsevier
7. Ghahfarokhi PS, Kallaste A, Belahcen A, Vaimann T (2019) Determination of heat transfer coefficient for the air forced cooling over a flat side of coil. 15(1):15–20. https://doi.org/10.2478/ecce-2019-0003
8. Rana SC, Bhanja D Effect of orientation angle in thermal performance analysis of a horizontal heat sink of perforated pin fins. https://doi.org/10.1063/1.5123925
9. https://www.electronics-cooling.com/2003/05/estimating-parallel-plate-fin-heat-sink-pressure-drop/
10. Culham JR, Muzychka YS (2001) Optimization of plate fin heat sinks using entropy generation minimization. IEEE 24(2)
11. Turgut O, Ozcan AC, Turkoglu H (2019) Laminar Forced Convection over an inclined Flat plate with unheated starting length. J Polytech. https://doi.org/10.2339/politeknik.403984

Numerical Modeling of Liquid Film Cooling Heat Transfer Coupled to Compressible Gas

Navaneethan Mansu, T. Sundararajan, and T. Jayachandran

Abstract Liquid film cooling is a widely adopted method for restricting the solid wall temperature within allowable limits in a combustor. In most of the earlier works, the hot gas and coolant is treated as incompressible, however, in many practical applications such as in rocket propulsion systems, hot gas is compressible. In this study, a numerical model for coupled heat transfer between the compressible hot combustion gas and the incompressible liquid film injected along the wall is presented. Interface energy balance and mass balance are modeled along with phase change of liquid and diffusion of vapor into the gas flow, to investigate the extent of cooling along the solid wall. Validation of turbulence-based in-house code with literature data for nozzle flow is also included in the study. Effects of initial liquid film thickness, film temperature, film velocity, and vapor mass fraction on wall-cooling are studied. The film cooled length obtained from the present numerical model is compared with experimental results available in the literature.

Keywords Film cooling · Hot gas · Coupling

1 Introduction

Rocket thrust chambers need to be protected from hot combustion gases with flame temperatures in the range of 3000 K, especially in the cases of reusable vehicles and engines with long burn times. Liquid film cooling is a preferred cooling technique adopted in liquid rocket engines, where the liquid coolant is caused to flow along the

N. Mansu (✉)
Thermal Design Division, VSSC/ISRO, Thiruvananthapuram 695022, Kerala, India
e-mail: mansu.999@gmail.com

T. Sundararajan
Department of Mechanical Engineering, IIT Madras, Chennai 600036, India

T. Jayachandran
Department of Aerospace Engineering, IIT Madras, Chennai 600036, India

S. Das et al. (eds.), *Proceedings of the 1st International Conference on Fluid, Thermal and Energy Systems*, Lecture Notes in Mechanical Engineering,
https://doi.org/10.1007/978-981-99-5990-7_47

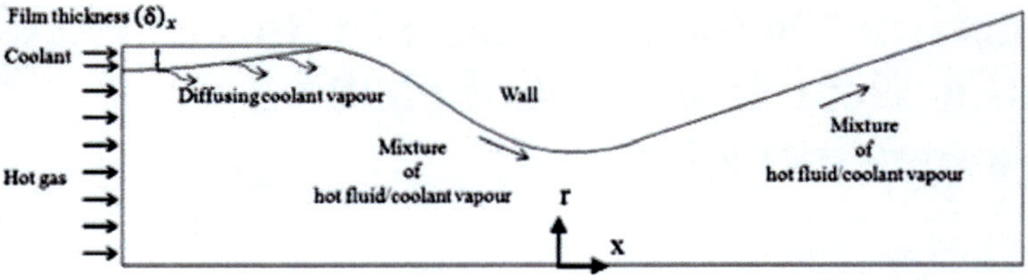

Fig. 1 Schematic of film cooling technique

inside surface of the thrust chamber wall. The high heat capacity of liquid and the latent heat needed for vaporization, render liquid film cooling as an attractive method, especially for high heat flux applications. Generally, the coolant is injected tangentially through discrete holes placed circumferentially on the combustion chamber. The liquid film insulates the chamber wall from direct contact with the hot combustion gas. Once the liquid film undergoes a phase change, the vapor formed also functions as a barrier for heat transfer to the wall. Figure 1 shows a schematic diagram of film cooling applied to the rocket chamber wall.

2 Literature Review and Objective

2.1 Literature Review

A proper design for film cooling system requires a good understanding of the interactions between compressible gas flow and incompressible liquid flow. Also, aspects such as phase change heat transfer, the influence of vapor shielding, and the variation of film thickness, along the wall, need to be considered. Shine and Nithi [1] have reviewed the theoretical and experimental studies carried out so far on film cooling of rocket engine thrust chambers. They also highlighted the aspects of film cooling that need to be further investigated. Advances in film cooling research were further reported in the latest article by Ludescher and Olivier [2], where film cooling efficiency was studied for coolant injection in the expansion section of a conical nozzle. In experimental and numerical works by Heufer and Olivier [3], the effects of major flow parameters on the heat transfer of gas film cooled surface in a laminar supersonic flow were studied. Liquid film cooling is still an important and active research area as it paves the way for supporting the design methods for liquid rocket combustion chambers and nozzles [4].

Early experiments by Kinney et al. [5] with air–water flow in pipes estimated the film cooled length using temperatures measured on the outer pipe surface. In studies by Stechman et al. [6], a modified Bartz equation was used to calculate the convective heat transfer coefficient. Shembharkar and Pai [7] solved incompressible flow equations for the hot gas, with Couette flow assumed for the liquid film. Zhang et al. [8] numerically solved the governing equations for the incompressible flow of both the liquid film and gas stream coupled through interfacial matching conditions; for an anon-evaporating film of constant thickness. Shine et al. [9] have reported an analytical study to predict the liquid film length by accounting for liquid entrainment through a correlation approach.

Wang [10] reported experimental data for a film cooled gaseous hydrogen/oxygen-based combustion chamber with hydrogen as the coolant. The effect of angular injection of coolant into the combustion chamber was examined by Xiang [11] for gas film cooled combustion chamber of a hydrogen–oxygen propellant system. In a recent paper by Strokacha et al. [12], an Eulerian–Lagrangian method was used to study film cooling performance in a kerosene–LOX combustion chamber using ANSYS CFX. In numerical simulations done by Yang and Sun [13], a liquid oxygen/RP1 propellant combination was considered with liquid water as film coolant.

2.2 *Objective*

A majority of the published papers on film cooling treat the gas flow as incompressible. In numerical studies [7, 8] done on liquid film cooling, the gas is assumed as incompressible. However, the combustion gas flow through the rocket thrust chamber and nozzle is definitely in the compressible regime. Correlations for film cooled length reported in the literature do not account for geometrical variations present in the combustion chamber and nozzle, with abrupt changes in the inner contour. At present, no studies are available in the literature that predict film thickness variation for realistic thrust chamber inner profiles. Very few papers address the variation of film thickness in the cooled portion of the wall and the effectiveness of cooling beyond the point where liquid film ceases to exist. In other words, the shielding aspects of vapor blankets are not considered. In the present work, a numerical methodology is presented, treating the gas flow as compressible and liquid flow as incompressible. The objective of the present study is to model the liquid film cooling mechanism by treating a hot medium as compressible and study the various flow features such as change in film thickness and heat transfer with phase change. However, the entrainment of liquid droplets by the high-speed gas flow is ignored for the sake of simplicity.

3 Methodology

This study employs a decoupled approach—the gas phase flow is solved with liquid film temperature as prescribed and the heat flux from hot gas to the liquid is predicted. Applying the calculated heat flux as the interface condition for the liquid, the temperature profile in the liquid and vapor flux (due to liquid evaporation and vapor diffusion into the hot gas) are predicted. The newly predicted liquid surface temperature is prescribed as the boundary condition for the hot gas flow model for the next time step. In a similar manner, the interfacial conditions of velocity and stress continuity, as well as temperature and heat flux continuity are applied to solve the combined gas and liquid flow in a decoupled manner, during the time-marching simulations of the two-phase flow problem.

Major assumptions applicable to the present study are listed below:-

(a) Coordinate system is cylindrical for axi-symmetric model
(b) Gas and liquid flows are treated as compressible and incompressible respectively
(c) Transport of vapor in the gas flow through species transport equation is modeled
(d) Temperature dependent properties are estimated for the gas phase, considering local gas and vapor mass fractions and gas temperature.
(e) Entrainment effects and radiation heat transfer are not considered. Also, surface tension effects are ignored with the assumption of a large radius of curvature for the liquid–gas interface

Gas Phase Equations

The axisymmetric governing equations of compressible in conservative form for hot gas flow is given below:

$$\frac{\partial U}{\partial t}+\frac{\partial F}{\partial x}+\frac{1}{r}\frac{\partial}{\partial r}(rG)=\frac{\partial F_v}{\partial x}+\frac{1}{r}\frac{\partial}{\partial r}(rG_v)+S \tag{1}$$

Turbulence Modeling

The turbulent effects are modeled in the present study using the two equation RNG *k*-epsilon model and the effective viscosity, thermal conductivity and mass diffusivity values are evaluated by accounting for both laminar and turbulent contributions to these properties. The equations for the conservation of turbulent kinetic energy (k) and the rate of turbulent dissipation (ϵ) are shown below.

$$\frac{\partial \rho k}{\partial t}+\frac{\partial \rho u k}{\partial x}+\frac{1}{r}\frac{\partial(r\rho v k)}{\partial r}=\frac{\partial}{\partial x}\left(\mu_k\frac{\partial k}{\partial x}\right)+\frac{1}{r}\frac{\partial}{\partial r}\left(r\mu_k\frac{\partial k}{\partial r}\right)+H_k-\rho\epsilon-H_c \tag{2}$$

$$\frac{\partial \rho\epsilon}{\partial t}+\frac{\partial \rho u\epsilon}{\partial x}+\frac{1}{r}\frac{\partial(r\rho v\varepsilon)}{\partial r}=\frac{\partial}{\partial x}\left(\mu_\varepsilon\frac{\partial \varepsilon}{\partial x}\right)+\frac{1}{r}\frac{\partial}{\partial r}\left(r\mu_\varepsilon\frac{\partial \epsilon}{\partial r}\right)+\mathrm{H}_\varepsilon \tag{3}$$

$$\begin{aligned}H_k=\mu_t\Bigg(2\left[\left(\frac{\partial u}{\partial x}\right)^2+\left(\frac{\partial v}{\partial r}\right)^2+\left(\frac{v}{r}\right)^2\right]+\left[\frac{\partial u}{\partial r}+\frac{\partial v}{\partial x}\right]^2\\-\frac{2}{3}\left[\frac{\partial u}{\partial x}+\frac{\partial v}{\partial r}+\frac{v}{r}\right]^2\Bigg)-\frac{2}{3}\rho k\left[\frac{\partial u}{\partial x}+\frac{\partial v}{\partial r}+\frac{v}{r}\right]\end{aligned} \tag{4}$$

$$H_\epsilon = C_{1\epsilon}\frac{\epsilon}{k}\mu_t * \left(2\left[\left(\frac{\partial u}{\partial x}\right)^2 + \left(\frac{\partial v}{\partial r}\right)^2 + \left(\frac{v}{r}\right)^2\right] + \left[\frac{\partial u}{\partial r} + \frac{\partial v}{\partial x}\right]^2\right) - C_{2\epsilon}\rho\frac{\epsilon^2}{k} \tag{5}$$

In the above equations, the values of constants are taken as follows:

$$\mu_k = 1.39(\mu_{\text{laminar}} + \mu_{\text{turbulent}})$$
$$\mu_e = 1.39(\mu_{\text{laminar}} + \mu_{\text{turbulent}})$$
$$\mu_{\text{turbulent}} = 0.0845\rho\frac{k^2}{\epsilon}$$
$$C_{1\epsilon} = 1.42 - \frac{\eta\left(1 - \frac{\eta}{4.377}\right)}{(1 + 0.012\eta^3)}$$
$$\eta = \sqrt{2(E_{ij}.E_{ij})}\frac{k}{\epsilon}$$
$$C_{2\epsilon} = 1.68$$

The compressibility effects are accounted by the inclusion of the term, $H_c = \frac{2\rho\epsilon k}{a^2}$.

Liquid Phase

As indicated in the introduction, the gas flow is externally coupled to the simplified form of the liquid model. The present section gives explanation of treatment of presence of liquid film on the gas flow boundary. A typical finite volume mesh for the gas domain near the boundary is shown in Fig. 2. The mesh near the wall boundary is shown along with the nodal and cell center locations. In the same figure, inlet conditions for liquid are at the location marked as 'A'. Conditions at location 'B' are computed by application of conservation of mass and energy using the upstream conditions of liquid and nodal/cell center values of nearby gas cells.

Mass conservation for liquid film, in terms of density, film thickness, average velocity at a given axial location, and evaporated mass through length x, is expressed as

$$(\rho_l\delta\overline{u_l})_{\text{in}} = (\rho_l\delta\overline{u_l})_{\text{out}} + \dot{m}_v x \tag{6}$$

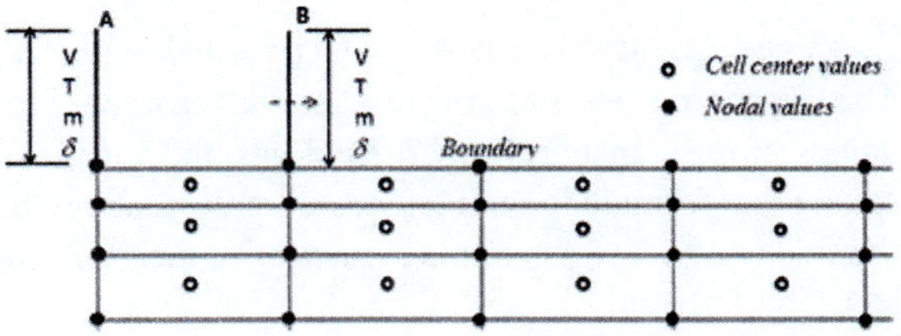

Fig. 2 Calculation of parameters for liquid

The evaporated mass flux is computed using the gradient of mass fraction as given below

$$\dot{m}_v = \left(\frac{-1}{1-m}\right)\left(\frac{\mu}{\sigma}\right)\frac{dm}{dn} \tag{7}$$

Computation of liquid exit temperature is performed through an energy equation where the heat balance is considered between the sensible heat along with heat conducted from hot gas and heat transported by evaporated mass through latent heat. For the liquid film, it is assumed that the film temperature is uniform across film thickness. In the energy equation, heat balance is considered between the sensible heat along with heat conducted from hot gas and heat transported by evaporated mass through latent heat L.

$$(\rho_l \delta \overline{u_l} C T_l)_{\text{in}} + Q_{\text{hotgas}} = (\rho_l \delta \overline{u_l} C T_l)_{\text{out}} + \dot{m}_v x L \tag{8}$$

From the temperature of the liquid, assuming thermodynamic equilibrium, the mass fraction of vapor at the inlet can be calculated through Antonio's equation and Daltons' law as given by the equation below

$$\log_{10} P_v = A - \frac{B}{T+C} \tag{9}$$

$$Y_{v,i} = \frac{P_v MW_v}{MW_g(P - P_v) + P_v MW_v} \tag{10}$$

Once vapor pressure is calculated using the Antonio equation using inlet temperature, vapor mass fraction can be computed using the pressure value and molecular weight values of gas and vapor. The pressure value in the nearby gas cell is denoted by P.

The unknows in the mass balance equation are δandu$_l$ at downstream location 'B'. To close the systems of equations to solve for these parameters, another expression is written in terms of shear stress at 'B'.

Considering the assumption of Couette flow with surface velocity u$_l$, the shear stress is written as

$$\frac{(u_2 - u_l)}{dy}\mu_g = \frac{(u_l)}{\delta}\mu_l \tag{11}$$

Values u_2 and u_1 are nodal values for velocities in gas cells on the location 'B'. The transport properties are based on the local mass fraction of vapor and individual values of the transport properties of gas and vapor. The expression is based on the square root formula for viscosity usingmolecularweight. The same is used to compute thermal conductivity in terms of respective thermal conductivity values for vapor and gas.

$$\mu = \frac{\left[\mu_v \frac{m}{\sqrt{MW_v}} + \mu_g \frac{(1-m)}{\sqrt{MW_g}}\right]}{\left[\frac{m}{\sqrt{MW_v}} + \frac{(1-m)}{\sqrt{MW_g}}\right]} \tag{12}$$

The liquid saturation temperature is computed assuming that the liquid is at the same pressure as that of the neighboring gas cell. This is justified considering the very fine thickness of the liquid film. Once the temperature of the film reaches the saturation value, there is no further change in liquid film temperature. At such a condition, heat transferred from gas is used for evaporation and there is no change in sensible enthalpy as shown by equation. Liquid reaching its saturation temperature corresponds to a maximum vapor fraction of 1.0. In the computation, a cut-off value of 0.98 is applied for vapor mass fraction. The velocities in imaginary cells for the gas boundary are expressed in terms of velocities of the liquid film using the interface normal, as

$$u_{\text{ghost}} = n_y u_l - n_x v_l \tag{13}$$

$$v_{\text{ghost}} = -n_x u_l - n_y v_l \tag{14}$$

A flow chart representing the methodology is given in Fig. 3.

4 Results and Discussion

4.1 *Validation with Published Nozzle Heat Transfer Results*

Validation of the predicted results obtained using in-house turbulence code with experimental data published by Back et al. [14] is discussed here. The geometry and boundary conditions are the same as those considered by Back et al. [14]. Stagnation pressure and temperature at the inlet are maintained at 1.752 MPa and 840 K, respectively. Mach number distribution obtained for these conditions is shown in Fig. 4, which illustrates a smooth transition from subsonic to supersonic flow across the throat. The heat transfer coefficient computed through the two equation RNG k-epsilon turbulence solver developed in-house is compared with the published values. The predictions give a better match in the locations of throat and divergent compared to the nozzle convergent region as shown in Fig. 5. This deviation is attributed to the isothermal wall boundary condition assumed in the present study for the liquid-cooled portion, whereas, in the actual experiment, the surface temperature is expected to vary along the axial length. The numerical simulations of the present work correctly predict the occurrence of maximum heat transfer coefficient at the nozzle throat, as demonstrated by Back et al. [14].

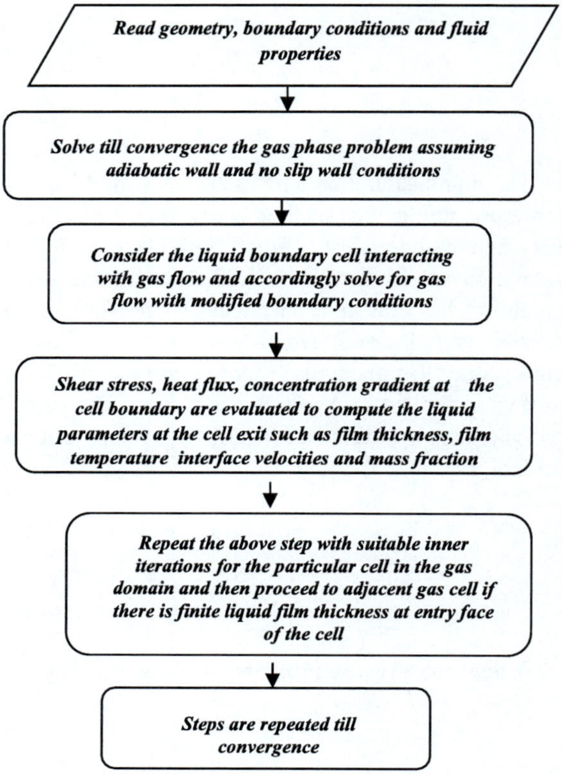

Fig. 3 Flow chart for methodology

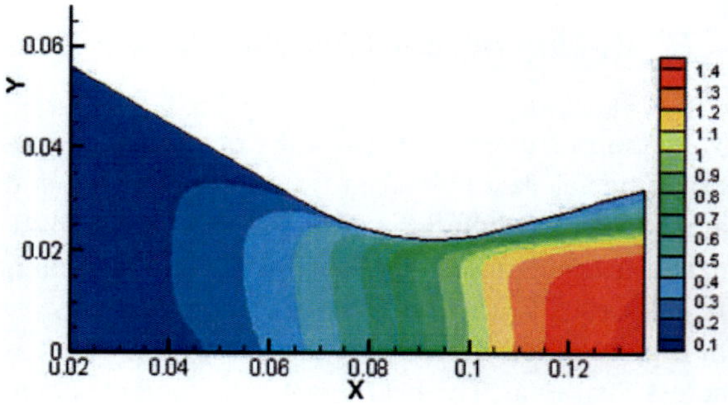

Fig. 4 Mach number distribution for the flow considered by Back et al. [14]

4.2 Simulation of Film Along Flow Boundary

In the present section, the effects of wall film injection in a nozzle with a hot gas as the core flow is discussed. Property of air is considered for hot gas and water is assumed as the coolant. The properties used are shown in the Appendix. The film cooled length is computed based on distance up to which the film just ceases to exist

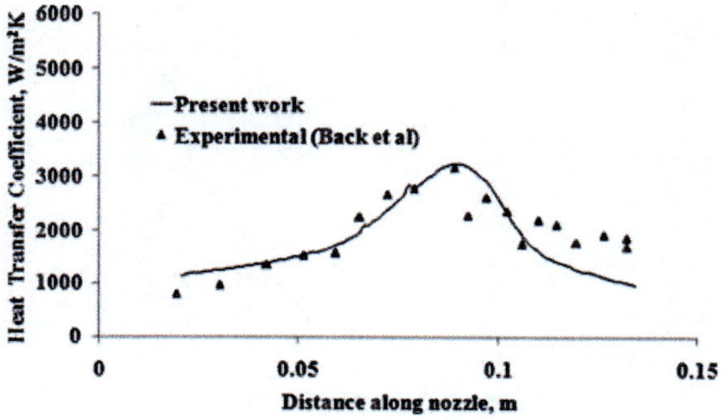

Fig. 5 Comparison of predicted heat transfer coefficient with literature data

along the wall. Initially, the gas flow field is solved with adiabatic wall assumption until convergence followed by injection of liquid along the wall. As film thickness is computed based on mass balance, value at a given location is governed by shear stress and mass evaporated from the film surface. The film will cease to exist at a location if the thickness drops to the minimum cut-off value or when negative values are obtained while solving for film thickness. The boundary conditions prescribed for the CD nozzle are shown in Table 1. Inlet conditions for the gas flow are shown in Table 1.

The distribution for temperature and vapor mass fraction in the nozzle is shown in Figs. 6 and 7, respectively. The performance of the liquid coolant with an inlet velocity of 1.0 m/s is considered. Figure 8 shows the variation of mass fraction of vapor and liquid temperature respectively, along the wall till the location where the film thickness is less than the imposed cut-off value. The bulk temperature of the film increases rapidly beyond a distance of 0.13 m which in turn increases the vapor mass fraction as it is governed by local temperature. Film thickness and evaporated mass from the liquid surface is plotted in Fig. 9. Neglecting the entry effects, the film thickness shows an overall decreasing trend except at the region from 0.1 to 0.13 m and also the near region where the film is fully consumed, the reason for the former being less increase in temperature at this location, thus affecting the mass evaporated from the surface. The heat flux from gas is plotted along with the temperature of the boundary cells occupied by the gas/vapor mixture in Fig. 10. An increase in heat flux from hot fluid to liquid causes an increase in liquid temperature and a decrease in the temperature of hot fluid.

The film injection velocity was varied with the slot height at the inlet fixed as 0.002 m. Figure 11 shows the variation in film thickness for different inlet velocities of the

Table 1 Application of boundary conditions

Axis	Center of symmetry
Outlet	Supersonic flow
Wall	Adiabatic
Inlet	Subsonic ($P0 = 40$ bar $T0 = 2500$ K)

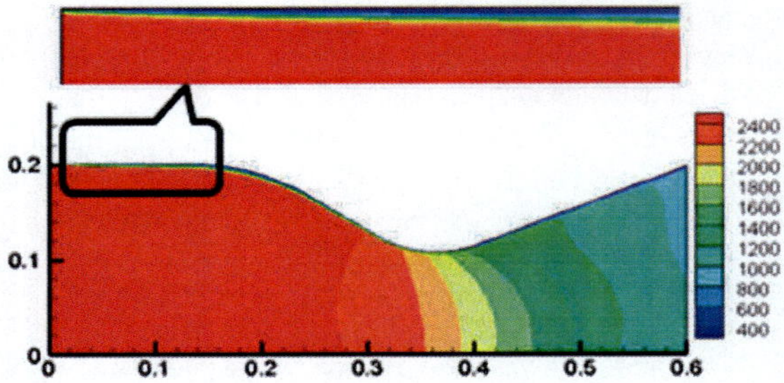

Fig. 6 Temperature distribution

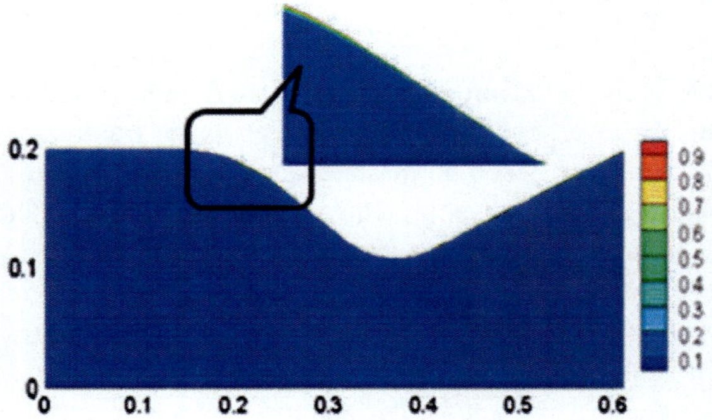

Fig. 7 Vapor mass fraction distribution

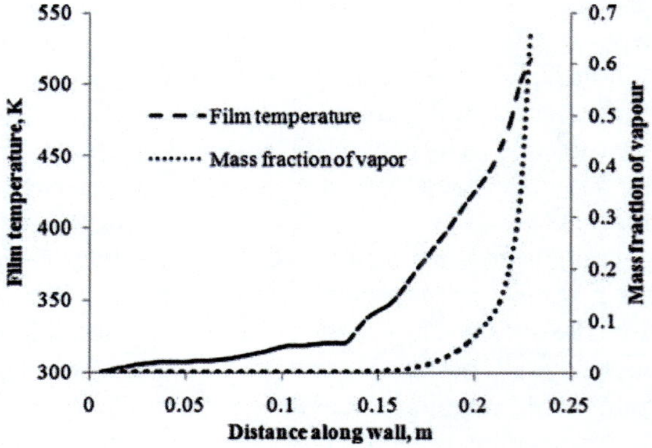

Fig. 8 Film temperature with mass fraction of vapor

film. At lower inlet velocities, the film thickness is lower compared to the values at higher velocities due to the higher shearing action of the surrounding fluid. The variation in film temperature, as in Fig. 12, indicates that in all cases of inlet velocities considered in the study, the film temperature reaches its saturation temperature at the given pressure. In most of the cases, there is a sharp rise in temperature at a distance of 0.13 m indicating higher heat flux from the gas. As the coolant velocity is increased, the film dry out point moves further downstream from the injection as in Table 2. At higher film inlet velocities, the increase in film dry out distance is lower

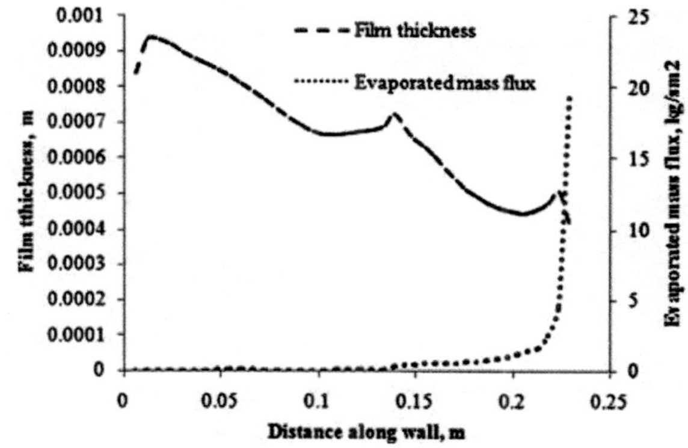

Fig. 9 Film thickness with evaporated mass flux

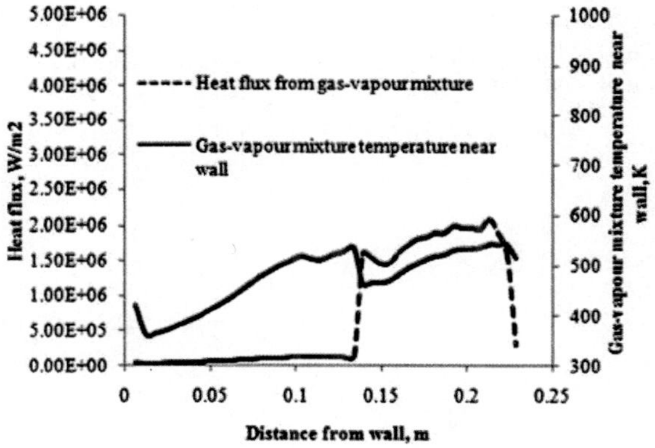

Fig. 10 Heat flux variation at the liquid boundary

compared to lower inlet velocities of the coolant. Also, the gradient of temperature in the flow direction is less for higher injection velocity. It should also be noted that due to changes in the inner contour of the wall, the variations in temperature and velocity gradients normal to the wall will increase resulting in higher shear stress and wall heat flux. Or in other words, there will not be any significant gain in film cooled length when we have a local change in surface curvature similar to the convergent in a typical convergent divergent nozzle.

Further studies were done by varying inlet slot width for a given initial coolant velocity whose results are shown in Fig. 13. For a given inlet velocity of 0.5 m/s, slot height was varied from 0.001 to 0.005 m to study the variation in film cooled length. The parameters were chosen so that the film cooled length includes a portion of the combustion chamber and nozzle convergent. The trend is linear for inlet slot height up to 0.3 mm, beyond which there is a change in the behavior of the curve; indicating a change in film cooled length. The shear stress distribution for a typical nozzle shows a decreasing trend in the cylindrical portion of the thrust chamber, whereas it increases in the convergent portion of the nozzle till the location of the throat is reached and again decreases in divergent. At a given location, higher shear stress

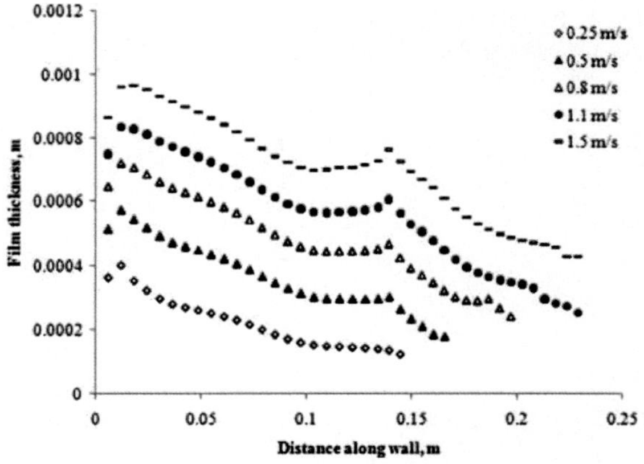

Fig. 11 Film thickness for different inlet velocities

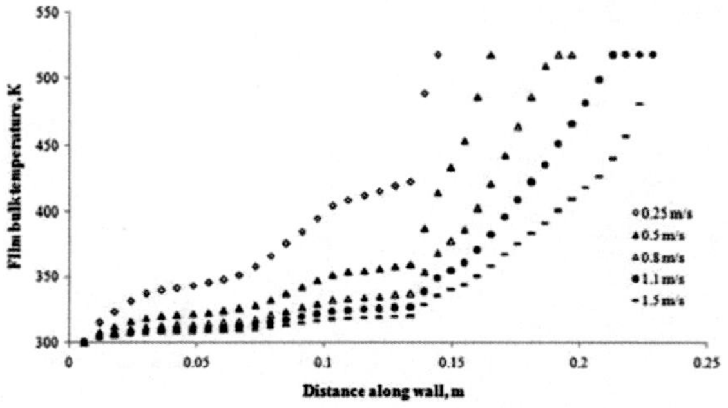

Fig. 12 Film temperature for different inlet velocities

Table 2 Film cooled length for different liquid inlet velocities

Liquid inlet velocity (m/s)	0.075	0.25	0.5	1.0	1.2	1.5
Film cooled length (m)	0.079	0.145	0.166	0.213	0.224	0.23

tends to squeeze the film resulting in the film extending further downstream. So, due to increased shear stress in convergent, film cooled length for a given set of conditions will be higher than the value of a straight duct with constant cross-sectional area.

4.3 *Literature Data*

As it is a complex task to measure film thickness for a hot gas flow, the present method is compared with available data in the literature. In literature, film cooled length is reported in terms of temperature recorded on the outer surface of the wall. Configuration mentioned in [5] was studied for comparison of numerical and experimental results. For validation, data reported for experimental evidence on a 2 in. smooth

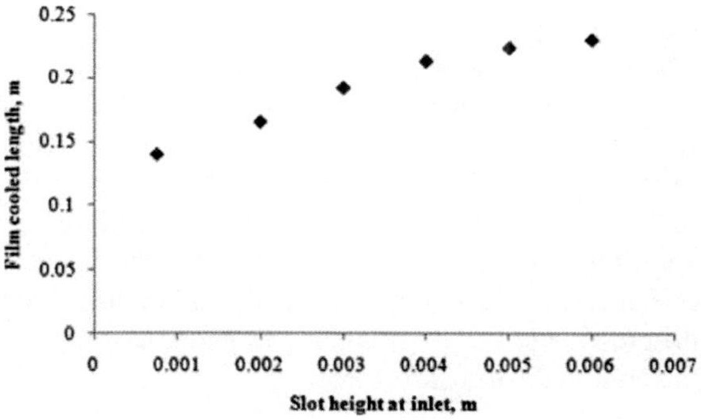

Fig. 13 Film cooled length for different slot heights

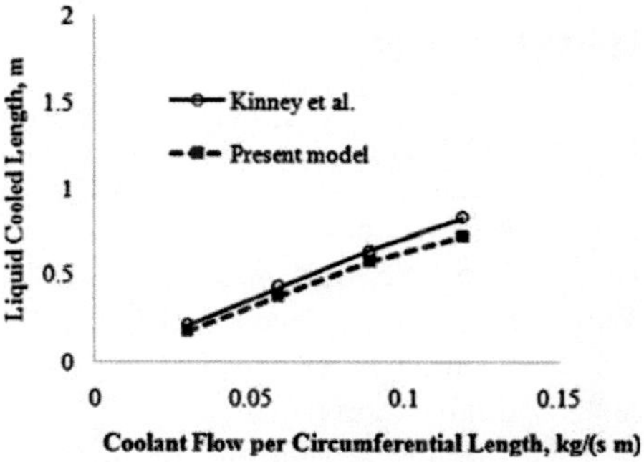

Fig. 14 Comparison with literature data

pipe with a length of 48 in. (4 ft) is considered. The film cooled length obtained from the present model is compared with that of the literature in Fig. 14. It is seen that the model is able to predict the literature data with reasonable accuracy. In the experiments, film cooled length determined based on skin temperature measured on the pipe is the distance where the temperature just reaches the boiling point of the liquid, whereas in the present model, the film cooled length is directly obtained from the numerical solution.

5 Conclusions

A numerical model for predicting film cooled length with heat transfer from hot combustion gas is presented. The model is suitable for investigating the distance to which the wall is expected to be protected from hot gas flow for arbitrary inner contour. The model based on the decoupling of liquid and gas flow gives a reasonable explanation for coupled solver which demands fine mesh near the interface and complicated models for incompressible to compressible fluid interaction within the domain. The approach followed here is justified in most of the applications, as the liquid is injected tangentially along the gas flow and thus there is less probability of liquid having a high transverse velocity component to flow across the gas interface. The present model is able to capture features of film cooling with a liquid undergoing phase change in the presence of gas treated as compressible. Results obtained from the model were compared with data reported in the literature and it was found to reasonably predict the measured data. The proposed method will be useful for internal flow applications where proper measurements or any intrusive technique are difficult to implement to characterize the liquid film. For studying the thermal effects downstream of film cooled length in the nozzle with ablative throat material, the present methodology is being extended to study the non-ablation of the composite material in the throat region.

Nomenclature

t	Time (s)
x	Axial coordinate (m)
r	Radial coordinate (m)
U	Conserved variables (–)
F/Fv	Convective, viscous flux in x (–)
G/Gv	Convective, viscous flux in r (–)
S	Source term (–)
k	Turbulent kinetic energy (m^2/s^2)
ϵ	Turbulent dissipation rate (m^2/s^3)
ρ_l	Liquid density (kg/m^3)
δ	Film thickness (m)
$\overline{u_l}$	Average liquid velocity (m/s)
$\dot{m}_v$	Evaporated mass flux (kg/sm^2)
x	Length of cell (m)
m	Vapor mass fraction (–)
μ	Viscosity (Ns/m^2)
σ	Schmidt Number (–)
T_l	Liquid temperature (K)
L	Latent heat (J/kg)
P_v	Vapor pressure (N/m^2)

Appendix

Properties used for coolant

(a) Density (kg/m^3) = 1000.0
(b) Specific heat (J/kg K) = 4179.0

(c) Viscosity (Ns/m^2) $= 0.00002414 \times 10^{\left(\frac{247.8}{T-140}\right)}$
(d) Latent heat (J/kg) $= (25 - 0.2274 \times (T - 273)) \times 10^5$

In (c) and (d), T is expressed in Kelvin.

References

1. Shine SR, Shri Nidhi S (2018) Review on film cooling of liquid rocket engines. Propuls Power Res 7(1):1–18
2. Ludescher, Olivier H (2018) Experimental investigations of film cooling in a conical nozzle under rocket-engine-like flow conditions. AIAA J 57(3)
3. Heufer KA, Olivier H (2008) Experimental and numerical study of cooling gas injection in laminar supersonic flow. AIAA J 46(11):2741–2751
4. (2021) Future space-transport-system components under high thermal and mechanical loads, notes on numerical fluid mechanics and multidisciplinary design
5. Kinney GR, Abramson A, Sloop JL (1952) Internal film cooling experiments with 2 and 4 inch smooth surface tubes and gas temperatures to 2000-F in 2 and 4 inch diameter horizontal tubes. Technical Report, NACA report 1087
6. Stechman RC, Oberstone J, Howell JC (1969) Design criteria for film cooling for small liquid-propellant rocket engines. J Spacecr Rockets 6:97–102
7. Shembharkar TR, Pai BR (1986) Prediction of film cooling with a liquid coolant. Int J Heat Mass Transf 29:899–908
8. Zhang HW, Tao WQ, He YL, Zhang W (2006) Numerical study of liquid film cooling in a rocket combustion chamber. Int J Heat Mass Transf 49:349–358
9. Shine SR, Sunil Kumar S, Suresh BN (2012) A new generalised model for liquid film cooling in rocket combustion chambers. Int J Heat Mass Transf 55:5065–5075
10. Wang T, Sun B, Liu D, Xiang J (2018) Experimental investigation of two-dimensional wall thermal loads in the near-injector region of a film-cooled combustion chamber. Appl Therm Eng 138:913–923
11. Xiang J, Sun B, Wang T, Yuan J (2020) Effects of angled film-cooling on cooling performance in a GO2/GH2 subscale thrust chamber. Appl Therm Eng 166:114627
12. Strokacha EA, Borovika IN, Bazarova VG, Haidnb OJ (2020) Numerical study of operational processes in a GOx-kerosene rocket engine with liquid film cooling. Propuls Power Res 1–10
13. Yang W, Sun B (2012) Numerical simulation of liquid film in a liquid oxygen/rocket propellant 1 liquid rocket. J Thermophys Heat Transf 26(2):328–336
14. Back LH, Massier PF, Gier HL (1964) Convective heat transfer in a convergent-divergent nozzle. Int J Heat Mass Transf 7:549–568

Study of Flow and Thermal Evolution in Zinc Bath During Galvanisation Using a 3D CFD Model

K. Muhammed Anas and Anirban Bhattacharya

Abstract Galvanisation, the coating of zinc on steel sheets, can increase the corrosion resistance of steel. For this purpose, the steel sheets from the furnace, after continuous casting, are passed through a molten zinc bath with the help of rollers. This is called a continuous galvanising bath. This process provides a surface finish almost similar to electroplating. While analysing the complex convection pattern during galvanisation, the convection current can be divided into forced and natural convection. The forced convection is due to strip movement and natural convection is due to temperature gradient. The heater (inductor), required for keeping the zinc in a molten state, also plays a significant role in flow patterns due to magnetic Lorentz force and Joule heating. The effect of strip velocity, natural convection, dross particle formation, ingot addition and inductor aspects have been studied in this paper. Based on the study, the chances of formation of dross particles in different regions of the zinc bath have been predicted. Also, a parametric study of strip angle and inductor power has been presented in the paper.

Keywords Galvanisation · Continuous galvanising bath · Zinc coating · Heat transfer · Convection effects

1 Introduction

The use of zinc-coated steel is increasing worldwide. Industries such as automobile industries heavily depend on these corrosion-resistant galvanised steel sheets. One of the widely used methods is electroplating to coat a metal over another metal. The main feature of electroplating is the excellent surface finish. But in terms of cost and time consumption while implementing on a large scale, it has disadvantages. So, at the industrial level, it is not a complete solution. After many production processes, the steel sheet coming through a continuous casting process has to be connected

K. Muhammed Anas · A. Bhattacharya (✉)
School of Mechanical Sciences, IIT Bhubaneswar, Bhubaneswar 752050, Odisha, India
e-mail: anirban@iitbbs.ac.in

S. Das et al. (eds.), *Proceedings of the 1st International Conference on Fluid, Thermal and Energy Systems*, Lecture Notes in Mechanical Engineering,
https://doi.org/10.1007/978-981-99-5990-7_48

with a galvanising system to get galvanised steel. The continuous galvanising bath provides a well-structured galvanising process with a surface quality almost near to that of electroplating.

A continuous galvanising bath is a molten zinc bath with rollers and additional structures. The steel sheets are guided to this molten zinc bath through rollers and taken out. During this time, the zinc is coated on the surfaces.

To get an optimal surface finish, an analysis has to be conducted to understand the flow pattern and heat transfer. Parameters such as strip velocity, strip thickness, strip temperature, etc., must be included in the study to get the flow pattern and temperature distribution. Moreover, the inductor heater performance plays a significant role in the flow pattern. As the heater is inductor-based, the magnetic Lorentz force and Joule heating considerably affect the flow pattern. While the process progresses, the zinc gets coated on the steel strips, and thus more zinc has to be added to the bath to compensate for this. For this purpose, zinc ingots are added to the bath. The resulting flow disturbance also influences the direction of flow and hence convection. The present study has carried out a detailed numerical analysis of the above-stated process.

2 Literature Review and Objective

In this section, a review of literature relevant to the present study is done. Ilinca et al. [1] analysed the flow pattern and thermal evolution for a galvanising bath. The study analysed the natural convection and induction heater flow redirection effects. Effects of important parameters such as line speed, strip width, strip temperature and inductor mixing were evaluated. Zhou et al. [2] explained the evolution of zinc transport in a continuous galvanising bath. This is important as the dross formation and the effect of those dross particles on the quality of steel strip gets influenced by the flow field. The flow of zinc inside the induction channel is important here. The flow inside the zinc bath was analysed and the effect of Lorentz force and Joule heating were studied. The detailed modelling of electro-hydrodynamics in the inductor channel was performed by Park et al. [3]. The effects of galvanising parameters on the galvanised product, such as surface morphology of coatings and resistance to corrosion were studied in a hot-dip process simulator. Similarly, the influences of galvanisation parameters on the properties of the final galvanised product were investigated by Shukla et al. [4]. A detailed analysis of inductor performance and current density relationship for magnetic Lorentz force was performed, and numerical studies were carried out. Zhou et al. [5] studied the effect of the inductor on the flow and heat transfer in the Zn bath. The trajectory of dross particles in two types of the layout was carried out. It was found that the course of dross particles was significantly changed by the position of the inductor heater [6].

The combined effect of natural convection and forced convection has to be analysed in detail to get a sound idea about the transport phenomena in a continuous galvanising bath. In this study, the above-mentioned analysis has been carried out.

Along with that, the inductor heater effect on zinc bath, which also has a major impact on the flow inside, has been studied with different inductor power values. The angle of insertion of steel sheets, called strip angle, has also been considered for the analysis and the effect of different strip angles has been analysed. The regions inside the zinc bath with very less velocity or zero velocity, are the regions where the chances of dross formation are more. Based on this, the probable regions for dross formation in a galvanising bath have been predicted.

3 Model Description

3.1 Problem Domain

The three-dimensional geometry is a fluid domain containing the zinc bath, one main roller and two guide rollers, an inductor heater, and the ingot adding surface. Due to the symmetric nature of the fluid domain, only half of the actual domain is considered for all simulations. Only the fluid domain was considered because the other outer compartments and additional structures including steel strips can be taken into account through boundary conditions. A 4114 × 1900 × 2440 mm fluid domain, as shown in Fig. 1, was considered for the analysis. The main roller and guide rollers have a diameter of 600 mm and 150 mm, respectively, and the thickness of the steel sheets was taken as 50 mm.

3.2 Mesh Structure

The fluid domain was discretised into the mesh structure as shown in Fig. 2. The three-dimensional mesh was created with an element size of 50 mm with 54,205 elements. The meshing method was multi-zone with a hexa-dominant cell type. The meshing was done by dividing the fluid domain into three parts in order to get proper mesh generation for the complex domain. The meshing was done using the commercial software ANSYS.

3.3 Mathematical Model

For the study, zinc is considered an incompressible fluid. The effect of radiation heat transfer is negligible throughout the study due to the low temperature of the fluid domain. Also, it was assumed that the physical properties of molten zinc do not change with temperature.

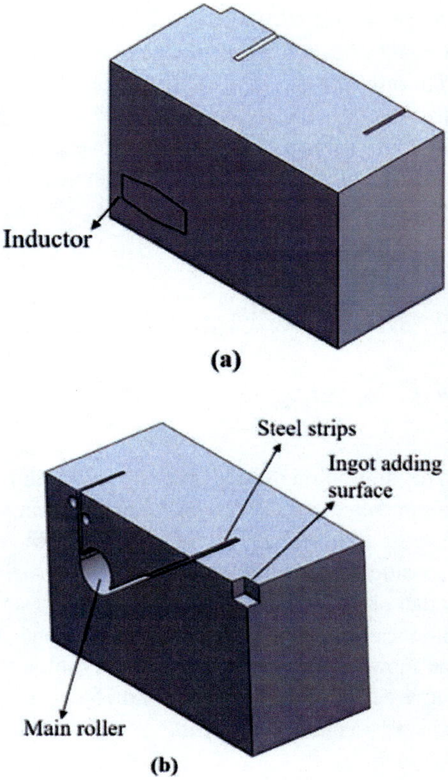

Fig. 1 Schematic diagram of the problem domain **a** view from the outside of the bath, **b** view from the central symmetric plane of the bath

Fig. 2 Generated mesh

Here the equations taken into account are the continuity equation, momentum equation and energy conservation equation. The continuity equation can be expressed as follows [1]:

$$\nabla \cdot \vec{u} = 0 \tag{1}$$

where $\vec{u}$ is the velocity of the fluid. The conservation of momentum equation is written as:

$$\rho\left(\frac{\partial \vec{u}}{\partial t} + \vec{u} \cdot \nabla \vec{u}\right) = -\nabla p + \nabla \cdot \left[2(\mu + \mu_T)\gamma\left(\vec{u}\right)\right] - \rho g_i \beta_T (T - T_0) + S_{mom} \tag{2}$$

where ρ is the density, β_T is the coefficient of thermal expansion, $\gamma(u)$ is the strain rate tensor, μ is the viscosity, p is the pressure, and S_{mom} is the source term for the Maxwell electromagnetic force. The conservation of energy equation is given by:

$$\rho C_p\left(\frac{\partial T}{\partial t} + \vec{u} \cdot \nabla T\right) = \nabla \cdot [(\lambda + \lambda_T)\nabla T] \tag{3}$$

where λ is the thermal conductivity, T is the temperature, and C_p is the specific heat capacity of zinc.

In the model, the presence of an inductor and the effect of ingot addition have also been considered. When the inductor is at its maximum power, then it is considered that the inductor outlet has 0.75 m/s velocity and 20 °C more temperature than the adjacent molten zinc temperature. And when the inductor is at 20% power, the velocity is taken as 0.45 m/s and the temperature is taken as 8 °C more than the adjacent molten zinc temperature. The inductor is assumed to work at maximum power when ingots are added to the bath and if not, the inductor works at 20% of its power.

Here, the temperature boundary condition is used to include the thermal aspect of the inductor heater as the inductor heater temperature will be higher than the bath temperature. Also, the inductor temperature will increase with the increase in inductor power. The velocity boundary condition is added to the inductor heater to include the effect of Lorentz force. As the heater is inductor-based, magnetic Lorentz force, along with Joule's heat, will influence the flow pattern near the inductor. When the ingot is added, the ingot surfaces are considered at 420 °C, which is the melting temperature of the ingot. As the zinc ingots are melting, the ingot adding surface will have an isothermal boundary condition. Here the magnetic Lorentz force is due to the interaction of the magnetic field of the inductor and the charged particles such as ions in the bath. These assumptions are taken from previous literature [1].

4 Results and Discussion

4.1 Effect of Forced Convection and Natural Convection

The flow pattern inside the zinc bath does not only depend upon forced convection, which is due to the movement of steel strips, but also on the natural convection effect due to the temperature gradient. To see the effect of natural convection and ingot addition, 2 cases are considered. In the first case, the forced convection due to the strip movement is only considered. Natural convection and ingot addition are not taken into account. Hence, the inductor works at 20% power, and the ingot surfaces are considered as zero heat flux faces. The strip velocity is 1.5 m/s. In the second case, both natural and forced convection are considered and ingot addition is also considered. As there are ingots, the inductor works at its maximum power, and ingot surfaces are taken at 420 °C. The strip velocity is kept the same equal to 1.5 m/s in this case.

Figure 3 shows the velocity vectors in the zinc bath for both cases. From the figure, it is clear that the flow pattern changes near the inductor surface when natural convection is taken into consideration. This is because the zinc near the inductor is at a high temperature, and hence the density will be low. Thus, it will move in an upward direction.

Another important aspect is the formation of dross particles inside the zinc bath. There is a high chance of the formation of dross particles where the velocity is

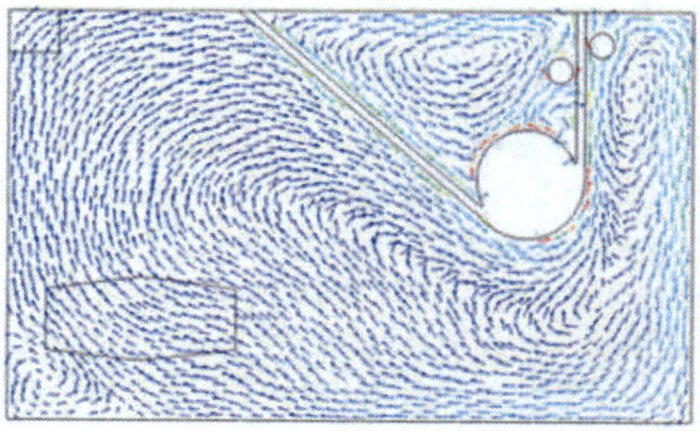

(a)

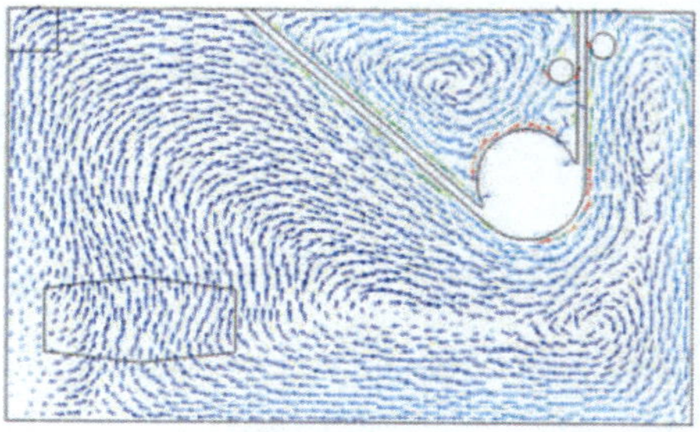

(b)

Fig. 3 The flow pattern parallel to the symmetry plane **a** with only forced convection **b** with both forced and natural convection

<0.01 m/s. The dross particles are intermetallic compounds (Fe–Al–Si intermetallic compound). Also, there will be aggregation consisting of dross particles and already existing metal oxide (CaO, SiO_2 and Fe_2O_3) impurities. Figure 4 shows the effect of natural convection in controlling the total velocity distribution inside the zinc bath compared to the case with only forced convection. The regions with a velocity higher than 0.01 m/s are shaded to show the regions where the chances of dross formation are low. It is seen that natural convection plays a major role in the flow patterns and hence reduction of dross formation in the zinc bath.

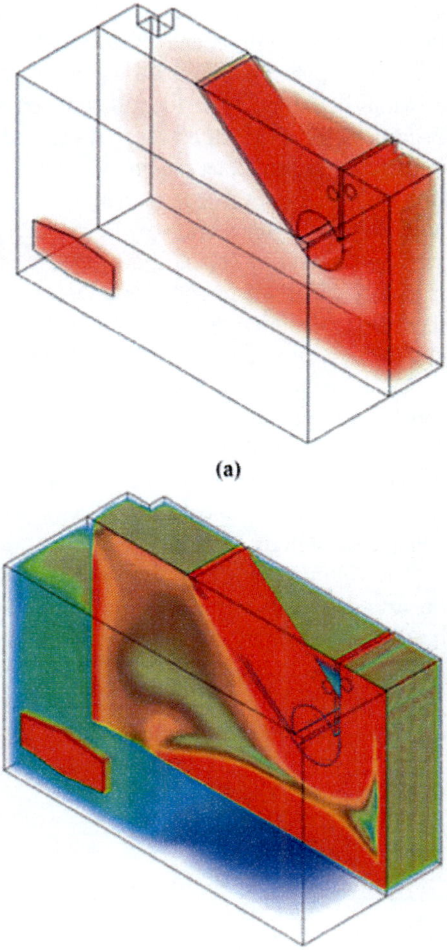

(a) (b)

Fig. 4 Regions with higher than 0.01 m/s velocity: **a** forced convection with no ingot addition **b** with both forced and natural convection and with ingot addition

4.2 Effect of Inductor Power

Inductor heater plays a major role in the velocity profile and temperature distribution of zinc bath, and hence it needs to be analysed. Here, as mentioned above, the inductor conditions were given with temperature boundary conditions and velocity boundary conditions. When the ingot is added, the inductor was assumed to be with a temperature of 20 °C more than the local bath temperature, and a velocity condition of 0.75 m/s was also given to account for magnetic Lorentz force. These conditions were taken as 100% power working conditions for the inductor. When ingots were not added, the inductor was assumed to be with a temperature of 8 °C more than the local bath temperature, and a velocity condition of 0.45 m/s was given. This is taken as the 20% power condition of the inductor. These assumptions were taken from the previous literature [1]. In the following parametric analysis of the inductor, the input conditions for different powers between 20% inductor power and 100% power were linearly interpolated, as shown in Table. 1.

The velocity of the zinc bath has an important role in the overall effectiveness of the bath. The regions with very low velocity or zero velocity are the regions of dross particle formation. This will affect the surface quality of steel sheets. The velocity is mainly due to the forced movements of steel strips. Other than forced convection, there are two more major effects. One is natural convection due to density difference and the other is due to the presence of an inductor. The heater is inductor-based and hence there will be magnetic Lorentz force on the fluid. Apart from the magnetic force, the Joule heating effect also creates movement inside the bath. As the inductor power increases, all these forces will also increase. This effect of inductor power on the volume averaged velocity in the domain is shown in Fig. 5. It shows that the velocity increases almost linearly with an increase of power but the magnitude of change is not high.

The volume averaged temperature of the zinc bath is also found to be increasing linearly with the increase of inductor power. As the power increases, heat flux increases. This increases the overall volume averaged temperature of the zinc bath. This is shown in Fig. 6.

The increase in volume averaged temperature and velocity is less because it is averaged over the entire fluid domain. The heat released from the inductor is mainly

Table 1 Temperature and velocity boundary conditions for different inductor power

Inductor power (%)	Temperature condition for inductor wall (K)	Velocity condition for inductor wall (m/s)
20	741	0.450
40	744	0.525
60	747	0.600
80	750	0.675
100	753	0.750

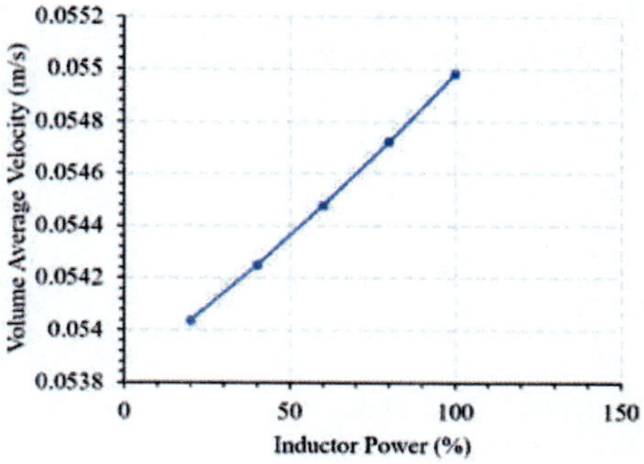

Fig. 5 Inductor power versus volume averaged velocity of zinc bath

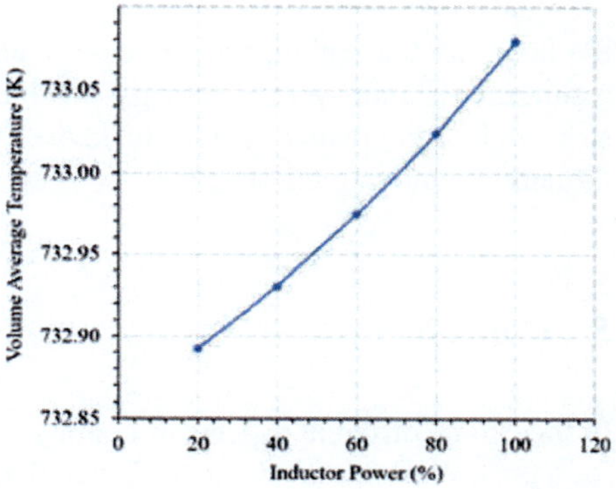

Fig. 6 Inductor power versus volume averaged temperature of zinc bath

used to melt the zinc ingots and used up as latent heat, thus having considerably less effect on the zinc bath temperature.

4.3 *Effect of Strip Angle*

Strip angle is the angle made by the steel strip with the horizontal while it starts moving into the zinc bath towards the primary roller. Here, the strip angle was varied from 40° to 70°. The maximum angle was taken based on the consideration that practically it may be difficult to carry the steel sheet into the zinc bath more steeply.

The relationship between the strip angle and volume averaged velocity is shown in Fig. 7. The convection pattern and volume averaged velocity of the zinc bath are found to be slightly affected by the strip angle with increasing velocity at higher strip angles. The flow pattern inside the bath is mainly driven by forced convection due to the steel sheet movement. When the angle is 40°, the flow from the left side is carried to the right side of the roller. But when the angle is increased, it was observed that the flow pattern changes with the formation of new vortices. This is due to the additional space present for the flow to redirect itself. Hence, it actually increased

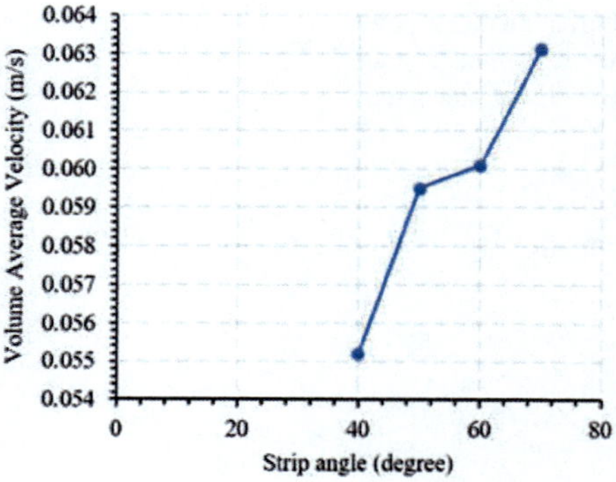

Fig. 7 Strip angle versus volume averaged velocity of zinc bath

the flow into the regions where velocity was very less, towards the inductor. As a result, the volume average velocity was found to be increasing with an increase in strip angle. But practically it is not advised to increase the strip angle steeper due to difficulty in moving the sheets.

5 Conclusions

In this study, different aspects of continuous galvanising baths have been analysed for a three-dimensional domain. Firstly, the effects of natural convection and forced convection have been studied. After that, parametric analysis was conducted for different inductor heater powers and for different strip angles. The change in the flow pattern and velocity has been analysed. The following conclusions can be made based on the simulations performed:

1. The natural convection has a great influence in determining the complex convection pattern in the zinc domain.
2. Dross particle formation is quite significant in the regions that are away from the roller. So, some mechanism has to be implemented to remove these particles.
3. Volume averaged velocity and volume averaed temperatures were found to be linearly increasing with an increase in inductor power.
4. The variation of strip angle shows that there is a change in the flow pattern with a slight increase in volume averaged velocity with an increase in strip angle.

References

1. Ilinca F, Hétu JF, Ajersch F (2005) Numerical simulation of 3D turbulent flow and heat transfer in a continuous galvanizing bath. In: 17th AIAA computational fluid dynamics conference, January, pp 463–482. https://doi.org/10.2514/6.2005-5100

2. Zhou X, Yuan S, Huang Y, Yang P, Zhou X (2015) Simulation of special flow affecting dross formation on steel strip in galvanising bath. Ironmak Steelmak 42(10):785–790. https://doi.org/10.1179/1743281215Y.0000000025
3. Park HS, Han KA, Lee J, Shim JW (2008) Numerical simulation of zinc flow and temperature distribution in a galvanizing zinc pot. ISIJ Int 48(2):224–229. https://doi.org/10.2355/isijinternational.48.224
4. Shukla SK, Saha BB, Triathi BD, Avtar R (2010) Effect of process parameters on the structure and properties of galvanized sheets. J Mater Eng Perform 19(5):650–655. https://doi.org/10.1007/s11665-009-9527-2
5. Zhou X, Yuan S, Liu C, Yang P, Qian C, Song B (2013) Performance of inductors attached to a galvanizing bath. Metall Mater Trans B Process Metall Mater Process Sci 44(6):1580–1585. https://doi.org/10.1007/s11663-013-9938-1
6. Vieira RR, Eleutério, HL De Oliveira TG, Bagatini MC, Tavares RP (2021) Numerical simulation of zinc flow in different layouts of galvanization pot. Mater Res 24(3). https://doi.org/10.1590/1980-5373-MR-2020-0477

Numerical Investigation into Effect of Sidewall Thermal Conductance in Darcy-Bénard Convection

P. Alam and U. Madanan

Abstract Coarse-grained porous-media thermal convection is crucial to the understanding of numerous natural and engineering applications. A survey of the literature reveals that most existing studies either disregard the effect of sidewall thermal conductance or apply a conduction-based model that assumes a linear temperature profile within the sidewalls when estimating the net convective heat transport. However, whether these approaches yield accurate results depends on several factors such as thermal conductivities of sidewalls and porous material and Rayleigh number. Therefore, the present study aims to conduct a three-dimensional numerical study to understand the effect of sidewall thermal conductance heat loss on heat transport for thermal convection of air in coarse-grained porous media. This effect is assessed by performing simulations using sidewalls of three different materials (styrofoam, acrylic, and high-density polyethylene) for a limited range of fluid-based Rayleigh numbers ($10^6 \leq Ra_f \leq 10^8$) and a fixed Darcy number ($Da = 2.91 \times 10^{-4}$). These simulations are repeated for four different porous materials to understand the effect of thermal conductivity of porous material on sidewall thermal conductance. Results show that sidewall thermal conductance can be modeled using a linear temperature profile with reasonable accuracy when Rayleigh numbers are smaller ($Ra_f \leq 10^7$) and the thermal conductivity of sidewalls is of the same order of magnitude as the stagnant fluid. A linear temperature profile assumption is also found to yield accurate results even at extremely high Rayleigh numbers, given that the porous matrix is highly conducting.

Keywords Porous-media thermal convection · Darcy-Bénard problem · Sidewall thermal conductance · Nusselt number · Rayleigh-Darcy number

P. Alam · U. Madanan (✉)
Experimental Thermo-Fluids Laboratory, Department of Mechanical Engineering, IIT Kanpur, Kanpur, Uttar Pradesh 208016, India
e-mail: umadanan@iitk.ac.in

S. Das et al. (eds.), *Proceedings of the 1st International Conference on Fluid, Thermal and Energy Systems*, Lecture Notes in Mechanical Engineering,
https://doi.org/10.1007/978-981-99-5990-7_49

1 Introduction

Thermal convection in fluid-saturated porous media is a widely studied research problem due to its vital role in understanding heat transport in applications such as geothermal energy systems, large thermal insulation systems in buildings, solar ponds, and solar aquifers. Modeling these applications in a laboratory setting involves two horizontal and differentially-heated confined plates (with the bottom plate warmer than the top) and four sidewalls to enclose the porous media and convecting fluid within. This configuration is typically termed as a Darcy-Bénard system. Many studies have been conducted to ascertain the influence of relevant parameters such as Rayleigh number, Prandtl number, size of the porous length scale, thermal conductivity, and location of porous media in an enclosure on the global heat transport. However, hardly any attempt has been made to understand the effect of sidewall thermal conductivity in the Darcy-Bénard problem. In consideration of the fact that most practical situations of the Darcy-Bénard problem do not possess sidewalls, it is crucial that the unwarranted effect sidewalls bring into this problem is understood and appropriately corrected for to accurately model such situations within a laboratory.

2 Literature Review

A review of existing literature shows that sidewall thermal conductance is either completely neglected or modeled using a linear temperature profile (LTP) approach in Darcy-Bénard systems. Kladias and Prasad [1] chose to ignore the effect of sidewall thermal conductance for their experimental study to ascertain the effect of flow and geometric parameters on Nusselt number for coarse-grained porous media ($2.9 \times 10^6 \leq Ra_f \leq 3.66 \times 10^9$; $1.16 \times 10^{-6} \leq Da \leq 2.12 \times 10^{-4}$). By contrast, Keene and Goldstein [2] applied an LTP model to correct for sidewall conductance. They have experimentally simulated Darcy-Bénard convection ($1.68 \times 10^9 \leq Ra_f \leq 3.86 \times 10^{11}$; $Da = 2.18 \times 10^{-5}$) in a test cell enclosed with acrylic sidewall and found the measured Nusselt numbers to be comparable with that for Rayleigh-Bénard convection. More recently, Ataei-Dadavi et al. [3] also used an LTP model in their experimental study on natural convection in coarse-grained porous media ($1.9 \times 10^7 \leq Ra_f \leq 1.7 \times 10^9$; $6.74 \times 10^{-6} \leq Da \leq 1.14 \times 10^{-4}$) enclosed within glass sidewalls. They have also communicated an observation similar to that of Keene and Goldstein [2]. However, Vadasz et al. [4], through their numerical study, reported that motionless solution might not yield accurate results for Darcy-Bénard systems involving conducting sidewalls.

In the case of Rayleigh-Bénard convection, wherein a confined fluid layer experiences heating from below and cooling from top, many investigators have examined the influence of sidewall thermal conductance heat loss on heat transport (or, Nu). One characteristic finding is that the thick thermal boundary layer that gets formed adjacent to the isothermal walls (for moderate Rayleigh numbers, i.e., 10^6–10^9)

offers significant resistance to heat transport, resulting in substantial heat leakage through the sidewalls. To estimate this heat loss, an empty-cell gradient method is commonly employed to correct for this heat loss, assuming convection (or, bulk motion) within the test cell to be absent and a linear temperature profile along the sidewalls. However, Ahlers [5] communicated that this method is inappropriate for estimating actual heat loss through the sidewalls, which can be significantly greater than the empty-cell estimates when the thermal conductivity of the sidewall material is at least an order of magnitude higher than that of the working fluid and when the Rayleigh number is below 10^9. He was the first to address this issue by proposing a model that takes into consideration the development of a thin thermal boundary layer on the sidewalls, particularly close to the bottom wall, where the thermal boundary layer plays an important role in enhancing heat transfer rate through the sidewalls. Later, Roche et al. [6] proposed an alternative approach, wherein they argued that sidewalls placed on the bottom surface experience steep temperature gradients as a result of the development of a slender thermal boundary layer, and thus these sidewalls can be treated as extended surfaces for the boundary layer to form on. More recently, Madanan and Goldstein [7] conducted experiments using test cells made with three different sidewall materials. They extrapolated the obtained results to estimate Nusselt numbers for the case of a perfectly insulated sidewall. They later used this data and proposed a semi-analytical model to estimate sidewall-corrected Nusselt numbers when sidewall thermal conductivity is an order of magnitude higher than that of the working fluid. They have also reported that Nusselt number estimates obtained using this model showed that heat flux through sidewalls can be up to 6 times that computed using an LTP model.

Based on these observations, the authors have attempted to assess the adequacy of an LTP approach and understand how thermal conductance heat loss varies with sidewall and porous media material (or, their thermal conductivities) in the present study. This is ascertained through a numerical study using a combination of three different sidewall materials (styrofoam, acrylic, and high-density polyethylene) and four different porous materials (glass, marble, stainless steel, and brass) for a range of fluid-based Rayleigh numbers ($10^6 \leq Ra_f \leq 10^8$) at a fixed Darcy number ($Da = 2.91 \times 10^{-4}$).

3 Materials and Methods

3.1 Geometry and Boundary Conditions

Computational domain for the present study is configured using a horizontal enclosure of dimension 76.2 mm × 76.2 mm × 76.2 mm filled with 27 spherical balls 25.4 mm in diameter and arranged in a simple cubic packing structure (refer to Fig. 1a). The convecting fluid is taken to be air and the limited range of Rayleigh numbers ($10^6 \leq Ra_f \leq 10^8$) investigated in the present study are realized using the concept of compressing gas outlined by Madanan and Goldstein [8]. The operating conditions and Rayleigh numbers for each of the numerical runs are listed in Table 1. Numerical simulations are carried out for a combination of three different sidewall materials, i.e., styrofoam ($k = 0.033\,\mathrm{W}/m \cdot K$), acrylic ($k = 0.2\,\mathrm{W}/m \cdot K$), and high-density polyethylene ($k = 0.49\,\mathrm{W}/m \cdot K$), and four different porous-media materials, i.e., glass ($k = 0.22\,\mathrm{W}/m \cdot K$), marble ($k = 3\,\mathrm{W}/m \cdot K$), stainless steel ($k = 16.2\,\mathrm{W}/m \cdot K$), and brass ($k = 120\,\mathrm{W}/m \cdot K$), to assess the effect of sidewall thermal conductivity on porous-media Nusselt number (Nu_m).

Thickness of the sidewalls is kept uniform and constant (3 mm) for all simulations. Temperature difference between bottom and top surfaces is maintained at 30 K for all simulations in order to ensure adherence to the Oberbeck-Boussinesq approximation ($\beta\Delta T \leq 0.2$) [9]. Due to the simple cubic packing structure of spheres within the enclosure, porosity of the geometry (ϕ) can be easily estimated and is found to be 0.476. Ergun formulation [10] is used ($K = \frac{\phi^3 D^2}{150(1-\phi)^2}$) to estimate permeability of the computational domain, where, D is the porous-media length scale. Based on the above, Darcy number ($Da = K/H^2$) is computed to be 2.91×10^{-4}.

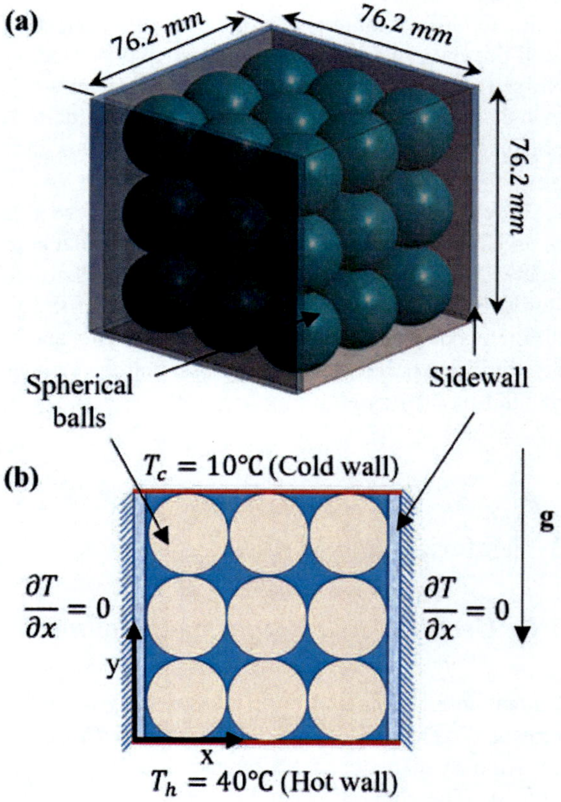

Fig. 1 Computational domain (**a**) schematic (**b**) cross-section with boundary conditions

Table 1 Operating conditions ($T_m = 298.15$ K and $\Delta T = 30$ K)

Run #	Operating pressure (kPa)	Ra_f
1	89.4	1×10^6
2	200.3	5×10^6
3	283.35	1×10^7
4	634.54	5×10^7
5	898.65	1×10^8

3.2 Governing Equations and Numerical Methods

Numerical simulation for the three-dimensional conjugate heat transfer problem is performed using ANSYS FLUENT 2021R1 based on the following assumptions:

- Fluid motion inside the cavity is incompressible, steady, and laminar
- Gravity is the only body force term in the negative y-direction
- Viscous dissipation and heat generation within the enclosure are negligible

Furthermore, the present problem is simulated by coupling together the enclosure containing the convective fluid and porous media. As per these assumptions, governing equations (i.e., continuity, momentum, and energy equation) for the present problem can be formulated as given below:

Fluid Phase:

$$\nabla.\boldsymbol{u} = 0 \tag{1}$$

$$\boldsymbol{u}.\nabla\boldsymbol{u} = -\frac{1}{\rho}\nabla \mathrm{p} + \nu(\nabla.\nabla\boldsymbol{u}) + g\beta(T_f - T_m) \tag{2}$$

$$\rho c_p(\boldsymbol{u}.\nabla T_f) = k_f\nabla^2 T_f \tag{3}$$

Solid Phase:

$$\nabla^2 T_S = 0 \tag{4}$$

Boundary conditions necessary to solve the above governing equations are: (i) bottom and top surfaces are maintained at 10 and 40 °C, respectively, (ii) convective fluid follows no-slip and no-temperature-jump boundary conditions along with Neumann condition at the solid–fluid interface, and (iii) outer surface of the sidewalls are adiabatic (refer to Fig. 1b for details). A pressure-based solver for incompressible flow is used to model the buoyant motion that adheres to the Oberbeck-Boussinesq approximation ($\beta\Delta T \leq 0.2$) [9]. The set of governing equations is then solved simultaneously by employing the numerical schemes listed in Table 2.

Table 2 Numerical schemes used for the present study

Parameter	Numerical scheme
Pressure	PREssure STaggering option
Momentum	Second-order upwind
Energy	Second-order upwind
Gradient scheme	Least square cell-based
Pressure velocity coupling	Semi-implicit method for pressure linked equations

3.3 Mesh Details and Grid Independence

Due to symmetry of the problem (refer to Fig. 1), only one quarter of the domain is considered for the analysis, as can be seen from Fig. 2. The reduced computational domain is meshed with unstructured tetrahedral mesh. A grid independence study is conducted using grids of 7.8 million, 8.7 million, 9.33 million, and 10.55 million cells to investigate the effect of grid number on the obtained results. Porous-media Nusselt number for the marble case with acrylic sidewalls obtained using these four grid configurations is plotted in Fig. 3. As can be inferred from the figure, the difference between Nusselt numbers estimated for 7.8 million cells and 10.55 million cells is found to be a meagre 0.28%. Therefore, owing to the obvious computational benefits, the relatively coarser grids (∼7.8 million cells) are utilized for the results presented in this paper.

Fig. 2 Mesh details

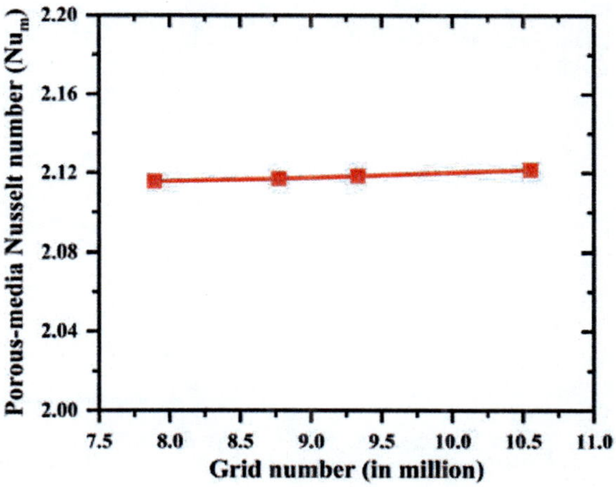

Fig. 3 Grid independence test (marble porous matrix with acrylic sidewalls at $Ra_f = 10^7$)

3.4 Data Validation

Porous-media Nusselt numbers for the glass case with acrylic sidewalls are estimated following two different approaches to correct for the effect of sidewall thermal conductance. In the first approach, a linear temperature profile is assumed in the sidewalls, whereas the second approach utilizes the sidewall thermal conductance predicted by ANSYS FLUENT. These results are corroborated by comparing with experimental results from the literature for $\frac{H}{D} = 3$ [1] (refer to Fig. 4). As can be inferred from the figure, they are in reasonable agreement with each other. Therefore, the numerical methodology adopted in the present study stands validated. However, Nusselt numbers computed using a linear temperature profile in the sidewalls yield a better comparison with the literature (mean absolute percentage error $\sim 7.6\%$) than that calculated using ANSYS FLUENT estimates (mean absolute percentage error $\sim 17\%$).

4 Results and Discussion

Figure 5 depicts the variation of normalized sidewall thermal conductance heat loss with normalized sidewall thermal conductivity for the range of investigated Rayleigh numbers. Here, sidewall thermal conductance is normalized by dividing it with that computed using an LTP-based approach and sidewall thermal conductivity is normalized by dividing it with the stagnant thermal conductivity of the convecting fluid. Here, only the results for marble and brass porous media are presented as representative cases. From both these plots, it is apparent that the normalized sidewall thermal conductance heat loss increases with increase in fluid-based Rayleigh number. This is thought to be a consequence of increased advection and formation

Fig. 4 Data validation (glass porous matrix with acrylic sidewalls)

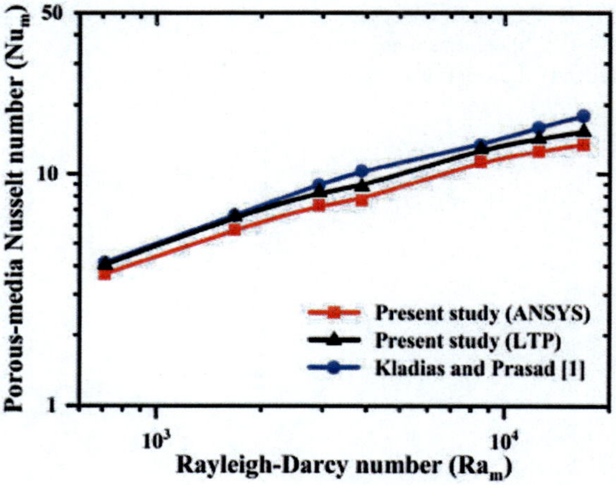

of thinner boundary layers near the vicinity of the hot surface at high Rayleigh numbers, which can expose sidewalls to a steeper temperature gradient and thereby induce a higher heat loss through them. However, for any given Rayleigh number, the normalized sidewall conductance is found to follow a decreasing trend with an increase in sidewall thermal conductivity. This indicates that when sidewalls are perfectly conducting, the normalized sidewall thermal conductance may approach a value of unity. Since thermal conductivity for a perfectly conducting material is infinitely large, and external convection can hardly dictate the temperature profile in the sidewall material, there will be a nearly linear temperature profile within side-walls. Therefore, this explains why sidewall conductance heat loss estimates in such cases will be very close to those obtained using an LTP-based approach.

Figure 6 shows porous-media Nusselt numbers calculated using both simulation-predicted sidewall conductance values and the LTP-based estimates plotted against the normalized sidewall thermal conductivity for all the studied Rayleigh numbers and porous materials. For this purpose, effective thermal conductivity of the porous matrix needs to be computed first to estimate porous-media Nusselt numbers. Effective thermal conductivity is the thermal conductivity of the overall porous matrix under pure diffusion conditions, i.e., in the absence of any bulk motion. In the present study, simulations are performed by turning off the body force term to establish the effective thermal conductivity. Effective thermal conductivities for each of the four different porous materials are determined using Fourier's law of thermal diffusion using the overall heat transfer rate obtained from the simulations, after correcting for the sidewall thermal conductance. The values thus obtained are tabulated in Table 3. After establishing effective thermal conductivities, the porous-media Nusselt number for each of these simulations can be computed as follows:

$$Nu_m = \frac{hH}{k_{eff}} = \frac{(q_o - q_{sw})H}{A_{cs} k_{eff} \Delta T} \tag{5}$$

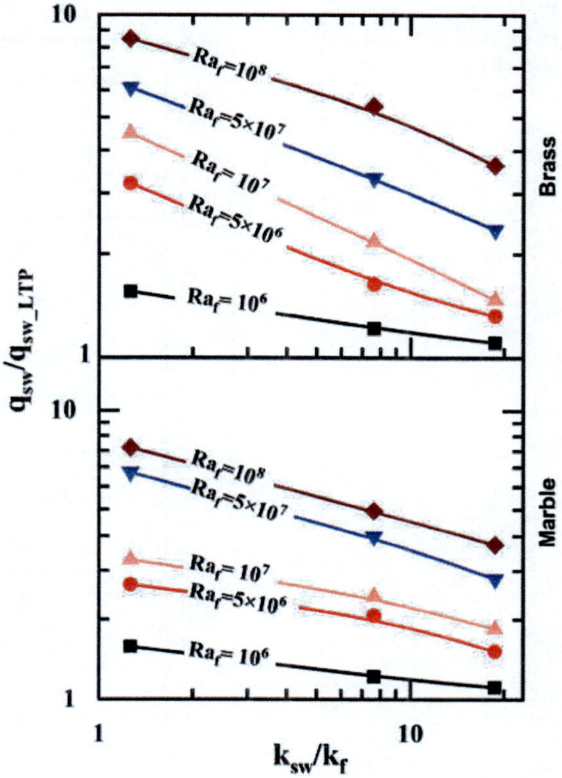

Fig. 5 Effect of Rayleigh number and sidewall thermal conductivity on sidewall thermal conductance

It is apparent from Fig. 6 that the numerically determined Nusselt numbers deviate significantly from the LTP-based estimates only when sidewall thermal conductivity is approximately an order of magnitude higher than that of the convecting fluid and fluid-based Rayleigh number is sufficiently high to produce notable buoyant flow within the porous matrix. It can also be understood from the figure that if the porous matrix is a higher thermal conductivity material such as brass, the LTP-based Nusselt numbers are well within 3.6% of the numerically predicted values even at the highest tested Rayleigh numbers. However, if thermal conductivity of the porous material is of the order of (or, a slightly higher) magnitude than the sidewall material (here, glass and marble), the LTP-based Nusselt number estimates are found to deviate significantly (as high as 18%) from the numerically predicted estimates.

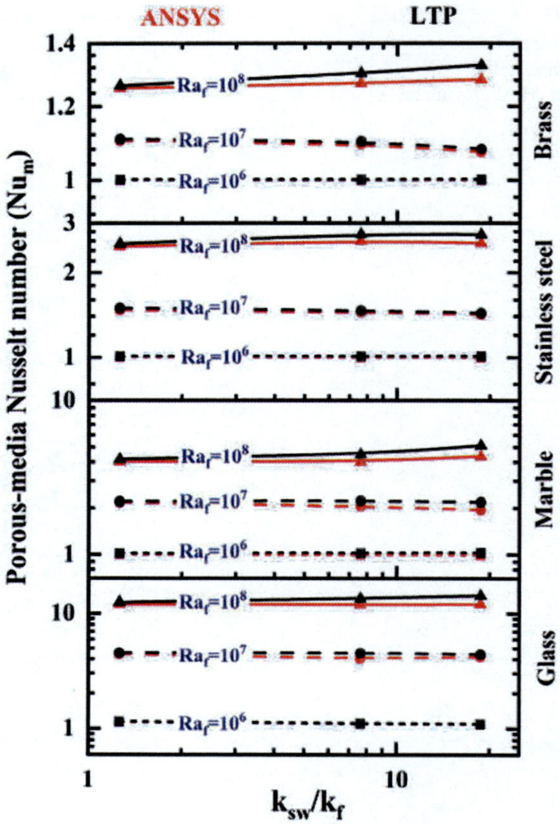

Fig. 6 Effect of sidewall thermal conductance on Nusselt number

Table 3 Ratio of effective and fluid thermal conductivities [$k_f = 0.026\,\mathrm{W}/m \cdot K$]

Porous material	$\mathrm{W}/m \cdot K \vert \mathrm{W}/mK$	$\frac{k_{eff}}{k_f}$
Glass	0.22	3.167
Marble	3	10.13
Stainless steel	16.2	29.16
Brass	120	173.28

5 Conclusions

The effect of sidewall heat conductance on Nusselt number for Darcy-Bénard convection is numerically studied for a range of fluid Rayleigh numbers $\left(10^6 \le Ra_f \le 10^8\right)$ and a fixed Darcy number ($Da = 2.9 \times 10^{-4}$). A linear temperature profile assumption is found to yield reasonably accurate results when:

(i) Rayleigh number is smaller ($Ra_f \le 10^7$) and the thermal conductivity of sidewalls is of the same order of magnitude as that of the stagnant fluid.

(ii) Rayleigh numbers are much higher, but the porous matrix is highly conducting.

Acknowledgements Author U.M. would like to thank DST SERB for funding this work through the Start-up Research Grant (SRG/2021/000194/ES).

Nomenclature

A_{cs}	Cross-sectional area of quarter domain [m^2]
c_p	Isobaric specific heat [J$/kg \cdot K$]
D	Pore diameter [m]
Da	Darcy Number ($= K/H^2$)
g	Acceleration due to gravity [m/s^2]
h	Heat transfer coefficient [W$/m^2 \cdot K$]
K	Permeability [m^2]
k_{eff}	Effective thermal conductivity [W$/m \cdot K$]
k_f	Fluid thermal conductivity [W$/m \cdot K$]
k_{sw}	Sidewall thermal conductivity [W$/m \cdot K$]
k_s	Pore material thermal conductivity [W$/m \cdot K$]
H	Porous layer height [m]
Nu_m	Porous-media Nusselt number
p	Operating pressure [Pa]
q_o	Overall heat transfer to test cell [W]
q_{sw}	ANSYS-based sidewall heat transfer [W]
q_{sw_LTP}	LTP sidewall heat transfer [W]
R	Specific gas constant [J$/kg \cdot K$]
Ra_f	Fluid-based Rayleigh number $\left(=\frac{p^2 c_p g \Delta T H^3}{R^2 T_m^3 k_f \mu_f}\right)$
Ra_m	Rayleigh-Darcy number $\left(=Ra_f Da \frac{k_f}{k_{eff}}\right)$
T_c	Cold wall temperature of an enclosure [°C]
T_h	Hot wall temperature of an enclosure [°C]
T_f	Fluid temperature [°C]
T_m	Mean temperature $\left(=\frac{T_h+T_c}{2}\right)$ [K]
T_s	Solid temperature [°C]
ΔT	Temperature difference ($=T_h - T_c$) [K]
u	Velocity vector [m/s]

Greek Symbols

β	Volumetric thermal expansion coefficient $\left(=\frac{1}{T_m}\right)$ [K^{-1}]
μ_f	Dynamic viscosity [Ns$/m^2$]
ρ	Density [kg/m^3]

References

1. Kladias N, Prasad V (1991) Experimental verification of Darcy-Brinkman-Forchheimer flow model for natural convection in porous media. J Thermophys Heat Transf 5(4):560–576
2. Keene DJ, Goldstein RJ (2015) Thermal convection in porous media at high Rayleigh numbers. ASME J Heat Transf 137(3):034503
3. Ataei-Dadavi I, Chakkingal M, Kenjeres S, Kleijn CR, Tummers MJ (2019) Flow and heat transfer measurements in natural convection in coarse-grained porous media. Int J Heat Mass Transf 130:575–584
4. Vadasz P, Braester C, Bear J (1993) The Effect of perfectly conducting sidewalls on natural convection in porous media. Int J Heat Mass Transf 36(5):1159–1170
5. Ahlers G (2000) Effect of sidewall conductance on heat-transport measurements for turbulent Rayleigh-Bénard convection. Phys Rev E 63(1):015303
6. Roche PE, Castaing B, Chabaud B, Hébral B, Sommeria J (2001) Sidewall effects in Rayleigh-Bénard experiments. Eur Phys J B 24(3):405–408
7. Madanan U, Goldstein RJ (2019) Effect of sidewall conductance on Nusselt number for Rayleigh-Bénard convection: a semi-analytical and experimental correction. ASME J Heat Transf 141(12):122504
8. Madanan U, Goldstein RJ (2020) High-Rayleigh-number thermal convection of compressed gases in inclined rectangular enclosures. Phys Fluids 32(1):017103
9. Gray DD, Giorgini A (1976) The validity of the Boussinesq approximation for liquids and gases. Int J Heat Mass Transf 19(5):545–551
10. Nield DA, Bejan A (2006) Convection in porous media. Springer, New York, USA

Combustion and Radiation Modeling of Non-premixed Turbulent DLR-A Flame

Naveen Kumar and Ankit Bansal

Abstract In this work, combustion and radiation models are employed to simulate a non-premixed turbulent DLR-A flame in OpenFOAM. In a DLR-A flame, the low-velocity air co-flow surrounds an axisymmetric methane flame. The mean flow was calculated using k-ε turbulence model. The Eddy Dissipation Concept (EDC) with GRI-Mech 3.0, to model turbulence-chemistry interactions (TCI), is applied. The flame temperature is validated against the experimental results. The discrete ordinates method (DOM) and the P_1 model are used to calculate the wall radiative heat flux for the gray gas mixture. The gas absorption constant has a significant impact on the radiative flux. H_2O and CO_2 are the primary radiative gases in the DLR-A flame. The weighted-sum-of-gray-gases (WSGG) and k-distribution approaches are implemented to evaluate the radiative flux.

Keywords Turbulent flame · Eddy dissipation concept · Radiative heat flux · Weighted-sum-of-gray-gases model · k-distribution

1 Introduction

Combustion and radiation studies are essential for various engineering applications such as furnaces, rockets, flames, and fire safety. There are two types of flames: premixed and non-premixed. A Bunsen burner is a common example of a premixed flame. An investigation carried out for the flame structure of diffusion flames disclosed that the concentration of the reactants decreases in the flame's vicinity and vanishes at the flame boundary [1]. Therefore, the product concentrations have maximum values at the boundary of the flame. Burning candles and matches in the air are typical examples of laminar diffusion flames. The easy control

N. Kumar (✉) · A. Bansal
Department of Mechanical and Industrial Engineering, IIT Roorkee, Roorkee 247667, Uttarakhand, India
e-mail: nkumar@me.iitr.ac.in

S. Das et al. (eds.), *Proceedings of the 1st International Conference on Fluid, Thermal and Energy Systems*, Lecture Notes in Mechanical Engineering,
https://doi.org/10.1007/978-981-99-5990-7_50

of a diffusion flame makes it more popular than a premixed flame [2]. Generally, industrial flames are turbulent.

Turbulent combustion combines turbulent and chemical kinetics. In turbulent reacting flows, the coupling between turbulence and chemistry occurs due to the random mixing of species and heat transport. Therefore, besides mass and heat transport (counting radiation), the turbulence-chemistry interaction (TCI) is also a crucial factor [3].

Large Eddy Simulation (LES) is the most powerful method to capture the TCI in chemically reactive flows. However, the LES method is more computationally expensive than Reynolds-Average Navier–Stokes (RANS). Therefore, RANS methods are still extensively used over LES because of their straightforward implementation, convergence, and lower computing cost. Among the various turbulence models of RANS, the k-ε model is widely used to simulate turbulent diffusion flames [3].

2 Literature Review and Objective

Many TCI models have been developed for chemically reactive flows, such as the Eddy Dissipation Concept (EDC), Partially Stirred Reactor model, Probability Density Functions, Steady Laminar Flamelet model, and the Conditional Moment Closure model [3–7]. However, the EDC has been extensively used to model the TCI in turbulent flames [3, 8]. In addition, GRI-Mech 3.0 [9] was specially designed for air-methane combustion.

Thermal radiation is vital in fires and combustion systems [10–15]. Radiation studies on flames require a thorough understanding of the physics of both, i.e., radiative heat transfer and radiation characteristics of combustion products. Radiation's negligence is responsible for inaccurate predictions of radiative fluxes and temperatures [1, 12].

The Zonal and the imaginary plane methods calculate the radiative fluxes in furnaces and boilers [16, 17]. However, the P_1 method and the discrete ordinates method (DOM) are widely used in radiation problems [18, 19].

The gray gas absorption coefficient assumes a constant value throughout the spectrum irrespective of wavelength. The combustion products, H_2O, CO_2, CO, SO_2, NO_2, and others, can emit and absorb radiation across a wide variety of narrow spectral lines and produce overlapped vibration–rotation bands. Within each band, the resulting absorption coefficient spectrum oscillates wildly, and across the bands, it is zero. A detailed and accurate radiation analysis of these gases is possible with the line-by-line (LBL) approach. However, it costs a lot of time to compute. Therefore, the research broadly employs the weighted-sum-of-gray-gases (WSGG) [18–20] and the k-distribution methods [18, 21–24].

Most radiation studies are limited to gray gas modelling for turbulent diffusion flames [15]. Only a few authors have paid attention to spectral models [2, 12–14]. However, we compared the radiative flux for gray gas and specifically used the k-distribution correlation based on HITEPM data for non-gray gas.

Therefore, the first objective of this work is to simulate the DLR-A flame and analyse the TCI effects on flame temperature and combustion products. The second objective compares the radiative heat flux with the DOM and P_1 methods using two different values of the gas absorption coefficient. Finally, the WSGG method [18] and full-spectrum k-distribution (FSK) correlations [21–24] are employed in conjunction with DOM to compute the wall radiative heat flux.

3 Methodology

Applying the Favre filter, the Reynolds-averaged equations for turbulent reactive flows with standard k-ε model [25] can be expressed as:

$$\frac{\partial \bar{\rho}}{\partial t} + \frac{\partial(\bar{\rho}\tilde{u}_j)}{\partial x_j} = 0 \tag{1}$$

$$\frac{\partial(\bar{\rho}\tilde{u}_i)}{\partial t} + \frac{\partial\left(\bar{\rho}\tilde{u}_i\tilde{u}_j\right)}{\partial x_j} = \frac{\left(\partial\bar{\tau}_{ij} - \bar{\rho}\widetilde{u_i''u_j''}\right)}{\partial x_j} - \frac{\partial \bar{p}}{\partial x_i} + \bar{\rho}\tilde{f}_i \tag{2}$$

$$\frac{\partial\left(\bar{\rho}\tilde{h}\right)}{\partial t} + \frac{\partial\left(\bar{\rho}\tilde{h}\tilde{u}_j\right)}{\partial x_j} = \frac{\partial}{\partial x_j}\left(\bar{\rho}\alpha\frac{\partial\tilde{h}}{\partial x_j}\right) - \frac{\partial}{\partial x_j}\left(\bar{\rho}\widetilde{u_j''h''}\right) + \bar{S}_h \tag{3}$$

$$\frac{\partial\left(\bar{\rho}\tilde{Y}_k\right)}{\partial t} + \frac{\partial\left(\bar{\rho}\tilde{Y}_k\tilde{u}_j\right)}{\partial x_j} = \frac{\partial}{\partial x_j}\left(\bar{\rho}D\frac{\partial\tilde{Y}_k}{\partial x_j}\right) - \frac{\partial(\bar{\rho}\widetilde{u_j''Y_k''})}{\partial x_j} + \bar{\omega}_k \tag{4}$$

where ρ, p, h, α, and D are the density, pressure, total enthalpy, thermal diffusivity, and mass diffusion coefficient, respectively. S_h, Y_k, and ω_k denotes the internal production of thermal energy, chemical source term, and mass fraction. u_j, $\bar{\tau}_{ij}$, and f_i represents the Cartesian velocity, viscous stress tensor, and body force in the x_j-direction and x_i-direction, respectively. The overbar and tilde refer to the Favre averages and Reynolds averages. The Reynolds stress and gradient model for the turbulence fluxes are expressed as follows:

$$\bar{\rho}\widetilde{u_i''u_j''} = -\mu_t\left(\frac{\partial\tilde{u}_i}{\partial x_j} + \frac{\partial\tilde{u}_j}{\partial x_i} - \frac{2}{3}\delta_{ij}\frac{\partial\tilde{u}_k}{\partial x_k}\right) + \frac{2}{3}\bar{\rho}\tilde{k}\delta_{ij} \tag{5}$$

$$\bar{\rho}\widetilde{u_j''Y_k''} = \frac{\mu_t}{\sigma_t}\frac{\partial\tilde{Y}_k}{\partial x_i}, \text{ and } \bar{\rho}\widetilde{u_i''h''} = \frac{\mu_t}{\sigma_t}\frac{\partial\tilde{h}}{\partial x_i} \tag{6}$$

where δ_{ij} is the Kronecker delta, σ_t and C_μ are the model constants, and $\mu_t = C_\mu\bar{\rho}\tilde{k}^2/\tilde{\varepsilon}$ is the eddy viscosity. The turbulence kinetic energy and its rate of dissipation can be expressed as:

$$\frac{\partial(\bar{\rho}\tilde{k})}{\partial t} = \frac{\partial(\bar{\rho}\tilde{u}_i\tilde{k})}{\partial x_i} + \frac{\partial}{\partial x_i}\left[\left(\mu + \frac{\mu_t}{\sigma_k}\right)\frac{\partial \tilde{k}}{\partial x_i}\right] + E - \bar{\rho}\tilde{\varepsilon} \tag{7}$$

$$\frac{\partial(\bar{\rho}\tilde{\varepsilon})}{\partial t} = \frac{\partial(\bar{\rho}\tilde{u}_i\tilde{\varepsilon})}{\partial x_i} + \frac{\partial}{\partial x_i}\left[\left(\mu + \frac{\mu_t}{\sigma_\varepsilon}\right)\frac{\partial \tilde{\varepsilon}}{\partial x_i}\right] + C_{\varepsilon 1}\frac{\tilde{\varepsilon}}{\tilde{k}}G - C_{\varepsilon 2}\bar{\rho}\frac{\tilde{\varepsilon}^2}{\tilde{k}} \tag{8}$$

The model constant values used by [25] are $C_\mu = 0.09$, $\sigma_k = 1$, $\sigma_\epsilon = 1.30$, $\sigma_t = 0.70$, $C_{\varepsilon 1} = 1.44$, and $C_{\varepsilon 2} = 1.92$.

3.1 Eddy Dissipation Concept (EDC)

EDC was originally invented by Magnussen and Hjertager [8]. In an EDC, the computational cells are divided into two regions (fine structures and surrounding structures). This model solves the chemical kinetics only in fine structures where cell size is defined by the Kolmogorov scale [26] and turbulence energy dissipates. The cascade's energy dissipation model [27] was used to obtain fine scales from large to small scales using turbulence model parameters (k and ε). The mass transfer rate $\dot{m}^*$ of the fine structure and its surroundings, as well as the fine structure mass to the total mass γ^*, are well understood. The average reaction rate of Eq. 4 is expressed as [8]:

$$\bar{\omega_k} = \frac{\bar{\rho}\gamma^*\dot{m}^*}{1-\gamma^*}\left(Y_k^* - \tilde{Y}_k\right) \tag{9}$$

The star ($*$) denotes the state of fine structures. The mass fraction of fine structures, Y_k^*, is evaluated by solving the ODEs that describe the perfectly stirred reactor.

3.2 Radiation Models

While radiation passes through a medium, it may be affected by emission, scattering, and absorption. These interactions are important to account for the radiation and are mathematically expressed in terms of Radiative Transfer Equation (RTE) [18]. An exact solution to RTE is impossible, even for a simple radiative enclosure. Therefore, the RTE for the turbulent diffusion flames is extensively solved by approximate radiation models such as P_1 and DOM.

3.2.1 P_1 Model

P_1 converts the RTE into a set of PDEs. It provides accurate results only for the media having isotropic radiation intensity. The accuracy can steadily improve for higher-order estimates with Mark's Boundary conditions [18]. The relationship between radiative heat flux and incident radiation is:

$$q = -\frac{1}{(3 - A_1\omega)}\nabla_\tau G \tag{10}$$

3.2.2 Discrete Ordinates Method (DOM)

DOM solves the RTE by discretizing both the spatial and angular domains of intensity dependence. Numerical quadrature is used to approximate the discrete representation of the directional intensity variation over a solid angle (4). The radiant heat flux, q, is expressed [18] as:

$$q(\mathbf{r}) = \int_{4\pi} I(\mathbf{r}, \hat{\mathbf{s}})\hat{\mathbf{s}}d\Omega \simeq \sum_{i=0}^{n} w_i I_i(\mathbf{r})\hat{\mathbf{s}}_i \tag{11}$$

3.2.3 Weighted-Sum-of-Gray-Gases (WSGG) Method

The WSGG method employed in this study is based on Hottel and Sarofim's work [19]. This method assumes three gray gases and one clear gas. The total radiative heat flux is evaluated using the summation of all gray gas fluxes [19]. The emissivity and absorptivity weights depend on surface and gas temperatures. The equations to calculate the total emissivity and absorptivity are given as [18, 19]:

$$\varepsilon = \sum_{i=0}^{I=3} a_{\varepsilon i}(T_g)\left[1 - e^{-k_i P s}\right] \tag{12}$$

$$\alpha = \sum_{i=0}^{I=3} a_{\alpha i}(T_g, T_s)\left[1 - e^{-k_i P s}\right] \tag{13}$$

where, a, k, s, and P denotes the emissivity weighting factor, absorption coefficient, path length and sum of partial pressure of absorbing gases.

The coefficients, $a_{\alpha i}$, which depend on the surface and gas temperatures, are expressed as:

$$a_{\alpha i}\left(T_g, T_s\right) = \sum_{j=1}^{J=4}\left[\sum_{k=1}^{K=4} c_{i,j,k} T_s^{k-1}\right] T_g^{j-1} \quad i = 1 \rightarrow 3 \tag{14}$$

$$a_{\alpha 0}\left(T_g, T_s\right) = 1 - \sum_{i=1}^{I=3} a_{\alpha i}\left(T_g, T_s\right) \tag{15}$$

3.2.4 Full-Spectrum *k*-Distribution Method

Only the gas absorption coefficient varies across the spectrum according to the *k*-distribution approach. The coefficients of spectral absorption are rearranged into monotonically increasing functions. These functions integrate radiation intensity across the spectrum efficiently, minimizing the number of RTEs that must be solved. To develop a FSK approach, it is necessary to assume the correlation or scaling among the spectral absorption coefficients [21].

3.2.5 Correlations for *k*-Distribution

The cumulative *k*-distribution correlations for water vapour, carbon dioxide, and gas mixtures are discussed in the literature [21–24]. We employed the FSK correlations to calculate the radiative properties of combustion products.

Correlations for H_2O, CO_2, and Gas Mixture

The cumulative *k*-distribution correlation for carbon dioxide is denoted by the following equation:

$$g\left(T_p, T_g, k\right) = \frac{1}{2}\tanh\left[P\left(T_p, T_g, k\right)\right] + \frac{1}{2} \tag{16}$$

where T_p and T_g are Planck function temperature and gas temperature, respectively. The P can be expressed as:

$$P\left(T_p, T_g, k\right) = \sum_{l=0}^{3}\sum_{m=0}^{3}\sum_{n=0}^{3} A_{lmn}\left[\frac{T_g}{T_{ref}}\right]^n\left[\frac{T_P}{T_{ref}}\right]^m\left[\log_{10}\left(\frac{k}{k_{ref}}\right)\right]^l \tag{17}$$

The coefficients of Eqs. 17 and 18 are needed to find the correlation for water vapour. At fixed cumulative *k*-distribution, *g*, the correlation between the k and k_0 are given by the following expression:

$$\log_{10}\left(\frac{k_0}{k_{ref}}\right) = \log_{10}\left(\frac{k}{k_{ref}}\right) + \sum_{l=0}^{2}\sum_{m=0}^{2}\sum_{n=0}^{1} B_{lmn}\left[\log_{10}\left(\frac{k}{k_{ref}}\right)\right]^n\left[\frac{T_g}{T_{ref}}\right]^l [x]^{m+1} \quad (18)$$

where A_{lmn} and B_{lmn} are the correlation coefficients. The correlation for the gas mixtures is expressed as:

$$g_k^{\text{mixture}} = g_k^{H_2O} + g_k^{CO_2} - 1 \quad (19)$$

4 Numerical Set-up

The simulation of the DLR-A flame is performed in the OpenFOAM platform. GRI-Mech 3.0 (53 species and 325 reactions) is applied to model the chemical kinetics. The reactingFoam solver (coupled RANS and EDC) is employed to capture the TCI in the DLR-A flame. The temperature and species concentration are calculated with the reactingFoam solver used as the radiation calculation inputs. The wall radiative heat flux is computed for both gray and non-gray gas mediums.

4.1 DLR-A Flame

The DLR-A flame [28] consists of a fuel mixture of 22.1% CH_4, 33.2% H_2, and 44.7% N_2. A pipe with an inner diameter of $D = 8$ mm and a nozzle of diameter, $d = 140$ mm, supplies the fuel mixture and air with a velocity of 42.15 m/s and 0.3 m/s, respectively. With respect to pipe diameter and fuel velocity, the corresponding Reynolds number is 15,200. The fuel is surrounded by low-velocity air.

4.2 Computational Domain and Boundary Conditions

The computational domain is extended to 125D and 24D in axial and radial directions. A 2D axisymmetric structured mesh with 3466 hexahedral cells is shown in Fig. 1. Fuel, oxidizer, and the ambient are kept at a pressure and a temperature of 292 K and 101,325 N/m^2, respectively. The outlet has a fixed-pressure boundary condition, while the zero-gradient boundary condition is assumed for all other quantities at the outlet and the wall.

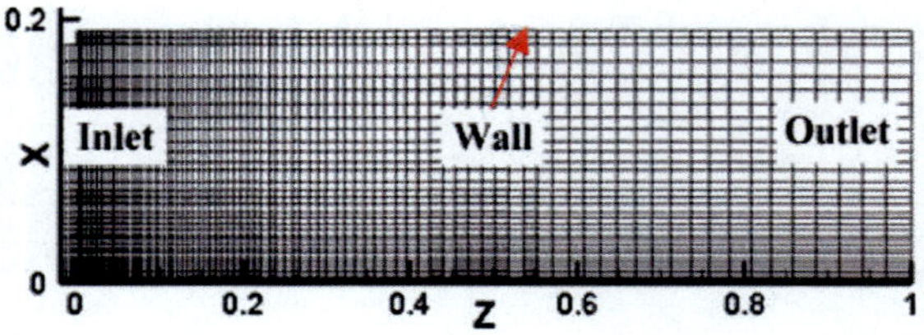

Fig. 1 Two-dimensional axisymmetric structured mesh of DLR-A flame

5 Results and Discussion

The simulation of a DLR-A flame is carried out in two sections. The flame temperature and species concertation were determined in the first section without considering the radiation effects. The temperature and species concentration are calculated and are used as the radiation calculation inputs. The second section calculates the radiative heat flux using the gray and non-gray gas radiation models.

5.1 *Validation*

As illustrated in Figs. 2 and 3, the flame temperature and the species concentrations are extracted along the axial direction and compared with the experimental outcome of the DLR-A flame [28]. The flame peak temperature is 1990 K, slightly higher than the 1846 K of the experimental peak temperature. The location of peak temperature is at $z/D = 61$ and $z/D = 67$ in the axial direction for experimental and numerical studies, respectively.

Figure 3 depicts that the concentrations of H_2O and CO_2 strongly depend on the flame temperature, and a little variation is observed between the present study and

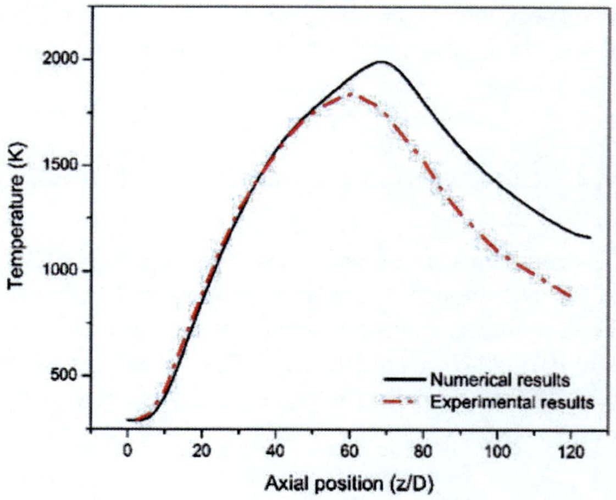

Fig. 2 Comparison of flame temperature with experimental result along the axial direction

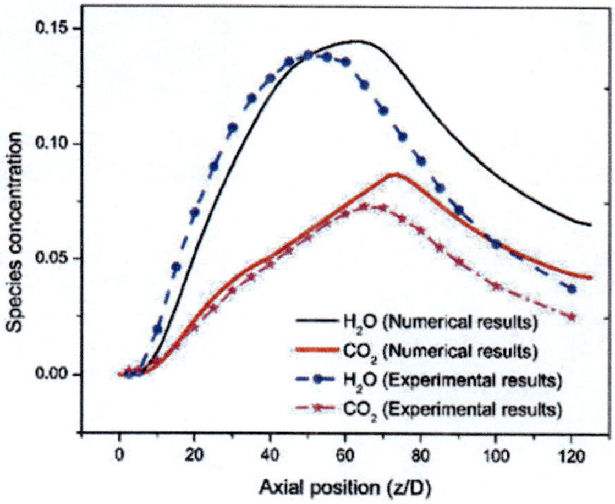

Fig. 3 Comparison of species concentration with experimental result along the axial direction

experimental data. Furthermore, the mass fraction of H_2O is higher than CO_2 due to the higher percentage of hydrogen atoms in the fuel mixtures.

5.2 Comparison of Radiative Heat Flux

5.2.1 Gray Gas Mixtures

For the gray gas mixtures, the DOM and P_1 calculate the wall radiative heat flux using two values, 0.1 and 0.5 m^{-1}, of the gas absorption coefficient. All surfaces in the computational domain are considered black and cold. The gas absorption coefficient is considered to be consistent across the whole computational domain.

As observed in Fig. 4, the flame thickness (species concentrations) increases from left to right. Consequently, the radiative heat flux gradually increases from left to right, as seen in Fig. 5. It indicates the radiative heat flux strongly depends on the flame temperature and gas absorption coefficient. The comparison of radiative heat flux over the computational domain's wall is shown in Fig. 5. For the gas absorption coefficient of 0.50 m^{-1}, the wall heat flux predicted by DOM and P_1 is very close for the entire computational domain, whereas it slightly differs for the absorption coefficient, 0.10 m^{-1}. In the extreme left region of the computational domain, the heat fluxes calculated by P_1 are almost the same for both absorption coefficients. This is because P_1 suffers from poor convergence for combustion applications compared to DOM.

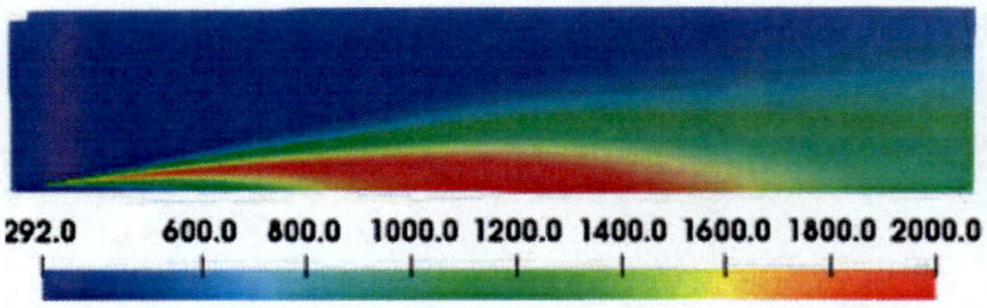

Fig. 4 Temperature contours of DLR-A flame

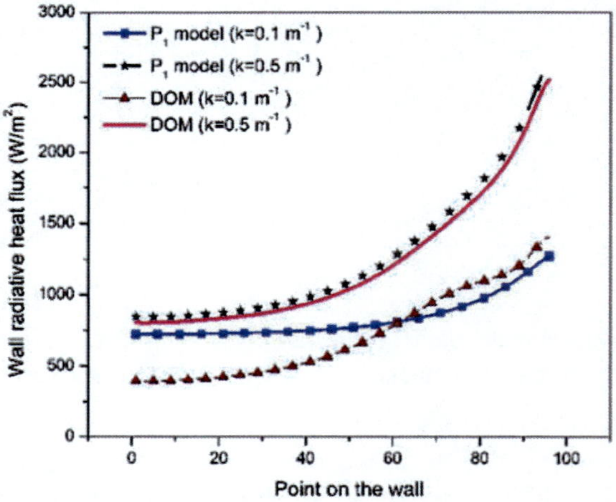

Fig. 5 Wall radiative heat flux comparison for gray gas mixtures

5.2.2 Non-gray Gas Mixtures

The major radiative gases in the flame are H_2O and CO_2. DOM solves the RTE of non-gray gas. The radiative fluxes computed by the WSGG method and FSK correlation follow the same trends for the entire computational domain, as shown in Fig. 6. The k-distribution method predicts 22% higher values of radiative heat flux than the WSGG method. The FSK employs monotonic functions to efficiently integrate radiation intensity across the spectrum, reducing the number of RTEs to be solved while minimizing accuracy losses. Hence, it provides more accurate results compared to the WSGG.

Figure 7 shows the logarithmic cumulative k-distribution for the individual species and gas mixture inside the grid cell.

Fig. 6 Wall radiative heat flux comparison for non-gray gas mixtures

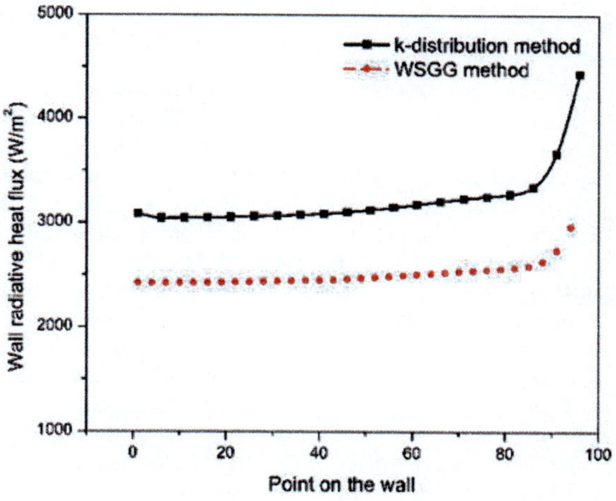

Fig. 7 Cumulative k-distribution for H_2O, CO_2, and non-gray gas mixtures

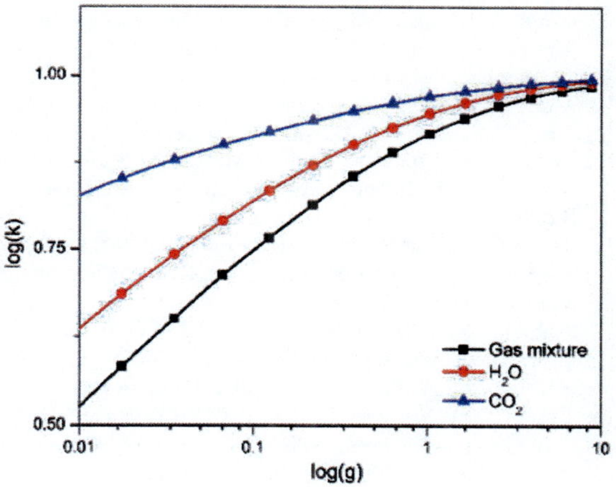

6 Conclusion

The DLR-A flame is simulated using the coupled RANS-EDC code in the open-source CFD-code OpenFOAM. The temperature profile along the axial direction agrees well with experimental results. The concentrations of H_2O and CO_2 strongly depend on the flame temperature. The wall heat flux predicted by P_1 and DOM is highly influenced by gas absorption coefficients, flame temperature, and species concentrations for gray gas mixtures. In comparison to DOM, P_1 has poor convergence for combustion applications. The k-distribution model predicts 22% higher values of radiative heat flux for non-gray gases than the WSGG model. The results indicated that the FSK with DOM is more accurate than the WSGG model.

References

1. Bradely JN (1969) Flame and combustion phenomena. Methuen & Co., Ltd., London
2. Lemos LD, Centeno FR, França FHR (2018) Effects of turbulence-radiation interactions in a non-premixed turbulent methane-air flame. Therm Eng 17(1):63–68
3. Lewandowski MT, Pozorski J (2016) Assessment of turbulence-chemistry interaction models computation of turbulent non-premixed flame. J Phys: Conf Ser 760(1):01215
4. Amani E, Nobari MRH (2010) An efficient PDF calculation of flame temperature and major species in turbulent non-premixed flames. Appl Math Model 34(8):2223–2241
5. Habibi A, Merci B, Roekaerts D (2007) Turbulence radiation interaction in Reynolds-averaged Navier-Stokes simulations of nonpremixed piloted turbulent laboratory-scale flames. Combust Flame 151(1–2):303–320
6. Haworth DC (2010) Progress in probability density function methods for turbulent reacting flows. Prog Energy Combust Sci 36(2):168–259
7. Kim JP, Schnell U, Scheffknecht G (2008) Comparison of different global reaction mechanisms for mild. Combust Nat Gas Combust Sci Technol 180(4):565–592
8. Magnussen BF, Hjertager BH (1977) On mathematical modeling of turbulent combustion with special emphasis on soot formation and combustion. In: Symposium international on combustion, vol 16, no 1, pp 719–729
9. Smith GP, Golden DM, Frenklach M, Moriarty NW, Eiteneer B, Goldenberg M, Bowman CT, Hanson RK, Song S, Gardiner JWC, Lissianski VV, Qin Z (1999) GRI-Mech 3.0. http://combustion.berkeley.edu/gri-mech/
10. Gray WA, Kilham JK, Mueller R (1976) Heat transfer from flames. Elek, London
11. Tien CL, Lee SC (1982) Flame radiation. Prog Energy Combust Sci 8(1):41–59
12. Gupta A, Haworth DC, Modest MF (2013) Turbulence-radiation interactions in large-eddy simulations of luminous and nonluminous nonpremixed flames. Proc Combust Inst 34(1):1281–1288
13. Pal G, Gupta A, Modest MF, Haworth DC (2015) Comparison of accuracy and computational expenses of radiation models in simulation of non-premixed turbulent-jet flames. Combust Flame 162(6):2487–2495
14. Habibi A, Merci B, Roekaerts DJEM (2006) Computational modeling of radiative heat transfers in non-premixed turbulent flames. In: 13th international heat transfer conference, Sydney, Australia. https://doi.org/10.1615/IHTC13.p26.130
15. Lysenko DA, Ertesvag IS, Rian KE (2014) Numerical simulation of non-premixed turbulent combustion using the eddy dissipation concept and comparing with the steady laminar flamelet model. Flow Turbul Combust 93(4):577–605
16. Hottel HC, Cohen ES (1958) Radiant heat exchange in a gas-filled enclosure: allowance for non-uniformity of gas temperature. AIChE J 4(1):3–14
17. Charette A, Erchiqui F, Kocaefe YS (1989) The imaginary planes method for the calculation of radiative heat transfer in industrial furnaces. Can J Chem Eng 67(3):378–384
18. Modest MF (2003) Radiative heat transfer. Academic Press, California
19. Hottel HC, Sarofim AF (1967) Radiative transfer. McGraw-Hill, New York
20. Denison MK, Webb BW (1995) The spectral-line weighted-sum-of-gray-gases model for H_2O/CO_2 mixtures. Trans ASME J Heat Transf 117(3):788–792
21. Zhang H, Modest MF (2003) Full-spectrum k-distribution correlations for carbon dioxide mixtures. J Thermophys Heat Transf 17(2):259–263
22. Modest MF, Singh V (2005) Engineering correlations for full spectrum k-distribution of H_2O from the HITEMP spectroscopic databank. J Quant Spectrosc Radiat Transf 93(1):263–271
23. Bansal A, Modest MF, Levin DA (2011) Multi-scale k-distribution model for gas mixtures in hypersonic nonequilibrium flows. J Quant Spectrosc Radiat Transf 112(7):1213–1221
24. Bansal A, Modest MF (2013) Spectral modelling of radiative heat transfer in carbonaceous atmospheres using k-distribution models. J Thermophys Heat Transf 27(2):217–225
25. Launder BE, Spalding DB (1974) The numerical computation of turbulent flows. Comput Methods Appl Mech Eng 3(2):269–289

26. Lysenko DA, Ertesvag IS, Rian KE (2014) Numerical simulations of the Sandia flame D using the eddy dissipation concept flow turbulence and combustion. 93(4):577–605
27. Ertesvag IS, Magnussen BF (2000) The eddy dissipation turbulence energy cascade model. Combust Sci Technol 159(1):213–235
28. TNF workshop, Proceedings of the third international workshop on measurement and computation of turbulent nonpremixed flames, Colorado, USA (1998)

Hydrogen Leakage Inside a Storage Warehouse and Local Combustion Threats: A Numerical Study

Amit Parmar, Mahesh Kumar Yadav, and Rakesh Dang

Abstract Hydrogen storage and safety are crucial for the envisaged hydrogen economy and sustainable development. In the present study, a CFD-based simulation is performed on a closed storage warehouse (12 m × 5 m × 3.2 m) model to analyze the accidental hydrogen release and subsequent combustion threats. The dispersion of accidental release of hydrogen, hydrogen stratification, and local combustion threats inside a storage warehouse are numerically analyzed. Two important parameters: leakage locations (total 6 locations) and leakage rates (0.91427, 1.82854 g/s) are examined. The dynamics of 4% volume hydrogen concentration iso-surface formation (minimum limit to start hydrogen combustion) and hydrogen distribution at a different time (up to 800s) in the storage warehouse are obtained. It is found that leakage locations near the wall, corners, and roof-top cause more serious safety issues than in the middle of the warehouse. Additionally, hydrogen stratification near the roof-top at the start of the leakage because of the momentum and buoyancy-driven flow increases the chance of hydrogen detonation whereas diffusion-dominated flow at the later time helps in hydrogen distribution. The time at which diffusion starts dominating is found to be a few hundred seconds for the parameter ranges considered in the present study.

Keywords Hydrogen storage · Warehouse · Hydrogen stratification · Combustion threats · CFD

1 Introduction

Hydrogen is actively researched for use as a fuel in different practical applications [1]. When hydrogen is burned, the combustion product is only water vapor. So, hydrogen is a clean source of energy. On the other hand, the combustion of fossil

A. Parmar · M. K. Yadav (✉) · R. Dang
Department of Mechanical Engineering, Punjab Engineering College (Deemed to be University), Chandigarh 160012, India
e-mail: maheshkyadav@pec.edu.in

S. Das et al. (eds.), *Proceedings of the 1st International Conference on Fluid, Thermal and Energy Systems*, Lecture Notes in Mechanical Engineering,
https://doi.org/10.1007/978-981-99-5990-7_51

fuels results in the production of oxides of carbon and nitrogen, and other pollutant gases. This poses a significant challenge to a clean environment and is the leading cause of the greenhouse effect. Therefore, the use of hydrogen as a fuel is an attractive alternative to fossil fuels for a clean environment and sustainable development. A deep understanding of the nature of hydrogen gas, its properties, storage, and safety issues are crucial to truly exploit the hydrogen energy potential.

Abdalla et al. [2] outlined the storage methods of hydrogen such as enveloping liquefied hydrogen, compressed hydrogen, carbon nanotubes, metal hydride, liquid organic hydrogen carrier, and metal–organic framework. The initial temperature and pressure, ambient temperature, and filling rate are identified as major challenges to the fast refueling of fuel cells from the hydrogen refilling station [3]. Hydrogen under high pressures (35–70 MPa) is generally used in the automotive industry [4]. Accordingly, the possibility of unexpected hydrogen leaks is significant. The leaked hydrogen may form flammable conditions and may lead to an explosion type of situation at local locations owing to the large combustibility range (4–75% by vol. concentration at room temperature) of hydrogen in air.

The hydrogen explosion depends upon local distribution, local transport behavior, accumulation of hydrogen, and the possibility of an ignition source. Therefore, there is a need to understand the transport dynamics and local-level stratification to mitigate the possibility of hydrogen combustion.

2 Literature Review and Objective

There is a finite chance of hydrogen leakage in practical situations and several studies have been conducted to examine the possibility of hydrogen deflagration, explosion, and jet flames [5, 6]. It is also accepted that even a minute leakage of hydrogen in the air can result in an explosion at local location(s) and pose severe safety challenges [7–10]. Accordingly, numerous studies have been conducted to examine local hydrogen transport and combustion threats. Vudumu and Koylu [11] examined local hydrogen flammability in straightforward geometries. Venetsanos et al. [12] utilized CFD tools to calculate the short- and long-term dispersion of hydrogen in confined areas. Li et al. [13] calculated the danger or hazard distances for unignited and ignited discharges in the outside environment.

In most of the studies, hydrogen leakage from a particular location is examined to find the hydrogen dispersion and formation of flammable clouds at local locations. In the present study, leakage from different locations are considered inside a hydrogen storage warehouse to simulate different accident scenario to evaluate the combustion threats/overall safety of the storage in a holistic manner. The hydrogen-air mixture compositions at six different leakage locations and two leakage rates are analyzed. The leakage locations and rates are chosen in such a way that different realistic accident scenarios can be analyzed. Accordingly, the study is of particular importance to identify the locations that will need particular attention in case of hydrogen leakage and will be helpful in designing/implementing the correct preventive measures.

3 Mathematical Model

3.1 Description of Problem

The warehouse model considered in the present study is 12 m in length, 5 m in width, and 3.2 m in height. In general, the leakage of hydrogen occurs from the hydrogen release-point distributor (having a typical size of 150 mm high and 305 mm square steel box with an open-top) [14–16]. Accordingly, a square-shaped cross-sectional area of the accidental release point of hydrogen with a hydraulic diameter of 0.305 m is taken in this study. The leakage of hydrogen inside the warehouse can occur at any location. To take into account the different accident scenarios, the hydrogen leakage from the leak points at the center, near the walls, and the corner of the warehouse are considered. Six different leakage locations L1, L2, L3, L4, L5, and L6, considered in the present study, are shown in Fig. 1. The center points of the leak locations are given in Table 1.

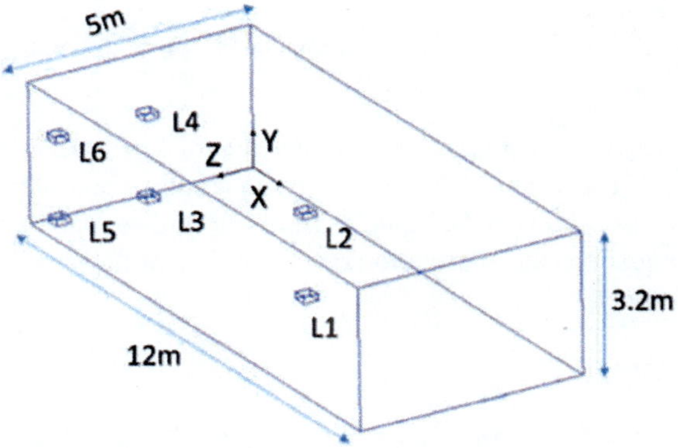

Fig. 1 Hydrogen storage warehouse schematic representation with six different leakage locations

Table 1 Coordinates of center points of the different leakage locations

Leakage locations	*x* (m)	*y* (m)	*z* (m)
L1	6.00	0.00	2.50
L2	6.00	2.00	2.50
L3	0.25	0.15	2.50
L4	0.25	2.00	2.50
L5	0.25	0.15	4.50
L6	0.25	2.00	4.50

3.2 Governing Equations

The mass conservation, momentum, and non-reacting transport equations (for hydrogen and air species) are given as follows:

Conservation of mass equation:

$$\frac{\partial \rho}{\partial t} + \nabla \cdot \left(\rho \vec{V}\right) = 0 \tag{1}$$

Conservation of momentum equation:

$$\frac{\partial}{\partial t}\left(\rho \vec{V}\right) + \nabla \cdot \left(\rho \vec{V} \vec{V}\right) = \nabla p + \nabla \cdot \overline{\overline{\tau}} + \rho \vec{g} \tag{2}$$

where $\overline{\overline{\tau}} = \mu\left[\left(\nabla \vec{V} + \nabla \vec{V}^T\right) - (2/3)\nabla \cdot \vec{V} I\right]$ (stress tensor)

Species Transport Equation: for *i*th species, i.e., Y_i, the mass fractions are evaluated using the equation:

$$\frac{\partial}{\partial t}(\rho Y_i) + \nabla \cdot \left(\rho \vec{V} Y_i\right) = -\nabla \cdot \vec{J}_i \tag{3}$$

where $\vec{J}_i$ is mass diffusion, given as $\vec{J}_i = -\rho D_i \nabla Y_i$

Please note that the energy equation is considered here so that the hydrogen combustion can be modeled in future studies. The turbulent kinetic energy transport equations and dissipation rate for turbulent flow are solved by using a realizable *k*-*ε* model.

3.3 Boundary Conditions and Solution of Governing Equations

The leakage area is set as a mass flow inlet boundary, and hydrogen is leaking toward the roof. The initial gauge pressure is set as 0 Pa. Two different leakage mass flow rates Q (0.91427 g/s) and 2Q (1.82854 g/s) are chosen in this study. The intensity and hydraulic methods are chosen as specification methods for turbulence. A realizable *k*-*ε* viscous model with a scalable wall function is imposed. The PISO scheme is preferred for pressure and velocity coupling. For spatial discretization, second-order upwind is used and also second-order implicit for transient formulation. The absolute convergence criteria for continuity, momentum, and species residuals are taken as 10^{-3} and 10^{-6} for energy residuals. The standard initialization method is chosen to initialize the problem with initial values of flow variables in ANSYS Fluent. For the

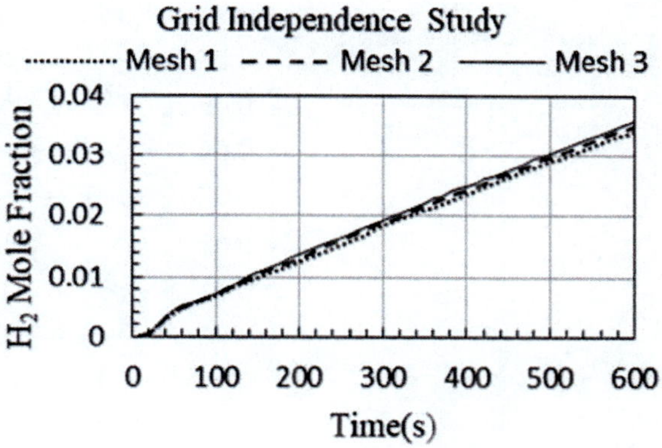

Fig. 2 Variation in hydrogen mole fraction with time for three different mesh sizes at a point (9, 2, 2.5)

calculation of results, the time step size is taken as 0.05 s to satisfy the Courant-Friedrichs-Lewy (CFL) stability criterion, and simulations are run for a total of 800s.

3.4 Grid Independence Study

High-quality hexahedral mesh is generated in ANSYS ICEM. The mesh near the hydrogen release area is refined to resolve the larger gradients. Three different mesh sizes with a varying number of hexahedral cells, namely, Mesh 1(240,128), Mesh 2(355,968), and Mesh 3(490,000) are chosen for the grid independence test. The hydrogen mole fraction obtained at a point (9, 2, 2.5) is plotted for 600 s to stamp out any calculation error. This is shown in Fig. 2. The relative error is calculated for Mesh 1 and Mesh 2 concerning Mesh 3 (finest mesh). The relative error has decreased from 5.5 to 2.36% when cells are increased from Mesh 1 to Mesh 2. Therefore, a Mesh 2 is selected in all the simulations to achieve good accuracy in reasonable computational resources.

4 Results and Discussion

4.1 Validation

To prove that the results of this study are free from numerical error, a validation of the model is carried out with a similar study conducted by Li et al. [12]. Figure 3 shows the comparison made in the numerical results at a point (0, 2.5, 0.38) for a total of 600 s, in the close warehouse of 12 m $\times$ 5 m $\times$ 3.2 m.

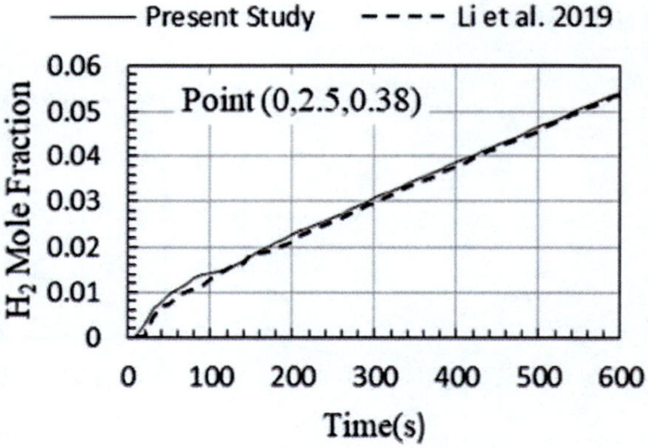

Fig. 3 Comparison of hydrogen mole fraction at the different times obtained from the present study and Li et al. [12] at a point (0, 2.5, 0.38)

A hydrogen mole concentration obtained in the present study and Li et al. [12] is found to be in good agreement with a relative error of 3%.

4.2 *Effect of Leakage Rate and Leakage Locations on Hydrogen Distribution*

Under the combustibility range of 4–75% by volume, the lower flammability limit (LFL) is 4%. The formation of a flammable atmosphere is seen through an iso-surface of 4% volume hydrogen concentration. Variation in the formation of the iso-surface of 4% volume hydrogen concentration at different durations is shown in Fig. 4. It is found that up to 300 s, 4% iso-surface has very small growth. However, after the 500 s there is a significant increase in the 4% iso-surface in the case of Q leakage rate. Whereas, for the case of 2Q leakage rate, the iso-surface formation takes place relatively faster. This shows that the formation of flammable regions will be faster as the leak mass flow rate increases. Further, at a large time (800 s), flammable clouds will develop in both cases.

There are different regions in the warehouse with different hydrogen mole fraction values. This is shown in Fig. 5. Most of the warehouse region is under 7% hydrogen mole fraction in the case of Q leakage rate at the end of 800 s, and theoretically, there is no possibility of detonation. On the other hand, the maximum region of the warehouse consists of up to 10% hydrogen mole fraction. Therefore, there is a high possibility of detonation in this case, as and when an ignition source is available.

In Fig. 6, the variation of iso-surface of 4% volume hydrogen concentration is shown for location 2. When the leakage rate is Q, the iso-surface strikes the roof in the 390 s which means the spreading of hydrogen cloud under the ceiling and after 630 s, the fast variation is there. By doubling the leakage rate, the rate of formation of iso-surfaces is faster and it strikes the roof in the 60 s. Even within 500 s, the whole warehouse has a concentration of more than 4% when the leakage rate is changed to

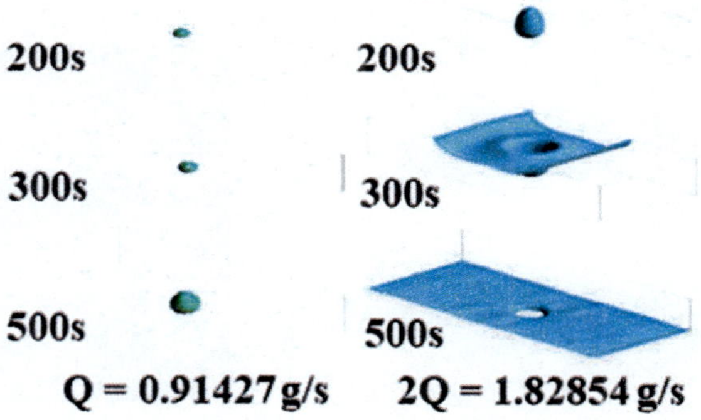

Fig. 4 Iso-surface of 4% volume hydrogen concentration for Q and 2Q mass flow rates at location 1

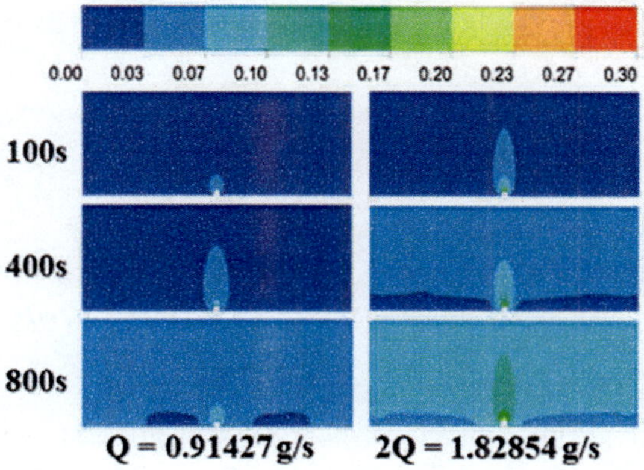

Fig. 5 Hydrogen mole fraction contours at location 1 for Q and 2Q mass flow rates

2Q. From the mole fraction contours in Fig. 7, there is a nearly uniform distribution of hydrogen across the storage warehouse as same as in location 1.

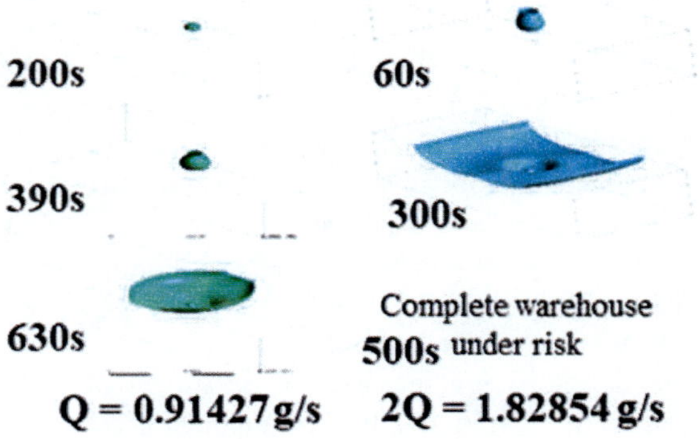

Fig. 6 Iso-surface of 4% volume hydrogen concentration for Q and 2Q mass flow rates at location 2

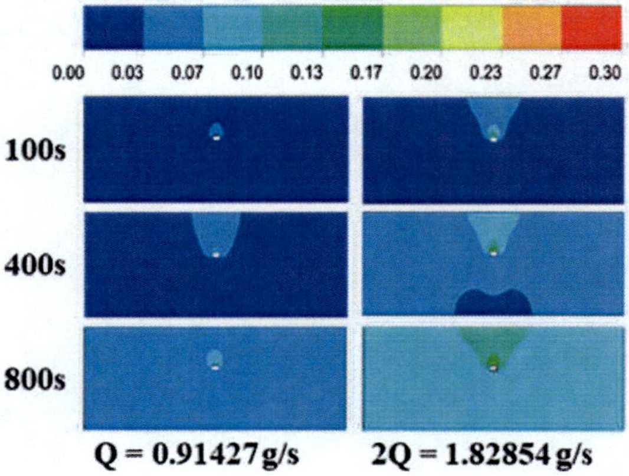

Fig. 7 Hydrogen mole fraction contours at location 2 for Q and 2Q mass flow rates

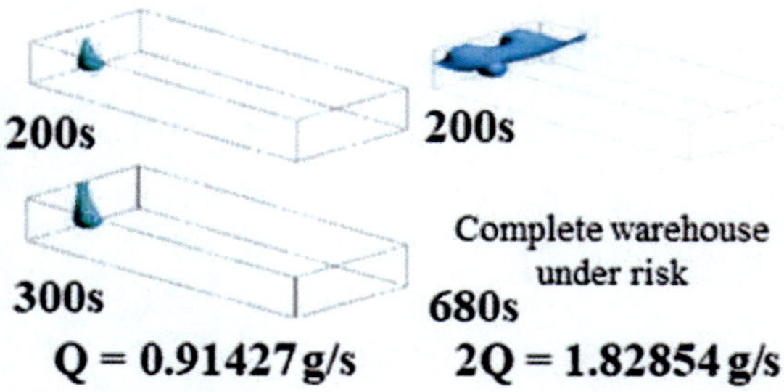

Fig. 8 Iso-surface of 4% volume hydrogen concentration for Q and 2Q mass flow rates at location 3

From Fig. 8, the propagation of hydrogen takes place non-uniformly, as the leakage starts along the wall until it hits the roof. After that dispersion process took place along the length.

Figure 9 depicts that a large amount of hydrogen gets accumulated near the corners just above the leakage location. At the end of the 800 s, for Q leakage rate, some areas are not in the flammability range and not in the detonation zone (18–59%). For the 2Q leakage rate, there is the possibility of detonation as the mole fraction of hydrogen is in the detonation zone. After doubling the leakage rate, at location 3, the complete warehouse is seen under risk at 680 s.

Figures 10 and 11 show the 4% hydrogen concentration iso-surface and mole fraction contours. As the leakage location is near the roof, therefore, up to 60 s, for Q leakage rate, very small growth in the 4% hydrogen concentration iso-surface, but as it touches the roof at the 90 s, the fast growth in the flammable atmosphere is seen. As location 4 is at a height above location 3, therefore, for a 1.82854 g/s leakage rate, the complete warehouse under risk atmosphere is seen in 610 s, i.e., 70 s before location 3.

Now, the leakage location is at the corner of the ground for location 5. From Figs. 12 and 13, it is seen that 4% iso-surface of hydrogen mole fraction gets developed up to the roof and this development is seen within the 30 s. After that, small

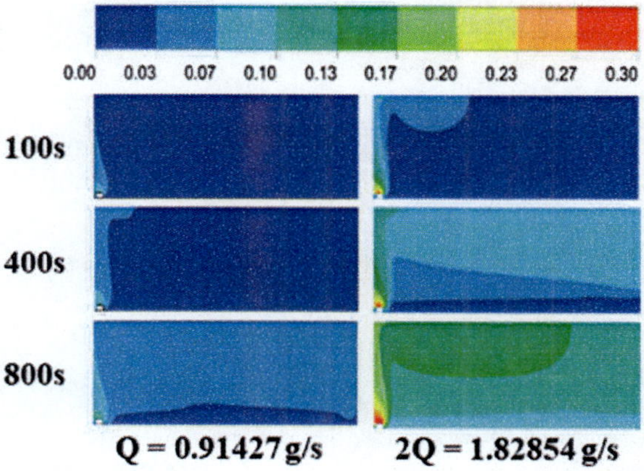

Fig. 9 Hydrogen mole fraction contours at location 3 for Q and 2Q mass flow rates

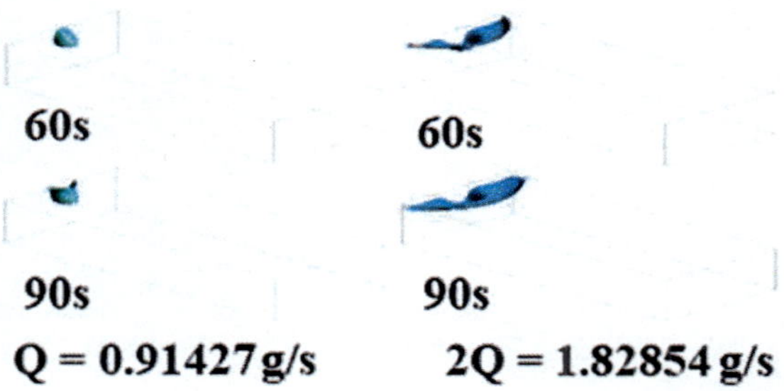

Fig. 10 Iso-surface of 4% volume hydrogen concentration for Q and 2Q mass flow rates at location 4

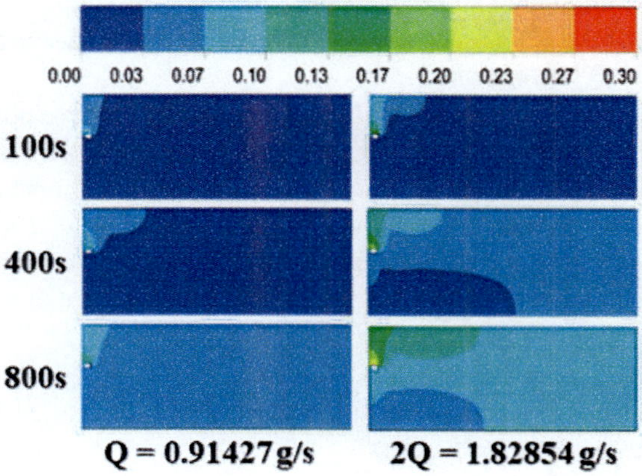

Fig. 11 Hydrogen mole fraction contours at location 4 for Q and 2Q mass flow rates

growth is there for the 380 s, which means flammable regions will start developing quickly in the case of Q leakage rate. For the 2Q leakage rate, a combustible atmosphere forms within 15 s just above the leakage source. The combustible mixtures increase rapidly after 190 s as hydrogen disperses along the ceiling.

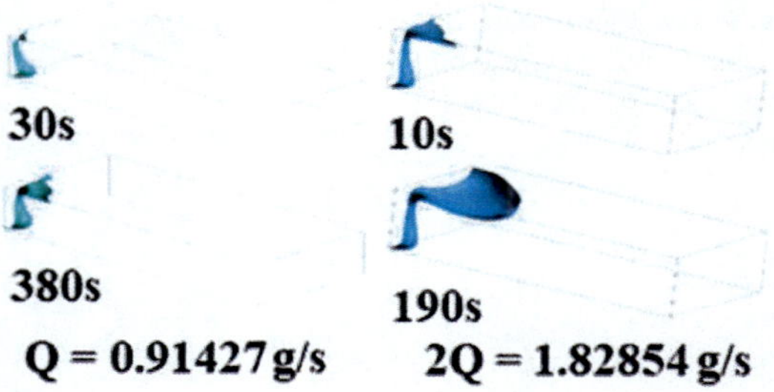

Fig. 12 Iso-surface of 4% volume hydrogen concentration for Q and 2Q mass flow rates at location 5

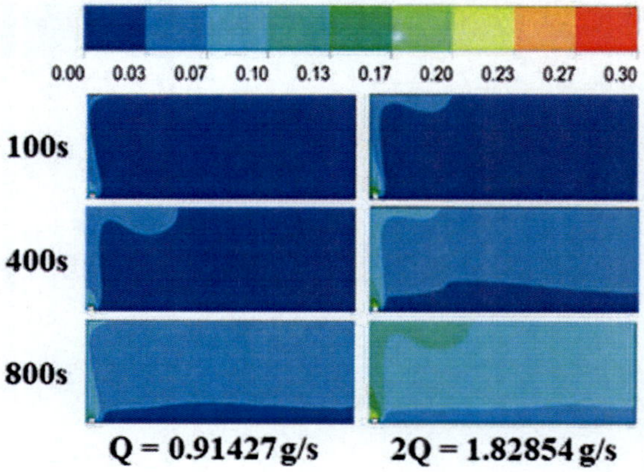

Fig. 13 Hydrogen mole fraction contours at location 5 for Q and 2Q mass flow rates

Figures 14 and 15 give the hydrogen distribution behavior for location 6. It is seen that the iso-surface of 4% hydrogen volume concentration hits the roof within 30 s for Q leakage rate.

For the initial 590 s, the flammable region develops at a constant rate. After this time, the flammable region expands significantly. As the leakage rate is doubled (2Q), the constant rate of generation of the flammable region is only up to 230 s then increasing that region both toward longer distance and ground surface. After the end of the 620 s, the whole region inside the warehouse comes in the range of flammability zone, i.e., greater than 4% volume concentration.

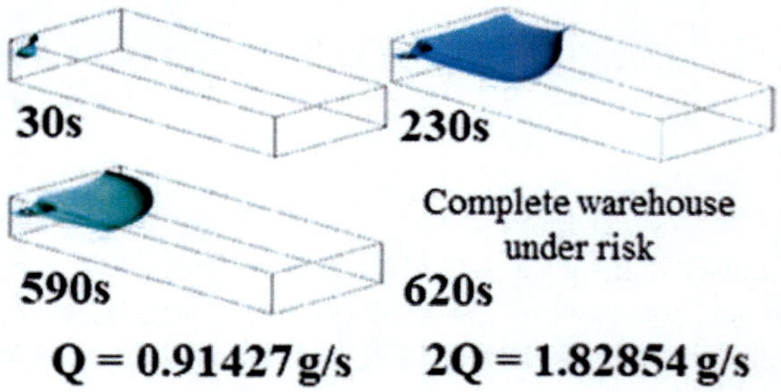

Fig. 14 Iso-surface of 4% volume hydrogen concentration for Q and 2Q mass flow rates at location 6

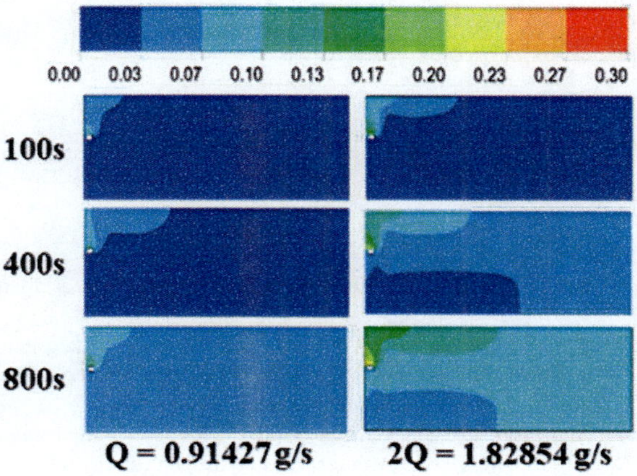

Fig. 15 Hydrogen mole fraction contours at location 6 for Q and 2Q mass flow rates

5 Conclusion

In this study, the accidental release of hydrogen in a confined storage warehouse is studied in the absence of any obstructions, and the spatial–temporal development of hydrogen flammable and detonation regions are investigated. Initially, due to momentum and buoyancy dominated flow, the leaked hydrogen moves upwards and strikes the roof of the warehouse. The diffusion of hydrogen takes place along and opposite the roof at a larger time scale. The time after which diffusion becomes significant comes out to be ~100–200 s, for the parameter values chosen in the present study. It is also observed that leakage nearer an obstacle such as the wall, corners, or roof-top increases the hydrogen stratification and thereby chances of a hydrogen explosion. The higher rate of the unexpected release of hydrogen also augments the possibility of the formation of a detonable zone. It is also observed that, even with minimal leakage, the possibility of hydrogen combustion is always present and the use of safety system(s) is essential to safeguard the hydrogen storage warehouse.

Nomenclature

Q	Leakage rate [g/s]
ρ	Density [kg/m^3]
t	Time [s]
α	Velocity vector [m/s]
∇	Unit vector operator
p	Static Pressure [Pa]
$\bar{\bar{\tau}}$	Stress Tensor [N/m^2]
g	Gravitational Acceleration [m/s^2]
μ	Dynamic viscosity [m^2/s]

I Unit Tensor
μ Diffusion flux [mol/m^{2}s]

References

1. Sasaki K et al (2016) Hydrogen energy engineering: a Japanese perspective. https://doi.org/10.1007/978-4-431-56042-5
2. Abdalla AM, Hossain S, Nis OB, Azad AT (2018) Hydrogen production, storage, transportation and key challenges with applications: a review. Energy Convers Manag 165:602–627. https://doi.org/10.1016/j.enconman.2018.03.088
3. Li M, Bai Y, Zhang C, Song Y (2019) Review on the research of hydrogen storage system fast refueling in fuel cell vehicle. Int J Hydrogen Energy 44:10677–10693. https://doi.org/10.1016/j.ijhydene.2019.02.208
4. Zheng J, Liu X, Xu P, Liu P, Zhao Y, Yang J (2012) Development of high pressure gaseous hydrogen storage technologies. Int J Hydrogen Energy 37(1):1048–1057. https://doi.org/10.1016/j.ijhydene.2011.02.125
5. Schefer RW, Houf WG, Williams TC, Bourne B, Colton J (2007) Characterization of high-pressure, underexpanded hydrogen-jet flames. Int J Hydrogen Energy 32(12):2081–2093. https://doi.org/10.1016/j.ijhydene.2006.08.037
6. Middha P, Hansen OR (2009) CFD simulation study to investigate the risk from hydrogen vehicles in tunnels. Int J Hydrogen Energy 34(14):5875–5886. https://doi.org/10.1016/j.ijhydene.2009.02.004
7. Gupta S, Brinster J, Studer E, Tkatschenko I (2009) Hydrogen related risks within a private garage: concentration measurements in a realistic full scale experimental facility. Int J Hydrogen Energy 34(14):5902–5911. https://doi.org/10.1016/j.ijhydene.2009.03.026
8. Merilo EG, Groethe MA, Colton JD, Chiba S (2011) Experimental study of hydrogen release accidents in a vehicle garage. Int J Hydrogen Energy 36(3):2436–2444. https://doi.org/10.1016/j.ijhydene.2010.04.056
9. Ekoto IW, Houf WG, Evans GH, Merilo EG, Groethe MA (2012) Experimental investigation of hydrogen release and ignition from fuel cell powered forklifts in enclosed spaces. Int J Hydrogen Energy 37(22):17446–17456. https://doi.org/10.1016/j.ijhydene.2012.03.161
10. Schiavetti M, Mattoli V, Lutzemberger G, Dario P, Carcassi M (2012) Experimental study of hydrogen releases in the passenger compartment of a Piaggio Porter. Int J Hydrogen Energy 37(22):17470–17477. https://doi.org/10.1016/j.ijhydene.2012.02.033
11. Vudumu SK, Koylu UO (2009) Detailed simulations of the transient hydrogen mixing, leakage and flammability in air in simple geometries. Int J Hydrogen Energy 34(6):2824–2833. https://doi.org/10.1016/j.ijhydene.2009.01.021
12. Venetsanos AG et al (2009) An inter-comparison exercise on the capabilities of CFD models to predict the short and long term distribution and mixing of hydrogen in a garage. Int J Hydrogen Energy 34(14):5912–5923. https://doi.org/10.1016/j.ijhydene.2009.01.055
13. Li ZY, Makarov D, Keenan J, Molkov V (2015) CFD study of unignited and ignlitedreelases from TPRD under a fuel cells car. In: 6th international conference on hydrogen safety (2015)
14. Li Y, Jiang J, Yu Y, Zhang Q (2019) Numerical simulation of dispersion and distribution behaviors of hydrogen leakage in the garage with a crossbeam. Simulation 95(12):1229–1238. https://doi.org/10.1177/0037549718825303
15. Pitts WM, Yang JC, Blais M, Joyce A (2012) Dispersion and burning behavior of hydrogen released in a full-scale residential garage in the presence and absence of conventional automobiles. Int J Hydrogen Energy 37(22):17457–17469. https://doi.org/10.1016/j.ijhydene.2012.03.074

16. Hajji Y, Bouteraa M, El Cafsi A, Belghith A, Bournot P, Kallel F (2014) Dispersion and behavior of hydrogen during a leak in a prismatic cavity. Int J Hydrogen Energy 39(11):6111–6119. https://doi.org/10.1016/j.ijhydene.2014.01.159

Numerical Study on Heat Transfer Through In-Phase, Out-Phase and Straight Microchannel

Zahoor Bhat, Yogesh M. Nimdeo, Harish Pothukuchi, and Divya Haridas

Abstract Thermal cooling is very important for the smooth functioning of electronic devices. The simulations were performed to obtain the pressure, velocity and temperature distributions for different flow conditions. Further, the surface Nusselt number is calculated to estimate and compare the thermal performance of straight and wavy microchannels. Water is used as a cooling medium and the inlet Reynolds number varies from 50–600 by varying the coolant inlet velocity. It was observed that as compared to the straight channel, in-phase and out-phase channel shows 49.53% and 45% enhancement in heat transfer, respectively. The comparison is made on the basis of the average Nusselt number obtained at Re = 300. However, the lower pressure drop is obtained in a straight channel followed by in-phase and out-phase wavy channels, respectively.

Keywords Wavy microchannel · Thermal cooling · In-phase channel · Out-phase channel · Microchannel heat sinks

1 Introduction

To design and enhance the thermal performance of microfluidic heat transfer devices, understanding the heat transfer characteristics in microchannels is important. Because of continuous functioning, the heat generated from such electronic components increases. The generation of high heat fluxes from miniature electronic equipment is a significant hurdle in the development of new-generation electronic gadgets. The performance of electronic elements is highly contingent on their working temperatures, and thus adequate heat dissipation is extremely important to detour their failure and to assure invariant performance. There exist numerous techniques embraced

Z. Bhat · Y. M. Nimdeo (✉) · H. Pothukuchi
Indian Institute of Technology, J & K, JammuNagrota, Jammu, Jagti, PO 181221, India
e-mail: yogeshnimdev@gmail.com

D. Haridas
Vellore Institute of Technology Bhopal University, Bhopal, M. P 466001, India

S. Das et al. (eds.), *Proceedings of the 1st International Conference on Fluid, Thermal and Energy Systems*, Lecture Notes in Mechanical Engineering,
https://doi.org/10.1007/978-981-99-5990-7_52

for the thermal management of electronic parts among which fluid flow via small dimensional channels has been ascertained to be an effective way of heat dissipation. Several researchers have pursued their work in this direction to enhance the thermal performance of such ongoing operating electronic gadgets by dispersing heat using microchannels. The heat transfer rates acquired from mini and microchannels are significantly higher as compared to conventional cooling techniques, viz. use of fins, fans, blowers, etc.

2 Literature Review and Objective

The era of microchannels started in 1981 when Tuckerman and Pease [1] used microchannels as high-performance heat sinks for the cooling of electronic devices. During the earlier phase of microchannel evolution, conventional microchannels were used. These were in the form of straight channels in which streamlines of the flowing fluid were nearly straight. The poor mixing of the fluid resulted in inefficient heat transfer. It is, therefore, very important to alter the structure of microchannels. When fluid flows through curved surfaces, there occurs a proper mixing of coolant and formation of vortices which inhibits the continuous growth of boundary layer and hence enhances heat transfer performance. Owing to the enhanced thermal performance of microchannels, many researchers have studied the heat transfer through wavy channels by both experimental and numerical approaches. Comini et al. [2] have performed numerical analysis to estimate heat transfer through a three-dimensional wavy channel. They noticed that decreasing the aspect ratio enhances the Nusselt Number. Ahmed et al. [3] numerically studied the heat transfer through wavy channel using nanofluids. The result showed that heat transfer improves due to nanoparticle volume fraction, Reynolds number and amplitude rather than the wavelength of channel. Sui et al. [4, 5] reported that the heat transfer performance of wavy channels is much superior to the straight channels through their experiments and numerical simulations. Jixiang et al. [6] studied the influence of wavy phase shifts on heat transfer through corrugated wavy channels. They noticed that the Nusselt number and friction factor decreases by increasing phase shifts. Rush et al. [7] experimentally observed the flow and heat transfer in sinusoidal wavy passages. They observed that relative amplitude is important in the stability of wavy channels at low Reynolds number. An amplitude larger than the related amplitude delayed the initial onset of instability. At higher Reynolds number, the effect of relative amplitude diminishes. Shuo et al. [8] numerically investigated the heat transfer enhancement of symmetric and parallel wavy microchannel heat sinks with a secondary branch design. As a result of the study, the in-phase configuration has a slightly higher Nusselt number in low Reynold ranges. But in case of moderate and high Reynold ranges, the heat transfer performance of the in-phase configuration is substantially less than that of the out-phase configuration. Lin et al. [9] investigated that thermal performance of wavy microchannels can be improved by changing amplitude and wavelength. Naphon et al. [10] studied the effect of various geometric configurations on flow and temperature distribution in wavy plate channels. They found that wavy plate with sharp edges significantly effect the heat transfer enhancement especially in V-shaped wavy channels.

According to the literature cited above, the geometry of the channel has a significant impact on heat transfer characteristics. Wavy microchannels possess a significant heat transfer advantage over straight microchannels. But in the wavy microchannels, the comparative thermal performance of in-phase and out-phase geometry of the wavy microchannels has not been addressed in detail as per our knowledge. Therefore, in this study, the performance characteristics of wavy sinusoidal channels (in-phase/out-phase) are compared to those of a straight channel. The analysis is carried out using velocity contours and variation of temperature, pressure, average Nusselt number and surface heat transfer coefficient as a function of distance along the axial length of the microchannel to get the thermal performances.

For all the simulations performed, water is used as the cooling medium and the inlet flow rate is varied to achieve different Reynolds numbers ranging from 50–600.

3 Materials and Methods

The numerical study has been conducted on two wavy channels of different phase shifts and a straight rectangular reference channel. In case of a wavy channel, the phase differences are introduced by varying the position of the top plate with that of the bottom plate. All the three channels are of rectangular cross section with specific dimensions (Length = 107 mm, Height = 2 mm, Width = 5 mm). The bottom plate made of copper is maintained at constant heat input of 5 W. Top plate which is also made of copper is maintained at isothermal condition. Rest two side walls are maintained at adiabatic condition. Figure 1 shows the schematic diagram of all the three channels used in the study. The inlet water is maintained at a temperature of 27 °C. Table 1 shows the description of variables used the numerical study.

Numerical Schemes:

The numerical simulations are performed using a finite volume based commercial solver ANSYS Fluent. The computational domain is discretized and SIMPLE Scheme is opted to solve pressure–velocity coupling. The momentum and energy

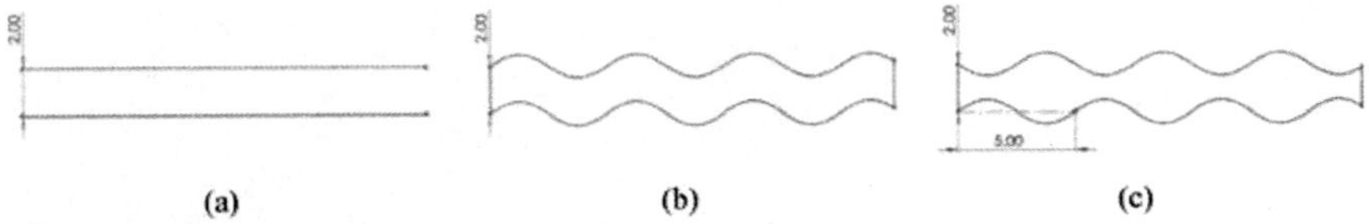

(a) (b) (c)

Fig. 1 Schematics of **a** straight rectangular channel (the reference for comparison), **b** in-phase and **c** out-phase wavy microchannel

Table 1 Numerical variables

Parameters	Values
Pitch of wavy channel (P)	5 mm
Amplitude of wavy channel (α)	0.5 mm
Inlet water temperature (T_{in})	27 °C
Heat input (Q)	5 W
Reynolds Range (Re)	50–600

equations solved by first order upwind scheme. The convergence criteria are chosen as residuals reduced to the order of 10^{-6}.

All simulations are conducted using hexahedra mapped meshes with 0.15 mm element size. The domain as a whole is discretized into 335,907 elements. Figure 2 illustrates how the chosen mesh distribution is organized.

Governing Equations:

The steady-state mass, momentum and energy conservation equations for the fluid are given as follows:

$$\text{Mass conservation}: \nabla \cdot (\rho U) = 0 \tag{1}$$

$$\text{Momentum conservation}: (U \cdot \nabla)\rho U = -\nabla P + \mu \nabla^2 U \tag{2}$$

$$\text{Energy conservation}: U \cdot \nabla T = \frac{k}{\rho c_p} \nabla^2 T \tag{3}$$

The heat transfer coefficient is measured by using the following formula,

$$h = \frac{Q'}{T_w - T_m} \tag{4}$$

where $Q\prime$ = applied wall heat flux

Tw = Area weighted average wall temperature.

Tm = Bulk fluid temperature ($\frac{T_i + T_o}{2}$).

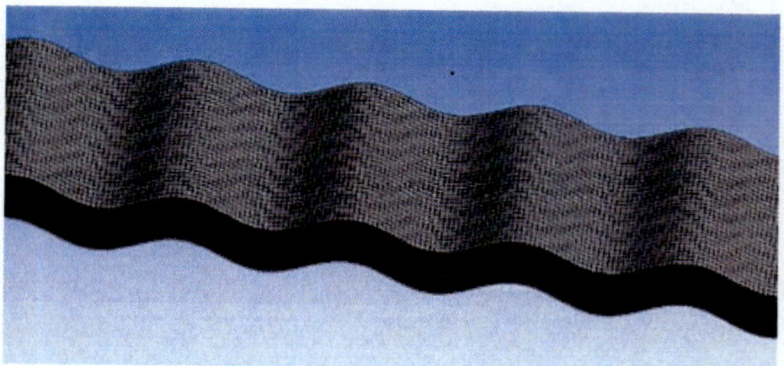

Fig. 2 Schematic of mesh distribution for in-phase channel

T_i = inlet fluid temperature and T_0= outlet fluid temperature.

Using the generated heat transfer coefficient data, the Nusselt number is calculated as

$$Nu = \frac{hD_h}{K} \quad (5)$$

Here, $D_h = \frac{4(H \times W)}{2(H+W)}$ is the hydraulic diameter of the channel and 'k' is the thermal conductivity of the fluid.

Boundary Conditions:

The following inlet and outlet boundary conditions are considered for the calculation of velocity, pressure and temperature.

Inlet velocity condition: $u = U_{in}$

The outlet pressure condition: $P = P_{atm}$

Inlet fluid temperature: 300 K

Top wall: Isothermal condition

Bottom wall: Constant heat flux condition

Side walls: Adiabatic condition.

Numerical Validation:

In the present work, the validation for verifying the performance prediction capability of the numerical model is carried out by comparison of measured Average Nusselt number as a function of Reynolds number with the reported study by Wang et al. [8] in the literature at similar boundary conditions. The validation is made on the basis of Nusselt number for in-phase channel and predictions of results are found to be in well agreement with that reported by Wang et al. as shown in Fig. 3.

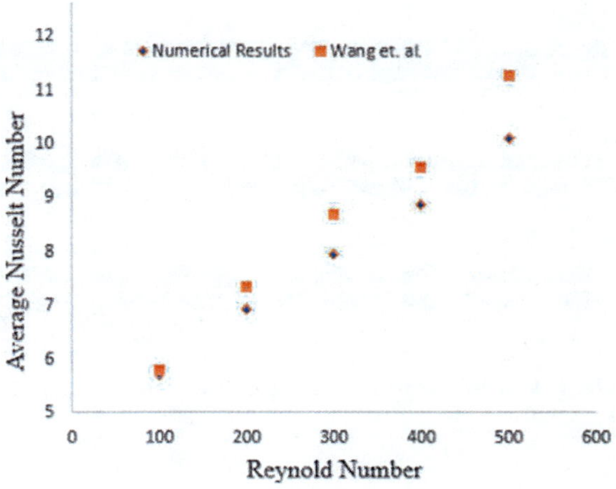

Fig. 3 Comparison of average Nusselt number obtained from present computation with results of Wang et al. [8]

4 Results and Discussions

4.1 Velocity Variation

The heat transfer performance of wavy channels with different phase shifts has been studied numerically for Reynolds number range of 50–600. Figure 4 illustrates the velocity contours at Re = 150. It can be seen that velocity is maximum towards the centre of the channel in all cases. However, overall velocity field is completely different in both the wavy channels. In case of in-phase wavy channel, the gap between the top and bottom wall remains constant, and therefore, velocity streamlines of the fluid towards the centre are almost continuous along the axial direction. While in case of out-phase channel, there is non-uniformity in the gap between the two walls due to formation of converging and diverging areas and it leads to the discontinuation of the velocity streamlines along flow direction. Therefore, in out-phase channel, there is maximum velocity in converging areas and minimum velocity in diverging areas. Based on discontinuity in the formation of velocity boundary layer, wavy microchannels exhibit enhancement in heat transfer over a straight microchannel.

From the velocity vectors depicted in Fig. 5, it is clearly evident that in straight channel (Fig. 5a), there is no formation of Dean Vortices in the flow. The fluid flows smoothly with improper mixing near the heated wall and results in poor convection. While in wavy channels (Fig. 5b, c), there is formation of Dean Vortices which leads to proper mixing of fluid and enhances the heat transfer performance of wavy channels.

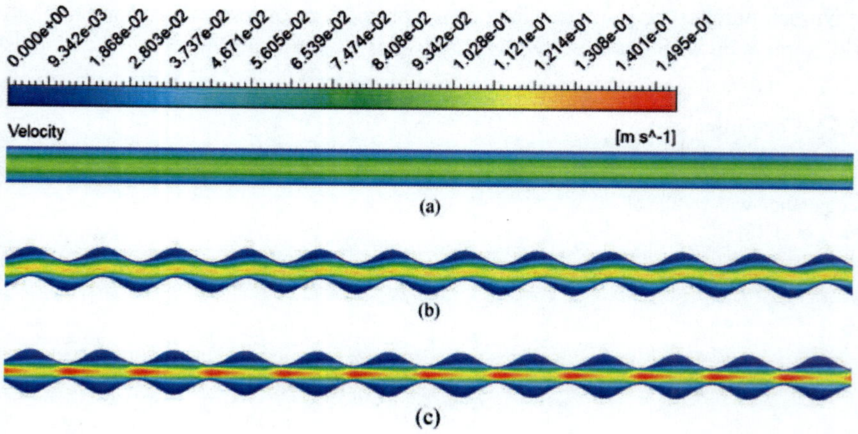

Fig. 4 Velocity contours of **a** straight channel, **b** in-phase and **c** out-phase wavy mini-channel at Re = 150

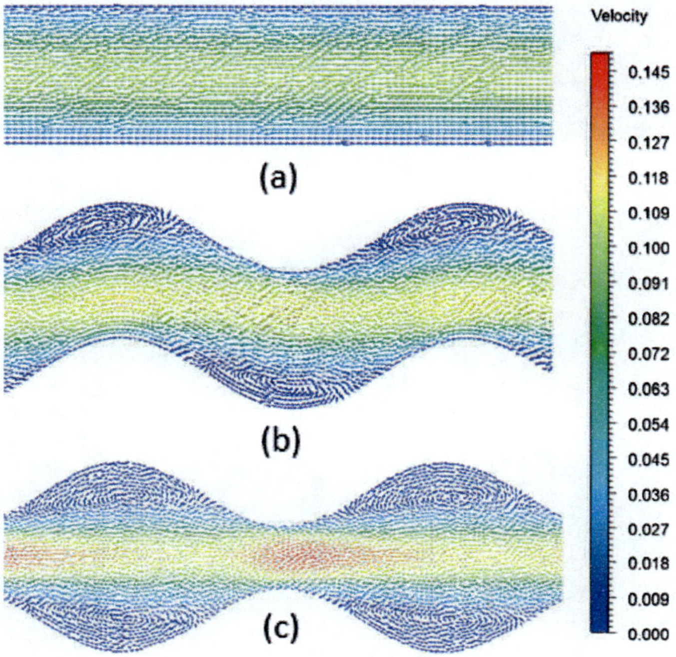

Fig. 5 Representation of velocity vectors along flow direction: **a** straight channel, **b** in-phase channel, **c** out-phase channel

4.2 Temperature Variation

In straight channel (Fig. 6a), the thickness of isotherms representing the thickness of thermal boundary layer is considerable and uniform along the flow direction. Therefore, the temperature gradients become weaken along the flow direction. However, in wavy channels (Fig. 6b, c), thermal boundary layer is not smooth along the flow direction. There arises disturbance in the formation of thermal boundary layer due to channel geometry and hence enhances the thermal performance of wavy channel.

The graphical variation of temperature along axial direction is shown in Fig. 7. Here also in wavy channels the temperature is not increasing gradually like in case of straight channel. The temperature starts increasing when fluid starts flowing over crest areas while it starts falling at trough areas. Which means destabilization of thermal boundary layer that results in enhancement in heat transfer in wavy channels than that of straight channel. However, in case of in-phase and out-phase channel, in-phase channel shows better heat performance than that of out-phase channel.

In Fig. 8, the variation of pressure along the flow direction is shown. Wavy microchannels have a greater pressure drop than straight microchannels. This is because wavy structure of the channel disturbs the entire flow field and there is rise of perturbation and vorticity inside the waves which causes the pressure to rise.

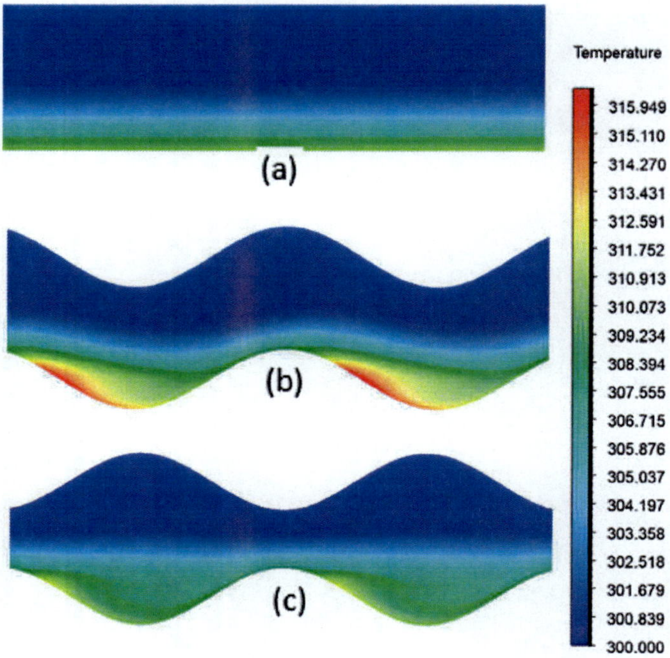

Fig. 6 The contours of temperature distribution at Re = 150: **a** straight channel, **b** in-phase channel, **c** out-phase channel

Fig. 7 Axial variation of temperature at Re = 150

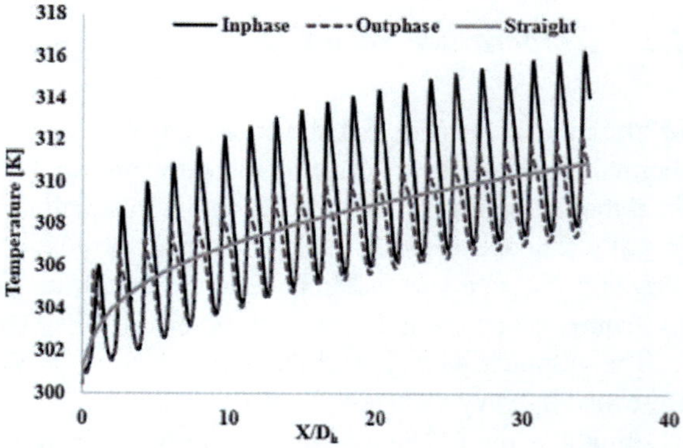

Further, it can be clearly seen that out-phase channel in comparison to in-phase channel attains higher pressure drop and requires higher input power to maintain the flow.

Figure 9 shows the variation of Surface Heat Transfer Coefficient in all the channels. In wavy channels, the trend follows the sinusoidal variation along axial direction with maximum peaks occurs at crest of the channel (corresponding to bottom wall) followed by lower peaks at trough portion of channel.

Fig. 8 Variation of Pressure in Axial direction at Re = 150

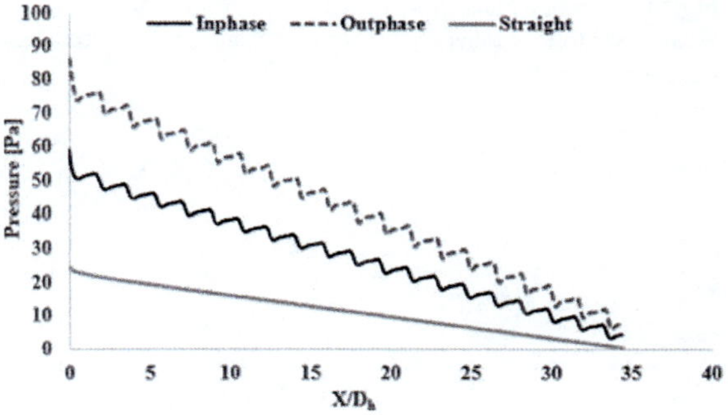

Fig. 9 Surface heat transfer coefficient at Re = 150

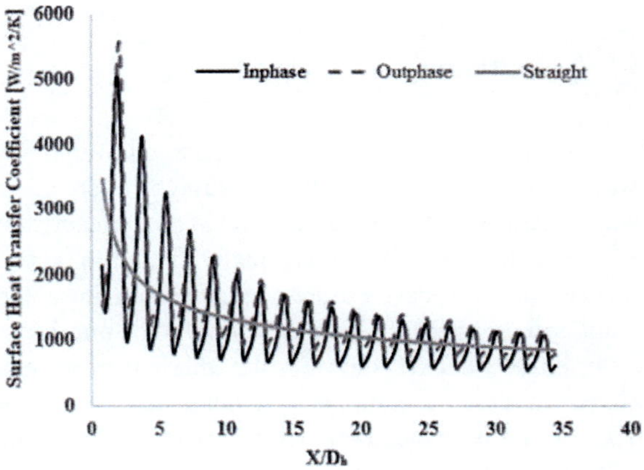

Figure 10 shows the average variation of Nusselt number with respect to Reynolds number. It is clearly seen that Nusselt number increases by increasing Reynolds number and wavy channels shows higher variation in average Nusselt number than that of straight channel. The reason for better thermal efficiency of wavy channels is because the thermal boundary layers are redistributed more due to the wave structure at the top and bottom walls. Also, the average Nusselt number comes out to be high for in-phase channel as compared to out-phase microchannel and for a given Reynolds number, the mixing of flow and the resulting instabilities occur further upstream in in-phase channel as compared to out-phase channel. Therefore, the mixing of the flow enhances the heat transfer in in-phase channel as compared to out-phase channel.

Moreover, for microchannel based heat sinks, in case of straight channels in laminar flow region, the flow becomes regular with considerable thermal boundary layer thickness. This led to the deterioration of heat transfer performance along the flow direction. However, for wavy channels Dean vortices quickly develop along the flow direction and disrupt the thermal boundary layer. It leads to an enhanced heat transfer performance of the wavy microchannel heat sinks.

Further experimental studies need to be carried out for better insight to investigate the heat transfer and fluid flow phenomena in great details.

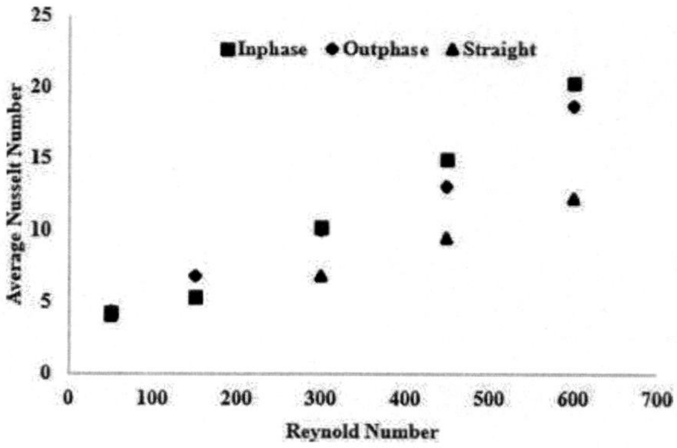

Fig. 10 Average Nusselt number w.r.to Reynolds number

5 Conclusion

Numerical study has been performed to analyse the heat transfer characteristics of wavy microchannels with that of straight microchannel. The results interpret that wavy channel shows higher heat transfer characteristics than that of straight channel. Wavy channel exhibits a sinusoidal variation in flow which results perturbation and formation of vortices, results an enhancement in heat transfer. Also, Average Nusselt number increases with increase in Reynolds number in all the cases.

In-phase channel shows better thermal performance than out-phase channel. However, higher pressure drop is obtained in wavy channels than straight channels. The overall study indicates that microchannel heat sinks based on wavy channel have advantage over those with straight channels.

Acknowledgements I express my sincere gratitude to Ministry of Human Resources and Development (MHRD) and Indian Institute of Technology (IIT) Jammu, for providing the financial support to carry out the project.

Nomenclature

P	Pitch of wavy channel (mm)
α	Amplitude of wavy channel (mm)
Re	Reynolds Number
T_{in}	Inlet water temperature (K)
Q	Heat input (W)
p	Pressure
U	Velocity
H	Height of channel
W	Width of channel
MCHS	Micro channel heat sinks

References

1. Tuckerman DP, Pease RFW (1981) High performance heat sinking for VLSI. IEEE Electron Dev Lett (2):126–129
2. Comini G, Nonino C, Savino S (2003) Effect of Aspect Ratio on convection enhancement in wavy channels. Numer Heat Transf, Part A 44:21–37
3. Vanaki SH, Mohammed HA, Abdullahi A, Wahid MA (2014) Effect of nanoparticle shapes on the heat transfer enhancement in a wavy channel with different phase shifts. J Mol Liq 196:32–42
4. Sui Y, Lee PS, Teo CJ (2011) An experimental study of flow friction and heat transfer in wavy microchannels with rectangular cross section. Int J Thermal Sci 50:2473e2482
5. Sui Y, Teo CJ, Lee PS, Chew YT, Shu C (2010) Fluid flow and heat transfer in wavy microchannels. Int J Heat Mass Transf 53:2760–2772
6. Yin J, Yang G, Li Y (2012) The effects of wavy plate phase shift on flow and heat transfer characteristics in corrugated channel. Energy Procedia 14:1566–1573
7. Rush TA, Newell TA, Jacobi AM (1999) An experimental study of flow and heat transfer in wavy passages. Int J Heat Mass Transf 42(9):1541–1553
8. Lin Wang S, Feng Zhu J (2022) Heat transfer enhancement of symmetric and parallel wavy microchannel heat sinks with secondary branch design. Int J Thermal Sci 171:107229
9. Lin L, Zhao J, Lu G, Wang XD, Yan WM (2017) Heat transfer enhancement in microchannel heat sink by wavy channel with changing wavelength/amplitude. Int J Thermal Sci 118:423e434
10. Naphon P (2009) Effect of wavy plate geometry configurations on the temperature and flow distributions. Int Commun Heat Mass Transfer 36:942–946

Implementation of Neural Networks for the Prediction of CHF Location

Kumar Vishnu, Rishika Kohli, Shaifu Gupta, and Harish Pothukuchi

Abstract Prediction of critical heat flux (CHF) location is a challenging issue in power generation industries. A variety of factors influence the occurrence of CHF and different mechanisms are proposed to understand the physical phenomena of CHF. Due to the nonlinear and complex behavior of CHF, these mechanisms cannot include all the factors simultaneously, as a result, some of them remain unresolved. Considering the above issues, this study employs a neural network (NN)-based machine learning method to model and predict CHF location. Becker et al. (An experimental investigation of post dryout heat transfer, 1983) data bank was selected for both training and testing. Hyperparameter tuning is applied to optimize the performance of the model. The model is found to be performing well with the training dataset as well as the testing dataset. The proposed model is helpful for the design engineers and analysts in the power plants for design optimization and to reduce the computational time.

Keywords Critical heat flux (CHF) · Machine learning (ML) · Neural network (NN)

Nomenclature

ANNs	Artificial neural networks
API	Application programming interface
ARE	Absolute relative error
CHF	Critical heat flux
MAPE	Mean absolute percentage error
MLPs	Multilayer perceptrons

K. Vishnu · H. Pothukuchi (✉)
Department of Mechanical Engineering, IIT Jammu, Jammu 181221, India
e-mail: pothukuchi.harish@iitjammu.ac.in

R. Kohli · S. Gupta
Department of Computer Science and Engineering, IIT Jammu, Jammu 181221, India

S. Das et al. (eds.), *Proceedings of the 1st International Conference on Fluid, Thermal and Energy Systems*, Lecture Notes in Mechanical Engineering,
https://doi.org/10.1007/978-981-99-5990-7_53

MSE	Mean square error
NN	Neural network
μ	Mean of the feature values
σ	Standard deviation of the feature values

1 Introduction

Boiling crisis or CHF location in a two-phase flow boiling system is detrimental to the design of thermal components in many engineering systems such as high-power microprocessor cooling, refrigeration industry, boilers, chemical plants, petroleum refineries, and most often in nuclear power plants. The heat transfer coefficient at the heater-coolant interface drops down sharply as the vapor phase completely covers the heated surface, restricting access to the liquid phase. Due to this reason, the maximum operating power is limited by the CHF which may cause a sudden rise in the heated surface temperature and may eventually result in catastrophic failure of the system [2]. The accurate design and modeling of the systems are necessary to ensure safety and operational efficiencies. The performance, reliability, and longevity of thermal systems significantly depend upon maintaining safe operating conditions [3].

Researchers and design engineers have extensively focused on the study of CHF to predict the safety limits during the system design and analyses. The failure of a system can be controlled/avoided by knowing the correct position of CHF occurrence and by regulating the operating power conditions. CHF and its location are the crucial parameters in the thermal–hydraulic system [4]. Therefore, an accurate prediction of CHF location is necessary to optimize the design and safety of nuclear reactors. CHF is a complex phenomenon that affects various parameters including the geometry and heating wall characteristics [4]. To the best of our knowledge, NN has been employed to predict the wall temperature at CHF [3] and CHF value. It has been seen in the literature that NN is a better-performing model. So, the present study explores NN to predict CHF location. The rest of the paper is organized as follows. Section 2 presents the literature review and objective; Sect. 3 and 4 describes the dataset and the methodology, while Sect. 5 presents the results and discussion followed by conclusions in Sect. 6.

2 Literature Review and Objective

Over the period of last 50 years, with the development of nuclear reactors, tremendous effort has been invested to understand the complexity of the CHF phenomena [5]. The methods developed so far for the prediction of CHF values include empirical correlations, analytical models [5], look-up tables [6], and detailed computational fluid dynamics (CFD) simulations [7]. Each method has its pros and cons owing to the highly complex and empirical relations, phenomenological complexity, and uncertainty of CHF occurrence [8]. The empirical correlations are easy to implement but limited due to highly complex correlations and often fail beyond their ranges of application ability. The analytical CHF models rely on too many assumptions and are empirical in nature. The CHF look-up table methods are based on the CHF data for an 8 mm diameter tube and require interpolation or extrapolation for the intermediate values in the table. The detailed computational fluid dynamics (CFD) simulations are computationally extensive, and the prediction accuracy depends on the choice of closure models [9].

The machine learning methodology is simple, fast, and reliable to predict CHF. While applying an ML model to predict CHF location, the effect of various parameters on the CHF location is analyzed adequately. Table 1 presents the brief literature on the application of ML to CHF prediction.

From the brief literature review, it is obvious that a generalized method that is robust and applicable to a wide range of operating conditions has not been developed so far for the prediction of CHF location. Therefore, it is important to devise an efficient method to overcome the above limitations and obtain accurate predictions of CHF location. With the advancements in computational capabilities and optimization techniques, artificial intelligence (AI) and machine learning (ML) have also received fast-growing interest in thermal-fluid systems and are promising solutions for more objectively predicting the CHF and its location [9]. Therefore, in the present study, an artificial neural network (ANN) model is employed to predict the location of CHF.

3 Dataset Description

In this section, Becker et al. [1] CHF data have been reviewed which covers a wide range of operating conditions with a total of 421 data points for a circular tube.

Multiple variables influence the occurrence of CHF and several mechanisms are intertwined. Furthermore, including the effects of various factors on CHF and its location in a single individual model is difficult. CHF location is a complex phenomenon that mostly depends upon the six input parameters: Wall Heat flux, Mass Velocity, Pressure, Pipe inner diameter, inlet sub-cooling, and heated length. So, the present model has been developed by choosing the above mentioned six parameters as input variables [10]. The target is the values of CHF location, i.e., output. Table 2 presents the operating parameters and their ranges in Becker et al. [1] dataset. The chosen benchmark dataset is divided into two parts, viz., training and testing sets. The training set includes 80% of the samples to train the machine learning model, whereas the testing set includes 20% of the samples to test the performance of the model.

Table 1 Brief literature review on the application of machine learning for the prediction of CHF

S. No	Authors	Description	Remarks
1	Ooi et al. [9]	Identification of flow regimes in boiling flows in a vertical annulus channel with NN and K-nearest neighbors (KNN) [9]	Flow regime identification
2	Kim et al. [10]	CHF prediction for narrow rectangular channels in a steady-state condition using NN [10]	CHF prediction
3	Tank et al. [11]	CHF Prediction on Microstructure Surfaces using Deep Belief Network [11]	Prediction of CHF on Micro surfaces
4	Gulli et al. [12]	CHF predictors in subcooled and low-quality flow boiling by NN and random forest [12]	CHF prediction
5	Jiang et al. [13]	Prediction of CHF using gaussian process regression and ant colony optimization [13]	CHF prediction
6	Almalki and Ahmed [8]	Application of classification Learner environment in MATLAB for the Prediction of Two-Phase Flow Patterns [8]	prediction of two-phase flow pattern
7	Park et al. [3]	Wall temperature prediction at critical heat flux using a NN [3]	CHF Wall Temperature
8	Zhao et al. [2]	CHF prediction Using a physics-informed Machine Learning-Aided NN Framework [2]	CHF prediction
9	He and Lee [4]	Application of ν-SVM for prediction of CHF [4]	CHF prediction

Table 2 Parameters and their ranges

S. No	Parameters	Range
1	Mass flux ($kgm^{-2}\ s^{-1}$)	496–3493
2	Wall heat flux (MWm^{-2})	0.147–60.5
3	Pressure (MPa)	2.98–119.8
4	Pipe inner diameter (m)	0.01–0.0247
5	ΔT_{sub} (K)	4.6–20.7
6	Heated pipe length (m)	7–7.1
7	CHF location (m)	0.63–6.75

4 Methodology

NN is useful for solving strong nonlinear and complex relations. Among the different types of ANNs [8], the feed-forward neural network is widely used in engineering applications [7]. It is also known as multi-layer perceptrons (MLPs), or simply neural networks (NN). It predicts the outcome with high accuracy when data is well sampled [11]. Further, the methodology has been elaborated into two sub-sections, i.e., Data pre-processing and Model explanation.

4.1 Data Pre-processing

In the present study, NN is implemented to predict the CHF location using Becker et al. [1] dataset comprising six input parameters (features) and one target feature. Input features include operating pressure, mass velocity, heat flux, inner diameter, heated length, and inlet sub-cooling [10]. The target is the values of CHF location (output). The input features are pre-processed by standard scaling as follows:

$$X\prime = \frac{X - \mu}{\sigma} \quad (1)$$

where μ and σ represent the mean and standard deviation of the feature values, respectively. X and $X\prime$ are the original and transformed values, respectively.

4.2 Model Explanation

The NN is a machine learning algorithm that recognizes the patterns and relations from the chosen dataset. The schematic of fully connected NN architecture is presented in Fig. 1. It consists of the input layer that accepts the input information from outside and passes it to the next layer called a hidden layer(s) which are either one or more in number depending upon the complexity of the data points, and the output layer that contains the problem result. Layers consist of neurons that are interlinked through connections having weights.

Each neuron has bias and it receives input from other neurons, processes it, and produces an output. Neurons are activated through an activation function. Intermediate values of weights and biases are calculated and stored in forward propagation.

$$Y = f(B + \sum_{i}^{n} W_i \times X_i) \quad (2)$$

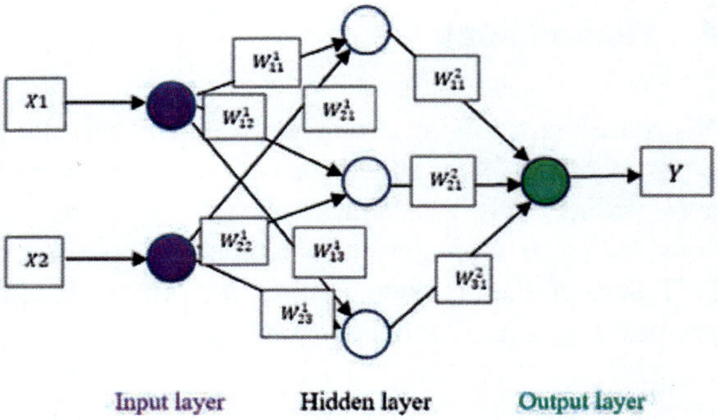

Fig. 1 Schematic of NN architecture

$$f(z) = \max(0, z) \tag{3}$$

where $f(z)$ is the rectified linear unit function activation function. B is the bias, W is the weight and X is the input feature. Y is the output.

The sequential model initiates the weights with random values and these weights are updated during backpropagation. the output error is calculated with respect to the expected output, and weights and biases are updated using RMSprop optimization algorithms to reduce the error. Different hyperparameters have been tuned such as the number of hidden layers and epochs, batch size, loss function, optimization algorithm, and activation function.

5 Results and Discussion

The NN model is implemented based on the experimental data of Becker et al. [1]. The NN is trained using the Keras application programming interface (API) with TensorFlow [12] backend and sci-kit-learn library in Python 3.6. The model framework is first trained and then the performance is validated with the testing dataset. It is seen empirically that the 6/50/50/50/1 architecture performs well for the data from the experiments. There is no specific formula to take the number of hidden layers and their neurons [14]. For the optimized architecture of the neural network, the trial-and-error method is used to find the number of hidden layers and their neurons. The NN architecture 6/50/50/50/1 consists of 6 neurons in the input layer, three hidden layers with 50 neurons, and 1 neuron in the output layer. The tuned optimization algorithm is the root mean square propagation (RMSProp). The loss function is the mean square error (MSE) [11] as follows:

$$MSE = \frac{1}{N}\sum_{i=1}^{N}(y_i - \hat{y}_i)^2 \tag{4}$$

Table 3 Quantitative performance of the NN model

Stages	MSE	MAPE	R^2
Training	0.0255	0.0311	0.9833
Testing	0.0609	0.0462	0.9663

where N is the total number of data points, $\hat{y}_i$ is the predicted and y_i is the experimental CHF location.

The mean absolute percentage error is used to evaluate the predicted model based on the experimental values as follows:

$$MAPE = \frac{1}{N}\sum_{i=1}^{N} |\frac{y_i - \hat{y}_i}{y_i}| \times 100 \tag{5}$$

Further, the measured MAPE for training and testing data points are listed in Table 3. R^2 score is the statistical measure of the goodness of fit for machine learning models and that is calculated as below equation

$$R^2 = 1 - \frac{\sum_{i=1}^{N}\left(y_i - \hat{y}_i\right)^2}{\sum_{i=1}^{N}(y_i - \overline{y})^2} \tag{6}$$

where $\overline{y}$ is the mean of the experimental CHF location.

The weights and biases are randomly initiated by their default values. The default value of weight is a random number and of bias is zero, and their best values are achieved by using the backpropagation method [9]. Rectified linear unit (ReLU) is applied as an activation function on hidden layers. The number of epochs and batch size are 1000 and 100, respectively.

The quantitative performance of the model is tabulated in Table 3. The performance of the model is checked by evaluating the MSE, MAPE, and R^2 scores. The MSE for training is 0.0255 and for testing data sets is 0.0609. MAPE of train and test data sets is 0.0311 and 0.0462 which shows the NN model is performing well for both labeled and unlabelled data points. R^2 score of the model for train and test data sets are 0.9833 and 0.9663 which indicates that the model is not over/underfitted.

Figure 2 compares the predicted results against the experimental values. It is seen that for different data points, the NN model predictions are in good agreement with the experimental values. Figure 3 shows the scatter plot of CHF location for experimental and predicted values. It can be noticed that the data points are close to the best fit line (red). The absolute relative error of all the test data points is within 0.25 (refer to Fig. 4).

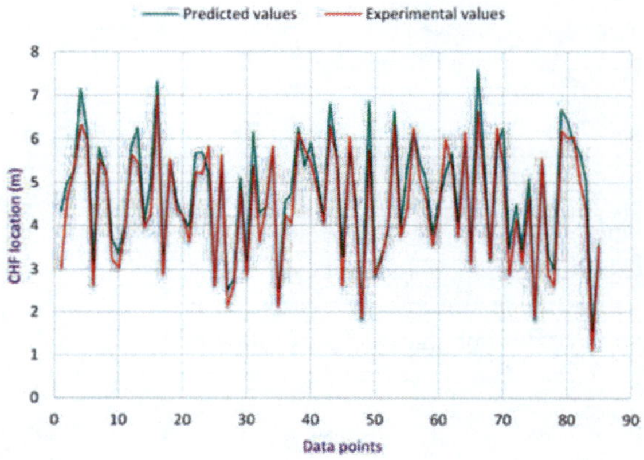

Fig. 2 Comparison of predicted and experimental values

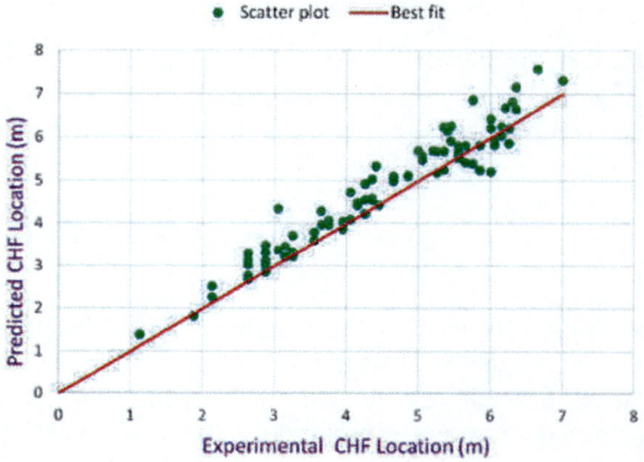

Fig. 3 Scatter plot of predicted and experimental values

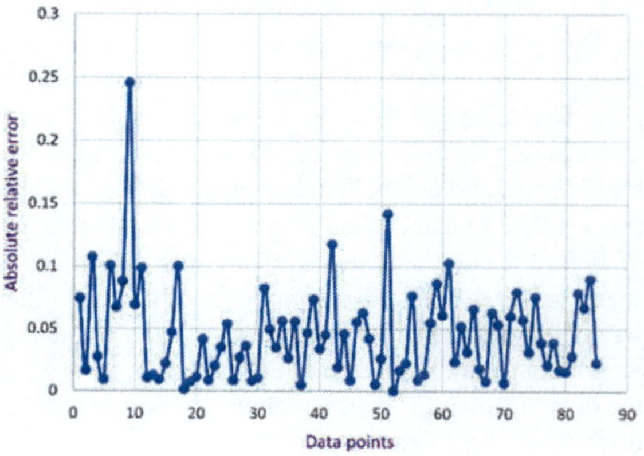

Fig. 4 Absolute relative error of the data points

6 Conclusions

The present study aims to employ an artificial neural network-based machine learning model to predict the CHF location covering a wide range of operating conditions. To address this objective, a NN model was implemented that has six input features and CHF location as the output feature. It is noticed that the model performed well, and the MAPE of training and testing sets are found to be 3.11 and 4.62%, respectively, which conforms a close agreement to the experimental values. Further, the prediction time is also much lower than the conventional CFD approach. From the present study, it can be concluded that the accurate CHF location predictions by the proposed model would result in optimal design and analysis of thermal systems which prevent their catastrophic failure.

Acknowledgements The authors Vishnu Kumar and Rishika Kohli would like to acknowledge the financial support received from the Ministry of Human Resources and Development (MHRD), the Government of India, to carry out this research.

References

1. Becker KM, Ling CH, Hedberg S, Strand G (1983) An experimental investigation of post dryout heat transfer. Department of Nuclear Reactor Engineering, KTHNEL-33, Stockholm, Sweden, Royal Institute of Technology
2. Zhao X, Shirvan K, Salko RK, Guo F (2020) On the prediction of critical heat flux using a physics-informed machine learning-aided framework. Appl Therm Eng 164:114540
3. Park HM, Lee JH, Kim KD (2020) Wall temperature prediction at critical heat flux using a machine learning model. Ann Nucl Energy 141:107334
4. He M, Lee Y (2018) Application of machine learning for prediction of critical heat flux: support vector machine for data-driven CHF look-up table construction based on sparingly distributed training data points. Nucl Eng Des 338:189–198
5. Pothukuchi H, Patnaik BSV, Prasad BVSSS (2020) A simple model to identify dryout location—validations and computational fluid dynamics predictions. Heat Transfer Eng 1–20
6. Groeneveld DC, Shan JQ, Vasic AZ, Leung LKH (2007) The 2006 CHF look-up table. Nucl Eng Des 237(15–17):1909–1922
7. He M, Lee Y (2020) Application of deep belief network for critical heat flux prediction on microstructure surfaces. Nucl Technol 206(2):358–374
8. Almalki N, Ahmed WH (2020) Prediction of two-phase flow patterns using machine learning algorithms. In *International Conference on Fluid Flow, Heat and Mass Transfer (FFHMT'20)*
9. Ooi ZJ, Zhu L, Bottini JL, Brooks CS (2022) Identification of flow regimes in boiling flows in a vertical annulus channel with machine learning techniques. Int J Heat Mass Transf 185:122439
10. Kim H, Moon J, Hong D, Cha E, Yun B (2021) Prediction of critical heat flux for narrow rectangular channels in a steady-state condition using machine learning. Nucl Eng Technol 53(6):1796–1809
11. Tang W, Hua G, Wang L (2017) How to train a compact binary neural network with high accuracy? In *Thirty-First AAAI Conference on Artificial Intelligence*
12. Gulli A, Kapoor A, Pal S (2019) Deep learning with TensorFlow 2 and Keras: regression, ConvNets, GANs, RNNs, NLP, and more with TensorFlow 2 and the Keras API. Packt Publishing Ltd.

13. Jiang BT, Zhou J, Huang XB, Wang PF (2020) Prediction of critical heat flux using Gaussian process regression and ant colony optimization. Ann Nucl Energy 149:107765
14. Nedal M (2019) Re: how to decide the number of hidden layers and nodes in the hidden layers? (2019)

Development of a Test Facility to Study the Agglomeration of Alkali Rich Biomass in Bubbling Fluidised Bed Gasification

P. Suraj, K. T. Abdul Azeez, P. Arun, and C. Muraleedharan

Abstract The thermochemical process like gasification increases the energy density of biomass by converting it into a mixture of fuel gases such as hydrogen (H_2), methane (CH_4) and carbon monoxide(CO). Despite the fact that the bubbling fluidised bed gasification is a proven technology with fuel flexibility, high carbon conversion efficiency, and uniform temperature distribution in the reactor, it is prone to bed agglomeration when alkali rich biomasses are used. The product gas composition depends on operating parameters like equivalence ratio, temperature, bed material, and gasification medium. The reported work illustrates the design approach for developing a test facility to study agglomeration behaviour in fluidized bed gasification of biomass. The bubbling fluidised bed gasifier design is such that the biomass feed rate is restricted to 1–4 kg/h with sand as bed material, coffee husk as feedstock, and air as gasification medium. The designed reactor has a diameter of 7.6 cm and a height of 117.2 cm. The components such as the distributor plate, cyclone separator, and wind box are also designed. The sudden fluctuation in temperature and pressure across the bed reveals the agglomeration tendency. The instrumentation and measurement part of the proposed system facilitates the monitoring of all the operating parameters. The quality of product gas may be enhanced by developing gas cleaning arrangements, which falls in the future scope of the work.

Keywords Gasifier · Agglomeration · Coffee husk · Biomass · Bubbling fluidised bed

1 Introduction

One-third of the world's renewable energy usage is met by bioenergy resources, among which more than half are fulfilled by conventional biomass [1]. In the year 2020, about 4.9% of primary energy utilisation of the U.S. was met by biomass,

P. Suraj (✉) · K. T. Abdul Azeez · P. Arun · C. Muraleedharan
Department of Mechanical Engineering, NIT Calicut, Calicut, Kerala 673601, India
e-mail: surajvvv121@gmail.com

S. Das et al. (eds.), *Proceedings of the 1st International Conference on Fluid, Thermal and Energy Systems*, Lecture Notes in Mechanical Engineering,
https://doi.org/10.1007/978-981-99-5990-7_54

which was around 4,532 trillion British thermal units [2]. Biomass gasification is a thermochemical conversion method in which the biomass is heated at a temperature of 700 °C with a limited supply of oxygen to produce syngas. The conversion of biomass into syngas makes it easy to transport and utilise by increasing its energy density. Bubbling fluidised bed (BFB) gasifiers are proven technology that has more heat transfer characteristics, better mixing, and fuel flexibility. As sand and air are inexpensive and readily available, they are the most widely utilised bed material and gasification medium. The gasification of alkali-rich biomass in the presence of sand and air often causes fouling, slagging, corrosion, and agglomeration in the bed. When the alkali elements in the biomass react with the bed materials, they bind together to create agglomerates. Unlike the other challenges, bed agglomeration leads to defluidization and shutdown of the plant. Prevention and timely prediction of agglomeration are essential for the smooth operation of the BFB gasifier.

2 Literature Review and Objectives

Several studies on biomass gasification in various gasifiers under different operating conditions have been reported [3–7]. In an air-stream gasification process, the influence of various operational parameters such as gasification temperature, equivalence ratio (ER), and steam-to-biomass ratio (S/B) on product gas yield and composition was reported [8]. The feedstock characteristics, pre-treatment, and gasification methods in different experimental studies were reviewed and consolidated [9]. Commonly used biomasses for gasification are wood, food crops, grass, agricultural residues, forestry leftovers, and organic substances of industrial and municipal waste. Out of these, herbaceous plants contain a substantial quantity of alkali in them [10]. The alkali component in biomass, such as potassium and sodium, has bi-functional activity as a catalyst and sorbent, which enhances H_2 production [11]. The presence of alkali metals in biomass aids in the reduction of tar and soot formation during woody biomass gasification [12]. During gasification, the alkali in biomass forms alkali carbonates. By trapping the sulphur in biomass, the generated alkali carbonates lower the sulphur concentration of syngas, making the gas cleaning process easy [13]. Despite the above mentioned benefits, high alkali content biomass has been identified as one of the primary causes of increased slagging, corrosion, fouling, and agglomeration in gasifiers [14–16]. The sticking together of bed particle materials is defined as agglomeration. Melt-induced agglomeration and coating-induced agglomeration are the two types of agglomeration [16].

A study with high potassium-content almond shells indicated that the gas phase potassium interacting with the solid phase silica induces agglomeration of the bed material, especially at higher temperatures. The degree of agglomeration in air gasification is significant compared to steam gasification due to the higher rate of kalsilite reactions in air gasification. In most cases, air gasification caused agglomeration in the bed, whereas steam gasification caused deposition on the gasifier walls [17]. The influence of bed materials on agglomeration was experimentally studied. There was no sign of agglomeration when dolomite was used as bed material, whereas a high agglomeration tendency was shown when magnesite was used. Silica sand exhibited the most intense agglomeration compared to other bed materials [18]. Gasification of cardoon in a 100 kW circulating fluidised bed gasifier was carried out to study the effects of agglomeration. Agglomeration was detected at a lower temperature within the first 30 min of operation due to the high potassium-rich ash. It was reported that the glue melt was high in sodium, potassium, calcium, magnesium, and silicon. The hollow appearance of the agglomerate is attributed to the presence of calcium [19]. During gasification, the potassium-rich biomass forms melt at a lower temperature which is responsible for agglomeration. During the process, the silicate ash completely melts and forms a viscous liquid composed of alkali silicates and other oxides or alkali salts in minor quantities. The sticky melt adhered to the bed particle grains, or a molten film on their surfaces necked them together [20].

In the present study, a bubbling fluidised gasifier test facility is designed to study the agglomeration of alkali-rich biomass like coffee husk. Coffee being the most traded commodity across the globe is accessible in large quantities. The alkali-rich nature of coffee husk makes it more prone to agglomeration, even though it has a high hydrogen yield during gasification. The developed design helps in performing parametric study as well as continuously monitors the pressure and temperature fluctuations in and across the bed, with the aid of the instrumentation facility provided. The design will help in the early prediction and prevention of agglomeration.

3 Materials and Methods

The main aim of the study is to design a laboratory-scale BFB gasifier to perform parametric and bed agglomeration behavioural studies. The system is designed to operate with a feed rate of 1–4 kg/h. The design is carried out considering coffee husk as the base fuel as it is rich in potassium and more prone to agglomeration. The design of the whole system can be subdivided into four, plenum/wind box, riser, distributor plate, and cyclone separator. The design approach makes use of the physical and chemical properties of the coffee husk. The calorific value is found using a bomb calorimeter. The gross and elemental composition of coffee husk are determined, using proximate analysis and ultimate analysis, respectively.

3.1 Characterisation of Biomass

The characteristics of biomass play a significant role in the design and operational behaviour of the reactor. The physical and chemical properties of biomass help in understanding the thermal behaviour and energy content associated with it. The proximate analysis provides the percentage composition of moisture, volatile matter, fixed carbon, and ash content in the fuel. The procedure is carried out with the aid of

Table 1 Testing standards used for proximate analysis

Composition	Testing standard used
Moisture	ASTM D3173M
Volatile matter	ASTM E872-82
Ash content	ASTM E1534
Fixed carbon	By subtraction

a muffle furnace and silica crucible following ASTM standards [21]. The proximate analysis is carried out on an air-dry basis and the standards used for analysis are tabulated in Table 1.

The composition of C, H N, S, and O in the coffee husk is determined by performing ultimate analysis with the aid of Elementar Vario EL-III ultimate analyser at Sophisticated Analytical Instruments Facility (SAIF), STIC, Kochi, India. The analysis was carried out on a dry ash-free basis. The calorific value of the coffee husk is determined using a digital bomb calorimeter according to ASTM standard D-2015 [22].

3.2 *Design of the Gasifier*

The design of a BFB gasifier is developed based on the characteristics of the fuel, which enables us to investigate the bed agglomeration behaviour and the effect of different parameters during the operation of the gasifier. The other important parameters involved in the design process are ambient temperature, ambient pressure, operating temperature, operating pressure, and ER, which are listed in Table 2.

Silica sand, a cheap and commonly available material, is taken as the bed material. The characteristics of bed material are taken as shown in Table 3.

The design of the gasifier is done based on a particular ER and by fixing the mass flow rate of air. ER and air mass flow rate are significant in calculating the fuel feed rate. Biomass feed rate can be determined by using Eqs. (1–3).

$$AFR_{\text{stoic}} = 0.115\text{C} + 0.345\text{H} - 0.043\text{O} \tag{1}$$

Table 2 Operating conditions of gasifier

Operating conditions	Value
Ambient temperature	303 K
Ambient pressure	1 atm
Gasification temperature	923 K
Gasification pressure	1 atm
Equivalence ratio	0.2

Table 3 Characteristics of bed material

Characteristics	Value
Particle size (d_p)	450 μm
Particle density (ρ_s)	2500 kg/m^3
Bulk density (ρ_b)	1500 kg/m^3
Sphericity (ϕ_s)	0.8

$$AFR_{\text{actual}} = ER \times AFR_{\text{stoic}} \tag{2}$$

$$\dot{m}_{\text{biomass}} = \frac{\dot{m}_{\text{air}}}{AFR_{\text{actual}}} \tag{3}$$

where C, H, and O are the percentage compositions of carbon, hydrogen, and oxygen, respectively.

In the present design, the air is considered as the gasification medium and the properties of air influence the design significantly. The properties of air like viscosity (μ_f) and density (ρ_f) at a particular temperature and pressure are calculated by the Eqs. (4) and (5).

$$\mu_f = \frac{1.458 \times 10^{-6} T^{1.5}}{T + 110.4} \tag{4}$$

$$\rho_f = \frac{P \times M_{wt}}{R \times T} \tag{5}$$

3.2.1 Design of Riser

The riser is a part of the gasifier where fluidisation and significant reactions occur. It includes the reactor portion and the freeboard section. The major dimensions to be determined are the diameter and height of the riser. The diameter of the riser (*D*) is selected based on the standard pipe dimensions of SS310. The selection of standard diameters helps in easy fabrication at minimum cost. The height of the gasifier is calculated based on the selected diameter and other fluidisation parameters. The different hydrodynamic parameters are calculated as given below.

Minimum fluidisation velocity

It is the velocity at which the drag force acting on the bed particle by the gasification medium gets balanced with the weight of the particle. The minimum fluidisation velocity (u_{mf}) for very small and very large particles are expressed in Eqs. (6) and (7), respectively [23].

$$u_{mf} = \frac{d_p^3(\rho_s - \rho_f)g}{150\mu} \frac{\varepsilon_{mf}^3 \phi_s^2}{1 - \varepsilon_{mf}} \quad Re_{p,mf} < 20 \tag{6}$$

$$u_{mf}^2 = \frac{d_p(\rho_s - \rho_f)g}{1.75\rho_f} \varepsilon_{mf}^3 \phi_s \quad Re_{p,mf} > 1000 \tag{7}$$

where $Re_{p,mf}$ and ε_{mf} are Reynolds number at minimum fluidisation and porosity or void fraction of the solid bed material at minimum fluidisation.

Terminal velocity

The free-fall velocity of a particle through a fluid is called terminal velocity (u_t). The velocity at which the buoyancy and drag force balance with the gravitational force for a freely falling particle is calculated using Eq. (8).

$$u_t^* = u_t \left[\frac{\rho_f^2}{\mu(\rho_s - \rho_f)g} \right]^{\frac{1}{3}} \tag{8}$$

where u_t^* is the dimensionless terminal gas velocity, which can be determined using dimensionless particle size [23].

Riser height

The total height of the reactor is the sum of the expanded bed height (H_{exp}) and freeboard height (H_{fb}). The expanded bed height has to be determined by the mass balance of the bed solids as given in Eq. (9).

$$\begin{aligned} W_s &= \rho_s A_c H_s (1 - \varepsilon_s) \\ &= \rho_s A_c H_{\exp} (1 - \varepsilon_{\exp}) \end{aligned} \tag{9}$$

where A_c: Area of cross-section, H_s : Static bed height, ε_s: Porosity of the static bed, and $\varepsilon_{\exp}$: Porosity at the expanded condition.

H_{fb} can be calculated by adding 0.3 to the transport disengaging height (TDH). The TDH is the height at which bed particles disconnect from the flowing gas and return to the bed. TDH is calculated using Eq. (10) [24].

$$TDH = 0.85 u_o^{1.2} (7.33 - 1.2 \log u_o) \tag{10}$$

where u_o: is the superficial gas velocity.

3.2.2 Design of Distributor Plate

The distributor plate is that part of the gasifier that supports the bed material and facilitates the passage of the gasification medium so that homogenous distribution occurs. The perforated plate type distributor plate is ideal for low-capacity laboratory settings. The bed pressure drop (ΔP_{bed}) and distributor plate pressure drop (ΔP_{dis}) is significant as it affects the homogenous flow of gasification medium and auxiliary power consumption. ΔP_{bed} and ΔP_{dis} can be found using Eqs. (11) and (12), respectively [25].

$$\Delta P_{bed} = \frac{W_s}{A_c} \tag{11}$$

$$\Delta P_{dis} = 0.3 \times \Delta P_{bed} \tag{12}$$

The orifice diameter (d_{orf}) is calculated as thrice of the particle diameter [26]. A mesh may be provided above the distributor plate to prevent the downflow of the particles. The thickness of the plate is assumed to be 5 mm to prevent material failure at higher operating temperatures. The major design parameter of the distributor plate is the number of orifices. The number of orifices can be determined using Eq. (13).

$$N_{orf} = \left(\frac{D}{d_{orf}}\right)^2 \left(\frac{u_o}{u_{orf}}\right) \tag{13}$$

where u_{orf}: is the gas velocity at the orifice, which is calculated using Eq. (14) [27].

$$u_{orf} = C_d \left(\frac{2 \times \Delta \mathrm{P_{dis}}}{\rho_{f,orf}}\right)^{0.5} \tag{14}$$

The centre-to-centre distance between the orifices is called pitch. The orifices in the distributor plates are arranged with a triangular pitch, as it has more holes per unit area. The pitch can be calculated from Eq. (15).

$$P_{pitch} = \sqrt{\left(\frac{2}{\sqrt{3}}\right) \times \frac{1}{N_{den}}} \tag{15}$$

where N_{den} is the number of orifices per unit area of cross-section.

3.2.3 Design of Cyclone Separator

The outlet gas from the riser will be mixed with the bed particle material. The separation of this particle from the syngas takes place in the cyclone separator. A design of a general purpose cyclone separator is adapted from [28]. The general purpose cyclone design serves the purpose of both higher efficiency and throughput. The dimensions of the cyclone separator are expressed in Table 4, and the schematic is illustrated in Fig. 1. The diameter of the cyclone is calculated according to the flow rate and inlet velocity of the cyclone.

Table 4 Design of cyclone separator

	Symbol	Relation
Body diameter of cyclone	D_o	D_o
Length of cyclone	L_a	2 D_o
Length of cone	L_b	2 D_o
Height of inlet	H	$D_o/2$
Width of inlet	W	$D_o/4$
Diameter of gas exit	D_g	$D_o/2$
Diameter of particle outlet	D_a	$D_o/4$
Length of vortex finder	S	$D_o/8$

Fig. 1 Cyclone separator

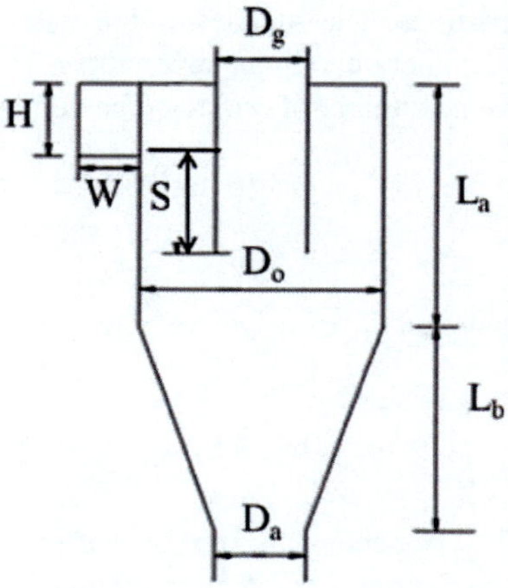

3.2.4 Design of Wind Box

The wind box ensures the uniform distribution of the gasification medium. The air entry can be made either through the side or bottom. During the bottom entry, there are chances that the jet directly hits the distributor plate, which results in uneven distribution of the bed. For a side entry wind box having an inlet pipe diameter (D_s), the position of the entry pipe from the distributor plate (H_b) can be calculated from Eqs. (16) and (17) [27].

$$H_b = 0.2D + 0.5D_s \quad D_s > \frac{D}{100} \tag{16}$$

$$H_b = 18D_s \quad D_s < \frac{D}{100} \tag{17}$$

3.2.5 Reactor Insulation

Gasification is an autothermal process in which partial combustion of the feedstock occurs. The generated heat is required to sustain the endothermic gasification reactions. The heat loss to the surroundings needs to be minimised for the safety and smooth operation of the plant. The insulation can be provided based on the critical insulation thickness. From the available insulating material, the double-layer ceramic wool with a density of 64/96 kg/m^3 is preferable. Fire clay and fire brick can be used along with ceramic wool if the stiffness of the reactor is also a concern.

3.3 Measurement and Instrumentation

The gasification process depends on several parameters like temperature pressure, ER, flow rates of different mediums, etc. To perform a parametric study in a BFB gasifier, these variables have to be monitored or measured continuously to determine an optimised operating condition. Different instrumentation facilities have to be provided to measure the parameters such as gas composition, temperature, flow rates, and pressure. K-type thermocouples are preferred for temperature measurement at different riser locations, and a differential pressure transducer is suitable to monitor the bed pressure drop. The flow rate of the gasification medium can be measured using a rotameter (0–200 lpm) or an orifice. A screw feeder with a variable frequency drive can be employed to control the fuel feed rate. The properties of syngas produced at different operating conditions can be analysed using a gas analyser which monitors the percentage composition of H_2, CH_4, CO, CO_2, O_2, NH_3, and H_2S. A tar sampling unit can also be attached to the reactor.

Gasification of alkali-rich biomass often leads to agglomeration of bed material. For the safe and proper operation of the BFB gasifier, provisions have to be provided for monitoring the bed behaviour under different operating conditions. A minimum of two thermocouples are inserted into the bed to capture the temperature variation precisely. The differential pressure transducer is also placed across the bed to measure the bed pressure drop. The output from temperature and pressure sensors is continuously monitored and logged with the aid of a data acquisition system. As there would be abrupt temperature and pressure variations in bed during agglomeration, temperature and pressure are two metrics that can be monitored to anticipate the agglomeration.

4 Results and Discussion

A detailed design of a BFB gasifier has been done, considering coffee husk as the feedstock. The key objective of the design is to conduct parametric research on gasification and anticipate and predict bed agglomeration behaviour. The elemental

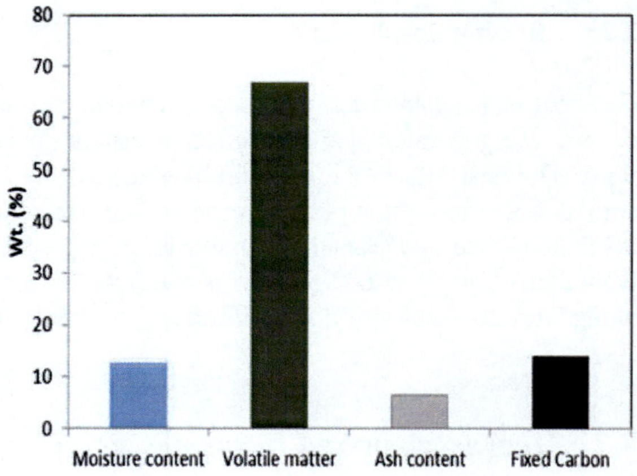

Fig. 2 Proximate analysis result of coffee husk

Table 5 Ultimate analysis results

	C	H	N	S	O
Wt. (%)	42.67	6.2	1.9	0.41	48.82

constitution of the coffee husk is an important factor in deciding the fuel feed rate for the system.

4.1 *Characteristics of Coffee Husk*

The gross composition of the coffee husk is determined by proximate analysis and is shown in Fig. 2.

The ultimate analysis result is shown in Table 5. The elemental composition is used to determine AFR_{stoic}. It can also be used to determine the lower heating value of the fuel [29]. The gross calorific value of feedstock obtained from the digital bomb calorimeter is 18.96 MJ/kg.

4.2 *Designed Dimensions of Gasifier*

The gasifier is designed for an ER of 0.2, and the diameter is limited to 7.6 cm to keep the fuel feed rate below 4 kg/h. For coffee husk as feedstock and air as gasification medium, Eqs. (1–3) gives the fuel feed rate as 2.9 kg/h. The hydrodynamic parameters like minimum fluidisation and terminal velocity are found using Eqs. (6) and (8), the values are 0.1367 and 3.127 m/s, respectively. The gasifier operates at a velocity between the minimum fluidisation velocity and terminal velocity, known as superficial velocity, and is taken as 0.15 m/s.

The riser height is the sum of the expanded bed height and freeboard height. The expanded bed height is determined from the static bed height. The value of H_s satisfies the condition $\frac{H_s}{D} \leq 2$ and is taken as 12 cm. From Eq. (9), H_{exp} is obtained as 14.66 cm. The TDH and H_{fb} are obtained as 72.57 and 102.57 cm, respectively. Therefore, the total height of the riser is determined as 117.2 cm.

The pressure drop across the bed and the distributor plate is critical for uniform distribution and efficient fluidisation. The total pressure drop is the sum of the pressure drop in the bed and the distributor plate, which is obtained as 2175 N/m^2. The diameter of the orifice (d_{orf}) is determined to be three times the particle diameter, which is 1.4 mm. The velocity of the gasification medium flowing through the orifice is determined using Eq. (14) as 37.5 m/s. On substituting the orifice velocity in Eq. (13), the number of orifices on the distributor plate is obtained as 31. The orifices in the distributor plate are arranged such that it has a triangular pitch so that it accommodates more holes per unit area and reduces the stagnant zones in the bed. According to Eq. (15), the triangular pitch of orifices on the distributor plate is 1.29 cm. The bubble diameter at the orifice should be less than the pitch to avoid the coalescence of the bubble at the initial stage itself.

The area of cross-section of the distributor plate is the same as the area of cross-section of the riser, which is 0.0045 m^2. On dividing the number of orifices by this area, the number of orifices per unit area (N_{den}) is determined to be 6834. According to [26], the fractional area of the opening should be greater than 0.5 and less than 3, as given in Eq. (18), where A_o is the area of orifice.

$$0.5 < N_{den} \times A_o \times 100 < 3 \tag{18}$$

The condition is satisfied and the calculated value is 1.0519. Another checkpoint in the distributor design is to satisfy the Eq. (19) [26] for stable operation.

$$\frac{16\rho_f u_o}{\left(\pi N_{den} C_D d_{orf}^2\right)^2} > 800 \tag{19}$$

The design satisfies the condition with a value of 967.39.

The outlet of the gasifier contains product gas along with ash and char particles. The gas–solid separation takes place in a cyclone separator. The geometrical dimensions of the cyclone separator are tabulated in Table 6.

The position of the inlet to the plenum is determined using Eq. (16). For an inlet diameter of 3.81 cm, the position of the inlet is calculated to be 3.425 cm from the distributor plate. For a lab-scale setup, side entry of the gasification medium is preferred. If the bottom entry of the gasification medium is to be used, either baffle plates or sufficient entry length must be provided.

Table 6 Dimensions of cyclone separator

	Symbol	Value (cm)
Body diameter of cyclone	D_o	6.53
Length of cyclone	L_a	13.06
Length of cone	L_b	13.06
Height of inlet	H	3.26
Width of inlet	W	1.63
Diameter of gas exit	D_g	3.26
Diameter of particle outlet	D_a	1.63
Length of vortex finder	S	0.82

5 Conclusion

A comprehensive design of a BFB gasifier with a biomass feed rate of 1–4 kg/h has been developed. The primary aim of the design is to study the agglomeration behaviour of alkali-rich biomass. Coffee husk being an alkali-rich biomass is selected for the design. The diameter of the reactor is selected such that the required feed rate can be achieved. The diameter is fixed as 7.6 cm and the total height of the riser is calculated as 1.172 m. A cyclone separator is designed following the available standards. Perforated distributor plate with 31 orifices and a wind box are also designed. The instrumentation requirements for an in-depth study on the bed behaviour during agglomeration and the insulation details are also discussed. Thus, the present design of the BFB gasifier will help in a detailed parametric study and timely prediction of bed agglomeration. The end products of gasification depend on the operating parameters and the gas quality can be significantly improved by designing and providing downstream gas cleaning arrangements that fall within the future scope of the work.

Acknowledgements Authors gratefully acknowledge the financial support provided by SERB, DST, Govt. of India through the research project on 'Agglomeration abatement in fluidised bed gasification of ligno-cellulosic biomasses' (Project File No. CRG/2021/00833).

References

1. International Renewable Energy Agency. [Online]. https://www.irena.org/
2. U.S. Energy Information Administration. [Online]. https://www.eia.gov/energyexplained/biomass/
3. Sharma P, Gupta B, Pandey M (2022) Hydrogen rich product gas from air–steam gasification of Indian biomasses with waste engine oil as binder. Waste Biomass Valor
4. Sitka A, Jodkowski W, Szulc P, Smykowski D, Szumiło B (2021) Study of the properties and particulate matter content of the gas from the innovative pilot-scale gasification installation with integrated ceramic filter. Energies 14(22)
5. Zhao S, Ding L, Ruan Y, Bai B, Qiu Z, Li Z (2021) Experimental and kinetic studies on steam gasification of a biomass char. Energies 14(21)

6. Ahmed II, Gupta AK (2010) Pyrolysis and gasification of food waste: syngas characteristics and char gasification kinetics. Appl Energy 87(1):101–108
7. George J, Arun P, Muraleedharan C (2018) Experimental investigation on co-gasification of coffee husk and sawdust in a bubbling fluidised bed gasifier. J Energy Inst
8. Turn S, Kinoshita C, Zhang Z, Ishimura D, Zhou J (1998) An experimental investigation of hydrogen production from biomass gasification. Int J Hydrog Energy 23(8):641–648
9. Widjaya ER, Chen G, Bowtell L, Hills C (2018) Gasification of non-woody biomass: a literature review. Renew Sustain Energy Rev 89:184–193
10. Narayan V, Jensen PA, Henriksen UB, Glarborg P, Lin W, Nielsen RG (2016) Defluidization in fluidised bed gasifiers using high-alkali content fuels. Biomass Bioenerg 91:160–174
11. Zhao M et al (2020) Alkali metal bifunctional catalyst-sorbents enabled biomass pyrolysis for enhanced hydrogen production. Renew Energy 148:168–175
12. Umeki K, Häggström G, Bach-Oller A, Kirtania K, Furusjö E (2017) Reduction of tar and soot formation from entrained-flow gasification of woody biomass by alkali impregnation. Energy Fuels 31(5):5104–5110
13. Furusjö E, Ma C, Ji X, Carvalho L, Lundgren J, Wetterlund E (2018) Alkali enhanced biomass gasification with in situ S capture and novel syngas cleaning. Part 1: gasifier performance. Energy 157:96–105
14. Kittivech T, Fukuda S (2020) Investigating agglomeration tendency of co-gasification between high alkali biomass and woody biomass in a bubbling fluidised bed system. Energies 13(1):56
15. Mac an Bhaird ST, Walsh E, Hemmingway P, Maglinao AL, Capareda SC, McDonnell KP (2014) Analysis of bed agglomeration during gasification of wheat straw in a bubbling fluidised bed gasifier using mullite as bed material. Powder Technol 254:448–459
16. Mettanant V, Basu P, Butler J (2009) Agglomeration of biomass fired fluidised bed gasifier and combustor. Can J Chem Eng 87(5):656–684
17. McCaffrey Z et al (2019) Air and steam gasification of almond biomass. Front Energy Res 7:1–11
18. Zhou C, Rosén C, Engvall K (2016) Biomass oxygen/steam gasification in a pressurised bubbling fluidised bed: agglomeration behavior. Appl Energy 172:230–250
19. Christodoulou C, Koytsoumpa EI, Panopoulos KD, Karellas S, Kakaras E (2014) Agglomeration problems during cardoon fluidised bed gasification. Therm Sci 18(2):645–656
20. Fryda LE, Panopoulos KD, Kakaras E (2008) Agglomeration in fluidised bed gasification of biomass. Powder Technol 181(3):307–320
21. Poyilil S, Palatel A, Chandrasekharan M (2021) Physico-chemical characterisation study of coffee husk for feasibility assessment in fluidised bed gasification process. Environ Sci Pollut Res 0123456789
22. Raveendravarrier B, Palatel A, Chandrasekharan M (2020) Physico-chemical characterisation of sewage sludge for thermochemical conversion processes. Energy Sources Part A Recover Util Environ Effects 1–25
23. Kunii D, Levenspiel O (1991) Fluidization engineering, 2nd edn. United States of America
24. Chan IH, Knowlton TM (1984) The effect of system pressure on the transport disengaging height (TDH) above bubbling gas-fluidised beds. In: AIChE Symp. Ser.; (United States), vol 80, p 241
25. Yang W-C (2003) Handbook of fluidisation and fluid-particle systems, vol 1, no 3
26. Basu P (1984) Fluidised bed boilers: design and application, vol 2
27. Basu P (2006) Combustion and gasification of fluidized beds
28. Bashir K (2015) Design and fabrication of cyclone separator. Tsinghua University
29. Rupesh S, Muraleedharan C, Arun P (2020) Energy and exergy analysis of syngas production from different biomasses through air-steam gasification. Front Energy 14(3):607–619

Performance Evaluation of Air-Cooled Condenser Coil

Rushil Patel, Anita Nene, Pankaj Anjankar, and Ajay Howal

Abstract In this investigation, an analytical study was carried out on the 15 models of air-cooled chillers that were operated using R134a refrigerant with capacities ranging from 0 to 246.18 kW. An innovative idea was introduced to redesign the air-cooled condenser with the concept of void fraction measuring technique inside the coils of the condenser. An equation has been formulated to predict the approximate void fraction along the length of the single circuit coil of the standard condensing tube of 8160 mm, after which for the remaining length of the single circuit coil, the best-suited diameter tube was examined using the Nusselt number correlation corresponding to the heat transfer coefficient calculated. Results showed that the new length calculated using void fraction was saving on an average 23.66728% of the original tube length, variation in the original performance of the single circuit coil and analytically calculated performance of the single circuit coil was around ±22.6%, and the heat transfer coefficient calculated using correlations and UNILAB coils software was ranging in ±40.1839. Hence, it was seen that a reduction in the dimensions of the condenser's coil with the same performance can be implemented for the air-cooled condenser.

Keywords Air-cooled condenser · Chiller system · Heat transfer coefficient · Heat exchanger · Two-phase flow

R. Patel (✉) · A. Nene
School of Mechanical Engineering, Dr. Vishwanath Karad MIT World Peace University, Pune, Maharashtra 411038, India
e-mail: rushil2805@gmail.com

P. Anjankar · A. Howal
Design and Development, Kirloskar Chillers Private Limited, Pune, Maharashtra 412801, India

S. Das et al. (eds.), *Proceedings of the 1st International Conference on Fluid, Thermal and Energy Systems*, Lecture Notes in Mechanical Engineering,
https://doi.org/10.1007/978-981-99-5990-7_55

1 Introduction

Each refrigeration and air-conditioning framework in view of a vapour compression cycle contains an evaporator and condenser. Since condensers and evaporators are heat exchangers, they share numerous things in common for all intents and purpose.

A condenser is a two-stage stream heat exchanger in which the heat is created from the change of fume into fluid and the heat produced is eliminated from the framework by a coolant. Condensers might be arranged into two fundamental sorts: those where the coolant and condensate stream are isolated by a strong surface, generally a cylinder divider, and those wherein the coolant and consolidating fume are brought into direct contact.

The immediate contact sort of condenser might comprise of a fume which is raised into a pool of fluid, a fluid which is splashed into a fume, or a pressed segment wherein the fluid streams downwards as a film over a pressing material against the vertical progression of fume. Condensers in which the streams are isolated might be partitioned into three primary sorts: air-cooled, shell-and-tube, and plate. An air-cooled condenser is just a pressure vessel which cools a circling liquid inside finned tubes by constraining encompassing air over the outside of the tubes. A typical illustration of an air-cooled condenser is a vehicle radiator. For a similar measure of heat transfer, the activity of air-cooled finned-tube condensers is more monetary as contrasted and water-cooled condensers; ordinarily, air-cooled condensers are of the round tube and blade type.

The void fraction is quite possibly the main boundary used to portray two-phase streams in tubes of the condenser. It is expected as contribution for deciding various other key stream boundaries, for example, the thickness of two-phase flowing and consistency of the typical speeds of the fluid and gas phases of the refrigerant separately. In addition, the void fraction assumes an essential part in the demonstration of two-phase stream design advances, heat movement, and pressure drop. The cross-sectional void fraction ε, which is a dimensionless mathematical stream boundary characterized as the proportion of the cylinder cross-sectional region Ag, was involved by the gas or fume phase to the absolute cross-sectional region A.

$$\varepsilon = \frac{Ag}{A} = \frac{Ag}{Al + Ag}[8] \tag{1}$$

Accordingly, the value of the void fraction is a stream boundary limited between 0 for the single-phase fluid stream, and 1 relating to single-phase gas stream. Upsides of the void fraction near 0 + are normal for an effervescent two-phase stream with the vast majority of the channel cross area involved by the fluid phase and a couple of gas bubbles entrained in the constant fluid phase. With the help of the void fraction technique inside the condensing tube, a visualization of the change of phase from one form to another can be seen. For condensing region value of void fraction less than 1 determines start of change of phase from vapour to liquid.

It is quite important that the void fraction ε term is a mathematical stream boundary, and is significantly unique in relation to the fume quality x, which is a vehicle stream boundary. While the neighbourhood worth of the fume quality is regularly known from the cycle or from energy or mass equilibrium, the void fraction rather should be anticipated from the stream factors.

2 Some Previous Work

H. Qian et. al., (2019) creators showed flow regimes and void fraction for level as well as vertical round smooth tubes which were having 7 mm inside width and R134a as refrigerant in the adiabatic circumstances for immersion temperature of 33 °C. Low mass flux going from 40 to 150 kg for each square meter second for flat tubes and mass flux going from 65 to 115 kg for every square meter second for vertical tubes were tested and are introduced in this paper. Flow regimes were caught by a fast camera, while void fraction was estimated by a speedy shutting valve strategy. Void fraction results for both even and vertical flows were contrasted with a few broadly utilized relationships. Results showed that when the vapour quality is kept steady, the void fraction of even tubes was bigger than that of vertical tubes and higher mass flux additionally brings about a bigger void fraction contrasted with that of lower mass flux. According to *J. Pabon et. al. (2019),* void fraction is a significant boundary to planning and mimicking thermal frameworks including two-phase streams. In this article, an exploratory examination of the void fraction in a two-phase flow of R1234yf refrigerant in even and smooth tubes which were having inner breadth of 4.8 mm was done. For the trial tests, the vapour quality was going from 0.1 to 1 while two immersion temperatures of 15 and 25 °C and two mass stream rates of 180 and 280 kg for each meter square second were explored. The quick-closing valve strategy was utilized to gauge the volumetric void fraction. Tests were additionally embraced with R134a refrigerant utilized as a kind of perspective in this review. The outcomes featured that the void fraction of R1234yf is 5% lower than that for R134a. Moreover, the trial information of R1234yf were analysed against seven connections from the writing from which the Baroczy and the Hughmark relationships were displayed to give the best forecast, with a mean outright mistake of 2 and 3.2%, individually. *R. Srisomba et. al., (2014)* in this article an experiment was done meaning to quantify the void fraction of R134a refrigerant moving through a level cylinder, utilizing a fast-shutting valve and optical perception procedures. The cylinder was straightforward, round, and 115 mm in length with an inside 8 mm thick, and has two actuator ball valves introduced at the two closures. The experiment was directed in conditions where fume quality went from 1 to 82%, immersed temperature was from 23 to 30 °C, and mass flux was from 644 to 1455 kg for each square meter second. The after-effects of the experiment show that raising the soaked temperature diminished the void fraction, though expanding mass flux significantly affected it. Besides, results additionally show three stream systems: intermittent, wavy, and annular. Each stream system prompted different anticipated void fractions. *J. Hajal et. al. (2003)* fostered another flow regime map and its heat model for condensation happening even inside plain tubes. In this article, another rendition of a two-stage flow pattern map, initially created by Kattan [3–5] writer for boiling, was introduced for inside even tubes. A few different adjustments were additionally made that were fitting for condensation instead of vanishing. Without any void fraction information at high decreased pressures for these circumstances, the new logarithmic mean void fraction strategy has been in a roundabout way approved utilizing the convective condensation model for annular flow and relating heat transfer test information at diminished pressures up to 0.8. *J. Thome et. al. (2003)* fostered another overall stream pattern of structure for condensation inside even, plain cylinders in light of improved on stream designs of the stream regimes-based heat transfer model, and furthermore remembered the impact of fluid fume which was interfacial roughness for the heat transfer rate was examined. These heat transfer models anticipated the neighbourhood heat transfer coefficients for the stream pattern specifically annular, irregular, delineated wavy, and completely separated and fog stream. The new model has been contrasted with test information for 15 refrigerants which were acquired in nine free examination research centres. In general, the heat transfer model anticipated/showed results of 85% of the heat transfer coefficients in the non-hydrocarbon information base to inside ±20% with almost uniform precision for each stream system and anticipated 75% of the whole data set to inside ±20% while including the hydrocarbons, the last option all from a solitary lab whose information had a few surprising exploratory patterns over piece of their test range.

3 Methodology

Out of the 15 models of air-cooled chiller [16], a model named 070.14 was taken into account for statistical data formation, where 070 defines capacity in tones of refrigeration for the chiller, 1 states the number of compressors used, and 4 defines the type of refrigerant. Since all these models were using R134a refrigerant [16], its respective refrigerant thermodynamic properties were taken into consideration. The data gathered for these models were of already sold chillers from year 2017 to 2022 by the company [16] to different customers over the globe. The first step forwarded was to gather some parameters like the capacity of the condenser in kilowatts, power consumption by fans and compressor in kilowatts, and ambient temperature of air in degree Celsius from the datasheets given by the customers of the chiller models. For ease in the statistical data formation, an MS Office Excel spreadsheet was used. As we were dealing with the air-cooled condenser, the main focus was on the parameters of the condenser. It was seen that for all the 15 models, on an average capacity of condenser was +219.607016 kW, ranging between +147.672 to +297.87552 kW considering sub cooling, total power consumption average was +80.398 kW, ranging between +58.58 kW to +92 kW, and ambient temperature of air was +42.24 °C, ranging between +35.6 to +50 °C. Further, saturation temperature for the evaporator and condenser in degree Celsius were calculated analytically using formulas. It was noted that the evaporator's saturation temperature was on an average (–0.6144) degree Celsius for the range between -5.914 to +7.9 °C, whereas the condenser's saturation temperature was +57.7186667 °C, varying between +51.57 to +65 °C. Next to that, sub-cooling in the condenser was also calculated and taken into account separately as the above gathered capacity of the condenser from all models was by considering sub-cooling. The above calculated parameters like saturation temperature of the evaporator and condenser in degree Celsius and sub-cooling in degree Celsius were inputted to FRASCOLD selection and simulation software. From FRASCOLD selection and simulation software, parameters like the coefficient of performance, suction gas temperature in degree Celsius, discharge temperature of the refrigerant in degree Celsius, mass flow of refrigerant in kilogram per second, etc., were obtained and noted for further analysis. Also, these parameters from FRASCOLD selection and simulation software for all the models were matched with our statistical data formed using the customer's requirement. Next to above, from GENETRON refrigerant modelling software, one of the thermodynamic properties, i.e., latent heat at condensing temperature in kilojoule per kilogram into kelvin of R134a refrigerant was noted for all 15 models. As above all data collected was of the condenser which had in total 8 numbers of coils for each of the models and our point of interest of void fraction was on a single circuit of the condenser coil which is 8160 mm, so further calculations were made considering single circuit coil and data was formed as needed. For the calculation of the condenser's capacity, the equation as shown below was used

$$Q = \dot{m}x\, h_{fg}[7] \tag{2}$$

where Q is the capacity of a single circuit coil in kilowatts, $\dot{m}$ is basically the mass flow rate of the refrigerant, and h_{fg} is the specific latent heat of the refrigerant in kilojoule per kilogram into kelvin. Using this equation capacity, all 15 models in a single circuit coil was calculated. After calculating the capacity of the single circuit coil in kilowatts, void fraction along the length of the single circuit coil was to be calculated for all 15 models.

$$\varepsilon\, length = \frac{Length\, of\, single\, circuit\, coil\, x\, latent\, heat\, lost\, by\, single\, circuit\, coil}{Capacity\, of\, single\, circuit\, coil\, from\, gathered\, data} \tag{3}$$

Here, the length of a single circuit coil of the condenser is in millimetres, and the capacity of a single circuit coil is in kilowatts. A new length in millimetres using void fraction was calculated, which highlighted that at this particular location along the length of the condensing tube, R134a refrigerant will change its phase from 100% vapour to 100% liquid. Next to this fulfilment, for the remaining length of the single circuit coil, it was assumed that 100% liquid would flow inside the tube and would now occupy a small portion of the tube to flow from the overall area. An idea to decrease the diameter of the tube from the standard tube been used for the condenser's coil was applied. For the simplicity of the calculation, decreasing diameter from 9.925 to 5.125 was taken with interval steps of 15 for each of the models. Corresponding to each model's condenser saturation temperature, some more thermodynamic properties like the density of the liquid in kilogram per cubic meter, the dynamic viscosity of the refrigerant in Pascal in seconds, the thermal conductivity of refrigerant in watts per meter into kelvin, and Prandtl number were gathered from GENETRON refrigerant modelling software. Firstly, Reynolds number for the remaining length of the tube was calculated such that which type of flow for liquid refrigerant is taking place can be visualized.

$$Reynolds\,number = \frac{Velocity\,of\,refrigerant\,in\,single\,circuit\,coil\,x\,Diameter\,of\,tube}{Kinematic\,viscosity\,of\,refrigerant}\,[9] \quad (4)$$

It was seen that for each of the models, the Reynolds number was greater than 4000 which indicates that the flow inside the tube was turbulent, and turbulent flow is beneficial as it increases the heat transfer rate. So as the three conditions were satisfying, i.e., the flow was fully developed turbulent, Prandtl number was between 1.5 to 500, and Reynolds number was between 2300 to 10^6 for all of the 15 models, below shown correlation was applied for calculating the Nusselt number.

$$Nu_d = 0.012\,x\left[Re_d^{0.87} - 280\right]x\,Pr^{0.4}x\left[1 + (Diameter/length)^{0.666}\right][7] \quad (5)$$

It was seen that for each of the models, the Nusselt number ranged from 31.92190273 to 96.5334738, and from the formula of Nusselt number as shown below, heat transfer coefficient for single circuit coil in watts per square meter into kelvin was calculated.

$$NuL = \frac{hL}{K}[7]. \quad (6)$$

Next, the heat transfer coefficient for a single circuit coil in watts per square meter into kelvin was calculated using UNILAB coils software. A UNILAB coil is the software used for designing and rating heat exchangers wherein condensing and direct expansion of refrigerants with phase change can be modelled. This tool also has an extensible archive with more than 50 geometries which are most used in air-conditioning and refrigeration industry. It is possible to insert a new geometry with staggered and aligned tubes, spheroidal or integral fins, tubes with different types of corrugation, and fins with different types of corrugation. For our investigation, we have condensing tool from the software and input parameters to this tool were condensing temperature of refrigerant, type of refrigerant used, superheating supplied to the refrigerant, sub cooling to be achieved, fouling factor at air side, as well as tube side, pressure drop at air side, as well as tube side, inlet dry bulb temperature of air to the condensing side, inlet relative humidity of air which affects the condensing tubes, air flow rate from fans, finned length of the condenser's coil, number of skipped tubes in a coil of the condenser, number of rows in a coil of the condenser, number of circuits in condenser coil, fins pitch, tube thickness, tube configuration, fins configuration, fins per inch, etc.,

4 Results and Discussion

4.1 Determining the Void Fraction Point Along the Length of the Condensing Tube

- As can be seen in Fig. 3a, the blue line represents the capacity of the single circuit coil at the original length, whereas the red line represents the newly calculated capacity of the single circuit coil analytically.

Fig. 1 Kirloskar's chiller having air-cooled condenser [16]

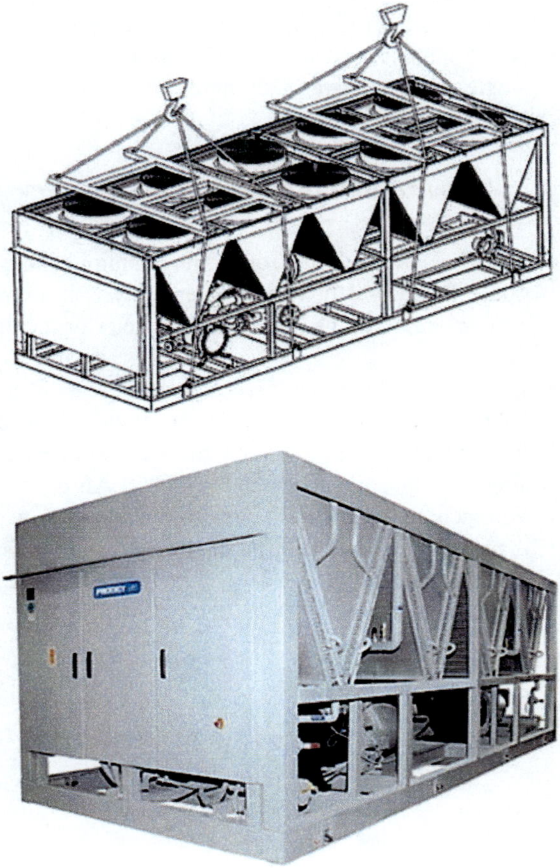

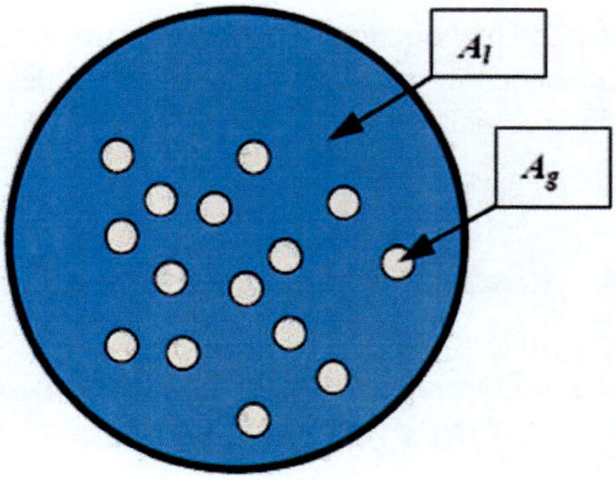

Fig. 2 Schematic of cross-sectional void fraction

- Variation in the capacities of each model's condenser was due to the required kilowatts of the customers as per their conditions of running the system.
- In Fig. 3b, blue line was representing original length of the single circuit coil which was the same as per standard [16] for all of the customers, whereas the red line represents the new length calculated using void fraction.
- All 15 models used standard dimensions condenser coil of which the length of a single circuit coil was 8160 mm.

4.2 Determining the Performance of New Condensing Coil

- As it was assumed that for the remaining length of the tube after void fraction, 100% liquid will flow, therefore, analysis in terms of heat transfer coefficient corresponding to the decreasing diameter of the tube for all the 15 models was done.
- In figure, the blue line represents the heat transfer coefficient for the remaining length of the single circuit coil calculated analytically using correlations, whereas the red line represents the heat transfer coefficient for the remaining length of the single circuit coil calculated from UNILAB coils software.
- For each and all models, separate calculations were carried out for decreasing diameter tubes, and the best fit corresponding to the requirement of the capacity was selected. The decreasing diameters of the tube were from 9.925 to 5.125.

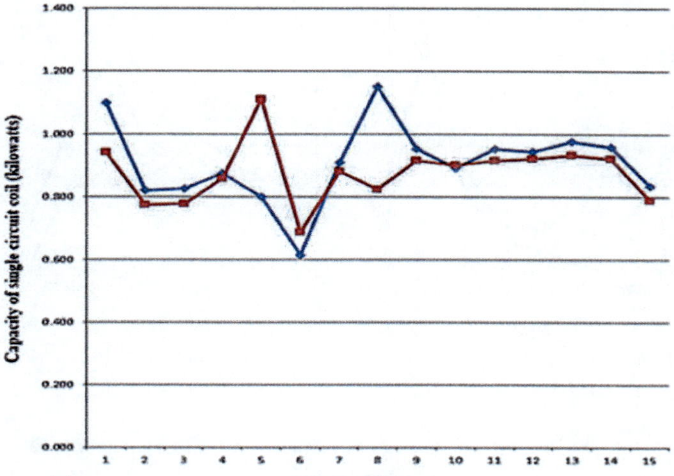

(a)

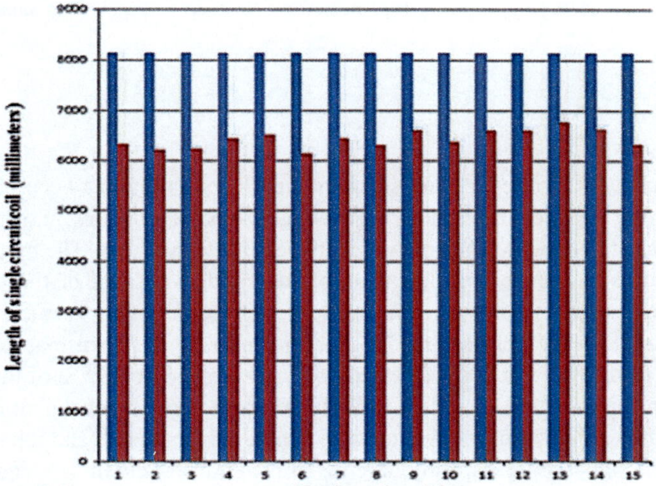

(b)

Fig. 3 **a** Original capacity of single circuit coil versus capacity of single circuit coil calculated analytically for all of the models. **b** Original length of the single circuit coil versus new length calculated using void fraction for all of the models

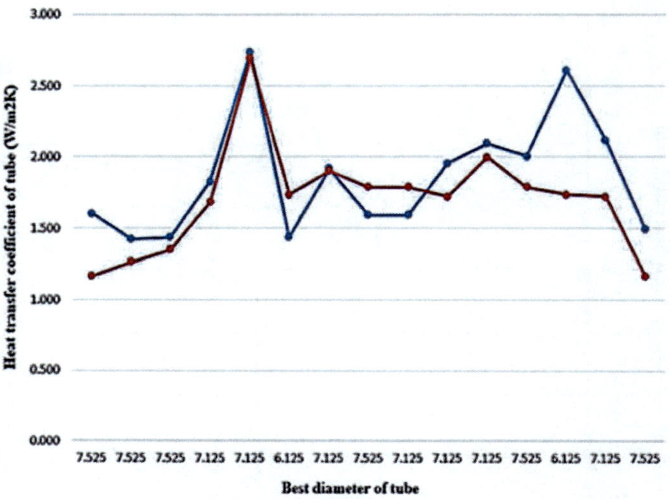

Fig. 4 Calculated heat transfer coefficient for the remaining length of the single circuit coil versus calculated heat transfer coefficient from UNILAB coils software for the remaining length of the single circuit coil corresponding to the best diameter of the tube for each of the model has been calculated

5 Conclusion

The analytical, experimental, and software simulation study has been carried out on a total of 15 models of chiller [16] having air-cooled condensers, each having different capacities in kilowatts ranging between 0 to 246.18 kW. The motto behind this study was to reduce the dimension of the condensing coil of the condenser using a decreasing diameter tube and void fraction technique by keeping steady or slight variation in the performance of the condenser. It was seen that new length calculated using void fraction was saving on an average 23.66728% of the original tube length for all of the models. Similarly, after the location of the void fraction along the tube length, a new diameter tube can be implemented, and the best fitted tube corresponding to the capacity in terms of heat transfer coefficient can be taken into practical use. It was seen that variation in the original capacity of the single circuit coil and capacity of the single circuit coil calculated analytically was around ±22.6%, whereas the heat transfer coefficient calculated using correlations and heat transfer coefficient calculated using UNILAB coils software was seen varying in ±40.1839 for all of the 15 models.

Acknowledgements I would like to thank Kirloskar Chillers Private Limited for their support and expertise in the experimentation and Dr. Vishwanath Karad MIT World Peace University for collaborating with the industry and their valuable support at every step.

References

1. Kays W, London A (1964) Compact heat exchangers, 2nd Edition, McGraw Hill
2. Arlington V (1972) Air-conditioning and refrigeration institute, standards of forced-circulation air-cooling and air-heating coils. Ari Standards, pp 410–472
3. Kattan N, Thome J, Favrat D (1998) Flow boiling in horizontal tubes: part-1 development of adiabatic two-phase flow pattern map. J Heat Transfer 120(1):140–147
4. Kattan N, Thome J, Favrat D (1998) Flow boiling in horizontal tubes: part-2 new heat transfer data for five refrigerants. J Heat Transfer 120:148–155
5. Kattan N, Thome J, Favrat D (1998) Flow boiling in horizontal tubes: part-3 development of a new heat transfer model based on flow pattern. J Heat Transfer 120:156–165
6. Cavallini A, Col D, Longo G, Rossetto L (2001) Experimental investigation on condensation heat transfer and pressure drop of new HFC refrigerants in a horizontal tube. Intern J Refrig 24:73–87
7. Kakac S, Liu H, Pramuanjaroenkij A (2002) Heat exchangers: selection, rating, and thermal design, 2nd edn. CRC Press, United States
8. Thome J (2006) Engineering data book III. Wolverine Tube Inc., United States
9. Lesson 22 (2008) Condensers and Evaporators, Version 1 Mechanical Engineering, IIT Kharagpur
10. Zhao H, Cao L, Wu H (2010) Study on the heat transfer coefficient of direct air-cooled condenser, International Conference On Advances in Energy Engineering
11. Hossain M, Afroz H, Talukder S, Miyara A (2016) Prediction of condensation heat transfer of low GWP refrigerants inside smooth horizontal tube. AIP Conf Proc 1754(1):050037
12. Prof. Das A (2017) Two phase flow and heat transfer, Mechanical Engineering, IITRoorkee
13. Khurmi R, Gupta J (2019) A textbook of refrigeration and air conditioning. S. Chand Publishing, India
14. Cengel Y (2019) Thermodynamics-an engineering approach, 9th edn. McGraw Hill, India
15. Ramgopal M, Rout S, Sarangi S (2021) Advances in air conditioning and refrigeration, select proceedings of RAAR 2019, 1st Edition, Springer
16. Kirloskar Chillers Homepage, https://www.kirloskarchillers.com/, Last Accessed 2022/05/15

A Quick Review on Experimental, Computational, and Optimization Studies of Coherent Structures in Swirling Flow

Prince Patel and Rohit Sharma

ABSTRACT Swirling jets are defined by a velocity component in the azimuthal direction (swirl). They can occur naturally, such as in tornadoes, or artificially, as in gas turbine combustors, swirl burners, etc. Two principal occurrences are observed when the swirl value exceeds a particular threshold: vortex breakdown (VB) and precessing vortex core (PVC). The emergence of a stagnation point in the jet, followed by a reverse flow zone, indicates vortex breakdown, whereas PVC arises after the onset of VB defined as the jet's precession about its geometrical axis. The coherent structures associated with PVC affect the performance of swirl flow thus it is crucial to examine and comprehend these phenomena. It is important for a researcher to become familiar with past investigations conducted on the relevant subject before beginning their research. The purpose of this study is to highlight recent experimental and Computational Fluid Dynamics, (CFD) work on coherent structures in swirling flow by different researchers. In addition, an overview of several optimization techniques such as Artificial Neural Network (ANN) is provided. The major purpose of this review is to discuss the potential of employing ANN in swirling flow along with making recommendations for future research.

Keywords Swirling flow · Vortex breakdown · PVC · CFD · ANN

1 Introduction

Swirling jets have a wide range of uses including swirl burners, gas turbine combustors, whirlpools, furnaces, cyclone separators, and spraying machines [1–3]. When the swirl value (S) imposed on a jet exceeds a specified level (i.e., $S > 0.6$), VB and PVC occur, where S represents the degree of rotation or swirl. There are several types of vortex breakdown which have been reported, as well as several processes for its

P. Patel · R. Sharma (✉)
School of Mechanical Engineering, VIT Bhopal University, Sehore, MP 466114, India
e-mail: rohit.sharma@vitbhopal.ac.in

S. Das et al. (eds.), *Proceedings of the 1st International Conference on Fluid, Thermal and Energy Systems*, Lecture Notes in Mechanical Engineering,
https://doi.org/10.1007/978-981-99-5990-7_56

onset [4]. The VB phenomena have recently been identified by various numerical and theoretical analyses of swirling flows [5–7]. These studies helped researchers learn more about the mechanisms that produce VB.

On the other hand, PVC is a well-defined flow structure that occurs frequently in swirling flows. PVC is coiled around the core vortex breakdown bubble, according to Tran et al. [8], with the windings facing in the opposite direction as the swirl. Because these phenomena are very unstable and three-dimensional, such events are difficult to explore experimentally, theoretically, or even statistically. Various computational and theoretical examinations of swirling flows have newly revealed the occurrence of PVC that can be reported in [8–10]. Even after having substantial research, these phenomena are still a topic of concern.

Nowadays, researchers have started employing ANN in swirling flow for the identification of dynamics of coherent structures after doing experimental, simulation, and numerical studies on swirling flow. Artificial neural networks are systems that mimic the functions of the human brain. Goodfellow et al. [11] explains how ANN can give a non-linear relationship between a set of easily seen input data and a set of associated output signals. ANN has recently gained popularity in the field of fluid mechanics due to their structural adaptability, which is required for reduced-order models [12]. The use of ANN in turbulence modeling has also been documented in [12, 13]. It is important for a researcher to become familiar with existing research in the relevant subject before beginning their research. As a result, the goal of this review is to describe the results of many researchers' experimental CFD work. In addition, a review of numerous optimization studies utilizing techniques such as ANN, which have previously received little attention. The main goal is to explore the application of ANN in the identification of coherent structures in swirling flows from recent studies. The majority of the articles in this study are recent. Figure 1 shows the year-wise distribution of articles.

Objectives of the Present Review Article

The purpose of this work is to describe key aspects and quality results from numerous experimental and CFD studies of coherent structures that have been conducted recently. Rather than focusing on several researches that have already been examined, excellent outcomes from a few current studies have been highlighted. Furthermore, optimization research has been highlighted, which has been rarely mentioned in the

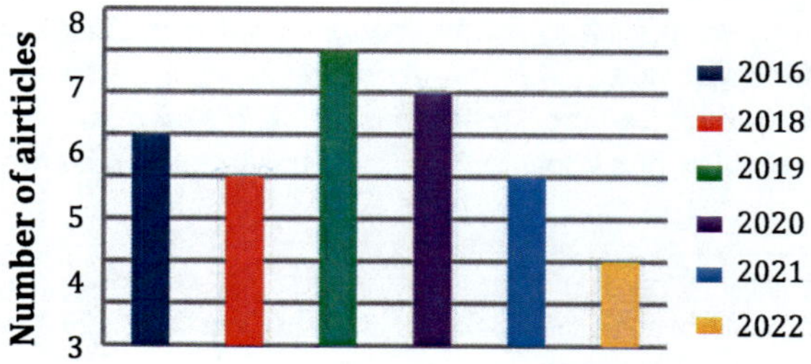

Fig. 1 Years of publication of the reviewed articles

published literature review. The major goal is to assist future researchers in determining the scope of their research by learning about earlier research in this field and bringing to light unpublished information. Section 2 summarizes key findings from experimental research, whereas Sect. 3 offers computational fluid dynamics work on VB and PVC. Section 4 discusses optimization studies on coherent structures that have been conducted recently.

2 Experimental Studies

In order to examine the coherent structures associated with swirling flow, many experimental and CFD simulations have been performed. Many studies have been conducted to determine the forms of VB and PVC [14–18].

PIV, OH*chemiluminescence, and OH planar laser-induced fluorescence (PLIF) techniques were utilized to analyze the lift-off process of swirl stabilized premixed flames [19], and it was observed that local flames disappeared surrounding the flame base, helical PVC emerged, and the flame base eventually vanished completely during the lift-off phase along with the vorticity field featured approximately balanced inner and outer shear layers when the flame was in the stable burner-attached state, as illustrated in Fig. 2a. Localized extinction occurrences around the flame base, and the inner shear layer showed a clear pattern of fragmentation and asymmetric displacement, as illustrated in Fig. 2b. The flame finally reached a stable elevated state, as shown in Fig. 2c by a visible zigzag pattern of vortices. This was the straight appearance of the helical revolving vortex phase, often known as the PVC.

In addition, Cozzi et al. [5] conducted an experimental investigation to analyze the coherent formations in the vicinity of an isothermal swirling jet. Stereoscopic PIV (S-PIV) was utilized to examine the flow field pattern under isothermal conditions. In both the swirling and non-swirling cases, the flow was investigated in both transverse and longitudinal directions. To highlight the essential flow features coupled with the precessing vortex core, a proper orthogonal decomposition (POD) was used for the

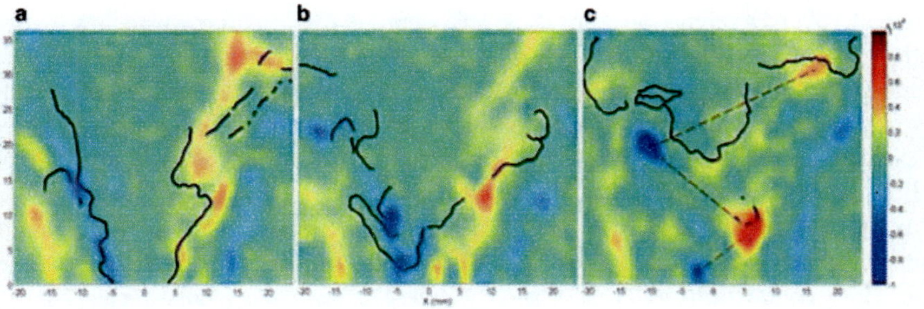

Fig. 2 Out-of-plane vorticity field (deep black line) placed on flame structure illustrating flow-field instability and PVC production [19]

instantaneous velocity vector field data. Vanierschot et al. [14] also looked at the flow patterns and characteristics of circular swirling jets in a transitional fluid flow. To extract the coherent structures from flow fields, time-resolved SPIV was coupled with spectral POD. The flow was examined using three different swirl values (Sw = 0.38, 0.47, and 0.57). With the lowest swirl number (Sw = 0.38), comparable to a Strouhal number St = 0.023, a vortex breakdown bubble developed and faded frequently. The breakdown bubble, together with the PVC, persisted in the flow field when the swirl was raised to Sw = 0.47. It was also found that the vortex breakdown bubble was wrapped with PVC, with the windings pointing in the reverse direction. This geometrical pattern, which had previously only been observed in completely laminar or turbulent swirling jets, was discovered the same in transitional jets found in that study. In addition, Clees et al. [15] carried out an experimental investigation to evaluate the occurrence of VB in a swirling jet in a range of strong swirls. Based on the examination of time-averaged velocity profiles and characteristics, three flow routes were identified: pre-breakdown, breakdown, and post-breakdown. The emergence of the recirculation zone suggestive of vortex breakdown, coupled with PVC, was observed in the post–breakdown region and its behavior was explored, employing dynamic mode decomposition, Rankine-vortex fitting, and POD to assess velocity fields in these regimes. Furthermore, in another experimental study, Using Tomographic PIV data, Vanierschot et al. [16] examined the 3D flow structures in a freely circular swirling jet flow undergoing vortex breakdown.

3 Computational Fluid Dynamics Studies

Apart from the experimental study X. Guo and B. Zhang [17] investigated PVC and energy isolation in a Ranque–Hilsch vortex tube (RHVT) using a CFD simulation, taking into account the impacts of cold mass fraction. In addition, a fresh examination of the energy separation method was also conducted. It was shown that raising the cold mass percentage below a particular threshold could lead to larger vortex cores, which improve energy extraction efficiency. When pressurized gas was fed tangentially in vortex chamber, it created a significant swirl, which resulted in the creation of an axial CRZ and a stagnation point as the VB occurred. Close to the reverse flow border, the vortex core deviated off the axis as a result of this, and the core precessed at a specific frequency. The fluid flow surrounding the PVC showed precession as well. The presence of extensive vortex forms of VB and PVC in the fluid flow was also used to characterize the energy separation mechanism. Wu et al. [18] also employed Large Eddy Simulation (LES) to examine isothermal and viscous turbulent swirling flows in a prototype gas turbine combustor design. With numerous VB flow topologies explored for contraction ratio in the range from 0.325 to 1.0, the influence of outlet topology contraction on VB shape and PVC were investigated. Houben et al. [20] also used CFD to determine the pressure and fluid flow inside the cyclone separator using the commercial CFD package (ANSYS) Fluent 13. The orientation of the vortex core was tracked in that investigation by looking for the lowest dynamic pressure point and the horizontal velocity field's center of moment as a function of the axial point. The obtained results were compared to the outcomes of the experiment and it was found that vortex was shown to be held in place when a stabilizer was employed. In addition, Gorbunova et al. [21] numerically odelled a non-stationary 3D swirling flow in an open tube with such a paraxial source of heat, demonstrating the presence of PVC in the flow. Increasing the heat-source power raised the PVC frequency while decreasing the amplitude of PVC oscillations, according to a variety of turbulence models such as Spalart–Allmaras, k–, and SST models.

4 Scope of Artificial Neural Network

Even after extensive experimental and computational fluid dynamics study on these coherent structures, especially vortex breakdown and precessing vortex core, these phenomena remain a subject of concern. In a recent study, Artificial Neural Networks were found to be one of the most effective data optimization approaches for recognizing and investigating the properties of coherent structures. A. Giannopoulos and J. L. Aider [22] investigated an experimental turbulence convective flow utilizing machine learning methods to find coherent structure dynamics and reduce system order. The flow was characterized using time-resolved optical flow PIV, which resulted in intense velocity profiles that could be utilized to track the flow's overall dynamics. After establishing a reduced-order system with POD, an optimal focused time-delay ANN was employed to identify a non-linear alignment between local input sensors and POD coefficients as outputs. Proper Orthogonal Decomposition (POD) is an analytical tool employed for extracting coherent structures from flow-field snapshots [5]. The various stages of the entire detection phase used in that investigation are depicted in Fig. 3.

A POD was utilized using the snapshot technique after collecting 6400 snapshots at a sample frequency of 150 Hz. A total of 17 POD modes were required to produce satisfactory results that have at minimum of 80% total energy. Large coherent structures could be noted in the instant field in Fig. 4, which are similar to the 3rd POD mode in Fig. 5.

In this turbulent flow, it was also shown that a deep ANN combined with the right sensor memory space could achieve good entire dynamics detection, coherent structure restoration, and overall order reduction. In addition, Deng et al. [23] stated that ANN was utilized to rebuild the flow field and also to detect the flow behavior of these coherent structures. In that paper, an LSTM-based POD model was presented to identify the relationship between velocity signals and time-varying POD coefficients acquired from non-TR-PIV observations. The LSTM model could reconstruct time series because it could accurately capture the temporal revolution of POD coefficients even in higher-level POD modes, according to the findings. The time-bounded flow fields might be well replicated using the suggested model's coefficients. In addition, Liu et al. [24] used deep learning approaches to reconstruct turbulent flows at super resolution. For super-resolution (SR), reconstruction of turbulent flows

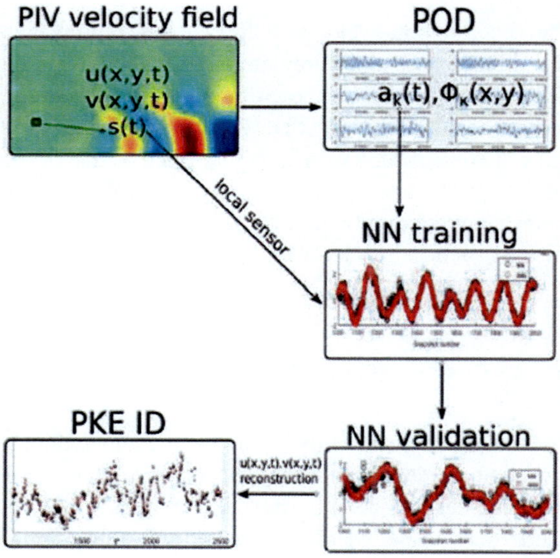

Fig. 3 The numerous procedures of detecting experimental pertubation energy via POD and training/validation are shown in a block diagram [22]

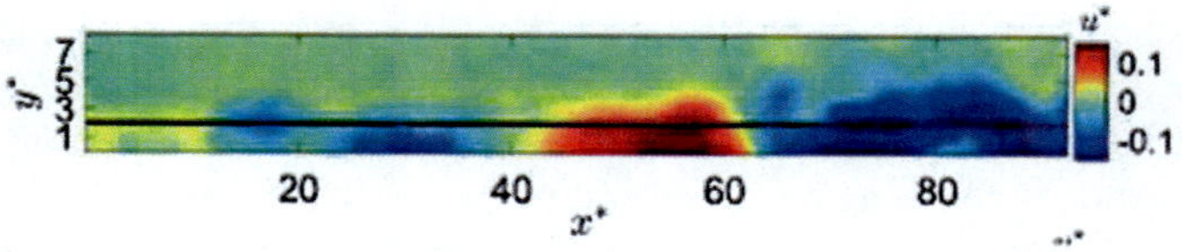

Fig. 4 The Turbulent Boundary Layer (TBL) experiment's instantaneous streamwise velocity fluctuation field [22]

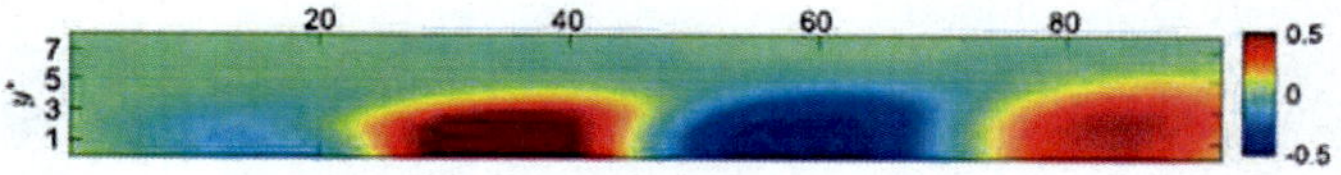

Fig. 5 3rd POD mode structure [22]

from poor rough flow fields, and two deep learning models were developed: the static convolutional neural network (SCNN) and the novel multiple temporal pathways convolutional neural network. The MTPC provided more information than that of the SCNN, and the accompanying high-resolution flow fields were rebuilt with outstanding accuracy. Reduced-order modeling and turbulence modeling in fluid mechanics could also be reported using Artificial Neural Networks [12, 13]. In another experiment, ANNs were used to forecast the dynamics of a Backward-Facing Step (BFS) flow and rebuild time-varying fields [25]. Abdurakipov et al. [26] used ANN to discover and assess the features of coherent structures in both non-swirling and swirling turbulence jets using data from the 3D PIV experiment and the DNS modeling database. The ability of a generative competitive neural network to recreate the form and composition of a turbulent flow with better spatial resolution, which is critical for analysis and evaluation, has been demonstrated. It was discovered that generative models were more effective than traditional approaches POD and DMD in reconstructing high moments of turbulent flow. It was also shown that the complex of established machine learning algorithms was capable of localizing coherent structures, detecting their trajectory, determining their characteristic size, and determining their phase propagation velocity. Additionally, Machine Learning was used by Kommalapati et al. [27] to interpret and analyze non- time-resolved PIV observations of turbulent flows. A huge PIV dataset was acquired with 50,000 non-time specified images of turbulent separated flow and neural network architecture was trained on it to outperform standard interpolation-based super resolution techniques. Finally, the current plan was shown to accurately replicate the features of turbulence-generated values, such as the streamwise deviation of velocity gradients, when compared to interpolation-based super resolution alternatives. Daniel et al. [28] also looked at the research work on ANN approaches, as well as current advancements in the usage of these modeling tools for thermal systems, as well as a new application for tackling thermal transfer uncertainties.

5 Conclusion

Swirling jets are used in a number of applications, such as swirl burners, gas turbine combustors, and so on. Wherever swirling jet applications are encountered, the two phenomena that come into play are VB and PVC. These occurrences can have a huge impact on the swirl flow system's overall performance. Because our current knowledge of swirling flow is limited, more empirical, theoretical, and statistical research is required to fully investigate and comprehend these phenomena. Although the majority of the work in this field has been experimental, several investigations have been recorded using CFD simulation as well. The goal of this paper was to interpret the findings of recent investigations on coherent structures that were not covered in prior review articles so that future research studies can profit from them and gain more understanding into this interesting zone of swirling flow. Furthermore, an overview of optimization studies with Artificial Neural Networks is offered. The possibilities of using ANN in swirling flow have been discussed as a promising topic and to propose research recommendations for the future.

6 Recommendation for Future Work

Without a question, the employment of ANN in fluid mechanics has proven to be a huge success. The following future work, on the other hand, is suggested:

1. Integration of artificial neural network modeling with CFD software for better reliable modeling practices.
2. The application of ANN to determine the dynamics of coherent structures in swirling flows should be investigated.
3. The use of ANN for flow field reconstruction in swirling flow should be researched further.

NOMENCLATURE

VB	Vortex Breakdown
PVC	Precessing Vortex Core
ANN	Artificial Neural Network
ML	Machine Learning
CFD	Computational Fluid Dynamics
PIV	Particle Image Velocimetry
POD	Proper Orthogonal Decomposition
LES	Large Eddy Simulation
3D	Three-dimensional

References

1. Jing L et al (2022) Numerical analysis of the effect of swirl angle and fuel equivalence ratio on the methanol combustion characteristics in a swirl burner. Process Saf Environ Prot 158:320–330. https://doi.org/10.1016/J.PSEP.2021.11.019
2. Alekseenko SV et al (2021) Investigation of transfer processes in swirling flows in application to vortex furnaces for coal fuel. Int J Therm Sci 161:106715. https://doi.org/10.1016/J.IJTHERMALSCI.2020.106715
3. Nassini PC, Pampaloni D, Meloni R, Andreini A (2021) Lean blow-out prediction in an industrial gas turbine combustor through a LES-based CFD analysis. Combust Flame 229:111391. https://doi.org/10.1016/J.COMBUSTFLAME.2021.02.037
4. Moise P, Mathew J (2019) Bubble and conical forms of vortex breakdown in swirling jets. J. Fluid Mech. pp 322–357, https://doi.org/10.1017/jfm.2019.401
5. Cozzi F, Sharma R, Solero G (2019) Analysis of coherent structures in the near- field region of an isothermal free swirling jet after vortex breakdown. Exp Therm Fluid Sci. 109:109860. https://doi.org/10.1016/J.EXPTHERMFLUSCI.2019.109860
6. Shen Y, Ghulam M, Zhang K (2020) Vortex breakdown of the swirling flow in a lean direct injection burner articles you may be interested in. Phys Fluids 32:125118. https://doi.org/10.1063/5.0028838
7. Naumov, IV, Sharifullin BR, Shtern VN (2020) Vortex breakdown in the lower fluid of two-fluid swirling flow. Phys. Fluids, 32(1), https://doi.org/10.1063/1.5132584
8. Tran CT, ping Long X, Ji B, Liu C (2020) Prediction of the precessing vortex core in the Francis-99 draft tube under off-design conditions by using Liutex/Rortex method. J Hydrodyn., 32(3), pp 623–628, https://doi.org/10.1007/s42241-020-0031-3

9. Manoharan K, Frederick M, Clees S, O'Connor J, Hemchandra S (2019) A weakly nonlinear analysis of the precessing vortex core oscillation in a variable swirl turbulent round jet. J. Fluid Mech. 884 https://doi.org/10.1017/jfm.2019.903
10. Lückoff F, Kaiser TL, Paschereit CO, Oberleithner K (2021) Mean field coupling mechanisms explaining the impact of the precessing vortex core on the flame transfer function. Combust Flame 223:254–266. https://doi.org/10.1016/J.COMBUSTFLAME.2020.09.019
11. Goodfellow I, Bengio Y, Courville A, Heaton J (2016) Ian Goodfellow, Yoshua Bengio, and Aaron Courville: Deep learning. Genet Program Evolvable Mach 19:262035618. https://doi.org/10.1007/s10710-017-9314-z
12. Maulik R, San O, Rasheed A, Vedula P (2018) Data-driven deconvolution for large eddy simulations of Kraichnan turbulence. Phys. Fluids, 30(12), https://doi.org/10.1063/1.5079582
13. Xie C, Wang J, Li H, Wan M, Chen S (2019) Artificial neural network mixed model for large eddy simulation of compressible isotropic turbulence. Phys. Fluids, 31(8), https://doi.org/10.1063/1.5110788
14. Vanierschot M, Ogus G (2019) Experimental investigation of the precessing vortex core in annular swirling jet flows in the transitional regime. Exp Therm Fluid Sci. 106:148–158. https://doi.org/10.1016/J.EXPTHERMFLUSCI.2019.05.001
15. Clees S, Lewalle J, Frederick M, O'Connor J (2018) Vortex core dynamics in a swirling jet near vortex breakdown. AIAA Aerosp Sci Meet. 2018(210059):1–18. https://doi.org/10.2514/6.2018-0052
16. Vanierschot M, Percin M, Van Oudheusden BW (2018) Double helix vortex breakdown in a turbulent swirling annular jet flow. Phys Rev Fluids 3(3):1–7. https://doi.org/10.1103/PhysRevFluids.3.034703
17. Guo X, Zhang B (2018) Computational investigation of precessing vortex breakdown and energy separation in a Ranque-Hilsch vortex tube. Int J Refrig 85:42–57. https://doi.org/10.1016/J.IJREFRIG.2017.09.010
18. Wu Y, Carlsson C, Szasz R, Peng L, Fuchs L, Bai XS (2016) Effect of geometrical contraction on vortex breakdown of swirling turbulent flow in a model combustor. Fuel 170:210–225. https://doi.org/10.1016/J.FUEL.2015.12.035
19. An Q, Kwong WY, Geraedts BD, Steinberg AM (2016) Coupled dynamics of lift-off and precessing vortex core formation in swirl flames. Combust Flame 168:228–239. https://doi.org/10.1016/J.COMBUSTFLAME.2016.03.011
20. Houben JJH, Weiss C, Brunnmair E, Pirker S (2016) CFD simulations of pressure drop and velocity field in a cyclone separator with central vortex stabilization rod. 9(1), pp 487–499, https://doi.org/10.18869/acadpub.jafm.68.224.23934
21. Gorbunova A et al (2016) Precessing vortex core in a swirling wake with heat release. Int J Heat Fluid Flow 59:100–108. https://doi.org/10.1016/J.IJHEATFLUIDFLOW.2016.03.002
22. Giannopoulos A, Aider JL (2020) Data- driven order reduction and velocity field reconstruction using neural networks: The case of a turbulent boundary layer. Phys. Fluids, 32(9), https://doi.org/10.1063/5.0015870
23. Deng Z, Chen Y, Liu Y, Kim KC (2019) Time-resolved turbulent velocity field reconstruction using a long short-term memory (LSTM)-based artificial intelligence framework, Phys Fluids, 31(7), https://doi.org/10.1063/1.5111558
24. Liu B, Tang J, Huang H, Lu XY (2020) Deep learning methods for super-resolution reconstruction of turbulent flows. Phys. Fluids, 32(2), https://doi.org/10.1063/1.5140772
25. Giannopoulos A, Aider JL (2020) Prediction of the dynamics of a backward-facing step flow using focused time-delay neural networks and particle image velocimetry data-sets. Int J Heat Fluid Flow 82:108533. https://doi.org/10.1016/J.IJHEATFLUIDFLOW.2019.108533
26. Abdurakipov S, Tokarev M, Butakov E, Dulin V (2019) Application of computer vision and neural network analysis to study the structure and dynamics of turbulent jets. 1421, pp 1–10, https://doi.org/10.1088/1742-6596/1421/1/012018

27. Machine learning for coherent structure identification and super resolution in turbulent flows, (2021)
28. Raphael D, Ewim E, Okwu MO, Onyiriuka EJ, Abiodun AS (2022) Engineering and applied science research A quick review of the applications of artificial neural networks (ANN) in the modelling of thermal systems. 49(3), pp 444–458, https://doi.org/10.14456/easr.2022.45.

Control of Trailing Edge Separation with the Use of Circular Endwall Slot in a Highly Loaded Turbine Blade Cascade

Devershi Mourya and P. M. V. Subbarao

Abstract The blade endwall slot was employed to effectively regulate boundary layer separation on the suction side and optimize the fluid properties following the blade's trailing edge. This investigation aims to determine the reason for trailing edge separation and how it may be suppressed using the converging blade endwall slot, with a focus on higher-turning stator blades. The circular blade endwall slot provides 20, 50, and 70% of the axial chord for evaluating endwall loss. The computational results are generated using the endwall slot blade and then compared to the datum blade and got a 5% reduction in both total pressure loss, vorticity magnitude, and 10% in turbulence kinetic energy. The findings reveal that the slot created at the endwall by adjusting the axial cords provides a self-adaptive jet that re-energizes and controls the low-velocity fluid downstream and regulates its movement towards midspan. Thus, trailing edge separation at a corner is eliminated, which enhances the fluid properties after trailing edge of the blade, improving the turbine's performance significantly.

Keywords Trailing edge separation · Blade endwall slot · High-turning compressor · Flow control

1 Introduction

Turbine blades for gas engines with excellent efficiency are in high demand due to the rising cost of fuel. It is necessary to lower the engine's size and weight without sacrificing performance to attain a high thrust-to-weight ratio in any gas engine. Therefore, high-loaded blades are required for gas engines. With a high-loaded blade, the secondary flow loss rises, including the formation of substantial corner separation on the suction side. In turbine blades, secondary losses comprise one-third of the total loss created during flow, while corner separations make up a significant portion.

D. Mourya (✉) · P. M. V. Subbarao
Department of Mechanical Engineering, IIT Delhi, New Delhi 110016, India
e-mail: Cdevershi05@gmail.com

S. Das et al. (eds.), *Proceedings of the 1st International Conference on Fluid, Thermal and Energy Systems*, Lecture Notes in Mechanical Engineering,
https://doi.org/10.1007/978-981-99-5990-7_57

Consequently, a variety of methods were used by researchers to manage the corner separation in heavily loaded axial turbine blades.

Active or passive flow control methods are used to regulate blade corner separation in heavily loaded blades. Some additional devices are utilized to blast fluid into the area where we wish to manage the secondary flow and corner separation in the cascade channel when using active control methods such as purge flow, plasma fluid actuators, vortex generator flow, boundary layer aspiration, synthetic purge flow. However, the formation of secondary and corner separations may be managed by a passive control technique, such as a modification in the design of the blade, hub, and. Passive control methods such as leaning, bowing, blade endwall slot, upstream cavity, blade endwall slot, endwall countering, and fencing may considerably reduce the formation and strength of secondary and corner separation. These techniques were favoured by researchers because of their simplicity and cheaper production costs. The substantial vortices generated by the vortex generator devices improve the mixing between the main flow and decline the boundary layer flow, lowering corner separation and overall pressure loss [1].

Endwall purge flow enhances boundary layer flow momentum by delaying or preventing separation [2, 3]. Lean is provided on the turbine blade to alter the stage reaction spanwise. Leaning may have beneficial effects when using blades with a high aspect ratio and a low hub-to-tip ratio [4]. Blade bowing enhances the aerodynamic performances in both turbines and compressors. A decrease in endwall flow may be achieved by rearranging and replacing vortex formations in cascade channel flow by employing bow blades [5]. Blade endwall slot minimizes the local loading at the endwall-blade junction and may increase fluid flow inside the cascade channel [6].

Slots in the endwall of a turbine blade may enhance the blade's efficiency. In a highly loaded blade, a novel blade end slots approach was presented to prevent 3D corner separation. NACA65 cascades with different incidence angles were studied by Sun [7]. Experiment and simulation reveal that a 4° incidence angle with a single endwall slot has substantial potential to diminish compressor losses and increase performance, as well as to significantly suppress corner separation. Tang [8] studied two endwall slots with an entire span in a highly loaded compressor blade profile with a wide incidence angle. They did both experimental and computational studies with high Mach Number flow and found that local low momentum fluid re-enrzies and minimizes corner separation because of the creation of self-adaptive jet flow. Wang [6] investigated two endwall slots in a blade with a solidity ratio of 1.66 and 1.36 and found that experimental with simulation resulted in a 23.2 and 27.7% decrease in total pressure loss. They found that by lowering the compressor's single-stage blade number, the endwall slot is more helpful with a low solidity ratio to boost the aero-engine thrust-to-weight ratio.

The slot in the blade's endwall suggests that there may be a way to reduce corner separation. The majority of prior investigations, on the other hand, utilize solely rectangular slots in front-loaded blade designs. As a result, a highly loaded blade may be used for research. Additionally, there is a need to examine the slot's perfect location, shape, and height of the slot relative to the endwall. This study employs a circular endwall slot with a diverging shape from the pressure side to the suction side at 30, 50, and 70% of the axial chord to reduce corner separation with trailing edge separation. Simulations and experiments are carried out at Mach 0.015 in low Reynolds number wind tunnels. The spanwise distributions of pitch averaged total pressure loss coefficient, vorticity magnitude, and turbulent kinetic energy with the three-dimensional flow field structures are examined to demonstrate the control of loss generation based on experimental results and the experimentally calibrated CFD results utilizing blade endwall slots.

2 Research Methodology

2.1 Computational Methodology

The geometry is created in SolidWorks 14, and due to the meshing of complicated geometry and dealing with local body mesh refinements, the unstructured mesh is created in Ansys Workbench 18. O-grid topology is used to discretize the boundary layer phenomena around the blade and tetrahedron topology for the rest of the domain. Fluent 18 is used to get a solution for simulation.

For the simulations, the fluent used the finite volume method (FVM) with a three-dimensional steady-state condition, which involves a pressure-based solver. Discretization is done under a second-order upwind scheme, where iteration stops when residual scales are less than 1×10^{-5}. Based on a literature survey for capturing the fluid behaviour near the endwall structure between the leading-edge and hub or shroud, a k-epsilon realizable model with a non-equilibrium wall treatment function is acceptable. Turbulent kinetic energy (T_{ke}) is shown in Fig. 1 for various grid numbers. After 928,319 nodes in the grid independence test, the effect of grid number on turbulent kinetic energy becomes asymptotic. The arrangement of the datum blade, a combination of enwall slot with datum blades, and its mesh details are shown in Fig. 2.

A domain is created with periodic boundary conditions in the pitch direction for carrying out simulations, shown in Fig. 3. Boundaries are kept at the no-slip position at the hub and blade wall, while the upper plane is considered symmetric. The inlet and outlet plane boundary are at 244 and 266% of C_{ax}, respectively. The iteration is started from the inlet plane. The outlet plane is kept as the atmosphere conduct at static pressure.

Datum blade and endwall slot blade simulations are performed according to the simulation methodology. The geometric parameters of the datum blade and endwall slot blade which are used in the computational domain for meshing are depicted in Table 1. The description of different sections of blade cascades is presented in Fig. 4. The computational domain is discretized with applied boundary conditions, and calculated results of the performance parameter of endwall slot blades are used later to compare with the Datum blade simulated result.

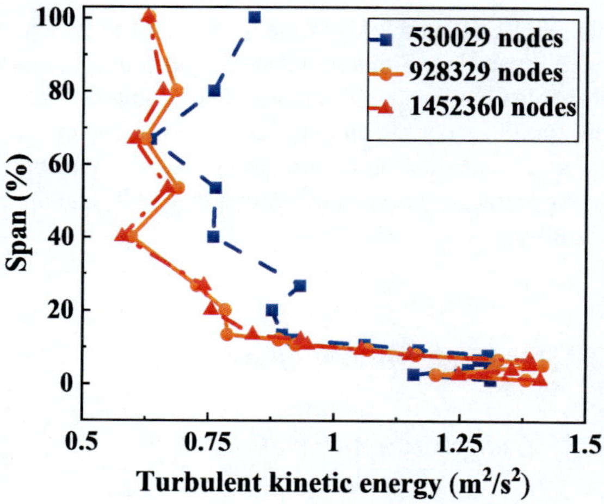

Fig. 1 Grid invariance test for turbulent kinetic energy along the spanwise direction at 107 Cax plane (Re-27106, M-0.015)

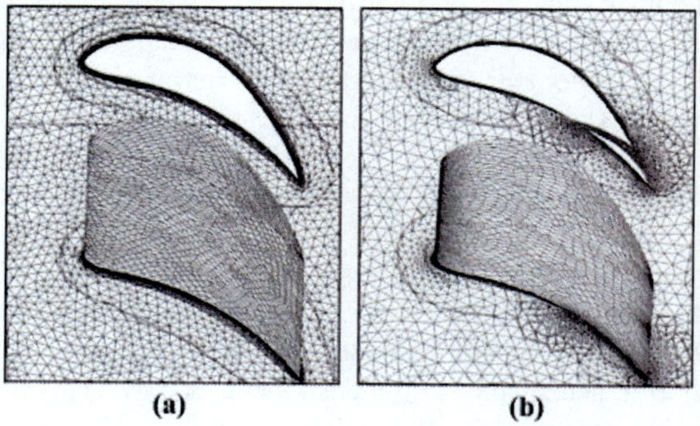

Fig. 2 Details of mesh **a** Datum blade, **b** Datum blade with endwall slot

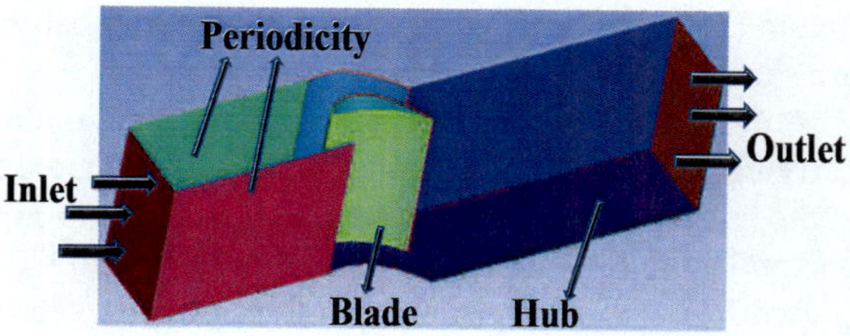

Fig. 3 Computational domain

2.2 Design and Optimization of Endwall Slot at Blade

The purpose of designing an optimized endwall slot at the turbine blade is to minimize the corner separation, measured by pitch averaged total pressure loss coefficient (TPLC) downstream of the trailing edge of the cascade. An endwall slot is created in the blade at 20, 50, and 70% of the chord at a height of 1 mm relative to endwall.

Table 1 Geometrical specifications of the blade

1	Blade	E-423
2	Blade axial chord length C_{ax} (mm)	82.48
3	Pitch s (mm)	91
4	Blade Span H (mm)	300
5	Inlet angle α_{in}	0°
6	Outlet angle α_{out}	21°
7	Stagger angle	35°
8	Endwall slot diameter D_s	1 mm
9	Endwall slot position	20, 50, and 70% of axial chord
10	Optimum endwall slot position	70% of axial chord
11	Slot height from endwall	1 mm
12	Blade Reynolds no. Re_b	27,106
13	Mach no. M	0.015

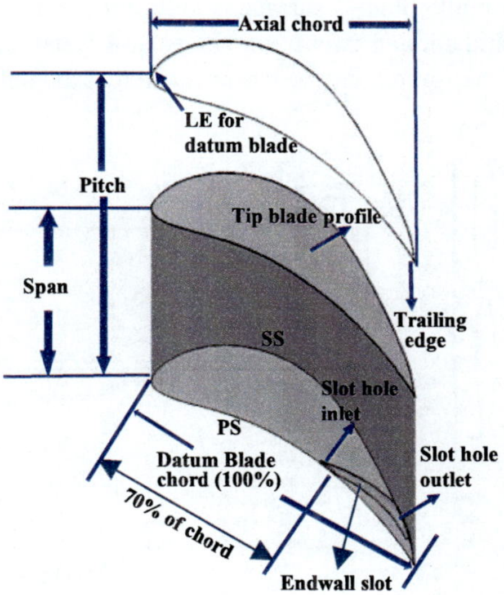

Fig. 4 Definitions of Datum and endwall slot blade

The dia and length of the slot are kept at 1mm and equivalent to 25% of chord, respectively.

2.3 Experimental Methodology and Validation

All the experiments are performed in a low-speed wind tunnel shown schematically in Fig. 5a. The cascade used in experiments is manufactured by using a 3D printing technique. The details of boundary layer properties are provided in Table 2. The wind tunnel dimensions are $300 \times 300 \times 2319\ \text{mm}^3$ with an axial fan operated by two HP DC motors. A two-hole pitot tube and a thermocouple are fixed at the inlet plane to measure the inlet air velocity and temperature, positioned at 2.44 C_{ax} upstream of the Leading edge (LE). After that, a cascade set-up is placed, as shown in Fig. 5a. An outlet plane is utilized at 2.66 C_{ax} downstream from the trailing edge, and the channel between 2.44 C_{ax} upstream from LE and 2.66 C_{ax} downstream from Trailing edge (TE) is considered a test section.

All measuring devices are mounted at this test section to calculate the parameters, so this section is simulated in computation. The fan pushes the air to a convergent channel where its velocity increases. After passing through the turbulence grid, the air velocity is kept constant at 5 m/s. Air at constant velocity flows into the considered test section consisting of three identical-shaped blades as shown in Fig. 5b. The middle blade's surface is split into twenty-one holes at a radial distance of 5 mm and another twenty-one holes at a radial distance of 15 mm from the hub endwall. The static pressure was measured at the outlet of the needle. Each measured values

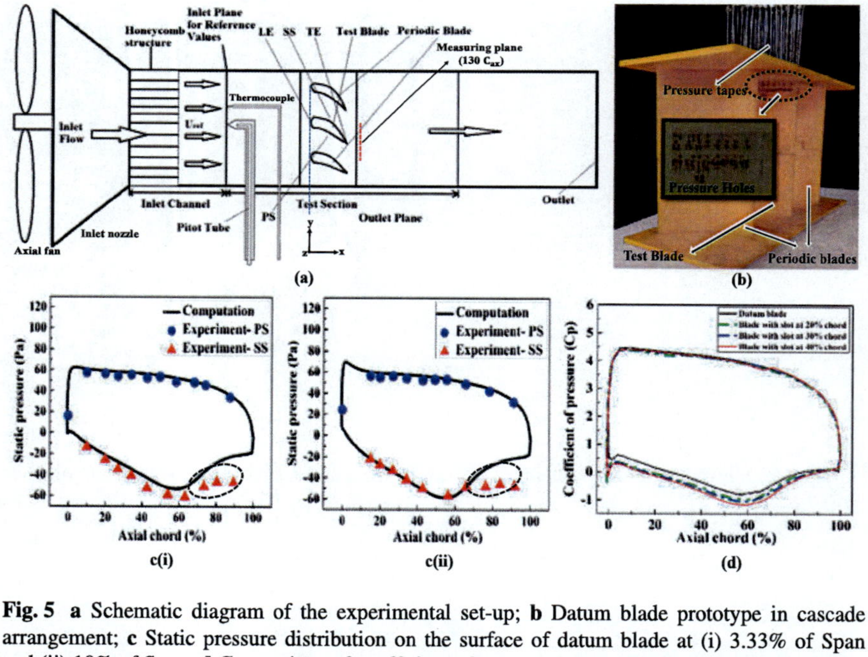

Fig. 5 **a** Schematic diagram of the experimental set-up; **b** Datum blade prototype in cascade arrangement; **c** Static pressure distribution on the surface of datum blade at (i) 3.33% of Span and (ii) 10% of Span; **d** Comparison of coefficient of pressure between Datum blade and endwall slot blades at 3.33% of span

Table 2 Reference properties at Upstream of experiment channel

1	Average velocity upstream U_{ref} (m/s)	5
2	Mach no. (M)	0.015
3	Stagnation pressure $P_{o, in}$ (Pa)	1.013×10^5
4	Boundary layer thickness δ(mm)	13.3
5	Displacement thickness $\dot{\delta}$ (mm)	1.5
6	Momentum thickness $\ddot{\delta}$ (mm)	1.12
7	Shape Factor ($\dot{\delta}/\ddot{\delta}$)	1.34
8	Turbulent intensity (%)	5
9	Turbulent viscosity ratio	10

are compared with simulation results for further understanding. The static pressure shows the condition of the fluid around the blade surface. The blue dots and red triangles indicated the measurement points at the blade's pressure side and suction side, respectively, in Figs. 5c(i), (ii). These experimental values are obtained using pressure holes at the blade surface. The produced data at 3.33 and 10% of the span are shown in Figs. 5c(i), (ii) with some divergence in the suction side. According to our investigation, the cause of these divergences of the flow is the condition of the boundary layer of the incoming flow, laminar in simulation but turbulent in experimental, and the early formation of blade wake and separation from the blade surface at the endwall.

3 Result and Discussion

3.1 Impact of Endwall Slot at Blade Loading

Figure 5d shows the comparison of the pressure coefficient at the blade with variation in end wall slot position. The endwall slot increases the removal of low-energy fluid from the boundary layer of the blade, which reduces the separation of flow at the trailing edge and enhances the blade loading, which is reflected in the suction side of the slotted blade in Fig. 5d. The endwall slot is unaffected on the pressure side, but on the suction side. The most significant blade loading occurs when the endwall slot is positioned at 40% of the chord.

3.2 Investigation of the Effect of Endwall Slot at TPLC on the Flow Field

Figure 6 compares the TPLC readings at 130% of the C_{ax} plane between the datum and the endwall slot blade. The TPLC value of the datum blade and end wall slot at 20 and 50% of chord blades is the highest, which may be obtained at 15% of blade span and 50% of blade pitch. A slot at these locations won't be able to decrease total loss, as shown in Fig. 6a–c. This is because Passage vortices have the highest TPLCs, which are enhanced by the interaction of boundary layer flow and crossflow with split branches of the Horshoe vortex. However, as shown in Fig. 6d, the maximum TPLC value drops by up to 10% when the slot is given at 70% of the chord. Lower trailing edge separation results in lower TPLC values over the whole span.

3.3 Investigation of the Effect of Endwall Slot at Vorticity Magnitude on the Flow Field

Vorticity magnitude (VM) readings are compared between the endwall slot blades at 130% of C_{ax} and the datum blade in Fig. 6. The VM values are greater in two areas, one having a bigger area and the other with a smaller area, as illustrated in Fig. 6. For all blades, the bigger area is achieved at 70% of blade pitch, whereas the smaller area is achieved at 20% of blade span and 35% of blade pitch. Figure 7a demonstrates that the datum blade has a significantly high VM value compared to other blades. This is the counter vortex, which has a lower Vorticity.

Magnitude in the smaller region. At the whole blade span, the trailing edge vortex has the maximum value of Vorticity Magnitude. As illustrated in Fig. 7b, c, the peak value of the endwall slot blade is 15% lower than the datum blade's highest value. There was a 22% drop in VM at 70% of the endwall slot, but the second region became negligible. As a result, the endwall flow losses are reduced, which may significantly diminish trailing edge separation. Because of this, including an optimally placed endwall slot may greatly decrease VM and improve downstream of the blade.

3.4 Investigation of the Effect of Endwall Slot at T_{ke} on the Flow Field

As can be seen in Fig. 8, the T_{ke} values at 130% of the C_{ax} plane are being compared between the datum and the slot blade on the endwall. Figure 8 shows that there are two distinct areas, one with a bigger size and the other with a smaller region, where the T_{ke} values are greater. For all blades, the bigger area is achieved at 15% of blade span and 57% of blade pitch, whereas the smaller area is achieved at 15% of blade span and 80% of blade pitch. Figure 8a shows that the datum blade has a very high

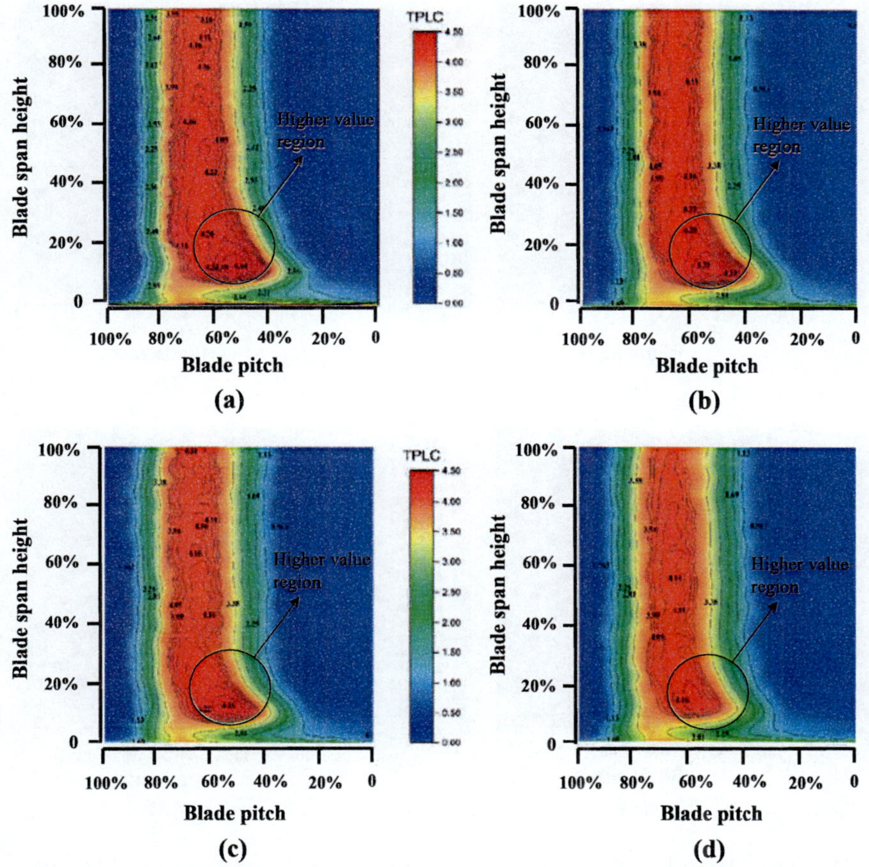

Fig. 6 Span-wise contour of TPLC at 30% C_{ax} downstream: **a** Datum Blade, **b** Endwall slot at 20% chord Blade, **c** Endwall slot at 50% chord Blade, **d** Endwall slot at 70% chord Blade (Re-27106, M-0.015)

T_{ke}. The counter vortex has a lower T_{ke} in the smaller region. The dominant passage vortex, which has the maximum T_{ke} on this plane, may be seen in the bigger region. Endwall slot blades, on the other hand, have a peak value that is lowered by 20% to the values of the datum blade, as illustrated in Fig. 8b, c. However, at 70% of the endwall slot, the first region displayed a lower value, and the second area vanished, as shown in Fig. 8d. As a result, the trailing edge separation is effectively suppressed by this reduction in endwall flow. Endwall slots lower T_{ke} considerably and function effectively even after trailing edge incorporation.

Table 3 lists the benefits of endwall slot blades over datum blades. When compared to other endwall slot blades, the endwall slot at 70% of the chord displays the greatest reduction in total pressure loss coefficient and vorticity magnitude, down to 4.76 and

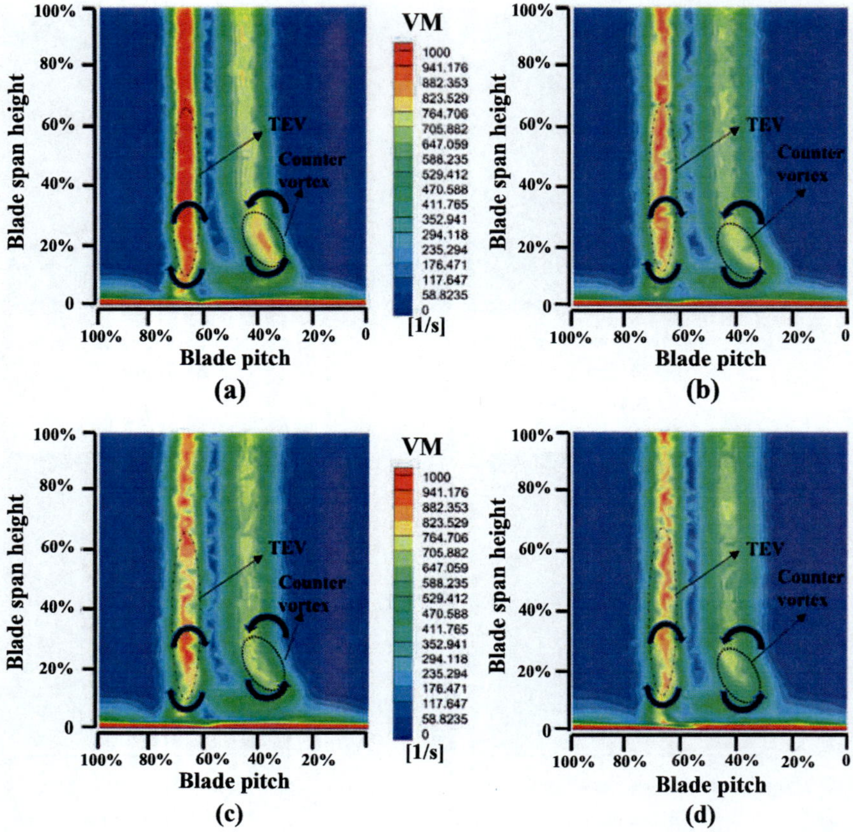

Fig. 7 Span-wise contour of Vorticity Magnitude (VM) at 30% C_{ax} downstream: **a** Datum Blade, **b** Endwall slot at 20% chord Blade, **c** Endwall slot at 50% chord Blade, **d** Endwall slot at 70% chord Blade (Re-27106, M-0.015)

5.5%, respectively, throughout the whole blade span. Compared to the datum blade, a 10.11% reduction in T_{ke} can be seen over the whole blade span.

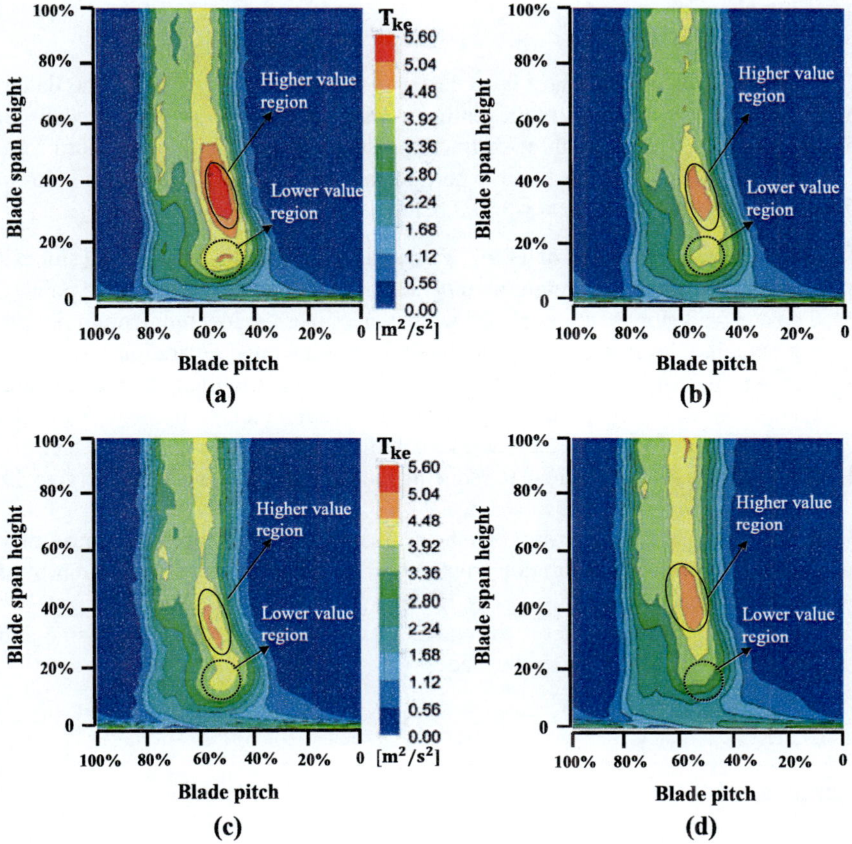

Fig. 8 Span-wise contour of Turbulence kinetic energy (Tke) at 30% C_{ax} downstream: **a** Datum Blade,**b** Endwall slot at 20% chord Blade, **c** Endwall slot at 50% chord Blade, **d** Endwall slot at 70% chord Blade (Re-27106, M-0.015)

Table 3 Comparison of pitch averaged parameters between datum blade and different endwall slot blades

	Datum	20% axial chord slot		50% axial chord slot		70% axial chord slot	
TPLC	1.815	1.822	*−0.399%*	1.773	*2.329%*	1.729	**4.76%**
VM	520.63	531.39	*−2.066%*	514.18	*1.239%*	491.76	**5.545%**
T_{ke}	1.563	1.468	*6.115%*	1.423	*8.981%*	1.406	**10.114%**

4 Conclusion

The primary goal of this study is to determine the best placement for the datum blade's endwall slot in terms of preventing trailing edge separation. An endwall slot of circular shape is used at three distinct positions on the datum blade chord for a comprehensive computational and experimental investigation. The endwall slot in the datum blade-related effects is listed below.

1. The maximum reduction of TPLC, VM, and T_{ke} is shown when endwall slot exit is located between minimum the separation point and the Trailing edge point.
2. The end wall slot generates a jet flow that energizes the low momentum fluid in the boundary layer which helps to delay the trailing edge separation.
3. Secondary losses from 3D vertical structure manipulation may be reduced and their strength can be reduced by using the endwall slot. In order to avoid the development of vortices, jet flow-like solutions are used in a specific place.
4. The jet flow is also beneficial to reducing the endwall flow due to re-energize the boundary layer fluid.
5. End wall slots are incredibly beneficial for highly loaded blades because they prevent suction-side separation from occurring too early, resulting in improved performance.
6. Also, the endwall slot helps to reduce axial vorticity on the suction side and downstream flow following the blades.

References

1. Hergt A, Meyer R, Engel K (2012) Effects of vortex generator application on the performance of a compressor cascade. J Turbomach 135(2)
2. Tao Z, Guo Z, Song L, Li J (2020) Uncertainty quantification of aero-thermal performance of a blade endwall considering slot geometry deviation and mainstream fluctuation
3. Schreiner BDJ, Wilson M, Li YS, Sangan CM (2020) Effect of purge on the secondary flow-field of a gas turbine blade-row. J Turbomach **142**(10)
4. Zhang S, MacManus DG, Luo J (2016) Parametric study of turbine NGV blade lean and vortex design. Chinese J Aeronaut 29(1):104–116
5. Xue X, Wang S, Luo L, Zhou X (2020) The compound bowing design in a highly loaded linear cascade with large turning angle. Proc Inst Mech Eng Part G J Aerosp Eng 234(16):2323–2336
6. Wang H, Qing Y, Liu B, Mao X (2021) Corner separation control using a new combined slotted configuration in a high-turning compressor cascade under different solidities. Energies 14(12)
7. Sun J, Ottavy X, Liu Y, Lu L (2021) Corner separation control by optimizing blade end slots in a linear compressor cascade. Aerosp Sci Technol 114:106737
8. Tang Y, Liu Y, Lu L, Lu H, Wang M (2019) Passive separation control with blade-end slots in a highly loaded compressor cascade. AIAA J 58(1):85–97

Characterization of Primary Atomization Processes of Like-on-Unlike Impinging Injectors

Bikash Mahato, Vivek Sahu, R. Vikram, P. Ganesh, and K. P. Shanmugadas

Abstract The present work investigates the primary atomization characteristics of like-on-unlike impinging injectors. Planar laser-induced fluorescence imaging experiments are conducted in both front and side view configurations to capture the sheet breakup processes. Four different breakup regimes are identified for a Weber number range of 187–2994. It is observed that the propagation of impact waves dictates the sheet breakup process and successive ligament formation, in higher Weber number cases. Different linear sheet parameters such as sheet breakup length, sheet width, and sheet thickness are quantified at low Weber number conditions.

Keywords Unlike impinging injectors · Planar laser-induced fluorescence imaging · Impact waves

1 Introduction

In many bipropellant liquid rocket engines, impinging injectors are utilized to atomize and mix the liquid fuel and oxidizer. Impinging injectors can be broadly divided into like-on-like and like-on-unlike types. Apart from the main thrust chamber, impinging injectors are also used in liquid rocket engines' gas generators and in tiny thrusters. They handle hypergolic-type propellants which undergo spontaneous combustion upon contact with each other. Rapid mixing, atomization, and combustion take place at the point of impingement [1]. The atomization processes of unlike impinging injector configurations are presented in the present work.

B. Mahato · V. Sahu · K. P. Shanmugadas (✉)
Department of Mechanical Engineering, IIT Jammu, Jagti 181221, Jammu, India
e-mail: shanmugadas.kp@iitjammu.ac.in

R. Vikram
National Centre of Combustion Research Centre (NCCRD), IIT Madras, Chennai 600036, India

P. Ganesh
ISRO Propulsion Complex, Mahendragiri, Tirunelveli 627133, India

S. Das et al. (eds.), *Proceedings of the 1st International Conference on Fluid, Thermal and Energy Systems*, Lecture Notes in Mechanical Engineering,
https://doi.org/10.1007/978-981-99-5990-7_58

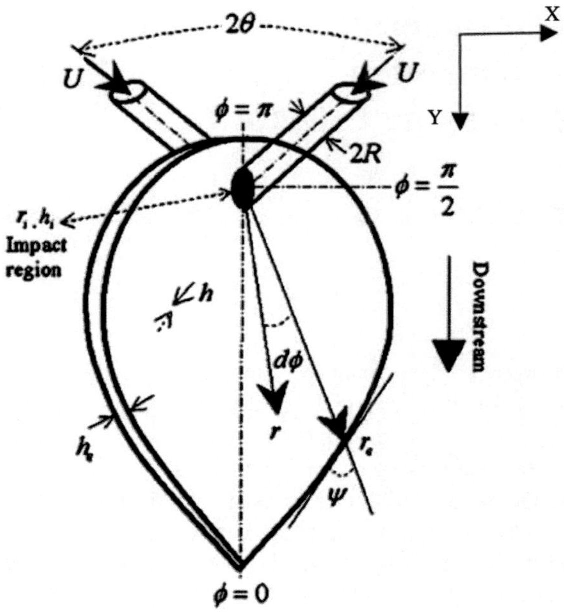

Fig. 1 Schematic of impinging atomization process [1]

In the case of like-on-like injectors, when two liquid jets impinge, a sheet is formed in a direction that is perpendicular to the plane of the jets. Waves emerge from the point of impingement and grow in amplitude till the liquid sheet breaks, forming ligaments and droplets. These waves are hydrodynamic in origin and are called impact waves [1]. A schematic of the atomization process is shown in Fig. 1. In the case of unlike impinging injectors, a jet of different diameters and momentum impinges each other and forms a liquid sheet with similar behavior but curved and inclined toward the lower momentum jet.

Investigation of the performance of rocket engine combustors is quite difficult considering the extreme operating conditions. Alternatively, experimental investigations are reported at scaled-down atmospheric ambient operating conditions. However, this approach is time-consuming and incurs considerable costs. Another approach is to investigate the atomization processes systematically and develop cold flow models which can be compared with flight test data to build correlations related to the injector performance. The development of both cold flow tests and theoretical understanding helps in a better way to understand the atomization and combustion phenomena, which also reduces the risk of handling hazardous liquid propellants [2]. The majority of earlier investigations concentrated on the like-on-like impinging injectors' atomization features [3]. The present work aims to conduct detailed investigations with unlike impinging injectors and to quantify different linear sheet parameters. The unsteady spray behavior due to sheet interaction with the atmosphere and breakup processes are also to be investigated, as only a limited number of investigations are conducted in this area, i.e., in the case of unlike doublet impinging injectors.

2 Literature Review and Objectives

The atomization processes of impinging injectors are investigated by a number of researchers. Heidmann et al. [3] noted that in the case of like-doublet injectors, four atomization regimes are formed based on his experimental observations, i.e., closed rim, periodic drop, open rim, and fully developed.

Dombrowski and Hooper [4] demonstrated that the breakdown features are significantly influenced by flow factors. For laminar flow cases below the critical jet velocity and above the critical jet velocity, impact waves are created at the point of impingement, resulting in closed rim or periodic drop spray. It is discovered that the breakup length is inversely proportional to the jet's velocity and impingement angle. However, the relationship between the sheet width and the jet velocity and impingement angle is straightforward. With an increase in sheet thickness and a drop in wave number, larger ligaments are created. Further beyond the injection plane, secondary atomization takes place.

Ryan et al. [5] studied the atomization characteristics of like-doublet injectors for different operating conditions, varying jet velocity and geometry of the injector. Results were consistent with the findings of Heidmann et al. [3], as well as Dombrowski and Hooper [4]. The wavelengths of impact waves and ligaments are directly proportional to the orifice diameter and are independent of jet velocity and the impingement angle.

Ibrahim and Przekwas [6] characterized the sheet breakup process for a wide range of Weber numbers, proposing detailed regime charts. They have also quantified the liquid sheet parameters. Sheet thickness is found to be inversely proportional to the distance from the impingement point. Anderson et al. [7] found that the breakup characteristics depend on the flow parameter. In the case of laminar jets, the sheet breakup length is dependent on the jet Weber number, whereas it is independent of the Weber number in the case of turbulent jets. It is also observed that the drop sizes are decreasing with the increase in Weber numbers.

Jung et al. [8] used water and kerosene for their cold flow experiments to characterize the breakup of liquid sheets and ligaments under atmospheric conditions. Their findings indicate that an increase in both turbulence strength and impact force shortened the breakup lengths and wavelengths of the liquid sheet and ligaments.

Inoue et al. [2] did both theoretical and experimental findings on doublet and triplet injectors for predicting the performance of bipropellant thrusters. The theoretical models to predict the C^* efficiency are in good agreement with the corresponding combustion test results with a 2% relative error.

Sweeney and Frederick [9] performed cold flow experiments under atmospheric ambient conditions for like-doublet injectors to study the effect of jet breakup length to impingement distance ratio. If the ratio is greater than one, a steady flat sheet is formed, and when the ratio is equal to one, an unsteady flat sheet is formed.

Even though detailed experimental characterization works are reported in the case of like-on-like impinging injectors, investigations with unlike impinging injectors are quite limited. In the case of unlike impinging injectors, the atomization mechanisms are mainly driven by the impact waves, and correlations for breakup lengths and sheet parameters are not available. The present work aims to investigate the primary atomization processes in unlike impinging injectors and to quantify different linear sheet parameters with respect to the variation in average Weber number.

3 Materials and Methods

3.1 *Test Conditions*

In the present work, experiments are conducted using an atmospheric test rig with water as the test fluid for both oxidizer and fuel. A brief schematic of the test rig is shown in Fig. 2. The injector has two manifolds, corresponding to fuel and oxidizer outlet lines. Two liquid lines (fuel and oxidizer) are connected to a storage tank where the flow is pressurized using nitrogen air and regulated using the flow control valve. Flow rate and injection pressure are controlled using needle valves and pressure regulators. The liquid is collected using a collection tank and pumped back into the storage tank.

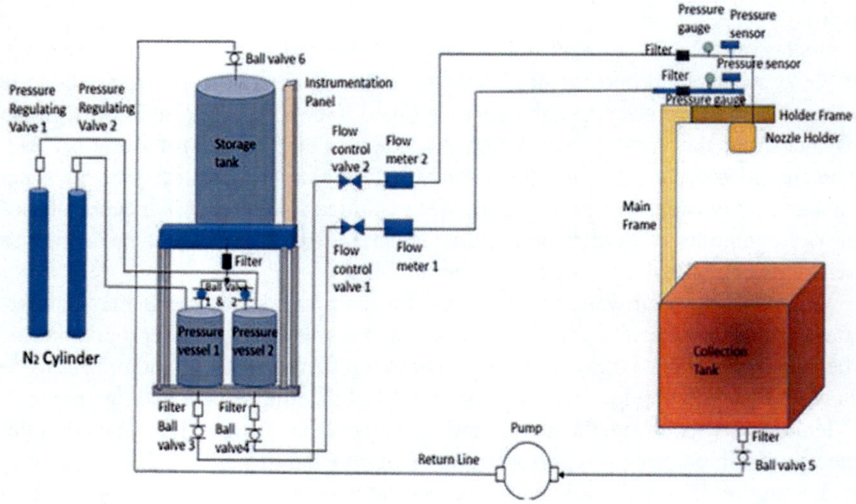

Fig. 2 Schematic diagram of the spray test rig

Table 1 Test conditions

Case no.	Fuel (Kg/s)	Oxidizer (Kg/s)	Fuel velocity (m/s)	Oxidizer velocity (m/s)	Average Weber number
1	0.005	0.0125	2.20	2.76	188
2	0.006	0.015	2.64	3.32	270
3	0.008	0.02	3.52	4.42	480
4	0.01	0.025	4.41	5.53	748
5	0.012	0.03	5.29	6.63	1078
6	0.014	0.035	6.17	7.74	1468
7	0.016	0.04	7.05	8.84	1916
8	0.018	0.045	7.93	9.95	2426
9	0.02	0.05	8.81	11.05	2994

For a constant mixture ratio of 2.5 and a momentum ratio of 3.14, the liquid flow rate for fuel is 0.005–0.02 kg/s, and for oxidizer, it is 0.0125–0.05 kg/s. The fuel jet velocity varied from 2.20 to 8.81 m/s and the oxidizer jet velocity from 2.76 to 11.05 m/s. A total of 9 cases are identified, and the average Weber number, We_{avg} (average of fuel and oxidizer Weber numbers), varied from 188 to 2994. Test conditions are given in Table 1.

3.2 Diagnostic Techniques

In the present work, the liquid sheet breakup characteristics are investigated using Planar laser induced fluorescence (PLIF) imaging. PLIF is a laser diagnostic technique in which a seeding dye (Rhodamine 6 G with a concentration of 5 mg/lit) is introduced to water. The mixture is excited by an ND:YAG laser at 532 nm (green) wavelength, and the fluorescence signal is collected at 556 nm using a narrow band-pass filter attached to the camera [10]. A CCD camera (PCO Pixelfly make) with a telescopic zoom lens (Sigma make) is used to capture the PLIF images. Both laser and camera are synchronized using a sequencer and the images are acquired at 10 Hz. Images are captured using a commercial image acquisition software 'Photron FASTCAM'. A field of view of 100 mm × 100 mm is selected and the camera resolution is 1024 × 1024 pixels. The scaling factor is 9.7 pixels/mm for breakup length, sheet width, and 10.95 pixels/mm for sheet thickness. The field of view is illuminated using a laser light sheet of thickness 1 mm, which is generated using a sheet optics set.

Two experimental configurations are used to capture the sheet breakup processes. Using the front view imaging, the planar sheet characteristics such as breakup length, sheet width, etc., are captured as shown in the schematic diagram Fig. 3. With the side view imaging, the sheet thickness is captured. Experimental arrangements are shown in Fig. 4.

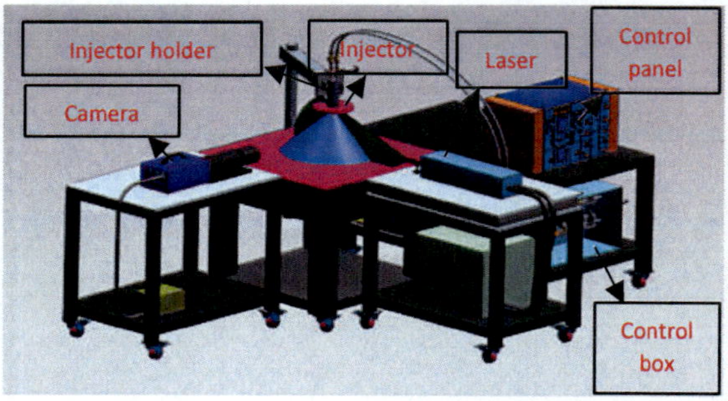

Fig. 3 Schematic representation of experimental arrangements

Fig. 4 Experimental arrangements

An image processing software is developed with a MATLAB GUI to process the images. The package contains options to do various image pre-processing tools such as edge detection, sharpening, contrast variations, etc. The software can measure linear parameters such as breakup length, sheet width, sheet thickness, etc. The codes are also validated with the commercial software DAVIS from LaVision.

Breakup length and sheet width are measured for Cases 1–6. The liquid sheets having locations of discontinuous intensities are identified for each local streamwise distance and the difference in these locations is measured to estimate the local breakup length. Similarly, the radial distance (perpendicular to the streamwise direction) is tracked and the maximum width is reported as the sheet width. The sheet thickness is measured by estimating the distance between the points in the row-wise direction where there is a sudden change in pixel intensity gradient. The variation of sheet thickness is tracked starting from the impinging point to the breakup point.

4 Results and Discussion

4.1 Sheet Breakup Regimes

The operating conditions are varied from Cases 1 to 9 by increasing the fuel and oxidizer line flow rates. The momentum ratio is held constant such that the liquid sheet makes a constant impingement angle of $2\,\theta = 87°$, similar to that in practical injectors. The nozzle holder is rotated, such that the sheet ensues exactly vertically and the laser sheet slices the sheet along the radial direction.

Front view and side view imaging provide details on the sheet propagation and primary atomization processes. Instantaneous images showing the liquid sheet breakup process are shown in Figs. 5 and 6. For a constant mixing ratio and momentum ratio, the atomization process depends on the average Weber number and four different sheet break up regimes are identified as listed below.

Flow regimes	We_{avg}
· Open rim breakup	188–480
· Wavy sheet breakup	480–1078
· Perforated sheet breakup	1078–1916
· Wavy and perforated breakup	1916–2994

At lower Weber numbers ($We_{avg} = 188–480$), a flat liquid sheet is formed with a closed rim structure. The liquid accumulates at the sheet tip, along the center line forming spherical liquid blobs and ligaments. These ligaments are shed periodically from the tip. As the Weber number is increased, an open rim structure is formed, with waves emerging from the point of impingement, due to the different momentum of fuel and oxidizer. Impact waves are observed at the impingement point and the amplitude of waves grows as the flow proceeds downstream. Ligaments are originated from the sheet tip and from the periphery as a result of the aerodynamic interaction with the ambient air. Regular shedding of ligaments along the edges of the sheet takes place, which breaks up into droplets.

As the We_{avg} is increased, the increase in jet momentum makes the sheet more unstable and the impact wave oscillations become more predominant.

This regime is characterized as the wavy sheet breakup regime ($480 < We_{avg} < 1078$).

Further increase in the flow rates ($1078 < We_{avg} < 1916$) results in a perforated sheet breakup. At this condition, the liquid sheet gets accelerated and spreads radially. The surface tension forces cannot hold the sheet surface locally which results in the perforation of the liquid sheet at multiple locations.

Further increase in the Weber numbers (>1916), results in rapid atomization near the injector exit plane. The impact wave amplitude increases and the wavelength decreases. This results in the breakup of the sheet in the crest region (Figs. 5 and 6 corresponding to Cases 7–9). Furthermore, the sheet breaks by the perforations

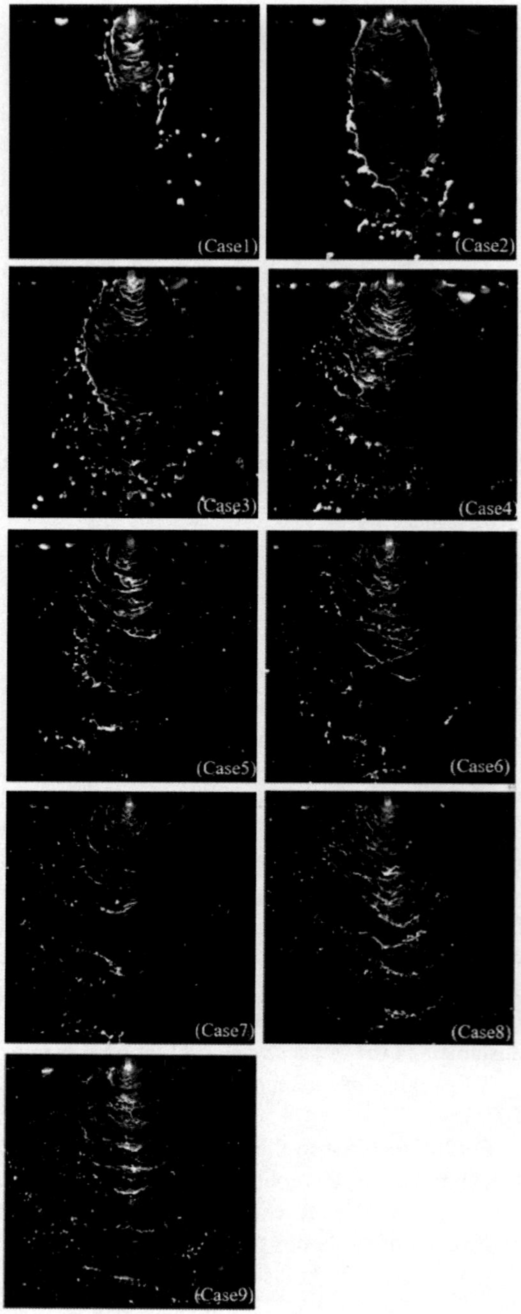

Fig. 5 Instantaneous PLIF images (front view imaging)

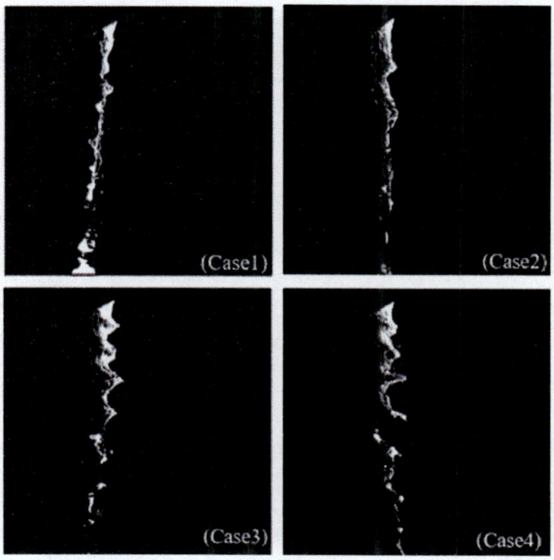

Fig. 6 Instantaneous PLIF images (side view imaging)

at multiple locations and also at the sheet periphery. This regime is identified as a combined perforation and wavy sheet breakup regime.

4.2 *Linear Sheet Parameters*

In the present work, linear sheet parameters, namely: sheet breakup length, sheet width, and sheet thickness are quantified at selected operating conditions. The sheet topology and sheet parameters are represented in Fig. 7.

The sheet breakup length is defined as the distance along the sheet from the point of impingement to the point where the sheet disintegrates forming ligaments and then droplets. Image processing is performed as explained in the previous Sect. 3.2. The

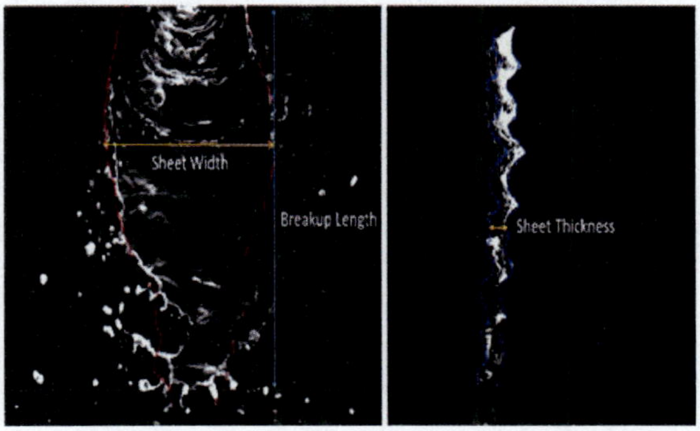

Fig. 7 Sheet topology

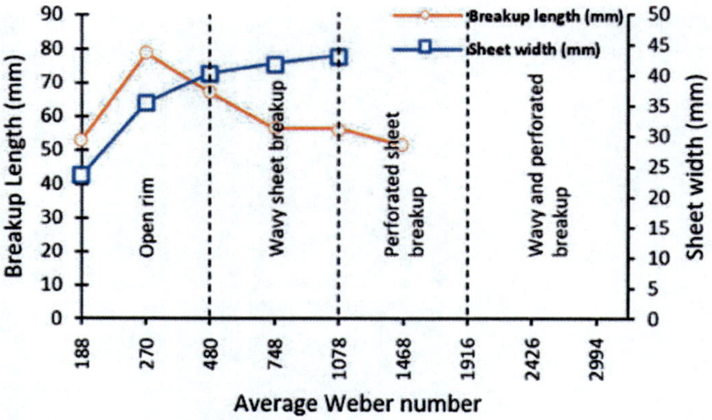

Fig. 8 Breakup length as a function of average Weber number

average breakup length for 500 images is estimated for We_{avg} ranging from 188 to 1468, i.e., Cases 1–6. The variation in breakup length with the average Weber number is plotted in Fig. 8. Different breakup regimes are also marked in the same figure. The sheet breakup length was found to increase up to Case 2 and then decreases as the We_{avg} increases. Up to the Case 2 condition, the surface tension force holds the liquid sheet and prevents its breakup, which showed an increase in the breakup length.

For higher Weber numbers (Cases 3–6), the higher momentum of the jets accelerates the liquid sheet and the impact waves become unstable resulting in a gradual reduction of breakup length. It becomes difficult to estimate the breakup length beyond Case 6 (in perforated and wavy sheet breakup regimes).

The width of the sheet is the maximum lateral spread of the sheet before breakup. An average of 500 images are taken for We_{avg} varying from 188 to 1078, i.e., cases 1–5. The sheet width increases linearly up to $We_{\text{avg}} = 1078$, as shown in Fig. 8. As the Weber number increases further, the sheet formation is difficult to identify since the disintegration of sheet takes place near to the point of impingement.

The liquid sheet thickness is identified as the thickness of the liquid sheet in the row-wise direction and is captured from impinging point along the flow direction. Side view imaging is used for measuring sheet thickness for different cases and the variation of sheet thickness along the flow direction is shown in Fig. 9. The sheet thickness is found to be decreasing along the flow direction. As the Weber number is increased, the liquid flow rate increases in both injectors and as a result thickness of the sheet increases near the impingement point. However, the higher momentum at higher Weber numbers results in rapid growth of impact waves with lateral spreading and dominance of perforated sheet breakup. This results in a faster reduction of liquid volume and lesser thickness as the flow proceeds downstream. In Case 1, an increase in sheet thickness is observed after 60 mm. The accumulation of the liquid blobs at the tip due to the closed rim type breakup appears as a higher sheet thickness in the side view imaging.

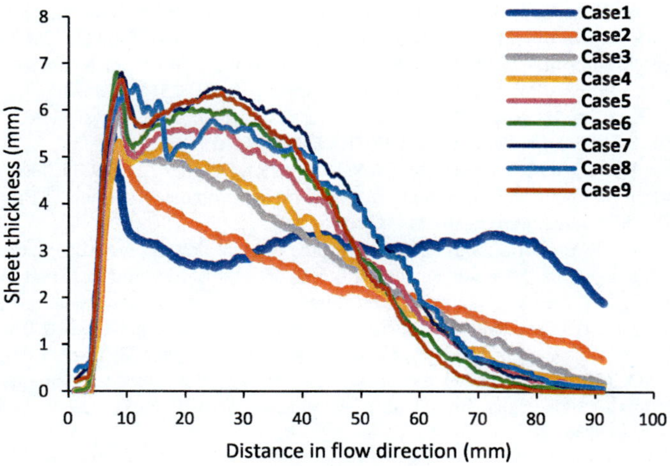

Fig. 9 Sheet thickness variation along the flow direction

5 Conclusion

The present work investigates the primary atomization processes in the case of like-on-unlike impinging injectors using the PLIF imaging technique. Different sheet breakup regimes and atomization processes are characterized using front and side view imaging and regime charts are presented. The propagation of impact waves dictates the sheet breakup process and successive ligament formation. Perforated and wavy mode of breakup becomes predominant at higher Weber numbers. Different linear sheet parameters such as breakup length, sheet width, and sheet thickness are quantified at low Weber number cases. Further works are planned to be performed for a detailed quantification of the spray parameters such as ligament ejection velocity, droplet size, etc., and also to investigate the unsteady impact wave dynamics.

Acknowledgements Authors wish to acknowledge IPRC and ISRO for the funding support. NCCRD is supported by the Department of Science and Technology, India.

References

1. Ashgriz N (ed) (2011) Handbook of atomization and sprays. Springer US, Boston, MA. https://doi.org/10.1007/978-1-4419-7264-4
2. Inoue C, Oishi Y, Daimon Y, Fujii G, Kawatsu K (2021) Direct formulation of bipropellant thruster performance for quantitative cold-flow diagnostic. J Propuls Power 37(6):842–849. https://doi.org/10.2514/1.B38310
3. Heidmann MF, Priem RJ, Humphrey JC (1957) A study of sprays formed by two impinging jets. UNT Digital Library, March 1957. https://digital.library.unt.edu/ark:/67531/metadc56084/m1/3/. Accessed 31 Mar 2022

4. Dombrowski ND, Hooper PC (1964) A study of the sprays formed by impinging jets in laminar and turbulent flow. J Fluid Mech 18(3):392. https://doi.org/10.1017/S0022112064000295
5. Ryan HM, Anderson WE, Pal S, Santoro RJ (1995) Atomization characteristics of impinging liquid jets. J Propuls Power 11(1):135–145. https://doi.org/10.2514/3.23851
6. Ibrahim EA, Przekwas AJ (1991) Impinging jets atomization. Phys Fluids Fluid Dyn 3(12):2981–2987. https://doi.org/10.1063/1.857840
7. Anderson WE, Miller KL, Ryan HM, Pal S, Santoro RJ, Dressler JL (1998) Effects of periodic atomization on combustion instability in liquid-fueled propulsion systems. J Propuls Power 14(5):818–825. https://doi.org/10.2514/2.5345
8. Jung K, Khil T, Yoon Y (2006) Effects of orifice internal flow on breakup characteristics of like-doublet injectors. J Propuls Power 22(3):653–660. https://doi.org/10.2514/1.20362
9. Sweeney BA, Frederick RA (2016) Jet breakup length to impingement distance ratio for like doublet injectors. J Propuls Power 32(6):1516–1530. https://doi.org/10.2514/1.B36137
10. Jardón-Pérez LE, Amaro-Villeda A, González-Rivera C, Trápaga G, Conejo AN, Ramírez-Argáez MA (2019) Introducing the planar laser-induced fluorescence technique (PLIF) to measure mixing time in gas-stirred ladles. Metall Mater Trans B 50(5):2121–2133. https://doi.org/10.1007/s11663-019-01631-y

Preliminary Analysis and Design of Turborocket Combustion Chamber

Sanjay Bhalekar and Rohit Sharma

Abstract The focus of this research is to provide a preliminary design of the combustion chamber of the turborocket engine. The present study delineates the specific steps for designing a combustion chamber. Firstly, the expected chemical composition was calculated by Gibbs Minimization. Knowing the composition of the mixture, the physical properties like density, the ratio of specific heats, etc., are calculated at temperatures that have been obtained from Glassman [2]. Then, the time frame of the reaction was calculated. The dimensions of the combustion chamber were calculated after obtaining all the above information. A parametric analysis was performed to characterize the behavior of the combustion chamber when its geometrical factors like flameholder positions and length are changed. The novelty of this study lies in the fact that there has not been any study that provides a complete framework for designing a combustion chamber and at the same time focuses on turborocket engine. The preliminary results show that the length leads to very significant changes in temperature wherein temperature increases up to a point after which it decreases as a result of completion of the reaction.

Keywords Combustion chamber · Flameholder · Parametric analysis · Geometry · Gibbs minimization

1 Introduction

The use of hydrogen for energy production is going to be the dominant way in the future as justified by Nazir et al. [1]. The propulsion industry can also benefit from this use, and turborocket combustion chamber design provides an effective way for utilizing this opportunity. The designing starts with knowing the fuel and getting its

S. Bhalekar · R. Sharma (✉)
School of Mechanical Engineering, VIT Bhopal University, Sehore 466114, Madhya Pradesh, India
e-mail: rohit.sharma@vitbhopal.ac.in

S. Das et al. (eds.), *Proceedings of the 1st International Conference on Fluid, Thermal and Energy Systems*, Lecture Notes in Mechanical Engineering,
https://doi.org/10.1007/978-981-99-5990-7_59

flame temperature, pressure, and equivalence ratio of the fuel-oxidizer which can be found by the standard table given by Glassman [2].

After this we use Mcbride [3] to find the chemical composition at equilibrium. Knowing the composition the properties of the resulting mixture are calculated which include density, ratio of specific heat, gas constant, etc. The length of the combustion chamber can be known by calculating the evaporation and reaction time. Zhang et al. [4] have provided a template for Hydrogen–Oxygen reaction time. After knowing all the previous values of reaction time, composition, and flame temperature and calculating the properties of the resulting mixture, we can calculate the combustion chamber geometries by Farokhi [5] and then simulations are performed on it. The parametric analysis is performed in order to characterize the behavior of the combustion chamber due to its change in some of the geometric features.

2 Literature Review and Objective

The objective of the following study is to provide a simple framework for designing a combustion chamber and demonstrating it. The process used here would be Gibbs Minimization to get an idea of the resulting composition which is obtained from Mcbride [3] where the method has been introduced including appropriate end conditions and steps for the iterative method they have provided. This is done at the temperature obtained from Glassman [2] where he gives results of different combustion mixtures and the conditions at which they achieve combustion. The calculation of time frame of the reaction is then achieved by. Zhang et al. [4] in which he forms a basic universal study of hydrogen- oxygen combustion and compares it with several distinct models. After that, geometry of combustion chamber from Farokhi [5] is calculated. The final design is taken from parametric analysis where lowest temperature design is selected as it simplifies turbine designing in turborocket engine.

3 Materials and Methods

The combustion chamber has the function of ignition of reactants. In the design of the turborocket, Hydrogen and Oxygen are made to react. The flame temperature and pressure are assumed considering the following Table 1.

This above information is obtained by Ref. [2]. This minimum $\Phi = 1.3$ and the maximum $\Phi = 1.7$.

$$\Phi = \frac{(\text{no. of moles of air/no. of moles of fuel})_{\text{Stoichiometry}}}{(\text{no. of moles of air/no. of moles of fuel})}$$

Φ here denotes the equivalence ratio where it equals the ratio of no. of moles of air by fuel. So, we first consider the reaction at stoichiometry.

Table 1 Maximum flame temperature of fuel–air and fuel–oxygen mixtures at 25 °C temperature and 1 or 20 atm pressure. Approximate flame temperatures of various stoichiometric mixtures, critical temperature 298 K

Fuel	Oxidizer	Pressure (atm)	*T*(K)
Acetylene	Air	1	2600
Acetylene	Oxygen	1	3410
Carbon monoxide	Air	1	2400
Carbon monoxide	Oxygen	1	3220
Heptane	Air	1	2290
Heptane	Oxygen	1	3100
Hydrogen	Air	1	2400
Hydrogen	Oxygen	1	3080
Methane	Air	1	2210
Methane	Air	20	2270
Methane	Oxygen	1	3030
Methane	Oxygen	20	3460

$$H_2 + 1/2O_2 \leftrightarrow n_{H_2O}H_2O + n_{H_2}H_2 + n_{O_2}O_2 + n_{OH}OH + n_OO + n_HH$$

The reaction is chosen to take place at $\Phi = 1.5$ due to the reason that it lies midway in the range and hence reduces the chances of a blowoff.

$$H_2 + 0.3333\ O_2 \leftrightarrow n_{H_2O}H_2O + n_{H_2}H_2 + n_{O_2}O_2 + n_{OH}OH + n_OO + n_HH$$

We will get the following value of products by Gibbs minimization. We will have the assumed temperature given in Table 1 as 3080 K at a pressure 1 atm pressure. According to Gibbs Minimization, 3 iterative equations govern the reaction at a given temperature and pressure.

$$\sum_{i=1}^{l}\sum_{j=1}^{l} a_{kj}a_{ij}n_j\pi_i + \sum_{j=1}^{NG+1} a_{kj}\Delta n_j\left(\sum_{j=1}^{NG} a_{kj}n_j\right)\Delta\ln n + \left(\sum_{j=1}^{NG}\frac{a_{kj}n_jH_j^0}{\text{RT}}\right)\Delta\ln T$$

$$= b_k^0 - b_k + \sum_{j=1}^{NG}\frac{a_{kj}n_j\mu_j}{\text{RT}}(k = 1,\ldots,l)$$

$$\sum_{i=1}^{NG} a_{ij}\pi_i + H_j^0\Delta\ln T = \frac{\mu_j}{\text{RT}}(j = NG+!,\ldots,NS)\sum_{i=1}^{l}\sum_{j=1}^{NG} a_{ij}n_j\pi_i$$

$$+\left(\sum_{j=1}^{NG} n_j - n\right)\Delta\ln n + \left(\sum_{j=1}^{NG}\frac{n_jH_j^0}{\text{RT}}\right)\Delta\ln T$$

$$= n - \sum_{j=1}^{NG} n_j + \sum_{j=1}^{NG}\frac{n_j\mu_j}{\text{RT}}$$

where:

$$\pi = \frac{\lambda}{\mathrm{RT}}$$

$$b = \sum_{j=1}^{NS} a_{ij} n_j$$

The above equations are obtained from Ref. [3]. From the properties given in Table 2, the heat transfer Q = 104202517.3 J and the ratio of specific heats are calculated.

The ratio of specific heats is obtained by,

$(n_1 + n_2 + n_3 + n_4)C_v = n_1 C_{v1} + n_2 C_{v2} + n_3 C_{v3} + n_4 C_{v4}$

$C_v = x_1 C_{v1} + x_2 C_{v2} + x_3 C_{v3} + x_4 C_{v4}$

$C_p = x_1 C_{v1} + x_2 C_{v2} + x_3 C_{v3} + x_4 C_{v4} + R$

$$\gamma = \frac{C_p}{C_v} = 1 + \frac{R}{x_1 C_{v1} + x_2 C_{v2} + x_3 C_{v3} + x_4 C_{v4}}$$

$$\gamma = 1 + \frac{R}{\frac{x_1}{\gamma_1 - 1} + \frac{x_2}{\gamma_2 - 1} + \frac{x_3}{\gamma_3 - 1} + \frac{x_4}{\gamma_4 - 1}}$$

$R_i = 6.472316$ atm · L/mol · K

$R_e = 3.25212439$ atm · L/mol · K

Table 2 Gibbs minimization table for Initial and final properties

	Initial state	Equilibrium state
Pressure (atm)	1.0000E+00	1.0000E+00
Temperature (K)	3.0800E+03	3.0800E+03
Volume (cm^3/g)	2.6573E+04	2.2404E+04
Enthalpy (erg/g)	9.8928E+10	1.5238E+10
Internal energy (erg/g)	7.2003E+10	−7.4630E+09
Entropy (erg/gK)	2.4063E+08	2.2659E+08

	Initial state		Equilibrium state	
	Mole fraction	Mass fraction	Mole fraction	Mass fraction
H_2	7.5002E-01	1.5897E-01	3.0580E-01	5.4652E-02
O_2	2.4998E-01	8.4103E-01	9.0477E-03	2.5666E-02
H_2O	0.0000E+00	0.0000E+00	1.3947E-02	1.9783E-02
OH	0.0000E+00	0.0000E+00	1.1029E-01	9.8555E-03
H	0.0000E+00	0.0000E+00	6.4682E-02	9.7523E-02
O	0.0000E+00	0.0000E+00	4.9623E-01	7.9252E-01

Substituting the values given below we get (Table 3)

$$\gamma_e = 2.369425 \text{ and } \gamma_i = 3.162759$$

After this for the calculation of the flame speed whether theoretical or experimental method can be used. The flame speed of the Hydrogen–Oxygen mixture is taken as 11.75 m/s as it won't lead to a flame-out scenario. The flame speed can be theoretically calculated by

$$U = \sqrt{\frac{2\lambda P Z a_{\text{eff}} b_{\text{eff}}}{a_0 \rho_0 \overline{c}_p (T_f - T_0)} \left(\frac{\lambda}{c_p D \rho}\right)_f^2 \left(\frac{n_T}{n_p}\right)^2 \left(\frac{R T_f^2}{E}\right) \exp\left(-\frac{E}{R T_s}\right)}$$

where, for stoichiometric air fuel-rich mixture ($\varnothing >= 1$), k_{eff} and b_{eff} are computed:

$$a_{\text{eff}} = a_0 \left(\frac{\mathrm{T}_0}{\mathrm{T}_\mathrm{f}}\right) \left\{1 - \left(\frac{1}{\varnothing}\left(1 - \frac{\frac{R T_f^2}{E}}{(T_f - T_0)}\right)\right)\right\}$$

$$b_{\text{eff}} = b_0 \left(\frac{T_0}{T_f}\right) \left(\frac{\frac{R T_f^2}{E}}{T_f - T_0}\right)$$

$$\lambda = \left(C_p + \frac{5}{4} R\right) \frac{\eta}{M}$$

$$\eta = \sum_i \eta_j N_j$$

$$C_p = \sum_j C_{pj} N_j$$

$$p = \frac{\rho}{\mathrm{MRT}}$$

$$Z = \left(\frac{\sigma_1 + \sigma_2}{2}\right)^2 \sqrt{8\pi R T \frac{M_1 + M_2}{M_1 M_2}}$$

Table 3 Table for ratio of specific heats of different chemicals

Temperature (°C)	Gas	The ratio of specific heat
2000	H_2	1.318
200	H_2O	1.310
400	O_2	1.394
20	OH	1.384
(1 + (2/*n*)) (*n*-degree of freedom) (*n* = 3)	H	1.666
(1 + (2/*n*))	O	1.666

The above equation is obtained from Dugger and Graab [6].

The remaining factor P, used herein as a semi-empirical factor, is found by averaging the values calculated by substituting the experimental values of maximum flame velocity for a given fuel and temperature into the laminar velocity equation.

To design the combustor boundaries the information related to residence time should be available to the designer. So, we have

$$t_i = t_e + t_{\text{reaction}}$$

The following equation is obtained from Farokhi [5].

The subscript "i" means stands for ignition delay, and "e" stands for evaporation timescale. Time of reaction can be found using Fig. 1.

Fig. 1 Arrhenius plot of overall rate constant data derived from the experimental data [symbols represent experimental data; the line is a least-squares fit of the form $k_G = A \exp(-Ea/RT)$]: **a** $C_{\text{hydrogen}} < 20\%$; **b** $C_{\text{hydrogen}} > 20\%$

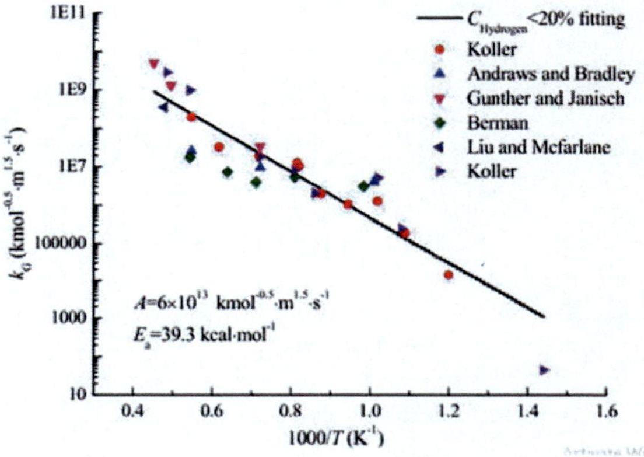

(a) $C_{\text{hydrogen}} < 20\%$

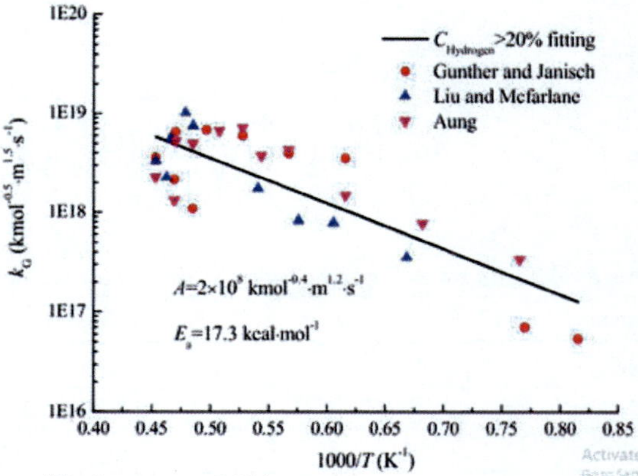

(b) $C_{\text{hydrogen}} > 20\%$

$$t_{\text{reaction}} = [x]/w = p/w\text{RT}$$

$$w_{c(\text{hydrogen})<20\%} = k_G[\text{H}_2]^1[\text{O}_2]^{0.5}$$

$$k_{\text{Gc(hydrogen)}<20\%} = 6 \times 10^{13}\exp(-1.65\text{E}8/\text{RT})$$

$$w_{c(\text{hydrogen}>20\%} = k_G[\text{H}_2]^{0.2}[\text{O}_2]^{1.2}$$

$$k_{\text{Gc(hydrogen)}>20\%} = 2.6 \times 10^{10}\exp(-0.72\text{E}8/\text{RT})$$

The above information is obtained from Zhang et al. [4].

Now after obtaining all the values we calculate combustion chamber area. Taking subscript *i* for "initial" and *e* for "exhaust". So the area is given by:

$$A = \frac{1 + \gamma_i M_i^2 \left(1 - \frac{C_D}{2}\right)}{\gamma_i M_i} \sqrt{\frac{\gamma_i - 1}{1 + \left(\frac{\gamma_i - 1}{2} M_i^2\right)} \left(\frac{1}{1 + \left(\frac{q}{c_{pi} T_{ti}}\right)}\right)}$$

The above equation is obtained from Farokhi [5].

The length of the combustion chamber would be obtained as

$$L_{CC} = \frac{\dot{m}_{gas} t_{gas}}{\rho_{cc} A_{CC}}$$

The above equation is obtained from Farokhi [5].

For viscosity, a method can be followed.

The following method is Gambill Method.

$$\vee^{1/3} = x_a \vee_a{}^{1/3} + x_b \vee_b{}^{1/3}$$

Gas constant of the mixture can be calculated as

$$R_{\text{mixture}} = \frac{R_{\text{H}_2} m_{\text{H}_2} + R_{\text{O}_2} m_{\text{O}_2}}{m_{\text{H}_2} + m_{\text{O}_2}}$$

Flameholder Simulation

One of the important values required for designing the combustion chamber is the coefficient of drag. The inlet velocity (Flame speed of hydrogen) is taken as 15 m/s and the diameter is taken as 1.128 m.

The viscosity is taken as 3.906292 e-6 and C_d is taken as 0.48 (Fig. 2).

Mesh Independence Study

See Fig. 3 and Table 4.

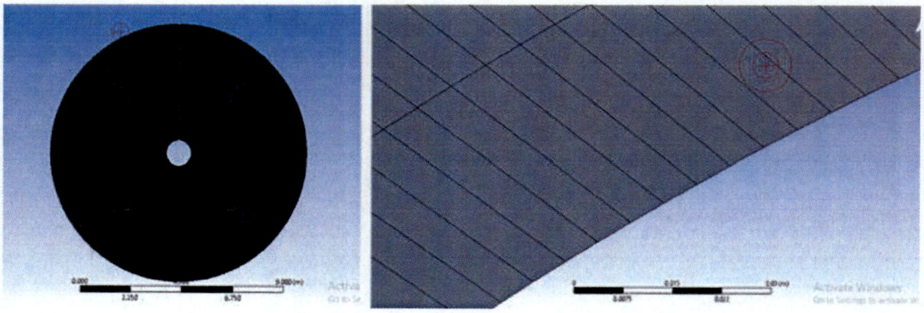

Fig. 2 Geometry and near boundary mesh

Fig. 3 Mesh independence analysis graph

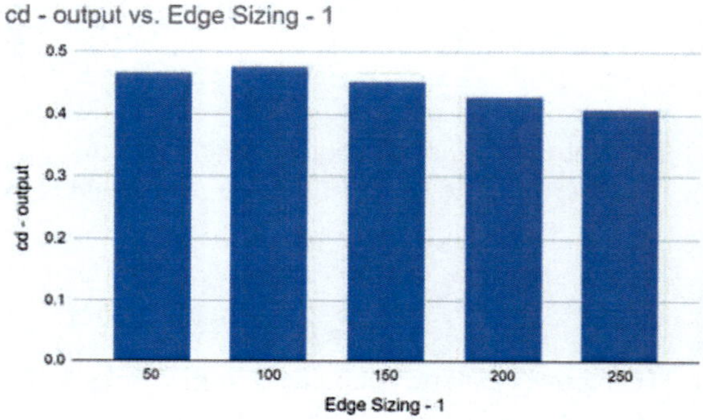

Table 4 Mesh independence analysis study table

Name	Edge sizing-2	Edge sizing-1	*cd*-output
DP1	150	50	0.467244
DP2	150	100	0.478466
DP3	150	150	0.453274
DP4	150	200	0.42893303
DP5	150	250	0.4088

Combustion Simulation

The flameholders are used as the inlets for Oxygen and Hydrogen inlets are taken at the extreme left. The fuel used are hypergolic and ignite directly on contact. The mesh is similar to the previous as it has proved to be accurate by mesh independence study (Table 5 and Figs. 4, 5, 6, 7).

Table 5 Parametric analysis study

#	XYPlane.H40 (m)	XYPlane.H21 (m)	XYPlane.H42 (m)	Pressure-op (Pa)	Density-op (kg m^{-3})	Velocity-op (m $s-^{1}$)	Temperature-op (K)
Name	P1	P3	P4	P5	P7	P8	P9
DP 0	2	8	2	−0.633	0.290	8.720	993.824
DP 1	2.5	8	2.5	−0.862	0.312	8.923	952.455
DP 2	3	8	3	−1.022	0.337	9.299	941.150
DP 3	2	9	2	−0.017	0.288	12.403	1521.135
DP 4	2.5	9	2.5	−0.922	0.252	9.141	997.745
DP 5	3	9	3	−0.021	0.257	12.902	1678.092
DP 6	2	10	2	−1.207	0.200	11.312	1387.370
DP 7	2.5	10	2.5	−1.006	0.199	10.987	1300.057
DP 8	3	10	3	−0.820	0.196	10.575	1262.237

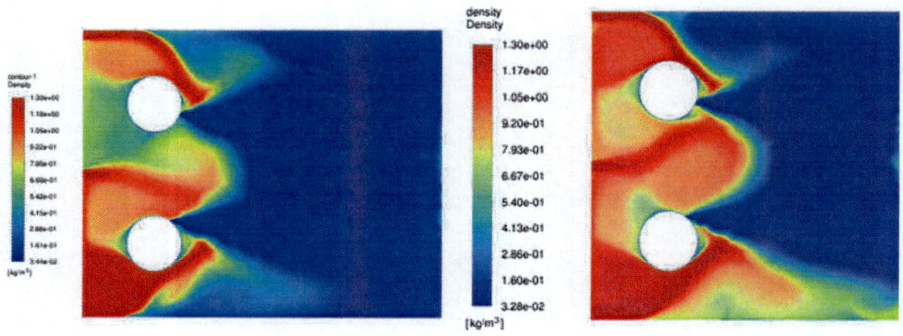

Fig. 4 Density contours (8 m left, 10 m right)

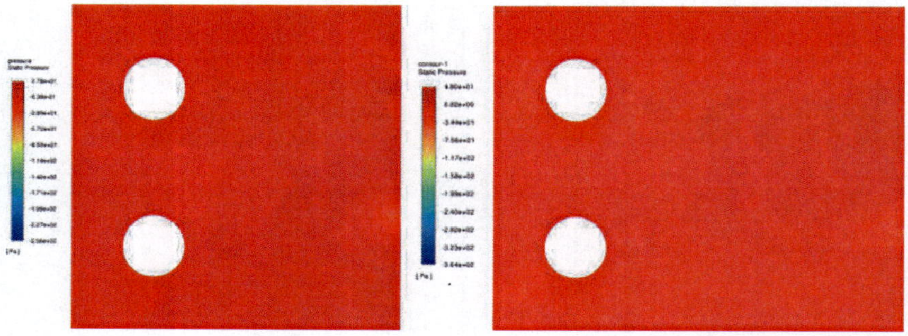

Fig. 5 Pressure contours (8 m left, 10 m right)

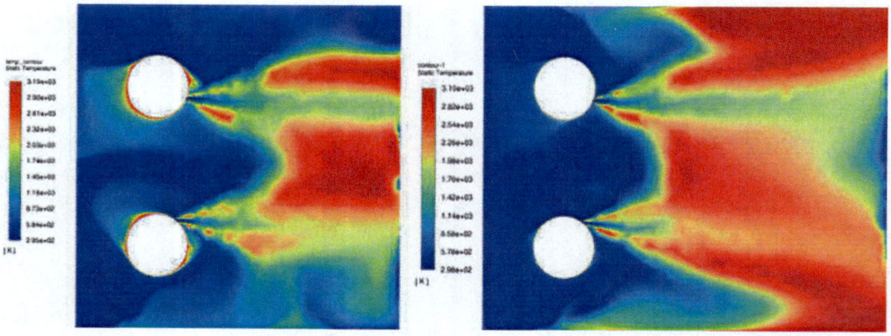

Fig. 6 Temperature contours (8 m left, 10 m right)

We can see a considerable reduction in the 10 m combustion chamber length and can conclude that at this point in time, the reactants and products react thoroughly as there is a significant decrease in temperature and hence can be sent to the turbine stage for power generation. The contours can also be compared to conclude the same

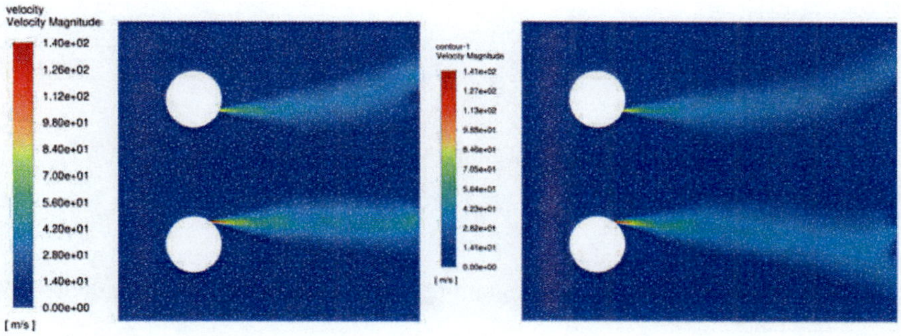

Fig. 7 Velocity contours (8 m left, 10 m right)

as it can be seen that there is no flame in temperature contour and right side in the middle indicating the completion of the reaction.

4 Results and Discussion

The design began by focusing on the resulting mixture which was obtained by Gibbs minimization. The geometry of the combustion chamber was calculated by calculating the physical properties. As it can be seen that there is a significant rise in the temperature due to flame spreading events. Also, there is a sharp change in temperature after combustion reaction ends. As can be observed in the parametric analysis, longer combustion chambers reduce heat load, however, this can cause a significant weight penalty.

5 Conclusion

This study focused on designing and analysis of the turborocket combustion chamber. The following framework is successfully used to design a combustion chamber. The method is thoroughly outlined and has been used exclusively. The calculation by Gibbs Minimization gives us the properties of the resulting mixture. Using this we calculate the properties of the resulting mixture to get the combustion chamber geometry. The results of parametric analysis have been produced by varying the geometry of the combustion chamber. The desirable values where there is a drop in temperature have been obtained by this method.

Acknowledgements The authors would like to express their sincere gratitude to VIT Bhopal University for providing essential support for the simulation work.

Nomenclature

Ø	Equivalence ratio
a	Stoichiometric coefficient
n	Number of moles of kilogram-moles
λ	Lagrangian multiplier
T	Temperature
H^0	Standard molar enthalpy
b^0	Is the assigned number of kilogram-atoms of an element per kilogram of total reactants
μ	Chemical potential per kilogram-mole of species
k	Fuel concentration, molecules/cm^3
b	Oxygen concentration, molecules/cm^3
C_p	Molar heat capacity at constant pressure, cal/(mole)(°K)
c_p	Specific heat, cal/(g)(°K)
$\underline{c_p}$	Mean specific heat, T_0 to T_f, cal/(g)(°K)
$\overline{E}$	Activation energy, kcal/g-mole
M	Molecular weight
N	Mole fraction
R	Gas constant, kcal/(g-mole)(°K) or ergs/(g-mole)(°K)
U	Flame velocity, cm/sec
η	Viscosity of the mixture, poise
σ	Collision diameter, cm
o	Initial condition
1, 2	Reactants in bimolecular reaction
eff	Effective mean reactant concentration
f	Condition at flame temperature
i	ith active particle
j	jth component in a mixture a - fuel concentration, molecules/cm^3
$\dot{m}$	mass flow rate, kg/sec
ρ	Density, kg/m^3
A	Area of the combustion chamber
t	Time (s)
۷	Viscosity
x	Mole fraction
ν	Volume fraction

References

1. Nazir H et al (2020) Is the H2 economy realizable in the foreseeable future? Part I: H2 production methods. Int J Hyd Energy 45(27):13777–13788
2. Glassman I (1987) Combustion, 2nd edn. Academic Press, New York

3. McBride BJ (1996) Computer program for calculation of complex chemical equilibrium compositions and applications, vol 2. NASA Lewis Research Center
4. Zhang Y et al (2017) Numerical simulation of hydrogen combustion: global reaction model and validation. Front Energy Res 5(November). https://doi.org/10.3389/fenrg.2017.00031
5. Farokhi S (2014) Aircraft propulsion. Wiley
6. Dugger GL, Graab DD (1953) Flame velocities of propane-and ethylene-oxygen-nitrogen mixtures. No. NACA-RM-E52J24

Gravitactic Bio-Thermal Convection Oscillates Vertically in a Porous Layer

Srikanth Kummari and Virendra Kumar

Abstract The onset of stationary convection is examined in a homogeneous suspension of gravitactic micro-swimmers in a thermally stratified porous layer subject to vertical oscillation. For the situation of high-frequency and low-amplitude vertical oscillation, the governed model-averaged differential equations are presented. The monotonic stability boundary for stationary convection was obtained using the Galerkin method. The influence of the altered thermal Rayleigh-Darcy number, R_a, on the modified bioconvection Rayleigh-Darcy number, R_b, is numerically examined. Due to porous media, the bioconvection strength is less in results than its non-existence.

Keywords Gravitactic · Stability · Time-averaging technique · Thermal · Porous

Nomenclature

$\tilde{b}$	Vibration amplitude [s]
c_a	Measure of acceleration [–]
c_p	Specific heat at constant pressure [J/kg.K]
D_a	Darcy number [–]
$\overline{D}_c$	Mean Diffusivity of gravitactic swimmers [m^2/s]
$\mathbf{g}$	Gravity [m/s^2]
h	Depth of the layer [m]

S. Kummari (✉) · V. Kumar
Department of Mathematics, School of Mathematics and Computer Sciences, Central University of Tamilnadu, Thiruvarur 610005, India
e-mail: srikanthehow@gmail.com

V. Kumar
e-mail: virendrakumar@cutn.ac.in

S. Kummari
Department of Mathematics, Rajiv Gandhi National Institute of Youth Development, Sriperumbudur, Tamilnadu 602105, India

S. Das et al. (eds.), *Proceedings of the 1st International Conference on Fluid, Thermal and Energy Systems*, Lecture Notes in Mechanical Engineering,
https://doi.org/10.1007/978-981-99-5990-7_60

$\overline{n}$	Mean concentration of the swimmers [mol/m^3]
$\overline{p}$	Averaged pressure [kg/m.s^2]
Pe	Péclet number [–]
$\overline{q}$	The mean fluid velocity [m/s]
q_c	Swimming velocity of the microorganism [m/s]
R_a	Thermal Rayleigh-Darcy number [–]
R_b	Bioconvection Rayleigh-Darcy number [–]
R_t	Thermal vibrational-Darcy parameter [–]
$\mathfrak{R}$	The medium permeability [m^2]
t	Time [s]
$\overline{T}$	The mean temperature [K]
ρ_f	Density of fluid [kg/m^3]
ρ_c	Density of cell [kg/m^3]
$\Delta\rho$	Density difference [kg/m^3]
θ	The average volume a gravitactic particle [m^3]
μ	The dynamic viscosity [kg/m.s]
β	The measure of volume expansion [1/K]
κ	The conductivity of thermal convection [W/m.K]
φ	The porosity of the layer [–]
ω	Frequency of vibration [1/s]
α	Wave number [1/m]

1 Introduction

Bioconvection is the phenomenon that describes the pattern development that occurs spontaneously in aqueous suspensions of microorganisms. Childes et al. proposed a mathematical model for collective response and pattern creation in a layer which contains gravitactic microbes (Tetrahymena pyriformis) [1]. The micro-swimmers respond to various stimuli called taxes, such as gyrotaxis, gravitaxis, phototaxis, and oxytaxis. A geometrical arrangement of gyrotactic swimmers and its theory is addressed by Kessler [2].

Biological applications have inspired many researchers to study the suspension of various micro-swimmers in fluid/porous media under different set-ups [(see References in) 3]. Kuznetsov and Avramenko reported the influence of critical permeability on thermo-bioconvection of micro-swimmers [4]. Gravitational bioconvection in a fluid layer (shallow) was analyzed by Nguyen and coworkers [5]. Kuznetsov et al. examined how vertical through-flow affected the commencement of bioconvection in a suspension of swimming organisms. They discovered that in a composite horizontal porous/fluid layer, vertical through-flow suppresses the onset of bioconvection [6]. Sharma and Kumar investigated the instability of gyrotaxis suspension in a porous layer due to heat effects [7]. Razi and coworkers investigated convection in porous materials under the vibrations impact and developed a solution analytically which describes the relationship for the mono-cellular domain [8]. Due to their uses in vibration production processes in specialized oil sectors, they have explored numerous researches on the suspension of motile micro-swimmers under the impact of vertical vibration. [3, 9, 10]. Researchers investigate vertically vibrated suspension swimming microorganisms in horizontal fluid/porous layers [11–16]. Nield and coworkers reported an analytical study using the stability theory of natural convection onset in the fluid layer containing nano-particles [17]. A numerical analysis of a non-Newtonian fluid containing nano and gyrotactic particles in a porous saturated vertical plate was addressed by Khan and coworkers [18]. They pointed out that nanofluid and bioconvection parameters strongly affect density, Sherwood, and Local Nusselt numbers. An investigation of a suspension of random swimmers (gyrotaxis) in a circular cylinder that rotates horizontally was reported by Pedley in [19]. A stochastic model is utilized to study a suspension of gyrotactic swimmers in a handled environment of uniform vorticity [20]. Experimental results were analyzed to compare the theory of micro-swimmers (gyrotactic) in a pipe flow [21].

An analytic study of porous saturated nanofluid gravitactic thermo-bioconvection subjected to a vertical through-flow, and with gyrotaxis in Hele-Shaw cell was studied by Saini and Sharma [22, 23]. Linear analysis and simulated results were used to present the bioconvection of gravitactic microbes [24]. In both our experimental and simulated studies, Kage et al. found that shape asymmetry contributed many times greater to the upward direction of swimmers (Chlamydomonas reinhardtii) than density asymmetric [25]. Kumar and Srikanth analyzed an oscillatory analysis of the gyrotactic thermo-vibrational convection in an infinite horizontal fluid layer [26]. Despite all these reports, there is no study on the suspension of gravitactic micro-swimmers in a thermally stratified porous medium (low porosity porous layer confined in $z = -h/2$ and $z = h/2$) under the influence of vertical vibration. The Galerkin approach is used to explore the numerical analysis of this problem.

2 Formulation and Stability Analysis:

Consider a homogeneous suspension of gravitactic swimmers trapped in an endless horizontal porous layer with a finite depth, as shown in Fig. 1. The fluid is considered to be incompressible, and the entire horizontal system is exposed to high-frequency, low-amplitude vertical oscillation [11, 26]. The fluid porous layer is supplied to uniform temperatures T_0 and $T_0 + \Delta T$ at the top ($z = h/2$) and no-slip lower ($z = -h/2$) walls, respectively. Then, the governed averaged conservation equations are:

Mass balance

$$\nabla \cdot \overline{q} = 0 \tag{1}$$

Momentum equation

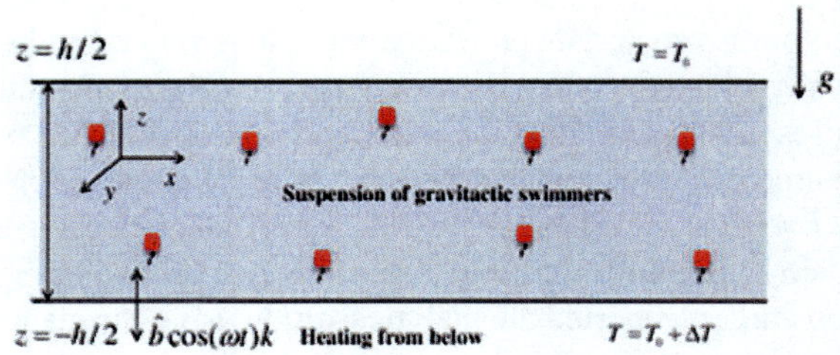

Fig. 1 Schematic diagram of the set-up

$$\begin{aligned}\rho_f(\partial\overline{q}/\partial t)c_a &= -\nabla p - (\mu/\Re)\overline{q} + \mathbf{g}[\overline{n}\Delta\rho\theta - \beta\rho_f\overline{T}]\\ &+ (\theta\beta\Delta\rho/2)\left(\tilde{\mathrm{b}}\omega\right)^2[(\overline{\eta}_1\cdot\nabla)(\overline{T}\mathbf{k}-\overline{\eta}_1) - (\overline{\eta}_2\cdot\nabla)(\overline{n}\mathbf{k}-\overline{\eta}_2)]\\ &+ (\rho_f/2)(\overline{\eta}_1\cdot\nabla)(\overline{n}\mathbf{k}-\overline{\eta}_1)\left[(\Delta\rho/\rho_f)\tilde{\mathrm{b}}\theta\omega\right]^2\\ &- (\rho_f/2)(\overline{\eta}_2\cdot\nabla)(\overline{T}\mathbf{k}-\overline{\eta}_2)\left(\tilde{\mathrm{b}}\beta\omega\right)^2\end{aligned} \quad (2)$$

Conservation of gravitactic swimmers

$$\varphi(\partial\overline{n}/\partial t) = -\nabla.\left[\overline{n}\,\overline{q} + \overline{n}\mathrm{q_c}\mathbf{k} - D_c\nabla\overline{n}\right] \quad (3)$$

Temperature balance

$$\rho_f\left[\partial\overline{T}/\partial t + (\overline{q}\cdot\nabla)\overline{T}\right]c_p = \kappa\nabla^2\overline{T} \quad (4)$$

Here, the quantities in Eq. (1) - (4) indicated by a $'-'$ sign represent the averaged variables, and the nomenclature defines these parameters. Suppose the oscillatory frequency is noticeably high enough, and the period of oscillation is sufficiently short compared to the symptomatic time scales. In that case, it is assumed that the above-averaged system of Eqs. (1)–(4) is valid., i.e. $\tau_{vibrational} << \min\left(\rho_f h^2/\mu, h^2/D_c, c_p\rho_f h^2/\kappa\right)$ [26, 27]. Furthermore, the terms corresponding to the fast oscillating velocity quantities can be disregarded because the vibrational amplitude is so small i.e. $\tilde{\mathrm{b}} << \min\left\{h/\theta(\overline{n}_1-\overline{n}_2)(\Delta\rho/\rho_f) - \beta\Delta\overline{T}\right\}$. The differences in mean concentration and thermal reference across the porous layer's borders are indicated by the variables $(\overline{n}_1-\overline{n}_2)$ and $\Delta\overline{T}$. $\overline{\eta}_1$ and $\overline{\eta}_2$ represent the mean volumetric and convective forces caused by the vertical oscillation, respectively. The $\overline{\eta}_1$ and $\overline{\eta}_2$ vectors are intended to satisfy the Helmholtz decomposition [16, 26]:

$$\left.\begin{aligned} curl\,\overline{\eta}_1 &= grad\,\overline{n}\times\mathbf{k}, \quad div\,\overline{\eta}_1 = 0\\ curl\,\overline{\eta}_2 &= grad\,\overline{T}\times\mathbf{k}, \quad div\,\overline{\eta}_2 = 0\end{aligned}\right\} \quad (5)$$

The fundamental state solution is obtained as [26]:

$$\left.\begin{array}{l}\overline{q}^{b}=\mathbf{0},\quad \overline{n}^{b}(z)=\nu\exp\left(\mathrm{q_c}z/D_c\right)\\ \Rightarrow \nu=\overline{n}^{av}Pe/\left(\exp(-Pe/2)-\exp(Pe/2)\right),\\ \overline{T}^{b}(z)=T_0+\Delta\overline{T}\left[(1/2)-(z/h)\right],\\ \overline{p}^{b}(z)=-\nu g\theta\Delta\rho\left(D_c/\mathrm{q_c}\right)\left[\exp\left(\mathrm{q_c}z/D_c\right)\right]\\ +c+\rho_0 g\Delta\overline{T}\beta\left[(z/2)-\left(z^2/2h\right)\right]\\ \overline{\eta}_1^{b}=\mathbf{0},\quad \overline{\eta}_2^{b}=\mathbf{0},\end{array}\right\} \tag{6}$$

A superscript $'b'$ is added to the quantities to indicate the quiescent stage and $Pe=\mathrm{q_c}h/D_c$ is the Péclet number and c is a constant. Here, $\overline{n}^{av}=1/h\int_{-h/2}^{h/2}\overline{n}^{b}(z)dz$ indicates the mean concentration of the micro-swimmers and constant ν is the particle's primary concentration at the lower wall [11].

To acquire linear stability results, perturbations are implemented to the fundamental state variables. The terms $\overline{T}^{*}\overline{p}^{*}$, $\overline{n}^{*}$ and $\overline{q}^{*}=(q_x,q_y,q_z)$ correspond to changes in temperature, pressure, number density, and fluid filtration velocity. $\overline{\eta}_1^{*}=(\eta_{1x},\eta_{1y},\eta_{1z})$ and $\overline{\eta}_2^{*}=(\eta_{2x},\eta_{2y},\eta_{2z})$ also represent the relative perturbations of the $\overline{\eta}_1$ and $\overline{\eta}_2$:

$$\nabla\cdot\overline{q}^{*}=0 \tag{7}$$

$$\begin{array}{l}\rho_f c_a\left(\partial\overline{q}^{*}/\partial t\right)=-\nabla\overline{p}^{*}-\left(\mu/\Re\right)\overline{q}^{*}+\mathbf{g}\left[\theta\Delta\rho\overline{n}^{*}-\rho_f\beta\overline{T}^{*}\right]\\ +\left(\rho_f/2\right)\left(\overline{\eta}_1^{*}\cdot\nabla\right)\left(\overline{n}^{*}\mathbf{k}-\overline{\eta}_1^{*}\right)\left[\left(\Delta\rho/\rho_f\right)\tilde{\mathrm{b}}\theta\omega\right]^2\\ +\left(\theta\beta\Delta\rho/2\right)\left(\tilde{\mathrm{b}}\omega\right)^2\left[\left(\overline{\eta}_1^{*}\cdot\nabla\right)\left(\overline{T}^{*}\mathbf{k}-\overline{\eta}_1^{*}\right)-\left(\overline{\eta}_2^{*}\cdot\nabla\right)\left(\overline{n}^{*}\mathbf{k}-\overline{\eta}_2^{*}\right)\right]\\ -\left(\rho_f/2\right)\left(\overline{\eta}_2^{*}\cdot\nabla\right)\left(\overline{T}^{*}\mathbf{k}-\overline{\eta}_2^{*}\right)\left(\tilde{\mathrm{b}}\beta\omega\right)^2\end{array} \tag{8}$$

$$\varphi\left(\partial\overline{n}^{*}/\partial t\right)=-div\left[\overline{n}^{b}\left(\overline{q}^{*}+\mathrm{q_c}k\right)+\overline{n}^{*}\mathrm{q_c}\mathbf{k}-D_c\nabla\overline{n}^{*}\right] \tag{9}$$

$$c_p\rho_f\left[\left(\partial\overline{T}^{*}/\partial t\right)-q_z\left(\Delta\overline{T}^{*}/h\right)\right]=\kappa\nabla^2\overline{T}^{*} \tag{10}$$

$$\left.\begin{array}{l}curl\,\overline{\eta}_1^{*}=grad\,\overline{n}^{*}\times\mathbf{k},\quad div\,\overline{\eta}_1^{*}=0\\ curl\,\overline{\eta}_2^{*}=grad\,\overline{T}^{*}\times\mathbf{k},\quad div\,\overline{\eta}_2^{*}=0\end{array}\right\} \tag{11}$$

Using Eqs. (11) and (8), respectively, perform $\mathbf{k}\cdot curl$ and $\mathbf{k}\cdot curl\,curl$:

$$\begin{aligned}&\rho_f c_a(\partial/\partial t)(\nabla^2 q_z) = \left(\rho_f\beta\nabla_1\overline{T}^* - \Delta\rho\theta\nabla_1\overline{n}^*\right)g - (\mu/\Re)\nabla^2 q_z\\&+(\rho_f/2)\left[(\Delta\rho/\rho_f)\tilde{\mathrm{b}}\theta\omega\right]^2\nabla_1\eta_{1z}(\partial\overline{n}^b/\partial z)\\&+(\theta\beta\Delta\rho/2)\left(\tilde{\mathrm{b}}\omega\right)^2\left[\nabla_1\eta_{1z}\left(\partial\overline{T}^b/\partial z\right) - \nabla_1\eta_{2z}(\partial\overline{n}^b/\partial z)\right]\\&-(\rho_f/2)\left(\tilde{\mathrm{b}}\beta\omega\right)^2\nabla_1\eta_{2z}\left(\partial\overline{T}^b/\partial z\right)\end{aligned} \quad (12)$$

$$\nabla^2\eta_{1z} = \nabla_1\overline{n}^*, \quad \nabla^2\eta_{2z} = \nabla_1\overline{T}^* \quad (13)$$

Here the quantities with '*' indicates the perturbed parameters, ∇_1 indicate the Laplacian operator (two dimensional). Analyzing the perturbed quantities using the normal mode:

$$\begin{aligned}&\left(q_z, \overline{n}^*, \overline{T}^*, \eta_{1z}, \eta_{2z}\right) = \left[U_z, N_z, \Theta_z, \Upsilon_{1z}, \Upsilon_{2z}\right](z)\\&\times\psi(x, y)\exp\left(\sigma_p t\right)\end{aligned} \quad (14)$$

Here, $\nabla_1\psi(x, y) = -\alpha^2\psi$, where $'\alpha'$ is the wave number [26]. Using the Eqs. (14) and (20), the non-dimensionalized equations are discovered at the onset of stationary convection (by allowing growth rate $\sigma_p = 0$) [28]:

$$\begin{aligned}&\left(\alpha^{\iota^2}U_z^\iota - D^2U_z^\iota\right) + \alpha^{\iota^2}R_b\left\{N_z^\iota - (\Theta_z^\iota/\mho)\right\} + \alpha^{\iota^2}\left[R_t\left\{\delta\Upsilon_{1z}^\iota - \Upsilon_{2z}^\iota\right\}\right]\\&\quad + \alpha^{\iota^2}R_v\exp(z^\iota Pe)\left\{(\Upsilon_{2z}^\iota/\mho) - \Upsilon_{1z}^\iota\right\} = 0\end{aligned} \quad (15)$$

$$D^2N_z^\iota - PeDN_z^\iota - \alpha^{\iota^2}N_z^\iota - U_z\exp(z^\iota Pe) = 0 \quad (16)$$

$$D^2\Theta_z^\iota - \alpha^{\iota^2}\Theta_z^\iota + (\mho R_a/R_b)\,U_z^\iota = 0 \quad (17)$$

$$D^2\Upsilon_{1z}^\iota + \alpha^{\iota^2}\left(N_z^\iota - \Upsilon_{1z}^\iota\right) = 0 \quad (18)$$

$$D^2\Upsilon_{2z}^\iota + \alpha^{\iota^2}\left(\Theta_z^\iota - \Upsilon_{2z}^\iota\right) = 0 \quad (19)$$

The non-dimensional values and quantities used in the aforementioned equations are:

$$\left.\begin{aligned}&z^\iota = zh, \quad \alpha^\iota = \alpha h, \quad U_z^\iota = \nu\theta q_c h^2 U_z/D_c^2,\\&\mho = \Delta\rho/\rho_f, \quad D_a = \Re/h^2, \quad \Upsilon_{1z}^\iota = \Upsilon_{1z}\theta,\\&\Upsilon_{2z}^\iota = \Upsilon_{2z}\beta, \quad N_z^\iota = N_z\theta, \quad \Theta_z^\iota = \Theta_z\beta,\\&D \equiv \partial/\partial z^\iota\end{aligned}\right\} \quad (20)$$

Here, $R_b = \Re q_c g v \theta \Delta\rho h^2 / \mu D_c^2$ is an altered bioconvection Rayleigh-Darcy number, $R_a = \rho_f^2 c_p g \beta \Delta T \Re h / \mu \kappa$ is an altered thermal Rayleigh-Darcy number, $\mho$ is the measure of density of gyrotactic particles, $R_v = \rho_f \Re\left[\theta \tilde{b} \omega v q_c h (\Delta\rho / \rho_f)\right]^2 / 2\mu D_c^3$ is the modified vibrational Rayleigh-Darcy number and $R_t = \rho_f \Re \theta q_c v h \beta \Delta T \left(\tilde{b}\omega\right)^2 / 2\mu D_c^2$ is the modified thermo–vibrational-Darcy parameter. To resolve the eigenvalue problem related to Eqs. (15–19), the Galerkin technique of weighted residuals is used [26]. The following are the dimensionless boundary restrictions of the flow:

$$\text{At } \left.\begin{array}{l} z^\iota = \pm 1/2;\quad U_z^\iota = 0,\quad DN_z^\iota = PeN_z^\iota, \\ \qquad \Theta_z^\iota = 0,\quad \Upsilon_{1z}^\iota = 0,\quad \Upsilon_{2z}^\iota = 0 \end{array}\right\} \tag{21}$$

The suitable functions meeting the boundary restrictions in Eq. (21) are:

$$\left.\begin{array}{l} N_z^\iota = Pe\cos(2\pi z^\iota) + \left(Pe^2/2\pi\right)\sin(2\pi z^\iota) \\ U_z^\iota = z^{\iota^2} - (1/4) \\ \Theta_z^\iota = \Upsilon_{1z}^\iota = \Upsilon_{2z}^\iota = (z^\iota - 1/2)\ (z^\iota + 1/2), \end{array}\right\} \tag{22}$$

Utilizing Eq. (22) into Eqs. (15–19), with respect to the Galerkin technique [29], the instability boundary for the stationary convection is discovered as follows:

$$\begin{aligned} &(120Pe)\alpha^{\iota^2}\xi_2\left(10+\alpha^{\iota^2}\right)^2 R_b \\ &+\alpha^{\iota^2}\xi_3\left(10+\alpha^{\iota^2}\right)R_a \\ &+120Pe\xi_2\alpha^{\iota^4}\left(10+\alpha^{\iota^2}\right)[\mho R_t - 30\xi_1 R_v] \\ &-\alpha^{\iota^4}\xi_3[30\xi_1 R_v - \mho R_t](R_a/R_b) \\ &= \left(10+\alpha^{\iota^2}\right)^3 \xi_3 \end{aligned} \tag{23}$$

Where, $\xi_1 = \sinh(Pe/2)\ \{4/(Pe)^3 + 48/(Pe)^5\} - \cosh(Pe/2)\ \{24/(Pe)^4\}$ $\xi_2 = \{8Pe^2\sinh(Pe/2) - \cosh(Pe/2)\left(8\pi^2 Pe + 2Pe^3\right)\}/\left(Pe^2 + 4\pi^2\right)^2$ and $\xi_3 = -\left(8\pi^2 + \alpha^{\iota^2}\right)Pe^4$. When there is no vibration and thermal influence, Eq. (23) collapses to:

$$(R_b)_{cr} = \underset{\alpha^\iota \geq 0}{Min}\left\{\left(10+\alpha^{\iota^2}\right)\xi_3 / 120\alpha^{\iota^2} Pe\xi_2\right\} \tag{24}$$

This is the representation for the critical bioconvection Rayleigh-Darcy number. In the limiting situation, the obtained $(R_a)_{cr} = 40$ for corresponding $(\alpha^\iota)_{cr} = 3.1622$, which is the same reported in [8, 16].

3 Results and Discussion

In this section, computations are being performed, and the results are schemed to explore the influences of the non-dimensional quantities such as thermal Rayleigh-Darcy number, R_a, the modified critical bioconvection Rayleigh-Darcy number, $(R_b)_{cr}$, for the fixed values Péclet number, Pe, wave number, α^ι, the vibrational Rayleigh-Darcy number, R_v, and thermo-vibrational-Darcy parameter, R_t [8, 26]. The bioconvection Rayleigh-Darcy number is a number that indicates and holds the suspension's stability criteria. The destabilizing or stabilizing nature of the suspension is revealed by the fall or rise in its magnitudes.

Figure 2 displays the effect of thermal Rayleigh-Darcy number, R_a, on bioconvection Rayleigh-Darcy number, $(R_b)_{cr}$, for distinct values of Pe. It observed that as Pe increases (from 0.1 to 1, 3), the magnitude of $(R_b)_{cr}$ enlarges. Hence Péclet number (which describes the micro-swimmers speed) shows a stabilizing effect [8, 11]. For the increased values of R_a the curves of $(R_b)_{cr}$ decrease in the $((R_b)_{cr},\ R_a)$ plane. The thermal influence across the Darcy porous layer makes the system destabilize [8, 16, 26].

Figure 3 shows the impact of R_a, on $(R_b)_{cr}$, for various values of vibrational Rayleigh-Darcy number, R_v. For increased values of R_v(from 0 to 50, 100), the magnitude of $(R_b)_{cr}$ increases. It was marked that the vibration's strength stabilizes the suspension. In the absence of thermal effect, this result entirely agrees with what was reported in [11]. In the specific range of values of $R_a (0 \leq R_a \leq 150)$, the curves of $(R_b)_{cr}$ decrease rapidly in the $((R_b)_{cr},\ R_a)$ plane.

Table 1 displays the influence of R_a on $\propto_{cr}^{\iota}$ for various values of Pe. It noted that for increased values of Pe (0.1 to 1), the magnitude of $\propto_{cr}^{\iota}$ decreases, later increases. These results agree with the reported works of [11, 15].

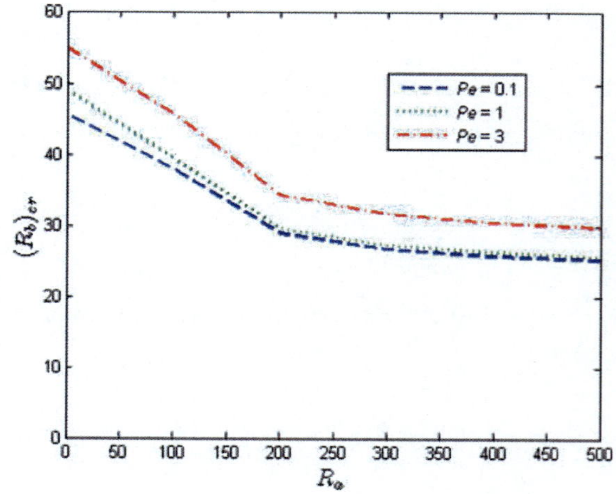

Fig. 2 Dependence of $(R_b)_{cr}$ on R_a for distinct values of Pe, $R_t = 10,\ (\alpha^\iota)_{cr} = 3.1,$ and $R_v = 50$

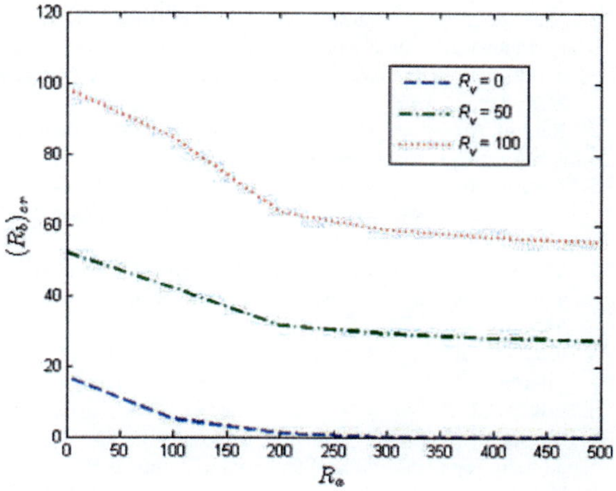

Fig. 3 Dependence of $(R_b)_{cr}$ on R_a for distinct values of R_v, $Pe = 1.5$, $R_t = 10$, and $\propto_{cr}^{\iota} = 3.1$

Table 1 Dependence of critical wave number $\propto_{cr}^{\iota}$ on R_a for distinct values of Pe, $R_b = 50$, $R_v = 50$, and $R_t = 10$

R_a	α_{cr}^{ι}		
	$Pe = 0.1$	$Pe = 1$	$Pe = 3$
0	3.1623	3.1623	3.1623
100	3.1621	2.9389	4.2703
200	3.1619	2.3715	5.1428
300	3.1617	2.5266	5.7695
400	3.1616	2.722	6.2903
500	3.1614	2.8788	6.7546

4 Conclusions

In this study, we explored the numerical investigation of the influence of the heat transfer phenomenon in gravitactic bioconvection in a fluid, porous medium. The monotonic instability boundary is obtained using the single-term Galerkin technique for stationary convection. It reported that the Péclet number, which describes the micro-swimmers speed, stabilizes the suspension. The vibration strength stabilizes the suspension; however, the system is destabilized by temperature variations across the Darcy porous layer. Compared to its absence, the bioconvection strength is suppressed by the action of the Darcy-porous matrix.

References

1. Childress S, Levandowsky M, Spiegel EA (1975) Pattern formation in a suspension of swimming microorganisms: equations and stability theory. J Fluid Mech 63:591–613
2. Kessler JO (1985) Hydrodynamic focusing of motile algal cells. Nature 313:218–220

3. Nield DA, Bejan A (2006) Convection in porous media, 3rd edn, Springer, New York
4. Kuznetsov AV, Avramenko AA (2002) A 2D analysis of stability of Bio-thermal convection in a fluid saturated porous medium-Estimation of the critical permeability value. Int Comm Heat Mass Trans 29:175–184
5. Bahloul A, Nguyen-Quang T, Nguyen TH (2005) Bioconvection of gravitactic microorganisms in a fluid layer. Int Commun Heat Mass Transfer 32:64–71
6. Avramenko AA, Kuznetsov AV (2006) The onset of convection in a suspension of gyrotactic microorganisms in superimposed fluid and porous layers: effect of vertical throughflow. Transp Porous Med 65:159
7. Sharma YD, Virendra Kumar (2011) Overstability analysis of Thermo-bioconvection saturating a porous medium in a suspension of gyrotactic microorganisms. Transp Porous Med 90:673–685
8. Razi YP, Mojtabi A, Charrier-Mojtabi MC (2009) A summary of new predictive high frequency Thermo-Vibrational models in porous media. Transp Porous Med 77:207–228
9. Eduardo J Gudiña, Jorge FB Pereira, Rita Costa, João AP Coutinho, José A Teixeira, Lígia R Rodrigues (2013) Biosurfactant-producing and Oil-degrading bacillus subtilis strains enhance oil recovery in laboratory Sand-pack columns. J Hazard Mater 261:106–113
10. Hong E, Jeong MS, Kim TH, Lee JH, Cho JH, Lee KS (2019) Development of coupled biokinetic and thermal model to optimize Cold-Water Microbial Enhanced Oil Recovery (MEOR) In Homog Reserv. Sustain 11:1652
11. Kuznetsov AV (2005) The onset of bioconvection in a Suspension of negatively geotactic microorganisms with High-frequency vertical vibration. Int Commun Heat Mass Transfer 32:1119–1127
12. Kuznetsov AV (2006) Investigation of the onset of bioconvection in a suspension of oxytactic microorganisms subjected to High-frequency vertical vibration. Theor Comput Fluid Dyn 20(2):73–87
13. Sharma YD, Virendra Kumar (2012) The effect of High-frequency vertical vibration in a suspension of gyrotactic microorganisms. Mech Res Commun 44:40–46
14. Virendra Kumar, Sharma YD (2014) Instability analysis of gyrotactic microorganisms: a combined effect of high frequency vertical vibration and porous media. Transp Porous Med 102:153–165
15. Srikanth K, Virendra Kumar (2022) Non-Darcian gravitactic bioconvection with a porous saturated vertical vibration. In: Ray SS, Jafari H, Sekhar TR, Kayal S (eds) Applied analysis, computation and mathematical modelling in engineering. AACMME 2021. Lect Notes Electr Eng 897, Springer, Singapore
16. Virendra Kumar, Srikanth K (2022) Vertically oscillated gyrotactic bio-thermal convection in a porous media, Forces in Mechanics 9:100136
17. Kuznetsov AV (2011) Non-oscillatory and oscillatory Nanofluid Bio-thermal convection in a horizontal layer of finite depth. Eur J Mech B/Fluids 30:156–165
18. Khan WA, Uddin MJ, Ismail AIM (2013) Free convection of Non-newtonian Nanofluids in porous media with gyrotactic microorganisms. Transp Porous Med 97:241–252
19. Pedley TJ (2015) Gyrotaxis in uniform vorticity. J Fluid Mech 762:R6
20. Cencini M, Franchino M, Santamaria F, Boffetta G (2016) Centripetal focusing of gyrotactic phytoplankton. J Theor Biol 399:62–70
21. Croze O, Bearon R, Bees M (2017) Gyrotactic swimmer dispersion in pipe flow: testing the theory. J Fluid Mech 816:481–506
22. Shivani Saini, Sharma YD (2018) Numerical study of nanofluid Thermo-bioconvection containing gravitactic microorganisms in porous media: effect of vertical through flow. Adv Powder Technol 29:2725–2732
23. Saini S, Sharma YD (2018) Numerical study of bioconvection saturated with nanofluid containing gyrotactic microorganisms confined within Hele-Shaw cell. Instrum Mes Métrologie 18(2018):573–591
24. Mil-Martínez SR, Ferrer VH, Turcio M, López-Serrano F, Ortega JA, R.O. (2019) Vargas, Stability analysis and numerical simulation of gravitactic bioconvection in a rectangular cavity. Comput Math Appl 77:222–236

25. Kage A, Omori T, Kikuchi K, Ishikawa T (2020) The shape effect of flagella is more important than bottom-heaviness on passive gravitactic orientation in Chlamydomonas reinhardtii. J Expl Biol 223, jeb205989
26. Virendra Kumar, Srikanth K (2021) An overstability analysis of vertically vibrated suspension of active swimmers subjected to thermal stratification. SN Appl Sci 3:612
27. Gershuni GZ, Lyubimov DU (1998) Thermal vibrational convection, Wiley, New York
28. Chandrasekhar S (1981) Hydrodynamic and hydromagnetic stability, Dover, New York
29. Finlayson BA (1972) The method of weighted residuals and variational principles, Academic Press, New York

Analysis of Different Heating Rates on the Thermal Degradation of *Chlorella Protothecoides* Microalgal Biodiesel Using Thermogravimetric Analysis (TGA)

Mukesh Kumar and Aritra Ganguly

Abstract Biodiesel is considered as a potential fuel for the diesel engine because of similar physiochemical properties compared to diesel. However, some fuel properties of biodiesel still require improvement to convert it into an ideal fuel such as biodiesel stability and cold flow properties. Thermal analysis is considered an effective method to analyze the thermal stabilities of any fuel. The present paper investigated the thermal degradation of *Chlorella protothecoides* microalgal biodiesel (CPMB) at varying heating rates. The Thermo-gravimetric analyzer (TGA) was conducted from 5 to 20 °C/min heating rates under both air and nitrogen atmosphere in a 25–700 °C temperature range.

Keywords Microalgae oil · TGA · DAM · CRM · Thermal stability

1 Introduction

Currently, biodiesel has received extensive focus worldwide because of the rapid demolishing of the crude oil reservoir, increasing cost, and exhaust emissions due to fossil fuel usage. Thus, it becomes necessary to search for alternative sources of fossil fuel which is renewable, sustainable, easy to process, and, most importantly, environment friendly [1, 2]. Transesterification process is most widely used to convert oil into biodiesel [3, 4]. Edible and non-edible oils are most widely used to convert into biodiesel production, however, these resources are not suitable to fulfill the global demand of fuel because of their slow growth rate and low lipid content [5]. Microalgae is considered as most promising feedstocks for biodiesel production as compared to 1st and 2nd second generation feedstocks due to higher oil content and faster growth rates [6–11]. Biodiesel stability and cold flow properties are two hurdles

M. Kumar (✉) · A. Ganguly
Department of Mechanical Engineering, IIEST Shibpur Howrah, Howrah, West Bengal 711103, India
e-mail: mukeshiitr13@gmail.com

S. Das et al. (eds.), *Proceedings of the 1st International Conference on Fluid, Thermal and Energy Systems*, Lecture Notes in Mechanical Engineering,
https://doi.org/10.1007/978-981-99-5990-7_61

to commercialize the biodiesel production from microalgae oil. To use microalgal biodiesel as engine fuel, it is necessary to study thermal behavior and kinetics.

Thermal analysis is an effective tool to estimate any fuel's properties like kinetic rate, frequency factor, enthalpy, boiling point, and activation energy [12–16]. Thermo-gravimetric analysis (TGA) is used to measure the weight loss of fuel concerning temperature [14, 17, 18].

A detailed investigation and comparison of thermal degradation of *Chlorella protothecoides* microalgal biodiesels at varying heating rates have not been reported to the best of the author's knowledge. In the present study, varying heating rates of 5, 10, 15, and 20 °C/min are employed to evaluate the thermal degradation of CPMB in air and nitrogen atmosphere. The activation energy of the biodiesel was calculated on the basis of both Coats & Redfern method (CRM) and Direct Arrhenius method (DAM).

2 Experiments and Methods

2.1 Materials

Biodiesel used in the present study was obtained from *Chlorella protothecoides* microalgal oil through a process known as transesterification. The detailed procedure for biodiesel production was discussed in our previous study [19].

2.2 TGA

TGA is employed to measure the thermal properties of fuel. It can be accomplished in the presence or absence of oxygen/nitrogen. Thermal degradation of CPMB is evaluated on a TGA (Perkin Elmer Pyris 6) using alumina pans. The sample is weighted to 10 ± 0.1 mg in an aluminum oxide ceramic crucible of 5 mm depth and 5 mm diameter. TGA experiments are conducted from 25 to 700 °C temperature at heating rates from 5–20 °C/min. Coats and Redfern method (CRM) and Direct Arrhenius method (DAM) are used to study the thermal stability of biodiesel.

Table 1 Properties of CPMB and diesel

Fuel properties	CPMB	ASTM D6751 limit	IS 15607 limit	EN 14214	Diesel
Density (kg/m^3)	862	–	860–900	860–900	813
Viscosity at 40 °C (cSt)	4.40	1.9–6.0	3.5–5.0	3.5–5.0	2.55
Flash point (°C)	165	130 min	120 min	120 min	65
Cloud point (°C)	0	–	–	–	–
Pour point (°C)	−3	–	–	–	–
CFPP (°C)	−10	–	–	–	−6
Oxidation stability, 110 °C (h)	4.6	3 h, min	6 h, min	6 h, min	17.3
Calorific value (MJ/kg)	39.48	–	–	–	42.83
Ester contents (%)	98.1	–	–	96.5 min	–
Activation energy (kJ/mole)	53.93	–	–	–	–
Cetane number	49.5	47 min	51 min	51 min	45

3 Results and Discussion

3.1 *Physiochemical Properties of Biodiesel*

The fuel properties of microalgae biodiesel were calculated using the standard method and listed in Table 1. These properties are compared with diesel and biodiesel specifications based on ASTM D6751, IS 15607 and EN 14,214 limits.

The density and kinematic viscosity of the CPMB were significantly higher than those of the diesel. The CPMB exhibits quite satisfactory fuel properties that meet standards upto the most biodiesel specifications. The CP and PP values of the CPMB were as low as 0 and -3 °C that indicated the use of CPMB in cold climates effectively. The Oxidation stability of CPMB, determined by using modified Rancimat method was 4.6 h and satisfies ASTM specification. This table concludes that the CPMB fuel properties are comparable to diesel and this fuel can be used directly or it can be used as a blend in diesel engine. The superior fuel properties of the CPMB demonstrated the suitability as a potential feedstock for biodiesel production. Therefore, the comparison containing validation of CPMB fuel properties indicates its suitability as a potential alternative to diesel.

3.2 *Thermal Stability of CPMB*

The experiments were conducted based on the above, and the corresponding results are reported in the subsequent paragraphs.

Figure 1a and b shows the TGA curves of CPMB in both air and Nitrogen medium at varying heating rates. It is shown in Fig. 1a that the maximum peak tends to be delayed towards the high temperature during increasing heating rates. A similar trend was reported by researchers for *Chlorella Vulgaris* decomposition [20–22]. The TGA trend is found to be similar for all heating rates, while a flat curve is obtained from 0–200 °C. The onset temperature (T_{onset}) is determined by extrapolating the horizontal baseline at 1% weight loss. The onset temperature represented the fuel's resistance to thermal degradation [22]. Once the sample completely burns, the curve becomes almost flat [22, 23].

Table 2 presents the onset and offset temperature of CPMB biodiesel at varying heating rates. It is pretty evident from Table 2 that the onset and offset temperature of the fuel increase with the increase in heating rates. The onset temperature increases from 196 to 251 °C in the air medium, while for Nitrogen, the same varies from 197 to 246 °C. The region where the significant weight loss occurred was from 196.09 to 248.4 °C, 214.64–263.39 °C/min, 248.59 to 297.41, and 251.61 to 302.44 °C at different heating rates, respectively. The study revealed that the residence time of the fuel inside the combustion furnace becomes shortened with the increase in heating rate [24]. The reason for the weight degradation of any fuel is the presence of saturated and mono-polyunsaturated acids [24]. TGA method can also used to find the different thermal properties of biodiesel in the presence and absence of both nitrogen and oxygen.

Figure 2a and b demonstrates the effect of heating rate on thermal degradation using DTG curve. Increasing heating rates tend to delay thermal degradation and this is because the material reaches a particular temperature (degradation temperature) in a shorter interval at a higher heating rate [24, 25]

A comparison between the slow and fast heating rates for weight loss is depicted in Fig. 3a and b. It very well may be closed from Fig. 3 that the weight reduction of material increments with an expansion in the heating rate. It is accepted that the high heating system diminishes the consistency of the sample while strengthening the responses framing volatiles [26].

4 Conclusions

In this work, the thermal degradation behavior of CPMB was considered. TGA of biodiesel was performed in an air as well as in a nitrogen atmosphere at various heating rates. The study revealed that an increase in heating rate tended to delay thermal degradation processes towards higher temperatures, most probably due to increased thermal lag at a given temperature. A higher heating rate implies that the material reaches that temperature in a shorter time, thereby degradation shifting towards a more elevated temperature, influencing the fuel combustion process.

Fig. 1 **a** TGA curves for *Chlorella protothecoides* in an air atmosphere for different heating rates, **b** TGA curves for *Chlorella protothecoides* in an air atmosphere for different heating rates

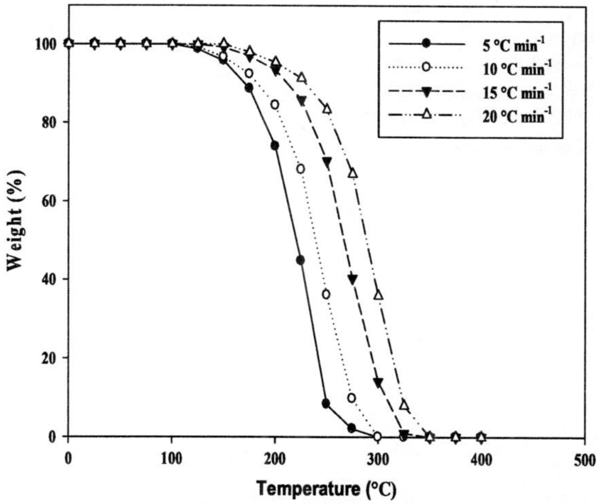

(a) TGA curves for *Chlorella protothecoides* in an air atmosphere for different heating rates

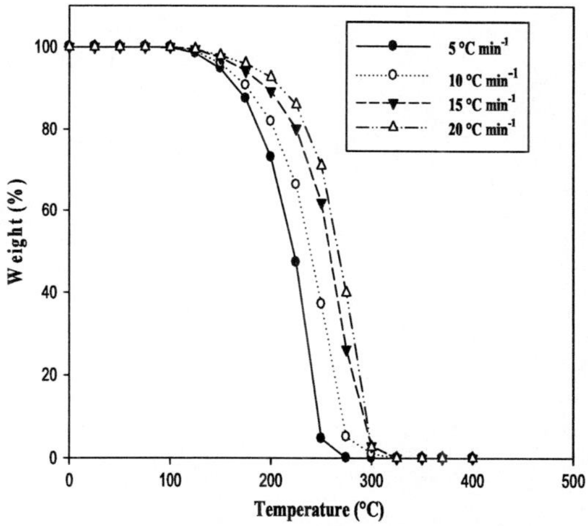

(b) TGA curves for *Chlorella protothecoides* in an air atmosphere for different heating rates

Table 2 Onset, offest and peak temperature of CPMB

Heating rate (°C/min)	*Chlorella protothecoides* biodiesel					
	Air atmosphere (°C)			Nitrogen atmosphere (°C)		
	Consent	Speak	Offset	Consent	Speak	Offset
5	196.09	237.16	248.38	197.21	244.20	251.88
10	214.64	258.56	263.39	213.07	263.58	272.45
15	248.59	292.43	297.41	232.60	281.16	289.99
20	251.61	293.31	302.44	246.73	287.87	295.93

Fig. 2 **a** Effect of heating rate on DTG curve for CPMB for air atmosphere, **b** Effect of heating rate on DTG curve for CPMB for a Nitrogen atmosphere

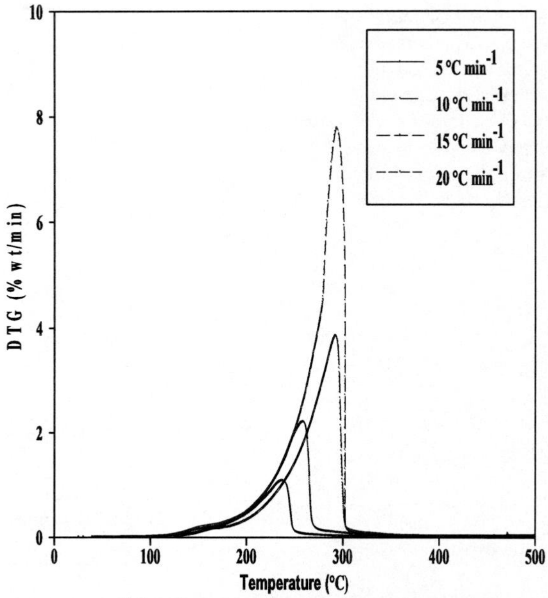

(a) Effect of heating rate on DTG curve for CPMB for air atmosphere

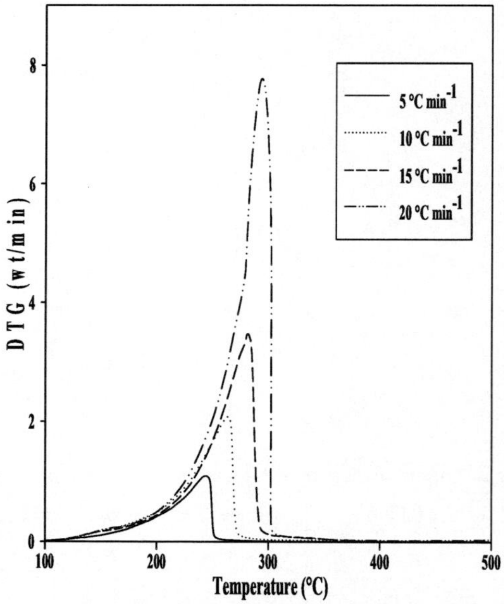

(b) Effect of heating rate on DTG curve for CPMB for a Nitrogen atmosphere

Fig. 3 **a** Degree of reaction versus temperature at varying heating rate for air atmosphere, **b** Degree of reaction versus temperature at varying heating rate for nitrogen atmosphere

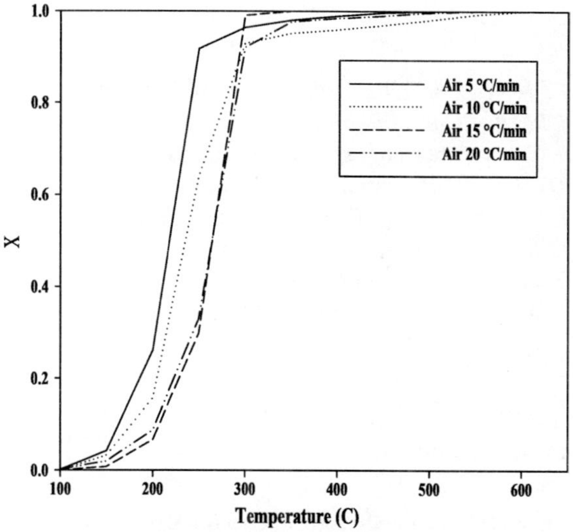

(a) Degree of reaction versus temperature at varying heating rate for air atmosphere

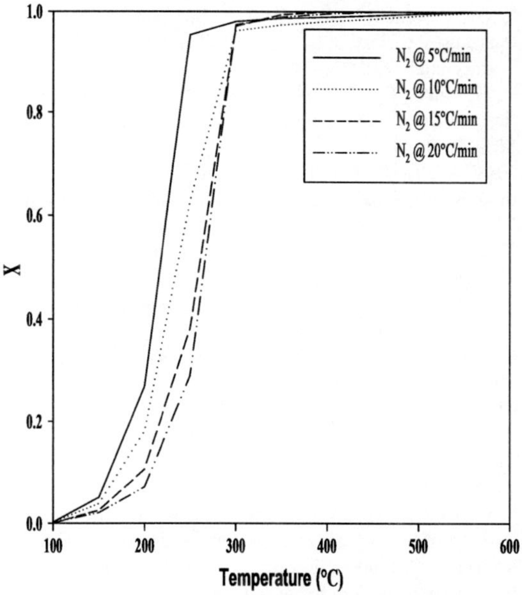

(b) Degree of reaction versus temperature at varying heating rate for nitrogen atmosphere

Nomenclature

CPM	*Chlorella protothecoides* Microalgal oil
CPMB	*Chlorella protothecoides* Microalgal oil
TAG	Thermogravimetric analysis
DTG	Differential thermal analysis
DAM	Direct Arrhenius method
CRM	Coats and Redfern method
T_{onsent}	Onset Temperature
T_{Offset}	Offset Temperature

References

1. Zhao H, Cao Y, Orndorff W, Cheng Y-H, Pan W-P (2012) Thermal behaviors of soy biodiesel. J Therm Anal Calorim 109:1145–1150
2. Chutia RS, Kataki R, Bhaskar T (2013) Thermogravemtric and decomposition kinetics studies of Mesua Ferrea L. deoiled cake. Bioresour Technol 139:66–72
3. Zhang F, Adachi D, Tamalampudi S, Kondo A, Tominaga K (2013) Real-time monitoring of the transesterification of soybean oil and methanol by fourier transform infrared spectroscopy. Energy Fuels 27:5957–5961
4. Li H, Niu S, Lu C, Liu M, Huo M (2014) Transesterification catalyzed by industrial waste-lime mud doped with potassium fluoride and the kinetic calculation. Energy Convers Manage 86:1110–1117
5. Nur ZAS, Taufiq-Yap YH, Nizah MFR, Teo SH, Syazwani ON, Islam A (2014) Production of biodiesel from palm oil using modified Malaysian natural dolomites. Energy Convers Manag 78:738–744
6. Kumar M, Sharma MP (2014) Potential assessment of micro-algal oils for biodiesel production: a review. J Mater Environ Sci 5:757–766
7. Kumar M, Sharma MP (2013) Production methodology of biodiesel from microalgae. Int J Appl Eng Res 8:1825–1832
8. Kumar M, Sharma MP (2015) Assessment of potential oils for biodiesel production. Renew Sustain Energy Rev 44:814–823
9. Chen YH, Tang TC, Chiang TH, Huang B-Y, Chang C-Y, Chiang P-C (2012) A complementary biodiesel blend from soap nut oil and free fatty acids. Energies 5(8):3137–3148
10. Tiwari P (2009) Isothermal and non-isothermal kinetic analyses of mahogany oil shale with TGA. Biotechnol Adv 27:583–587
11. Silva SA, Conceio MM, Souza AG, Prasad S, Silva Marta CD, Fernandes VJ (2004) Thermal analysis of the powder and the bran of algaroba. J Therm Anal Calorim 75(2):411–417
12. Rodríguez R, Sierens R, Verhelst S (2009) Thermal and kinetic evaluation of biodiesel derived from soybean oil and higuereta oil. J Therm Anal Calorim 96:897–901
13. Oliveira LE, Giordani DS, Paiva EM, De Castro HF, Da Silva MLCP (2013) Kinetic and thermodynamic parameters of volatilization of biodiesel from babassu, palm oil and mineral diesel by thermogravimetric analysis (TG). J Therm Anal Calorim 111:155–160
14. Niu S, Han K, Zhou F, Lu C (2011) Thermogravimetric analysis of the decomposition characteristics of two kinds of calcium based organic compounds. Powder Technol 209:46–52
15. Goodrum JW (2002) Volatility and boiling points of biodiesel from vegetable oils and tallow. Biomass Bioenerg 22:205–211

16. Ramalho EFSM, Santos IMG, Maia AS, Souza AL, Souza AG (2011) Thermal characterization of the poultry fat biodiesel. J Therm Anal Calorim 106:825–829
17. Tutunea D (2013) Thermal investigation of biodiesel blends derived from rapeseed oil. J Therm Anal Calorim 111:869–875
18. Kumar M, Sharma MP (2016) Optimization of transesterification of *Chlorella protothecoides* oil to biodiesel using box-behnken design method. Waste Biomass Valoris 016:9485–9495
19. Chouhan APS, Singh N, Sarma AK (2013) A Comparative analysis of kinetic parameters from TGDTA of Jatropha curcas oil, biodiesel, petroleum diesel and B50 using different methods. Fuel 109:217–224
20. Chen C, Lu Z, Ma X, Long J, Peng Y, Hu L, Lu Q (2013) Oxy-fuel combustion characteristics and kinetics of microalgae Chlorella vulgaris by thermogravimetric analysis. Biores Technol 144:563–571
21. Chen C, Ma X, Liu K (2011) Thermogravimetric analysis of microalgae combustion under different oxygen supply concentrations. Appl Energy 88:3189–3196
22. Jain S, Sharma MP (2012) Application of thermogravimetric analysis for thermal stability of Jatropha curcas biodiesel. Fuel 93:252–257
23. Ali SA, Razzak SA, Hossain MM (2015) Apparent kinetics of high temperature oxidative decomposition of microalgal biomass. Bioresoure Technol 175:569–577
24. Volli V, Purkait MK (2014) Physico-chemical properties and thermal degradation studies of commercial oils in nitrogen atmosphere. Fuel 117:1010–1019
25. Jeguirim M, Trouvé G (2009) Pyrolysis characteristics and kinetics of Arundo donax using thermogravimetric analysis. Biores Technol 100:4026–4031
26. Meesri C, Moghtaderi B (2002) Lack of synergetic effects in the pyrolytic characteristics of woody biomass/coal blends under low and high heating rate regimes. Biomass Bioenerg 23:55–56

Analysis on Solid Rocket Motor Using Computational Method for Uniform Flow Field

Shruti Dipak Jadhav, Ankit Kumar Mishra, Aswin MR, and Athul Krishna S

Abstract The purpose of this paper is to get a complete study of flow parameters over a typical solid rocket motor and further analyze it according to the desired particular needs. Solid rocket motor consists of key inert components, namely the motor case, igniter case, and nozzle. The motor attached to the thrust vector control mechanism in a large-scale rocket can also be considered to be an important component. The paper follows the trait of a numerical study of the uniform flow equations, motor design, and computational fluid dynamics analysis over the motor using ANSYS. The analysis is done on the solid rocket motor for various parameters like dynamic and static pressure over the surface, surface shear stress and mainly focusing on the flow velocity. The detailed study of flow velocity includes the streamlined flow attached to the surface and the possible vortex formation areas over the motor case surface. Here, the calculations are done at a flow field velocity of mach 0.6.

Keywords CFD · Nozzle · Rocket motor · Structural · Uniform flow

1 Introduction

Solid Rocket Motors is mainly used in launch vehicles to provide high thrust during the initial launch sequence. Solid rocket motor design is based on the high energy solid propellant and high propellant mass fraction. In a solid propellant rocket motor, the propellant is burnt in the combustion chamber by mixing oxidizer and propellant

S. D. Jadhav (✉)
Department of Aerospace Engineering, Sandip University, Nashik 422001, India
e-mail: jadhavshruti743@gmail.com

A. K. Mishra
Department of Research and Development, ASTROEX RESEARCH ASSOCIATION, Deoria 274001, India
e-mail: ankitkumarm1998@gmail.com

A. MR · A. K. S
Department of Aerospace Engineering, Lovely Professional University, Phagwara 144411, India

S. Das et al. (eds.), *Proceedings of the 1st International Conference on Fluid, Thermal and Energy Systems*, Lecture Notes in Mechanical Engineering,
https://doi.org/10.1007/978-981-99-5990-7_62

with the help of the igniter [1, 2]. These further results in a downward force giving high thrust to the rocket. The problem here arises in the case of performance and stability in the combustion chamber. Alongside the sub-category, pressure oscillations at a high burn rate can cause the rocket motor to have large-scale instabilities. Here one more factor that needs to be taken care of is the heat transfer rate at which the metal or composite case can withstand [3, 4].

For the computational study of uniform flow over the solid rocket motor, the desired parameters needed to consider the pressure variations over the surface and the flow circulation at the rear end of the solid rocket motor. These are forms of hydrostatic instabilities. The computational fluid dynamics process and structural models help to study the motor's internal aerodynamics [5–8].

Usually, the solid propellant combustion produces gaseous endproducts which can induce low axial Mach number, large Reynolds number and the internal shear flow inside the combustion chamber. These are measured on the basis of applying time-dependent boundary conditions [9–11].

Anthoine et al. on different fuel grain burn velocities came to a conclusion that the pressure oscillations are reduced when the upstream grain burn velocity is higher than the downstream grain burn velocity. So, increasing the upstream burn grain velocity can significantly reduce the pressure oscillations that the solid rocket motor is experiencing [12].

From Kirkkopru, Kassoy, and Zhao, the computational study shows induced acoustic waves planar to the sidewall planes interact to sidewall induced flow generating unsteady vortices. Here the maximum vorticity intensity approximates to 1/M, where M is the solid rocket motor surface mach number. These transient gradients can cause shear stresses on the walls and the high flow pressure regimes [13].

Ankit et al. studied about the theoretical concept required for designing solid rocket motor in order to achieve higher amount of thrust. The software used for this research are; Fusion 360 as well as ANSYS [14].

Mishra et al. investigated a computational study on two stage sounding rockets to be demonstrated with CD nozzle has been considered. The theoretical and numerical calculations were compared with the computational analysis to achieve the ideal condition [15, 16].

The main purpose of the following research paper is to analyze the uniform flow field around the solid rocket motor at mach 0.6 and to conclude on the basis of that.

2 Materials and Methods

The CFD analysis is done on Ansys fluent with a step-by-step solver process. The geometry of the solid rocket motor is made in Fusion 360 with the assembly of three components, nozzle, motor case, and the nose cone. This is further imported to the Ansys design modeler where the model is enclosed at a ratio of 3:2 X-axis box dimensions. Further Boolean operator is being operated on the rocket motor to

Fig. 1 Section analysis of the designed solid rocket motor

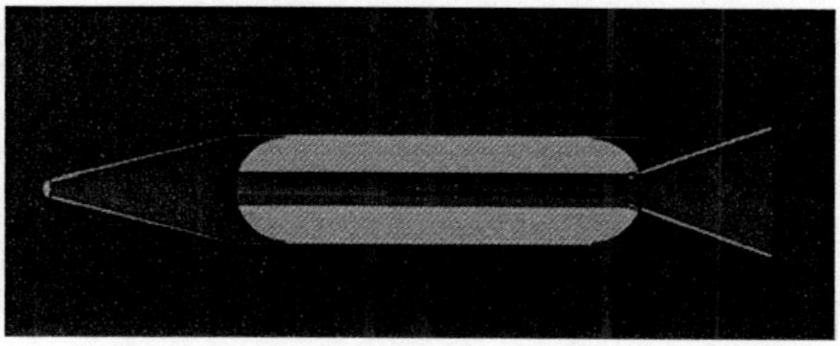

perform the subtract process. This is done in order to get a fine mesh around the rocket motor case.

The next process is the meshing where the enclosed box is fully meshed with adding an inflation layer. The inflation layer gives a greater number of congested layers with a small layer spacing which tends to give more precise flow analysis over the surface of the rocket motor, the element order is set to the quadratic and the total number of nodes being meshed is coming under 50000.

Next is the fluent solver. Here the energy equation is turned on along with the reference values, boundary conditions are being set for the calculation process. When the values are entered, the initialization process is activated and thus the calculation is started. The final procedure was to obtain the results in the form of contour maps, vector flow, and plots for the data analysis.

3 Flow Analysis Over Solid Rocket Motor

3.1 Geometry Design

The solid rocket motor is made in Fusion 360 with an assembly feature. The rocket motor is divided into three components for a smooth transition of modeling. The length to diameter ratio of the motor case is set to 6 which would be having a higher propellant mass ratio. The designed nozzle is of fixed type. The nozzle inlet valve is put under a fixed constraint on the end of the motor case. Then the nose cone is being decided and placed under fixed constraint at the front of the motor case [4] (Fig. 1).

3.2 Meshing

A linear element order meshing has been done over the enclosed box having a cut section of the solid rocket motor. Here an additional program-controlled inflation layer with a first-layer thickness of 2.5 mm has been added. Then the inflation layer over the solid rocket motor is set to include. This provides a narrow and steeper small-scale mesh element around the rocket motor thus increasing the flow rate over

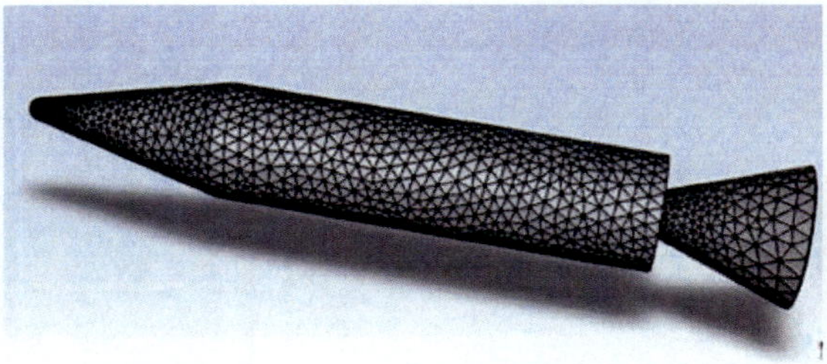

Fig. 2 Exterior mesh structure over the solid rocket motor

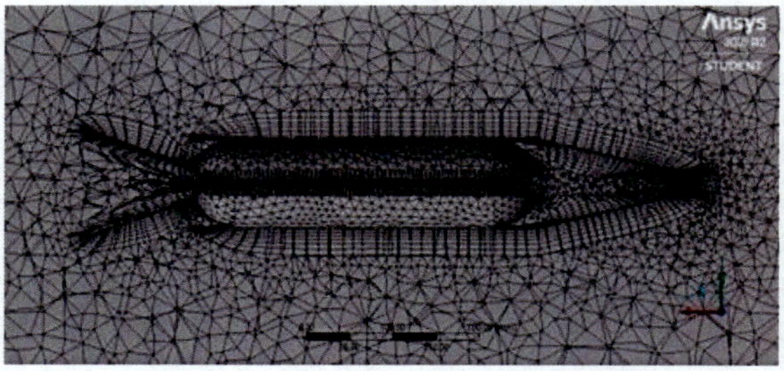

Fig. 3 Interior mesh structure with inflation layers over the body surface

the rocket motor to be more precise with more elements. The finalized mesh structure will result in the following Fig. 2.

3.3 Dynamic Pressure Distribution

The dynamic pressure distribution along the surface of the solid rocket motor can be seen from the pressure contour in the Fig. 3. Here, the pressure increases due to the force acting on the surface by the flow element throughout the entire motion. And the compressible dynamic pressure over a fluid element can be calculated by,

$$q_c = \left(\left[1 + \frac{(\gamma - 1)}{2} M^2\right]^{\frac{\gamma}{\gamma - 1}} - 1\right) \tag{1}$$

3.4 Static Pressure Distribution

The static pressure at the surfaces is the normal drag force acting on the elemental mesh nodes. The variation of the static pressure can be seen from the contour as well from the s t a t i c p r e s su r e versus position graph. The decrease in the static pressure at the component joint/fixed section is due to the change in the flow vector.

Fig. 4 Static pressure variation on the rocket motor

The static pressure can be measured in terms of mach number and the specific weight of air. Since the specific weight is almost constant throughout the body, the variation can be seen in the mach number on the body surfaces (Figs. 4, 5, 6 and 7).

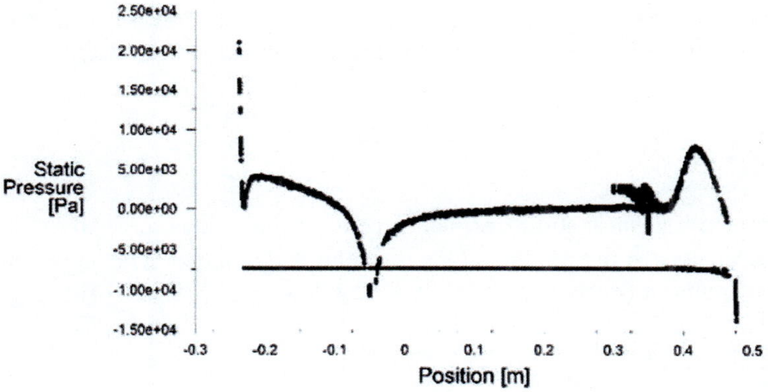

Fig. 5 Static pressure versus position on rocket motor

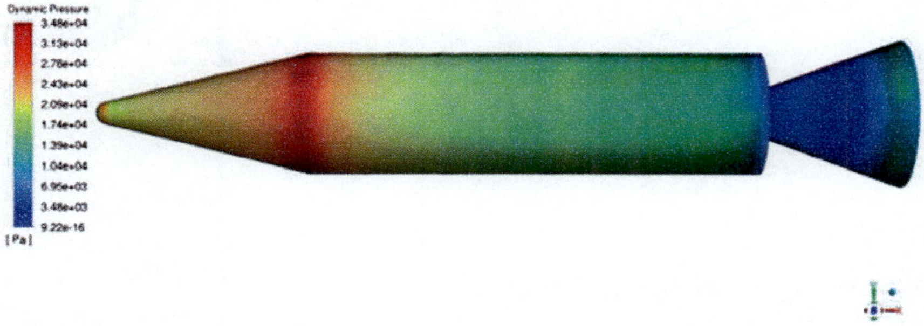

Fig. 6 Dynamic pressure variation on the rocket motor

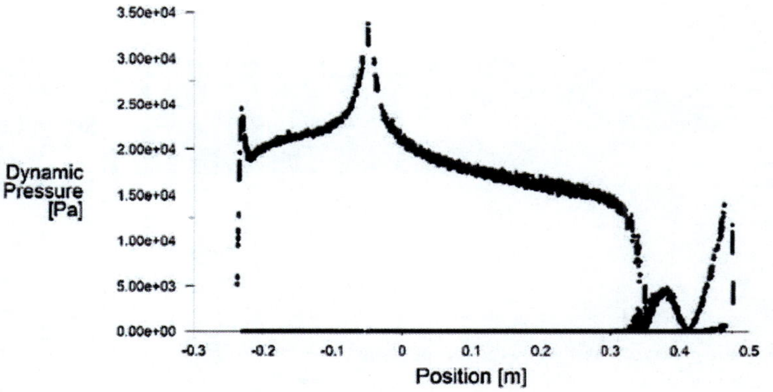

Fig. 7 Dynamic pressure versus position along the solid rocket motor

3.5 *Total Pressure Variation*

The total pressure is the summation of static and dynamic pressure. Therefore, the total pressure over a surface element for compressible flow is calculated by,

$$P_0 = p + p\left(\left[1 + \frac{(\gamma - 1)}{2} M^2\right]^{\frac{\gamma}{\gamma - 1}} - 1\right) \equiv p + q_c \qquad (2)$$

From the wall shear stress distribution contour, it can be observed that the shear stress decreases as the position along the solid rocket motor. This is because the velocity is almost kept constant along with the flow (Figs. 8, 9 and 10).

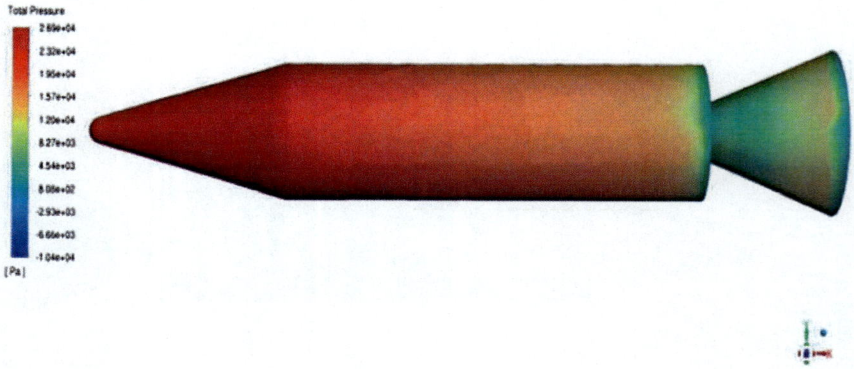

Fig. 8 Total pressure variation on the rocket motor

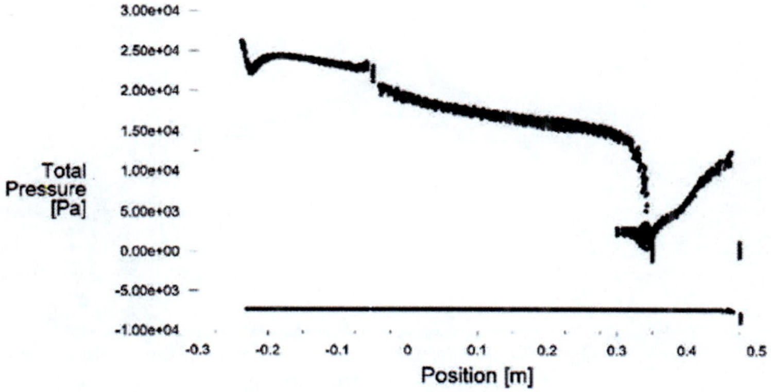

Fig. 9 Total pressure versus position along the rocket motor

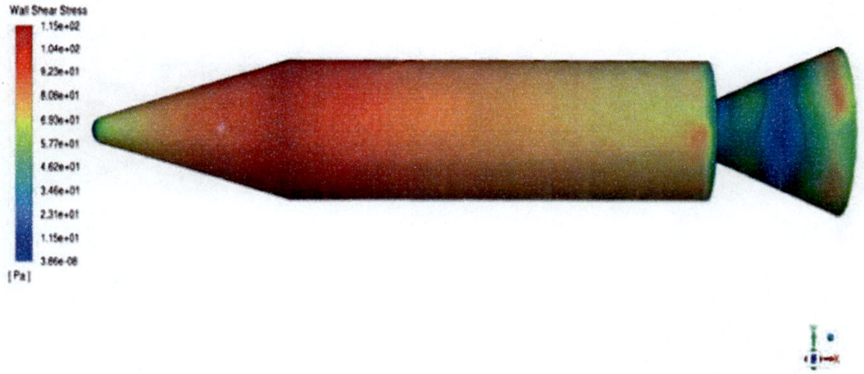

Fig. 10 Wall shear stress variation on the rocket motor

3.6 Wall Shear Stress Distribution

The maximum shear stress along the solid rocket motor can be seen on the angular variance from the nose cone to the motor case. This is because of the shear force acting against the flow vector. Here, for a viscous flow, as the velocity increases along with the position of the solid rocket motor, the wall shear stress increases. The wall shear stress can be calculated using (Fig. 11)

3.7 Velocity Angle Variation

The variation along the angle at mach 0.6 which the flow hits the solid rocket motor is opposite over the upper and lower surface of the body. The velocity angle increases

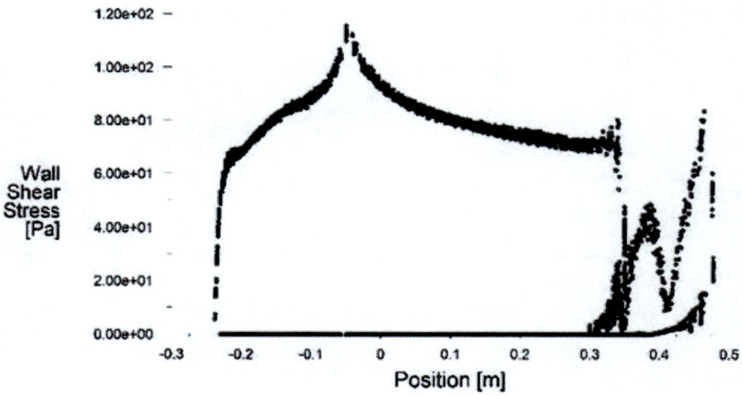

Fig. 11 Wall shear stress versus position along with the rocket motor

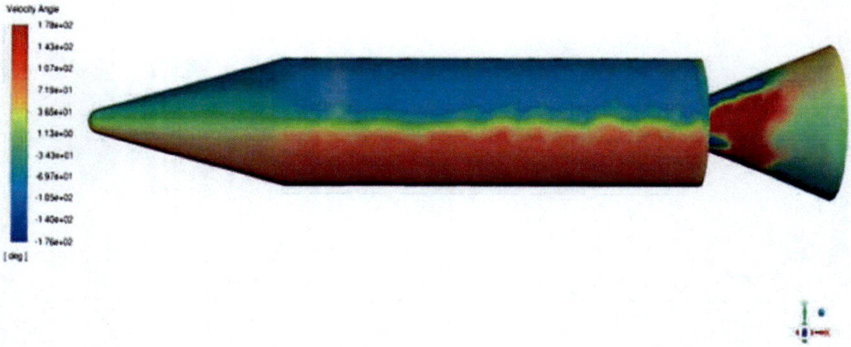

Fig. 12 Velocity angle variation

in both directions as the end surfaces of the rocket motor reaches. This can be seen in the variation contour for the velocity angle given below (Fig. 12).

$$r_w = \mu\left(\frac{\partial u}{\partial x}\right)_{x=0} \tag{3}$$

where, μ = Dynamic viscosity.

3.8 Velocity Variation

From the velocity contour, it is visible that, the flow is attached to the surface till the end of the motor case. This provides a smooth, non-turbulent launch of the solid rocket motor. At the end of the motor case, it is seen a small flow circulation is being

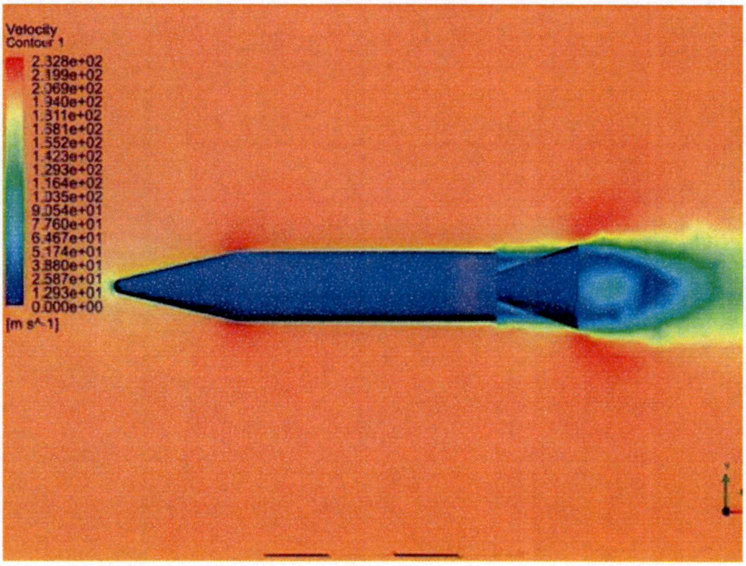

Fig. 13 Velocity profile on the solid rocket motor

developed. This causes a small amount of drag on the rocket motor. The velocity profile at the nozzle section will not be the same when the thrust provided by the oxidizer and the propellant grain are activated. So, the drag due to that is significantly lower when the thrust due to the rocket motor is turned on (Fig. 13).

From the velocity streamline figure, it is clearly observed that the flow circulation has clearly lower velocity profile. This can decrease the bypass ratio and even create small amounts of vortices at the edge of the motor case. It may be reduced by surface flow attaching methods and such (Fig. 14).

4 Conclusions

The main objective of this paper was to determine and identify the uniform compressible flow analytics and structural analytics over the solid rocket motor by the computational method. This is achieved by the results from the pressure distributions, wall shear stress distribution, and velocity profile. Calculations over the mesh model with minimum inflation layer thickness provided near to the accurate results over the theoretical approximations. It can be increased based on higher nodal elements at the mesh period. The high pressure developed area over the rocket motor is where the wall shear stresses act the most intending to create deformation on the body which can lead to high drag and failure of the mission objective. This can be resolved by using higher tensile and Young's modulus material such as magnesium composites

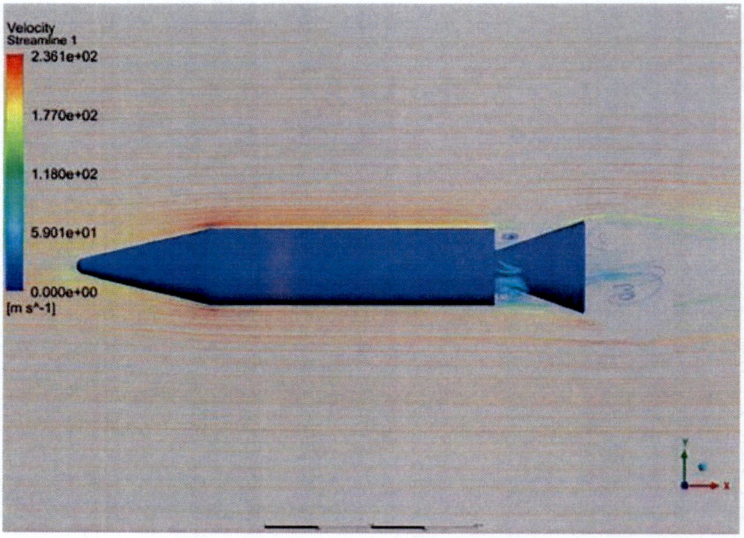

Fig. 14 Velocity streamline

or any titanium alloys. The results from the velocity profile give a more significant advantage to attached surface flow which is better during the launch sequence. Also, the flow circulation at the end of the motor case causes a reduction in flow velocity creating drag. This can also be resolved by the addition of vortex generators or any flow attaching surface models which can be implemented.

References

1. Mejia GL, Rocha RJ, Iha K, Rocco JAFF (2017) Multiperforated solid rocket motor burn simulation. 53rd AIAA/SAE/ASEE Jt Propuls. Conf. 2017(July):1–6. https://doi.org/10.2514/6.2017-4693
2. Navarrete-Martin L, Krus P (2018) Sounding Rockets: Analysis, simulation and optimization of a solid propellant motor using Hopsan. Transp. Res. Procedia 29(2017):255–267. https://doi.org/10.1016/j.trpro.2018.02.023
3. Rajesh S, Suresh G, Mohan RC (2017) A Review on Material Selection and Fabrication of Composite Solid Rocket Motor (SRM) Casing. Int J Mech Solids 9(1):973–1881
4. Kumar D, Nayana B (2016) Design and structural analysis of solid rocket motor casing hardware used in aerospace applications. J Aeronaut Aerosp Eng. 5(2). https://doi.org/10.4172/2168-9792.1000166
5. Khan SA, Ibrahim OM, Aabid A (2021) CFD analysis of compressible flows in a convergent-divergent nozzle. Mater Today Proc 46:2835–2842. https://doi.org/10.1016/j.matpr.2021.03.074
6. Anthoine J, Lema MR (2009) Passive control of pressure oscillations in solid rocket motors: Cold-flow experiments. J Propuls Power 25(3):792–800. https://doi.org/10.2514/1.39794

7. French AD,Panelli M, Di Lorenzo G, Schettino A, Paglia F (2017) Combustion instability and Pressure oscillation numerical simulation in a solid rocket motor. In: 53rd AIAA/SAE/ASEE Jt. Propuls. Conf. 2017, no. July, pp. 10–12. https://doi.org/10.2514/6.2017-4952
8. Ji S, Wang B (2019) Modeling and analysis of triggering pulse to thermoacoustic instability in an end-burning-grain model solid rocket motor. Aerosp Sci Technol 95:105409. https://doi.org/10.1016/j.ast.2019.105409
9. Gallier S, Godfroy F (2009) Aluminum combustion driven instabilities in solid rocket motors. J Propuls Power 25(2):509–521. https://doi.org/10.2514/1.37664
10. Simulation N, Motors R (1992) Shedding phenomenon in 2D Test
11. Suksila T (2018) Experimental investigation of solid rocket motors for small sounding rockets. IOP Conf Ser Mater Sci Eng 297(1). https://doi.org/10.1088/1757-899X/297/1/012009
12. Anthoine J, Jézéquel P, Prévost M, Prévot P, Casalis G (2015) Variable fuel grains burn velocities to reduce solid- rocket-motor pressure oscillations. J Propuls Power 31(1):342–351. https://doi.org/10.2514/1.B35206
13. Kirkkoprn K, Kassoy DR, Zhao Q (1995) Unsteady vorticity generation and evolution in a model of a solid rocket motor: Sidewall mass addition transients. In: 33rd Aerosp. Sci. Meet. Exhib., vol. 12, no. 4. https://doi.org/10.2514/6.1995-603
14. Mishra AK, Jadhav S (2022)Theoretical aspects on design and performance characteristics for solid rocket motor. Int J All Res Educ Sci Method 10(2):894–898.
15. Mishra AK, Gandhi K,Sharma K, Sumanth N (2021) Conceptual design and analysis of two stage sounding rocket. Int J Univ Sci Eng 7(1):52–72
16. Mishra AK, Madhumitha M, Shenoy V (2021) Numerical and Computational Analysis on Two Stage Sounding Rocket. Int J Sci Eng Technol 9(4):01–06

Effects of Confinement and Reynolds Number Variation on the Flow Field of Swirling Jets

Rohit Sharma and Fabio Cozzi

Abstract Swirling jets are commonly used in the burners of gas turbines to enhance mixing or to stabilize the flames. In this manuscript, a stereo (3C2D) particle image velocimetry (SPIV) was applied to examine the near exit region of an isothermal swirling jet inside an octagonal-shaped combustion chamber with a focus on the effects of confinement and Reynolds number (Re) variation on the swirling flow field. Measurements were performed on an axial plus tangential entry swirl burner with a geometric swirl number (S_g) of 1.8 and two different Re, corresponding to 10,900 and 21,800. The results observed in our experimental work are scaled appropriately with the swirl number based on the flux of the axial momentum. The contours of the mean axial velocity field reveal the occurrence of vortex breakdown (VB) for the confined jets compared to the unconfined jets for both Reynolds numbers. Upon confinement, the flow field is dominated by the existence of VB with a wider central recirculation zone (CRZ) and with enhanced axial velocity fluctuations. The enhancement in the Re further increased the CRZ and enhanced the magnitudes of the mean axial velocity and its fluctuations. The outcome obtained from these results includes a better knowledge of the swirl jet in the swirl burner's near region. In addition, the experimental data can be useful for validating computational fluid dynamics (CFD) modeling.

Keywords Swirling jets · SPIV (Stereoscopic particle image velocimetry) · Confinement · Vortex breakdown

Nomenclature

ATSB	Axial plus tangential entry swirl burner

R. Sharma (✉)
School of Mechanical Engineering, VIT Bhopal University, 466114, Bhopal, India
e-mail: rohit.sharma@vitbhopal.ac.in

F. Cozzi
Department of Energy, Politecnico Di Milano, 20156 Milano, Italy

S. Das et al. (eds.), *Proceedings of the 1st International Conference on Fluid, Thermal and Energy Systems*, Lecture Notes in Mechanical Engineering,
https://doi.org/10.1007/978-981-99-5990-7_63

A_t	Total area of tangential air inlets
CC	Combustion chamber
CoRZ	Corner Recirculation zone
CRZ	Central Recirculation zone
D	Nozzle diameter
G_z	Axial flux of axial momentum
G_θ	Axial flux of angular momentum
IRZ	Inner Recirculation zone
$\dot{m}$	Mass flow rate
$\dot{m}_{axial}$	Mass flow rate of axial air inlet
$\dot{m}_{tangential}$	Mass flow rate of tangential air inlet
PIV	Particle image velocimetry
PVC	Precessing vortex core
R	Nozzle radius
R_o	Radial distance of tangential inlets
frr,z.y	Coordinates
RB	Recirculation bubble
Re	Reynolds number
S	Swirl number
S_{cr}	Critical swirl number
S_g	Geometric swirl number
SJ	Swirling Jet
SP	Stagnation Point
SR	Split ratio
U	Mean Axial velocity
u'	Axial velocity fluctuation
U_{bulk}	Bulk velocity
V	Mean radial velocity
v'	Radial velocity fluctuation
VB	Vortex breakdown
W	Mean tangential velocity
w'	Tangential velocity fluctuation
ρ	Density of air
μ	Dynamic viscosity of air

1 Introduction

Swirling flows are widely used in gas turbines, including both aviation and non-aviation, for controlled mixing of species to increase flame stability and reduce emissions [1–4]. These considerable advantages along with the distribution of swirl in other technical applications have prompted both experimental and numerical studies, to describe the swirl flow under various conditions such as aspect ratio, confinement, non-reacting and reacting conditions, etc. [1]. The behavior of the swirl jet is known to depend strongly on a non-dimensional parameter called the swirl number, S [5, 6]. Swirling flows have been investigated in depth over the past decades and continue to be a topic of interest both in industry and academia which can be attributed to the fact that the SJs exhibit a rich variety of phenomena such as recirculation, VB, processing vortex core (PVC) and its associated instability [5, 7–9]. The VB corresponds to a sudden change in flow structure evidenced by the presence of a stagnation point (SP) in the jet and the presence of a region of reverse flow; PVC on the other hand ordinarily visible after the inception of VB, characterized by the precession of the jet around its geometrical axis. Even after substantial research has been done, these phenomena are still a matter of concern.

2 Literature Review and Objective

The importance of SJs in combustion applications has accelerated a significant amount of work on confined swirling jets [1, 10, 11]. Syred and Dahman [10] showed that the effects of the confinement would be more noticeable when the confinement ratio is less than 9 for high swirl numbers [10, 12]. A study on the different confinement levels indicated the existence of a distinct number of recirculation zones (RZ) for larger and smaller ducts [13]. Stohr et al. [14] examined the effect of the PVC on the flow field, mixing, ignition and the shape of the flame zone of a turbulent swirling jet. Vashahi et al. [11] conducted measurements in a model rectangular gas turbine combustor to evaluate the effect of rectangular confinement on the generated flow field. Fu et al. [15] demonstrated the aerodynamic features of the isothermal swirling flow field connected with a counter-rotating radial-inlet swirler and the effects of Reynolds number on the flow shape were also explored. They found the effect of the confinement on turbulent flow field and concluded that the size of CRZ for the confined flow was almost twice as compared to an unconfined case.

Studies [1] have shown that the RZ was significantly enlarged upon confinement and with the increase in Re for both non-reacting and reacting cases. Additionally, the length of CRZ was found to be closely related to the geometry of the CC.

There is yet a paucity of studies addressing flow behavior in confined arrangements compared with other key highlights of swirling flows. However, experimental data of the confined flow regimes on the axial plus tangential type swirl burner are still scarce. In the present investigation, the confinement is due to the presence of an octagonal combustion chamber and the confinement ratio (the area ratio is about 29) is much greater than in previous works. To this extent, it is quite interesting to understand flow behavior and predict its modes. The current study's main goal is to quantitatively describe the Re and confinement variation effects on the swirling jet flow field.

In the context of this investigation, this research paper is organized as follows. Section experimental setup and measuring techniques provide an overview of the axial plus tangential entry swirl burner, octagonal confinement and details of the

SPIV configuration. Section characterization parameters describe the various non-dimensional numbers used to characterize the swirling flow field. Section results and discussion emphasize the experimental findings and discussion on the present investigation. Finally, in the conclusions section, our main findings are summarized.

3 Experimental Setup and Measuring Techniques

The experiments were conducted using a vertically mounted swirl combustor under various configurations. More details on the ATSB can be found in Cozzi et al. [16] and Sharma and Cozzi [17]. The ATSB was confined using an octagonal combustion chamber with optical access as shown in Fig. 1 and having a confinement ratio, $\frac{D_{confinement}}{D} = 5.4$, where $D_{confinement} = 193$ mm is the apothem of the octagonal combustion chamber and $D = 36$ mm is the nozzle diameter. The dimension of each window of the CC is 290 mm in height and 80 mm in width.

SPIV system and Dantec's software were used to acquire the flow field and to process the snapshots respectively. 700 PIV snapshots were acquired for each case to get statistical information on the flow field. The relative error of the mean velocity components is 4% of the bulk velocity. The relative statistical uncertainty in the estimation of r.m.s. value of velocity fluctuations ($\varepsilon_{r.m.s} = \frac{1.96}{\sqrt{2N}}$, where$N = 700$) is 5% of the measured value, with a confidence level of 95%. For more details about the PIV system, cameras, errors, etc. can be found in Cozzi et al. [5].

3.1 Characterization Parameters

The swirl intensity for swirling flows is usually characterized by the non-dimensional swirl number, defined as the ratio of axial flux of azimuthal momentum (G_θ) to axial flux of axial momentum (G_z) times the nozzle radius (R), can be written as

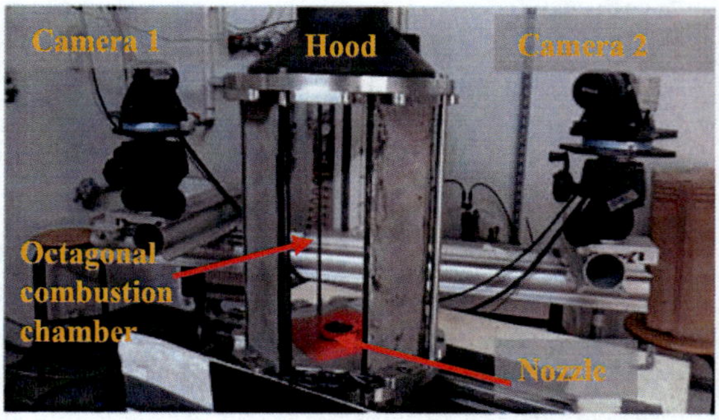

Fig. 1 Photograph of the SPIV setup for confined flow in the longitudinal plane

Table 1 Experimental conditions

Confined	Re	S_g	S	VB	PVC
No	10,900	1.8	0.80	No	No
No	21,800	1.8	0.78	No	No
Yes	10,900	1.8	1.03	Yes	No
Yes	21,800	1.8	1.10	Yes	No

$$S = \frac{G_\theta}{G_z R} \tag{1}$$

Assuming the boundary layer approximations and negligibly small turbulent stresses w.r.t the mean velocity terms, the swirl number is simplified as [6, 18]

$$S = \frac{2\pi\rho\int_0^R r^2 UW dr}{2\pi\rho\left[\int_0^R r(U^2 - \frac{1}{2}W^2)dr\right] \times R} \tag{2}$$

and the geometric swirl number is defined as [19–21]

$$S_g = \frac{\pi R_o R}{A_T}\left(\frac{\dot{m}_{tangential}}{\dot{m}_{tangential} + \dot{m}_{axial}}\right)^2 \tag{3}$$

The Re is defined as the ratio between the inertial and the viscous forces in the flow and is written as.

$$Re = (\rho U_{bulk} D)/\mu \tag{4}$$

see the nomenclature for terms shown in Eqs. (2), (3), and (4). Table 1 shows the experimental parameters for each experimental case.

4 Results and Discussion

The flow field under unconfined and confined conditions is obtained and characterized by means of Reynold numbers (Re) and geometric swirl number (S_g). This section describes the effects of Re and confinement variation swirling jets flow field.

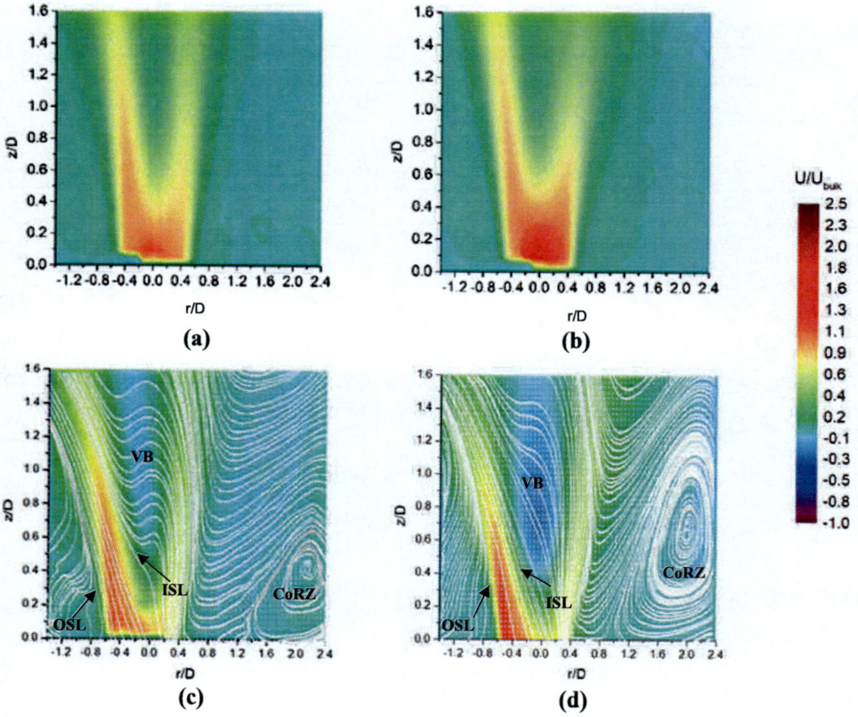

Fig. 2 Contour plots of mean axial velocity, U/U_{bulk} in a vertical plane for $S_g = 1.8$, **a** unconfined case 1, Re = 10,900, **b** unconfined case 2, Re = 21,800, **c** confined case 1, Re = 10,900, and **d** confined case 2, Re = 21,800. Streamlines superimposed on contour plots of confined cases

4.1 Unconfined Flow Field

With the same experimental configuration, a series of tests were conducted at similar swirling conditions with two different Re, corresponding to 10,900 and 21,800. The contours of mean axial velocity (U) which refer to cases related to these Reynolds number is shown in Fig. 2a and b. For brevity, we will compare only the contour map of the U on the same scale. The structure of the jet does not appear different for the two different Re for the unconfined jet at $S_g = 1.8$. However, no negative mean axial velocity was observed, so no recirculation shows up in the average flow field.

4.2 Confined Flow Field

The axial velocity contour for the confined jet ($S_g = 1.8$ and Re = 21,800) is shown in Fig. 2d. The location of the inner shear layer (ISL), corner recirculation zone (CoRZ), vortex breakdown (VB), and outer shear layer (OSL) can be noticed in the confined mean axial velocity field cases (see Fig. 2c and d). One can notice the presence of the reverse flow region upon confinement as compared to the unconfined cases for both Reynolds numbers. It is worth noting that the stagnation points for Re = 10,900 and Re = 21,800 confined swirling jets are located on the axis of the jet at approximately z/D = 0.63 and z/D = 0.43, respectively. Both the reverse flow and the stagnation point presence on the jet axis are a feature of the vortex breakdown phenomena, and they only show up in the confined mean flow field. The confinement of the jet thus affects the vortex breakdown process as it makes the axial velocities (compare Fig. 2a and b with Fig. 2c and d) and tangential velocities (not shown here) decay faster in the axial direction so as to develop a stronger axial pressure gradient [22] which eventually promotes the vortex breakdown formation.

For the unconfined and confined SJ case with the same Re and Sg, confined SJ undergoing VB due to further increase of the adverse pressure gradient due to hood and combustion chamber and flow reverse its direction in the central region of the jet. An increase in Re under confined isothermal conditions further strengthened the RZ downstream of the nozzle. Although the CRZ with the lowest Re has a shorter length and shorter width (see Fig. 2c and d). An increase in Re under the confined configurations significantly enhanced the RZ width as shown in Fig. 2d.

The presence of combustion chamber/confinement results in the appearance of a corner recirculation zone (CoRZ) as shown in Fig. 2c and 2d. The streamlines are also superimposed on the mean axial velocity contour to outline the CoRZ. The presence of the confined wall has a strong effect on the apparent flow structure. The confinement restricts the entrainment phenomena, and the position of the confined wall controls the strength of the pressure gradient in the corner recirculation zone as a result of the Coanda effect [23]. Furthermore, the increased Reynolds number enhanced the magnitude of the CoRZ and led to the downstream displacement of the CoRZ, as shown by the streamlines for both confined cases (see Fig. 2c and d).

4.3 *Axial Velocity Fluctuations*

The contour plots of the axial velocity fluctuations (u′) for the unconfined jet are shown in Fig. 3a and b (notice the same scale). High u′ was noticed in the center, the area where the deficit of axial velocity is observed, confirming the strong instability present in the transition between jet-like shape and wake-like in the axial velocity. The high RMS could be due to the movement of the vortex bubble from one position to another and back again which is visible in the instantaneous axial velocity maps (as shown in Fig. 4).

The contour maps of the u′ for the confined jet are shown in Fig. 3c and d. The confinement increases the axial velocity fluctuations near the nozzle exit (z = 0.08D – 0.56D for Re = 10,900 and z = 0.02D – 0.38D for Re = 21,800) and within the shear layers (ISL and OSL), but it decreases the u′ further downstream (z/D > 1.11 for Re = 10,900 and z/D > 0.78 for Re = 21,800) in the vicinity of the jet axis compared to the unconfined jet. It is noteworthy that for the confined jet, lower

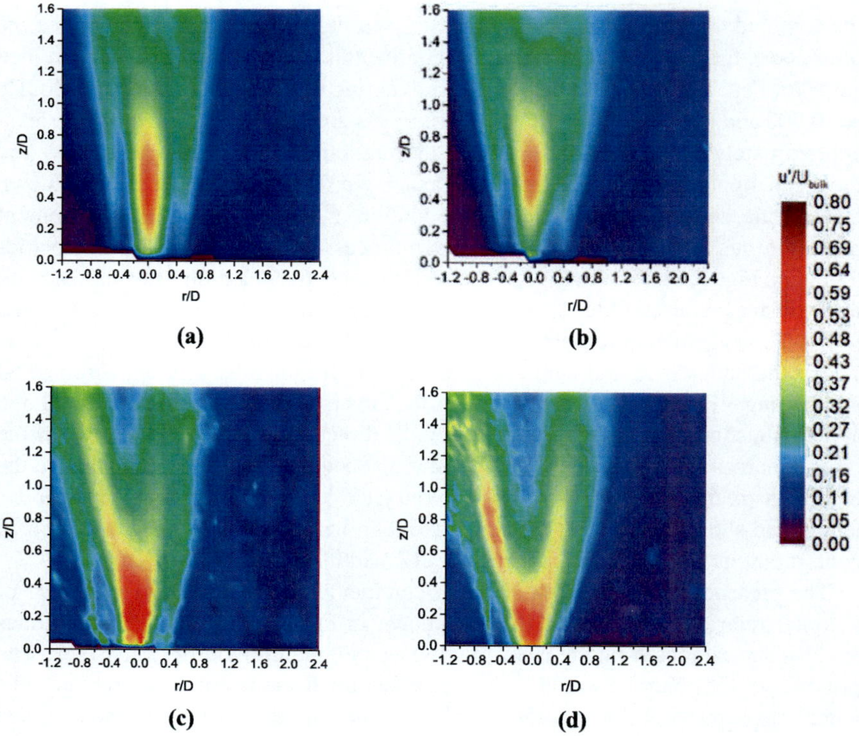

Fig. 3 Contour plots of axial velocity fluctuations, u'/U_{bulk} in a vertical plane for $S_g = 1.8$, **a** unconfined case 1, Re = 10,900, **b** unconfined case 2, Re = 21,800, **c** confined case 1, Re = 10,900, and **d** confined case 2, Re = 21,800

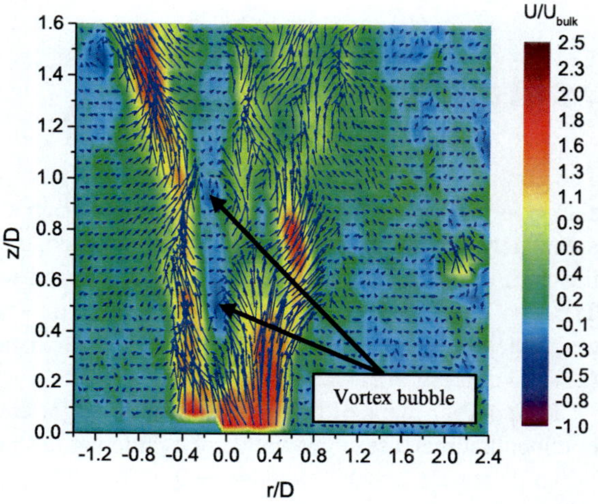

Fig. 4 Contour plot representation of instantaneous axial velocity and vectors superimposed on it for $S_g = 1.8$ and Re = 21,800

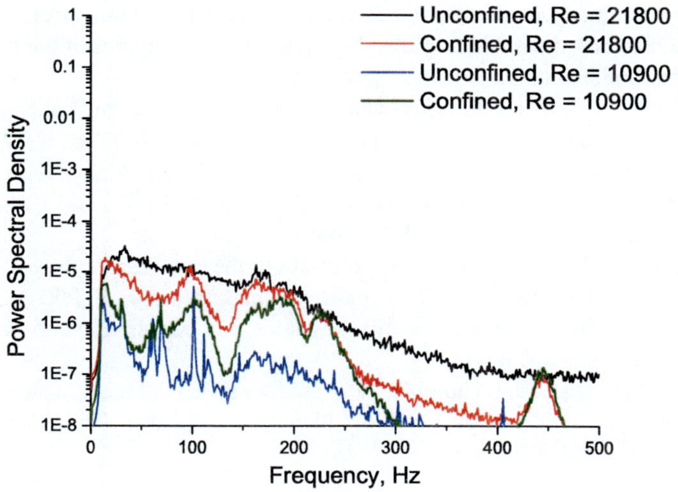

Fig. 5 Comparison of power spectral density of capacitive microphone signals for the unconfined and confined swirling jets

axial velocity fluctuations occur in the reverse flow region downstream of the axial stagnation point.

4.4 Frequency Spectra of Pressure Fluctuations

To determine the occurrence of PVC under unconfined and confined conditions for the cases discussed in the present manuscript, the pressure fluctuations signal was recorded using a capacitive microphone. For Re = 10,900 and 21,800, no distinct sharp peak is detected in the spectra of pressure fluctuations as shown in Fig. 5, indicating that no PVC was found at $S_g = 1.8$.

5 Conclusions

The flow field of the ATSB has been described under unconfined and confined circumstances, with a focus on demonstrating the effect of Reynold's number variation and octagonal shape confinement on the swirling flow field. The main findings are summarized below:

1. The results reveal the occurrence of vortex breakdown earlier due to confinement. Also, confinement has a significant effect on the strength of the CRZ.

2. Under confined conditions, increasing the Reynolds number from 10,900 to 21,800 increased the width of the CRZ. Also, the magnitudes of the mean axial velocity and its fluctuations increased.
3. Confinement increased the axial velocity fluctuations near the nozzle exit (z = 0.08D − 0.56D for Re = 10,900 and z = 0.02D − 0.38D for Re = 21,800) and within the shear layers, but decreased the velocity fluctuations in the vicinity of the jet axis towards downstream (z/D > 1.11 for Re = 10,900 and z/D > 0.78 for Re = 21,800) as compared to the unconfined jet.
4. For confined flow, the stagnation point along the nozzle axis moves upstream from z/D = 0.63 to z/D = 0.43 for Reynolds number = 10,900 and 21,800, respectively. This revealed that the Reynolds number significantly affects the position of the CRZ at a fixed S_g.
5. For the investigated swirl number, the frequency spectra of the measured pressure fluctuations didn't evidence any significant sharp peaks, thus evidencing the absence of the PVC phenomenon in the flow.

The outcome of the experimental results includes a better knowledge of the swirling flow field in the near exit region of the swirl burner. Furthermore, the experimental data may be useful for validating CFD modeling.

Acknowledgements The authors would like to thank Prof. Aldo Coghe for the fruitful discussion and invaluable assistance in carrying out the experiments.

References

1. Khalil AEE, Brooks JM, Gupta AK (2016) Impact of confinement on flowfield of swirl flow burners. Fuel 184:1–9. https://doi.org/10.1016/j.fuel.2016.06.098
2. Cozzi F, Coghe A, Sharma R (2018) Analysis of local entrainment rate in the initial region of isothermal free swirling jets by Stereo PIV. Exp Therm Fluid Sci 94:281–294. https://doi.org/10.1016/j.expthermflusci.2018.01.013
3. Barakat S, Wang H, Jin T, Tao W, Wang G (2021) Isothermal swirling flow characteristics and pressure drop analysis of a novel double swirl burner. AIP Adv 11. https://doi.org/10.1063/5.0041361
4. Litvinov IV, Suslov DA, Gorelikov EU, Shtork SI (2021) Swirl number and nozzle confinement effects in a flat-vane axial swirler. Int J Heat Fluid Flow 91:108812. https://doi.org/10.1016/j.ijheatfluidflow.2021.108812
5. Cozzi F, Sharma R, Solero G (2019) Analysis of coherent structures in the near-field region of an isothermal free swirling jet after vortex breakdown. Exp Therm Fluid Sci 109:67–74. https://doi.org/10.1016/j.expthermflusci.2019.109860
6. Vignat G, Durox D, Candel S (2022) The suitability of different swirl number definitions for describing swirl flows: accurate, common and (over-) simplified formulations. Prog Energy Combust Sci 89:100969. https://doi.org/10.1016/J.PECS.2021.100969
7. Santhosh R, Miglani A, Basua S (2014) Transition in vortex breakdown modes in a coaxial isothermal unconfined swirling jet. Phys Fluids 26. https://doi.org/10.1063/1.4870016
8. Oberleithner K, Paschereit CO, Seele R, Wygnanski I (2012) Formation of turbulent vortex breakdown: intermittency, criticality, and global instability. AIAA J 50:1437–1452. https://doi.org/10.2514/1.J050642

9. Alekseenko SV, Abdurakipov SS, Hrebtov MY, Tokarev MP, Dulin VM, Markovich DM (2018) Coherent structures in the near-field of swirling turbulent jets: A tomographic PIV study. Int J Heat Fluid Flow 70:363–379. https://doi.org/10.1016/j.ijheatfluidflow.2017.12.009
10. Syred N, Dahman DR (1978) Effect of high levels of confinement upon the aerodynamics of swirl burners. J Energy 2:8–15. https://doi.org/10.2514/3.47950
11. Vashahi F, Lee S, Lee J (2016) Experimental analysis of the swirling flow in a model rectangular gas turbine combustor. Exp Therm Fluid Sci 76:287–295. https://doi.org/10.1016/j.expthermflusci.2016.03.032
12. Syred N (2006) A review of oscillation mechanisms and the role of the precessing vortex core (PVC) in swirl combustion systems. Prog Energy Combust Sci 32:93–161. https://doi.org/10.1016/j.pecs.2005.10.002
13. Farokhi S, Taghavi RRE (1989) Effect of initial swirl distribution on the evolution of a turbulent jet. AIAA J 27:700–706
14. Stöhr M, Sadanandan R, Meier W (2011) Phase-resolved characterization of vortex-flame interaction in a turbulent swirl flame. Exp Fluids 51:1153–1167. https://doi.org/10.1007/s00348-011-1134-y
15. Yongqiang Fu, Jun Cai S-MJ and HM (2007) Characteristics of the swirling flow generated by a counter-rotating swirler. In: 43rd AIAA/ASME/SAE/ASEE Jt. Propuls. Conf. Exhib., American Institute of Aeronautics and Astronautics. https://doi.org/10.2514/6.2007-5690
16. Cozzi F, Sharma R, Coghe A, Arzuffi F (2015) An Experimental Investigation on Isothermal Free Swirling Jet. XXXVIII Meet Ital Sect Combust Inst 3–8
17. Sharma R, Cozzi F (2017) Experimental study of unconfined and confined isothermal swirling jets. Int J Mech Mechatronics Eng 11:386–396. https://doi.org/10.1999/1307-6892/10006459
18. Panda J, McLaughlin DK (1994) Experiments on the instabilities of a swirling jet. Phys Fluids 6:263–276. https://doi.org/10.1063/1.868074
19. Feikema D, Chen RH, Driscoll JF (1990) Enhancement of flame blowout limits by the use of swirl. Combust Flame 80:183–195. https://doi.org/10.1016/0010-2180(90)90126-C
20. Claypole T, Syred N (1981) The effect of swirl burner aerodynamics on NO_x formation. In: Eighteenth Symp Combust Combust Inst 81–9
21. Sharma R, Cozzi F, Coghe A (2016) Phase-averaged characterization of turbulent isothermal free swirling jet after vortex breakdown 4–7
22. Lucca-Negro O, O'Doherty T (2001) Vortex breakdown: a review. Prog Energy Combust Sci 27:431–481. https://doi.org/10.1016/S0360-1285(00)00022-8
23. Douglas CM, Lesshafft L (2022) Confinement effects in laminar swirling jets. J Fluid Mech 945:27. https://doi.org/10.1017/jfm.2022.589

Effect of Corner Curvature of Square Cylinder on Flow Transition and Heat Transfer

Prashant Kumar, Saurabh Singh Chauhan, and Prasenjit Dey

Abstract Fluid flow and heat transfer over a square cylinder with modified corners are studied numerically at a very low Reynolds number (*Re*). The present study aims to investigate the flow transition from a creep flow to a steady separated flow and the corresponding effect on the heat transfer characteristic. The corners of the square cylinder are modified by changing the corner curvature and are varied from 0 to 1. The effect of corner curvature radius on the critical Reynolds number (Re_{cr}) where the flow transition occurs is investigated along with the flow features, and heat transfer characteristics at the Re_{cr}. It is found that Re_{cr} increases approximately exponentially with the increasing of corner curvature radius and the Re_{cr} is found always less than 7 for all the considered cases. In addition, the drag coefficient decreases with the increment of corner curvature radius. Corner curvature radius (*CCR*) also influences the Nusselt number (*Nu*). It is perceived that the *Nu* increases with increase in corner curvature radius (*CCR*) and as well as with increasing *Re*.

Keywords Square cylinder · Curved corner · Flow transition · Critical Reynolds number · Heat transfer

1 Introduction

Fluid flow and heat transfer over a cylindrical body is one of the most fundamental research subjects in external flow due to its enormous practical importance in almost every field of engineering such as cooling systems of electronic devices, heat exchangers, chimney stacks, ventilation systems, etc. [1]. Fluid flow and heat transfer both are greatly influenced by flow transitions such as flow separation, and onset of periodicity [2]. A critical Reynolds number (Re_{cr}) is defined as at which *Re* the flow transitions take place and the Re_{cr} depends on the geometrical parameters

P. Kumar · S. S. Chauhan · P. Dey (✉)
Department of Mechanical Engineering, National Institute of Technology Goa, Goa 403401, India
e-mail: prasenjit.dey@nitgoa.ac.in

S. Das et al. (eds.), *Proceedings of the 1st International Conference on Fluid, Thermal and Energy Systems*, Lecture Notes in Mechanical Engineering,
https://doi.org/10.1007/978-981-99-5990-7_64

of the body. Dependency of Re_{cr} on the geometry of a cylindrical body is studied by various techniques [3].

Sharma and Eswaran [4] numerically studied a two-dimensional laminar flow and heat transfer around a square cylinder and observed that the flow separation for the square cylinder occurs in between $Re = 1$ to 2. Sen et al. [5] studied the flow over a circular cylinder at low Reynolds number and noticed that the flow separation occurs at $Re = 6.29$ for blockage ratio = 0.05. They also found that if the blockage ratio is less than 0.01 then the blockage ratio doesn't have any significant effect on the characteristic of flow parameters. The effect of cylinder corner curvature along with the angle of attack (α) was studied numerically by Alam et al. [6] and found that corner curvature radius (*r/R*) essentially influences the heat transfer at $\alpha < 30°$ and fluid forces i.e. drag and lift at $\alpha > 10°$. Clearly, there are a lot of foregoing studies of flow transition, but most of them are mainly dedicated to square and circular cylinders. Only a handful of studies are available that explore the fluid flow and heat transfer over a square cylinder with modified corners at low *Re*. However, as per author's best knowledge, no study is available on the critical *Re* of a square cylinder with rounded corners in the steady laminar flow regime when the heat transfer is considered. In addition, the behavior of heat transfer around the cylinder in the vicinity of critical *Re* is not explored yet. The present study thus efforts to fill this gap in the literature for the transition of the flow regime around a square cylinder with rounded corners at low *Re*.

2 Geometric Configuration

Schematic representation of the present computational domain of fluid flow and heat transfer over a cylinder is shown in Fig. 1. Cylinder is placed in a rectangular domain of dimension, 20*D* (Height, *H*) × 55*D* (length, *L*), where *D* is the cylinder diameter (side of square). These dimensions yield a blockage ratio $(D/H) = 5\%$. The 5% of blockage is considered as there is no blockage effect on the flow over a square cylinder at 5% of blockage, as suggested in the earlier studies [7, 8]. Distance of the inlet boundary and the outlet boundary from the cylinder center is 15*D* and 40*D*, respectively. These distances are quite enough to adequately capture the flow separation in the downstream [9].

Roundness of the corners is defined by the corner curvature radius (*CCR*) which is the ratio of corner radius and the radius of the cylinder. $CCR = r/R$ (Fig. 2).

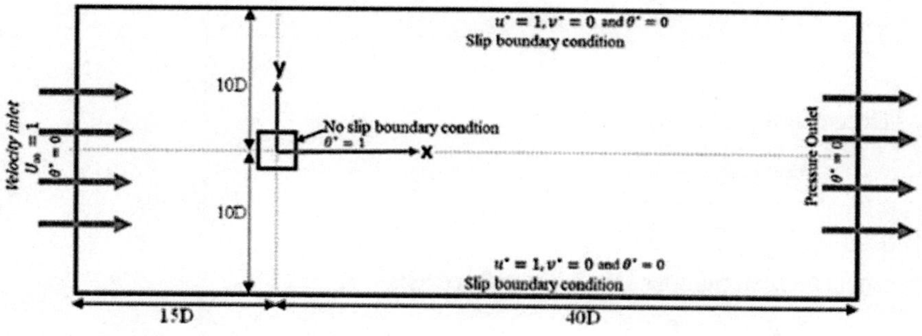

Fig. 1 Computational domain

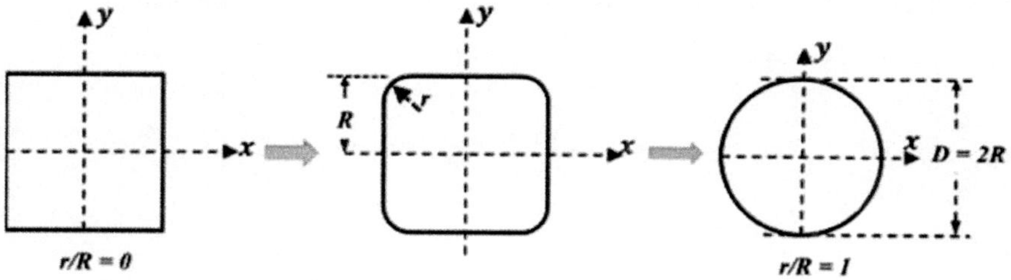

Fig. 2 Corner curvature radius

3 Governing Equations and Boundary Conditions

The Fluid flow problems with heat transfer are governed by Navier–Stokes and energy equations. For an incompressible, steady, laminar flow with heat transfer in two-dimensional coordinate, these equations can be written in non-dimensional form as follows:

The momentum equations:

$$u^*\frac{\partial u^*}{\partial x^*} + v^*\frac{\partial u^*}{\partial y^*} = -\frac{\partial p^*}{\partial x^*} + \frac{1}{Re}\left(\frac{\partial^2 u^*}{\partial x^{*2}} + \frac{\partial^2 u^*}{\partial y^{*2}}\right) \tag{1}$$

$$u^*\frac{\partial v^*}{\partial x^*} + v^*\frac{\partial v^*}{\partial y^*} = -\frac{\partial p^*}{\partial y^*} + \frac{1}{Re}\left(\frac{\partial^2 v^*}{\partial x^{*2}} + \frac{\partial^2 v^*}{\partial y^{*2}}\right) \tag{2}$$

The continuity equation:

$$\frac{\partial u^*}{\partial x^*} + \frac{\partial v^*}{\partial y^*} = 0 \tag{3}$$

The energy equation:

$$u^* \frac{\partial \theta^*}{\partial x^*} + v^* \frac{\partial \theta^*}{\partial y^*} = \frac{1}{RePr} \left(\frac{\partial^2 \theta^*}{\partial x^{*2}} + \frac{\partial^2 \theta^*}{\partial y^{*2}} \right) \tag{4}$$

Drag coefficient (C_D) is defined as:

$$C_D = \frac{F_D}{0.5 \rho {U_\infty}^2 D} \tag{5}$$

Local Nusselt number (Nu_{loc}) is defined as:

$$Nu_{loc} = \frac{h_{loc} D}{K} \tag{6}$$

Surface average Nusselt number is calculated by the following equation,

$$Nu_{avg} = \frac{\int_0^{A_s} Nu_{loc} dA_s}{A_s} \tag{7}$$

3.1 Boundary Conditions

Boundary conditions used in present study are as follows (Fig. 1):

(a) Inlet is given as 'velocity inlet' with $u^* = 1$, $v^* = 0$ and constant temperature, $\theta^* = 0$.
(b) Pressure outlet ($\nabla p^* = 0$) and constant temperature $\theta^* = 0$ condition is applied at the outlet.
(c) Cylinder walls are assigned as no-slip boundary condition i.e., $u^* = 0$, $v^* = 0$ and constant temperature $\theta^* = 1$.
(d) Two lateral boundaries of domain are considered as slip boundary condition, $u^* = 1$, $v^* = 0$ and constant temperature $\theta^* = 0$.

4 Numerical Methodology and Validation

After computational domain is generated, it is divided into 9 sub-blocks. Square cylinder is in the central block. A non-uniform grid structure with 45,000 quadrilateral elements is generated after a detailed grid independency study to ensure an efficient solution in less computational time. Four sub-blocks, each of which shares a common edge with central sub-block, are having finer mesh in order to capture shear layer topology efficiently. Central block (marked with 5) has even more fine mesh. It has the smallest element of size 0.01D. A thorough grid independency study is conducted for all the cylinders and it is observed that a total cell of 45,000 is adequate to achieve the necessities of the both mesh independence and calculation time limit. The detailed study of the grid independency study for the square cylinder at $Re = 10$ is presented in Table 1 (Fig. 3).

Present numerical analysis is accomplished by using a Finite Volume Method based commercial Computational Fluid Dynamics solver, Ansys Fluent 2021 R1. A steady solver with laminar viscous model is used and Standard pressure discretization scheme is adopted for interpolating the pressure on the faces of cells. For pressure–velocity coupling, Semi-Implicit Method for Pressure Linked Equations (SIMPLE) algorithm is used. Second order upwind scheme is implemented for the discretization of momentum and energy equations.

In order to check the authenticity of results obtained in the present numerical study, the present results are authenticated with the available literature values for square ($r/R = 0$) and circular cylinder ($r/R = 1$) at $Re = 5$, 10 and 20, blockage ratio of 5%, and Prandtl number, $Pr = 0.71$. Recirculation length (W_r) and surface mean drag coefficient ($C_{d,mean}$) are calculated for the validation study. The discrepancies in the results are tabulated in Table 2 from which it can be concluded that the results of the present study are in great agreement with those in available literature.

Table 1 Grid independency study

No. of cells	Parameters			% Change		
	W_r	$C_{D,mean}$	Nu_{avg}	W_r	$C_{D,mean}$	Nu_{avg}
30,000	0.52	3.69	1.57			
45,000	0.50	3.55	1.51	3.24	3.88	4.01
60,000	0.50	3.52	1.50	0.8	0.91	0.98

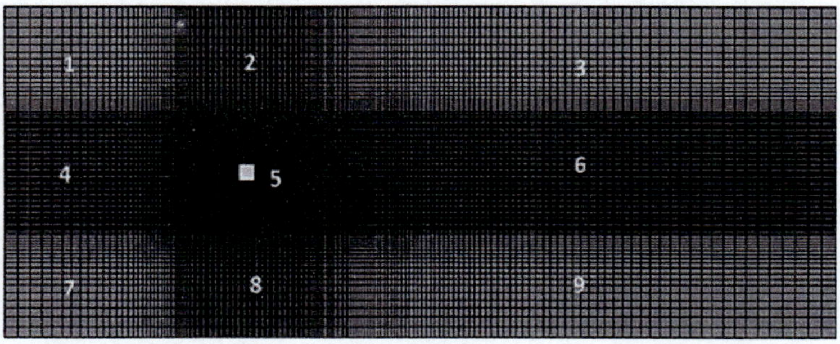

Fig. 3 Grid structure of computational domain

Table 2 Validation for a square and circular cylinder for $Pr = 0.71$

Literature	Re = 5		Re = 10		Re = 20	
	W_r	$C_{d,mean}$	W_r	$C_{d,mean}$	W_r	$C_{d,mean}$
r/R = 0.0						
Gupta et al. [10]	–	4.840	0.49	3.63	1.05	2.44
Dey and Das [11]	–	4.716	0.50	3.52	1.10	2.42
Present study	–	4.760	0.50	3.55	1.08	2.42
r/R = 1.0						
Tritton [12]	–	–	–		–	2.22
Dey and Das [11]	–	–	–	3.22	–	2.24
Present study	–	–	–	3.21	–	2.24

5 Results and Discussion

5.1 Streamlines

Figure 4a shows the flow characteristics for *r/R* = 0.4 at three different *Re* cases based on the observed flow separation from the cylinder surface. Flow transition from the creep flow to the steady separated flow from the rear surface can be observed in Fig. 4a with zoomed view. At a certain *Re*, the shear layers start to separate from the rear surface resulting in a steady symmetric recirculation bubble as seen in Fig. 4a (middle) and the corresponding minimum *Re* is called as Re_{cr}. The shear layers are found completely attached around the cylinder surface at $Re < Re_{cr}$ [Fig. 4a (top)] and when the *Re* is increased to $Re > Re_{cr}$, the size of the recirculation bubble increases [Fig. 4a (bottom)]. Further, the steady symmetric recirculation bubble for all the considered *r/R* cases at $Re = Re_{cr}$ is shown in Fig. 4b. It is found that the height of the recirculation bubble (H_r) decreases with increasing *r/R*, whereas the width (W_r) increases with increasing *r/R*. The values of Re_{cr} for various values of CRR are shown in Table 3. It is noticed that the Re_{cr} is significantly dependent on the corner curvature radius and can be fitted by an exponential equation as shown in Fig. 5.

5.2 Velocity Magnitude

Velocity magnitude contours (Fig. 6) help in evaluating the shear layer topology. Shear layer possess high velocity gradients hence larger shear stresses. It is observed that as the corner radius ratio increases, low velocity region near the front stagnation point shrinks because the cylinder's shape is becoming continuous as CRR increases.

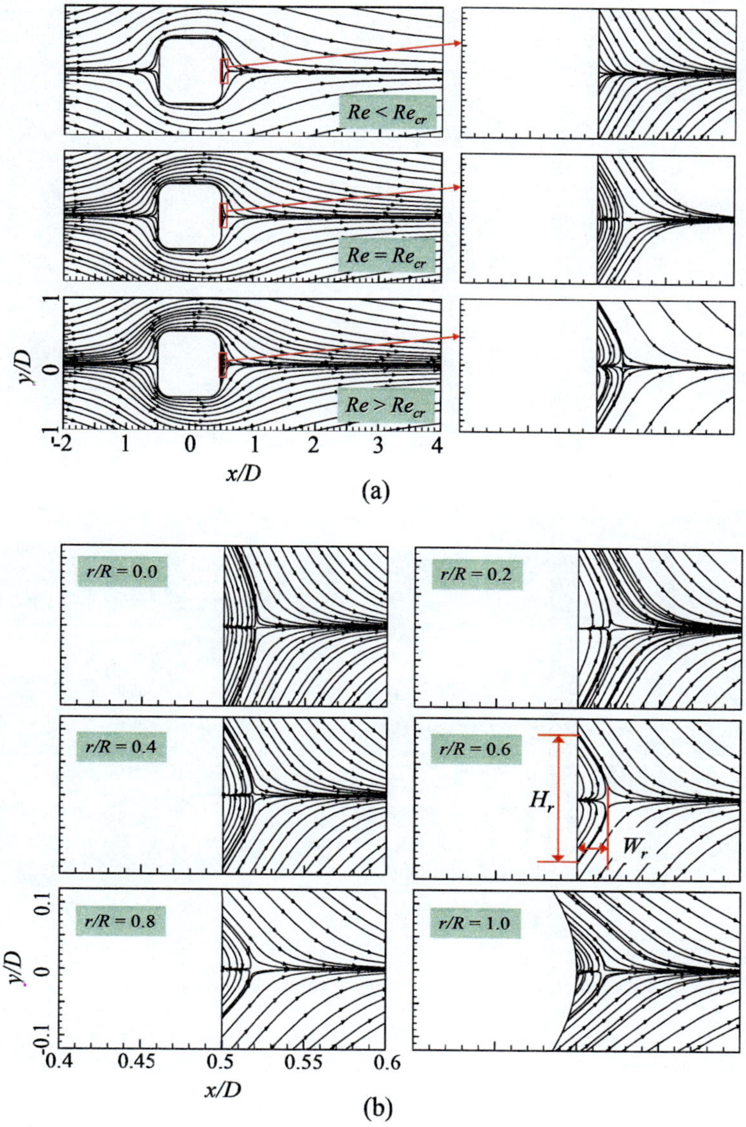

Fig. 4 Streamline pattern **a** around a rounded corner Square Cylinder, *r/R* = 0.4 at three different conditions and **b** around all the rounded corner Square Cylinder at $Re = Re_{cr}$

Table 3 Summary of Re_{cr} for various CRR

CRR	0	0.2	0.4	0.6	0.8	1
Re_{cr}	1.52	2.07	2.8	3.7	5	6.37

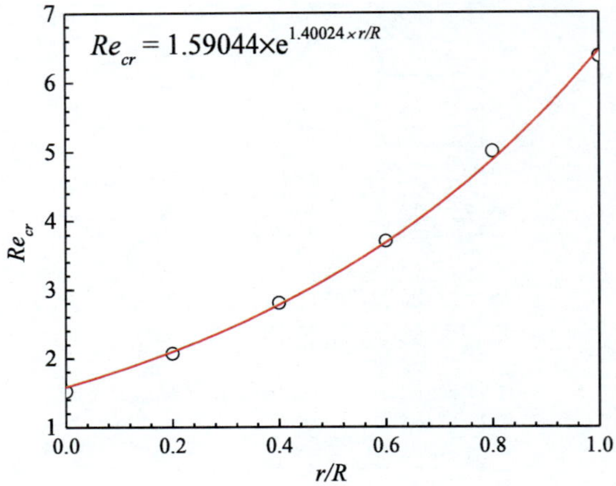

Fig. 5 Variation of critical *Re* with *r/R*

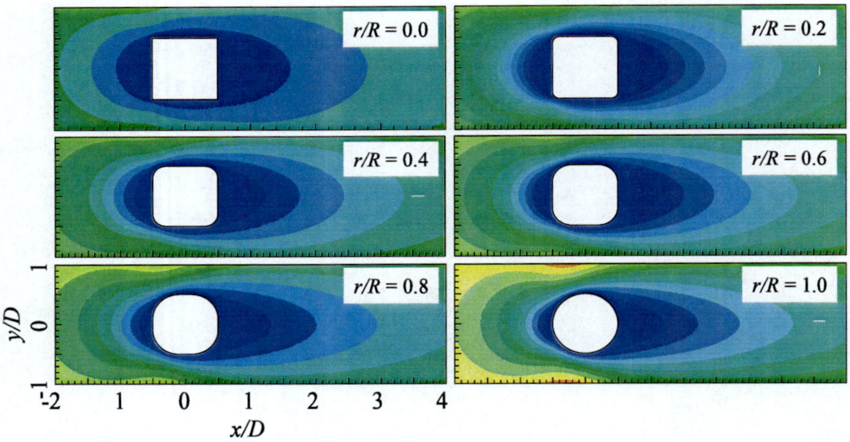

Fig. 6 Velocity distribution around different rounded corner square cylinders

5.3 *Pressure*

Coefficient of pressure distribution in the vicinity of different corner curvature square cylinders is shown in Fig. 7 at $Re = Re_{cr}$. It is observed that maximum pressure region is shifted from the leading edge to the front stagnation point. It is observed that the C_p is maximum at the leading edge and minimum at the trailing edge. In case of square cylinder, high pressure region is spread in larger area as compared to the circular cylinder. This leads to high pressure difference on the front and rear surfaces. This difference in pressure results to higher drag on the cylinders with low *r/R* (shaper corner).

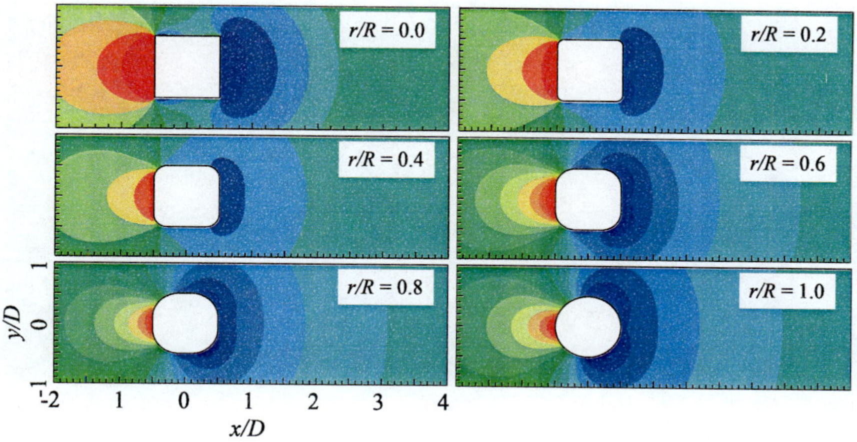

Fig. 7 Pressure coefficient distribution around different rounded corner square cylinders

5.4 *Temperature*

Figure 8 shows the isotherm contour in the vicinity of the cylinder $Re = Re_{cr}$. It is evident from the figure that the thermal field is steady and symmetric along the central axis passes through the cylinder center. It is observed that the temperature gradient is maximum at the front face of the cylinder resulting thinner thermal boundary layer. As a result, maximum heat is transferred from the cylinder to the fluid by the front surface compared to the other surfaces. It is noticed that the temperature gradient increases or the high temperature zone around the cylinder shrinks with increase in *r/R*. Consequently, heat transfer rate increases with increasing the *r/R*.

6 Drag Coefficient and Nusselt Number

Effect of the corner curvature ratio (*r/R*) on the fluid force and heat transfer at three different cases of *Re*, $Re < Re_{cr}$, $Re = Re_{cr}$ and $Re > Re_{cr}$. It is found that the surface average drag coefficient ($C_{d,mean}$) decreases with increasing of *r/R* and *Re* and is shown in Fig. 7. As discussed earlier, the high pressure region at the leading edge of the cylinder reduces with increasing *r/R* resulting reduction in the pressure drag. Whereas, the adverse pressure increases at the trailing edge of the cylinder with increasing the *Re* which further reduces the pressure drag on the cylinders. As a result, $C_{d,mean}$ decreases with increasing of *r/R* and *Re*. It is also evident that the change in $C_{d,mean}$ with increasing of *Re* is significant for lower values of *r/R* ($\leq$0.4) (Tab).

The overall heat transfer rate is calculated by the dimensionless parameter called Nusselt number (*Nu*) and it is observed that the surface average *Nu* (Nu_{avg}) increases

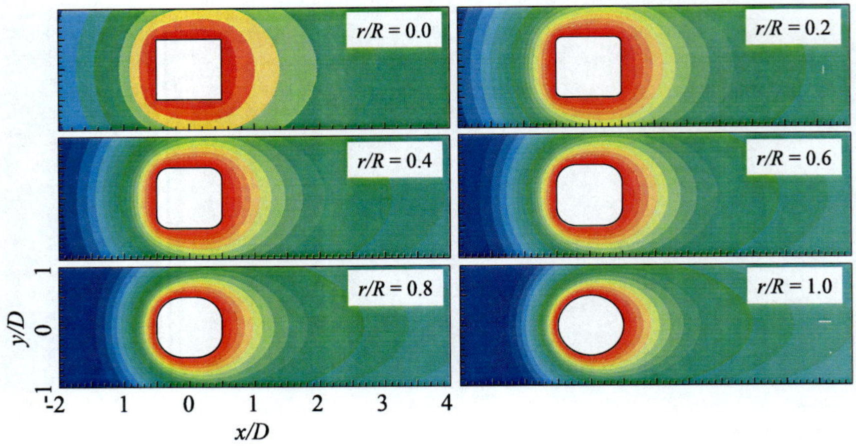

Fig. 8 Temperature distribution around different rounded corner square cylinders

Table 4 Correlation of drag coefficient and Nusselt number with *r/R*

Re	$C_{d,mean}$	Nu_{avg}
$Re < Re_{cr}$	$C_{d,mean} = 10.95147 \times e^{0.33627 \times r/R}$	$Nu_{avg} = 0.81757 \times e^{1.97987 \times r/R}$
$Re = Re_{cr}$	$C_{d,mean} = 10.84154 \times e^{0.33826 \times r/R}$	$Nu_{avg} = 0.82187 \times e^{1.97349 \times r/R}$
$Re > Re_{cr}$	$C_{d,mean} = 10.45954 \times e^{0.3533 \times r/R}$	$Nu_{avg} = 0.83222 \times e^{1.95326 \times r/R}$

with increasing of *r/R* and as well as *Re*. The variation of $C_{d,mean}$ and Nu_{avg} with the change in *r/R* is fitted exponentially and is shown in Table 4 for all the three cases of *Re*, $Re < Re_{cr}$, $Re = Re_{cr}$ and $Re > Re_{cr}$ with the constraint of *Re*, $0.1 \leq Re \leq 40$ (Fig. 9).

7 Conclusions

Effect of the corner curvature radius (*r/R*) on the critical Reynolds number (Re_{cr}) dictating the transition from creep flow to separated steady laminar flow is studied. Critical Reynolds number is found always less than 7 for all the considered corner radius ratio, $r/R = 0$ to 1. Critical Reynolds number for flow separation is found to be increasing (exponentially) with corner radius ratio. It is observed that as the corner radius ratio increases, low velocity region near the front stagnation point shrinks in size. The maximum pressure region is observed to be shifting from the corners to the center of the front face of the cylinder as *r/R* increases from 0 to 1. Drag coefficient is found to be decreasing with the increasing of *r/R* and the Nusselt number is observed to be increasing with the increasing of *r/R*.

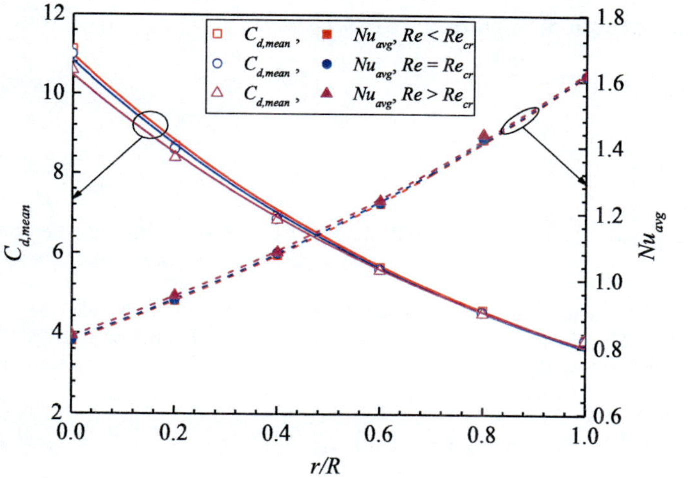

Fig. 9 Variation of $C_{d,mean}$ and Nu_{avg} with r/R along with fitting line

Nomenclature

A_s	Cylinder surface area [m^2]
D	Diameter of cylinder [m]
x^*	Dimensionless form of x coordinate; $(\frac{x}{D})$
y^*	Dimensionless form of y coordinate; $(\frac{y}{D})$
U_∞	Free stream velocity [m/s]
u^*	Dimensionless form of x component of velocity; $(\frac{u}{U_\infty})$
v^*	Dimensionless form of y component of velocity; $(\frac{v}{U_\infty})$
$C_{d,mean}$	Surface average drag coefficient
F_D	Drag force [N]
p_∞	Inlet pressure [Pa]
p^*	Dimensionless form of pressure; $\left(\frac{p-p_\infty}{\rho U_\infty{}^2 D}\right)$
T_∞	Inlet temperature [°C]
T_w	Temperature of cylinder wall [°C]
θ^*	Dimensionless form of temperature; $\left(\frac{T-T_\infty}{T_w-T_\infty}\right)$
ρ	Density of air [kg/m^3]
K	Thermal conductivity of the air [W/m.K]
h_{loc}n	Local heat transfer coefficient [W/m^2.k]
μ	Viscosity of the airPa.s
Nu	Nusselt number
r	Corner radius [m]
R	Cylinder radius [m]
Re	Reynolds number
Pr	Prandtl's number

$P*$ Dimensionless pressure $\left(\frac{p-p_\infty}{\rho U_\infty{}^2}\right)$

References

1. Kumar B, Srivastava GP, Kumar M, Patil AK (2018) A review of heat transfer and fluid flow mechanism in heat exchanger tube with inserts. Chem Eng Process Process Intensif 123:126–137
2. Talaat Abdelhamid, Md. Mahbub Alam, Islam Md. (2021) Heat transfer and flow around cylinder: effect of corner radius and Reynolds number. Int J Heat Mass Transfer 171
3. Bai H, Mahbub Alam Md. (2018) Dependence of square cylinder wake on Reynolds number. Phys Fluids 30:015102
4. Sharma A, Eswaran V (2004) Heat and fluid flow across a square cylinder in the two-dimensional laminar flow regime. Numer Heat Transfer, A: Appl: Int J Comput Methodol 45:3:247–269
5. Sen S, Mittal S, Biswas G (2009) Steady separated flow past a circular cylinder at low Reynolds numbers. J Fluid Mechanics 620:89–119
6. Alam Md.M, Abdelhamid T, Sohankar A (2020) Effect of cylinder corner radius and attack angle on heat transfer and flow topology. Int J Mech Sci 175
7. Sohankar A, Norberg C, Davidson L (1998) Low-Reynolds-number flow around a square cylinder at incidence: study of blockage, onset of vortex shedding and outlet boundary condition. Int J Numer Methods Fluids 26:39–56
8. Zheng Q, Alam MM (2017) Intrinsic features of flow past three square prisms in side-byside arrangement. J Fluid Mech 826:996–1033
9. Dey P, Das AK (2016) A utilization of GEP (gene expression programming) metamodel and PSO (particle swarm optimization) tool to predict and optimize the forced convection around a cylinder. Energy 95:447–458
10. Gupta AK, Sharma A, Chhabra RP, Eswaran V (2003) Two-dimensional steady flow of a power-law fluid past a square cylinder in a plane channel: momentum and heat-transfer characteristics. Ind Eng Chem Res 42(22):5674–5686
11. Dey P, Das AK (2017) A numerical study on effect of corner radius and Reynolds number on fluid flow over a square cylinder. Sādhanā 42(7):1155–1165
12. Tritton DJ (1959) Experiments on the flow past a circular cylinder at low Reynolds numbers. J Fluid Mech 6(4):547–567

Influence of Base Fluid on the Thermo-Physical Properties of Hybrid (Fe_3O_4 + SiC) Nanofluids

T. Kanthimathi and P. Bhramara

Abstract In the present work, experiments are performed to determine the effect of the base fluid when spiked with very low volume fraction of Hybrid (mixture of Fe_3O_4 and SiC) nanoparticles on the thermo-physical properties like specific heat, density, viscosity and thermal conductivity of nanofluids. The volume percentage of nanoparticles suspended in the base fluids are 0.02, 0.04, 0.06 and 0.08. Three different base fluids, viz., Distilled Water (DW), a mixture of Ethylene Glycol–Water in the volume ratios of 20:80 and 40:60 (20:80 EG-W and 40:60 EG-W) are considered in the analysis. The effect of temperature on the properties is also studied. For the range of volume concentration considered, there was observed to be a negligible variation in the density and specific heat of the nanofluids compared to that of respective base fluids. However, there is significant variation in the thermal conductivity and viscosity of the nanofluids. 20:80 EG-W based nanofluids exhibited higher enhancements in thermal conductivity and viscosity. Prandtl number of Hybrid nanofluids decreased as the nanoparticle concentration increased indicating the increase in heat transfer penetration compared to momentum penetration which is a favorable condition for enhancement of heat transfer rate.

Keywords Base fluid · Nanofluid · Thermos-physical properties · Concentration · Nanoparticles

T. Kanthimathi (✉)
Department of Mechanical Engineering, JNTU Hyderabad, Hyderabad, Telangana 500085, India
e-mail: pkanthi1978@gamil.com

Department of Mechanical Engineering, K L Deemed to Be University, Guntur Dt., AP 522302, India

P. Bhramara
Department of Mechanical Engineering, JNTUHCEH, Hyderabad, Telangana 500085, India

S. Das et al. (eds.), *Proceedings of the 1st International Conference on Fluid, Thermal and Energy Systems*, Lecture Notes in Mechanical Engineering,
https://doi.org/10.1007/978-981-99-5990-7_65

Abbreviations

C_p	Specific Heat, J/kg K
d	Diameterm
k	Thermal conductivity, W/mK
Pr	Prandtl number
Re	Reynolds number
T	Temperature, °C
W	Mass, kg

Greek symbols

ρ	Densitykg/m^3
Ø	Particle concentrationμ
α	Viscosity
Thermal diffusivity	Centi poise(cp), m^2/s

Subscripts

bf	Base fluid
nf	Nanofluid
np	Nanoparticle
p	Particle
hnf	Hybrid nano particle
w	Water

1 Introduction

The research on nanofluids is at its peak. However, there are no generalized correlations to predict the properties (thermo-physical) of the nanofluids, mainly due to their dependence on the type of nanoparticles as well as on the type of base fluid, size and volume fraction of nanoparticles, operating temperature, and combination of these parameters.

2 Literature Review and Objective

Esfe et al. [1] determined experimentally the thermal conductivity of Ag-MgO/Water nanofluid for 40 nm and 25nm sized MgO and Ag nanoparticles respectively, for the volume fraction ranging from 0 to 2%. New correlations for viscosity and thermal conductivity of nanofluids were proposed. Moldoveanu et al. [2] experimentally estimated the viscosity of Al_2O_3, SiO_2 and their hybrid nanofluids for the concentrations ranging from 1 to 5% and in the temperature range of 285 K to 330 K. It was noticed from the results that relative viscosity of alumina nanofluids is higher than SiO_2, but by replacing 0.5% of SiO_2 with alumina, the viscosity of the hybrid mixture was observed to increase compared to that of SiO_2. Suresh et al. [3] determined thermal conductivity and viscosity of Al_2O_3 + Cu(90:10) nanocomposite with water as the base fluid for the concentrations ranging from 0.1 to 2%. An enhancement of 12.11% in thermal conductivity and 115% in viscosity is observed. Leong et al. [4] experimentally determined thermal conductivity of Cu + TiO_2 nanocomposite suspended in 50:50 EG-W base fluid for the concentration range of 0.2 to 0.8%. Maximum of 9.8% enhancement in thermal conductivity is obtained at 0.8% volume concentration. Xian et al. [5] estimated the thermal conductivity of 50%COOH-Gnp + 50% TiO_2/40:60 EG-W hybrid nanofluid for the volume concentrations of 0.025, 0.05, 0.075 & 0.1% experimentally. Maximum of 23.74% and 21.59% enhancement in thermal conductivity is obtained for COOH-Gnp and Hybrid nanofluids. Tahat et al. [6] experimentally estimated the thermal conductivity and viscosity of (70%)Al_2O_3 + (30%)CuO/25:75 EG-W Hybrid nanofluid for the concentrations range of 0.5 to 2%. 95% enhancement in thermal conductivity and 159% enhancement in viscosity was observed.

Babar et al. [7] conducted a review on the preparation, stability and thermo-physical properties of hybrid nanofluids. Based on their extensive review, it was reported that the hybrid nanofluids offer better thermal characteristics than single component nanofluids. It was also stated by the authors that the effect of different parameters like particle concentration, particle size, base fluid type, the temperature on the thermal properties of hybrid nanofluids is not understood clearly so they insisted on the necessity of more research in this context.

Nabil et al. [8] conducted a comprehensive review on the thermo-physical properties of hybrid nanofluids and hybrid nano lubricants. They emphasized the need for extensive research on hybrid nanofluids to make them available for commercial applications.

In the present work, the effect of three different base fluids, viz., DW, 20:80 EG-W and 40:60 EG-W on the thermo-physical properties of and the Hybrid mixture of (Fe_3O_4 + SiC) nanofluids is determined for the particle concentrations ranging from 0.02 to 0.08 vol% at three different temperatures of 40, 45 and 50 °C.

3 Materials and Methods

The nanoparticles Fe_3O_4 of 30nm size and SiC of 50nm size are procured from Nano Amor USA Texas with 99% purity. The TEM images of Fe_3O_4 and SiC nanoparticles is shown in Figures 1 and 2.

The hybrid mixture is prepared by two step method, i.e., suspending Fe_3O_4 and SiC nanoparticles in the ratio of 50:50 by volume in three different base fluids. The hybrid nanoparticle concentration is determined using Eq. (1) [12].

$$\phi = \frac{\frac{WSiC}{\rho SiC} + \frac{WFe3O4}{\rho Fe3O4}}{\left(\frac{WSiC}{\rho SiC} + \frac{WFe3O4}{\rho Fe3O4} + \frac{Wbf}{\rho bf}\right)} \times 100 \quad (1)$$

To prepare stable nanofluids, continuous stirring of the solution is carried out for about 4–6 h. The stability of nanofluids is up to 15 days and their stability is tested by determining Zeta potential. It is implied that if the nanofluids are stable at high

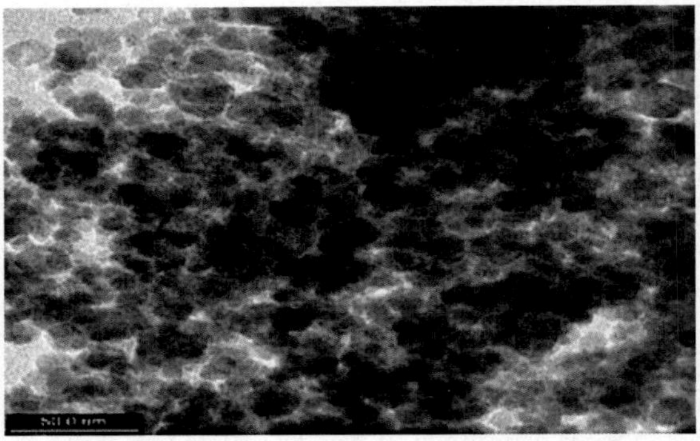

Fig. 1 TEM image of Fe_3O_4 nanoparticles

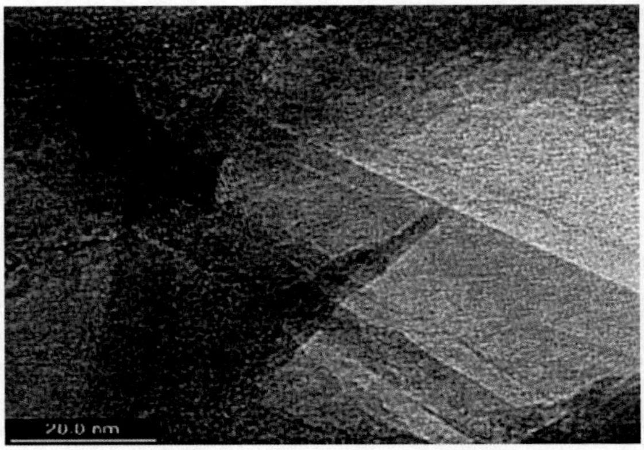

Fig. 2 TEM image of SiC nanoparticles

Table 1 Nanofluids zeta potential at 0.08% Concentration by Volume

Base fluid	Zeta potential of hybrid nanofluids
DW	−36.3
20:80 EG-W	−46.8
40:60 EG-W	−39

Table 2 Range and accuracies of nanofluid property measuring instruments

Instrument	Range	Accuracy
Density Measuring instrument	0.0011–0.9999 g/cm^3	±0.001%
Specific heat apparatus	0.5–4 MJ/m^3 °C	±0.005%
Viscosity-DV2T Viscometer	0.2–3000 cp	±0.1%
Thermal Conductivity	0.2–2 W/mK	±10%

concentrations then they will be stable at low concentrations as well, hence the Zeta potential of Hybrid nanofluids at 0.08% concentrations is shown in Table 1.

Zeta Potential for the colloidal solutions considered in the analysis is above ±30 mV, indicating stable suspensions [15].

The density of the nanofluids is tested using the Antonpaar density measuring instrument. Mittal Nanofluid specific heat capacity apparatus is used for measuring nanofluids heat capacity. The viscosity is measured using a DV2T viscometer and nanofluids thermal conductivity is determined using Tempos thermal property analyzer. The range of measurement and accuracy of the instruments is presented in Table 2.

4 Results and Discussion

4.1 Density

The variation in the density with the particle concentration is shown in Fig. 3a–c for hybrid nanoparticles dispersed in DW, 20:80 EG-W, and 40:60 EG-W base fluids. The experimental values of density are compared with the Pak and Cho [9] correlation given by Eq. (2).

$$\rho_{hnf} = \phi_{np1}\rho_{np1} + \phi_{np2}\rho_{np2} + \left(1 - \phi_{np1} - \phi_{np2}\right)\rho_{bf} \ldots\ldots(1.2) \quad (2)$$

Hybrid nanofluids density decreased with the rise in temperature and increased as the particle concentration increased. 20:80 EG-W based Hybrid nanofluid exhibited a maximum increment of 2.22% at 0.08 vol% and 50 °C. Maximum deviation of less than 2% is obtained with the correlation.

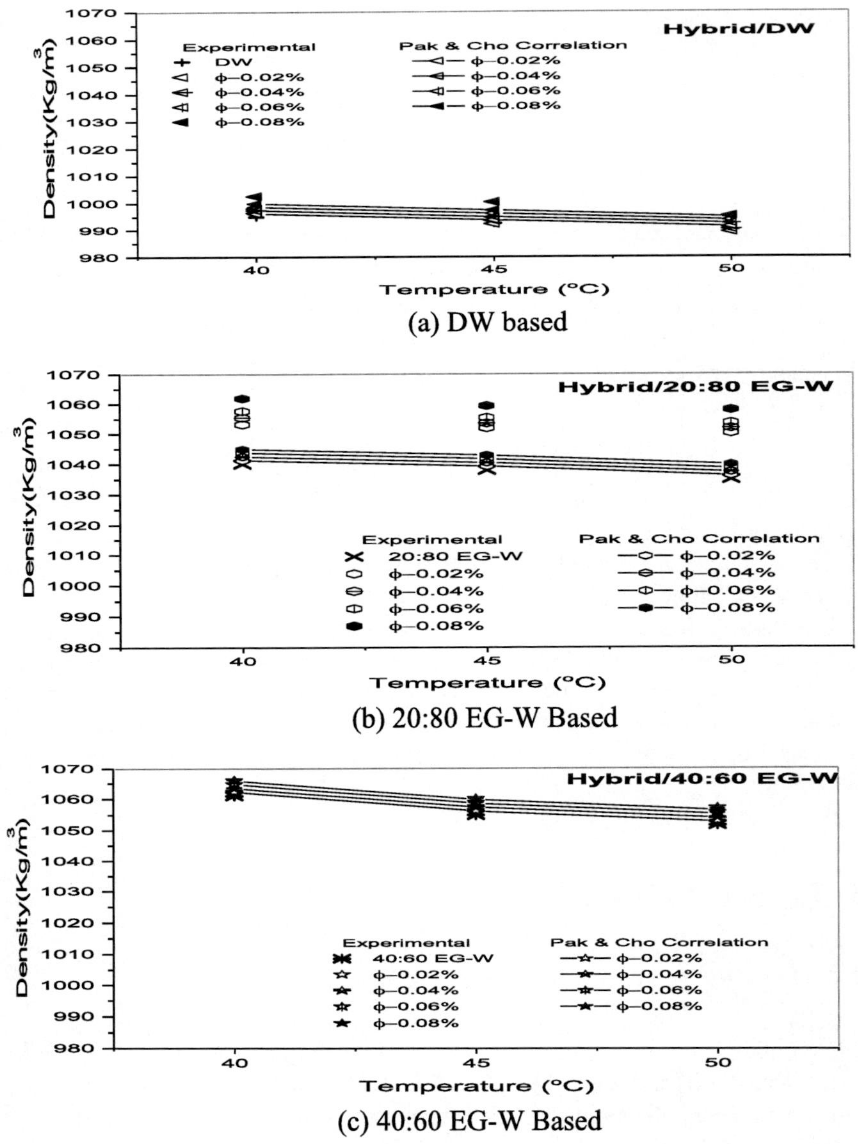

Fig. 3 Density of hybrid nanofluids

4.2 *Specific Heat*

The variation in the specific heat of Hybrid nanoparticles dispersed in DW, 20:80 EG-W, and 40:60 EG-W base fluids with the particle concentration is shown in Fig. 4a–c. The experimental values of specific heat obtained are compared with the Pak and

Cho [9] correlation given by Eq. (9) for a hybrid nanofluid.

$$(\rho C_p)_{hnf} = \phi(\rho C_p)_{np1} + \phi_{p2}(\rho C_p)_{np2} + (1 - \phi_{p1} - \phi_{p2})\rho_{bf} Cpbf_{hnf} \quad (3)$$

Specific heat of hybrid nanofluids increased with the increase in temperature and decreased with the increase in concentration. No significant variation in specific heat with temperature is observed, due to low concentrations of nanoparticles considered. The deviation of experimental values with the correlation is less than 0.2%.

4.3 Viscosity

The various factors which affect viscosity of nanofluids include temperature, particle concentration, base fluid etc. The experimental values of viscosity are correlated with theoretical viscosity values obtained by the correlations of Sharma et al. [10] and Corcoine et al. [11] which are given by Eqs. (4) and (6).

$$\frac{\mu_{nf}}{\mu_{bf}} = \frac{1}{1 - 34.87(\frac{d_p}{d_{bf}})^{-0.3}(\frac{\phi}{100})^{1.03}} \quad (4)$$

where d_{bf} is the base fluid molecule's equivalent diameter given by Eq. (5)

$$d_{bf} = 0.1\left[\frac{6M}{N\pi\rho_{bfo}}\right]^{1/3} \quad (5)$$

in which M is the molecular weight of the base fluid, N is the Avogadro number = 6.023×10^{23} and ρ_{bfo} is the base fluid's density at 20 °C.

$$\frac{\mu_{nf}}{\mu_{bf}} = \left[(1 + \frac{\phi}{100})^{11.3}(1 + \frac{T_{nf}}{70})^{-0.038}(1 + \frac{d_p}{170})^{-0.061}\right] \quad (6)$$

Figure 5a–c show the variation of Hybrid nanofluids viscosity with temperature and particle concentration also the variation between experimental and theoretical values of viscosity are shown.

The percentage increment in the viscosity of Hybrid nanofluids increased from 10.15 to 23.92%, 33.99 to 61.29% and 31.6 to 51.06% for DW, 20:80 EG-W and 40:60 EG-W based nanofluids respectively for the particle concentrations ranging from 0.02 to 0.08 vol% at 50 °C. The correlations predicted the experimental values with higher deviation of 28.69%, 65.26% and 56.63% with Sharma et al. [10] correlation and 25.8%, 61.28% and 51.06% with Corcoine et al. [11] correlation at 0.08 vol% and 50 °C.

Highest enhancement in viscosity is exhibited by EG-W based nanofluids when compared with DW based nanofluids due to the higher values of viscosity for EG

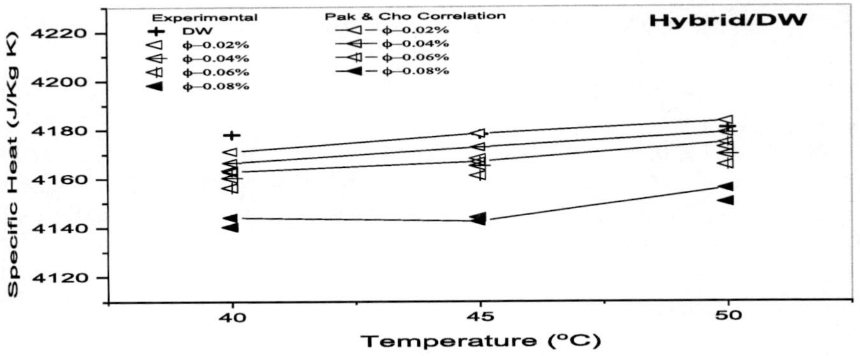

(a) DW Based

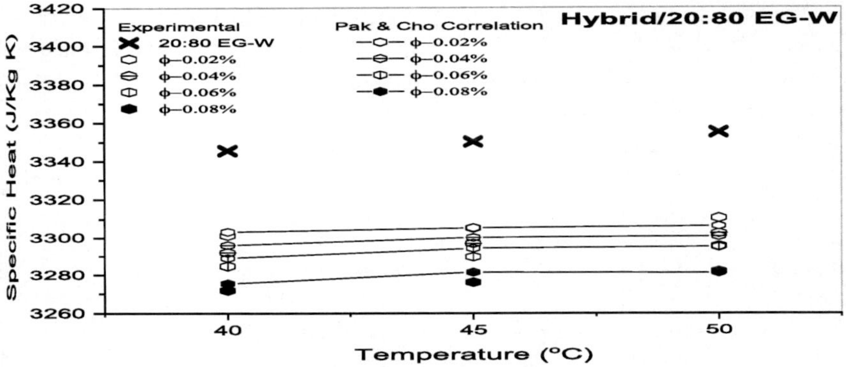

(b) 20:80 EG-W Based

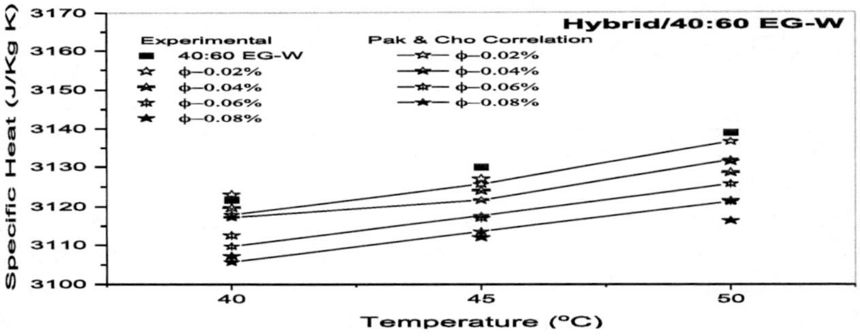

(c)40:60 EG-W Based

Fig. 4 Specific heat of hybrid nanofluids

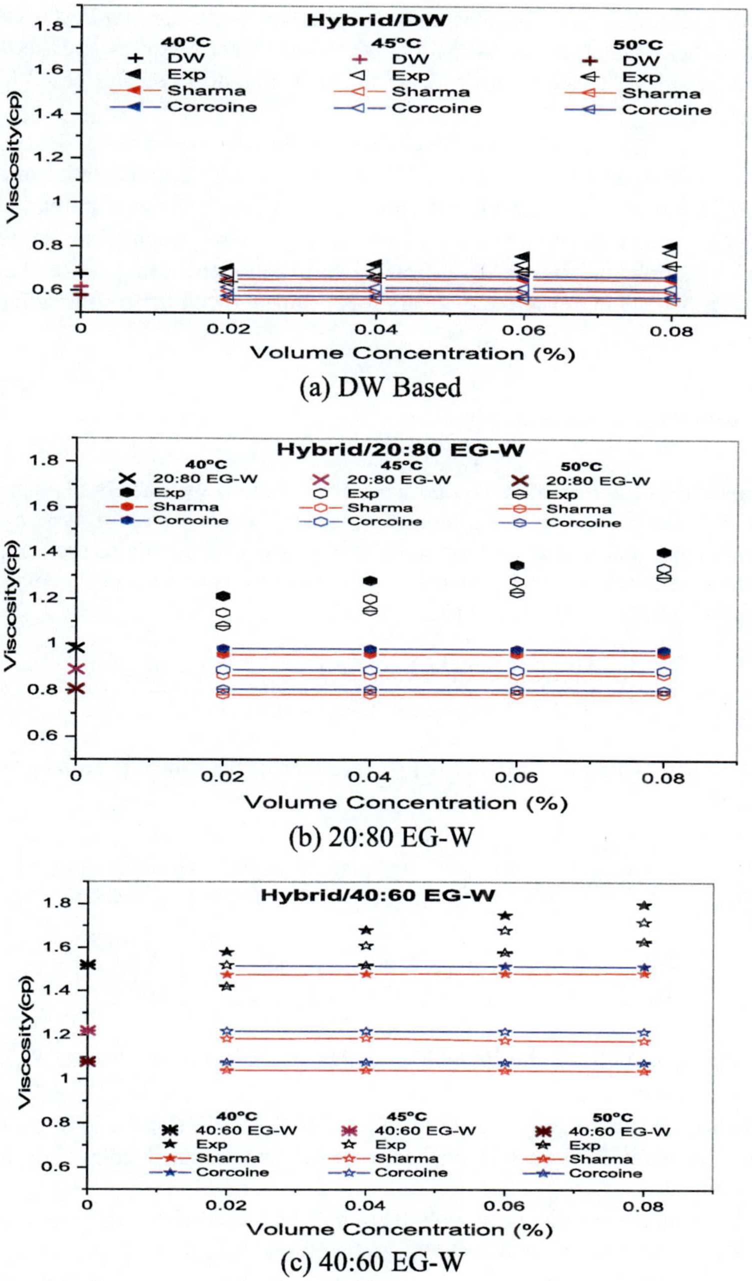

(a) DW Based

(b) 20:80 EG-W

(c) 40:60 EG-W

Fig. 5 Viscosity of hybrid nanofluids

than Water. Sundar et al. [14] also presented similar results for ND-Fe_3O_4 Hybrid nanofluids with water, 20:80, 40:60 and 60:40 EG-Water mixtures as base fluids. The enhancements in the viscosity were higher for 20:80 EG-water based hybrid nanofluids.

In general, the correlations are observed to predict the experimental data of DW based nanofluid better than the EG-W based nanofluids. This deviation may be attributed to various parameters like the preparation method, pile up of the particles in the base fluid and sonication time etc,. Giwa et al. [12] also presented similar results for their experimentations on Al_2O_3-Fe_2O_3 hybrid nanofluids using DW and EG-W base fluids, when their experimental data was compared with these correlations.

4.4 Thermal Conductivity

The nanofluid's thermal conductivity depends on various parameters like nanoparticle's size and shape, operating temperature, particle concentration type of base fluid, etc. The experimental thermal conductivity values of nanofluids are compared with the correlations of Hamilton Crosser [13], Sharma et al. [10] and Corcoine et al. [11] given by the Eq. (7), (8) and (9).

$$k_{nf} = k_{bf}\left[\frac{k_p + (n-1)k_{bf} - \phi(n-1)(k_{bf} - k_p)}{k_p + (n-1)k_{bf} + \phi(k_{bf} - k_p)}\right] \tag{7}$$

Here n = 3/Ψ, where Ψ is the ratio of surface area of the sphere to surface area of nanoparticle.

$$\frac{k_{nf}}{k_{bf}} = \left[0.8938(1+\frac{\phi}{100})^{1.37}(1+\frac{T_{nf}}{70})^{0.277}(1+\frac{d_p}{150})^{-0.0336}(\frac{\alpha_p}{\alpha_w})^{0.01737}\right] \tag{8}$$

$$\frac{k_{nf}}{k_{bf}} = 1 + 4.4(\mathrm{Re})^{0.4}(\mathrm{Pr})_{bf}^{0.66}\left[\frac{T}{T_{fr}}\right]^{10}\left[\frac{k_p}{k_{bf}}\right]^{0.03}\left[\frac{\phi}{100}\right]^{0.66} \tag{9}$$

where $\mathrm{Re} = \frac{2\rho_{bf} K_b T}{\pi(\mu_{bf})^2 d_p}$ is the Reynolds number of the nanoparticle, and K_b is the Boltzmann constant.

The thermal conductivity of Hybrid nanofluids with DW, 20:80 EG-W and 40:60 EG-W is shown in Figs. 6a–c for the temperatures and particle loadings considered in the analysis.

The percentage enhancement of 40.04 to 43.65%, 60.47 to 64.33% and 60.09 to 63.94% in thermal conductivity is obtained for Hybrid nanofluids as the temperature varies from 40 to 50 °C at 0.08 vol% of particle concentration for DW, 20:80 EG-W and 40:60 EG-W based nanofluids.

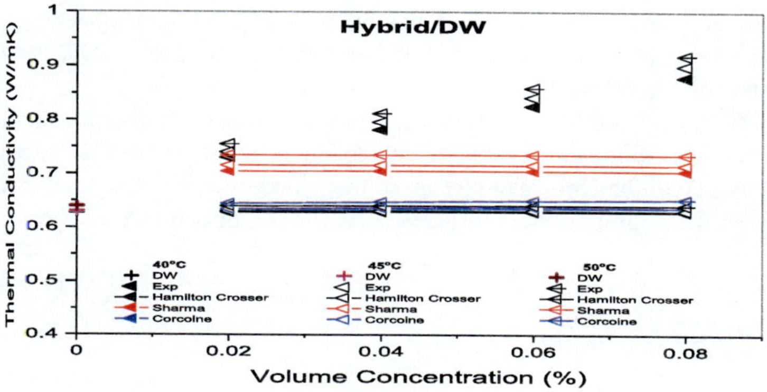

(a) DW Based

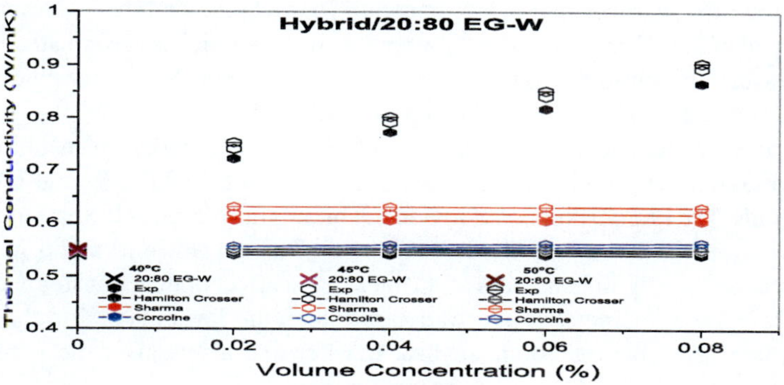

(b) 20:80 EG-W

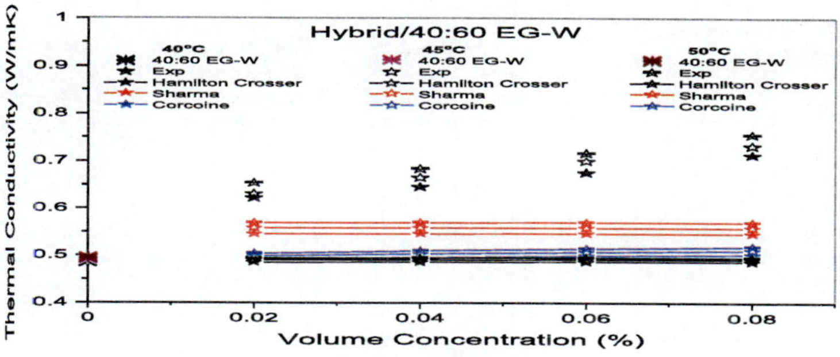

(c) 40:60 EG-W

Fig. 6 Hybrid nanofluids thermal conductivity

Thermal conductivity of Hybrid nanofluids is in good agreement with Sharma et al. [10] correlation with and an average deviation of 13.82 to 31.93% for the three base fluids in the analysis.

EG-W based nanofluids exhibited higher augmentation in thermal conductivity even with the addition of very low particle loadings of 0.02 vol%. This phenomenon is mainly due to the Brownian motion of the nanoparticles in the base fluid and in particular, the stability of nanoparticles in EG-W is higher than in water.

4.5 Prandtl Number

The properties (thermo-physical) of the nanofluids depend on parameters such as base fluid properties, particle loadings, temperature, particle size and shape. To analyze the combined effect of thermophysical properties, Prandtl number is evaluated based on the measured thermo-physical properties. The Prandtl Number of Hybrid nanofluids for the concentrations considered is represented in Fig. 7.

It can be noted from Fig. 7, that the Prandtl number of hybrid nanofluid with 40:60 EG-W base fluid has higher values than that of 20:80 EG-W and DW based nanofluids. The Prandtl number of hybrid nanofluids decreased as the particle loading increased for the three base fluids considered. The decreasing Prandtl number of hybrid nanofluids is an indication of the increasing effect of heat transfer penetration compared to the momentum penetration into the fluid layers lateral to the surface, which is a favorable condition for heat transfer and also causes the reduction in pressure drop.

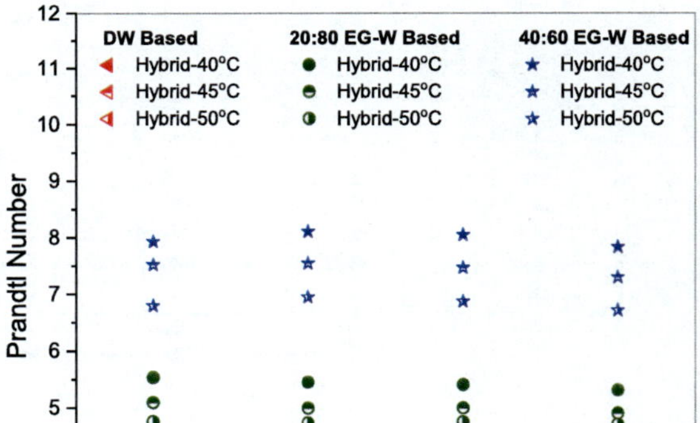

Fig. 7 Hybrid nanofluids thermal conductivity

5 Conclusion

The thermo-physical properties of Hybrid (Fe_3O_4 + SiC) nanoparticles suspended in DW, 20:80 EG-W and 40:60 EG-W base fluids are determined experimentally for the particle concentration ranging from 0.02 to 0.08% at the temperatures of 40, 45 and 50 °C. The following are the conclusions drawn based on the experimental analysis of the effect of base fluid, on the thermo-physical properties of the nanofluids.

- Stability of Hybrid nanofluids is more in EG-W compared with water as base fluid.
- No significant enhancement in density and specific heat is observed for the Hybrid nanofluids considered and also the Pak and Cho Correlation was in excellent agreement with the experimental values
- Enhancement in Viscosity and thermal conductivity of hybrid nanofluids is found to be higher for 20:80 EG-W than Water and 40:60 EG-W base fluids.
- Effect of base fluid and particle loading on the thermos-physical properties was found to be greater than temperature.
- Considered correlations for viscosity and thermal conductivity were found to underestimate the experimental values due to the variation in the type of nanoparticle, temperature and base fluid considered during the experimentation and the correlations used. This has necessitated the need of more experimentation.
- Prandtl number of the Hybrid nanofluid decreased with the increase in particle concentration indicating the increase in the heat transfer penetration compared to momentum penetration resulting in the favorable condition for heat transfer.
- 40:60 EG-W based hybrid nanofluids were found to be more suitable for heat transfer experimentations among the three different Hybrid nanofluids considered.

References

1. Esfe MH, Arani AAA, Rezaie M, Yan W-M, Karimipour A (2015) Experimental determination of thermal conductivity and dynamic viscosity of Ag–MgO/water hybrid nanofluid. Int Commun Heat Mass Transfer 66:189–195
2. Moldoveanu GM, Huminic G, Minea AA, Huminic A (2018) Experimental study on thermal conductivity of stabilized Al2O3 and SiO_2 nanofluids and their hybrid. Int J Heat Mass Transfer 127: 450–457
3. Suresh S, Venkitaraj KP, Selvakumar P, Chandrasekar M (2011) Synthesis of Al_2O_3–Cu/water hybrid nanofluids using two step method and its thermo physical properties. Colloids Surf A Physicochem Eng Aspects 388(1–3):41–48
4. Leong KY, Razali I, Ku KZ, Ahmad HC, Ong MJ Ghazali MR, Rahman A (2018) Thermal conductivity of an ethylene glycol/water-based nanofluid with copper-titanium dioxide nanoparticles: an experimental approach. Int Commun Heat Mass Transf 90:23–28
5. Xian, H.W., Sidik, N.A.C. and Saidur, R., (2020); "Impact of different surfactants and ultrasonication time on the stability and thermophysical properties of hybrid nanofluids"; *International Communications in Heat and Mass Transfer*, vol.110, pg no..104389.

6. Tahat MS, Benim AC (2017) Experimental analysis on thermophysical properties of Al2O3/CuO hybrid nano fluid with its effects on flat plate solar collector. In: *Defect and diffusion forum*, Vol. 374, pp 148–156. Trans Tech Publications Ltd
7. Babar H, Ali HM (2019) Towards hybrid nanofluids: preparation, thermophysical properties, applications, and challenges. J Mol Liquids 281:598–633
8. Nabil MF, Azmi WH, Hamid KA, Zawawi NNM, Priyandoko G, Mamat R (2017) Thermophysical properties of hybrid nanofluids and hybrid nanolubricants: a comprehensive review on performance. Int Commun Heat Mass Transfer 83:30–39
9. Pak BC, Cho YI (1998) Hydrodynamic and heat transfer study of dispersed fluids with submicron metallic oxide particles. Exp Heat Transfer Int J 11(2):151–170
10. Korada SV, Vandrangi SK, Kamal S, Minea AA (2017) Experimental studies on the influence of metal and metal oxide nanofluid properties on forced convection heat transfer and fluid flow. Adv New Heat Transfer Fluids. CRC Press, pp 1–28
11. Corcione M (2011) Empirical correlating equations for predicting the effective thermal conductivity and dynamic viscosity of nanofluids. Energy Conver Manag 52(1):789–793
12. Giwa SO, Sharifpur M, Goodarzi M, Alsulami H, Meyer JP (2021) Influence of base fluid, temperature, and concentration on the thermophysical properties of hybrid nanofluids of alumina–ferrofluid: experimental data, modeling through enhanced ANN, ANFIS, and curve fitting. J Thermal Anal Calor 143(6): 4149–4167
13. Hamilton RL, Crosser OK (1962) I&EC Fund 1:187
14. Sundar LS, Ramana EV, Graça MPF, Singh MK, Sousa AC (2016) Nanodiamond-Fe3O4 nanofluids: preparation and measurement of viscosity, electrical and thermal conductivities. Int Commun Heat Mass Transfer 73: 62–74
15. Ajeet K, Dixit CK (2017) Methods for characterization of nanoparticles. Adv Nanomed Deliv Therap Nucleic Acids 43–58. doi:https://doi.org/10.1016/B978-0-08-100557-6.00003-1. ISBN 9780081005576

Effect of an Air Cavity Appended to the Solar Still: A Computational Approach

Y. S. Prasanna and Sandip S. Deshmukh

Abstract In recent years, saline water distillation using solar energy has been identified as a high renewable energy need. Despite the fact that numerous researchers have taken varied techniques in this area, this work presents a novel design and its computational analysis which is an essential criterion in improving the performance of a solar still. An air cavity is added to the passive still in this investigation to prevent heat loss to the atmosphere. To explore the effect of the air cavity in the heat transfer phenomena within the solar still, a finite volume technique based on the SIMPLE algorithm is used in Ansys Fluent software to solve the governing equations. The findings of this process include Nusselt number (Nu), Reynolds number (Re), and variation of total heat flux with respect to the variation of the height of an air cavity while holding Prandtl ($Pr = 0.712$). The findings showed that reducing the height of an air cavity resulted in a significant increase in mass and heat transfer rates inside the air cavity. The temperature gradient between the glass cover and the saline water plays a critical role in the optimisation of the design parameters of the solar still. From the CFD simulations, a height of 20 cm of an air cavity is determined to be the optimum height for improving the solar still performance in terms of the evaporation rate inside the still.

Keywords Air cavity · CFD · Nusselt number · Heat transfer · Solar still

1 Introduction

Even though water covers 71% of the planet's surface, ensuring that it is safe to drink is becoming very difficult. Most of the world's water is too salty to be consumed. Further, polluted water negatively impacts the human body, causing harmful diseases, and atleast a thousand children die each day from water-related illnesses caused by

Y. S. Prasanna (✉) · S. S. Deshmukh
Department of Mechanical Engineering, BITS Pilani Hyderabad Campus, Hyderabad, Telangana 500072, India
e-mail: p20190441@hyderabad.bits-pilani.ac.in

S. Das et al. (eds.), *Proceedings of the 1st International Conference on Fluid, Thermal and Energy Systems*, Lecture Notes in Mechanical Engineering,
https://doi.org/10.1007/978-981-99-5990-7_66

dirty water, according to the United Nations Children's Fund (UNICEF). Fortunately, solar energy, the elixir of power production [1], being abundantly available throughout the world, will be the best renewable source for water distillation. Water purification will be done cheaply and sustainably using solar energy.

Solar distillation is a method for distilling brackish water using its energy as a source of heat. It is possible to clean water of non-volatile pollutants and germs by using solar stills. This technique produces pure water with a 30 PPM or below TDS level, which is safe for human consumption [2]. Solar energy, a natural form of energy, doesn't emit any greenhouse gases and hence is included in the low carbon economy list. Solar still works similarly to a natural hydrological cycle, accompanying the evaporation and condensation process. Solar stills are classified into passive and active types. The most common type of solar still is a passive one that performs heat absorption and distillation processes. The active solar still requires an additional heat source (solar collectors, solar ponds, industrial and power plant waste heat) to boost the evaporation rate of the salty water in the basin. Active stills are expensive and impractical for residential usage; however, passive stills, which are simple in design, easy to fabricate, and inexpensive, are predominantly used for domestic purposes.

2 Literature Review and Objective

The distillate production from a passive solar still Fig. 1 is comparatively much less than the active still, whereas increasing the productivity of the still while keeping its price low is a difficult task. Therefore, to compensate for the need, we produced a new design, simulated it and produced the results obtained in this research work. In evaluating the performance of the still, the initial significant step is modelling evaporation and condensation processes using CFD tools which may greatly aid in optimising the parameters of solar stills. Simulating before experimenting will drastically cut down on the time and money needed to run many experiments.

Khare et al. [3] created a CFD model and used it to improve the solar still's performance via a series of parametric studies. The researchers concluded that the

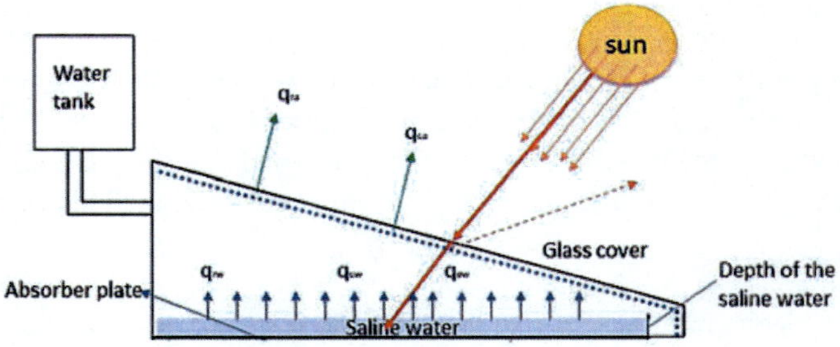

Fig. 1 Passive solar still

simulated findings matched the experimental results well. Natural convection in a solar still unit was studied by using a three-dimensional CFD modelling system. At 300 and 350 K, the researchers set the absorber and condensation surfaces of the solar still. The other surfaces were kept at adiabatic temperatures. They came to the conclusion that there was a significant flow recirculation area with high velocity zones in both the top and bottom regions. Due to the high velocity zones, the shear stress on the top and bottom walls rose significantly. Experimental and CFD data were compared and found to be in good agreement [4]. The hourly productivity of a single slope in the solar still was studied in a two-dimensional CFD simulation. The Chilton–Colburn comparison also provided a new equation for evaluating the solar still's production. The findings showed that the data from well-known models (particularly the Chilton–Colburn connection) are in good agreement with each other [5].

The primary goal of this study is to offer a thorough understanding of the heat transfer phenomenon inside a simple-to-build novel solar still under different operating and boundary conditions. There are very few investigations performed on the simulation aspects, to optimise and enhance the performance of a solar still. This research presents a novel design of solar still with an air cavity. The air cavity beneath the absorber plate functions as a perfect insulator, impeding heat loss to the atmosphere. The effect of air cavity height on the performance of the solar still is investigated in this study using the volume of fluid method in a multi-phase flow model from Ansys fluent workbench. The multi-phase model is selected as the heat transfer phenomenon inside the still is evaporation and condensation, which constitutes a three phase fluid flow.

3 Methodology

The methodology adopted for the current research is presented as a flow chart in Fig. 2. An extensive literature review has helped to identify the research gap for the proposed work.

4 Modelling

4.1 Model and Its Configuration

Figure 3 shows the thermal modelling of the proposed solar still setup. The temperature of the basin bottom $T_b = 283$ K; glass cover outside temperature is $T_0 = 298$ K as the glass cover is exposed to atmospheric temperature; and convection heat transfer coefficient h_0 is used to simulate the proposed still. The still is well insulated throughout its circumference to avoid heat losses to the atmosphere. Accordingly,

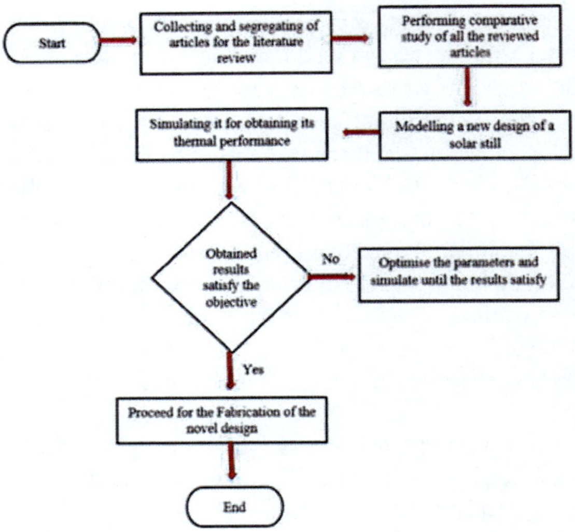

Fig. 2 Footsteps in conducting the current study

the simulation is performed by solving the conservation equations numerically by means of the finite volume method. The still basically is a two-dimensional trapezoidal space with area of absorber plate being 1.23m^2 and angle of inclination of the glass cover being 17.55^0. filled with a mixture of water–air–vapour which is further attached to an air cavity underneath.

Gravity feeds water to the solar still from an overhead tank that is higher than the still. A thin film of tap water flows through the absorber plate in the middle of the solar still, which is covered with glass at the top. The heated plate evaporates water and Condenses at the glass after passing through the air. The condensate water is collected in the collecting trough on the right side. The bottom, left and right side

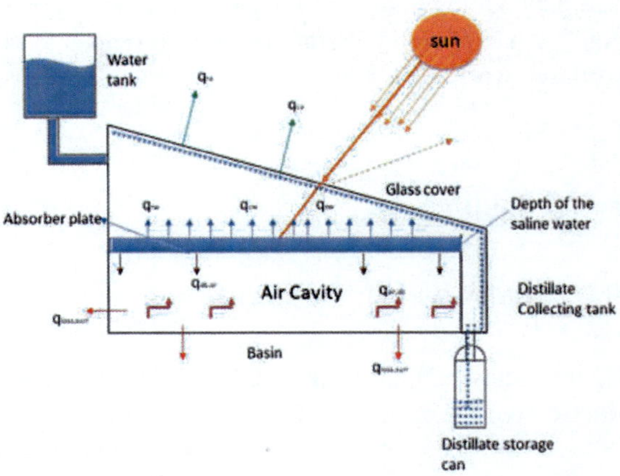

Fig. 3 Thermal Model of proposed solar still

walls are completely airtight and well insulated with 50 mm thick polystyrene and 8 mm thick water resistant plywood to avoid heat losses to the environment.

In this case, the flow is described as both transient and turbulent. Except for density, which obeys the Boussinesq approximation, the working fluid is incompressible and has constant characteristics. The thermophysical characteristics determined at T_0 and a relative humidity of 50% are considered from [6].

4.2 *Grid Independence Test*

The grid convergence test was carried out with the element size being the input parameter and the average surface Nusselt number of the basin bottom as the output parameter in the simulation of a solar still. Table 1 illustrates the convergence criteria observed in the simulation of a solar still with an air cavity. The simulations were performed for the 2D linear meshed solar still with element size starting from 0.009 m to 0.005 m and the deviation of output parameter for each of the element sizes is tabulated in Table 1. The error percentage for the 5 × 5 element size is observed to be very low, therefore the final simulations have been carried out for the solar still meshed with 5 × 5 sized elements. While Fig. 4 shows the orthogonal quality and skewness of the meshed solar still model.

Table 1 Grid Convergence

Element size (mm)	Elements	Nodes	Nu_m	%Error
9 × 9	6601	6898	8.727	–
8 × 8	8473	8807	8.728	0.01
7 × 7	10,932	11,315	8.7302	0.02
6 × 6	14,782	15,229	8.7305	0.005
5 × 5	21,383	21,918	8.7307	0.02

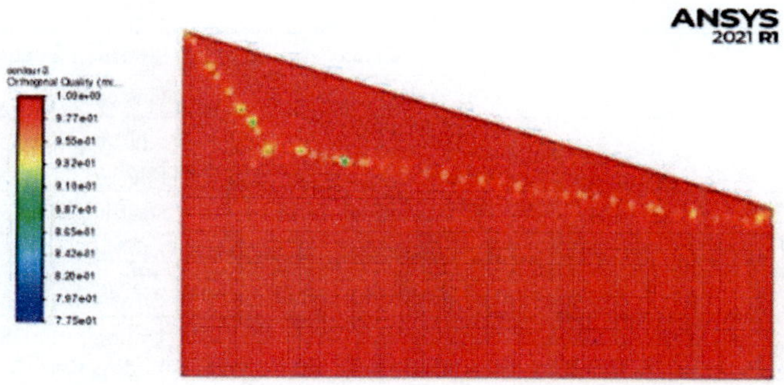

Fig. 4 Orthogonal quality of mesh

5 Results and Discussion

The buoyancy force is responsible for the system's heat transfer. In the gaseous phase, the density difference is caused by the temperature difference between the water and glass cover causing the motion of the vapour to be circular.

The convection from water to glass cover is given as seen in Eq. (1),

$$\dot{q}_{cw} = h(T_w - T_g) \tag{1}$$

whereas, the heat transfer coefficient h_{cw} is obtained from Eq. (2).

$$Nu = \frac{h_{cw} \cdot d}{k} = C(Gr \cdot \mathrm{Pr})^n \tag{2}$$

Nusselt and Sherwood numbers will be used to determine the heat and mass transfer in a solar still. From the given input parameters see Table 2 The nusselt number obtained in the proposed simulation is observed to be slightly higher than that of the previous studies [10]. For the Nusselt number to be high, either grashoff or prandtl number should be more for the effective convective heat transfer phenomena. Also, the saline water depth and height of the air cavity affect the evaporation rate inside the solar still. The difference in temperatures between the glass cover and the water over the absorber plate increases the evaporation rate causing good turbulence inside the still.

The molecular prandtl number is a critical parameter for understanding heat and mass transfer inside the solar still. According to Fig. 5, at a saline water depth of 1.8 cm and an air cavity height of 26 cm, the molecular prandtl number is high near the top surface of the absorber plate in contact with the saline water. And the molecular Pr falls considerably from the absorber plate to the glass cover. The molecular Pr

Table 2 Input Parameters

Parameter	Specification	
Solver	Space	2D
	Turbulence	k-epsilon 2 equation
	Time	Unsteady
	Multi-Phase Model	3 phase, VOF, Implicit
	Radiation Model	Discrete Ordinate
Materials	Solid	Aluminium
	Fluid	Air, water
Method and operating conditions	Simple Gravity Temperature Pressure	1st order transient Y- direction 293 K 1.01 bar

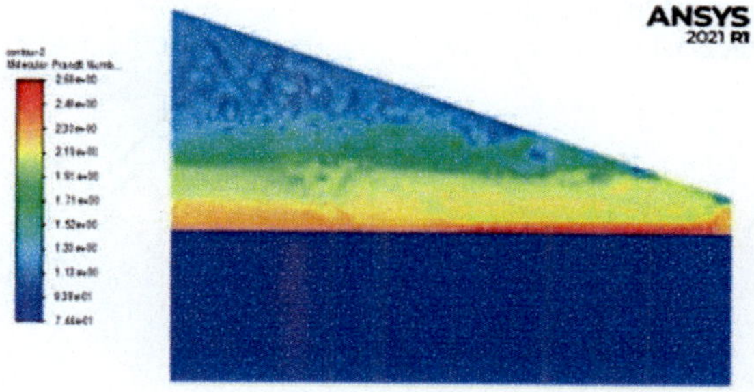

Fig. 5 Variation of molecular Pr number

is reduced when there is turbulence, indicating that the Pr is inversely related to the Re number. The Reynolds number of the still alone and still with an air cavity is found to be 7314 and 12,848. Thus Re influencing the flow to be turbulent in nature indirectly explains the evaporation rate inside the solar still.

5.1 *Effect of wall temperature*

The temperature of water and glass cover are the essential parameters to be investigated in the performance evaluation of the solar still. As the temperature gradient between the glass cover and the water increases, the exergy and energy efficiencies of the still also increases. This means the absorber plate should have a temperature always greater than that of water stored over it. Figure 7 depicts the temperature variation along the walls of the absorber plate of the still.

From Fig. 7, we can observe clearly the maximum temperature of the plate is observed to be at the lower wall of the still. This is due to the fact that there is less clearance between the glass cover and the water over the absorber plate, allowing the plate to collect the greatest amount of solar irradiation.

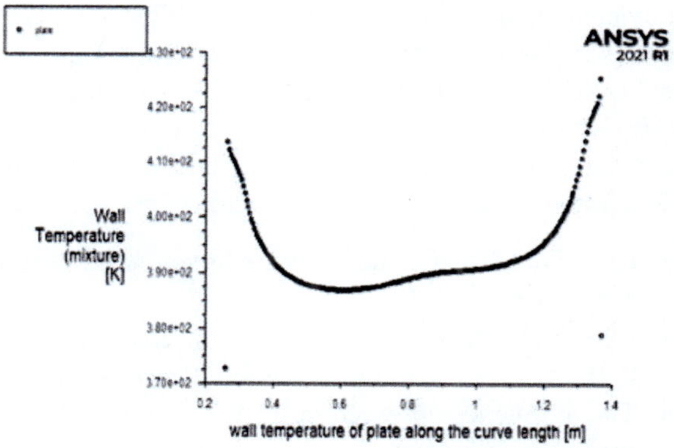

Fig. 7 Temperature variation in the absorber plate along its length

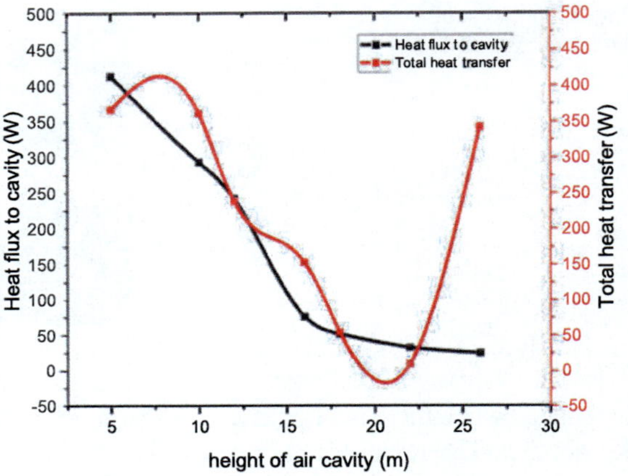

Fig. 8 Effect of variation of the height of an air cavity

5.2 *Effect of Air cavity height of air cavity (m)*

The air cavity appended to the solar still prevents the heat loss to the atmosphere acting as a perfect insulation. The total heat transfer rate from still to the air cavity varies with change in height of the cavity. It is observed from Fig. 8 that more amount of heat is transferred inside the still with a lower height of the air cavity. Also, the heat transfer to the air cavity from the absorber plate decreases as the height of the air cavity increases. From Fig. 8, the optimum height of an air cavity is observed to be 20 cm respectively as it encourages to lower the heat transfer rate from the still to the air cavity.

6 Conclusions

The solar still attached with an air cavity filled with an air–water-vapour mixture is computationally studied from the top wall, a glass cover to the bottom of the cavity and is exposed to external free stream temperature. It has been determined that the effect of critical factors, which include Reynolds number (Re_∞), Nusselt number (Nu) and height of the cavity (H), turbulence inside the solar still is pictorially explained. Before solving for the results, a mesh convergence test was performed. The following are the short conclusions drawn from the current study.

- Heat and mass transfer rates are larger on the bottom wall of the still than on the glass cover.
- The height of the air cavity is optimised and discovered to be a crucial parameter in the study of heat transfer within the solar still.
- The turbulence observed inside the solar still directly affects the evaporation rate inside the still. A lower molecular Prandtl number is observed near the glass cover.

NOMENCLATURE

CFD	Computational fluid dynamics
Gr	Grashof number
Nu	Nusselt number
Pr	Prandtl number
Sh	Sherwood number
Q	Heat flux
T	Temperature

suffix

b	Basin
s	Surface
0	Ambient
cw	Convenience from water
g	Glass cover
w	Water

References S

1. Prasanna YS, Deshmukh SS (2020) Significance of nanomaterials in solar energy storage applications. Mater Today Proc. https://doi.org/10.1016/j.matpr.2020.08.218
2. Katekar VP, Deshmukh SS, Vasan A (2020) Energy, drinking water and health nexus in India and its effects on environment and economy. J Water Clim Chang. https://doi.org/10.2166/wcc.2020.340
3. Khare VR, Singh AP, Kumar H, Khatri R (2017) Modelling and performance enhancement of single slope solar still using CFD. Energy Proc 109:447–455. https://doi.org/10.1016/j.egypro.2017.03.064
4. Edalatpour M, Aryana K, Kianifar A, Tiwari GN, Mahian O, Wongwises S (2016) Solar stills: a review of the latest developments in numerical simulations. Sol Energy 135:897–922. https://doi.org/10.1016/j.solener.2016.03.005
5. Tsilingiris PT (2013) The application and experimental validation of a heat and mass transfer analogy model for the prediction of mass transfer in solar distillation systems. Appl Therm Eng 50:422–428. https://doi.org/10.1016/j.applthermaleng.2012.07.007
6. Poling BE, Prausnitz JM, O'connell JP (2001) Properties of gases and liquids. McGrawHill Education
7. Koufi L, Cherif Y, Younsi Z, Naji H (2019) Double-diffusive natural convection in a mixture-filled cavity with walls' opposite temperatures and concentrations. Heat Transf Eng 40:1268–1285. https://doi.org/10.1080/01457632.2018.1460928
8. de Paula ACO, Ismail KAR (2021) Comprehensive investigation of water film thickness effects on the heat and mass transfer of an inclined solar still. Desalination 500:114895. https://doi.org/10.1016/j.desal.2020.114895

9. Saleem KB, Koufi L, Alshara AK, Kolsi L (2020) Double-diffusive natural convection in a solar distiller with external fluid stream cooling. Int J Mech Sci 181:105728. https://doi.org/10.1016/J.IJMECSCI.2020.105728
10. Ana Carolina Oliveira de Paula, Kamal Abdel Radi Ismail (2021) Comprehensive investigation of water film thickness effects in the heat and mass transfer of an inclined solar still. Desalination 500:114895, ISSN 0011-9164. https://doi.org/10.1016/j.desal.2020.114895

Enhancement of Energy Storage Using Phase Change Material and Nano Materials in Advancement

Hiranmoy Samanta, Joydip Paul, Soumyodeep Mukherjee, Somnath Mitra, and Sreejit Roy Chowdhury

Abstract In recent years the deterioration of the energy crisis and environmental pollution, a lot of research on the utilization of alternative energy sources and methodologies were found. Thermal energy storage plays a greater role in the era of energy. Conventional energy sources are pollutive in many ways and as well the storage is shallow. The PCMs (Phase change Materials) based thermal energy storages are investigated based on different parameters during the Melting and solidification process. Due to the high energy storage density and constant phase change temperature, phase change thermal energy storage (PCTES) has gradually become one of the preferred thermal energy storage systems over the last three decades. Analysis of PCMs has been done for thermal energy storage because of their high thermal energy densities per unit volume mass and their availability in wide temperature ranges. PCMs store the energy in the form of latent heat during phase change and this energy is being used afterwards for different and several different purposes PCM thermal storage plays a key role in improving energy efficiency and in limiting the discrepancy between the energy supply and energy demand of solar thermal energy applications (STEAs). This paper deals with the case study of different PCMs used in energy storage. The case study focuses on the application of PCM in Ice storage and the building materials have been presented.

Keywords PCM · Energy storage · Performance

1 Introduction

Thermal energy storage process is achieved through mainly three ways namely (a) sensible heat, (b) Latent Heat, (c) Chemical Energy. The phase change Process mainly accounted for high latent heat density for the phase change material over

H. Samanta (✉) · J. Paul · S. Mukherjee · S. Mitra · S. R. Chowdhury
Department of Mechanical Engineering, Gargi Memorial Institute of Technology, Kolkata, India
e-mail: hiranmoy.me_gmit@jisgroup.org

S. Das et al. (eds.), *Proceedings of the 1st International Conference on Fluid, Thermal and Energy Systems*, Lecture Notes in Mechanical Engineering,
https://doi.org/10.1007/978-981-99-5990-7_67

other conventional energy systems. By combining two or more PCM the energy storage is accompanied by the required temperature range.

Modeling of Thermal Energy Storage using Phase Change Materials.

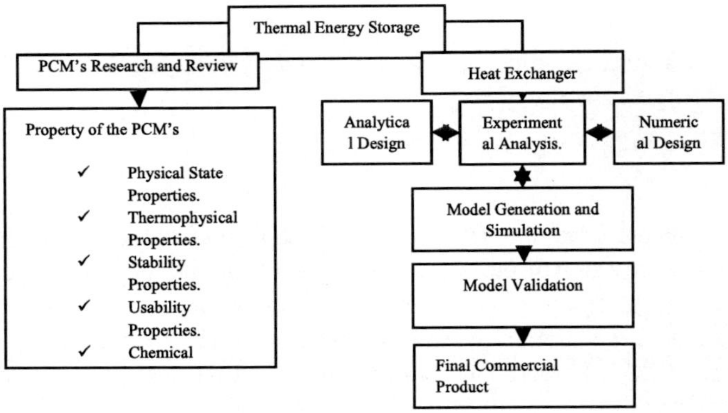

2 Literature Review and Objective

Soares et al. [22] examined how and where to use Phase Change Material (PCM) in a passive latent heat storage system (LHTES) and provided an overview of how these building solutions relate to the energy efficiency of the building. It is found that the potential for passive building solutions using PCM is to reduce heating and cooling energy consumption by reducing/shifting loads and improving indoor thermal comfort by reducing indoor temperature fluctuations is showing. Whiffen and Riffat [23] have reviewed PCM technology by keeping applications of Thermal Energy Storage in mind to meet the goal of reducing CO_2 emission and preserving the environment. Akeiber et al. [24] reviewed current research on the use of PCM for passive cooling of buildings. It has been proved that full-scale testing and numerical modeling are the most common research methods for experimental and theoretical analysis of PCM. Thamaraikannn et al. [21] have introduced the examination and investigation of stage change materials utilized in various sorts of capacity frameworks for various applications. Kaviarasu and Prakash [18] provided perspectives on different PCM types, selection of nanomaterials for embedding, and PCM applications that incorporate nanomaterials. It is observed that the most usually utilized kinds of PCM are water, paraffin, hydrated salts, and bio-based PCM. Notwithstanding the high advantages of dormant intensity stockpiling, the low warm properties require the fuse of nanomaterials. The thermal properties of the base PCM are optimized by the ratio of high surface area to volume of the nanomaterials. Mohamad and Sidik [19] presented an overview of an experimental approach to phase change materials (PCM) to improve thermal properties. Graphene-based nanoparticle has been discussed in it and found that graphene surface area is a key property used in determining the best thermophysical properties of nano-reinforced phase change materials. Tofani and Tiari [20] reviewed nano-enhanced PCM (NePCM) only and additional enhancements. The review is categorized based on improvements: NP only, NP and fins, NP and heat pipes, NP using highly conductive porous materials, NP and multiple PCMs, and nano-encapsulated PCM. Both experimental and numerical methods are considered, focusing on how much NP improves the system.

3 Materials and Methods

Phase Change Materials (PCMs) are used to store and release thermal energy during the process of melting & freezing (changing from one phase to another) (FIg. 1).

Whenever such a material freezes, it delivers a lot of energy as inert intensity of combination, or energy of crystallization. On the other hand, when the material is liquefied, an equivalent measure of energy is consumed from the prompt climate as it changes from strong to fluid. This property of PCMs can be utilized in various ways, for example, in nuclear power stockpiling by which intensity or coolness can be put away from one cycle or period in time, and utilized sometime in the future area (Fig. 2).

The selections of PCMs are done for a specific application based on the following properties (Table 1).

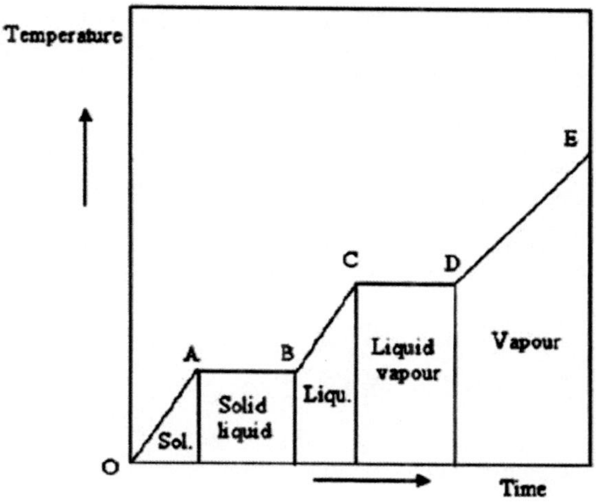

Fig. 1 Phase transition curve [23]

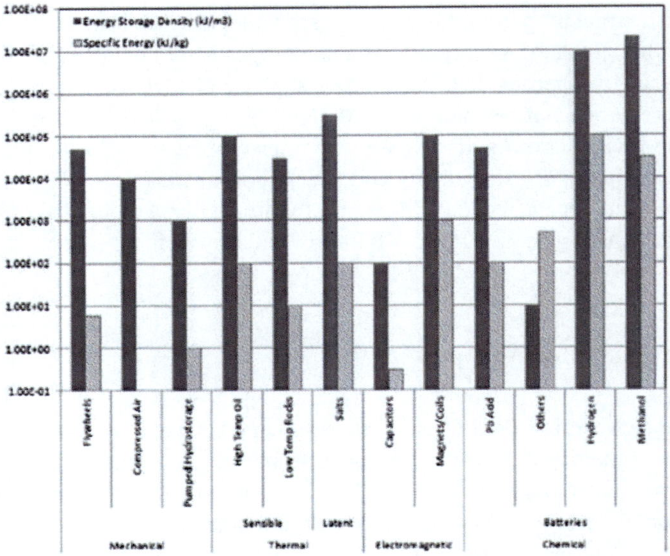

Fig. 2 Comparison between different energy storage technologies [23]

Table 1 Different types of PCMs

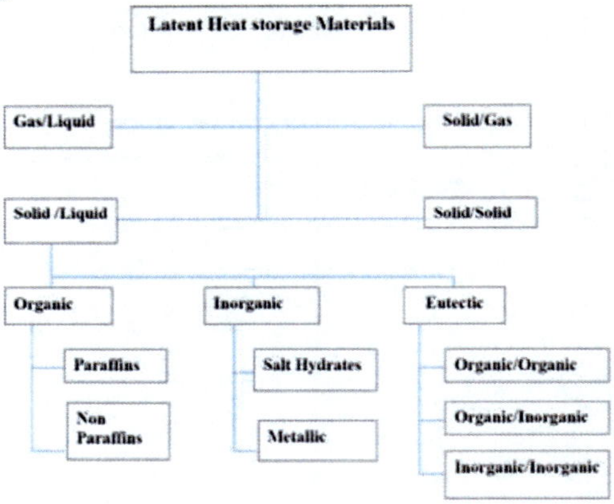

3.1 Thermodynamic Properties

The phase change material should possess,

- Temperature of melting within the intended operational temperature range.
- High fusion latent heat per unit volume.
- High thermal conductivity, high density, and high specific heat.

- To decrease the confinement challenge, minimal volume changes on phase transformation and small vapor pressure at working temperatures are used.
- Melting that is consistent.

3.2 *Kinetic Properties*

- High nucleation rate to prevent liquid phase supercooling.
- High crystal growth rate, such that the system can satisfy heat recovery demands from the storage system

3.3 *Chemical Properties*

Chemical Resistance

- Completely reversible freezing and melting cycle.
- After a high number of freeze/melt cycles, there is no deterioration
- Materials that are non-corrosive, non-toxic, non-flammable, and non-explosive.
- Economic characteristics
- Low cost

3.4 *Applications of Phase Change Materials Include, but Are not Limited to:*

- Storage of thermal energy
- Building conditioning, for example, 'ice storage'
- Heat and electrical engines cooling.
- Food, drinks, coffee, wine, milk products, and greenhouses all require cooling.
- Medical uses include blood transfer, operating tables, and hot–cold therapy.
- Under heavy garments or costumes, the human body cools.
- Recovery of waste heat
- Off-peak electricity usage: hot water heating and cooling
- Systems with heat pumps
- In bioclimatic architecture, passive storage is used (HDPE, paraffin)
- In chemical processes, smoothing exothermic temperature peaks
- Solar power plants
- Thermal systems in Spacecraft
- Vehicles thermal comfort
- Electronic devices' thermal protection
- Transport, hotel commerce, ice cream, and other food thermal protection

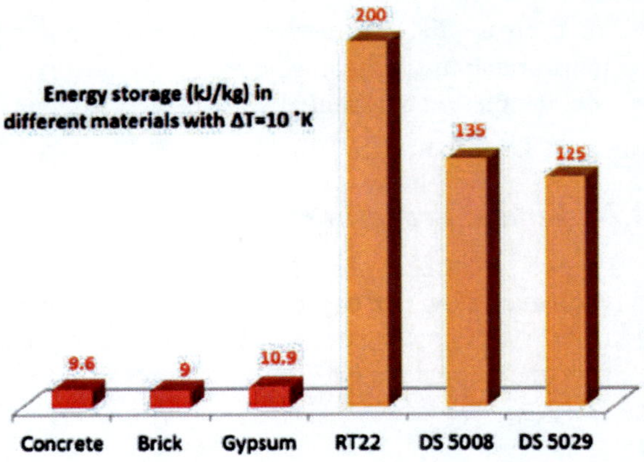

Fig. 3 Heat storage capacity comparison between commercial PCMs and conventional building materials [24]

- Textiles that are utilized in apparel
- Cooling of computers
- Thermal energy storage and turbine inlet chilling
- In tropical locations, telecom shelters are used to safeguard high-value equipment by maintaining the internal air temperature below the maximum allowable by absorbing heat generated by heavy equipment such as a base station Subsystem. When conventional cooling systems lose power, PCMs reduce the need for diesel generators, resulting in significant cost reductions across thousands of telecom sites in the tropics.*[1] (Fig. 3).

4 Nanomaterials: Their Advantages and Limitations and Its Application in PCM

Nanomaterials offer significantly improved ion transport and electron conductivity compared to conventional materials for batteries and supercapacitors. They also allow the particles to occupy all available interstitial sites in the space they occupy, resulting in elevated specific volume and fast ion diffusion. These properties permit nanomaterial-based terminals to endure high flows. This is a promising answer for high energy, high limit energy capacity. Aside from multiwalled carbon nanotube-added substances and carbon covering on silicon particles in lithium particle battery cathodes, the utilization of nanomaterials in business gadgets is exceptionally restricted. There is presently a library of nanomaterials with assorted synthetic creations and structures, from oxides, chalcogenides and carbides to carbon and lithium alloying components. The success of nanomaterials in energy storage applications is multifaceted. Nanostructures have become essential to control electrochemical performance and exploit different charge-storage mechanisms. The restrictions of nanomaterials in energy capacity gadgets are connected with their huge surface

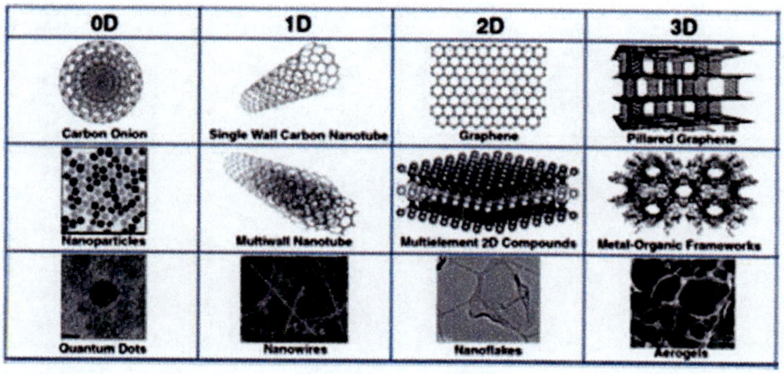

Fig. 4 0D, 1D, 2D, and 3D nanomaterials [25]

region, which causes parasitic responses with electrolytes. It is important to integrate the high electron conductivity of graphene or MXene with nanomaterials with additional capacitance, such as the high operating voltage of the oxide and the high redox motion. The high surface-area-to-volume ratio of nanomaterials and the characteristic short dispersion path provide the answer to achieving high power density and energy density at the same time. Different types of nanomaterials are available from 0 to 3D (Fig. 4).

Carbon is important for energy capacity because of its low intrinsic gravity and high abundance, coupled with the high electron conductivity of graphitic carbon. However, nanostructured carbon often offers limited redox potential. And it is often used as a two-layer capacitor material or conductive support, not as a dynamic material for energy capacity devices. The exception is graphite, which consists of a regular stack of graphene sheets and has a specific limitation of 372 mAh g1 for inserting lithium particles between the sheets. The fundamental cycles controlling electrochemical execution might be somewhat not quite the same as those of mass battery materials. For example, ion transport in nanoscale systems often occurs within the confines of the mid or upper surface of the electrode material, such as in a supercapacitor with a porous carbon electrode, rather than in a solid block electrolyte and solid electrodes in conventional batteries. For large applications, we want to build batteries and supercapacitors in a normal structure, but digging up nanomaterials considers faster activity, higher limits, and longer life than current innovations. For example, replacing graphite with nanostructured silicon can radically increase energy density compared to traditional batteries. One of the main advantages of nanoscale materials is that they can be used to create cathodes of any size, shape, or structure (Fig. 5).

Nanoparticles (NPs) are little materials with sizes from 1 to 100 nm and can be grouped into various classes in light of their properties, shape, or size. The different gatherings incorporate fullerenes, metal NPs, clay NPs, and polymer NPs. Because of its huge surface region and nanoscale size, NPs have exceptional physical and compound properties [25]. Their optical properties are size subordinate and have been accounted for to give various shadings because of retention in the noticeable reach.

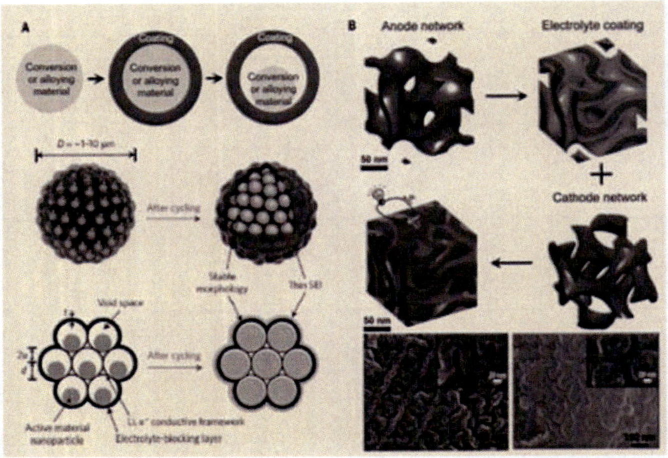

Fig. 5 Ways of preventing execution constraints of nanomaterials in energy capacity applications [25]

Their reactivity, strength, and different properties additionally rely upon their exceptional size, shape, and design. These properties make them reasonable contenders for an assortment of business and homegrown applications, for example, catalysis, imaging, clinical applications, energy-based research, and ecological applications. In this way, it is enthusiastically prescribed to involve NP in PCM as it works on the warm physical science of hotness stockpiling. A significant impact in working on the warm properties of nano PCM is the impact of the surface region and size of the nanoparticles. On account of graphene, an extremely enormous surface region uses the volume part (%) of the filler. The filler is uniformly dispersed all through the composite. The enormous surface region and size impact of the nanoparticles give a huge point of interaction region between the nanoparticles and the framework. Lessening the size and volume proportion of nanoparticles appears to attempt to work on the thermophysical properties of the PCM. The conductivity of nanofluids increments with expanding temperature of the nanofluids. Accordingly, high surface region and size impact is one of the critical standards for getting the best warm properties from improved nano PCM (FIg. 6).

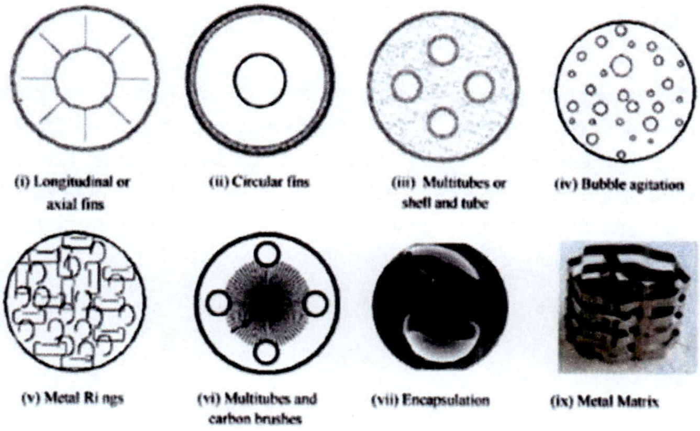

Fig. 6 Different kinds of heat transfer improvements fins [21]

5 Case Study

5.1 Ice Storage

In subsequent years commercial as well as industrial ice storage systems found in ASHRAE Handbook HVAC application based on analysis of pure substances, often water. London, Goodman, and Saitoh's studies based on pure substances are cited here. Different Geometrical designs are being used in the study of random Crystallization and under-cooling. Ismail et al. have studied ice storage in a bank with finned tubes; different parametric studies are also done.

5.2 Building Application

Mainly two types of building applications are found in the literature.

- Building material.
- Room heating and cooling procedure

Salyer studied the reaction to fire and the possible method of safety by adding fire retardant additives, with advantages and disadvantages review of materials, etc. It has been seen that Paraffin type PCM's have a greater advantage over other PCM's.

6 Conclusions

To replace the conventional energy storage systems, PCMs (Phase change Materials) based thermal energy storages are investigated based on different parameters during Melting and solidification process. Their classifications and properties are studied. Later energy storage using nanomaterials has been discussed. Their structures and utilization have been discussed. At last, a brief case study of the application of PCM in ice storage and building application is presented.

References

1. Zalba B, Marin J, Cabeza L, Mehling H Review on thermal Storage with Phase Change Materials, heat Transfer Analysis and Applications
2. London AL, Seban RA (1943) Rate of ice formation. Trans ASME 65:771–778
3. Goodman TR (1958) The heat-balance integral and its application to problems involving a change of phase. Trans ASME 335–342
4. Lazaridis A (1970) A numerical solution of the multidimensional solidification (or melting) problem. Int J Heat Mass Transfer 13:1459–1477
5. Agyenim F, Eame P, Smyth M (2011) Experimental study on the melting and solidification behavior of a medium temperature phase change storage material (Erythritol) system augmented with fins to power a LiBr/H2O absorption cooling system. Renew Energy 36(1):108–117
6. Agyenim F, Eames P, Smyth M (2009) A comparison of heat transfer enhancement in a medium temperature thermal energy storage heat exchanger using fins. Sol Energy 83(9):1509–1520
7. Agyenim F, Eames P, Smyth M (2010) Heat transfer enhancement in medium temperature thermal energy storage system using a multi-tube heat transfer array. Renew Energy 35(1):198–207
8. Al-Abidi AA, Mat S, Sopian K, Sulaiman MY, Mohammad A (2013) Internal and external fin heat transfer enhancement technique for latent heat thermal energy storage in triplex tube heat exchangers. Appl Therm Eng 53(1):147–156
9. Fan LX, Khodadadi JM (2011) Thermal conductivity enhancement of phase change materials for thermal energy storage: a review. Renew Sustain Energy Rev 15(1):24–46
10. Farid MM, Khudhair AM, Siddique AKR, Hallaj S (2004) A review on phase change energy storage: materials and applications. Energy Convers Manage 45(9–10):1597–1615
11. Fernández I, Renedo CJ, Pérez S, Carcedo J, Mañana M (2010) Advances in phase change materials for thermal solar power plants Quality. In: International Conference on Renewable Energies and Power Quality (ICREPQ'11) Las Palmas de Gran Canaria (Spain), 13th to 15th April, 2010
12. Jaworski M (2012) Thermal performance of heat spreader for electronics cooling with incorporated phase change material. Appl Therm Eng 35:212–219
13. Jegadheeswaran S, Pohekar SD, Kousksou T (2010) Exergy-based performance evaluation of latent heat thermal storage system: A review. Renew Sustain Energy Rev 14(9):2580–2595
14. Lingayat AB, Suple YR (2013) "Review on phase change material as the thermal energy storage medium: Materials Applications. Int J Eng Res Appl 3(4):916–921
15. Sharma A, Tyagi VV, Chen CR, Buddhi D (2009) Review on thermal energy storage with phase change materials and applications. Renew Sustain Energy Rev 13(2):318–345
16. Trp A, Lenic K, Frankovic B (2006) Analysis of the influence of operating conditions and geometric parameters on heat transfer in water-paraffin shell-and-tube latent thermal energy storage unit. Appl Therm Eng 26(16):1830–1839

17. Waqas A, Din ZU (2013) Phase change Material (PCM) storage for free cooling of buildings—A review. Renew Sustain Energy Rev 18:607–625
18. Kaviarasu C, Prakash D (2016) Review on phase change materials with nanoparticle in engineering applications. J Eng Sci Technol Rev 9(4):26–36
19. Mohamad AT, Sidik NAC (2019) Nano-enhanced phase change material effects on the supercooling degree improvement: a review. IOP Conf. Series: Mater Sci Eng 469:012036. IOP Publishing
20. Tofani K, Tiari S (2021) Nano-enhanced phase change materials in latent heat thermal energy storage systems: a review. Energies 14:3821. https://doi.org/10.3390/en14133821
21. Thamaraikannn R, Kanimozhi B, Anish M, Jayaprabakar J, Saravanan P, Rohan Nicholas A (2017) Review of phase change materials based on an energy storage system with applications. IOP Conf Ser: Mater Sci Eng 197:012034
22. Soaresa N, Costa JJ, Gaspar AR, Santos P (2013) Review of passive PCM latent heat thermal energy storage systems towards buildings' energy efficiency. Energy Buildings 59:82–103
23. Whiffen TR, Riffat SB (2013) A review of PCM technology for thermal energy storage in the built environment: Part I. Int J Low-Carbon Technol 8:147–158
24. Akeiber H, Nejat P, Majid MZA, Wahid MA, Jomehzadeh F, Famileh IZ, Calautit JK, Hughes BR, Zaki ZA (2016) A review on phase change material (PCM) for sustainable passive cooling in building envelopes. Renew Sustain Energy Rev 60:1470–1497
26. Pomerantseva E, Bonaccorso F, Feng X, Cui Y, Gogotsi Y (2019) Energy storage: the future enabled by nanomaterial. Science 366:969s

Numerical Study of a Microchannel Thermocycler with Sequential PCM Array for RNA Amplification

B. Indulakshmi, Nikhil Prasad, and Ranjith S. Kumar

Abstract Numerical studies are performed for a microchannel embedded in three different phase changing material (PCM) encapsulations for achieving isothermal conditions required for denaturation, annealing, and extension phases in a sequential manner for thermocycling. PCM confinements for the denaturation and extension phases are preheated such that they transfer latent heat to the fluid passing through the microchannel. Whereas, the solid phase PCM kept in the annealing section absorbs latent heat of fusion from the fluid and thereby reduces its temperature congenial for annealing. Two-dimensional unsteady fluid flow and heat transfer in a microchannel with melting/solidification boundaries are solved using a finite volume solver based on SIMPLEC algorithm. A detailed parametric study is carried out by varying the size of the PCM storage and fluid flow rate ($0.05 < Re < 1$). Fluid temperatures in the microchannel and thermal performance aspects have been assessed for all the configurations. An accurate control over successive isothermal fluid conditions is found to attain. Hence the present configuration is a suitable candidate for miniature thermocycler configuration for RNA amplification.

Keywords Microchannel convection · Phase changing material (PCM) · Thermocycler · Heat transfer in melting/solidification · Finite volume method (FVM) · RNA amplification

1 Introduction

The recent surge in communicable diseases necessitates the quick detection of infectious microorganisms through genome sequencing. Polymerize chain reaction (PCR) process has become the backbone of biomedical research that involves creating a large number of copies of nucleic acid sequences through sequential changes

B. Indulakshmi (✉) · N. Prasad · R. S. Kumar
Micro/Nano Fluidics Lab, Department of Mechanical Engineering, College of Engineering Trivandrum-695016, Trivandrum, India
e-mail: indulakshmi.d@gmail.com

S. Das et al. (eds.), *Proceedings of the 1st International Conference on Fluid, Thermal and Energy Systems*, Lecture Notes in Mechanical Engineering,
https://doi.org/10.1007/978-981-99-5990-7_68

in temperature. Thermocycler is a device used to artificially facilitate exponential nucleic acid amplification by exposing the sample to various prescribed temperature levels. A continuous flow polymerase chain reaction (CF-PCR) method is gaining research interest due to its suitability in implementing thermocycling in compact microfluidic platforms [1, 2]. It is well known that microfluidic paths have the inherent advantage of large surface-to-volume ratio that facilitates enhanced transport phenomena. Active control of temperature sequences in the microfluidic device is a challenge due to the limited space implementation of externally powered heaters. PCMs offer passive thermal control in microfluidic passages and are ideally suitable for mass production through microfabrication methods. A microchannel thermocycler for facilitating three temperature levels for the nucleic acid amplification of *Choanephora cucurbitarum* [3] is envisaged in the present work. Basic architecture of the PCM based microchannel thermocycler is shown in Fig. 1. A sample containing the target gene for amplification is fed through a microchannel inlet which sequentially fed the sample through three isothermal fluid path segments. The denaturation segment is to be maintained at 95 °C. This is achieved by passing the sample fluid through an encapsulation filled with preheated sorbitol ($C_6H_{14}O_6$) in a molten condition. The second segment has a primer-specific annealing temperature of 52 °C. This is attained in the second segment by melting solid phase Tetracosane ($C_{24}H_{50}$). The third microfluidic segment in this sequence for the extension phase is surrounded by molten Tritriacontane ($C_{33}H_{68}$) which sets an isothermal condition of 72 °C. A numerical study is performed to check the sequential variation of the sample fluid while it is carried continuously through the aforementioned microfluidic segments in tandem. A unit of the thermocycler to perform only one amplification cycle consisting of denaturation, annealing, and extension region each is considered for simulation. Further, the study is also extended to a detailed thermal response characterization of the microfluidic convection process with phase changing boundaries to facilitate thermocycler design.

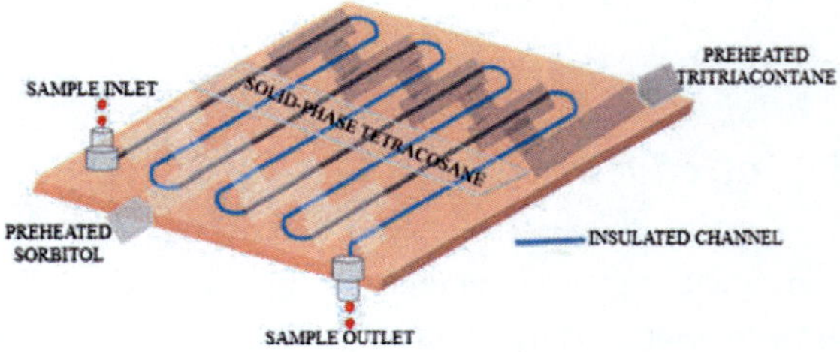

Fig. 1 PCM based Microfluidic thermocycler

2 Literature Review and Objective

Polymerase chain reaction (PCR) has become an accepted practice in pathogen detection in various facets of our day-to-day life especially in light of the current insurgency due to the outbreak of Covid-19 pandemic [5]. Food safety, forensic investigations, genetic studies, bio-molecular research, etc. are some of the other industrial and scientific applications of PCR. This process involves multiple temperature sequences to facilitate reactions that lead to exponential amplification of the target nucleic acid sequence.

Microfluidic platforms are ideal for continuous flow polymerase chain reaction (CF-PCR) devices due to its inherent advantages of low sample volume, quick thermal and mass transfer response, minimal reagent requirement, multifunctional integration, etc. [6]. This also facilitates the development of ultra-miniature process equipment Micro-Total-Analysis Systems (μTAS) and Lab-on-a-Chip. Recent advances in microfabrication also facilitated this highly poised technological development for achieving cost-effective instrumentation ideal for mass testing to curb communicable diseases [7].

Various types of phase changing material (PCM) encapsulations [8] are gaining research interest recently. PCMs are suitable to integrate in microfluidic platforms with modern microfabrication methods such as ultra-rapid prototyping (URP) [9] and have been successfully used in micro-thermoregulation applications [11]. Heat generation and control for maintaining isothermal conditions in CF-PCR microfluidic passages is quite challenging. Exothermic reactions [12] and Peltier heaters [13] are conventionally used for microfluidic thermocycler applications. This necessitates the support of additional accessories which makes the system further complex. A precise and passive temperature control is possible with microchannel laden in PCM [14]. Passive control of microfluidic thermocycler is not explored much for medical instrumentation and hence it is envisaged in the present study. This makes use of molten PCMs for denaturation and extension phases. These can be conveniently fed to the identified imprinted microfluidic passages and fed to the premises of the respective microchannel boundary. Whereas, the solid phase PCM used for the annealing phase can be prefabricated and assembled along with microfluidic circuit of the thermocycler.

The salient objective of the present work is to numerically simulate the PCM based microfluidic thermocycler and establish its suitability for CF-PCR applications. A detailed thermal response assessment is carried out for various flow and geometric conditions of a base unit of the thermocycler.

3 Governing Equations and Numerical Methods

The physics of the present case study include two-dimensional incompressible fluid flow and unsteady heat transfer to a phase changing medium across an immiscible boundary. Conservation of mass, momentum, and energy of the carrier fluid (water) is modeled as

$$\frac{\partial u_i}{\partial x_i} = 0 \tag{1}$$

$$\frac{\partial u_i}{\partial t} + u_j \frac{\partial u_i}{\partial x_j} = -\frac{1}{\rho}\frac{\partial p}{\partial x_i} + \nu \frac{\partial^2 u_i}{\partial x_{j^2}} \tag{2}$$

$$\frac{\partial T_f}{\partial t} + \frac{\partial \left(u_i T_f\right)}{\partial x_i} = \alpha_f \frac{\partial^2 T_f}{\partial x_i^2} + \frac{\nu}{c_f}\frac{\partial u_i}{\partial x_j}\left(\frac{\partial u_i}{\partial x_j} + \frac{\partial u_j}{\partial x_i}\right) \tag{3}$$

Thermal convection in PCM is modeled such that the Boussinesq approximation is assumed to be valid. Hence the momentum equation for PCM domain is

$$\frac{\partial u_i}{\partial t} + u_j \frac{\partial u_i}{\partial x_j} = -\frac{1}{\rho}\frac{\partial p}{\partial x_i} + \nu \frac{\partial^2 u_i}{\partial x_{j^2}} - B\left(u_i - u_p\right) + g\beta(T - T_m) \tag{4}$$

Energy conservation model of the PCM domain involves treating mushy region as a single component porous medium.

$$\frac{\partial T_p}{\partial t} + \frac{\partial \left(u_i T_p\right)}{\partial x_i} = \alpha_p \frac{\partial^2 T_p}{\partial x_i^2} + \frac{\nu_p}{c_p}\frac{\partial u_i}{\partial x_j}\left(\frac{\partial u_i}{\partial x_j} + \frac{\partial u_j}{\partial x_i}\right) - L_p \frac{\partial f}{\partial t} \tag{5}$$

Liquid fraction is estimated as

$$f = \begin{cases} 1 \text{ if } T_i > T_l \\ \frac{T_i - T_s}{T_l - T_s} \text{if } T_s < T_i < T_l \\ 0 \text{ if } T_i < T_s \end{cases} \tag{6}$$

The local heat transfer coefficient is

$$h_x = q_s / (T_{sx} - T_x) \tag{7}$$

where T_s is the mean temperature of the fluid at an axial location in the channel calculated based on mean velocity of the fluid U as

$$T_x = \frac{1}{UA}\int_A u T_f dA_c \tag{8}$$

The following data are deduced for the thermal performance characterization. Average heat transfer coefficient in PCM laden microfluidic channel is estimated based on respective melting temperature as

$$\overline{h} = mc(|T_i - T_e|) \bigg/ \left(\frac{|(T_m - T_i) - (T_m - T_e)|}{\text{In}|(T_m - T_i)/(T_m - T_e)|} \right) \quad (9)$$

Surface average of Nusselt number

$$\overline{Nu} = \frac{\overline{h}H}{k} \quad (10)$$

where the characteristic dimension is chosen as the height of the channel H.

The aforementioned conservation equations are numerically solved using a finite volume solver (Ansys Fluent [15]). Semi-Implicit Method for Pressure Linked Equations-Consistent (SIMPLEC) algorithm is used for the sequential solution of the momentum equation. Pressure staggering option (PRESTO) scheme is used for pressure term discretization. The energy equation is solved using a second-order upwind scheme. A second-order implicit scheme has been chosen for temporal discretization. Convergence criteria of all conservation variables are set as 10^{-6} residual values for each time step. Thermo-physical properties of the phase changing materials are given in Table 1.

Aforementioned computational procedure is validated based on the experimental results reported by Steinke et al. [16]. Temperature rise measured in a 250 μm wide microchannel with heat flux conditions for Re = 61 and 91 have been compared (Figs. 2, 3).

A schematic of the computational domain and boundary conditions used for the present work is shown in Fig. 2. Four configurations are analyzed in this work by varying the dimensions of the PCM storages in three successive process chambers,

Table 1 Thermo-physical properties of PCMs

Property [unit]	PCM		
	Sorbitol ($C_6H_{14}O_6$)	Tetracosane ($C_{24}H_{50}$)	Tritriacontane ($C_{33}H_{68}$)
Melting Temperature [K]	368	323.6	345
Conductivity [W/mK]	0.4	0.37	0.2
Solidus temperature [K]	367	321.25	344
Liquidus temperature [K]	369	323.75	346
Specific heat [kJ/kgK]	2500	2100	1110
Density [Kg/m^3]		779	782
Latent heat [kJ/kg]	165	259	256

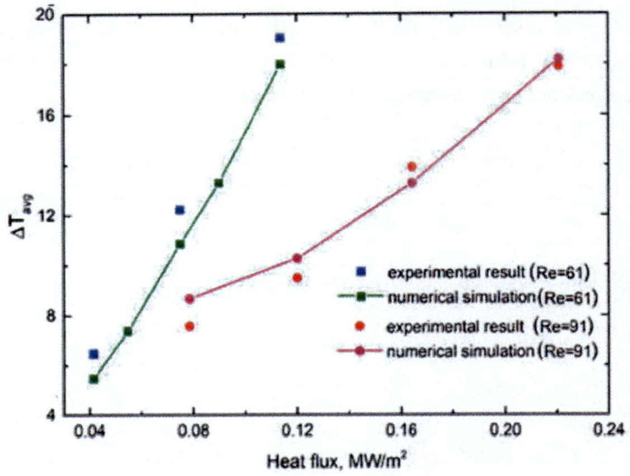

Fig. 2 Validation results

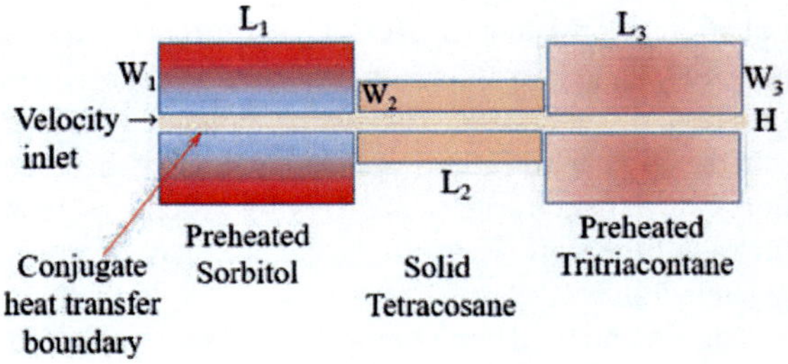

Fig. 3 Computational domain and boundary conditions

viz; denaturation, annealing, and extension as given in Table 2. Flow rate through the channel is also varied ($0.05 < Re < 1$) to study the thermal response of the PCM with an objective to ensure the maintenance of isothermal condition in these three phases of the thermocycling process. Length of the PCM confinements is chosen as 12 mm by maintaining a gap of 750 μm between the sections. Width of the microchannel is $H = 100$ μm. Velocity at the inlet of the channel is prescribed to realize four different $Re = 0.05$, 0.1, 0.5 and 1.0).

Table 2 Chamber dimensions of various cases considered in the study

Case	Section width (μm)		
	Denaturation section (W_1)	Annealing section (W_2)	Extension section (W_3)
I	1000	1000	1000
II	500	500	500
III	1000	500	1000
IV	1000	250	1000

4 Results and Discussion

Results of the detailed parametric study by varying the size of the PCM storage (Table 2) and fluid flow rate ($0.05 < Re < 1$) are presented. Flow rate is selected based on the thermal response of the system for the efficient maintenance of isothermal conditions in microfluidic thermocycler. Initial conditions of PCM are set such that it mimics the molten state. Initial conditions of PCM in the denaturation and extension sections are respectively 95 °C and 72 °C. Whereas, solid PCM is maintained at ambient conditions (27 °C) in the annealing section. Isothermal condition maintenance of fluid passing in annealing section is crucial as it can be achieved only by a well-balanced latent energy transfer from the fluid to PCM. A comparison of time elapsed to attain annealing temperature and extent of isothermal region observed in annealing (Fig. 4) indicates that quick transfer of adequate heat to PCM (to maintain isothermal conditions in the fluid) is not possible in lower flow rate conditions. Hence, further detailed studies of thermal performance comparison are carried out for $Re = 0.5$ and 1.0 only. Extent of isothermal region obtained in type IV at $Re = 1.0$ is attaining the largest extent of isothermal region in minimum time (5 s).

Temperature and liquid fraction contours at various instances for case IV at $Re = 0.5$ and 1.0 are shown in Fig. 5. Fluid admitted at ambient conditions (27 °C) is absorbing energy from the molten sorbitol in denaturation section. It quickly attains the desired denaturation temperature (95 °C) and maintains the same for a considerable extent of the section. As a result, sorbitol gets solidified in the initial region due to large latent energy transfer. A similar observation can be made for the extension region as well. A comparison of the axial temperature profile of various configurations for $Re = 0.5$ and 1.0 conditions is given in Figs. 6, 7. This clearly indicates that type IV configuration is outperforming others in quick maintenance of the required isothermal conditions in respective sections.

Thermal response characteristics of two better performing cases are compared in Fig. 8. The increase in fluid velocity enhances the rate of melting of PCM in annealing section (as observed by the higher rate of liquid fraction). This results in a sudden reduction of temperature from denaturation section to annealing section. However, the increase in velocity causes a sluggish rise of temperature from the annealing to the extension section. This reduces the available isothermal region in the extension region.

The degree of latent energy transfer in the current problem is restricted by the mass of PCMs taken in respective confinements. Hence latent energy transfer is highly influenced by the flow rate in the channel. A comparison of melt progression in annealing section (Fig. 8) and solidification trends in denaturation and extension sections (Figs. 10, 11) clearly evidences the mass flow rate dependence of the latent energy transfer. Time needed for response (4–16 s) is found comparable to the operation of similar microfluidic based detection device (Fig. 9).

A detailed assessment of heat transfer performance of PCM laden microfluidic channels is essential for its design and implementation. Since lower flow rates ($Re =$

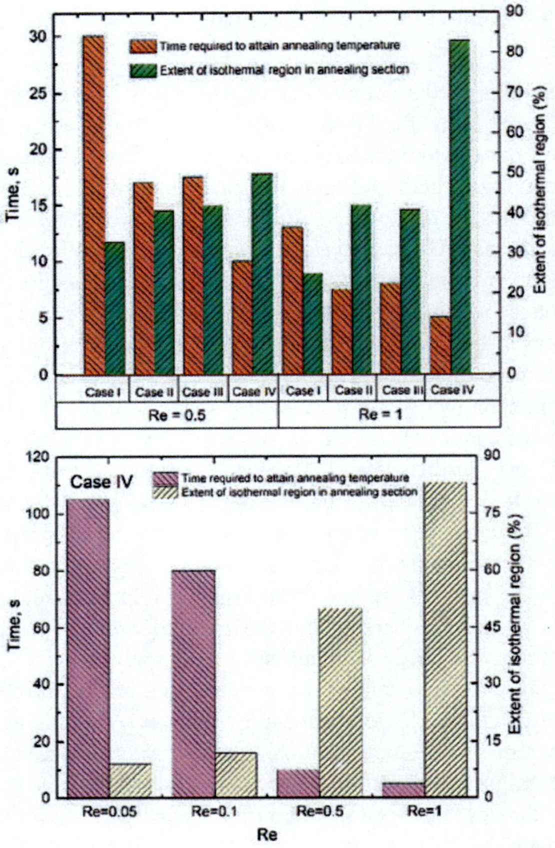

Fig. 4 A comparison of time elapsed to attain annealing temperature and extent of isothermal region observed in annealing

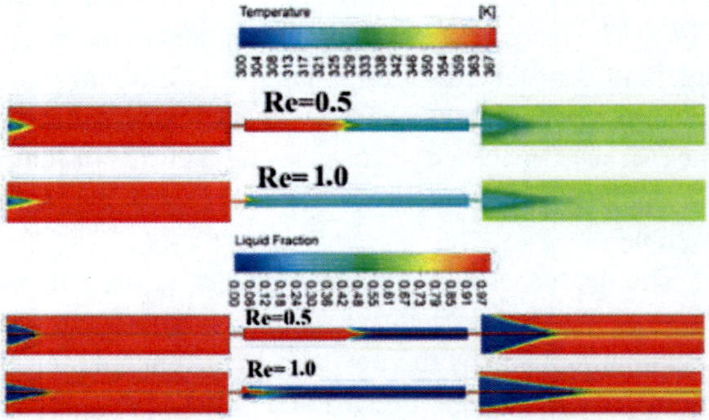

Fig. 5 Temperature and liquid fraction contours at various instances for case IV for Re = 0.5 and 1.0

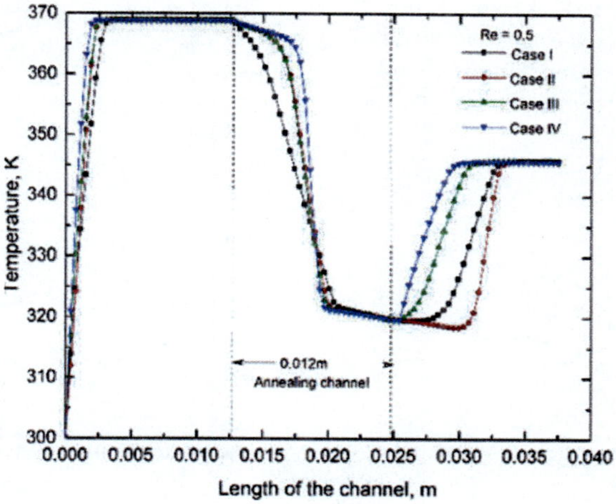

Fig. 6 A comparison of the axial temperature profile of various configurations with Re = 0.5

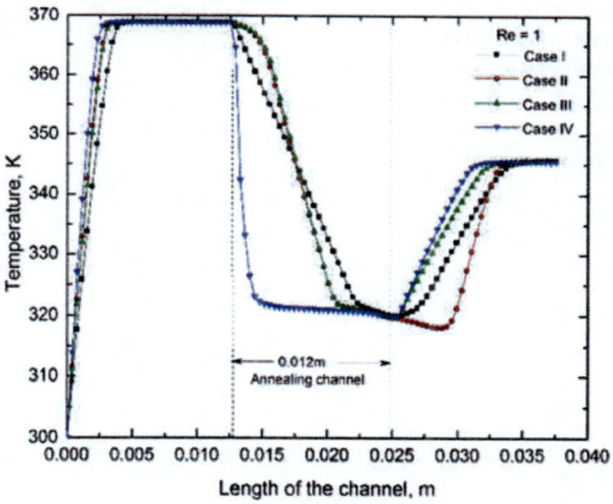

Fig. 7 A comparison of the axial temperature profile of various configurations with Re = 1.0

0.05 and 0.1) are not providing adequate response, higher flow rates (Re = 0.5 and 1.0) are only considered for further comparison of the heat transfer performance.

A comparison of the thermal performance of various configurations with Re = 0.5.and 1.0 is given in Fig. 12. Heat transfer coefficient increases with flow rate. Though all phase-change heat transfer rates are of similar order, the heat transfer coefficient in extension region is comparatively higher than the other two regions. This is due to larger temperature difference existing at the entry of the extension region. This observation agrees well with the shorter isothermal region observed in the extension region for most of the cases considered here. This essentially call for a varying area PCM storage to allocate an ideal amount of latent energy transfer towards the working fluid. Present study could establish the use of micro-PCM encapsulations for the maintenance of isothermal region in miniscule thermocyclers.

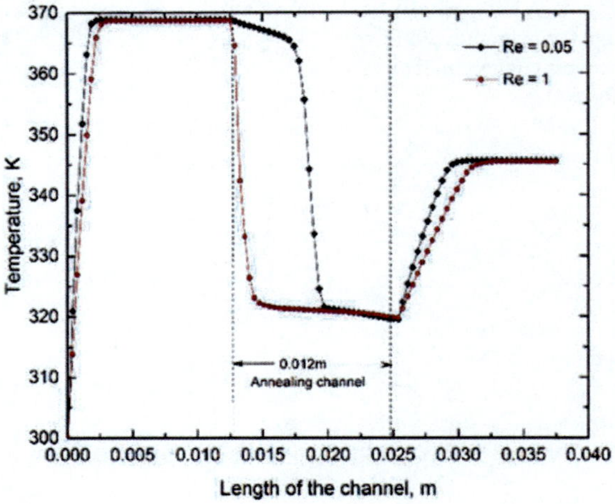

Fig. 8 A comparison of the axial temperature profile of case 4 configuration with Re = 0.5 and 1

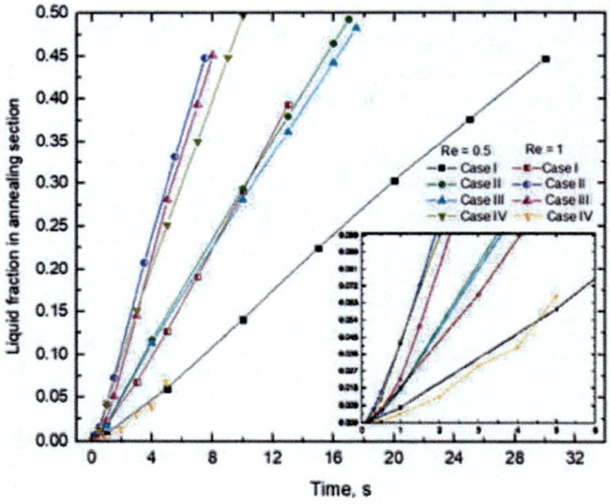

Fig. 9 A comparison of melt progression in annealing section

5 Conclusions

A numerical study of PCM based thermocycler has been carried out using an experimentally validated SIMPLEC algorithm based finite volume solver. Computations could portray the spatio-temporal advancements of the melt front in PCM encapsulations associated with denaturation, annealing, and extension sections of the thermocycler. Among the various flow rates ($0.05 < Re < 1$), $Re = 0.5$ and 1.0 could realize both sufficient isothermal region and minimal melting time. Among the various configurations analyzed in the present study, type IV is found to provide adequate isothermal regions in all the three sections with minimal melt time when sample fluid is admitted at $Re = 1.0$. A quick reduction to the annealing temperature is

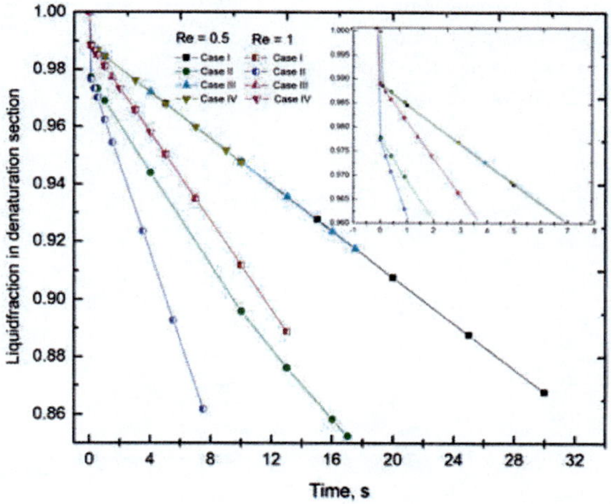

Fig. 10 A comparison of solidification trends in denaturation section

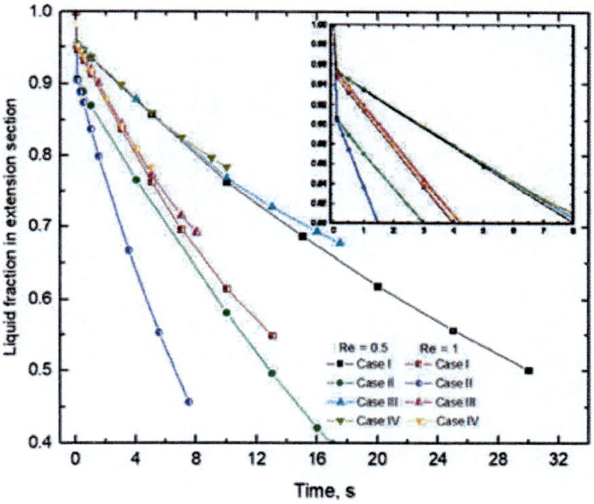

Fig. 11 A comparison of solidification trends in extension section

also observed for type IV configuration when fluid is admitted at Re = 1.0. Simulations could open up an avenue using PCM laden microfluidic path for thermocycling requirements in CF-PCR applications.

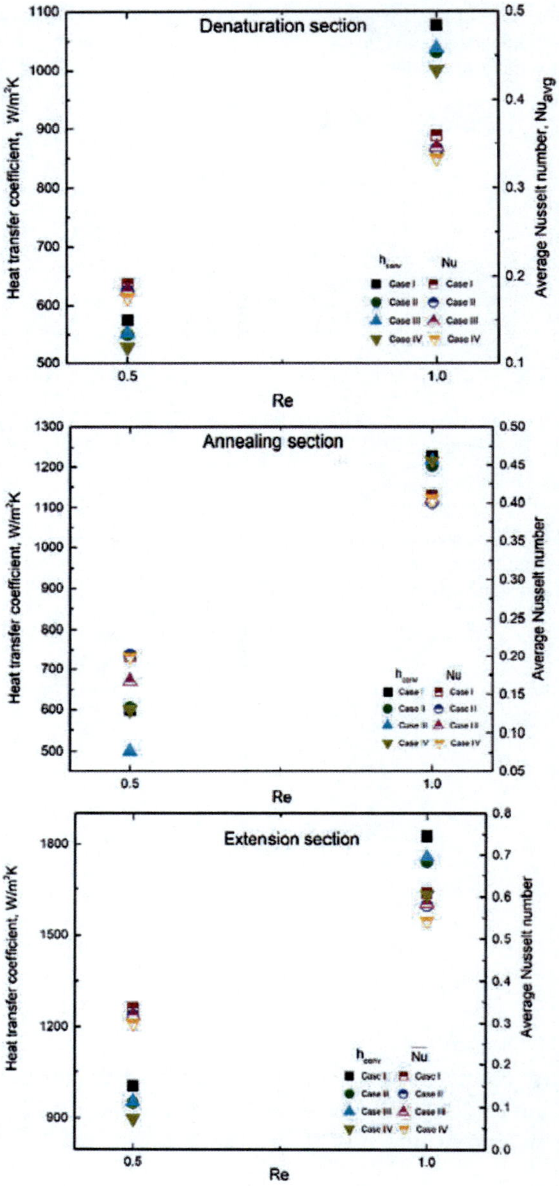

Fig. 12 A comparison of the thermal performance of various configurations with Re = 0.5.and 1.0

Acknowledgements Authors acknowledge the financial support from Kerala State Council for Science, Technology and Environment (KSCSTE) for this work.

Nomenclature

c	Specific heat [J/kgK]
F	Liquid fraction [–]
h	Heat transfer coefficient [W/m^{2}K]
k	Thermal conductivity [W/mK]
L	Latent heat of fusion [J/kgK]
p	Pressure [Pa]
u	Velocity [m/s]
T	Temperature [K]
α	Thermal diffusivity [m^2/s]
β	Thermal expansion coefficient [K^{-1}]
ρ	Density [kg/m^3]
ν	Kinematic viscosity [m^2/s]

References

1. Hashimoto M, Chen PC, Mitchell MW, Nikitopoulos DE, Soper SA, Murphy MC (2004) Rapid PCR in a continuous flow device. Lab Chip 4(6):638–645
2. Yang B, Wang P, Li Z, Tao C, You Q, Sekine S, Zhuang S, Zhang D, Yamaguchi DY (2022) A continuous flow PCR array microfluidic chip applied for simultaneous amplification of target genes of periodontal pathogens, Lab on a Chip, 22, pp 733–737
3. Liu PQ, Wei MY, Zhang JZ, Wang RB, Li BJ, Chen QH, Weng QY (2019) Leaf blight on okra caused by Choanephora cucurbitarum in China. Can J Plant Path 41(3):366–371
4. Dong X, Liu L, Tu Y, Zhang J, Miao G, Zhang L, Ge S, Xia N, Yu D, Qiu X (2021) Rapid PCR powered by microfluidics: A quick review under the background of COVID-19 pandemic. TrAC, Trends Anal Chem 143:116377
5. Zhang Y, Ozdemir P (2009) Microfluidic DNA amplification—A review. Anal Chim Acta 638(2):115–125
6. Burki T (2020) Mass testing for COVID-19. The Lancet. Microbe 1(8):e317
7. Waldbaur A, Rapp H, Länge K, Rapp BE (2011) Let there be chip—towards rapid prototyping of microfluidic devices: one-step manufacturing processes. Anal Methods 3(12):2681–2716
8. Han X, Kong T, Zhu P, Wang L (2020) Microfluidic encapsulation of phase-change materials for high thermal performance. Langmuir 36(28):8165–8173
9. o'Neill PF, Ben Azouz A, Vazquez M, Liu J, Marczak S, Slouka Z, Chang HC, Diamond D, Brabazon D (2014) Advances in three-dimensional rapid prototyping of microfluidic devices for biological applications. Biomicrofluidics, 8(5), p 52112.
10. Shi T, Hu P, Wang J (2020) Preparation of polyurea microcapsules containing phase change materials using microfluidics. Chem Sel 5(7):2342–2347
11. Sözmen AB, Arslan Yildiz A (2021) Cost-effective and rapid prototyping of PMMA microfluidic device via polymer-assisted bonding. Microfluid Nanofluidics, 25(8), pp 1–11
12. Singleton J, Zentner C, Buser J, Yager P, LaBarre P, Weigl BH (2013) Instrument-free exothermic heating with phase change temperature control for paper microfluidic devices. In Microfluidics, BioMEMS, and Medical Microsystems 8615:136–149
13. Miralles V, Huerre A, Malloggi F, Jullien MC (2013) A review of heating and temperature control in microfluidic systems: techniques and applications. Diagnostics 3(1):33–67

14. Yan WM, Ho CJ, Tseng YT, Qin C, Rashidi S (2020) Numerical study on convective heat transfer of nanofluid in a minichannel heat sink with micro-encapsulated PCM-cooled ceiling. Int J Heat Mass Transf 153:119589
15. ANSYS Fluent, Ansys fluent. Academic Research. Release, 14, 2015
16. Steinke ME, Kandlikar SG, Magerlein JH, Colgan E, Raisanen AD (2005) Development of an experimental facility for investigating single-phase liquid flow in microchannels. Nanochannels, Microchannels, and Minichannels 41855:233–243

Experimental and Numerical Studies of N–Heptane Pool Fires

S. Raja, B. Ashutosh, and V. Raghavan

Abstract Pool fires may occur in industries and facilities handling flammable fuels and solvents triggering losses to life and property. This work reports experimental and numerical investigations of n–heptane pool fires considering three square pools of sizes 8, 12 and 16 cm. Average mass loss rates have been measured in experiments using three trials. Fire Dynamics Simulator (FDS) is used to simulate these pool fires to compliment the experimental data. Time averaged mass loss rates predicted numerically are validated against the experimental results. The temporal behaviour of these flames is analyzed systematically using contours of oxygen mass fraction and temperature, and velocity vectors. In this study, the instantaneous flame heights are determined by analyzing the contours of n–heptane mass fractions across the entire burning duration. Further, the heights are also determined by locating the points where the flames are 50% intermittent.

Keywords Pool fires · Mass loss rate · Large eddy simulation · Fire dynamics simulator · Flame intermittency

1 Introduction

Pool fires consist of a liquid fuel kept in a container over which a buoyant diffusion flame is established upon ignition. These can occur in many of the scenarios including liquid fuel storage and leakage [1]. It is, therefore, essential to unveil the dynamics of pool fires and help us understand the risks associated with it [2]. Experimental studies of pool fires across scales were conducted by Lock and Hamins [3], where averaged temperature profiles were obtained using thermocouples after correcting for losses and species profiles reported using gas chromatography. Oliveria et al. [4] compared the temperature fields and flame heights as predicted numerically for

S. Raja · B. Ashutosh · V. Raghavan (✉)
Department of Mechanical Engineering, Indian Institute of Technology Madras, Chennai–600036, Tamil Nadu, India
e-mail: raghavan@iitm.ac.in

S. Das et al. (eds.), *Proceedings of the 1st International Conference on Fluid, Thermal and Energy Systems*, Lecture Notes in Mechanical Engineering,
https://doi.org/10.1007/978-981-99-5990-7_69

different liquid fueled pool fires against the experimental results. However, in this, the burning rate was not predicted but rather provided as the pool boundary condition. Chen et al. [5] determined the flame height using the intermittency of the visible flame at different heights and explained the pulsating behaviour of the flame. Kumar et al. [6] investigated the effects of fuel depth, size of the pan, and wall material on the mass–burning rates and temperature fields. Numerical studies on the pool fires using FDS carried out by several researchers [7–9] were focused mainly on determining the mass loss rates, heat release rates, temperature, species and velocity fields.

In this work, experimental mass loss rates from three square pool n–heptane fires are carefully measured and FDS is used to complement the analysis of flame dynamics. The numerically obtained mass loss rates for pools of square cross section with varying side lengths are validated against the experimental data. Further, the transient behaviour of pool fires has been studied by means of instantaneous flow, species and temperature fields. Analyzing the time dependent extents of n-heptane mass fractions and recording the location on the axis at which 99% of n–heptane is consumed, flame heights have been determined numerically. Flame heights have also been located as the axial locations, where intermittency of the flame is 50%. Study of transient behaviour of pool fires deepens the understanding of flame dynamics and facilitates the design of suppression mechanisms.

2 Experimental Methodology

The experiments are conducted at Fire Test Enclosure (FTE), a dedicated facility for fire testing at National Center for Combustion Research and Development (NCCRD) at Indian Institute of Technology Madras (IITM). The setup consists of square shaped stainless steel pools of side lengths varied as 8, 12 and 16 cm. Liquid n–heptane is filled up to 1.5 cm depth in pans, which are 4 cm deep, such that an ullage of 2.5 cm is maintained initially. The fuel pan is enclosed with black fire retardant curtains, which are hanged from all the four walls in order to suppress the sudden air entrainment. This also facilitates the capturing of high-quality images of the flame. The pool container, kept on top of a fireproof insulation board, is placed on a weighing balance. The weighing balance can weigh a maximum mass of 6 kg with an accuracy of 0.1 g. The distance of the top of the pool container from the floor is 21 cm. A digital camera is mounted on a tripod and kept parallel to the pool surface so that it captures a two-dimensional high-definition video of the flame at 60 fps. The schematic diagram in Fig. 1 shows the experimental setup.

The mass of n–heptane fuel required to fill in the container of the side length 8 cm for the ullage of 1.5 cm is calculated based on its density and the volume of the filled portion. This mass of n–heptane fuel is poured carefully into the container and ignited using the butane torch. The same procedure is repeated for the pools of side lengths 12 and 16 cm. Also, each of these experiments is repeated 3 times to determine the repeatability and uncertainty of the experimental data. Adequate time

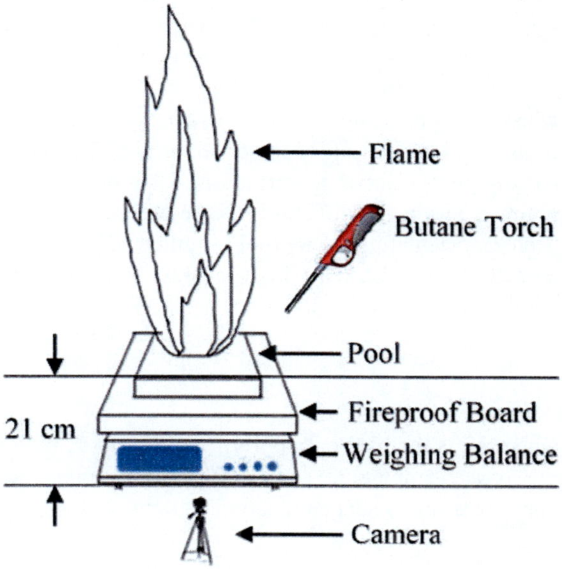

Fig. 1 Schematic diagram of the experimental setup

gap is provided between the two experiments so that the fuel pan cools to ambient temperature before beginning the next experiment.

3 Numerical Model

The numerical studies are carried out in FDS 6.7.6 [10], an open source fire simulation software, developed by NIST. FDS solves Navier–Stokes equations in a form appropriate for low-speed thermally driven flow for smoke and heat transport from fires. FDS utilizes Large Eddy Simulation (LES) based solver or Direct Numerical Simulation (DNS). In this work, LES based solver is used. The governing equations for momentum conservation are discretized by second order finite difference scheme on three-dimensional staggered rectilinear grid. A second order explicit Runge–Kutta method is used for time marching. Combustion in the gas phase is modelled by a single-step mixing controlled chemical reaction using appropriate fuel, oxidizer and product species. For n–heptane, the single-step reaction is written as:

$$C_7H_{16} + 11(O_2 + 3.76N_2) \rightarrow 7CO_2 + 8H_2O + 41.36N_2 \quad (1)$$

Radiative heat transfer is included by solving Radiation Transport Equation considering a gas medium as gray. RTE is solved using finite volume method. The rate of evaporation of the liquid fuel is obtained using Stefan's ordinary diffusion, written as,

$$\dot{m}'' = h_m \frac{\overline{p}_m W_F}{RT_g} ln\left(\frac{X_{F,g} - 1}{X_{F,l} - 1}\right); h_m = \frac{\mathrm{Sh} D_{l,g}}{L}, \quad (2)$$

where $\overline{p}_m$ is pressure, T_g is temperature, $X_{F,g}$ is the volume fraction of fuel vapour in the first cell on top of the pool surface, $X_{F,l}$ is the volume fraction of the fuel vapour coming out of the surface, h_m is mass transfer coefficient, Sh is the Sherwood number, $D_{l,g}$ is the diffusivity of the fuel in the gas phase and L is the length scale used for calculating the Reynolds number. The volume fraction of fuel vapour at the surface, $X_{F,l}$ is defined using Clausius–Clapeyron relation as,

$$X_{F,l} = exp\left[-\frac{h_v W_F}{R}\left(\frac{1}{T_s} - \frac{1}{T_b}\right)\right] \quad (3)$$

where h_v is the heat of vaporisation, W_F is the molecular weight of the fuel, T_s is the surface temperature and T_b is the boiling temperature of the fuel. For liquid pool, one-dimensional heat conduction equation is solved.

$$\rho c \frac{\partial T}{\partial t} = \frac{\partial}{\partial x}\left(k \frac{\partial T}{\partial x}\right) + \dot{q}_r''' \quad (4)$$

Here x = 0 represents the liquid surface and the direction of x is downwards towards the bottom of the pool. The heat absorbed in the pool due to in-depth thermal radiation is added as a source term, $\dot{q}_r'''$. The variables k, ρ and c are thermal conductivity, density and specific heat, respectively. No internal convection within the pool is solved. The thermal boundary condition on the top of the liquid surface is calculated using the expression given as,

$$-k \frac{\partial T}{\partial x}(0, t) = \dot{q}_c'' - \Delta h_v \dot{m}'' \quad (5)$$

Here, $\dot{q}_c''$ is the convective heat flux and Δh_v is the latent heat of vaporization. Further, $\dot{q}_c''$ is defined as:

$$\dot{q}_c'' = h\left(T_g - T_w\right)\frac{W}{m^2}. \quad (6)$$

Here,

$$h = max\left[C\left|T_g - T_w\right|^{1/3}, \frac{k}{L}\mathrm{Nu}\right]\frac{W}{m^2 K} \quad (7)$$

In Eq. (7), C is an empirical coefficient for free convection (taken as 1.52 for horizontal plate), L is a characteristic length based on pool size and k is the thermal conductivity of the gas. The Nusselt number (Nu) is found as,

Table 1 Thermophysical properties of n-Heptane

Properties	Values (Units)
Density	675 kg/m^3 [11]
Specific heat	2.2 kJ/kg-K [11]
Conductivity	0.12 W/m–K [11]
Emissivity	0.9 [assumed]
Absorption coefficient, κ	187.5 m^{-1} [10]
Boiling temperature	98.0°C [11]
Latent heat of vaporisation	365 kJ/kg [11]
Radiative fraction, χ_r	0.40 [11]
CO Yield, y_{CO}	0.01 [11]
Soot Yield, y_s	0.037 [11]

Table 2 Thermophysical properties of stainless steel

Properties	Values (Units)
Density	7820 kg/m^3 [12]
Specific heat	0.46 kJ/kgK [12]
Conductivity	45 W/mK [12]
Emissivity, ε	0.95 [assumed]

$$\mathrm{Nu} = 0.037\mathrm{Re}^{4/5}\mathrm{Pr}^{1/3};\ \mathrm{Re} = \frac{\rho|\boldsymbol{u}|L}{\mu};\ \mathrm{Pr} = 0.7 \quad (8)$$

The heat of combustion is 45000 kJ/kg [11]. Since the flame is transport controlled, infinite rate chemistry is employed. The convective heat flux to the surface is evaluated from the gas phase temperature field and surface temperature. The thermo-physical properties of gas phase species and the thermo-physical properties of stainless steel pool containers are listed along with appropriate references in Tables 1 and 2, respectively.

Only one-quarter of the domain is considered in simulations owing to the symmetry of the flame. As shown in Fig. 2, the top and side boundaries of the computational domain are specifed with "OPEN" condition, through which combustion products leave the domain or atmospheric air enters. Pressure is specified as ambient pressure and gradients in normal direction for all variables are set to zero in the case of outflow. For incoming flow, oxygen and nitrogen mass fractions are set to 0.23 and 0.77, respectively, and the temperature to the ambient temperature of 30 °C.

The dimensions of the computational domains used for the three cases simulated are listed in Table 3.

The domain is discretized using a multi–block structured mesh. In order to capture the gradients along x, y and z directions, cells of sizes 5 mm are used up to 1.5 times the side lengths of the pool and up to a height of 0.5 m. After that, the cell sizes are increased progressively. The maximum cell size is 20 mm.

Grid resolution for LES is often quantified using the plume resolution index, RI [9]. It has been defined as, $RI = D^*/\Delta x$, where D^* is the characteristic diameter of the fire and Δx is the cell size. Characteristic fire diameter, D* is obtained as,

Fig. 2 Computational domain

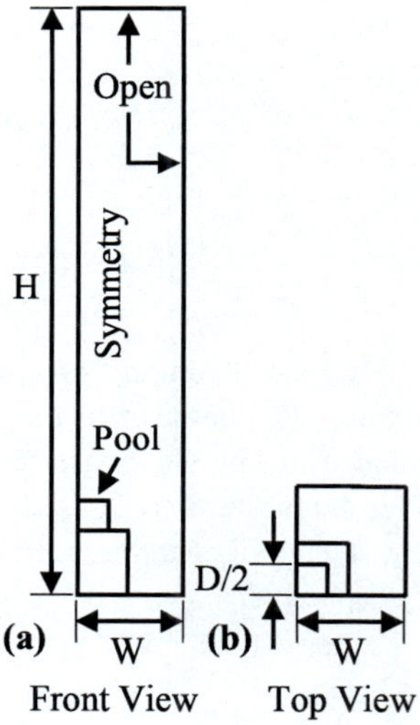

Table 3 Dimensions of the computational domain

Dimensions	Values (m)		
	Case 1	Case 2	Case 3
H	2.0	2.5	3.0
W	0.2	0.3	0.4
D	0.08	0.12	0.16

Table 4 RI and number of cells using $\Delta x = 5$ mm

D (m)	D^* (m)	$RI = D^*/\Delta x$	No. of cells
0.08	0.1389	27.79	44,468
0.12	0.1634	32.68	92,300
0.16	0.1833	36.67	184,252

$$D^* = \left(\frac{\dot{Q}}{\rho_\infty C_p T_\infty \sqrt{g}} \right)^{2/5} \tag{9}$$

Table 4 reports the values of pool diameter (D) characteristic fire diameter (D*), RI and the total number of cells used to discretize the three domains.

4 Results and Discussion

4.1 *Mass Loss Rates*

The predicted average mass loss rates (MLR) are validated against the experimental results. Figure 3 shows the numerical values of mass loss rates and the experimental values along with error bars. For the pools of side lengths 8 cm, 12 and 16 cm, the numerically obtained MLR values are 0.071 g/s, 0.226 g/s and 0.483 g/s, respectively, as compared to the experimental values of 0.077 g/s, 0.201 g/s and 0.456 g/s, respectively. The resultant errors are 6.73%, 12.09% and 5.87%, respectively. The numerical model predicts the experimental MLR values quite accurately.

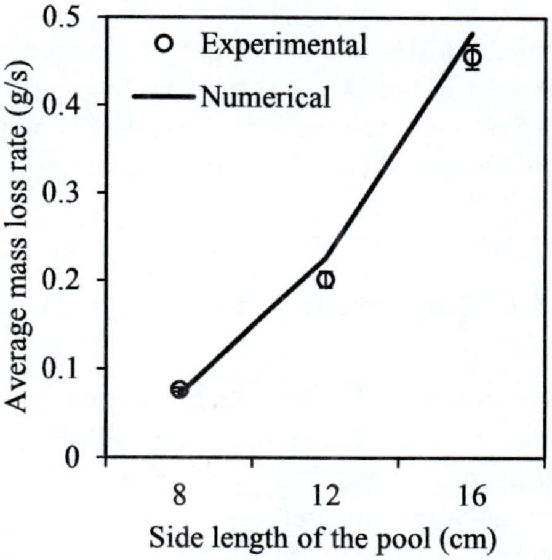

Fig. 3 Variation of **a** average mass–loss rates and **b** average HRR with side length of the pool

Fig. 4 Variation of flame intermittency with height up the pool centerline for pools of side– lengths **a** 8 cm **b** 12 cm **c** 16 cm

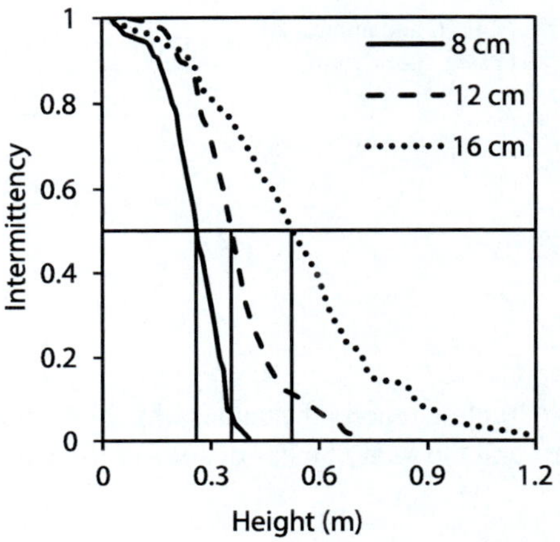

4.2 *Flame Intermittency and Heights*

The temporal behaviour of pool fire is explained by analysing the flame intermittency, I. It is defined as the fraction of total burning duration in which the flame zone exists at a given axial location. The instantaneous flame height is defined as the distance above the pan surface, where 99% of the fuel is consumed [11]. The overall flame height, H_f, is also defined as the height at which the intermittency is 0.5 [13]. The contour lines with n–heptane mass fraction of 0.01 are analysed over the entire burning duration and the time for which the flame is present at various heights along the pool centerline is obtained. Figure 4 shows the intermittency, I = 0.5 line and locations, where the vertical lines are drawn to meet the horizontal axis. Figure 5 shows the comparison among flame heights for the three pools as determined by time-averaging the instantaneous flame heights from n–heptane mass fraction criterion (method M1) and the height estimated using the criterion of intermittency = 0.5 (method M2). Flame heights, as determined using method M1 are 0.219, 0.385 and 0.528 m as compared to 0.261, 0.357 and 0.525 m, got using M2, for the pools of side lengths 8 cm, 12 cm and 16 cm, respectively.

4.3 *Temperature, Velocity and Oxygen Profiles*

Instantaneous flow, temperature and oxygen fields for pool fires of side lengths 12 cm and 16 cm are shown in Figs. 6 and 7, respectively. In both cases, the first two time instants represent growing fire. The third time instant corresponds to the instant of maximum heat release rate. Line contours for oxygen mass–fraction (left side), grayscale contours for temperature and velocity vectors (right side) at three instants for both these cases demonstrate the transient behaviour of pool fires. In Fig. 6, the fire fields at 21, 80 and 345 s for the pool of 12 cm side length are shown. In this, high

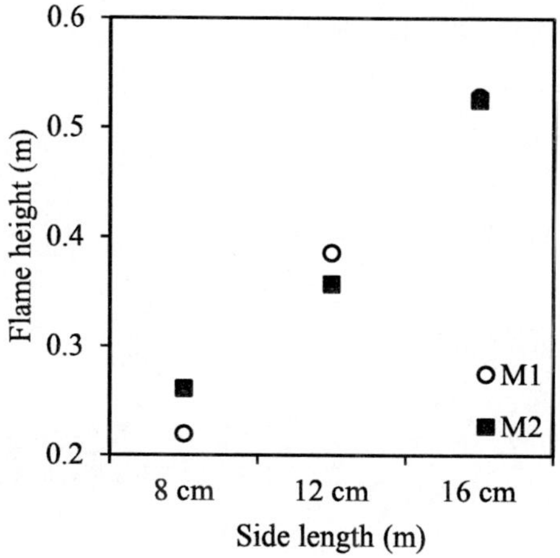

Fig. 5 Comparison among the time-averaged and 50% intermittent flame heights for pools of side–lengths **a** 8 cm **b** 12 cm **c** 16 cm

temperature of the order of 1200 °C confirms the presence of reaction zone near the pool surface. Temperature reduces downstream due to air entrainment. An increase in the extent of flame zone with time is observed from temperature contours, as shown by 600 °C isotherm extending to 0.6, 0.8 and 1.0 m along the axis for these three instants. The oxygen mass fractions range from 0.02–0.04 kg/kg near the centerline to 0.22 kg/kg at the flame peripheries at near ambient conditions. An increase in velocity along the height is because of buoyancy. In Fig. 7, the corresponding fields at 47, 92 and 274 s for the pool of 16 cm are shown, which shows similar features.

5 Conclusions

The experimental and numerical studies are carried out to analyse the temporal behaviour of the flames for square shaped n–heptane pools with side lengths of 8 cm, 12 cm and 16 cm. FDS is used to simulate the flames using the LES solver. Numerically predicted mass loss rates are seen to compare well with the experimental results. From numerical predictions, thorough analysis of temperature, velocity and oxygen fields systematically at three time instants. Further, the flame heights are obtained using the axial location of contour of fuel mass fraction equal to 0.01 as well as from the location of intermittency equal to 0.5. The flame heights are estimated as 0.261 m, 0.357 m and 0.525 m, from intermittency method, as compared to the average flame heights 0.219 m, 0.385 m and 0.528 m got from the n–heptane mass fraction method, for the pools of side lengths 8 cm, 12 cm and 16 cm, respectively.

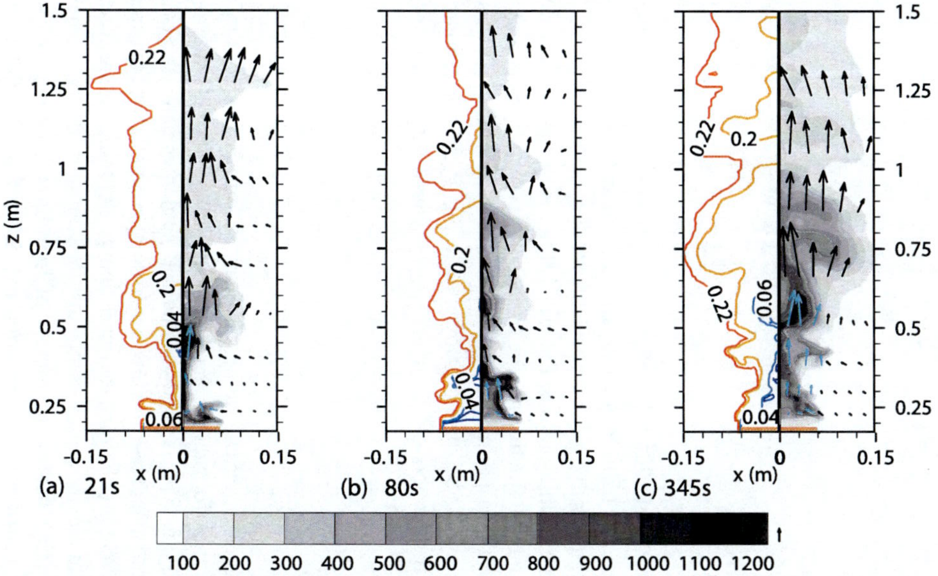

Fig. 6 Instantaneous contours of temperature (right) and oxygen mass fraction (left), and velocity vectors (right) at **a** 21 s **b** 80 s and **c** 345 s for the pool of side length 12 cm

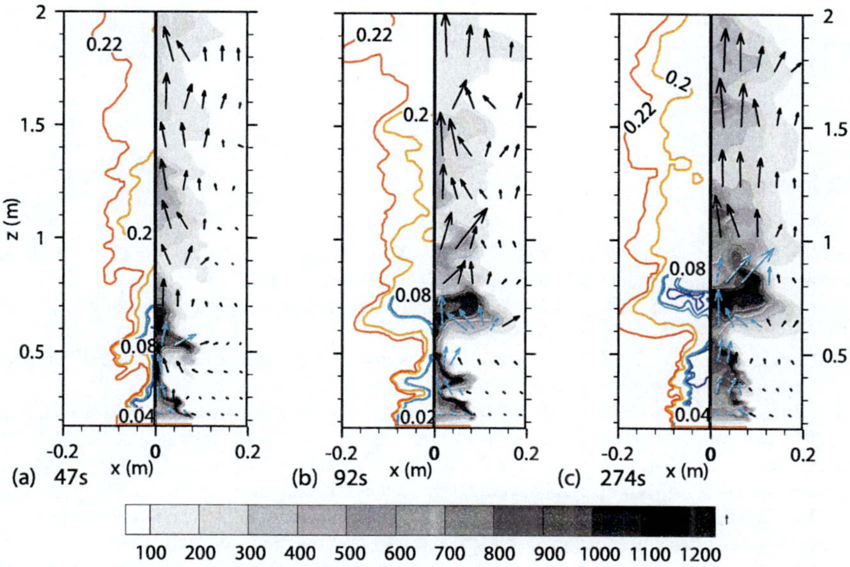

Fig. 7 Instantaneous contours of temperature (right) and oxygen mass fraction (left), and velocity vectors (right) at **a** 47 s **b** 92 s and **c** 274 s for the pool of side length 16 cm

Acknowledgements The authors sincerely acknowledge the research grant received through Science and Education Research Board (SERB) National Post-Doctoral Fellowship (N–PDF), Department of Science and Technology, Government of India, the fire testing facility at National Center of Combustion Research and Development (NCCRD) and high performance computing facility at P.G. Senapathy Center for Computing Resource, Indian Institute of Technology Madras.

Nomenclature

$\dot{m}''$	Rate of evaporation of fuel [kg/m^{2}s]
W_F	Molecular weight of fuel [kg/kmol]
ρ	Density [kg/m^3]
R	Specific gas constant [J/kgK]
T_g	Gas–phase temperature [K]
T_s	Surface temperature [K]
T_b	Boiling temperature [K]
c	Specific heat [kJ/kgK]
k	Thermal conductivity [W/mK]
$\dot{q}_r'''$	Rate of heat absorption per unit volume due to in–depth thermal radiation [kW/m^3]
I	Intermittency [–]

D Side of the square pool [m]
*D** Characteristic fire diameter [m]

References

1. Liu C, Ding L, Jangi M, Ji J, Yu L, Wan H (2020) Experimental study of the effect of ullage height on flame characteristics of pool fires. Combust Flame 216:245–255
2. Miao Z, Wenhua S, Ji W, Zhen C (2014) Accident consequence simulation analysis of pool fire in fire dike. Procedia Eng 84:565–577
3. Lock A, Hamins A (2016) The structure of a Moderate—Scale methanol pool fire, NIST report, https://nvlpubs.nist.gov/nistpubs/TechnicalNotes/NIST.TN.1928.pdf
4. Oliveira RLF, Doubek G, Vianna SV (2019) On the behaviour of temperature field around pool fires in controlled experiment and numerical modelling. Process Saf Environ Prot 123:358–369
5. Chen X, Liew KM, Wang X, Zhang J (2016) Pulsation behaviour of pool fires in a confined compartment with a horizontal opening. Fire Technol 52:515–531
6. Kumar AS, Sowriraajan AVE, Bhaskar Dixit CS, Mukunda HS (2021) Experiments on unsteady pool fires—effects of fuel depth, pan size and wall material. Sadhana, 46, pp 1–11
7. Choudhary A, Gupta A, Kumar S (2018) Studies on jatropha oil pool fire. Therm Sci Eng Prog 6:104–127
8. Stewart JR, Phylaktou HN, Andrews GE, Burns AD (2021) Evaluation of CFD simulations of transient pool fire burning rates. J Loss Prev Process Ind 71:104495
9. Sikanen T, Hostika S (2016) Modeling and simulation of liquid pool fires with in-depth radiation and absorption and heat transfer. Fire Saf J 80:95–109
10. McGrattan K, McDermott R, Marcoss V, Hostikka S, Floyd J. (2020) Fire dynamics simulator—Technical reference guide Volume 1: Mathematical Model. Gaithersburg: Natl Inst Stand Technol
11. Yip A, Haelssig JB, Pegg MG (2021) Simulating fire dynamics in multicomponent pool fires. Fire Saf J 125:103402
12. Karlsson B, Quintiere JG (2000) Enclosure fire dynamics, CRC Press
13. Maynard TV, Butta JW (2018) A physical model for flame height intermittency. Fire Technol 54:135–161

Studies on Fire Spread Over Centrally Ignited Forest Fuel Bed Using Fire Dynamics Simulator

B. Ashutosh and V. Raghavan

Abstract This work reports simulations of flame spread over a square region of forest biomass fuel that is centrally ignited. The dependence of the rate of spread (ROS) of fire on the angle of inclination of the fuel surface with respect to horizontal is studied. Flame spread over the solid fuel surface is modelled using appropriate sub-models for moisture evaporation, pyrolysis of fuel, char oxidation and gas phase combustion in Fire Dynamics Simulator (FDS). The numerical model is able to predict a flame spread rate of 0.279 m/min for the horizontal fuel bed, which compares well with the experimental value reported in the range of 0.17–0.28 m/min. The numerical model is then used to investigate the effect of inclination of the fuel bed with respect to the horizontal, by varying the inclination angle as 30°, 45°, 60°, 75° and 90°. Results show that when the angle of inclination is increased, ROS increases in the upslope direction and decreases in the downslope direction. Along the lateral direction, ROS increases with an increase in the angle of inclination. The fire dynamics are explained using instantaneous contours of temperature and the fields of velocity vectors, for several inclination cases.

Keywords Forest fuel · Fire dynamics · Large Eddy simulation · Fire spread · Rate of spread

1 Introduction

Wildland (forest) fires are disastrous and have a significant impact on the atmosphere [1]. Perry et al. [2] studied fire spread over various terrains of forest material by using flux equations for physical and chemical processes and studied ignition of subsequent strips along the spread direction. Vakalis et al. [3, 4] used contour propagation model and neural network to characterize the fire consequences and developed a software

B. Ashutosh · V. Raghavan (✉)
Department of Mechanical Engineering, Indian Institute of Technology Madras, Chennai 600036, Tamil Nadu, India
e-mail: raghavan@iitm.ac.in

S. Das et al. (eds.), *Proceedings of the 1st International Conference on Fluid, Thermal and Energy Systems*, Lecture Notes in Mechanical Engineering,
https://doi.org/10.1007/978-981-99-5990-7_70

package for studying the fire spread. Cancellieri et al. [5] used a hybrid technique to study the thermal degradation kinetics of Erica Arborea, in which the apparent activation energy was obtained using a model free approach and the reaction order and pre-exponential factor were estimated using a model fitting approach.

So far, to describe wildland fire propagation five models, namely empirical, semiempirical, reduced physical, fully physical, and simplified fully physical have been employed [6]. Empirical methods involve simple approaches for determining the rate of spread (ROS) in terms of wind velocity, ground topography and fuel moisture content. In this, the modelling parameters must be recalibrated when transferring from lab scale to field scale investigations. Semiempirical models involve the computation of at least one of the physical conservation laws, such as energy conservation of the vegetative fuel, in addition to using the empirical sub-models. This approach too has an operational range, but since the conservation law(s) handle some of the relevant components of fire physics, it has a higher ability to convert from one scale to another. Reduced physical models compute the outputs using the relevant physical laws. In fully physical models, a set of coupled nonlinear partial differential equations across a three-dimensional grid is solved numerically.

Adou et al. [7] numerically investigated the surface fire spread over white birch considering bed inclination ranging from $-15°$ to $+15°$ and compared the rate of spread with the lab scale experiment in a wind tunnel. This work considered modelling of both gas and solid phases to bring out the coupled heat and mass transfer processes. The flame spread rates were observed to be strongly dependent on the model parameters. Simpson et al. [8] performed numerical studies on fire spread using Weather Research Forecasting (WRF) and WRF-Fire coupled models considering wind speed, wind direction and the terrain slope as parameters. In this study, the slope angle was varied in the range of 0° to 40°.

Based on the review of literature, it has been understood that there are limited numerical studies on wildland fire spread and limited studies analyzing flame spread over surfaces with different inclinations. Study with slope angle ranging from 0° (horizontal) to near $\approx$90° (near vertical) can provide interesting results, since topography plays an important role in fire spread. This forms the motivation of this study. In addition, this study also serves the purpose of understanding the capability of FDS 6.7.7 to simulate the surface fire spread as compared to a previous study [9]. Further, a correlation is proposed to estimate ROS as a function of inclination angle.

2 Numerical Model

FDS 6.7.7 is capable of simulating wildland fire spread. There are three methodologies incorporated for this. They are particle method, boundary fuel method and level set method. In particle method, the ground vegetation, such as leaves and grass can be represented as multiple Lagrangian particles having rectangular, cylindrical or spherical shapes. Thin vegetation, for example, surface litter, can be introduced in the computational domain using boundary fuel method. Here, a porous boundary with specified moisture fraction, surface to volume ratio, packing ratio and thickness is considered to be lying over a hard ground. Level set method is based on experimentally observed rate of spread, which is given as an input to the simulation to analyze the smoke transport alone. In this study, boundary fuel method is used.

A computational domain of length 1 m, width 1 m and height 0.5 m is used as shown in Fig. 1. Ambient humidity is taken as 47% and the ambient temperature is taken as 33.7 °C. The moisture content in the biomass fuel (vegetation) is taken as 10%. Packing ratio (0.02028) is calculated as the ratio of bulk density (13.18 kg/m^3) [9] to the density of dry vegetation (650 kg/m^3) and given as an input. Surface to volume ratio is estimated as 4000 (1/m) [9]. The biomass fuel, positioned at the bottom, is represented as porous solid packed to the size of $0.96 \times 0.96 \times 0.12$ m^3. Total mass of the fuel used is 1.458 kg, calculated as the product of bed volume, packing ratio and the density of the fuel elements.

The top and side boundaries of the computational domain are specifed with "OPEN" condition, through which combustion products leave the domain or the atmospheric air enters. In these boundaries, pressure is specified as ambient pressure and gradients in the normal direction for all the variables are set to zero in case of outflow. For inflow (occurring based on pressure gradient), the mass fractions of oxygen and nitrogen are set as 0.23 and 0.77, respectively, and the temperature is set to ambient temperature. A uniform cell size of 10 mm is taken throughout the domain and the total number of cells comes out to be 500,000.

The solid phase pyrolysis process is expressed using the following three steps:

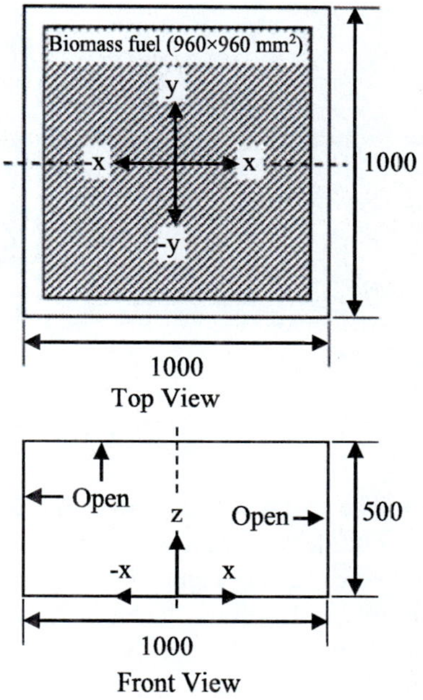

Fig. 1 Computational domain. All dimensions are in mm

1. Moisture evaporation (endothermic):

$$\text{Wet Fuel} \rightarrow \nu_{H_2O}H_2O + (1 - \nu_{H_2O})\ \text{Dry Fuel} \tag{1}$$

Here,

$$\nu_{H_2O} = \frac{M}{1+M} \tag{2}$$

2. Pyrolysis of dry fuel (endothermic):

$$\text{Dry Fuel} \rightarrow \nu_{char}\text{Char} + (1 - \nu_{char})\ \text{GPP} \tag{3}$$

3. Char oxidation (exothermic):

$$\text{Char} + \nu_{O_2,Char}O_2 \rightarrow (1 + \nu_{O_2,char} - \nu_{ash})\text{Products} + \nu_{ash}\text{Ash} \tag{4}$$

Here, GPP represents gas-phase pyrolysis products. Sum of the component densities is equal to the density of the composite solid.

$$\rho_s = \rho_{s,dry} + \rho_{s,H_2O} + \rho_{s,char} + \rho_{s,ash} \tag{5}$$

The rate of endothermic moisture evaporation is expressed in terms of pre-exponential factor (A) and activation energy (E) as:

$$r_{H_2O} = \left(\frac{\rho_{s,H_2O}}{\rho_s(0)}\right) A_{H_2O}T^{-\frac{1}{2}}\exp\left(-\frac{E_{H_2O}}{RT}\right) \tag{6}$$

Similarly, the rate of endothermic pyrolysis of the dry fuel is expressed as:

$$r_{pyr} = \left(\frac{\rho_{s,dry}}{\rho_s(0)}\right) A_{pyr}T^{-\frac{1}{2}}\exp\left(-\frac{E_{pyr}}{RT}\right) \tag{7}$$

The rate of exothermic char oxidation is written as:

$$r_{char} = \left(\frac{\rho_{s,char}}{\rho_s(0)}\right) A_{char}\exp\left(-\frac{E_{char}}{RT}\right)\frac{\rho Y_{O_2\sigma}\left(1 + \beta_{char}\sqrt{Re}\right)}{\nu_{O_2,char}\rho_s(0)} \tag{8}$$

The heat of combustion is 17700 kJ/kg [9]. Since the flame is transport controlled, an infinite rate chemistry is employed. The convective heat flux to the surface is evaluated from the immediate gas phase temperature field and the surface temperature. Radiative fraction is taken as 0.01 and soot yield is taken as 0.01 [9]. Thermophysical properties and pyrolysis kinetic parameters of dry fuel are listed in Table 1. Ten percent of the dry vegetation converts to char, which in turn oxidizes based on the input of oxygen. leaving ash behind. A constant, β_{char} (0.2) is used to account for

Table 1 Thermophysical properties and pyrolysis parameters of dry vegetation

Properties	Values (References)
Density, ρ_{dry}	650 kg/m^3 [9]
Conductivity, λ_{dry}	0.1 W/m–K [10]
Specific heat, $c_{p,dry}$	0.01 + 0.0037 T (K) kJ/kg-K [10]
Vegetation moisture, M	10% [9]
Packing ratio, β_e	0.020277 [9]
Surface to volume ratio, σ_e	4000 m^{-1} [9]
Pre-exponential factor, A_{dry}	36,300 s^{-1} [10]
Activation energy, E_{dry}	60,300 J/mol [10]
Heat of pyrolysis, Δh_{pyr}	416 kJ/kg [9]
Char fraction, χ_{char}	0.1 [9]

the blowing effect on char oxidation. Table 2 lists all the thermophysical properties and oxidation parameters for char. The thermophysical properties of ash are listed in Table 3.

Ignition is provided using a hot patch over an area of 0.02 × 0.02 m^2 located centrally over the fuel bed, for 5 s. The hot patch is specified as heat release rate per unit area equal to 1500 kW/m^2.

Table 2 Thermophysical properties and oxidation parameters for char

Properties	Values (References)
Density, ρ_{char}	300 kg/m^3 [10]
Conductivity, λ_{char}	0.05 W/m–K [10]
Specific heat, $c_{p,char}$	0.42 + 0.002 T + 6.85 × 10^{-7} T^2 kJ/kg-K [10]
Exponent, N_S	0 [10]
$\nu_{O_2,Char}$	1.65 kg/kg-char [11]
β_{char}	0.2 [11]
Pre-exponential factor, A_{char}	430 s^{-1} [11]
Activation energy, E_{char}	74,800 J/mol [11]
Heat of oxidation, ΔH_{char}	−32,000 kJ/kg [10]

Table 3 Thermophysical properties of ash

Properties	Values (References)
Density, ρ_{char}	67 [10]
Conductivity, λ_{char}	0.1 [10]
Specific heat, $c_{p,char}$	1.244 (T(K)/300)$^{0.315}$ [10]

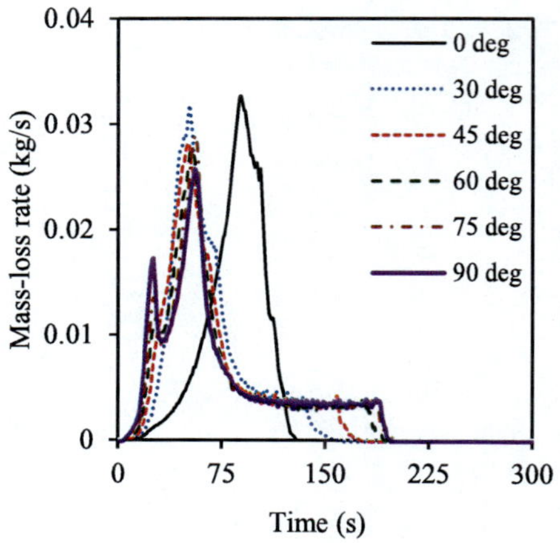

Fig. 2 Variation of mass loss rate with time for different inclinations of the fuel surface

3 Results and discussion

3.1 Mass Loss Rates

Considering six inclinations (by rotating *x*-axis in the front view of Fig. 1, in an anti-clockwise direction) of the fuel surface (0°, 30°, 45°, 60°, 75° and 90°) with respect to the horizontal, the variations of spatially-averaged mass loss rates with time are shown in Fig. 2. For the horizontal fuel surface (0° inclination), the maximum mass loss rate is around 0.0327 kg/s, which is the highest among all the cases. For the fuel surface inclined at 30° with the horizontal, the maximum mass loss rate is around 0.0318 kg/s. The maximum mass loss rate values slightly decrease as the angle of inclination of the fuel surface increases. As the slope is increased, the peak in the mass loss rate is attained earlier. This is due to a faster flame spread rate in the upslope direction and more area being ignited by the spreading fire.

3.2 Fire Spread Rates

Average rate of spread (ROS) of the fire is estimated as the ratio of the length through which the fire has spread to the time it takes to spread through that distance. For validation of the numerical model, the average ROS for the fuel surface is compared with the experimentally measured ROS reported by Buffachi et al. [9]. In this study, for horizontal fuel surface (0° inclination), the predicted average ROS is around 0.279 m/min. This lies within the range of experimentally measured values (0.17–0.28 m/min) [9].

Table 4 Rate of spread, ROS (m/min) as a function of inclination of the fuel surface

Angles (°)	ROS (m/min)		
	Along $+x$	Along $-x$	Along $\pm y$
0	0.279	0.279	0.279
30	0.363	0.211	0.353
45	0.421	0.180	0.384
60	0.480	0.161	0.397
75	0.500	0.151	0.407
90	0.511	0.153	0.411

Using the validated numerical model, a parametric study varying the inclination angle of the fuel surface is carried out and predicted average ROS are reported in Table 4. In Table 4, the upslope flame spread rates (along $+x$) are found to increase with an increase in the inclination angles due to the buoyancy driven flow (Fig. 3a). Further, the flame stand-off is closer to the downstream fuel surface, increasing the heating rate. On the other hand, the downslope spread rates (along $-x$) decrease with an increasing inclination, as it corresponds to flame spread opposed by buoyancy driven flow (Fig. 3a). Along the directions perpendicular to $+x$ and $-x$, (y direction in top view of Fig. 1), the rate of spread is found to increase with an increase in the angle of inclination (Fig. 3a). These rates are comparable to corresponding cases in $+x$ direction for inclinations up to 30°. After that, a further increase in the angle of inclination leads to a lower spread rate in y direction as compared to that in $+x$ direction. Also, there is a negligible difference in the spread rates in the y direction for inclination angles of 75° and 90°. Second order polynomial fits for the rate of flame spread (m/min), as a function of inclination angle (°) are shown in Fig. 3b–d.

3.3 *Temperature and Velocity Profiles*

Grayscale contours of temperature and velocity vectors in Fig. 4 are used to explain the flame characteristics. In all the cases, the time instants are chosen when the fire has the maximum mass loss rates, where a fully developed fire scenario exists, showing induced air flow into the fire at that time instant. Figures in the right column show the field in front view (as in Fig. 1). Figures in the left column show the field in side view (y–z plane). At 0° inclination, a pool type fire is seen, having a closer stand-off near the edges and higher stand-off at the middle. Air entrainment occurs from sides and buoyancy causes an acceleration of the hot gases in upward direction. For inclination greater than zero, the flame stands closer to the fuel surface due to the resultant flow field. The flow field as shown in Fig. 4, exhibits the relative dominance of air entrainment from the bottom side as compared to the top, thus compressing the flame even further towards the unburnt fuel surface. This causes an enhancement in heat and mass transfers at the fuel surface. Flame stand-off in the lateral direction

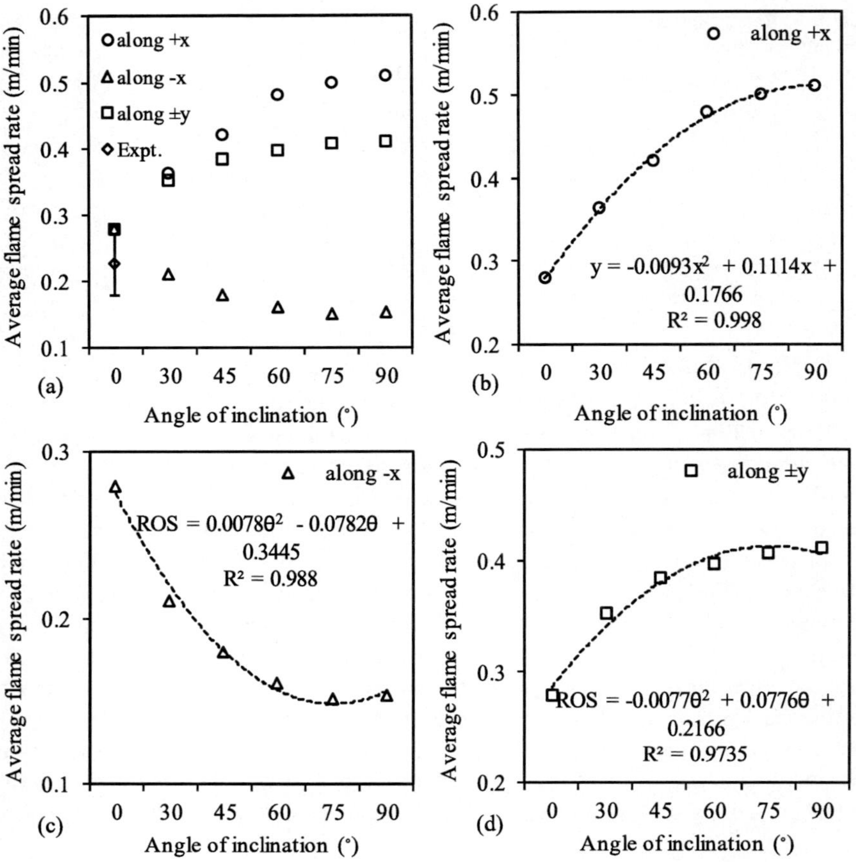

Fig. 3 Variation of fire spread as a function of inclination along **a** all the directions **b** $+x$ **c** $-x$, and **d** $\pm y$

(y direction) is almost uniform for inclination greater than zero. Air flow is almost parallel to the flame surface. However, in $-x$ direction, as the inclination is increased, the flame is seen to spread at a much slower rate, as indicated by its limited presence to the left of $x = 0.5$ m (right side figures in Fig. 4).

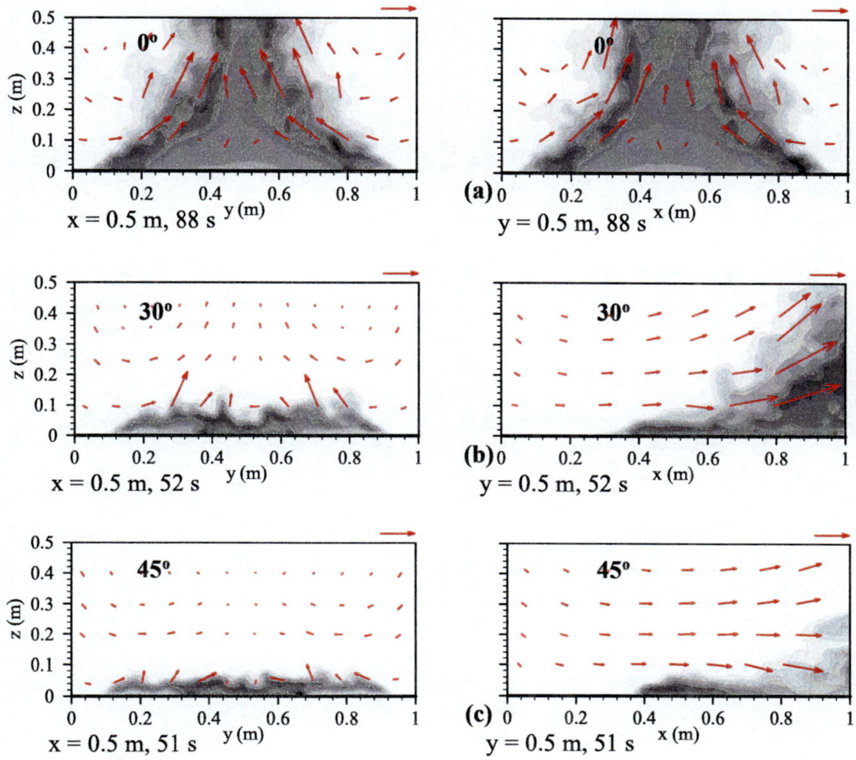

Fig. 4 Grayscale temperature contours with velocity vectors at time instants where the mass loss rates are maximum, along $x = 0.5$ m (left side) and $y = 0.5$ (right side) for six inclination angles: **a** 0°, **b** 30°, **c** 45°, **d** 60°, **e** 75° and **f** 90°

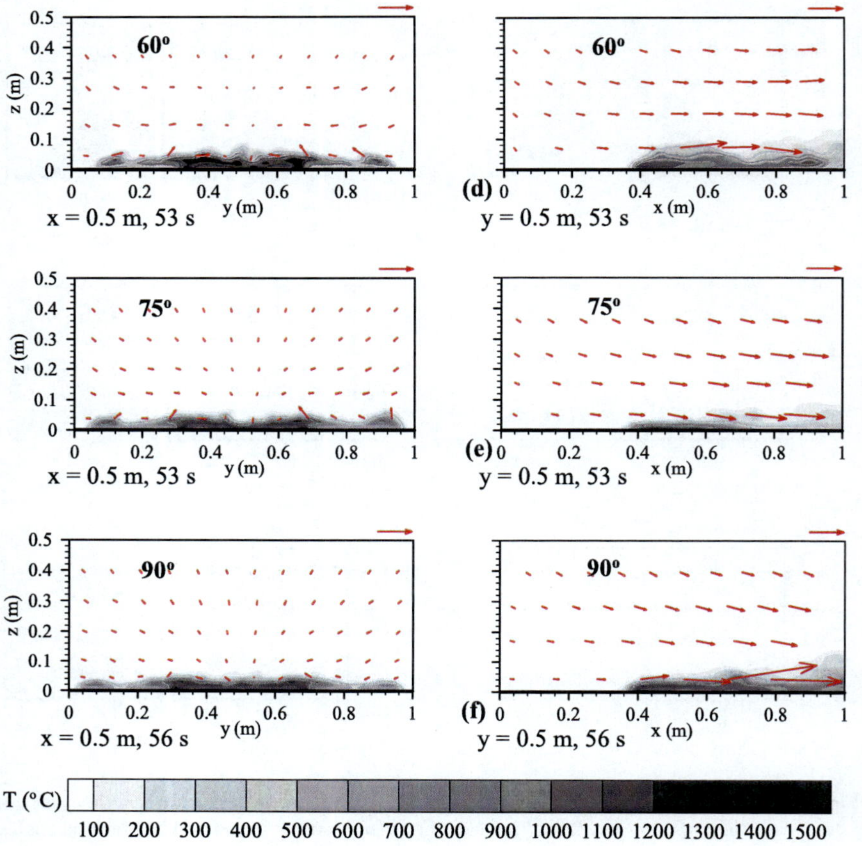

Fig. 4 (continued)

4 Conclusions

Numerical studies are carried out using FDS 6.7.7 to understand the behaviour of fire spreading over a square boundary of wildland fuel bed, centrally ignited and kept inclined at 0°, 30°, 45°, 60°, 75° and 90° with respect to the horizontal. For validation of the numerical model, the numerically predicted flame spread rate for horizontal fuel bed (0.279 m/min) is compared with the corresponding experimental data (0.17–0.28 m/min). Parametric studies for positive inclination angles are conducted and the rate of flame spread have been analysed to explain the involved physics. With the increase in the angle of inclination, in upslope (+*x*) direction, the flame spread rate increases and in downslope (−*x*) direction, it decreases, due to aiding and opposing effects of buoyancy, respectively. In the lateral direction, flame spread rate increases and asymptotically reaches a maximum value, which is less than that obtained for +*x* direction. For 0° inclination, a pool type fire scenario exists, which has air entrainment only from side boundaries. Flame stand-off is the maximum at the centre. As the inclination increases, flame stands closer to fuel surface causes an enhancement in heat and mass transfers. Air flow is seen to be parallel to the flame surface. Flames spread in −*x* direction only after they completely spread in other directions.

Nomenclature

ρ	Density	(kg/m^3)
Λ	Thermal conductivity	(W/m-K)
c_p	Specific heat	(kJ/kg-K)
M	Moisture content	(%)
β_e	Packing ratio	(–)
σ_e	Surface to volume ratio	(–)
r	Rate of the reaction	(kmol/m^{3}s)
A	Pre-exponential factor	(kg/m^3)
E	Activation energy	(J/kgK)
Δh_{pyr}	Heat of pyrolysis	(kJ/kg)
χ_{char}	Char fraction	(–)
ΔH_{char}	Heat of char oxidation	(kJ/kg)
$\nu_{\text{O2,char}}$	Coefficient of O_2in char oxidation reaction	(–)
Re	Reynolds number	(–)
R	Specific gas constant	(kJ/kg-K + 6)
ROS	Rate of spread	(m/min)
θ	Angle of inclination	(°)
T	Temperature	(K)

References

1. Schultz MG, Heil A, Hoelzemann JJ, Spessa A, Thonicke K, Goldammer JG, Held AC, Pereira JMC, Bolscher MVH (2008) Global wildland fire emissions from 1960 to 2000. Global Biogeochem Cycles 22:1–17
2. Perry GLW, Sparrow AD, Owens IF (2004) A GIS-supported model for the simulation of the spatial structure of wildland fire, Cass Basin, New Zealand. J Appl Ecol 36:502–518
3. Vakalis D, Sarimveis H, Kiranoudis C, Alexandridis A, Bafas G (2004) A GIS based operational system for wildland fire crisis management I. Mathematical modelling and simulation. Appl Math Model 28:389–410
4. Vakalis D, Sarimveis H, Kiranoudis C, Alexandridis A, Bafas G (2004) A GIS based operational system for wildland fire crisis management II. System architecture and case studies. Appl Math Model 28:411–425
5. Cancellieri D, Leoni E, Rossi JL (2005) Kinetics of the thermal degradation of Erica arborea by DSC: hybrid kinetic method. Thermochimica Acta 438:41–50
6. Balbi JH, Morandini F, Silvani X, Filippi JB, Rinieri F (2009) A physical model for wildland fires. Combust Flame 156:2217–2230
7. Adou JK, Brou ADV, Porterie B (2015) Modeling wildland fire propagation using a semi-physical network model. Case Stud Fire Saf 4:11–18
8. Simpson CC, Sharples JJ, Evans JP (2016) Sensitivity of atypical lateral fire spread to wind and slope. Geophys Res Lett 43:1744–1751
9. Buffachi P, Krieger GC, Mell W, Alvarado E, Santos JC, Carvalho JA Jr (2016) Numerical simulation of surface forest fire in Brazilian Amazon. Fire Saf J 79:44–56
10. McGrattan K, Hostikka S, McDermott R, Floyd J, Weinschenk C, Overholt K (2013) Fire dynamics simulator user's guide. NIST Spec Publ 1019(6):1–339
11. Porterie B, Consalvi JL, Loraud JC, Giroud F, Picard C (2007) Dynamics of wildland fires and their impact on structures. Combust Flame 149:314–328

Numerical Modeling of Liquid Film Cooling Heat Transfer Coupled to Compressible Gas

Navaneethan Mansu, T. Sundararajan, and T. Jayachandran

Abstract Liquid film cooling is a widely adopted method for restricting the solid wall temperature within allowable limits, in a combustor. In most of the earlier works, the hot gas and coolant are treated as incompressible however in many practical applications such as in rocket propulsion systems, hot gas is compressible. In this study, a numerical model for coupled heat transfer between the compressible hot combustion gas and the incompressible liquid film injected along the wall is presented. Interface energy balance and mass balance are modeled along with phase change of liquid and diffusion of vapor into the gas flow, to investigate the extent of cooling along the solid wall. Validation of turbulence based in-house code with literature data for nozzle flow is also included in the study. Effects of initial liquid film thickness, film temperature, film velocity and vapor mass fraction on wall-cooling are studied. The film-cooled length obtained from the present numerical model is compared with experimental results available in the literature.

Keywords Film cooling · Hot gas · Coupling

1 Introduction

Rocket thrust chambers need to be protected from hot combustion gases with flame temperatures in the range of 3000 K, especially in the cases of reusable vehicles and engines with long burn times. Liquid film cooling is a preferred cooling technique adopted in liquid rocket engines, where the liquid coolant is caused to flow along the

N. Mansu (✉)
Thermal Design Division, VSSC/ISRO, Thiruvananthapuram, Kerala 695022, India
e-mail: mansu.999@gmail.com

T. Sundararajan
Department of Mechanical Engineering, IIT Madras, Chennai 600036, India

T. Jayachandran
Department of Aerospace Engineering, IIT Madras, Chennai 600036, India

S. Das et al. (eds.), *Proceedings of the 1st International Conference on Fluid, Thermal and Energy Systems*, Lecture Notes in Mechanical Engineering,
https://doi.org/10.1007/978-981-99-5990-7_71

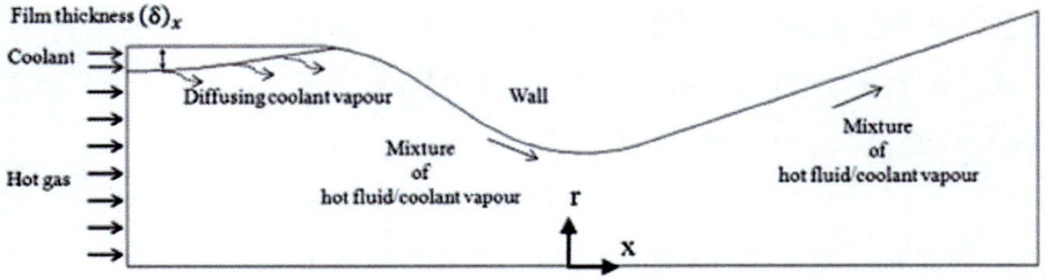

Fig. 1 Schematic of film cooling technique

inside surface of the thrust chamber wall. The high heat capacity of liquid and the latent heat needed for vaporization, render liquid film cooling as an attractive method, especially for high heat flux applications. Generally, the coolant is injected tangentially through discrete holes placed circumferentially on the combustion chamber. The liquid film insulates the chamber wall from direct contact with the hot combustion gas. Once the liquid film undergoes a phase change, the vapor formed also functions as a barrier for heat transfer to the wall. Figure 1 shows a schematic diagram of film cooling applied to the rocket chamber wall.

2 Literature Review and Objective

2.1 Literature Review

A proper design for film cooling system requires a good understanding of the interactions between compressible gas flow with the incompressible liquid flow. Also, aspects such as phase change heat transfer, influence of vapor shielding and the variation of film thickness along the wall, need to be considered. Shine and Nithi [1] have reviewed the theoretical and experimental studies carried out so far on film cooling of rocket engine thrust chambers. They also highlighted the aspects of film cooling which need to be further investigated. Advances in film cooling research were further reported in the latest article by Ludescher and Olivier [2] where film cooling efficiency was studied for coolant injection in the expansion section of a conical nozzle. In experimental and numerical works by Heufer and Olivier [3], the effects of major flow parameters on the heat transfer of gas film-cooled surface in a laminar supersonic flow were studied. Liquid film cooling is still an important and active research area as it paves the way for supporting the design methods for liquid rocket combustion chambers and nozzles [4].

Early experiments by Kinney et al. [5] with air–water flow in pipes estimated the film-cooled length using temperatures measured on the outer pipe surface. In studies by Stechman et al. [6], a modified Bartz equation was used to calculate the convective heat transfer coefficient. Shembharkar and Pai [7] solved incompressible flow equations for the hot gas, with Couette flow assumed for the liquid film. Zhang et al. [8] numerically solved the governing equations for incompressible flow of both the liquid film and gas stream coupled through interfacial matching conditions, for anon-evaporating film of constant thickness. Shine et al. [9] have reported an analytical study to predict the liquid film length by accounting for liquid entrainment through a correlation approach.

Wang [10] reported experimental data for a film-cooled gaseous hydrogen/oxygen based combustion chamber with hydrogen as the coolant. The effect of angular injection of coolant into the combustion chamber was examined by Xiang [11] for gas film-cooled combustion chamber of a hydrogen–oxygen propellant system. In a recent paper by Strokacha et al. [12], an Eulerian–Lagrangian method was used to study film cooling performance in a kerosene–LOX combustion chamber using ANSYS CFX. In numerical simulations done by Yang and Sun [13], liquid oxygen/RP1 propellant combination was considered with liquid water as film coolant.

2.2 *Objective*

A majority of the published papers on film cooling treat the gas flow as incompressible. In numerical studies [7, 8] done on liquid film cooling, the gas is assumed as incompressible. However, the combustion gas flow through the rocket thrust chamber and nozzle is definitely in the compressible regime. Correlations for film-cooled length reported in literature do not account for geometrical variations present in the combustion chamber and nozzle, with abrupt changes in the inner contour. At present, no studies are available in the literature which predict film thickness variation for realistic thrust chamber inner profiles. Very few papers address the variation of film thickness in the cooled portion of the wall and the effectiveness of cooling beyond the point where liquid film ceases to exist. In other words, the shielding aspects of vapor blanket are not considered. In the present work, a numerical methodology is presented, treating the gas flow as compressible and liquid flow as incompressible. The objective of the present study is to model the liquid film cooling mechanism by treating hot medium as compressible and study the various flow features such as change in film thickness and heat transfer with phase change. However, the entrainment of liquid droplets by the high-speed gas flow is ignored for the sake of simplicity.

3 Methodology

This study employs a decoupled approach—the gas phase flow is solved with liquid film temperature as prescribed and the heat flux from hot gas to the liquid is predicted. Applying the calculated heat flux as the interface condition for the liquid, the temperature profile in the liquid and vapor flux (due to liquid evaporation and vapor diffusion into the hot gas) is predicted. The newly predicted liquid surface temperature is prescribed as the boundary condition for the hot gas flow model for the next time step. In a similar manner, the interfacial conditions of velocity and stress continuity, as well as temperature and heat flux continuity are applied to solve the combined gas and liquid flow in a decoupled manner, during the time-marching simulations of the two-phase flow problem.

Major assumptions applicable to the present study are listed below:-

(a) Coordinate system is cylindrical for axisymmetric model.
(b) Gas and liquid flows are treated as compressible and incompressible respectively.
(c) Transport of vapor in the gas flow through species transport equation is modeled.
(d) Temperature dependent properties are estimated for the gas phase, considering local gas and vapor mass fractions and gas temperature.
(e) Entrainment effects and radiation heat transfer are not considered. Also, surface tension effects are ignored with the assumption of a large radius of curvature for the liquid–gas interface.

Gas Phase Equations

The axisymmetric governing equations of compressible in conservative form for hot gas flow are given below:

$$\frac{\partial U}{\partial t}+\frac{\partial F}{\partial x}+\frac{1}{r}\frac{\partial}{\partial r}(rG)=\frac{\partial F_v}{\partial x}+\frac{1}{r}\frac{\partial}{\partial r}(rG_v)+S \tag{1}$$

Turbulence Modeling

The turbulent effects are modeled in the present study using the two equation RNG k-epsilon model and the effective viscosity, thermal conductivity and mass diffusivity values are evaluated by accounting for both laminar and turbulent contributions to these properties. The equations for the conservation of turbulent kinetic energy (k) and the rate of turbulent dissipation (ϵ) are shown below.

$$\frac{\partial \rho k}{\partial t}+\frac{\partial \rho u k}{\partial x}+\frac{1}{r}\frac{\partial (r\rho v k)}{\partial r}=\frac{\partial}{\partial x}\left(\mu_k\frac{\partial k}{\partial x}\right)+\frac{1}{r}\frac{\partial}{\partial r}\left(r\mu_k\frac{\partial k}{\partial r}\right)+H_k-\rho\epsilon-H_c \tag{2}$$

$$\frac{\partial \rho \epsilon}{\partial t}+\frac{\partial \rho u \epsilon}{\partial x}+\frac{1}{r}\frac{\partial (r\rho v \varepsilon)}{\partial r}=\frac{\partial}{\partial x}\left(\mu_\varepsilon\frac{\partial \varepsilon}{\partial x}\right)+\frac{1}{r}\frac{\partial}{\partial r}\left(r\mu_\varepsilon\frac{\partial \epsilon}{\partial r}\right)+H_\varepsilon$$

$$H_k=\mu_t\left(2\left[\left(\frac{\partial u}{\partial x}\right)^2+\left(\frac{\partial v}{\partial r}\right)^2+\left(\frac{v}{r}\right)^2\right]+\left[\frac{\partial u}{\partial r}+\frac{\partial v}{\partial x}\right]^2-\frac{2}{3}\left[\frac{\partial u}{\partial x}+\frac{\partial v}{\partial r}+\frac{v}{r}\right]^2\right)$$
$$-\frac{2}{3}\rho k\left[\frac{\partial u}{\partial x}+\frac{\partial v}{\partial r}+\frac{v}{r}\right] \tag{3}$$

$$H_{\in} = C_{1\in}\frac{\in}{k}\mu_t * \left(2\left[\left(\frac{\partial u}{\partial x}\right)^2 + \left(\frac{\partial v}{\partial r}\right)^2 + \left(\frac{v}{r}\right)^2\right] + \left[\frac{\partial u}{\partial r} + \frac{\partial v}{\partial x}\right]^2\right) - C_{2\in}\rho\frac{\epsilon^2}{k} \quad (4)$$

In the above equations, the values of constants are taken as follows:

$$\mu_k = 1.39(\mu_{laminar} + \mu_{turbulent})$$

$$\mu_e = 1.39(\mu_{laminar} + \mu_{turbulent})$$

$$\mu_{turbulent} = 0.0845\rho\frac{k^2}{\in}$$

$$C_{1\in} = 1.42 - \frac{\eta\left(1 - \frac{\eta}{4.377}\right)}{\left(1 + 0.012\eta^3\right)}$$

$$\eta = \sqrt{2(E_{ij} \cdot E_{ij})}\frac{k}{\in}$$

$$C_{2\in} = 1.68$$

The compressibility effects are accounted by inclusion of the term, $H_c = \frac{2\rho\epsilon k}{a^2}$

Liquid Phase

As indicated in the introduction, the gas flow is externally coupled to the simplified form of liquid model. The present section gives an explanation of the treatment of presence of liquid film on the gas flow boundary. A typical finite volume mesh for gas domain near boundary is shown in Fig. 2. The mesh near the wall boundary is shown along with the nodal and cell center locations. In the same figure, inlet conditions for liquid at the location marked as 'A'. Conditions at location 'B' are computed by application of conservation of mass and energy using the upstream conditions of liquid and nodal/cell center values of nearby gas cells.

Mass conservation for liquid film, in terms of density, film thickness, average velocity at a given axial location and evaporated mass through length x, is expressed as

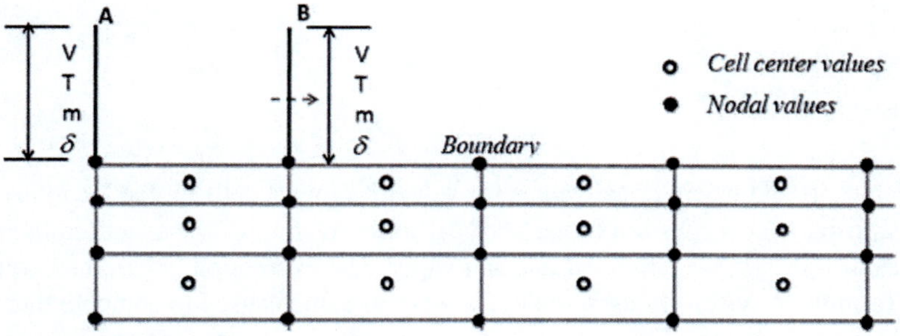

Fig. 2 Calculation of parameters for liquid

$$(\rho_l \delta \overline{u_l})_{in} = (\rho_l \delta \overline{u_l})_{out} + \dot{m}_v x \quad (5)$$

The evaporated mass flux is computed using gradient of mass fraction as given below

$$\dot{m}_v = \left(\frac{-1}{1-m}\right)\left(\frac{\mu}{\sigma}\right)\frac{dm}{dn} \quad (6)$$

Computation of liquid exit temperature is performed through an energy equation where the heat balance is considered between the sensible heat along with heat conducted from hot gas and heat transported by evaporated mass through latent heat. For the liquid film, it is assumed that the film temperature is uniform across film thickness. In the energy equation, heat balance is considered between the sensible heat along with heat conducted from hot gas and heat transported by evaporated mass through latent heat L.

$$(\rho_l \delta \overline{u_l} C T_l)_{in} + Q_{hotgas} = (\rho_l \delta \overline{u_l} C T_l)_{out} + \dot{m}_v x L \quad (7)$$

From the temperature of liquid, assuming thermodynamic equilibrium, the mass fraction of vapor at inlet can be calculated through Antonio equation and Daltons law as given by the equation below

$$\log_{10} p_v = A - \frac{B}{T+C} \quad (8)$$

$$Y_{v,i} = \frac{P_v MW_v}{MW_g(P - P_v) + P_v MW_v} \quad (9)$$

Once vapor pressure is calculated using Antonio equation using inlet temperature, vapor mass fraction can be computed using the pressure value and molecular weight values of gas and vapor. Pressure value in the nearby gas cell is denoted by P.

The unknows in mass balance equation are δandu_l at downstream location 'B'. To close the systems of equations to solve for these parameters, another expression is written in terms of shear stress at 'B'.

Considering the assumption of Couette flow with surface velocity u_l, the shear stress is written as

$$\frac{(u_2 - u_l)}{dy}\mu_g = \frac{(u_l)}{\delta}\mu_l \quad (10)$$

Values u_2 and u_1 are nodal values for velocities in gas cells on the location 'B'. The transport properties are based on local mass fraction of vapor and individual values of transport properties of gas and vapor. The expression is based on square root formula for viscosity usingmolecularweight. Same is used to compute thermal conductivity in terms of respective thermal conductivity values for vapor and gas.

$$\mu = \frac{\left[\mu_v \frac{m}{\sqrt{MW_v}} + \mu_g \frac{(1-m)}{\sqrt{MW_g}}\right]}{\left[\frac{m}{\sqrt{MW_v}} + \frac{(1-m)}{\sqrt{MW_g}}\right]} \tag{11}$$

Liquid saturation temperature is computed assuming that liquid is at the same pressure as that of the neighboring gas cell. This is justified considering very fine thickness of the liquid film. Once the temperature of film reaches the saturation value, there is no further change in liquid film temperature. At such a condition, heat transferred from gas is used for evaporation and there is no change in sensible enthalpy as shown by equation. Liquid reaching its saturation temperature corresponds to a maximum vapor fraction of 1.0 is achieved. In the computation, a cut-off value of 0.98 is applied for vapor mass fraction. The velocities in imaginary cells for gas boundary is expressed in terms of velocities of the liquid film using the interface normal, as

$$u_{ghost} = n_y u_l - n_x v_l \tag{12}$$

$$v_{ghost} = -n_x u_l - n_y v_l \tag{13}$$

A flow chart representing the methodology is given in Fig. 3.

4 Results and Discussion

4.1 Validation with Published Nozzle Heat Transfer Results

Validation of the predicted results obtained using in-house turbulence code with experimental data published by Back et al. [14] is discussed here. The geometry and boundary conditions are same as those considered by Back et al. [14]. Stagnation pressure and temperature at inlet are maintained at 1.752 MPa and 840 K respectively. Mach number distribution obtained for these conditions is shown in Fig. 4, which illustrates a smooth transition from subsonic to supersonic flow across the throat. The heat transfer coefficient computed through the two equation RNG k-epsilon turbulence solver developed in-house is compared with the published values. The predictions give a better match in the locations of throat and divergent compared to the nozzle convergent region as shown in Fig. 5. This deviation is attributed to the isothermal wall boundary condition assumed in the present study for the liquid-cooled portion whereas, in the actual experiment, surface temperature is expected to vary along the axial length. The numerical simulations of the present work correctly predict the occurrence of maximum heat transfer coefficient at the nozzle throat, as demonstrated by Back et al. [14].

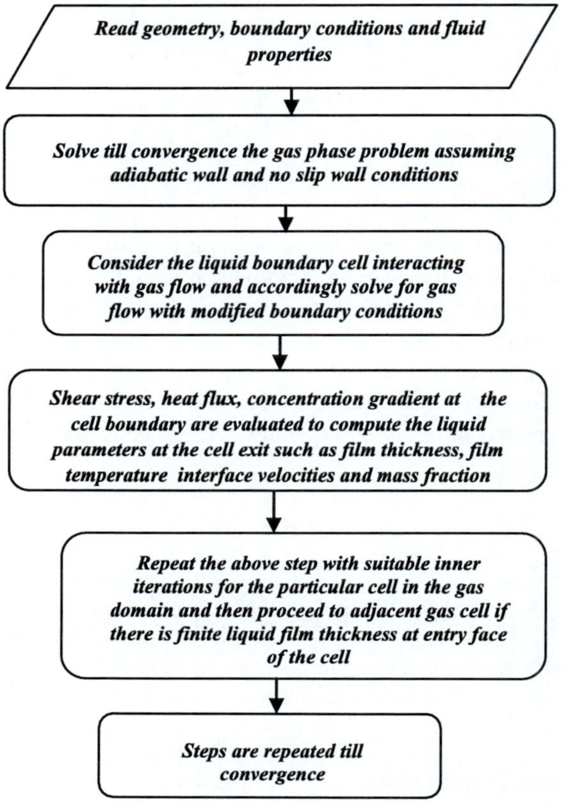

Fig. 3 Flow chart for methodology

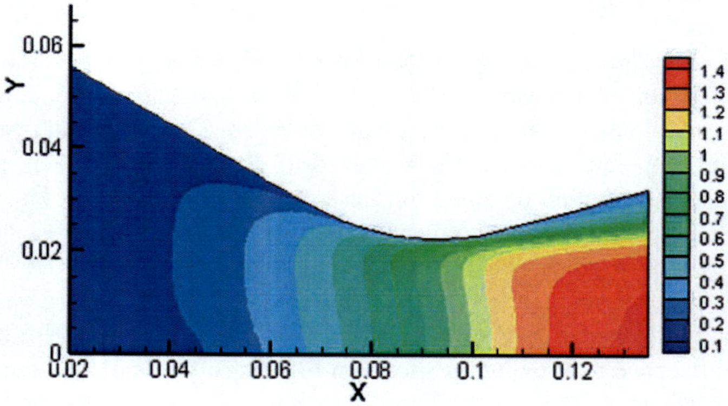

Fig. 4 Mach number distribution for the flow considered by Back et al. [14]

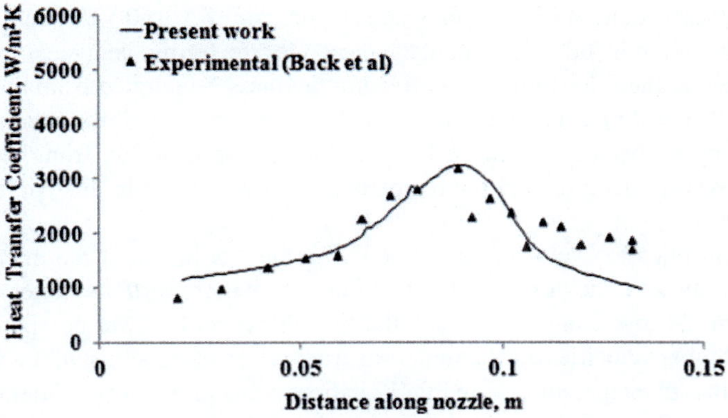

Fig. 5 Comparison of predicted heat transfer coefficient with literature data

4.2 Simulation of Film Along Flow Boundary

In the present section, the effects of wall film injection in a nozzle with a hot gas as the core flow are discussed. Property of air is considered for hot gas and water is assumed as the coolant. The properties used are shown in the Appendix. The film-cooled length is computed based on the distance up to which the film just ceases to exist along the wall. Initially, the gas flow field is solved with adiabatic wall assumption until convergence followed by injection of liquid along the wall. As film thickness is computed based on mass balance, the value at a given location is governed by shear stress and mass evaporated from the film surface. Film will cease to exist at a location if the thickness drops to the minimum cut-off value or when negative values are obtained while solving for film thickness. The boundary conditions prescribed for CD nozzle are shown in the table. Inlet conditions for the gas flow are shown in Table 1.

The distribution for temperature and vapor mass fraction in the nozzle is shown in Figs. 6 and 7 respectively. Performance of the liquid coolant with an inlet velocity of 1.0 m/s is considered. Figure 8 shows the variation of mass fraction of vapor and liquid temperature respectively, along the wall till the location where film thickness is less than the imposed cut-off value. The bulk temperature of the film increases rapidly beyond a distance of 0.13 m which in turn increases the vapor mass fraction as it is governed by local temperature. Film thickness and evaporated mass from the liquid surface is plotted in Fig. 9. Neglecting the entry effects, the film thickness shows

Table 1 Application of boundary conditions

Axis	Center of symmetry
Outlet	Supersonic flow
Wall	Adiabatic
Inlet	Subsonic (P0 = 40 bar T0 = 2500K)

an overall decreasing trend except at the region from 0.1 to 0.13 m and also near region where film is fully consumed, the reason for the former being less increase in temperature at these locations thus affecting the mass evaporated from the surface. The heat flux from gas is plotted along with the temperature of the boundary cells occupied by gas/vapor mixture in Fig. 10. Increase in heat flux from hot fluid to liquid causes an increase in liquid temperature and a decrease in the temperature of hot fluid.

The film injection velocity was varied with the slot height at the inlet fixed as 0.002 m. Figure 11 shows the variation in film thickness for different inlet velocities of the film. At lower inlet velocities, the film thickness is lower compared to the values at higher velocities due to higher shearing action of the surrounding fluid. The variation in film temperature, as in Fig. 12 indicates that in all cases of inlet velocities considered in the study, the film temperature reaches its saturation temperature at the given pressure. In most of the cases, there is a sharp rise in temperature at a

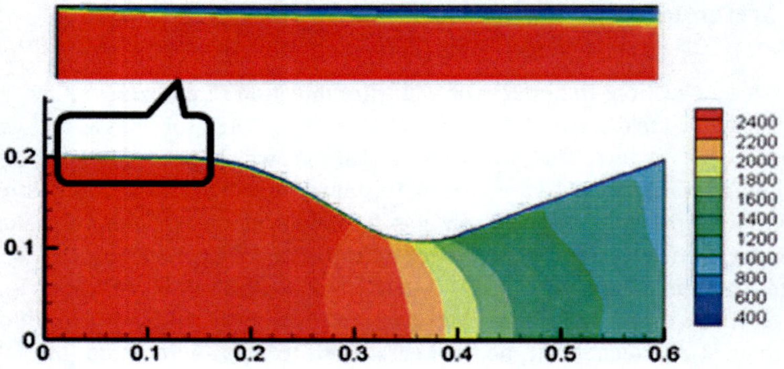

Fig. 6 Temperature distribution

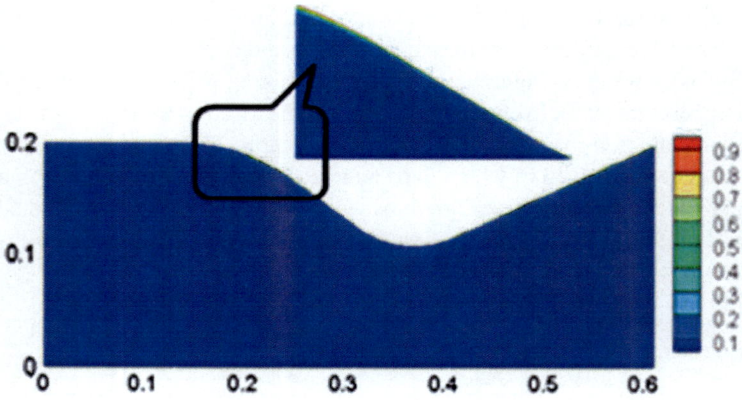

Fig. 7 Vapor mass fraction distribution

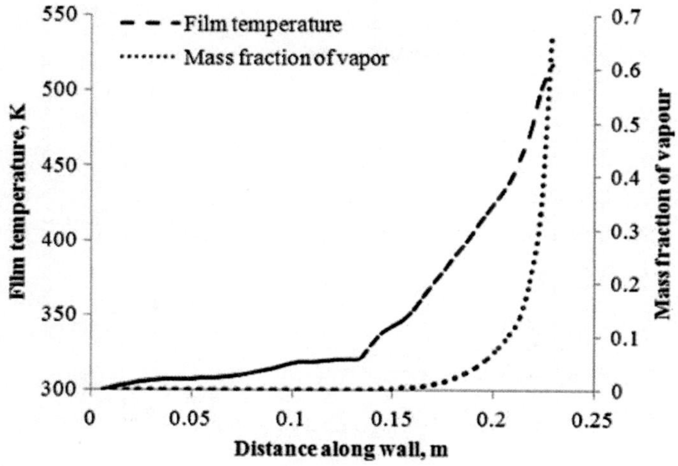

Fig. 8 Film temperature, mass fraction of vapor

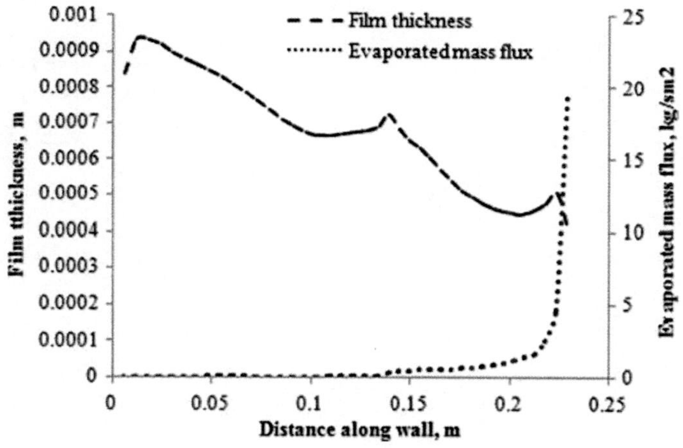

Fig. 9 Film thickness with evaporated mass flux

distance of 0.13 m indicating higher heat flux from the gas. As the coolant velocity is increased, the film dry out point moves further downstream from the injection as in Table 2. At higher film inlet velocities, increase in film dry out distance is lower compared to lower inlet velocities of the coolant. Also, the gradient of temperature in the flow direction is less for higher injection velocity. It should also be noted that due to changes in inner contour of the wall, the variations in temperature and velocity gradients normal to the wall will increase resulting in higher shear stress and wall heat flux. Or in other words, there will not be any significant gain in film-cooled length when we have a local change in surface curvature similar to the convergent in a typical convergent divergent nozzle.

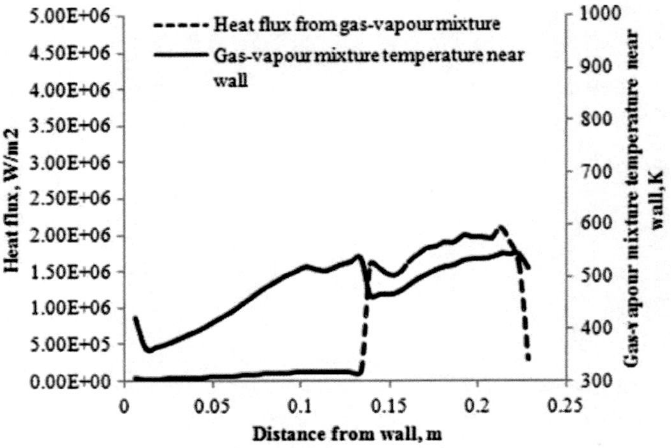

Fig. 10 Heat flux variation at liquid boundary

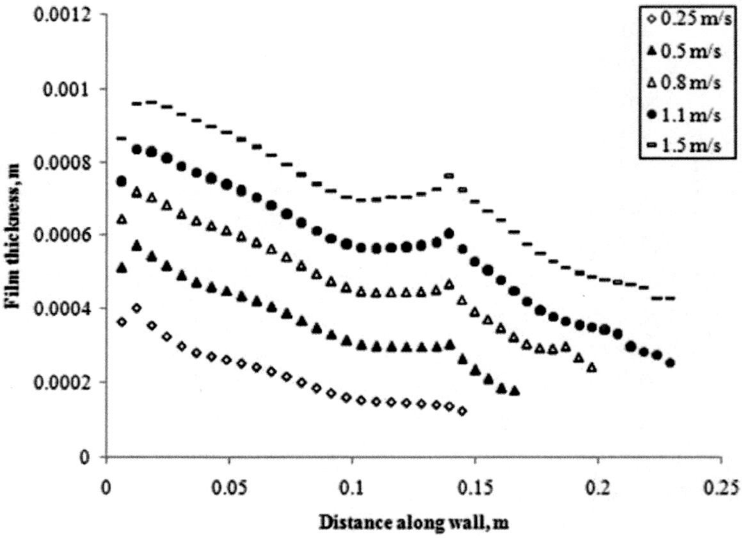

Fig. 11 Film thickness for different inlet velocities

Further studies were done by varying inlet slot width for a given initial coolant velocity whose results are shown in Fig. 13. For a given inlet velocity of 0.5 m/s, slot height was varied from 0.001 to 0.005 m to study the variation in film-cooled length. The parameters were chosen so that the film-cooled length includes a portion of the combustion chamber and nozzle convergent. Trend is linear for inlet slot height up to 0.3 mm, beyond which there is a change in behavior of the curve, indicating a change in film-cooled length. The shear stress distribution for a typical nozzle

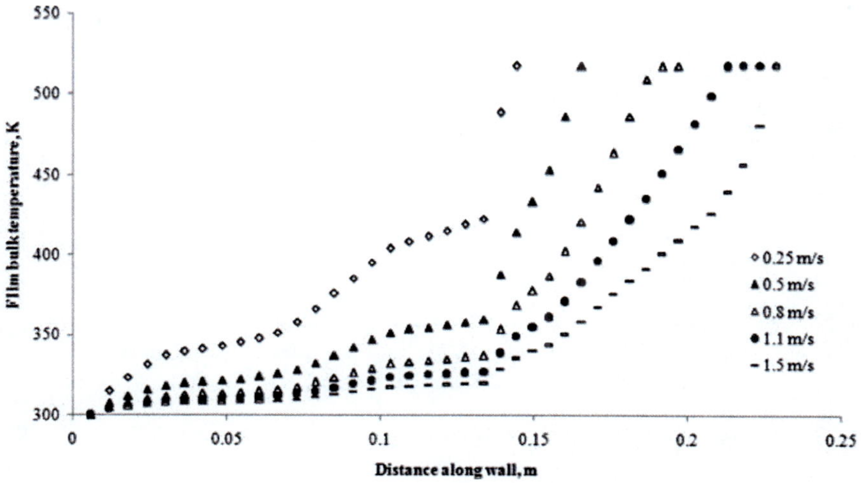

Fig. 12 Film temperature for different inlet velocities

Table 2 Film-cooled length for different liquid inlet velocities

Liquid inlet velocity, m/s	0.075	0.25	0.5	1.0	1.2	1.5
Film-cooled length, m	0.079	0.145	0.166	0.213	0.224	0.23

shows a decreasing trend in the cylindrical portion of thrust chamber whereas it increases in convergent portion of the nozzle till location of throat is reached and again decreases in divergent. At a given location, higher shear stress tends to squeeze the film resulting in film extending to further downstream. So, due to increased shear stress in convergent, film-cooled length for a given set of conditions will be higher than the value of a straight duct with constant cross sectional area is considered.

4.3 Literature Data

As it is a complex task to measure film thickness for a hot gas flow, the present method is compared with available data in the literature. In literature, film-cooled length is reported in terms of temperature recorded on the outer surface of the wall. Configuration mentioned in [5] was studied for comparison of numerical and experimental results. For validation, data reported for experimental evidence on a 2-inch smooth pipe with a length of 48 inches (4 ft) is considered. The film-cooled length obtained from the present model is compared with that of the literature in Fig. 14. It is seen that the model is able to predict the literature data with reasonable accuracy. In the experiments, film-cooled length determined based on skin temperature measured on pipe, is the distance where the temperature just reaches the boiling point of the

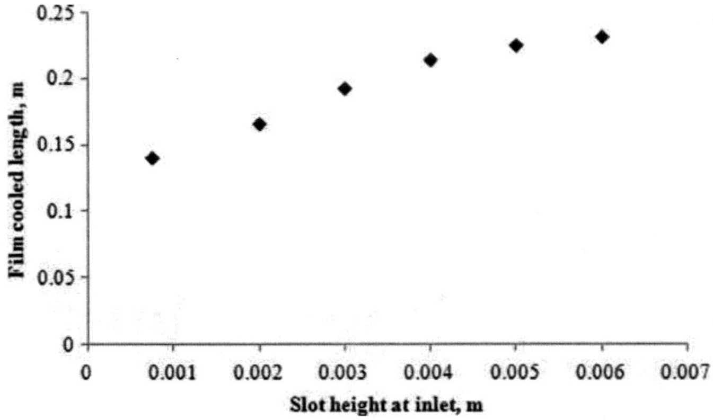

Fig. 13 Film-cooled length for different slot height

Fig. 14 Comparison with literature data

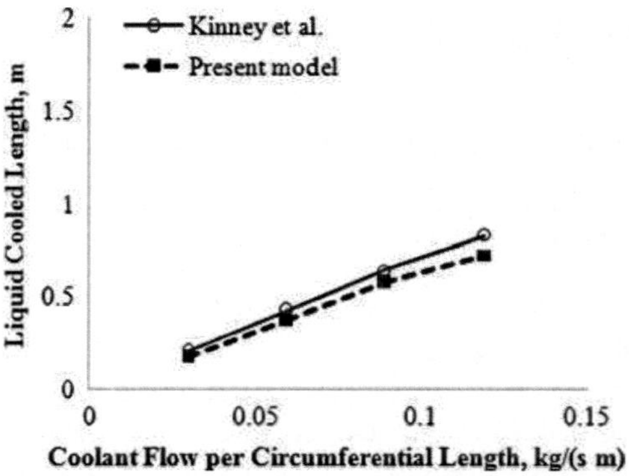

liquid, whereas, in the present model, the film-cooled length is directly obtained from numerical solution.

5 Conclusion

A numerical model for predicting film-cooled length with heat transfer from hot combustion gas is presented. The model is suitable for investigating the distance to which the wall is expected to be protected from hot gas flow for arbitrary inner contour. The model based on decoupling of liquid and gas flow gives a reasonable explanation to a coupled solver which demands fine mesh near the interface and complicated models for incompressible to compressible fluid interaction within the domain. The approach followed here is justified in most of the applications, as the liquid is injected tangentially along the gas flow and thus there is less probability of liquid having a high transverse velocity component to flow across the gas interface. The present model is able to capture features of film cooling with a liquid undergoing phase change in the presence of gas treated as compressible. Results obtained from the model were compared with data reported in the literature and it was found to reasonably predict the measured data. The proposed method will be useful for internal flow applications where proper measurements or any intrusive technique are difficult to implement to characterize the liquid film. For studying the thermal effects downstream of film-cooled length in a nozzle with ablative throat material, the present methodology is being extended to study the non-ablation of the composite material in the throat region.

Nomenclature

t	Time (s)
x	Axial coordinate (m)
r	Radial coordinate (m)
U	Conserved variables
F/Fv	Convective, viscous flux in x
G/Gv	Convective, viscous flux in r
S	Source term
k	Turbulent kinetic energy [m^2/s^2]
ϵ	Turbulent dissipation rate [m^2/s^3]
ρ_l	Liquid density [kg/m^3]
δ	Film thickness [m]
$\overline{u_l}$	Average liquid velocity [m/s]
$\dot{m}_v$	Evaporated mass flux [kg/sm^2]
x	Length of cell [m]
m	Vapor mass fraction
μ	Viscosity [Ns/m^2]
σ	Schmidt Number
T_l	Liquid temperature [K]
L	Latent heat [J/kg]
P_v	Vapor pressure [N/m^2]

Appendix

Properties used for coolant.

(a) Density (kg/m^3) = 1000.0
(b) Specific heat (J/kg K) = 4179.0

(c) Viscosity $(Ns/m^2) = 0.00002414 \times 10^{\left(\frac{247.8}{T-140}\right)}$
(d) Latent heat $(J/kg) = (25 - 0.2274 \times (T - 273)) \times 10^5$

In (c) and (d), T is expressed in Kelvin.

References

1. Shine SR, Nidhi SS (2018) Review on film cooling of liquid rocket engines. Propul Power Res 7(1):1–18
2. Ludescher S, Olivier H (2018) Experimental investigations of film cooling in a conical nozzle under rocket-engine-like flow conditions. AIAA J 57(3) (2018)
3. Heufer KA, Olivier H (2008) Experimental and numerical study of cooling gas injection in laminar supersonic flow. AIAA J 46(11):2741–2751
4. Future space-transport-system components under high thermal and mechanical loads, notes on numerical fluid mechanics and multidisciplinary design, 2021
5. Kinney GR, Abramson A, Sloop JL (1952) Internal film cooling experiments with 2 and 4 inch smooth surface tubes and gas temperatures to 2000- F in 2 and4 inch diameter horizontal tubes. Technical Report, NACA report 1087
6. Stechman RC, Oberstone J, Howell JC (1969) Design criteria for film cooling for small liquid-propellant rocket engines. J Spacecraft Rockets 6:97–102
7. Shembharkar TR, Pai BR (1986) Prediction of film cooling with a liquid coolant. Int J Heat Mass Transfer 29:899–908
8. Zhang HW, Tao WQ, He YL, Zhang W (2006) Numerical study of liquid film cooling in a rocket combustion chamber. Int J Heat Mass Transfer 49:349–358
9. Shine SR, Sunil Kumar S, Suresh BN (2012) A new generalised model for liquid film cooling in rocket combustion chambers. Int J Heat Mass Transfer 55:5065–5075
10. Wang T, Sun B, Liu D, Xiang J (2018) Experimental investigation of two-dimensional wall thermal loads in the near-injector region of a film-cooled combustion chamber. Appl Therm Eng 138:913–923
11. Xiang J, Sun B, Wang T, Yuan J (2020) Effects of angled film-cooling on cooling performance in a GO2/GH2 subscale thrust chamber. Appl Therm Eng 166:114627
12. Strokacha EA, Borovika IN, Bazarova VG, Haidnb OJ (2020) Numerical study of operational processes in a GOx-kerosene rocket engine with liquid film cooling. Propul Power Res 1–10
13. Yang W, Sun B (2012) Numerical simulation of liquid film in a liquid oxygen/rocket propellant 1 liquid rocket. J Thermophys Heat Transfer 26(2):328–336
14. Back LH, Massier PF, Gier HL (1964) Convective heat transfer in a convergent-divergent nozzle. Int J Heat Mass Transf 7:549–568